PHYSICAL PROPERTIES OF PLANT AND ANIMAL MATERIALS

PHYSICAL PROPERTIES OF PLANT AND ANIMAL MATERIALS

STRUCTURE, PHYSICAL CHARACTERISTICS AND MECHANICAL PROPERTIES

Second Revised
and
Updated Edition

NURI N. MOHSENIN

Formerly Professor of Agricultural Engineering
The Pennsylvania State University

Presently Food Physics Information Systems
State College, Pennsylvania

GORDON AND BREACH PUBLISHERS
Australia Canada China France Germany India Japan Luxembourg
Malaysia The Netherlands Russia Singapore Switzerland Thailand
United Kingdom

First Edition published 1970
Second Edition published 1986
Fourth printing 1996

Emmaplein 5
1075 AW Amsterdam
The Netherlands

Library of Congress Cataloging in Publication Data

Mohsenin, Nuri N.
Physical properties of plant and animal materials.

Bibliography: p. 841
Includes index.
1. Food. I. Title.

TX531.M582 1986 664 85-24940
ISBN 0-677-21370-0

CONTENTS

PREFACE TO THE SECOND EDITION

MORE THAN A DECADE has passed since the first edition of this book was made available. During this period the number of investigations on the physical characteristics and mechanical properties of foods and agricultural products has increased at a phenomenal rate. To incorporate even some of the new developments into this volume would have required doubling the size of this book. On the other hand, the consulting and in-house teaching experiences of the author with the food industry in recent years have clearly indicated the need for some revision of this book in order to incorporate materials which are essential to the study of physical properties of food materials.

For these reasons, the revision in this second edition were limited mainly to the addition of Chapter 2a "Water Retention," to introduce the concept and implications of water activity; compaction of agricultural materials; failure strength and characterization of solid foods; kinematics and dynamics of impacting fruits; and additional illustrated problems. Other subjects expanded or added were measurement of the surface area of granular materials, grain interporosity, particle size measurement and particle size reduction, pipe friction of fluid foods, and the application of rheological parameters to the design of food extruders.

As for new developments, to keep the new information and data on the physical properties of foods and agricultural products current and updated, the author has chosen to develop a machine-readable data base that can be updated and be used by computer-based retrieval systems.

It is gratifying to note that the original plan of the author to produce three books on the subject of physical properties of foods and agricultural products has become a reality. The first book, this volume, had three printings in English and one in Japanese before this edition. The second book, published in 1980 and currently being revised for a second edition, dealt with the thermal properties of foods and agricultural materials. The third book, entitled *Electromagnetic Radiation Properties of Foods and Agricultural Products,* was published in 1984. The next book will be a handbook on the physical properties of foods and agricultural products that will be published both in the conventional hardcopy form and as a computer version that can be edited and updated by the user periodically.

The author appreciates the corrections, criticisms and suggestions received from numerous colleagues both in academia and industry. The author is the first person to recognize the limitations of his books on the subject of the physical properties of foods and agricultural products. He is satisfied, however, that the publication of these books has generated sufficient interest to inspire researchers to look more seriously into the physical principles involved in the handling, processing, storage, preservation, quality evaluation, and quality control of raw and processed food materials.

State College, Pennsylvania NURI N. MOHSENIN

PREFACE TO THE FIRST EDITION

THE INCREASING economic importance of food materials, together with the complexity of modern technology for their production, handling, storage, processing, preservation, quality evaluation, distribution and marketing, and utilization, demands a better knowledge of the significant physical properties of these materials. From the production units on the farm to the consumer, food materials are subjected to various physical treatments involving mechanical, thermal, electrical, optical, and sonic techniques and devices. It is essential to understand the physical laws governing the response of these biological materials so that the machines, processes, and handling operations can be designed for maximum efficiency and the highest quality of the end products. Maintenance of quality under adverse conditions of handling, storage, and distribution, savings in weight and bulk in some cases, reduced costs in handling and processing operations, and finding new ways for utilization may result from an understanding of the basic physical properties of economic plant and animal materials.

Despite the importance of the subject, intensive research and study in this area have only started during recent years. In the past, however, considerable research has been reported in various scientific journals on physical properties and characteristics of biological materials, particularly those of interest to the food and agricultural industries. This book, which is the outgrowth of class notes prepared for teaching a course in this area, is an attempt to place the most significant research reports on physical characteristics and properties of plant and animal materials under one cover. The contents should be of value as text material for students and teachers as well as researchers in any branch of science and technology concerned with the physical behavior of biological materials. Agricultural and food engineers, bioengineers, and food, plant and animal scientists should find the book useful as a reference material.

The book is divided into two volumes. This first volume contains topics of gross structure and terminology of the economic plant and animal materials, physical characteristics such as shape, size, volume, density, porosity, etc., and mechanical and rheological properties. The second volume will contain topics on thermal, electrical, and optical properties.

Emphasis is placed on the fundamentals of engineering sciences as applied to biological materials for their characterization and determination of physical properties. Once the material has been characterized and a range for its various physical properties has been established, it is necessary to interpret the results in a form that would have engineering utility and implications. For this reason, when possible, examples have been given to illustrate the application of a certain property in design and analysis, utilization, and quality control.

When tabulated data are needed for discussion, they are given in the main body of the text. Otherwise, the data on physical properties are given in tables in the Appendix. The symbols used are usually defined in the text. However, for ease of reference, a definition of symbols used in each chapter is also found at the end of that chapter.

Acknowledgment is due to Professor F. W. Peikert, Head, Department of Agricultural Engineering, and other colleagues for their encouragement in preparation of the manuscript. The writer sincerely appreciates the painstaking work of George Kann for preparing many of the ink drawings and Shirley Brungart, Dorothy Olbricht, and Sue Brocail for typing the multilith masters for the preliminary copies of this book printed in 1966 and 1968. The author is particularly indebted to C. T. Morrow for his continuous assistance in supplying references, conducting experiments, helping with class and laboratory instructions covering materials from this book, and reading the first draft of the manuscript. Other associates, in particular, R. K. White, J. R. Hammerle, D. R. Bittner, and Leora Shelef, have also contributed to the compilation, checking and analysis of the data reported as unpublished work. The graduate students who have been associated with the author in research on physical properties of agricultural products over the past nine years have all contributed to this work through their theses, special assignments, or working of the problems. They are D. R. Bittner, H. E. Cooper, D. G. Cowart, R. S. Devnani, R. G. Diener, J. Duru, E. E. Finney, W. F. Fletcher, J. J. Gaffney, J. R. Graham, J. R. Hammerle, C. T. Morrow, C. W. Nelson, R. K. White, J. D. Whitney and Y. M. Yang.

It would be impossible to name and acknowledge all the organizations and individuals for their permission to make use of their published work and their cooperation in supplying the requested information and illustrative material. Throughout the book, however, an attempt has been made to indicate the sources of all materials used. The bibliographical references to these sources are given at the end of this volume.

Obviously, a preliminary work of this nature cannot be without error and could stand considerable improvement. The author will much appreciate receiving any suggestions and criticisms concerning the organization or condensation and notification of any inaccuracies in statements or illustrations or any typographical errors.

University Park, Pennsylvania NURI N. MOHSENIN

1

IMPORTANCE

MODERN AGRICULTURE has brought about the handling and processing of plant and animal materials by various means such as mechanical, thermal, electrical, optical and even sonic techniques and devices. Despite these ever increasing applications, little is known about the basic physical characteristics and properties of these materials. Specific heat and other thermal characteristics, electrical conductivity and dielectric constants, light transmittance characteristics, and such mechanical properties as stress-strain behavior, resistance to compression, impact and shear, and coefficient of friction are a few examples of these unknown properties. A knowledge of these properties should constitute important and essential engineering data in design of machines, structures, processes and controls; in analyzing and determining the efficiency of a machine or an operation; in developing new consumer products of plant or animal origins; and in evaluating and retaining the quality of the final product. Such basic information should be of value not only to engineers but also to food scientists and processors, plant and animal breeders, and other scientists who may exploit these properties and find new uses. To understand and appreciate the need for information on physical properties of plant and animal products, a few examples are given in the following (Fig. 1.1).

1.1 PHYSICAL CHARACTERISTICS

Shape, size, volume, surface area, density, porosity, color and appearance are some of the physical characteristics which are important in many problems associated with design of a specific machine or analysis of the behavior of the product in handling of the material.

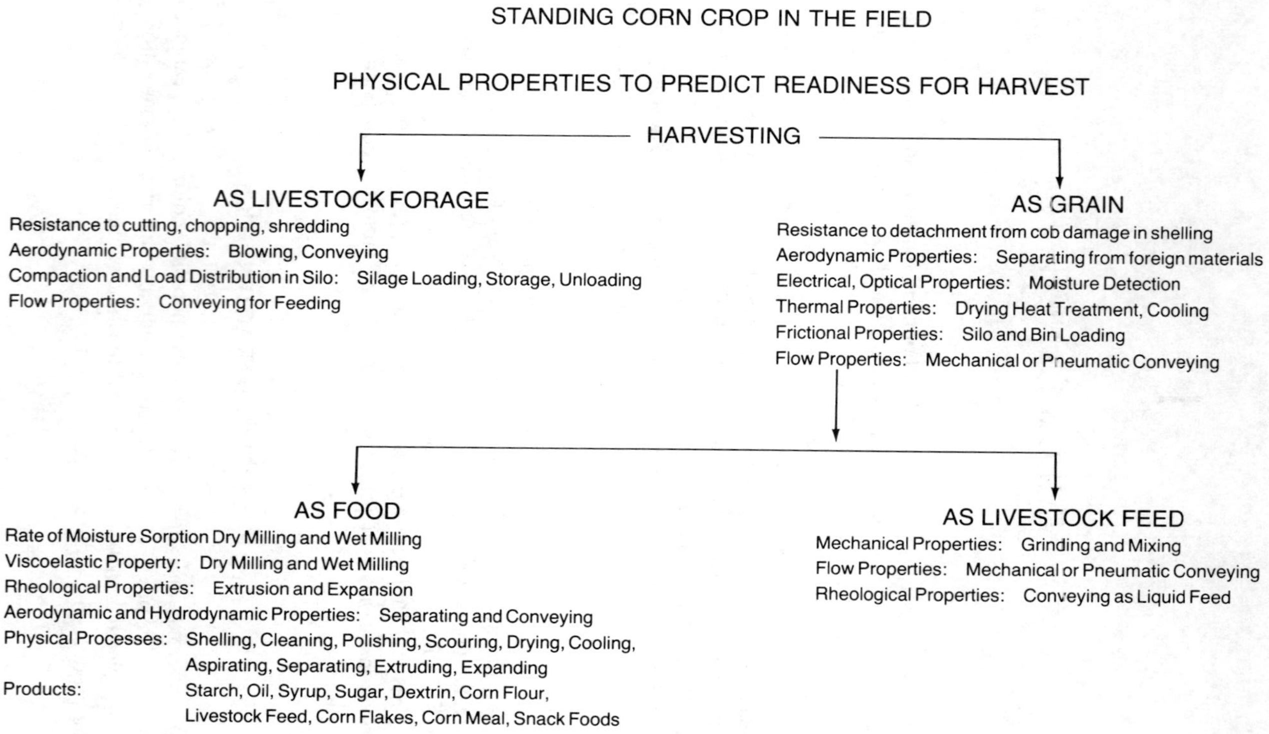

Figure 1.1 Importance of physical properties in the case of a standing corn crop in the field

What shape is to be assumed for the material and which dimension is to be employed in calculations are two first questions which one must answer before analyzing the cooling curve of a fruit or understanding the problem of separation of seeds and grains from undesirable materials by pneumatic or electrostatic devices.

There are many charts available for solving the problems of transient heat flow in engineering materials. A glance at these charts shows that accurate estimates of the shape and the related dimensions of the material are necessary before the chart can be used in solution of the problem of heat transfer in the body in its natural state. As will be seen later the problem has been solved by either assuming a spherical shape for the product or confining a bulk of the material in a cylindrical container or a slab-shaped box before placing the product in the heating or cooling medium.

One of the important design parameters in conveying of solid materials by air or water is the assumption for the shape of the material. Accurate estimates of the frontal area and the related diameters are essential for determination of terminal velocity, drag coefficient, and Reynolds number.

Table 1.1 shows the assumptions made by several investigators in calculating aerodynamic characteristics of some agricultural products.

The question of shape and size is also important in problems of stress distribution in the material under load, in electrostatic separation of seeds and grains, in light reflectance and color evaluation, and in development of sizing and grading machinery.

A knowledge of density and specific gravity of agricultural products is needed in calculating thermal diffusivity in heat transfer problems, in determining Reynolds number in pneumatic and hydraulic handling of the material, in separating the product from undesirable materials and in predicting physical structure and chemical composition.

The irregular shape and porous nature of many agricultural products present difficult problems in volume and density measurements. Such simple techniques as water displacement can result in appreciable errors if the water can penetrate into the material or if the material is very small in size, such as small seeds. A difficult example in this area has been the density evaluation of an expanded forage wafer.

Surface color and appearance of agricultural products are valuable physical characteristics for selective separation in the field as subsequent handling and processing (Fig. 1.1a). In selective harvesting of fruits and vegetables and in sorting and grading at post harvest and during storage, desirable

Table 1.1 Frontal area and diameter assumed in investigating the aerodynamic characteristics of agricultural products

Material	Shape chosen	Diameter chosen	Frontal area	Reference
Seed grains	Ellipse	$\frac{L_1 + L_2 + L_3}{3}$	$\pi L_1 L_2/4$	Bilanski (1962)
Grains	Sphere	D	$\pi D^2/4$	Garrett and Brooker (1965)
Grass seeds	Sphere	d	$\pi d^2/4$	Keck and Goss (1965)
Fruits	Sphere	d_{ave}	$\pi d^2_{ave}/4$	Schmidt and Levin (1963)
Potatoes	—	—	measured (max.)	Gilfillan and Growther (1959)

$L_1 L_2 L_3$ = length, width, and depth, respectively

D = diameter of a sphere having volume equal to that of the grain (volume found from weight divided by specific weight)

d = diameter of a sphere found by taking the geometric mean of the 3 mutually perpendicular measured seed dimensions

d_{ave} = average of maximum and minimum diameters of the fruit

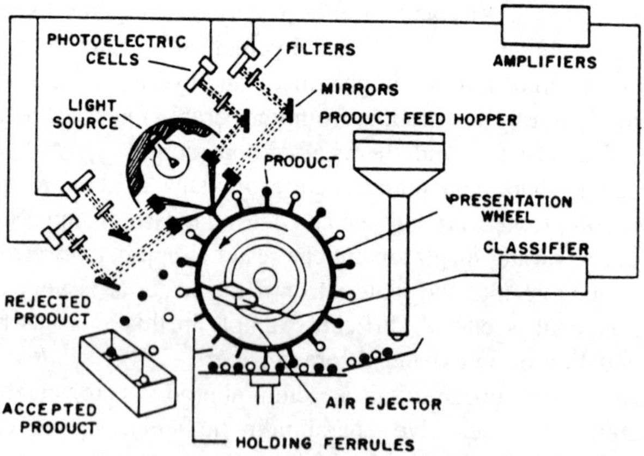

Figure 1.1a Application of light reflectance characteristics of agricultural Products in sorting and grading machines (courtesy Mandrel Industries, Inc.)

products can be selected on the basis of color and appearance. Here the light reflectance characteristics of the product is the important physical property which must be known before equipment can be designed for separation and sorting or grading. However, irregular shape and nonuniform color of the object with certain reflectance characteristics which occur at only one or two narrow regions of the spectrum present problems which require special techniques and instrumentation.

1.2 MECHANICAL PROPERTIES

Mechanical Damage to seeds and grains which occur in harvesting, threshing, and handling can seriously affect viability and germination power, growth vigor, insects and fungi attack, and quality of the final product (Table 1.2). Depression of viability is due to mechanical damage to the embryo of the seed. The depression of growth vigor of the damaged seed has been demonstrated by the decrease of the size of shoots and of the weight of the plants (Fig. 1.2).

Pea bean crackage has been a problem during handling in elevators. The amount of damage is apparently affected by temperature, moisture content and impact loading.

Hardness of grains has been a subject of interest to millers, livestock feeders, breeders and other agricultural scientists. Biting or cutting the grain has provided a qualitative evaluation of grain hardness. A number of attempts have been made to find an objective and a quantitative measure of the hardness of individual kernels or the average of a collection of kernels. The data are used to ascertain the relationship between hardness and certain physical and chemical properties, feeding value, and size reduction and milling characteristics of the grain.

Table 1.2 Mechanical damage to wheat grain in threshing (King and Riddolls, 1959)

	Concave clearance, in.				Drum speed, rpm			
	1/8	5/32	3/16	7/32	1100	1000	900	800
Visible damage %	7.7	8.3	8.5	7.2	19.9	10.0	8.1	5.0
Germination of visibly undamaged seed %	90.4	89.9	90.6	90.9	78.7	88.0	92.6	92.6
Wastage %	16.0	16.9	16.8	15.2	36.9	20.9	14.9	11.9

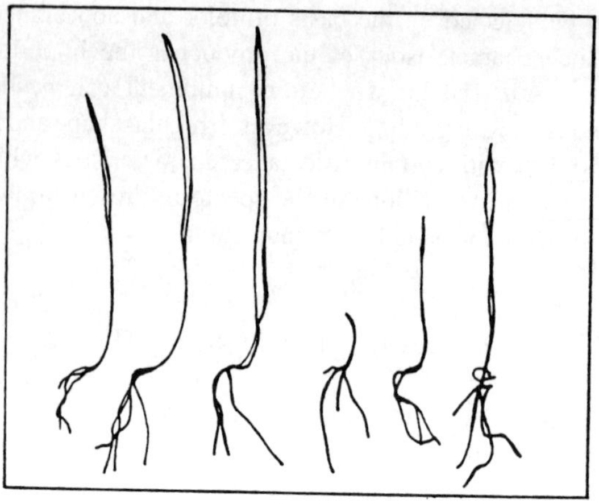

Figure 1.2 Mechanical damage to seeds causes depressed germination and growth vigour (Usenko, 1952)

Mechanical properties such as compressive strength, impact and shear resistance, are important and in some cases necessary engineering data in studying size reduction of cereal grains as well as seed resistance to cracking under harvesting and handling conditions (Fig. 1.3). From an energy standpoint, this information can be used to determine the best method (shear, impact or static crushing) to breakup or grind grain.

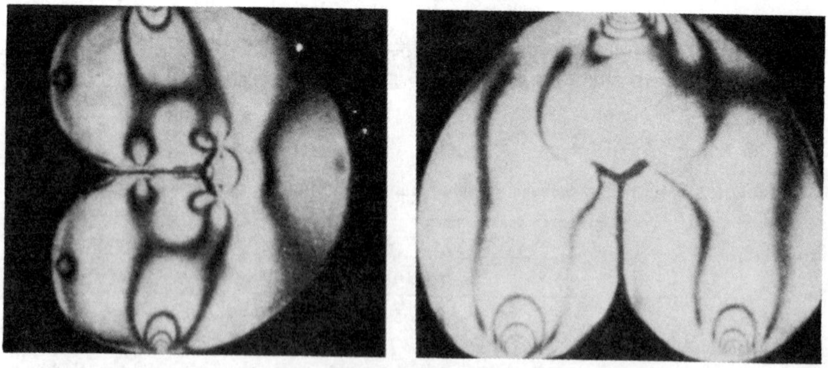

Figure 1.3 Photo-elastic techniques are employed to study mechanical damage in wheat grains (Arnold and Roberts, 1966)

Static and sliding coefficients of friction of grains, forage materials, and some other farm products on metals, wood and other materials are needed by design engineers for rational design and predicting motion of the material in harvesting and handling equipment. Coefficient of friction is also important in determining the pressure of grain and silage against bin walls and silos.

Compressibility, expansion characteristics, coefficients of internal friction and cohesion, and elasticity of forage or silage mass are important in studying compressibility of the material and determining methods of compressing and packaging (Fig. 1.4). Shearing resistance and bending strength of forage crops as they are cut are also important mechanical properties for understanding the nature of the cutting process and energy requirements in mowing machines.

Aerodynamic and hydrodynamic properties of agricultural products are needed for air and water conveying and separation of foreign materials. Density size, shape, and drag coefficient are the physical properties needed in calculating the terminal velocity of an object in the fluid. In air conveying or pneumatic separation, an air velocity greater than terminal velocity would lift the particle. To allow the particle to fall gently, the air velocity is adjusted to a point just below the terminal velocity.

Mechanical harvesting, bulk handling, transporting, and storage of fruit and vegetable products have also indicated a need for basic information

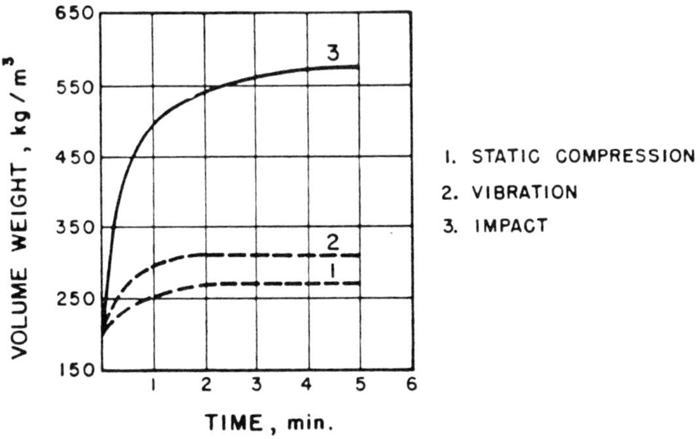

Figure 1.4 Under otherwise equal conditions, the greatest compression of silage has been obtained by impact compression (Yaremenko, 1956)

in mechanical properties. Bruising and skinning of mechanically harvested potatoes, distortion of onion bulbs in bottom of the storage pile, and mechanical damage to fruits and vegetables by compression, impact, and vibration have lowered the grade of these products, with consequent loss to the grower (Figs. 1.5, 1.6).

If mechanical handling of agricultural products result in deformation and flow of the materials, the mechanical properties involved will be referred to in this work as rheological properties. On the basis of this definition, many of the preceding examples involve deformation and flow and thus concern the rheological properties. Additional examples of rheological properties are such textural attributes of food products (Fig. 1.7) as firmness, yielding quality, crispness, fibrousness and such flow characteristics as viscosity, consistency, and fluidity of liquid feed and waste materials handled in slurries.

1.3 THERMAL PROPERTIES

Many of the agricultural products of plant or animal origin are subjected to various types of thermal processing before they are placed at the access of the consumer. The thermal processing may include heating, cooling, drying, and freezing. It is upon the thermal properties of the product that any change of temperature will largely depend.

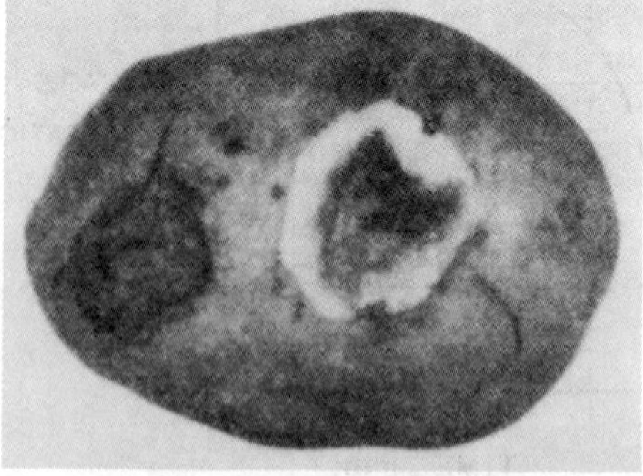

Figure 1.5 Dead loads experienced by potato in storage piles causes mechanical damage (courtesy C. H. Green, NIAE, England)

Figure 1.6 Mechanical damage to pears due to in-transit vibration (O'Brien, 1965)

Hard seed of alfalfa and red clover can be made permeable by application of properly regulated heat. Heating of these seeds for four minutes at 220°F. has reduced the number of hard seeds by as much as 80 percent. Specific heat of seeds is an important physical property in heat treating applications where germination and viability of seeds may be endangered if critical temperature or time of heating is exceeded.

Heat treatment of wheat, corn, and legumes has shown some promise in stimulating germination. Enzymes in cotton seed can be completely inactivated by dielectric heating, thus preventing seed spoilage while in storage. The fungus causing the decay of onions in storage vanishes in the temperature range of 40° to 60°C (Gasnikow, 1960). Heat treatment of peaches has shown promise in delaying the decay of the fruit in storage and improving the quality retention.

In fruits and vegetables the action of enzymes and microorganisms causing deterioration can be controlled by low temperature. It is said that fresh produce deteriorate as much in an hour at 90°F. as in a day at 50°F. or in a week at 32°F. (Guillou, 1958). To cool the fresh fruit as rapidly as possible, a "portable harvest cooler" has been developed which can be hauled right out to the orchard.

Heating or cooling of agricultural products may be accomplished by the methods of convection, conduction and radiation. A knowledge of such thermal characteristics as specific heat, thermal conductivity, thermal diffusivity surface conductance, and emissivity as well as such physical charac-

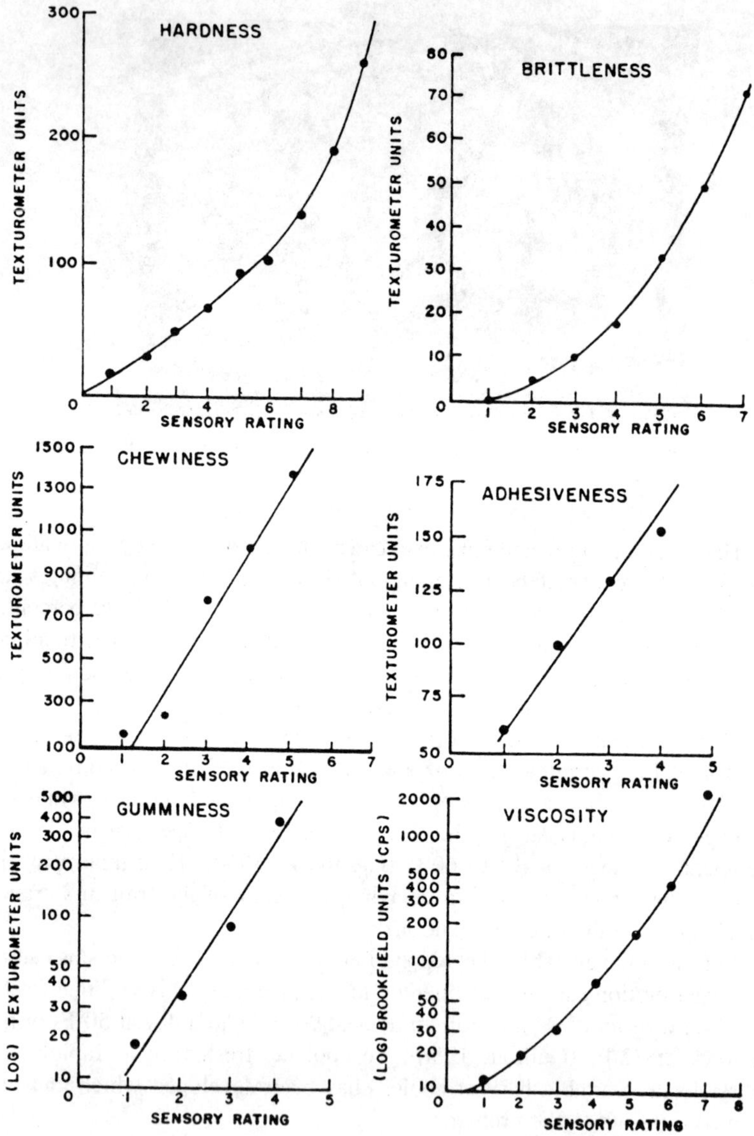

Figure 1.7 Mechanical parameters are the most important attributes in texture evaluation of food products (Szczesniak *et al.*, 1963)

teristics as density, shape and size is essential for design of the equipment and predicting of the processes. A heat balance for a heating or cooling system cannot be attempted without a knowledge of the heat capacity of the material. To define the magnitude and location of a temperature that denotes the heat content of the material at any time during a heating or cooling process, a knowledge of the thermal characteristics of the material is required. In heat treatment of steel, a desired microstructure can be obtained by controlling the time and temperature of heating. In heat treatment of biological materials, time and temperature are equally important if viability, nutrients and quality of the material are to be preserved.

1.4 ELECTRICAL PROPERTIES

Some electrical properties of agricultural products which are important in handling and processing are electrical conductance and capacitance, dielectric properties, and reaction to electromagnetic radiation.

Electrical conductance or capacitance properties have been used in moisture content determination of products such as cereal grains. Electrical resistance methods have been employed for precise measurement of cotton fiber length distribution and fineness of wool fiber.

The principle of electrostatic separation which has been known for centuries is being investigated for separation and cleaning of agricultural seeds (Fig. 1.8). With small seeds, it has been found that electrostatic separation is essentially independent of size, shape, weight, and surface texture. When devices depending upon these physical characteristics fail to separate similar seed varieties, the seed's ability to hold electrostatic charge can be used for separation. Conductivity of the seed is the property which would determine, basically, the ability of the seed to hold surface charge.

An impedance technique can be used to determine the extent of injury to plant tissues due to frost, poisoning by spray, or other means of damage. This technique is based on the fact that when a tissue is dead it has no capacitance and when it is uninjured and healthy, has reistance and capacitance comprising an impedance. Thus the ratio of low-frequency impedance to high-frequency impedance, measured with a wide-range a.c. bridge, is an indication of the degree of injury to the plant tissue.

Dielectric heating, which is the heating of the material due to its own dielectric losses when placed in an electrodynamic field, has been used quite extensively in drying plastics, ceramics and other nonconducting

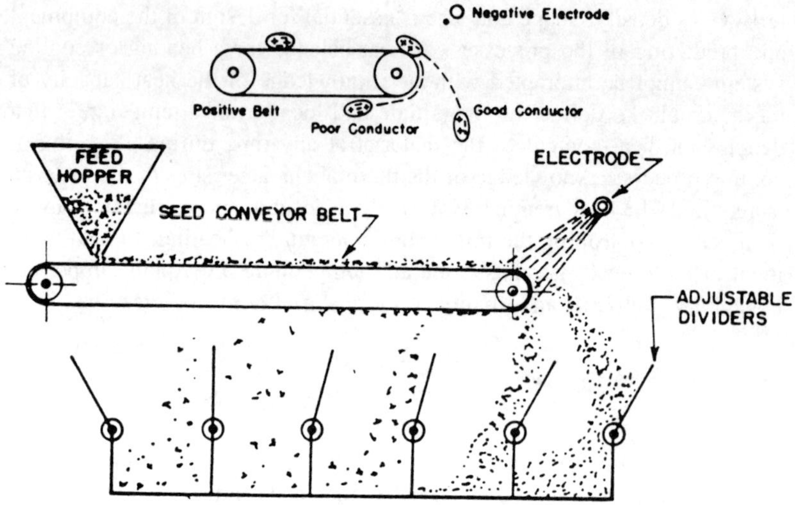

Figure 1.8 Ability of seeds to conduct an electrical charge can be used for seed cleaning by electrostatic separation (Harmond *et al.*, 1961)

materials. Uniformity of heating and the high rates of temperature rise are two principal advantages which make dielectric heating attractive for heat treatment and drying of agricultural products (Fig. 1.9). The possibilities of dielectric heating have been investigated for inactivating the enzymes in cotton seed, controlling insects and fungi in seeds, and drying rice and other grains. The use of high-frequency electric field to cotton seeds has shown that the enzymes in the seed could be completely inactivated in six minutes. Further, it was found that during and after the treatment, drying of the cotton seed took place which further improved the quality for storage. Dielectric heating has also been used in heat treatment of seeds. Using the correct amount of exposure to radio-frequency electric fields, germination and early growth was stimulated while some dormant or hard seeds were made immediately germinative.

Electromagnetic radiation has considerable potential for processing of agricultural products. Scientists have found many practical applications for any of the several ranges of radiation in the electromagnetic spectrum (radio-frequency, infrared, visible, ultraviolet, x-rays, and gamma rays). The principal effects of radiation can be divided into 1) the heating effects (longer wave length and lower energy range) and 2) chemical effects

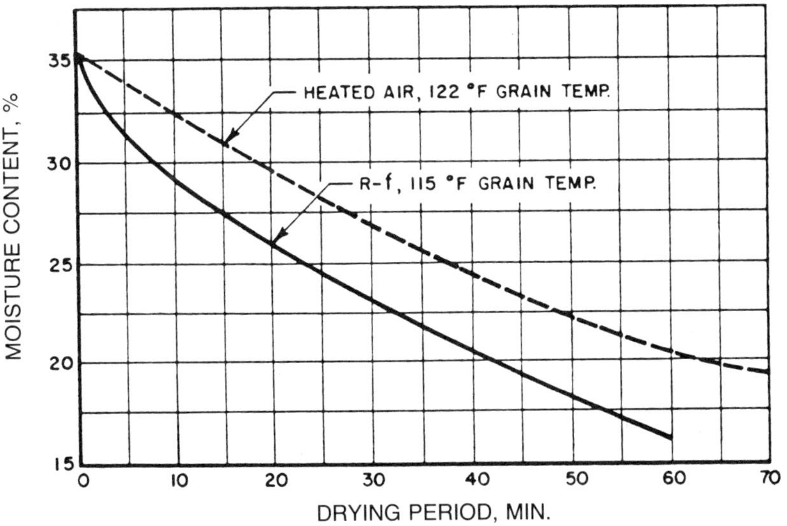

Figure 1.9 Grain drying by radio-frequency (R-f) heating results in faster and more uniform heating of the grain. Comparison with heated air drying requires a knowledge of dielectric properties of the grain (Knipper, 1956)

(shorter wave length and higher energy range) causing ionization of the atoms. Certain physical properties of agricultural products would probably be important in determining their reactions to electromagnetic radiations.

1.5 OPTICAL PROPERTIES

Light transmittance and reflectance properties of agricultural products have been explored in recent years for electronic sorting and grading, maturity, and surface color determinations, and study of the interior characteristics of fruits and vegetables.

From an optical point of view intact agricultural products are dense, light scattering materials which require a highly sensitive and specially designed spectrophotometer for measuring their spectral transmittance characteristics.

An instrument has been developed which allows the transmittance and measurement of monochromatic light through intact biological specimens

(Norris, 1958). The technique, which is based on light transmittance characteristics and absorption spectra of the material has been exploited to determine internal color of tomatoes, smut content of wheat, fruit maturity, degree of milling of rice, internal discoloration of potatoes, blood spots and green rot in eggs, water core in apples, insect infestation in wheat, moisture content of seeds, scald damage in cherries, and damage in yellow corn (Figs. 1.10, 1.11).

Several workers have considered the light reflectance characteristics of agricultural products for selecting, grading, and separation of desirable products from foreign materials. The reflectance properties of potatoes were found sufficiently different from those of soil clods to suggest that reflectance may be used to differentiate between the two materials electronically (Palmer, 1961). The sorting efficiency of workers in cherry processing plants as affected by the spectral distribution of the illuminating light as well as the reflectance of cherries and their defects has been investigated (Parker and Wiant, 1954). Light reflectance curves of lemons have shown distinct variation in various stages of fruit maturity. The use of this principle has been explored for the design of a color sorting machine (Powers, *et al.,* 1953).

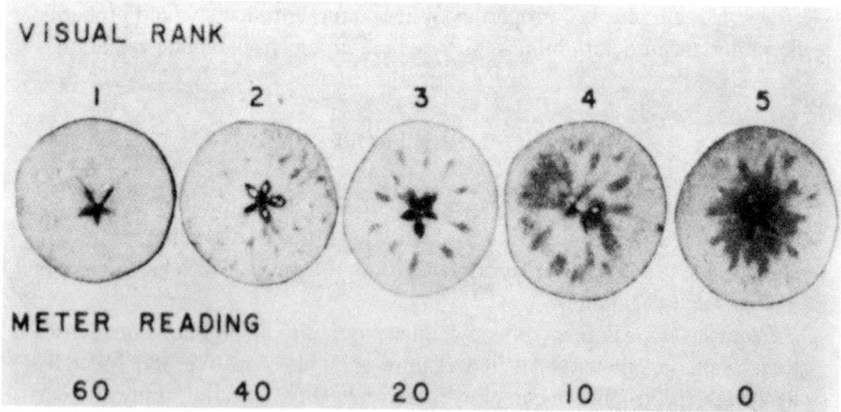

Figure 1.10 Presence of water core in apples can be found non-destructively by light transmittance (courtesy USDA Instrumentation Research Laboratory, Beltsville, Maryland)

Figure 1.11 An experimental fruit sorting machine employing principles of light transmittance (Yeatman and Norris, 1965)

1.6 GLOSSARY OF TERMS RELATED TO PHYSICAL PROPERTIES OF FOODS AND AGRICULTURAL MATERIALS

Following is a compilation of terms related to structure and physical properties of food materials of plant or animal origins. It shows the extent to which biological and physical principles are involved in handling, processing, storage, preservation, and quality evaluation and quality control of foods and agricultural products. This book covers the basic principles and applications related to those terms related to structure, physical state, and mechanical properties. A second book will cover the thermal and diffusional properties. The last, being the third book in the series, will cover the electromagnetic radiation properties.

ANATOMY OF PLANTS AND ANIMAL MATERIALS

Cells
Tissues
Protective Tissues
Epidermis and periderm
Conducting and vascular tissues
Phloem
Xylem
Supporting tissues
Parenchyma tissues (edible)
Collenchyma tissues (mechanical strength)
Cell wall
Cellulose
Semicellulose
Lignin
Pectic substances
Turgor pressure
Relationship of Turgor pressure and rigidity
Dimensional characteristics
Cell volume, size
Air spaces
Surface area
Cells/unit volume
Fiber cells
Constituents of plant and animal products

RETENTION OF WATER

Free water (capillary absorption)
Bound water (molecular adsorption)
Polar sites
Adsorption compression
Swelling of hygroscopic solids
Water activity, equilibrium relative humidity
Equilibrium moisture content
Raoult's law
Activity coefficient
Sorption-desorption isotherm
Sorption-desorption hysteresis
Regions of isotherms

PHYSICAL STATE

Sphericity
Roundness
Frontal area
Geometric mean diameter
Porosity
Bulk density
Solid density
Unit density
Permeability
Specific surface
Specific gravity
Convex body
Average projected area
Prolate spheroid
Oblate spheroid

RHEOLOGICAL PROPERTIES OF SOLID AGRICULTURAL MATERIALS

Rheology
Axial stress
Axial strain
Bulk stress
Bulk strain
Triaxial stress
Triaxial strain
Shear stress
Shear strain
Shear deformation
Bulk deformation
Deviatoric stress
Isotropic stress
Modulus of elasticity
Bulk modulus
Compressibility
Shear modulus or rigidity
Poisson's ratio
Strength (shear, compressive, tensile, bulk)
Yield point
Bioyield point
Rupture point
Pressure (axial, bulk)
Toughness
Stiffness
Resilience
Mechanical hysteresis
Elasticity
Plasticity
Viscoelasticity (linear and non-linear
Stress relaxation
Relaxation time
Creep
Retardation time
Viscosity
Viscosity coefficient (absolute)

RHEOLOGICAL PROPERTIES OF ORGANIC FLUIDS

Flow diagram
Shear stress
Shear rate or velocity gradient
Flow behavior index (power law constant)
Viscosity index
Newtonian flow
Non-Newtonian flow
Plastic flow
Yield stress
Pseudoplastic
Dilatant
Thixotropic fluid
Rheopectic fluid
Absolute viscosity
Plastic viscosity
Apparent viscosity
Kinematic viscosity
Power law equation to describe flow
Rheometry

MECHANICAL PROPERTIES OTHER THAN RHEOLOGICAL PROPERTIES

Texture of foods
Readiness for harvest
Brinell hardness
Meyer hardness
Meyer's law for hardness
Four phases of impact
Dynamic yield pressure
Coefficient of restitution
Height of rebound
(how to determine?)
Fragility factor
Energy balance in an impact
Stress cracking
Maximum allowable load
(strength of food materials)
Drag coefficient
Reynold's number
Terminal velocity
Coefficient of surface friction
Surface tension
Adhesion, surface tension
Rolling resistance
or maximum stability angle
Coefficient of internal friction
Angle of repose
Pressure ratio in silos and bins
Cohesion of granular materials
Yield locus in flow
of granular materials
Flow function of granular
materials
Fluidized conveying
Asperity contact

THERMAL AND DIFFUSIONAL PROPERTIES

Heat capacity
Specific heat
Latent heat
Eutectic mixtures
Sensible heat
Heat of adsorption
Heat of wetting
Heat of respiration
Thermal conductivity
Thermal diffusivity
Unit surface conductance
surface transfer coefficient)
Diffusion coefficient thermal
diffusivity kinematic viscosity
Fick's law of mass diffusion,
Fourier's law of heat
conduction, Newton's law
of viscosity
Mass transfer coefficient
convection heat transfer
coefficient
Dimensionless numbers
Biot number (heat transfer)
Biot number (mass transfer)

Thermal emissivity
Heat transfer by
  conduction
  convection
  radiation
Steady-state heat transfer
Unsteady-state heat transfer
Mass transfer by
  ordinary diffusion
  pressure diffusion
  thermal diffusion
  forced diffusion

Fourier number (heat transfer)
Fourier number (mass transfer)
Temperature ratio (heat transfer)
Concentration ratio
  (mass transfer)
Permeability coefficient
Solubility coefficient,
  Henry's law
Coefficient of thermal expansion
Cooling rate
Half-cooling time
Mass average temperature

ELECTROMAGNETIC RADIATION PROPERTIES

Spectrum of electromagnetic
  radiation
Spectral response of human eye
Geometric optics
Physical optics
Frequency of radiation
Energy of radiation
Irradiation
Illumination
Irradiance
Radiance
Candle, lumen, luminance
Foot candle
Luminosity
Refraction and refraction index
Snell's law
Diffraction
Interference
Diffraction grating
Regular reflectance
Body reflectance

Absorption coefficient
Kubeika-Munk function
Absorptivity and emissivity
Black body
Gray body
Collimator
Monochrometer
Spectral band width
Integrating sphere
Spectrophotometer
Optical system geometry
  45°/0°, d/0°, 0°/d
Spectral reflectance
Spectral transmittance
Holography interframetry
Physical attributes of color
  radiant energy
  dominant wave length
  purity
  reflectance
  directional reflectance

Total reflectance
Fresnel's law
Transmittance
Beer-Lambert law
Absorbance
Law of conservation of energy in optic
Scattering coefficient
ICI color system
Standard illuminants
Tri-stimulus values of color
Chromaticity coordinates
Chromaticity diagram
Non-spectral colors
Complementary wave length
Munsell color system
Hunter color system
Delayed light emission, DLE
Electrical resistivity
Conductance
Electrical conductivity
Dielectric polarization
Complex dielectric constant or relative complex permitivity
Psychological attributes of color
lightness
hue
chroma
saturation
value
gloss
Dielectric constant or relative permitivity
Dielectric loss factor or relative dielectric loss factor
Power factor
Dielectric loss tangent or dissipation factor
Magnetic permeability
Dielectric heating
Power absorption
Mechanical analogs of dielectric properties
Electrostatic force
Electrostatic charge

1.7 SELECTED BOOKS AND BIBLIOGRAPHIES ON PHYSICAL PROPERTIES OF FOOD MATERIALS

Adam, M. 1969. Bibliography of Physical Properties of Foodstuffs. Praug.

Adam, M. 1978. Bibliography of Physical Properties of Foodstuffs-Texture Properties. Praug.

Anon. 1980. Physical Properties of Agricultural Materials and Their Influence on Technological Processes. Proceedings of the 2nd Int. Congress. Godollo, Hungary, 26–28 August.

Anon. 1976. Physical Properties of Plant Materials and Their Influence on Technological Processes. Transactions of the International Conference by the Institute of Agrophysics of the Polish Academy of Sciences. Polish Academy of Sciences, Warsaw.

Burmistrova, M. F. 1956. *Physicomechanical Properties of Agricultural Crops*. Translated from Russian and published for the National Science Foundation by the Israel Program for Scientific Translations. U. S. Dept. of Commerce, Washington, D.C.

Finney, E. E., Jr. 1973. *Measurement Techniques for Quality Control of Agricultural Products*. Am. Soc. Agr. Engrs., St. Joseph, Michigan, and Optical Society of America, Pittsburgh, Pennsylvania.

Gaffney, J. J. (ed.). 1976. *Quality Detection in Foods*. Am. Soc. Agr. Engrs., St. Joseph, Michigan.

Linko, P., V. Malkki, J. Olkku, and J. Larinkari (eds.), 1980. *Food Process Engineering Vol. 1, Food Processing Systems*. Proceedings of the 2nd Int. Congress on Engineering and Food, Helsinki, Finland.

Mohsenin, M. and J. P. Mittal. 1976. Bibliography of Ph.D. and M. S. Theses Related to Physical Properties of Agricultural Materials. Dept. of Agr. Engng., Penn State University, University Park, Pennsylvania.

Peleg, M. and E. Bageley. (eds.). 1983. *Physical Properties of Food*. AVI Publishing Company, Inc., Westport, Conn.

Rha, C. (ed.). 1975. *Theory, Determination, and Control of Physical Properties of Food Materials*. Reidel Publishing Co., Dordrecht, Holland.

1.8 SELECTED BOOKS IN FOOD PROCESS ENGINEERING WITH PHYSICAL PROPERTIES DATA AND APPLICATIONS

Blakebrough, N. (ed.). 1968. *Biochemical and Biological Engineering Science*. Vol. 2, Academic Press, New York.

Brennan, J. G., J. R. Butters, M. D. Cowell and A. E. V. Lilly. 1976. 2nd Ed. *Food Engineering Operations*. Elsevier Pub. Co., New York.

Charm, S. E. 1978. *Elements of Food Engineering*. 3rd Ed. AVI, Westport, Conn.

Farral, A. W. 1979. *Food Engineering Systems*. AVI Pub. Co., Westport, Conn.

Harper, J. H. 1976. *Elements of Food Engineering*. AVI Pub. Co., Westport, Conn.

Henderson, S. M. 1976. *Agricultural Process Engineering*. 3rd Ed. AVI Pub. Co., Westport, Conn.

Heldman, D. R. and R. P. Singh. 1981. *Food Process Engineering*. 2nd Ed. AVI Pub. Co., Westport, Conn.

Karel, M., O. R. Rennema and D. B. Lund. 1925. *Physical Principles of Food Preservation*. Marcel Decker, Inc., New York.

Leniger, H. A. and W. A. Beverloo. 1975. *Food Process Engineering*. Reidel Pub. Co., Dordrecht, Holland, Boston, Mass.

Loncin, M. and R. L. Merson. 1979. *Food Engineering and Selected Applications*. Academic Press, New York.

Toledo, R. T. 1980. *Fundamentals of Food Process Engineering*. AVI Pub. Co., Westport, Conn.

Zenz, F. A. and D. F. Othmer. 1960. *Fluidization and Fluid Particle Systems*. Reinhold Publishing Co., New York.

2

STRUCTURE AND CHEMICAL COMPOSITION

STUDY OF PHYSICAL properties of materials from plant or animal origin requires some knowledge of their structure as well as certain physiological activities influencing these properties. In this chapter we will discuss briefly only those aspects of structure which have direct bearing on the physical properties of selected economic plant and animal materials. For further information in the structure and physiology of these materials the reader is referred to the selected list of references compiled for this chapter. A portion of this discussion is devoted to the absorption and desorption of moisture because the physical properties are highly dependent on the moisture content of the material.

2.1 ORGANIZATION OF THE PLANT BODY

The plant body consists of structural units called cells. Each cell is enclosed in its own cell wall and united with other cells by means of a cementing substance. Cells are then grouped together to form *tissues* which may be classified as protective tissues, conductive tissues, and ground tissues.

The *protective tissues* are composed of guard cells to protect the organ from mechanical injury, insects, fungi, microorganisms and control the transpiration and aeration in the tissue system. They include the *epidermis* and *periderm* tissues which are the primary and the secondary outer protective covering of the plant body (Fig. 2.1). The protective cells are closely pressed together and are usually quite tough. Epidermis of fresh produce usually contain minute valves, called stomata, for exchange of gases. Also cutinization of the outer walls of these cells make them impervious to water. Cutin is a layer of waxy material which usually presents problems

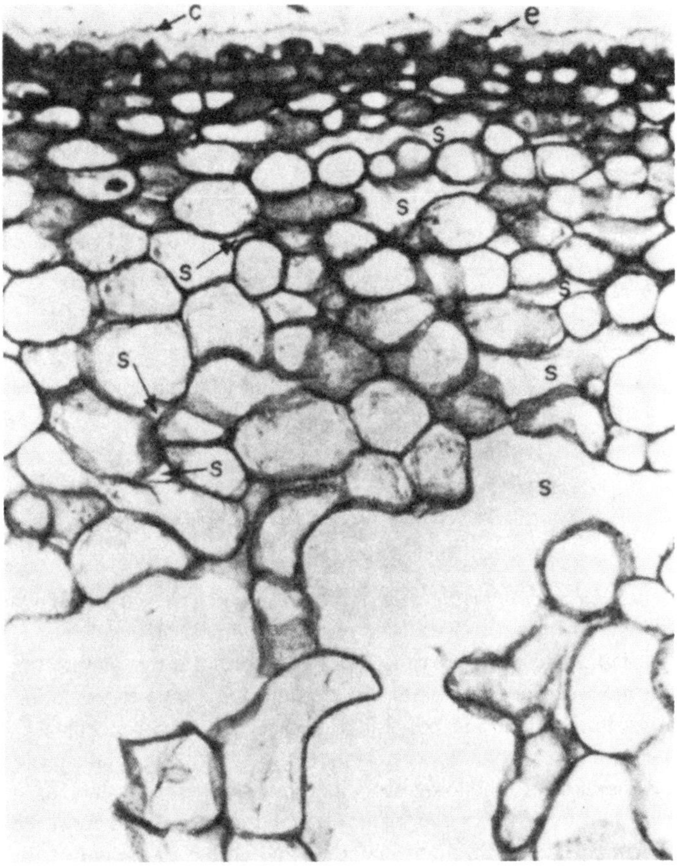

Figure 2.1 Transition from skin to flesh parenchyma in Granny Smith apples, ×150: c—cuticle, e—epidermis, s—intercellular space (Reeve, 1953)

in mechanical testing, when one attempts to use epoxy resins or other bonding agents on the surface of the intact specimen.

The *conducting or vascular tissues* contain the *phloem* (food conduction) and the *xylem* (water conduction) tissues (Fig. 2.2). The cells are composed of long tubes with their walls made of primarily cellulose and sometimes lignin forming the fibrous material which contributes to stringiness and toughness in some food materials. This part of the structure of conducting tissues may also be classified as supporting tissues.

STRUCTURE AND RETENTION OF WATER

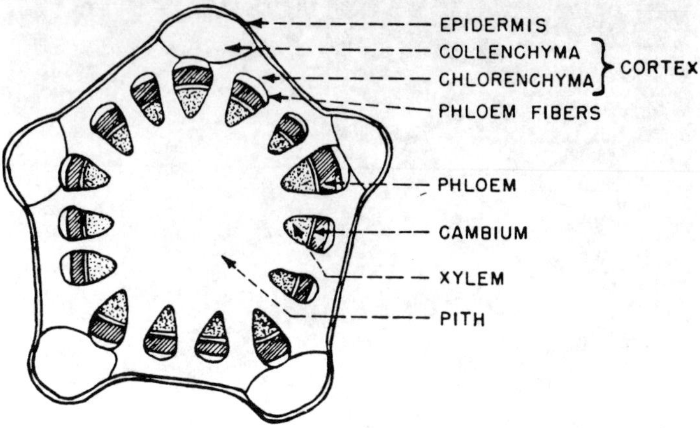

Figure 2.2 Schematic diagram of a young alfalfa stem

The *ground or supporting tissues* are the parenchyma, the collenchyma, and the sclerenchyma tissues. The *parenchyma* cells are the chief type of cells in plant materials appearing in various special forms as the structural units in most of the other parts of the plant body. The parenchyma cells are large, thin-walled and polygonal in shape as shown in Fig. 2.3. The internal structure of parenchyma cells (Fig. 2.4) may contain plastids or may be adapted for the storage of water and reserve foods such as starch. They are living cells, capable of growth and division, and form the bulk of the primary tissues in plant materials and edible portions in fruits and vegetables. The parenchyma cells do not fit tightly together and are often separated by intercellular spaces which may be filled with air or water and constitute as much as 25 percent of total volume of the tissue in such fruits as the apple (Table 2.3). The cementing agent, which may be pectic substances, lignin or other compounds at the middle lamella, hold these cells together to form the parenchyma tissues.

Collenchyma and *sclerenchyma* tissues provide strength and mechanical support for the plant body (Fig. 2.5). The cells in the former are a modified version of parenchyma cells (uneven thickened wall) and are the strengthening tissues in young plant materials. The latter are strengthening elements in mature plant parts with long slender cells distinguished as fiber cells (Fig. 2.6).

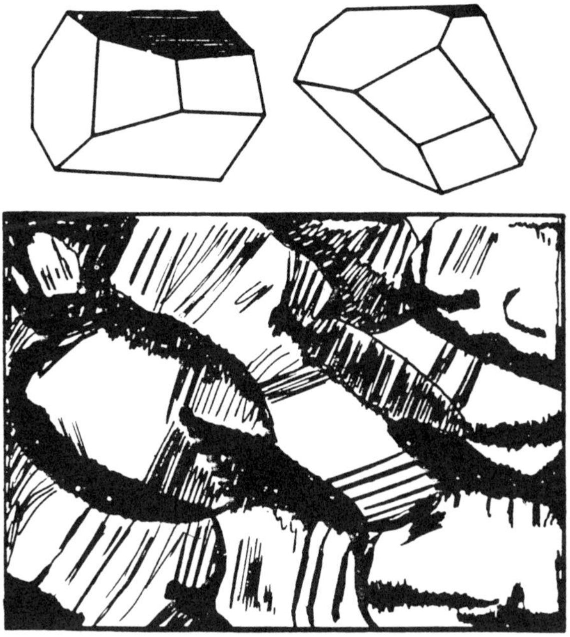

Figure 2.3 Parenchyma cells in edible portion of fruits and vegetables. Above: polygonal shape of the cells. Below: intercellular air shown in dark shadows (Meyer, 1960)

2.2 THE CELL WALL

It is generally agreed that mechanical properties of cell walls in plant materials reflect the mechanical properties of the plant tissues (Falk *et al.*, 1958; Frey-Wyssling, 1952). The elasticity, strength and rigidity of plant tissues, for example, are due to rheological properties of the cell wall (Frey-Wyssling, 1952). These cell walls are composed essentially of *cellulose microfibrils* embedded in an amorphous matrix. The microfibrils are relatively inert but their number and arrangement are largely responsible for the form and structural mechanics of the cell wall. Matrix materials also contribute to structure but are generally considered more reactive than the microfibrils and have been assumed to control rigidity in the cell wall.

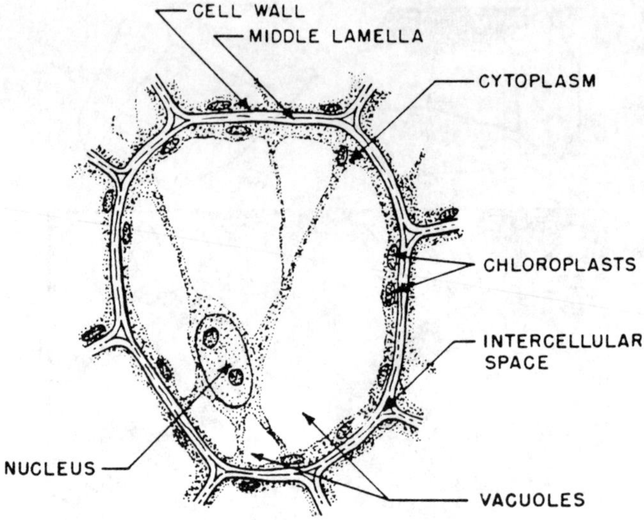

Figure 2.4 Structure and contents of a parenchyma cell (Brook, 1964)

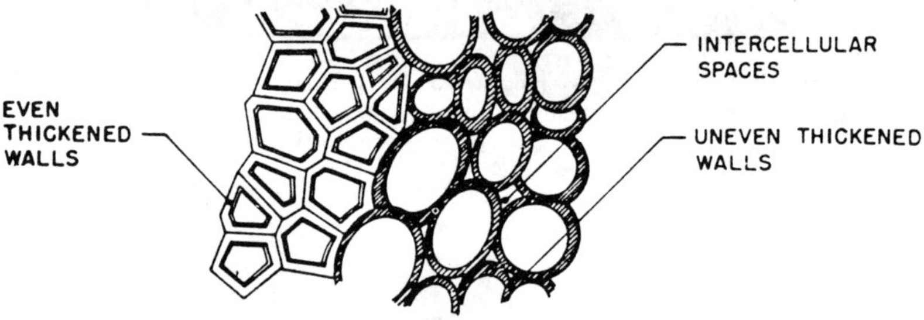

Figure 2.5 Collenchyma tissue (right) and sclerenchyma tissue (left) provide strength and mechanical support (Braungart and Arnett, 1962)

The microfibrils are the real morphological units in the structure of a plant material. Similar fibrillar elements have been found in proteins. Figure 2.7 shows the fibrillar structure of natural cellulose fibre. Greater details of microfibrils are shown in Fig. 2.8 for a ramie fibre. The microfibrils of all cellulosic cell walls have about the same diameter (250–300 Å) and structure (Frey-Wyssling, 1952). They consist of some 25 *micellar strands* with about 2500 cellulose chains. However, the arrangement of the microfibrils is not equal in all cell walls. Furthermore, the crystallinity

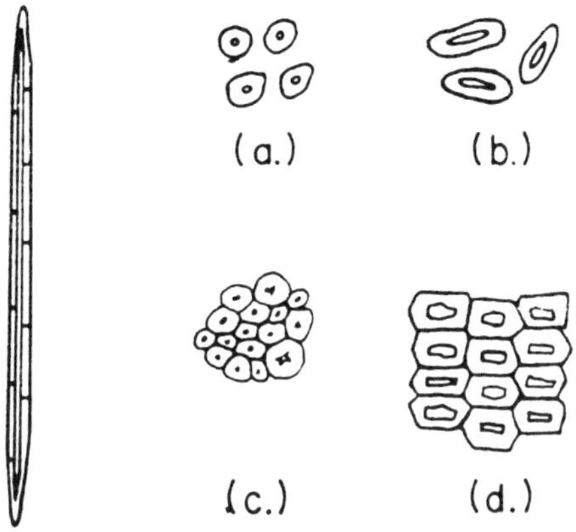

Figure 2.6 Single fibre cell (left) and cross-section of fibre cells in cotton (a), ramie (b), flax (c) and spruce wood (d) (a, b, c, d from Frey-Wyssling, 1952)

Figure 2.7 Fibrillar structure of natural cellulose fibre; dark regions represent intercellular spaces (Meredith, 1956)

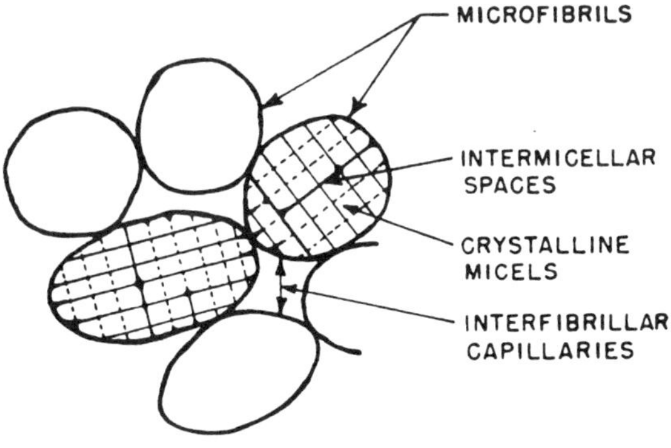

Figure 2.8 Cross-section of microfibrils in ramie fibre (Frey-Wyssling, 1952)

of the cell walls may be limited to the homogeneous crystal lattice within the micellar strands. The fraction of the amorphous cellulose holding these regions together may reach as much as 30 percent (Frey-Wyssling, 1952). These areas of crystallization and amorphous cellulose have been studied by x-ray techniques. It is in the amorphous portions of the microfibrils that water absorption and swelling takes place. Also this part of the cellulose is believed to be flexible and capable of distortion without breaking. This property gives strength to plant tissues without rendering them rigid.

While the microfibrilar component of the cell wall consists basically of cellulose, the amorphous matrix component of the wall is composed predominantly of semicellulose and pectic substances (Setterfield and Bayley, 1961).

Semicellulose is an alkali-soluble, nonfibrous compound which is not very well defined. It is found mostly in wood but its presence in alfalfa hay, sugar cane, cornstalk, oat hulls and other agricultural materials has been reported (Meyer, 1960).

Pectic substances are basically straight chain polymers of considerable interest in foods of plant origin, particularly fruits and vegetables. In young plant materials pectic substances are in the form of protopectin which is water-insoluble. Upon aging, the water-insoluble protopectin changes to water-soluble pectin capable of forming gels with sugar and acid under suitable conditions. The "mealiness" of over-ripe fruits is said to be partly attributed to this change in pectic substances. Figure 2.9 shows the changes in hardness of apples accompanying the changes of water-insoluble protopectin to water-soluble pectin. Note the fall of protopectin and the rise of water-soluble pectin until late in the storage period when a reversal of this trend occurs.

Although the exact location of pectic substances in a tissue is still a subject of dispute, it is agreed that they occur in the *middle lamella* between cells, acting as part of the cementing agent, as well as in the cell wall. In older plant materials, the cementing agent may, in addition to pectic substances, contain *lignin*. The presence of lignin together with thickening of cellulose layer in cell walls render the plant material woody and tough.

2.3 THE CELL CONTENTS

Within the cell wall is the *protoplasm* differentiated into various regions of cytoplasm, nucleus, and various inclusions (Fig. 2.4). The *cytoplasm*

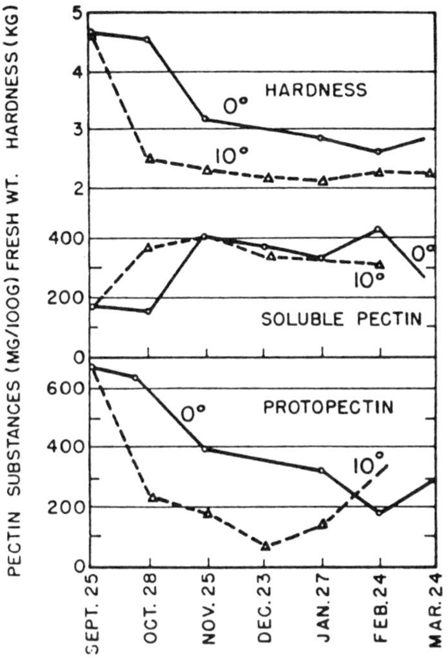

Figure 2.9 Changes in tissue firmness of canada apple as influenced by pectic substances (Meyer, 1960)

constitutes the main mass of the protoplasm with such rheological properties as non-Newtonian viscous behavior, elasticity, swelling and shrinkage, a measurable rigidity even though a fluid, and tensile strength. The *nucleus* directs much of the activity of the living cell including cell division, cellular metabolism and transmitting the heritable characters of the organism. The various inclusions in the protoplasm include numerous small bodies called *plastids,* such as chloroplasts important in the process of photosynthesis, and *vacuoles,* made up of droplets of solutions referred to as the "*cell sap*." In young plants, the vacuoles are small and numerous. As the cell grows older, the size of the vacuoles increase and join together to form often only one large vacuole in a mature cell. While the elasticity of the cell walls is recognized as the main factor responsible for elasticity of tissues, the cell sap is responsible for exerting a pressure called "*turgor pressure*" on the cell walls and keeping them in a state of elastic stress.

The combined effects of this hydrostatic pressure (turgor pressure) of the cell contents with the elastic cell walls determine the viscoelastic properties of the tissues in the biological material. The effect of turgor pressure on stress and strain in the walls of the cells with simple geometric shape is discussed by Frey-Wyssling.

2.4 RELATIONSHIP BETWEEN TURGOR PRESSURE AND TISSUE RIGIDITY

The cellular structure of biological materials is such that it would be difficult to deduce the rheological properties of the whole material from the properties of the structural units, namely the cells. Even a simple uni-directional stress has to be transmitted through an irregular network of intercellular spaces and a random cellular arrangement. However, by assuming some simple and idealistic cellular models, it has been possible to explain certain relationships related to mechanical properties of tissues.

Nilsson and co-workers (Nilsson *et al.*, 1958) showed that the dependence of elastic modulus in potato tissue upon turgor pressure in the cells, found experimentally, can be explained by a simple model in which the liquid-filled cells are assumed to be bounded by their elastic membranes. If all cells were spherical with radius r, the cell structure could be represented by Fig. 2.10 where in the cubical packing each sphere is assumed to be in contact with six neighboring cells. It is further assumed that the turgor pressure is equal in all cells for the model; the cell fluid cannot penetrate the cell walls while the tissue is being stretched; the cell walls are homogeneous, isotropic and elastic following the Hooke's law; and in the absence of external forces, all cell walls are free from internal stresses when the turgor pressure of the cell fluid is zero.

With these assumptions, the theory of elastic membrane was applied to derive the following linear relationship between the elastic modulus, E, of the plant tissue and the turgor pressure, p, in the cell fluid.

$$E=3.6p+2.5\times 10^7 \text{ dynes/cm}^2$$

When the more realistic cell structure (isodiametric polyhedra) was considered in the analysis, it was found that the elastic modulus was only slightly sensitive to the assumed cell form and to the direction of the applied force (Fig. 2.10).

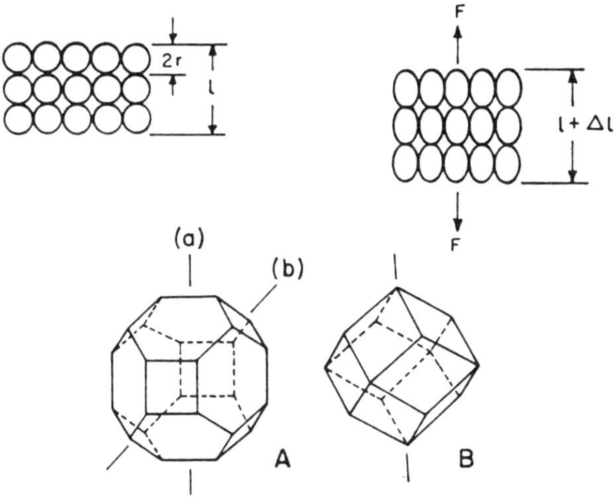

Figure 2.10 Top: The simplified model of potato parenchyma assumed in deriving the relationship between rigidity and turgor pressure. The more realistic form of the cell structure shown below showed little difference in the calculated elastic modulus (Nilsson, 1958)

Experimental data (Falk *et al.*, 1958) has shown good agreement with theoretical results. Elastic modulus of the tissue was obtained by determining the elastic stretching of a specimen under a given force. The specimens were brought to osmotic equilibrium in solutions of different concentrations of mannitol and then taken out and dipped in paraffin oil to prevent osmotic changes during the modulus determinations. Turgor pressure was obtained by taking the difference between the osmotic value of the cell contents and that of the surrounding solution. Since the change in length of the cell is the same in all directions (isotropic cell wall), the change of volume can be calculated from the change in length and the osmotic value of the cell contents calculated using the Boyle-van't Hoff law (Tamiya, 1938).

2.5 DIMENSIONAL CHARACTERISTICS OF CELLS AND INTERCELLULAR SPACES

Information on size, shape and volume of the cells and the physical nature and characteristics of the intercellular spaces is often necessary for understanding the mechanism of flow and deformation and for interpreting the experimental data on mechanical and rheological properties. Such information can be found in source materials on plant physiology or periodicals concerned with specific commodities such as "Cereal Chemistry," "Tobacco Science," "American Potato Journal," etc.

The following information (Tables 2.1 through 2.5 and Figs. 2.11 and 2.12) gathered for apple fruits is presented here as examples of the type of data which may be found in the literature. Cell measurements were made on parenchyma tissues histologically prepared from the flesh of the fruit. Estimation of volume of average cell, and cell wall surface were based on average diameter of cells measured microscopically with a calibrated eyepiece micrometer. The estimation was based on the assumption that the 14-sided parenchyma cells are nearly spherical. The total intercellular space per unit volume of tissue was made by vacuum infiltration of the tissue submerged in water. Knowing the percentage of intercellular space, the volume of the tissue actually occupied by cells was obtained as a percentage of total tissue volume. In determining this volume it was assumed that spheres of cells packed into the volume have no intercellular spaces. Models approaching this condition have been constructed with foam bubbles or by compressing soft lead shot in a steel cylinder (Marvin, 1939).

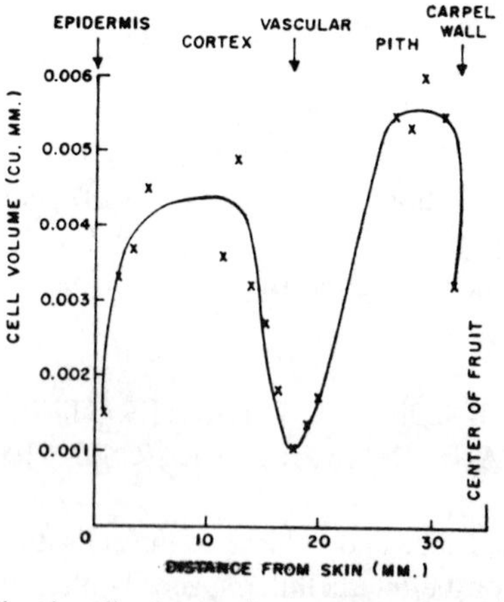

Figure 2.11 Gradient in cell volumes along an equatorial radius in a mature apple (Bain, 1951)

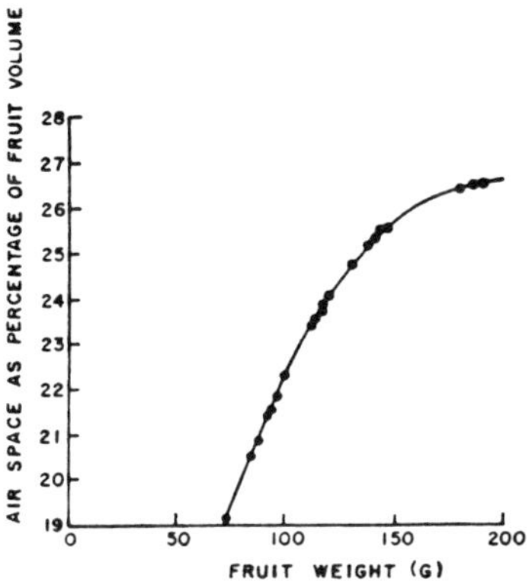

Figure 2.12 Relationship between air space and fruit weight in Granny Smith apples (Bain, 1951)

Table 2.1 Dimensions (in microns) of parenchyma cells in different tissue regions of apples (Reeve, 1953)

Variety	Skin	Outer flesh 1/8″–1/4″ (depth)	Inner flesh 3/8″–5/8″ (depth)	Core (length)	Core (width)
	Av SD	AV SD	Av SD	Av SD	Av SD
Newtown Pippin	76±16	182±42	210±39	261±58	136±29
Winesap	67±21	194±42	214±35	266±42	142±33
Delicious	67±18	161±27	170±30	266±57	124±25
Rome Beauty	67±24	193±34	207±29	327±72	142±21
Gravenstein	85±27	221±43	235±58	494±95	157±32

Av=average
SD=standard deviation

Table 2.2 Dimensions (in microns) of intercellular spaces of middle flesh region of different apples (Reeve, 1953)

Variety	Width Av SD	Length Av SD	Greatest lengths observed
Delicious	269 ± 93	487 ± 154	1,000
Newtown Pippin	257 ± 72	485 ± 192	1,200
Winesap	275 ± 74	480 ± 157	1,200
Ripe Gravenstein	350 ± 108	665 ± 274	1,500
Rome Beauty	257 ± 75	562 ± 231	1,800
McIntosh	300 ± 75	590 ± 250	2,000

Table 2.3 Amount of intercellular space in different tissue regions of apples (Reeve, 1953)

Variety and character of fruit	Percent of tissue occupied by intercellular space as determined by vacuum infiltration with water		
	Calyx-end flesh	Stem-end flesh	Middle portion of flesh
Delicious, ripe, not mealy	15	18.0	21.0—22.0
Gravenstein, ripe	17.6	20.1	25.4—27.0
Newtown Pippin, ripe	13.8—14.2	16.5	21.5—23.6
Rome Beauty, ripe	13.8	17.9	22.0—23.6
Winesap, ripe	15.3	17.7	19.8—20.7

Table 2.4 Size and number of cells and cell wall surface per unit volume of apple flesh parenchyma (Reeve, 1953)

Variety and tissue	Average cell diameter (microns)	Average wall area per cell (microns)^2	Average cell volume (microns)^3	Approx. per cent occupied by cells	Cells per unit volume (1 cc.)	Total surface of cell wall area per unit volume (1 cc.) (microns)^2
Newtown Pippin						
Outer flesh	182	1.04×10^5	3.16×10^6	80	2.53×10^5	2.64×10^{10}
Inner flesh	210	1.39×10^5	4.85×10^6	80	1.65×10^5	2.3×10^{10}
Delicious						
Outer flesh	161	0.814×10^5	2.19×10^6	80	3.66×10^5	2.99×10^{20}
Inner flesh	170	0.91×10^5	2.57×10^6	80	3.11×10^5	2.83×10^{10}
Gravenstein						
Outer flesh	221	1.53×10^5	5.65×10^6	75	1.32×10^5	2.03×10^{10}
Inner flesh	235	1.73×10^5	6.8×10^6	75	1.1×10^5	1.9×10^{10}
Rome Beauty						
Outer flesh	193	1.17×10^5	3.71×10^6	78	2.11×10^5	2.47×10^{10}
Inner flesh	207	1.35×10^5	4.6×10^6	78	1.69×10^5	2.29×10^{10}
Winesap						
Outer flesh	194	1.19×10^5	3.8×106	80	2.10×10^5	2.50×10^{10}
Inner flesh	214	1.44×10^5	5.13×10^6	80	1.56×10^5	2.24×10^{10}

Table 2.5 Relationship between cell number, cell volume and weight of Australian apple variety Granny Smith 149 days after full blossom (Bain, 1951)

Weight of fruit (g.)	Volume of tissue (cc.)	Mean cell volume (cu.mm. $\times 10^4$	Cell number ($\times 10^{-6}$)
142.02	129	36	35.9
161.51	147	48	30.6
161.41	146	38	38.5
163.10	148	39	38.0
167.46	152	39	39.0
172.31	156	37	42.2
174.15	158	39	40.5
177.50	162	36	45.0
180.20	164	47	34.9
185.14	168	34	49.5
188.20	171	40	42.7
189.50	173	36	48.0
200.10	182	39	46.7
201.07	183	45	40.7
202.85	185	32	57.9
205.72	187	43	43.5
222.10	202	34	59.5
229.16	217	33	65.7
249.91	297	40	56.5

2.6 GROSS STRUCTURE AND CHEMICAL COMPOSITION OF SELECTED AGRICULTURAL PRODUCTS

Many food and agricultural products are multicomponents, multiphase systems, and partially solved in gel condition. The fibrous components constitute the solid phase, holding a liquid phase, dispersed or continuous. When oil and fat droplets are present, even more liquid phases exist. The internal structure of these systems is very difficult to characterize. From a morphological point of view, organization of the plant and animal materials as related to physical handling and processing is given by illustrations in this section.

From an engineering point of view, we are concerned with the gross structure and the role that each structural component plays in response to internal and external forces of mechanical, thermal, diffusional, or electromagnetic sources. Consider the list of terms and definitions related to the anatomy of raw materials of plant or animal origins. We can see that a physical function can be assigned to each structural component. For example, the mechanical strength of tissues can be affected by structural as well as physical factors such as water, heat, light, and electric current.

For the benefit of the readers who are unfamiliar with the morphology and nomenclature of natural food materials, the gross structure and nomenclature for selected agricultural products are presented at the end of this chapter.

This descriptive nomenclature is followed by a table summarizing the constituents of selected food materials which should be helpful in understanding the physical behavior of the product. In this table water content values are on the wet basis, obtained by drying in vacuum at 100°C. The term "protein" is generally considered to be total nitrogen times the factor 6.25. Fat and oil content determination is based on the ether extract method. Nitrogen-free extract represents the so called "carbohydrates" such as starch, sugars, dextrins, and some unknown substances. Sugar and starch have also been reported separately. Acids, though widely removed from carbohydrates, are also included in nitrogen-free extracts. Ash is assumed to contain all of the mineral constituents.

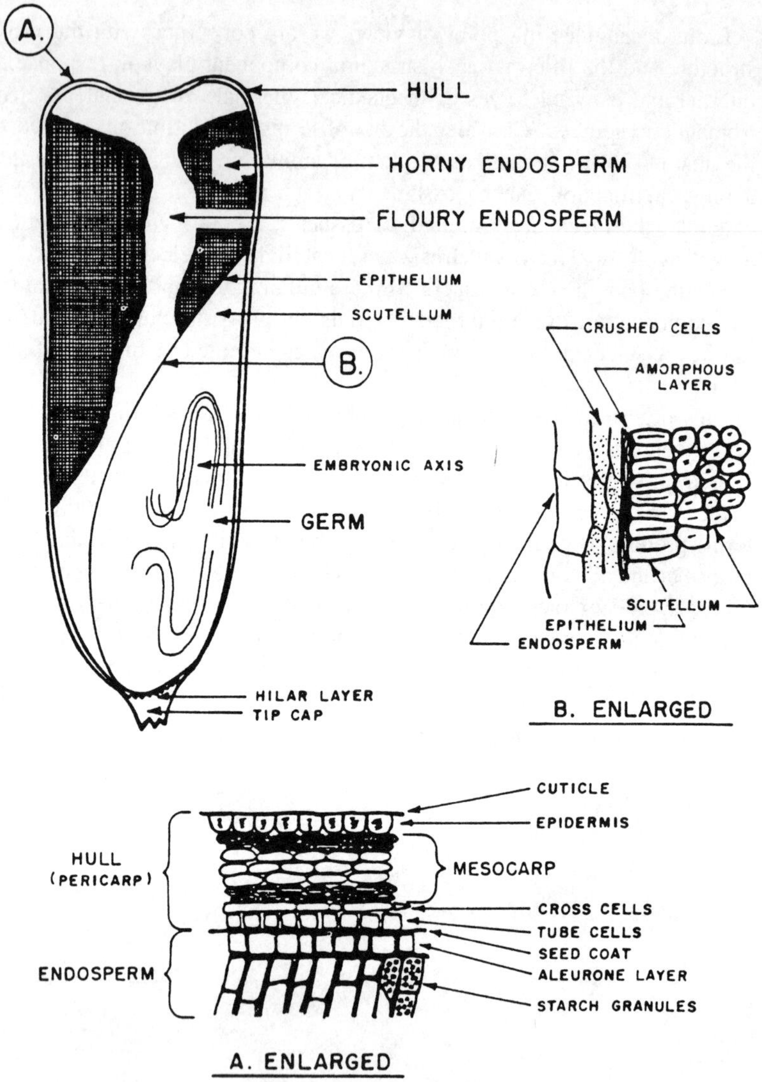

Figure 2.13 Structural elements of the corn kernel (White, 1966)

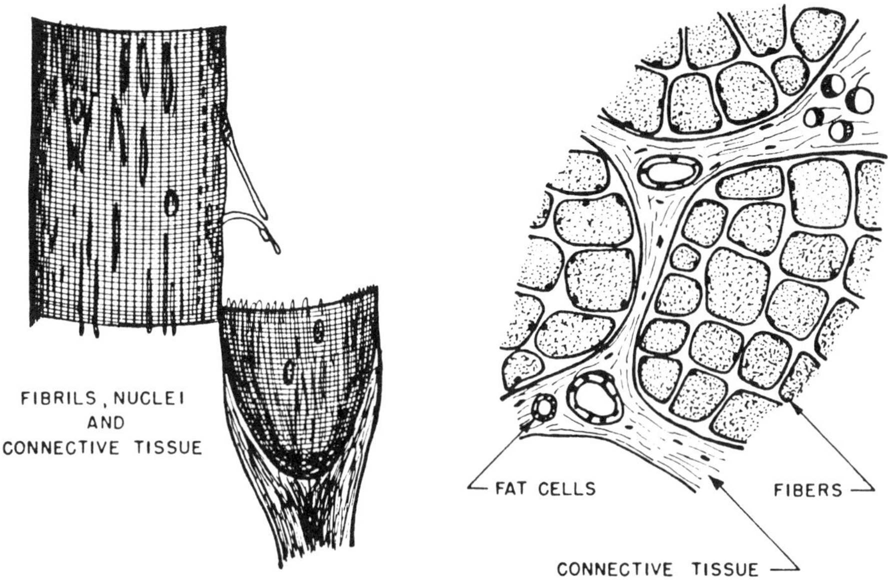

Figure 2.14 Microscopic structure of beef muscle × 160 (A typical fiber is about 50 μ in diameter and many times as long) (Winton and Winton, 1937)

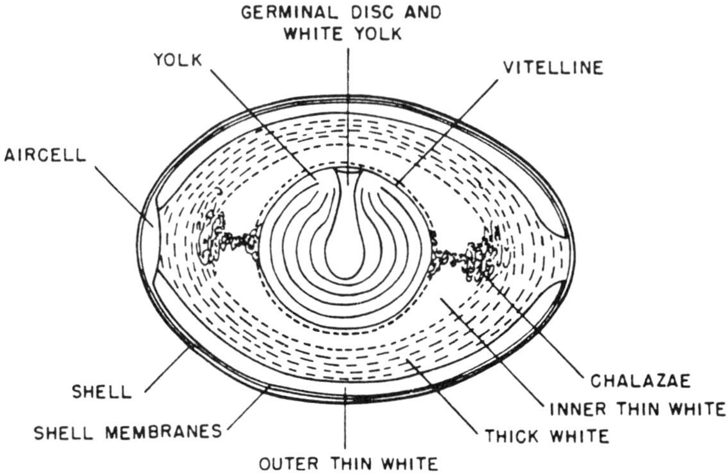

Figure 2.15 Structure of an avian egg (Romanoff and Romanoff, 1949)

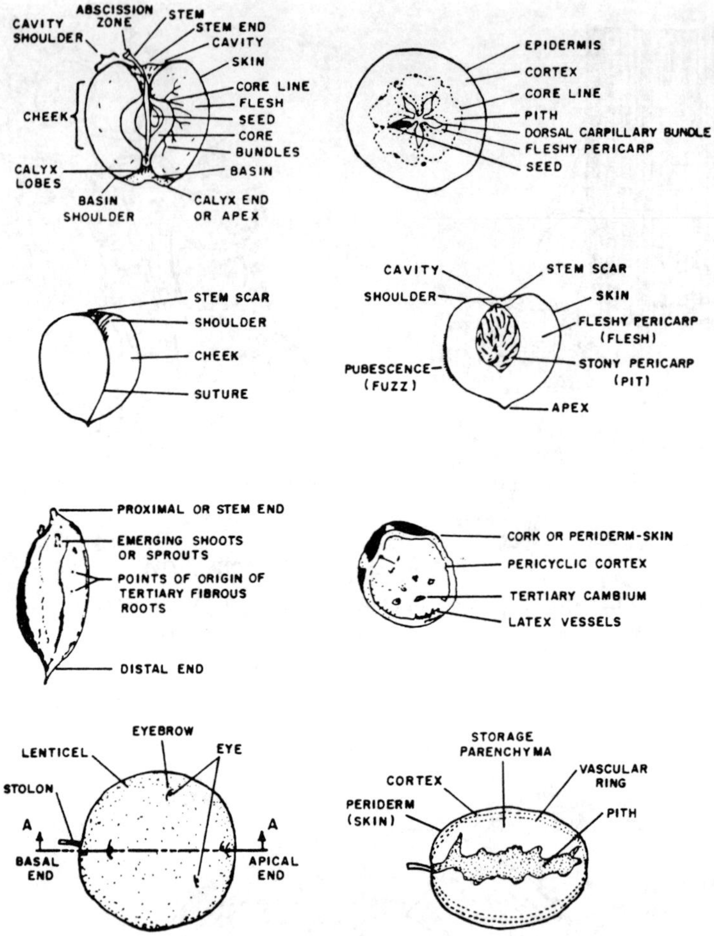

Figure 2.16 Morphological parts of apple, peach, sweet potato, white potato (Mohsenin, ed., 1965)

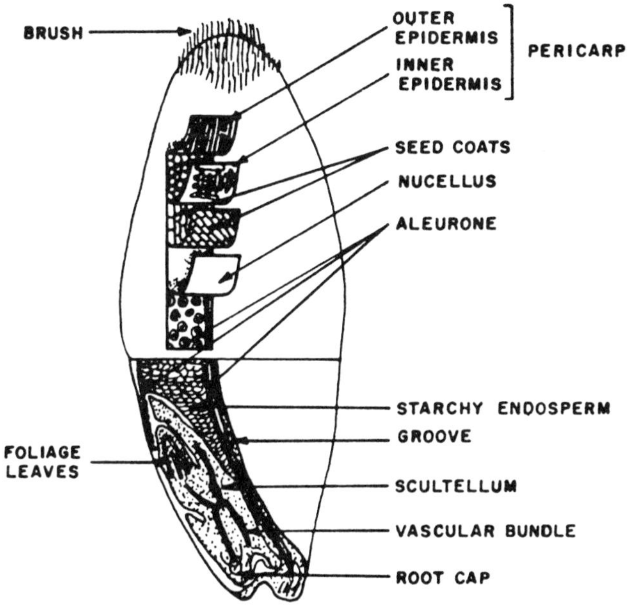

Figure 2.17 Structure of the wheat grain (Robbins, Weier, Stocking, 1957)

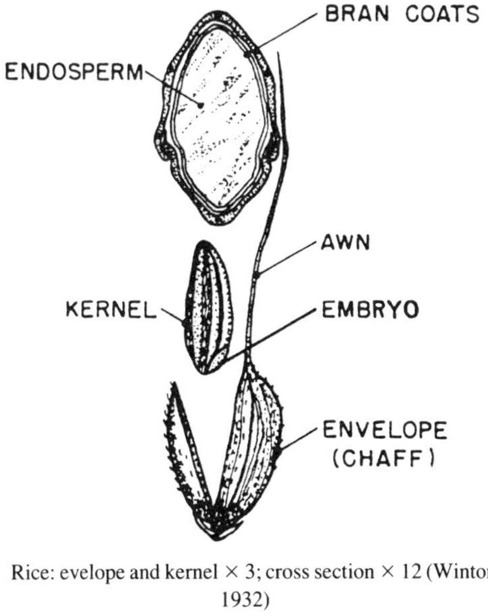

Figure 2.18 Rice: evelope and kernel × 3; cross section × 12 (Winton and Winton, 1932)

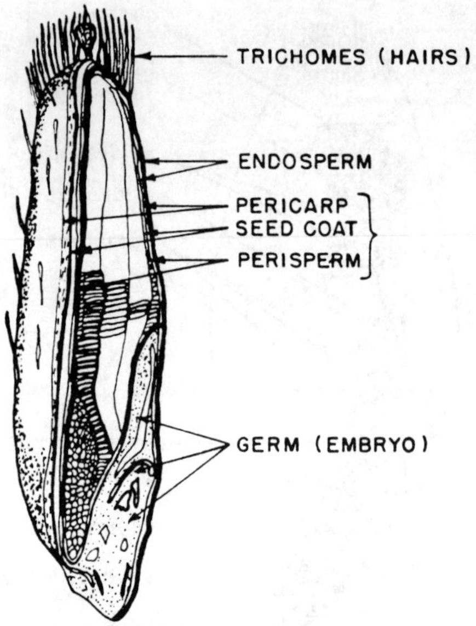

Figure 2.19 Oats kernel with hull removed (Parker *et al.*, 1952)

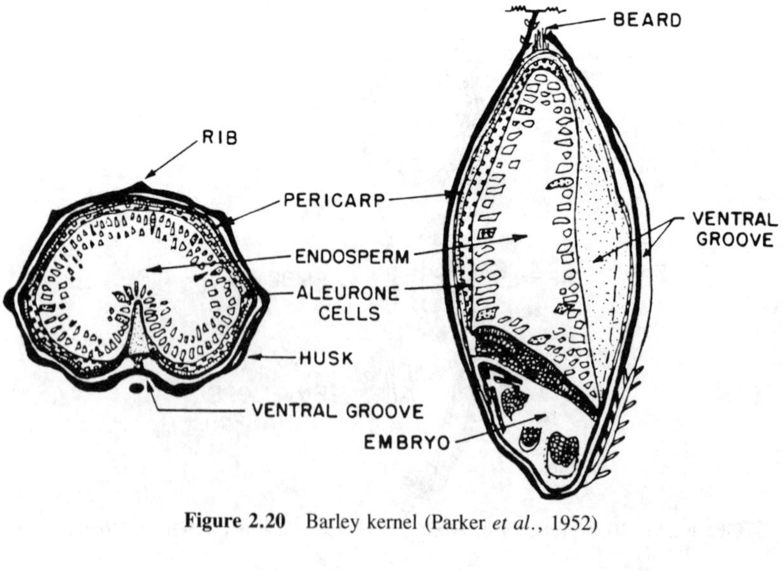

Figure 2.20 Barley kernel (Parker *et al.*, 1952)

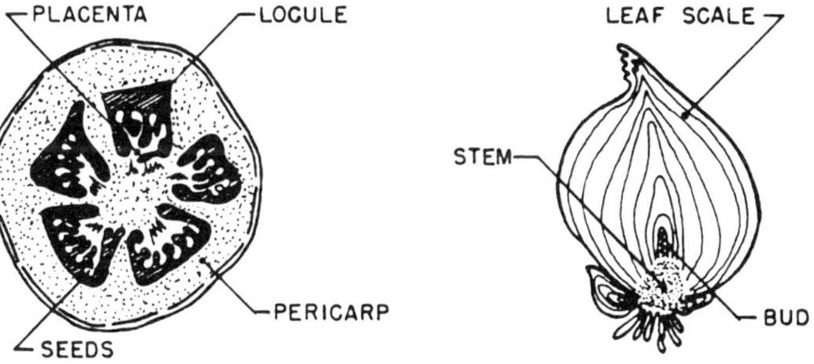

Figure 2.21 Tomato (left), onion (right) (Braungart and Arnett, 1962)

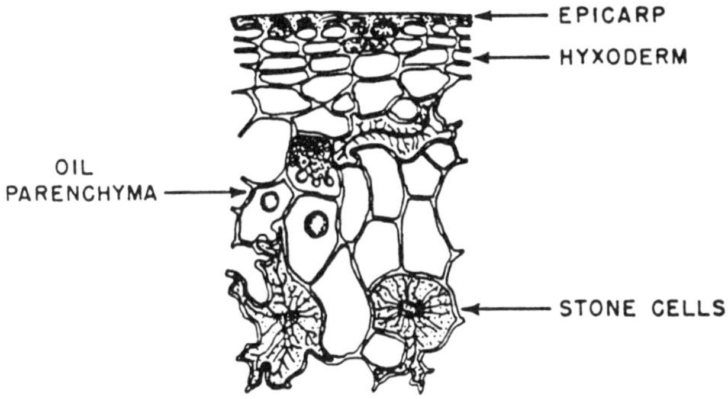

Figure 2.22 Olive: outer pericarp in cross section × 160 (Winton and Winton, 1962)

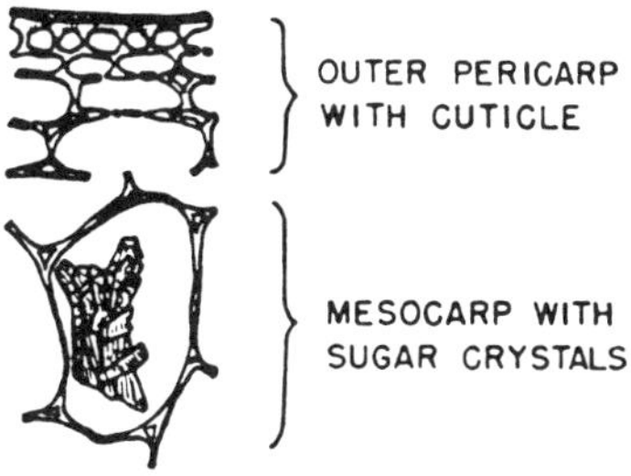

Figure 2.23 American grape × 160 (Winton and Winton, 1935)

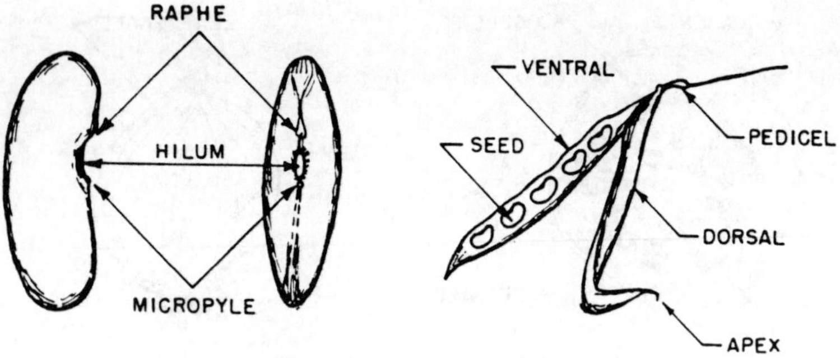

Figure 2.24 Bean seed (Robbins, Weier, Stockings, 1957); bean pod (Mohsenin, ed., 1965)

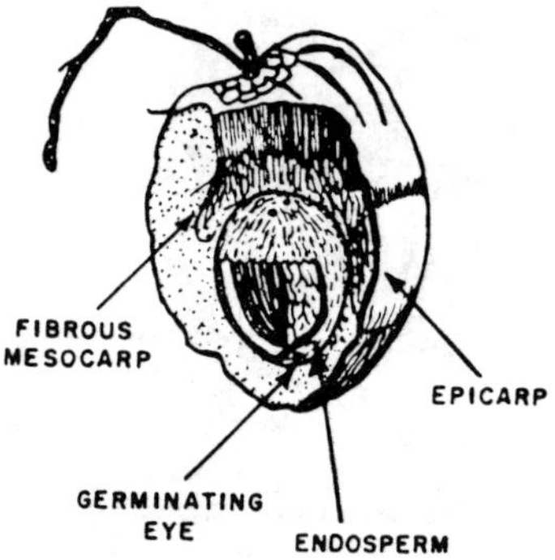

Figure 2.25 Cocoanut fruit × 1/5 (Winton and Winton, 1932)

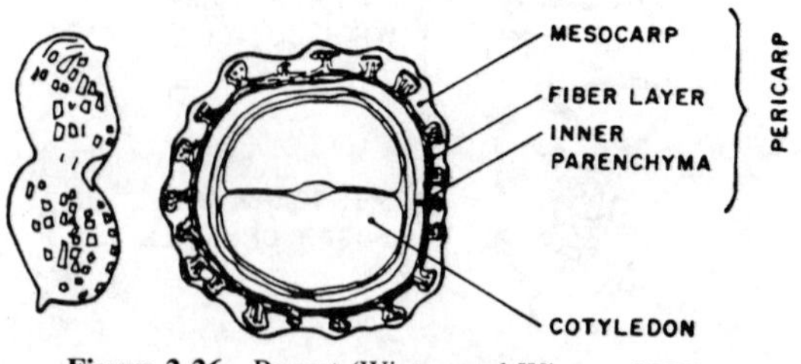

Figure 2.26 Peanut (Winton and Winton, 1932)

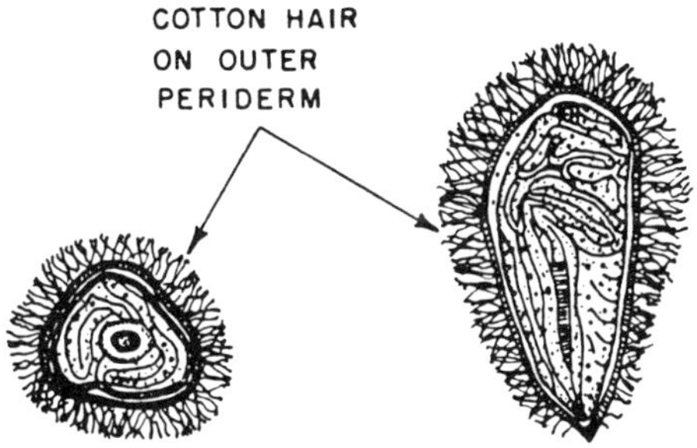

Figure 2.27 Cottonseed: cross-section (left) and longitudinal section (right) (Winton and Winton, 1932)

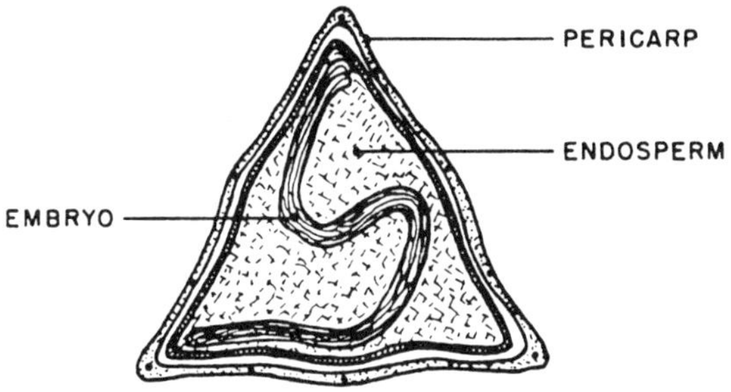

Figure 2.28 Buckwheat × 16 (Winton and Winton, 1932)

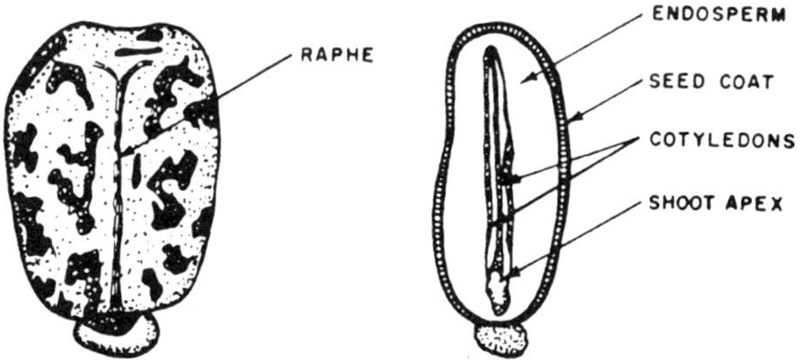

Figure 2.29 Castor bean seed (Robbins, Weier, Stocking, 1957)

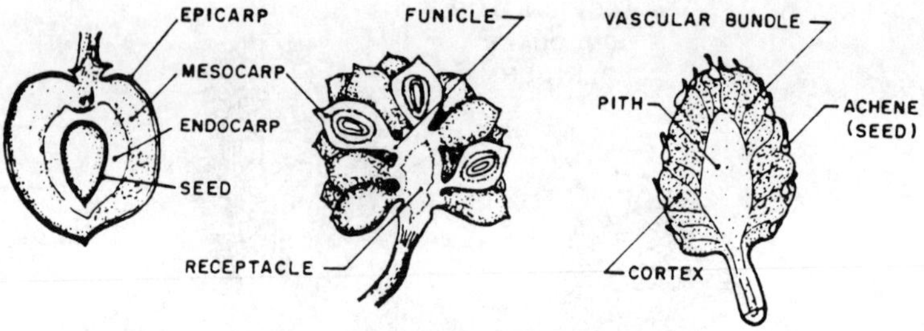

Figure 2.30 Cherry and blackberry (Brook, 1964) strawberry (Braungart and Arnett, 1962)

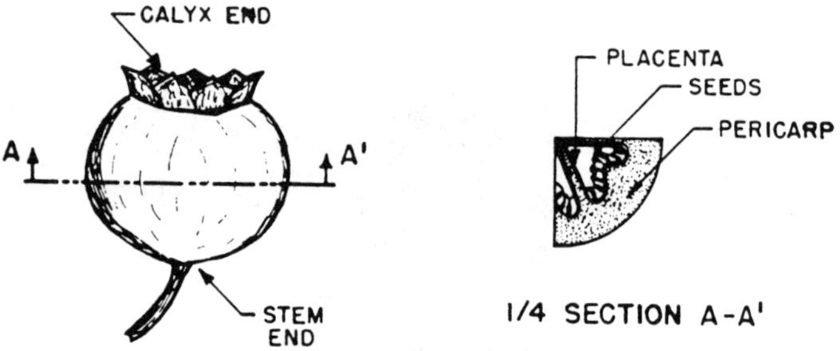

Figure 2.31 Blueberry (Mohsenin, ed., 1965)

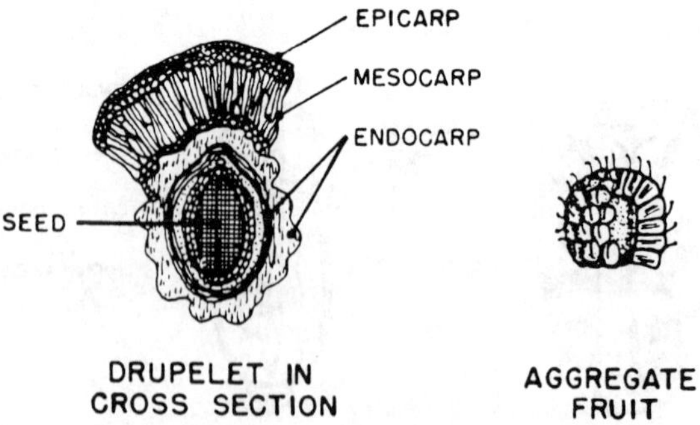

Figure 2.32 Raspberry (Winton and Winton, 1932)

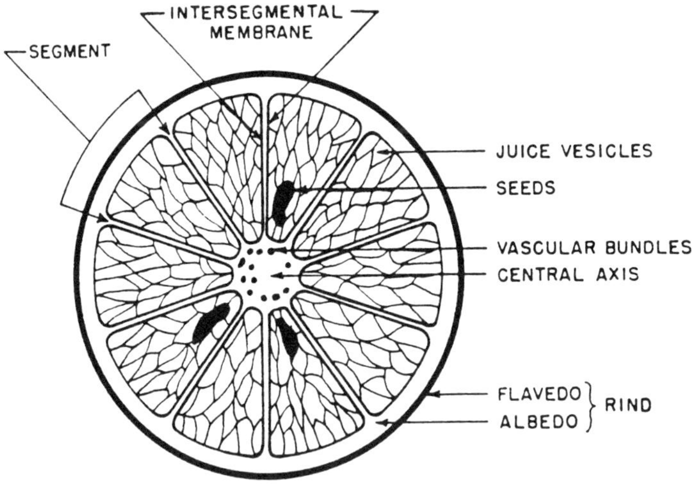

Figure 2.33 Orange (Agriculture Handbook No. 98, 1956 and courtesy James Suole, University of Florida)

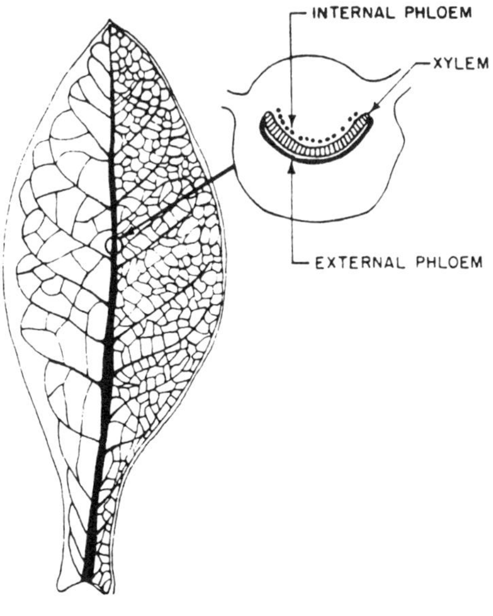

Figure 2.34 Tobacco leaf (Avery, 1933)

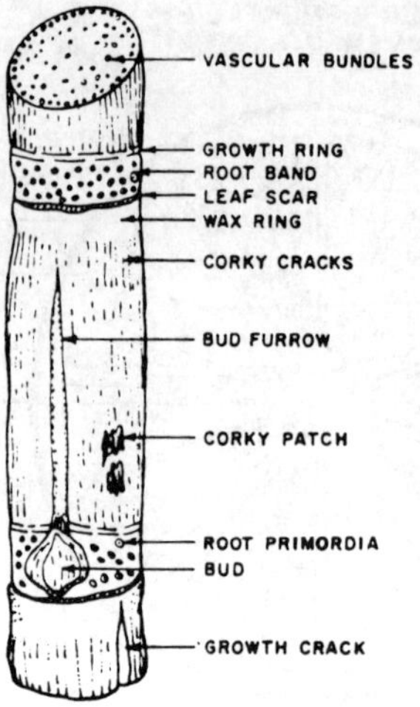

Figure 2.35 Sugar cane (Dillewign, 1952)

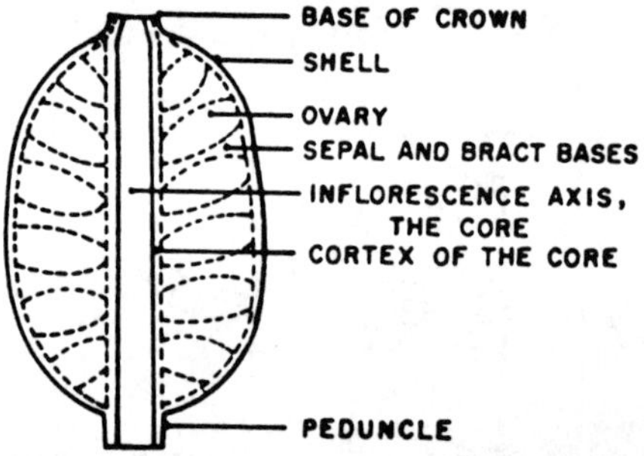

Figure 2.36 Pineapple inflorescence (Okimoto, 1948)

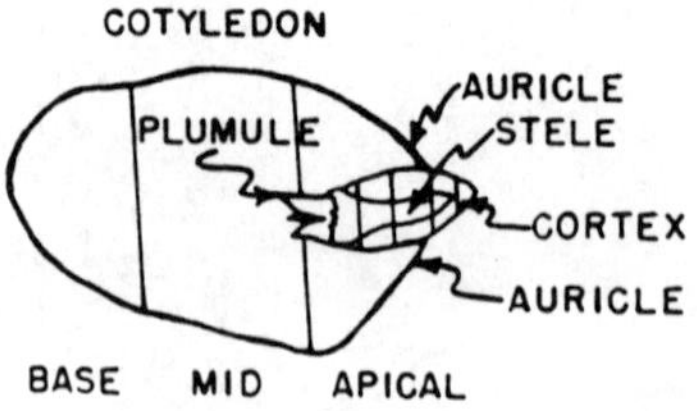

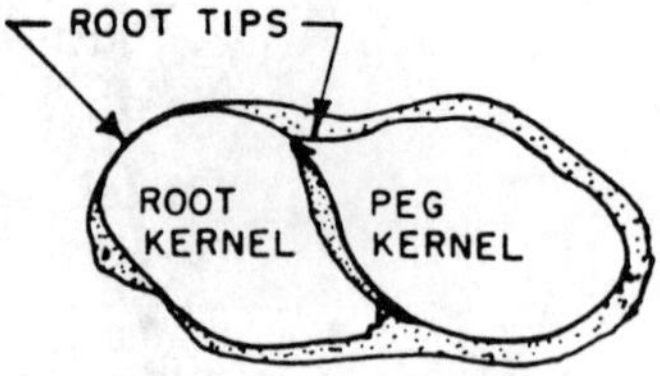

Figure 2.37 Peanut structure. Left: Parts of the embryo. Right: Kernels cradled in the hull (Turner *et al.*, 1965)

Figure 2.38 Photomicrograph (× 240) of cross section of egg shell 20 μ thick. Egg shell membrane has been stripped away (Terepka, 1963)

Table 2.6 Proximate constituent of some plant products[3]

Product	Water	Protein	Fat	N-Fext[1]	Fiber	Ash	Starch	Sugar	Solids[2]	Sucrose	Acids
	(Average values in percent)										
Alfalfa (Hay)	8.36	15.59	1.83	33.38	30.18	10.66					
Alfalfa (green)	71.75	4.84	0.97	12.39	7.39	2.66					
Almond	4.8	21.0	54.9	14.3	3.0	2.0					
Apple	85	0.10				0.27		11.5	16.7	4.00	0.61
Apricot	89.0	0.65				0.51		13.3	12.4	4.3	1.23
Asparagus	93.9	1.83	0.25	2.55	0.74	0.67					
Banana		1.6	0.2		0.50	0.90	3.0	20.2	27.54	8.27	0.41
Barley[3]	9.32	13.39	1.87	76.05	5.64	3.04					
Beets	86.5	1.75	0.10	9.89	0.88	0.88					
Blackberry		0.92				0.59			12.7	0.16	0.91
Black Raspberry		1.02				0.55			14.7	0.10	1.60
Black Walnut	2.5	30.3	57.8	5.8	1.6	2.0					
Blueberry		0.5	3.0	13.0	3.2	0.11			15.0	0.19	0.9
Buckwheat	12.62	10.02	2.24	64.43	8.67	2.02					
Cashew Nut	3.80	9.70	47.15		1.27	2.59	8.90				
Celery	94.5	1.1	0.1	3.3 Total		1.0					
Chick Pea	10.5	22.6	5.0	56.1	3.1	2.9	49.3				
Cocoanut											
(Kernel)	20	6.8	50.6	31.5 Total		1.3					
(Milk)	92.7	0.4	0.15	4.6	0.0	0.80					
Common Bean	89.2	2.3	0.3	7.4	1.9	0.8					
Corn (Cob)	10.68	2.37	0.52	54.89	30.13	1.41					
Corn (Hard)	10.90	10.35	5.00	70.20	1.89	1.49					

Table 2.6 (continued)

	(Average values in percent)										
Product	Water	Protein	Fat	N-Fext[1]	Fiber	Ash	Starch	Sugar	Solids[2]	Sucrose	Acids
Corn (Pop)	10.71	11.22	5.18	69.66	1.76	1.47					
Corn (Sweet)	8.82	11.62	8.13	66.72	2.79	1.92					
Cottonseed		30.29	25.76	33.64	14.37	3.24					
Cranberry		0.35	0.51	10.7	1.2	0.17		3.7	12.2		2.3
Cucumber	95.8	0.70	0.09	1.6	0.65	0.42					
Date	15.4	2.1	2.8	78.4		1.3					
English Walnut	2.5	16.6	63.4	16.1	2.6	1.4					
Grapefruit	88	0.6	0.14		1.9	0.5		2.8	11.0	3.1	1.3
Lemon	80	1.04	1.0	15.0	0.08	0.51	0.20	2.0	10.0	0.58	7.8
Lettuce	91.5	1.82	0.6	3.77	1.09	1.22					
Lima Bean	10.4	18.1	1.5	65.9 Total		4.1					
Lime	85	0.78	3.5		0.07	0.66		1.6	9.7	0.00	6.6
Linseed	7.09	24.75	37.28	27.62 Total			3.27				
Maize	12.36	12.11	3.63	68.08	2.39	1.43					
Olive (Pickled)	58.0	1.1	25.9	11.6 Total			1.7				
Onion	85.3	2.3	0.22	10.80	0.76	0.68					
Orange	86.1	1.2	0.4		2.00	0.41		7.1	10.0	4.0	1.6
Peach		0.60				0.65			13.1	7.0	0.6
Pear	82	0.56	0.60	11.5	2.7	0.20			3.1	4.0	0.26
Pineapple		0.50	1.1	13.5	0.20	0.53	5.0		12.5	7.59	1.06
Plum	76.6	1.14				0.77		11.5	12.0	0.11	0.35
Podded Pea	88.6	3.3	0.10	6.10	1.06	0.75					
Red Clover	70.8	4.4	1.1	13.5	8.1	2.1					
Red Clover (Hay)		15.3	3.9	45.8	27.8	7.2					

Table 2.6 (continued)

Product	Water	Protein	Fat	N-F ext[1]	Fiber	Ash	Starch	Sugar	Solids[2]	Sucrose	Acids
	(Average values in percent)										
Red Raspberry		1.02				0.55			14.7	0.10	1.60
Rice	12.6	10.3	1.4	73.3	1.0	1.0					
Rice Straw	8.97	4.72	1.87	32.21	32.25	20.00					
Rye	11.59	12.03	1.84	72.64	1.67	1.86					
Shelled Peanut	4.9	29.1	48.8	18.1	2.00	2.3					
Sorghum	12.36	12.11	3.63	68.08	2.39	1.43					
Sour Cherry		1.26				0.6		11.0	20.5	0.46	0.73
Soybean	10.80	33.98	16.85	28.89	4.79	4.69					
Strawberry	86.5	0.80				0.66		5	10.9	0.46	1.21
Sunflower	6.88	15.19	28.29	17.36	28.54	3.20					
Sweet Cherry		1.26				0.6		11.0	20.5	0.46	0.73
Sweet Corn	75.4	3.1	1.1	19.7 Total		0.7					
Timothy (Green)	61.6	3.1	1.2	20.3	11.8	2.1					
Timothy (Hay)		8.0	3.1	52.8	30.7	5.4					
Tomato	93.0	0.85	0.05		0.32	0.60					
Wheat	10.52	11.87	2.09	71.90	1.79	1.83					
Whole Oats	10.00	13.76	4.38	66.29	12.20	3.96					
Winter Squash	88.3	1.4	0.5	9.0	0.8	0.8					

[1]N-F ext = nitrogen free extract

[2]Some values are for juice

[3]Winton and Winton, 1935; Jacobs, 1951, 1958

Table 2.7 Proximate constituents of some animal products

	(Average values in percent)							
Product	Water	Protein	Fat	Ash	Total solids	Solids not fat	Lactose	Salt
Dairy: (Jacobs, 1958)								
Milk	87.3	3.42	3.67	0.73	12.69	8.77	4.78	—
Butter	13.90	1.18	82.41	—	—	—	—	2.51
Cheese (cheddar)	36.8	23.7	33.8	5.6	—	—	—	—
Meat:								
Beef (Meyer, 1960)	75	22		5.1	1.1			
Pork (Meyer, 1960)	41.9	10.8		44.8	2.1			
Lamb (Winton and Winton, 1937	58.6	17.8		22.6	1.0			
Chicken (Winton and Winton, 1937)	74.8	21.6		2.5	1.1			
Fish:								
Cod	82.6	16.5		0.4	1.2			
Salmon	64.6	22		12.8	1.4			
White fish	69.8	22.9		6.5	1.6			
Egg (chicken): (Meyer, 1960)								
Whole	73.7	14.8	10.5	1.6				
White	86.2	13.0	0.2	0.6				
Yolk	49.5	16.1	33.3	1.1				

	(Average values in percent)			
	Organic matter	Calcium carbonate	Magnesium carbonate	Calcium phosphate
Eggshell (chicken) (Winton and Winton, 1937)	4.2	93.7	1.3	0.8

2.7 SELECTED BIBLIOGRAPHY ON STRUCTURE OF PLANT AND ANIMAL PRODUCTS

Bourne, G. H. ed. 1960. *The Structure and Function of Muscle, Vol. 1, Structure*. Academic Press, New York.

Bower, F. O. and C. W. Wardlaw. 1947. *Botany of the Living Plant*. Macmillan and Company, London.

Bradbury, O., I. M. Cull and M. M. MacMasters. 1956. Structure of the Mature Corn Kernel. *Cereal Chemistry* 33:329–391.

Briskey, E. J., R. G. Cassens, and J. C. Troutman, ed. 1966. *The Physiology and Biochemistry of Muscle as a Food*. The University of Wisconsin Press. Madison.

Esau, K. *Plant Anatomy*. 1953. John Wiley and Sons, Inc., New York.

Esau, K. *Anatomy of Seed Plants*. 1960. John Wiley and Sons, Inc., New York.

Galston, A. W. 1961. *The Life of the Green Plant*. Prentice-Hall, Inc. Englewood Cliffs, New Jersey.

Giese, A. C. 1957. *Cell Physiology*. W. B. Saunders Company, Philadelphia.

Hayward, H. E. 1938. *The Structure of Economic Plants*. The Macmillan Company, New York.

MacMasters, M. M. 1953. Grain Structure and Grain Storage. USDA, AIC Bull. 348.

MacMasters, M. M. 1962. Important Aspects of Kernel Structure. *Trans. of the ASAE* 5(2):247–249.

Miller, E. C. 1938. *Plant Physiology*. McGraw-Hill Book Company, Inc., New York.

Porter, C. L. 1959. *Taxonomy of Flowering Plants*. W. H. Freeman and Company, San Francisco.

Stiles, W. and W. Leach. 1952. *Respiration in Plants*. Methuen's Monographs on Biological Subjects. John Wiley and Sons, Inc., New York.

Terepka, A. R. 1963. Structure and Classification in Avian Egg Shell. *Experimental Cell Research* 30:171–182.

Tukey, H. B. and J. O. Young. 1939. Histological Study of the Developing Fruit of the Sour Cherry. *Botanical Gazette*. 100:723–749.

Tukey, H. B. and J. O. Young. 1942. Gross Morphology and Histology of Development Fruit of the Apple. *Botanical Gazette* 104(1):3–25.

Uphof, J. C. 1959. *The Dictionary of Economic Plants*. Hafner Publishing Company, New York.

Wolf, M. J., C. L. Buzen, M. M. MacMasters and C. T. Rist. 1952. Structure of the Mature Wheat Kernel. Cereal Chemistry 29:321–382.

2a

RETENTION OF WATER IN FOOD AND AGRICULTURAL MATERIALS

2a.1 FORMS OF WATER RETENTION

Water is one of the most important components affecting all the physical properties of food materials. The way water affects the physical nature and properties of food materials is complicated because of the interaction between water and the medium, which involves the physical structure as well as the chemical components such as various solutes including polymers and colloidally dispersed particles.

Until moisture content of foods is lowered to substantially below 50 percent (wet basis), water in food is said to exist as *free water*. In this form, the water behaves physically as pure water with properties such as vapor pressure equal to those of pure water. When water content of food is lowered, a point is reached at which the water becomes less active in that it cannot act physically or chemically as pure water. For example, it cannot freeze or act as a solvent or reactant. In this state the water in food is usually referred to as *bound water*.

Another way to describe water in foods is by the mechanisms of molecular adsorption, corresponding to "bound water" and capillary absorption corresponding to "free water." Molecular adsorption occurs when the water molecules adhere to specific points in the molecular structure of the material. When the distance between the water molecule and the cell wall becomes small enough (of the order of 10^{-7} cm), the force of attraction is large enough to draw the water into the micellar network of the cell wall (Barkas, 1953). The force of attraction at low moisture contents is so high that an "adsorption compression" results in a net decrease in volume of the solid-water aggregate.

As the moisture content increases, the molecular attraction becomes smaller and there is a volume increase which is roughly equal to the volume of water added. Due to the initial adsorption compression, however, the total volume of the aggregate remains smaller than that of constituents.

Example

Figure 2a.1 shows the relationship between specific volume of corn kernels and their moisture content. Using these data show that the rate of change of specific volume of the corn is approximately equal to the specific volume of free water.

From the slope of the straight line in Fig. 2a.1

$$\frac{0.53\times10^{-2}\text{ cu ft/lb dry matter}}{32\times10^{-2}\text{ lb }H_2O\text{/lb dry matter}}=0.0166\text{ cu ft/lb}$$

Specific volume of water at 70°F=1/62.3=0.0161 cu ft/lb

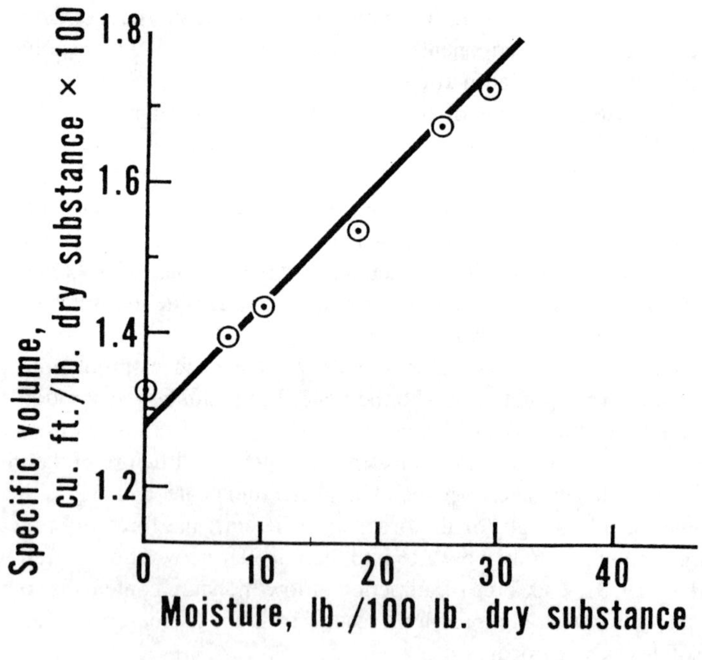

Figure 2a.1 Relationship between moisture content and specific volume of corn kernels (Lancaster, USDA, ARS, Peoria, Ill.)

The extent and the nature of the surface on which adsorption compression can take place appear to be the prime factor in molecular adsorption (Stamm and Seborg, 1935). Molecular attraction can be due to electronic and van der Wall's attraction but mostly due to hydrogen bonding. Thus the greater the number of ionic or polar type molecules, the more water is held in the food in this form. Molecular adsorption is the primary cause of swelling in hygroscopic food materials. The starch in corn kernel, for example, contains more polar sites for attraction of water molecules than does the cellulose (Chung, 1966). This should allow a larger amount of moisture to be adsorbed by starchy materials than cellulosic material.

At still higher moisture contents, where the vapor pressure has not yet reached the saturation point, most of the available points of attraction have been filled with water and further holding of water molecules is possible only through the formation of chains of water molecules or "water bridges" extending between those molecules which have been directly adsorbed. A direct tensile stress within the elastic limit of the material tends to rupture these water bridges. Removal of the stress can cause the water bridges to reform without any sign of plastic deformation. In the case of wood, it is claimed that when a shear stress is acting between parallel cellulose chains, the water molecules can jump from one point of attraction to another, resulting in some energy loss and plastic deformation in the material (Barkas, 1953). As there is no evidence that plasticity in wood has its origin in a slipping of entire cells over one another, this theory of inelastic jump of water molecules has been offered as the most plausible explanation for the increasing of plasticity in cellulosic materials with increasing moisture content (Glasstone *et al.*, 1941).

Capillary absorption occurs when voids in the cellular structure are of the size to hold water in liquid form by forces of surface tension. The size of capillaries that will fill with water under different relative humidities can be calculated by the Kelvin equation given below

$$r = \frac{2\sigma M}{\rho RT \ln(P_0/P)} \qquad (2a.1)$$

where r = capillary radius
σ = surface tension
M = molecular weight
ρ = density of the liquid
R = gas constant

T = absolute temperature
P_0/P = relative pressure

Based on this equation, the sizes of capillaries for various relative vapor pressures are given in Table 2a.1

Many of the familiar physical phenomena such as the spherical shape of rain drops or the rise of water in a capillary are due to surface tension. The surface tension of a liquid is defined as the force per unit length on the surface of a liquid which opposes the expansion of the surface area. When a liquid, like water, wets the surface of a solid surface like a tube, it pulls the body of liquid up forming a concave surface. When a liquid, like mercury, does not wet the surface of the tube, the liquid does not adhere to the wall and forms a convex surface to the horizontal plane. In a capillary tube, the surface tension is given by (Daniels, 1949)

$$T_s = \frac{1}{2} h \rho g r \qquad (2a.1a)$$

where h is the height of rise of liquid, ρ is mass density of the liquid, g is gravitational force, r is the radius of the capillary tube, and T_s is force/unit length. For mercury, T_s is about 5 times as that of water.

In the drying process while most of the capillary water will disappear first, there is simultaneous evaporation of both capillary water and molecularly-held water right from the saturation pressure to the lowest vapor pressure at which capillary water can be held (Barkas, 1953). This makes it difficult to determine precisely what percentage of moisture content is

Table 2a.1 Size of capillaries that will fill with water under different relative vapor pressures (from Stamm, 1964)

Relative vapor pressure	Capillary radius (microns)
0.9	0.010
0.95	0.020
0.97	0.035
0.98	0.053
0.99	0.106
0.995	0.210
0.999	0.060
0.9999	10.60

held by capillary absorption and that by molecular adsorption. In a work on equilibrium moisture content of shelled corn, however, it was concluded that in the range of 5 to 90 percent relative humidity the moisture binding mechanism is predominately multimolecular adsorption (Hall and Rodriguez-Arias, 1958). In the case of cellulosic materials it has been stated that capillary adsorption occurs at relative humidities exceeding about 90 percent (Stamm, 1962).

There are several methods for determining the bond water in food materials. Among these methods are (1) determining the amount of unfrozen water, (2) determination of dielectric properties of foods, (3) measurement of nuclear magnetic resonance, and (4) measurement of vapor pressure of foods.

There is always a certain amount of "unfreezable water" present in food. If food samples of different moisture contents were frozen at temperature well below the freezing point of water (−50°C and lower) and then heated in a differential thermal analyzer (DTA), the presence of no peaks on the DTA recorder indicates that only "unfreezable water" is available. Upon heating, the latent heat of diffusion produces DTA peaks. From the samples of a food at different moisture contents, the point at which all the remaining water is unfreezable can be determined (Duckworth, 1971).

Nuclear magnetic resonance method (NMR) for determination of bound water is based on the principle that hydrogen protons provide a very strong absorption-signal and thus show clearly the condition of resonance when placed in a magnetic field upon which is imposed an external radio-frequency (r-f) magnetic field. The forced orientation of hydrogen nuclei absorbs energy from the magnetic field in direct proportion to the number of hydrogen nuclei present.

The dielectric method of bond water determination is based on the fact that the method differentiates between dielectric properties of rigidly-held dipoles of ice and that of the much more mobile liquid water molecules.

The most common and least costly method for studying the properties of water in foods is the sorption and desorption method which will be discussed later. Fig. 2a.2 shows the major types of water in foods.

Water Activity

Water activity is the equilibrium relative humidity (ERH) at which the food neither gains nor loses moisture. Water activity is also defined as the ratio of vapor pressure exerted by the water in the food (P) to the vapor

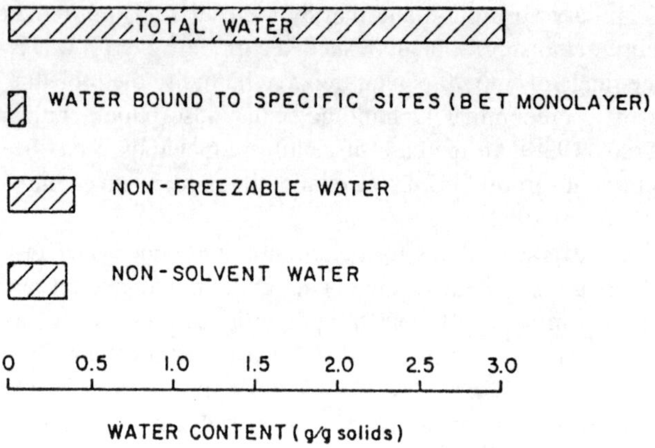

Figure 2a.2 The types of bond water in foods (Karel *et al.* 1975)

pressure of pure water (P_0) at the same temperature. Denoting water activity by a_w, in equation form

$$a_w \times 100 = \%ERH = P/P_0 \times 100 \tag{2a.2}$$

Since the a_w for pure water at any temperature above its freezing point is 1.00, the a_w for any food has to be in the range of 0–1.00. For example, the a_w for high moisture food such as fruits, vegetables, meat, etc. is more than 0.97, for intermediate moisture food between 0.6 and 0.9 and for dry food such as crackers and non-fat dry milk is below 0.6 (see Table 2a.2).

According to *Raoult's law*, for pure solutions the ratio of vapor pressure exerted by the solvent, P, to the vapor pressure of the pure solvent, P_0, is equal to the mole fraction of the solvent, N.

$$P/P_0 = N = n_1/(n_1 + n_2) \tag{2a.3}$$

where n_1 and n_2 are, respectively, mole of solvent and mole of solute. It follows then that for water being the solvent, there exists an equality between equations (2a.2) and (2a.3). Accordingly, water activity in terms of mole fraction for ideal solutions is

$$a_w = n_1/(n_1 + n_2) \tag{2a.4}$$

For non-ideal solutions, a coefficient γ, called activity coefficient, is added to Eq. (2a.4) to account for deviation from Raoult's law so that

$$a_w = \gamma(n_1/n_1 + n_2) \tag{2a.5}$$

whereas in an ideal solution, Raoult's law expresses the water vapor pressure in equilibrium with the solution, in foods, which are non-ideal, the vapor pressure can be predicted only by empirical methods (Ross, 1975).

2a.2 EFFECT OF TEMPERATURE ON WATER ACTIVITY

As temperature varies, the excitation of molecules as well as the distance and thus attraction between molecules varies. It has been shown that activity

Table 2a.2 Approximate water activity a_w values of some foods, sodium chloride, sucrose solutions (Troller and Christian, 1978)

Foods	Water Activity a_w	NaCl (%)	Sucrose (%)
Dried milk, dried vegetables, chopped walnuts	0.2	—	—
Dried potato flakes, crackers, cake mixes, pecan halves	0.3	—	—
Dried egg, cocoa	0.4	—	—
Chocolate, confectionary, honey, noodles	0.5–0.6	—	—
Parmesan cheese, dried fruit, corn syrup, licorice	0.6–0.7		
Molasses, soft dried figs, heavily salted fish	0.7–0.8	19% to saturation	—
Cheddar cheese (aged), jams, margarine, salami, condensed milk	0.8–0.9	14–19	59% to saturation
Processed cheese, bakery goods, high moisture prunes, raw ham, dry sausage, high salt bacon, orange juice concentrate	0.9–0.95	8–14	44–59
Fresh meat, fruits, vegetables, canned fruit in syrup, canned vegetables in brine, frankfurter, butter, low salt bacon	0.95–1.00	0–8	0–44

coefficient γ, is dependent on temperature leading to the following expression relating a_w and T, absolute temperature (van den Berg and Bruin, 1978):

$$\ln \frac{(a_w)T_1}{(a_w)T_2} = C\ (1/T_1 - 1/T_2) \tag{2a.6}$$

where C is a function of heat of mixing, which has a negative value for nearly all moist food systems. It follows that for these materials a_w increases with increase of temperature. In the ambient temperature range, the observed change in water activity per °C is for common sugar solutions of the order of 0.0001 (Ross, 1975) and for some cereals about 0.0023 (van der Berg and Leniger, 1976).

Water activity at subfreezing temperatures is given by the ratio of vapor pressure of ice and vapor pressure of supercooled water, P_{ws}, at the system's temperature.

$$a_w = P_{ice}/P_{ws} \tag{2a.7}$$

For example, at -18°C an aqueous system has a water activity of 0.839 (Fennema and Berny, 1974).

Sorption-Desorption Isotherms

The plot of equilibrium moisture content of a material versus relative humidity of the environment at a given temperature is referred to as sorption or desorption isotherms, depending whether the material is being wetted or dried. It has been established that in gels and other hygroscopic materials the sorption-desorption isotherms are sigmoidal in shape and show marked hysteresis as illustrated in the case of corn kernel in Fig. 2a.3. Figures 2a.4 and 2a.5 show isotherms of other food materials. In these curves, at any given relative humidity the moisture content reached from a drier state is lower than that reached from a wetter one. In other words, the amount of water held by these materials is not only dependent upon the equilibrium relative vapor pressure, but is also dependent upon the direction from which equilibrium is approached. Since relative humidity is the equilibrium relative humidity, according to Eq. (2a.2), the x-axis in the isotherm can also be represented as water activity a_w.

Several explanations have been proposed for this hysteresis phenomenon. The one favored most is that for the case of cellulosic materials. It is based

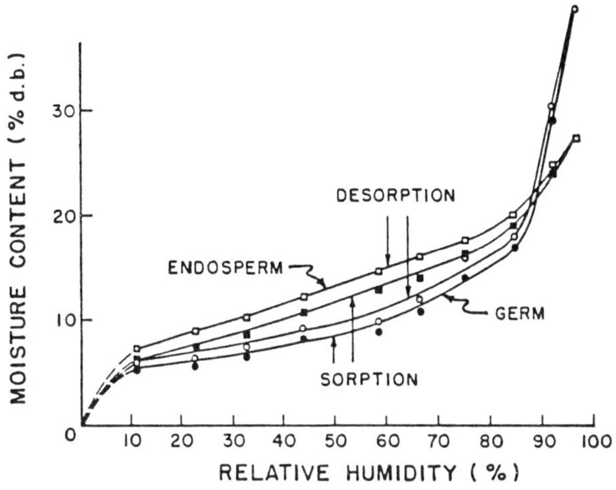

Figure 2a.3 Sorption-desorption isotherms for germ and endosperm in corn kernel at 74°F (Shelef and Mohsenin, 1966)

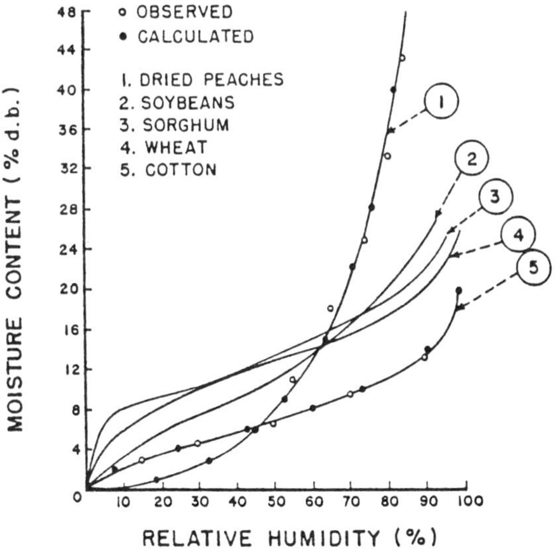

Figure 2a.4 Equilibrium moisture curves for several agricultural products (Henderson, 1952—Similar curves are given by Pichler, 1956; Day and Nelson, 1963; Young and Nelson, 1965, Chung and Pfost, 1967.)

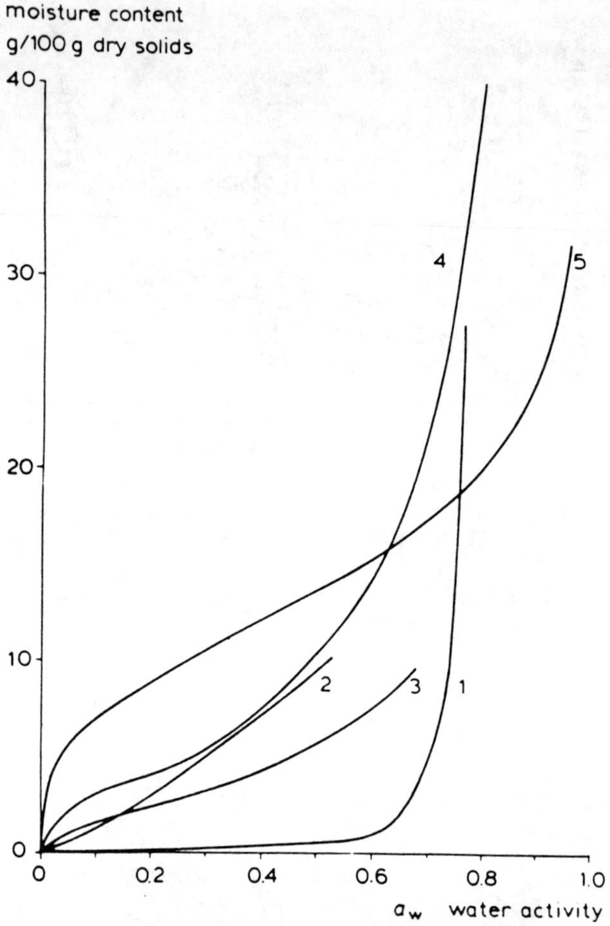

Figure 2a.5 Resorption isotherms for (1) Composed sweet, (2) Spray-dried chicory extract, (3) Roasted Colombian coffee, (4) Pig pancreas extract powder, and (5) Native rice starch (van den Berg and Bruin, 1978)

on the theory of the change in availability of active polar sites for the bonding of water molecules (Stamm, 1964, p. 147). Under this theory, in the original wet condition, the polar sites in the molecular structure of the material are almost entirely satisfied by adsorbed water. Upon drying and shrinkage, the molecules and their water-holding sites are drawn closely enough together to satisfy each other. This reduces the water holding capacity of the material upon subsequent adsorption. In other words, due to the fact that rehydration is never as complete as the original hydration, for any given relative vapor pressure, the material shows a higher moisture content along the desorption curve than it does along the adsorption curve.

Regions of Isotherms

The sorption-desorption isotherms can be divided into three regions (Fig. 2a.6):

I Bound water or monolayer region which corresponds to absorption of up to a monmolecular layer of water. The completion of this region occurs very close to the inflection point of the curve.

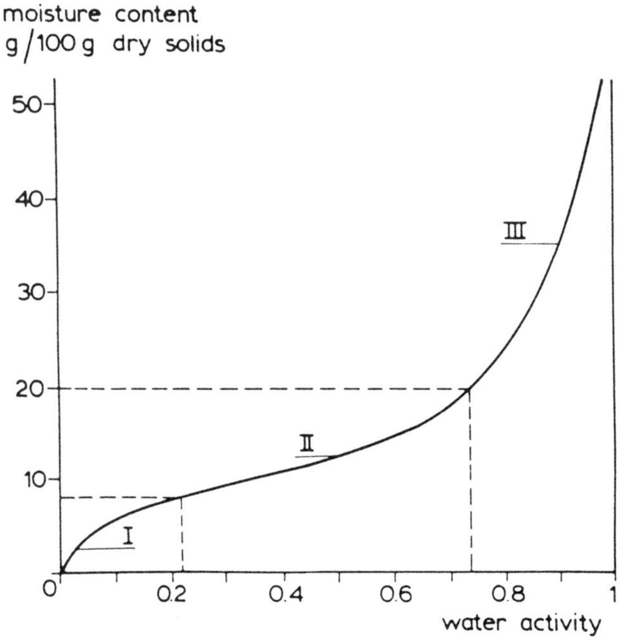

Figure 2a.6 Regions of an isotherm for hygroscopic materials (van den Berg and Bruin, 1978)

II Intermediate water or multilayer region which corresponds to adsorption of additional layers of water over the monolayer.

III Free water or capillary absorption region which refers to the water held in the pores of the material.

Equilibrium Moisture Content

Water activity is relatively a new term used mostly in the food processing industry. Equilibrium moisture content is the term used commonly in the agricultural engineering and processing industry. For example, one of the classical equations used for estimating equilibrium moisture content is the following equation given by Henderson and Perry (1966):

$$1-rh=e^{-KTM_e^n} \qquad (2a.7a)$$

where rh = relative humidity, decimal
T = absolute temperature, °R
M_e = equilibrium moisture content, % d.b.
k and n are are constants as specified in the following table.

The values of k and n for some agricultural grains are given in the following table.

Material	k	n
Shelled corn	1.10×10^{-5}	1.90
Sorghum	3.40×10^{-6}	2.31
Soybeans	3.20×10^{-5}	1.52
Wheat	5.59×10^{-7}	3.03

Example

A sealed box is filled with soybeans at 20% moisture (w.b.). Estimate the relative humidity of the interseed air. The temperature of the grain in the box is 60°F.

From the above table and the given data,
$k = 3.20\times10^{-5}, n=1.52$
$T = 60+460 = 520°R$
$M_e = 20\% \text{ (w.b.)} = 25\% \text{ (d.b.)}$

Substituting the Eq. 2a.7a,

$$1-rh = e^{-3.2\times10^{-5}(520)(25)^{1.52}}$$

$$rh = 89\%$$

Adsorption Equations

Attempts to express the adsorption process mathematically have led to several empirical and derived equations (Freundlich, 1922; Smith, 1947; Brunauer, Emmett and Teller, 1938; Young and Nelson, 1967). The best known and widely used mathematical representation of the adsorption phenomenon in organic materials is, however, given by the so called B.E.T. equation (Brunauer, Emmett, Teller, 1938). A general form of the B.E.T. equation for n molecular layers of adsorbed water is

$$V = \frac{V_mC\ (P/P_0)}{1-(P/P_0)}\ \frac{1-(n+1)(P/P_0)^n+n(P/P_0)^{n+1}}{1+(C-1)(P/P_0)-C(P/P_0)^{n+1}} \tag{2a.8}$$

where V = volume of gas or vapor adsorbed at vapor pressure, P, per unit weight of adsorbent when the saturation pressure is P_0

V_m = constant representing the volume adsorbed when a monomolecular layer has just been formed

C = constant depending on the heat of adsorption E_a, the heat of condensation E_c, absolute temperature, T, and gas constant, R, such that $C=e^{-(E_a-E_c)/RT}$

If in place of the volume, V, the weight of adsorbate, W, per unit weight of adsorbent is used in the equation, the volume, V_m, will also be replaced by weight, W_m, per unit weight of adsorbent. In the case of sigmoidal types of adsorption curves, the volume, V_m, or weight, W_m, corresponding to the completion of a monomolecular layer, occurs very close to the inflection point in the curve. When n=1, the B.E.T. equation reduces to an expression valid for monomolecular layer adsorption. When $n=\infty$, the equation is valid for multimolecular layer adsorption.

Kunze and Hall (1964, 1966), introduced the method of plotting rice equilibrium moisture content lines directly on a psychrometric chart to illustrate grain and ambient environmental conditions relationships in moisture adsorption and desorption processes.

A review of the literature on the mechanism of adsorption and a discussion of the several adsorption equations are given by Young and Nelson (1967).

These authors have also made an attempt to explain the hysteresis effect in the wetting and drying process of biological materials. Among other work appearing in the literature on development of equilibrium moisture content equations, theories for sorption-desorption phenomena, on the explanation of the hysteresis loop, and data on food and agricultural materials are Coppens and Wei (1954), Chung and Pfost (1967), Strohman and Yoerger (1967), Chen and Clayton (1970), Ngoddy and Bakker-Arkema (1970, 1972), Gustafson and Hall (1973), Friesen (1971, 1974), and Meddings and Potter (1974). In addition, considering the diffusion theory and the mass transfer phenomena, many other published works appear in the literature. The references for these works are found in the revised edition of *Thermal Properties of Food and Agricultural Materials* (Mohsenin).

In terms of water activity, a simplified B.E.T. equation can be presented in the following form for use in food materials:

$$a_w/(1-a_w)m = 1/m_0C + (C-1)a_w/m_0C \qquad (2a.9)$$

where m is moisture content at given a_w, m_0 is the monolayer moisture value, and C is a constant related to temperature, energy of sorption, and other constants. If the quantity $a_w/(1-a_w)m$ were plotted against water activity a_w, a straight line is obtained. The monolayer moisture value m_0 is then obtained from the slope and intercept of the line. The value of m_0 is usually the moisture content at which the water is tightly bound and cannot participate as a solvent. Thus, m_0 is the moisture content for maximum stability of dehydrated foods.

Example

Using the data in Fig. 2a.7 determine the monolayer value for this food material. From the intercept of Fig. 2a.7

$$m_0C = 1/0.022 = 45.45$$

and from the slope of Fig. 2a.7

$$C-1 = 0.89\ m_0C$$

Therefore C=41.45 and

$$m_0 = 45.45/41.45 = 1.1\text{g } H_2O/\text{g solid.}$$

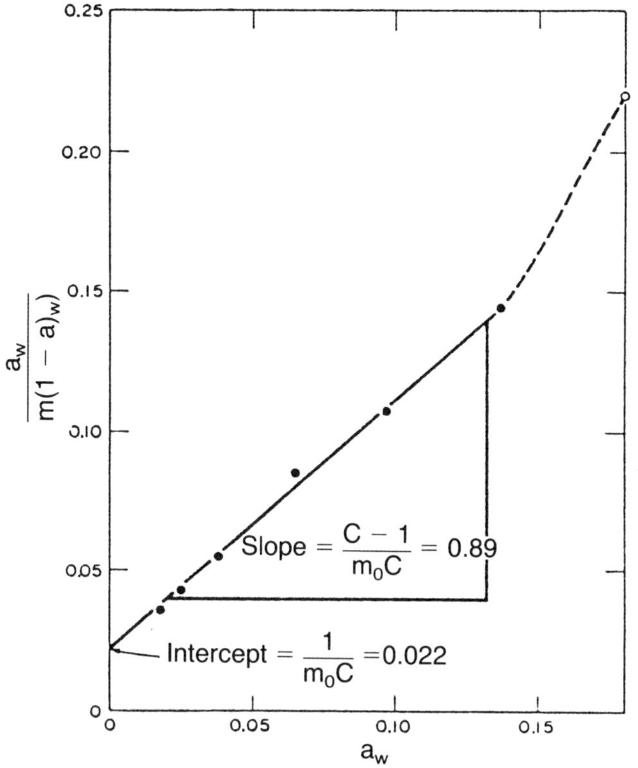

Figure 2a.7 B.E.T. plot for dehydrated orange crystals at 25°C (Karel and Nickerson, 1964)

Example

The following data were obtained for ground, precooked, freeze-dried beef at 50°F. Using the B.E.T. equation, determine the monolayer moisture value m_0 and the C constant.

a_w	m (d.b.)	a_w	m (d.b.)
0.0	0.0	0.10	0.077
0.1	0.011	0.15	0.093
0.02	0.019	0.20	0.106
0.03	0.027	0.30	0.121
0.04	0.041	0.40	0.137

Solution

Since B.E.T. equation given as Eq. (2.9) represents the equation of a straight line, the plot of $a_w/(1-a_w)m$ versus a_w yields almost a straight line with the following slope and intercept

$$\text{slope} = (C-1)/m_0C = 8.7$$
$$\text{intercept} = 1/m_0C = 0.7.$$

From the above expressions

$$C = 13.4 \text{ and } m_0 = 0.1066 \text{ g } H_2O/\text{g solid}.$$

Availability of Water for Microbial Growth

Since microbial growth is directly related to the availability of water in food, water can be made unavailable by drying, freezing, or addition of solutes such as salt or sugar, which will tie up the water needed to support microbial growth (Kaplow, 1970 and Potter, 1970). Keeping the foods' water activity below a certain value, they can be protected against microbial deterioration. For example (Fig. 2a.8) the lower limits of water activity at which mold can grow is 0.80, while for bacteria growth is 0.91 (Bone, 1969). Fig. 2a.8 is a generalized illustration of the effect of water activity on food stability, frequently seen in the literature which has been designed based on the literature data (Labuza, 1970). Starting from $a_w = 1$, a decrease of water activity slows down and finally stops all types of chemical deteriorations and microbial growth except for lipid oxidation.

Microbial growth, like other types of food deterioration, is based on chemical and biochemical reactions. Except for diffusion limitations, the speed of these chemical or biochemical reactions is given by the following relationship:

$$R_A = -K_A C_A^n \tag{2a.10}$$

where R_A = rate of conversion of component A per unit volume

K_A = first order reaction rate constant being a function of pressure, P, temperature, T, and composition C_A, C_B, C_C, etc.

C_A = concentration of component A

n = reaction order

If water is a reactant itself, it acts more directly through its concentration together with its influence on K_A. If water is not a reactant, water

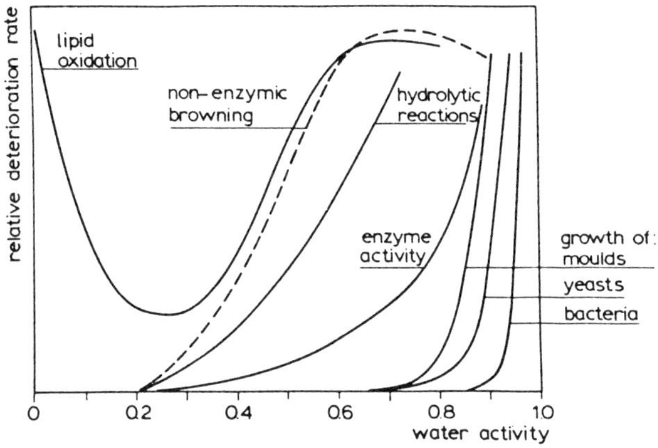

Figure 2a.8 Availability of water for microbial activity (van den Berg and Bruin, 1978)

activity, a_w, acts only through K_A. Only when the water component is the rate limiting factor, can a_w be expected to have a direct influence on deterioration reaction in a food system.

Water also exerts a significant influence on the chemical reactions affecting the stability, shelf life, color and appearance, taste, texture, and many other physical and chemical properties of foods (Labuza and Chou, 1974; Loncine *et al.*, 1968).

Theoretical Estimation of Water Activity

For theoretical estimation of a_w, sometimes the role of water activity in a food system is expressed in terms of a basic thermodynamic concept. However, because of the complexity of food systems and the fact that a_w is only one factor in the deterioration reaction not acting solely by means of a gas phase as a vapor pressure, the analysis of the system becomes quite complex. Van den Berg and Bruin (1978) have discussed the theoretical estimation of a_w based on the following approaches:

1 Theoretical models of Gibb's free energy for solutions-one phase systems.
2 Theories of physical sorption if a surface can be distinguished.
3 Combination of 1 and 2.

Applying this approach, they have shown that with multiphase food systems physical sorption theories apply better at low a_w while solution theories are more successful at higher a_w. Examining the various theories and formulas given in the literature, these investigators concluded that it is difficult to derive practical recommendations for use of these theories and equations to estimate water activity in food systems.

Another approach for estimating the water activity in food material is that given by Loncin and Merson (1979). A liquid present in a capillary is subjected to a suction pressure ΔP such that

$$\Delta P = 2\,T_s/r \qquad (2a.11)$$

where r is the radius of the capillary. In the boiling process, the above expression is used to explain why boiling is not produced even though the liquid is brought to a temperature with its vapor pressure greater than the ambient pressure (Glassmann, 1970).

Capillary action and its applications to porous products has been researched and discussed by several investigators (Loncin, 1975). As water content increases beyond an a_w of 0.6 to 0.7, a sharp increase of moisture content is observed with a small increase of a_w. This indicated the beginning of the region of capillary condensation. In this region, the so called Kelvin-Thompson equation can be used for estimating the water activity as follows:

$$a_w = e^{-(M\Delta p/\rho RT)} \qquad (2a.12)$$

where M and ρ are molecular weight and density of water while R and T are gas constant and absolute temperature.

Table 2a.3 Surface tension of some liquids in N m^{-1} (Barrow, 1973)

Liquid	Temperature °C 20	60	100
H_2O	0.07275	0.06618	0.05885
Hg	0.48 at 0°C	—	—
C_2H_5OH	0.0223	0.0223	0.0190
C_6H_6	0.0289	0.0237	—

Example

A porous solid food material has pore size diameter of 2×10^{-8} m containing water of 20°C. Estimate the water activity of this food.

From Table 2a.3, surface tension of water at 20°C=0.07275 N/M. From Eq. (2a.11), suction pressure P is

$$
\begin{aligned}
P &= 2T_s/r \\
&= 2\times0.07275/2\times10^{-8} \\
&= 7.275\times10^{6}\ \text{Pa} \\
&= 0.152\times10^{5}\ \text{lbf/ft}^2
\end{aligned}
$$

Using Eq. (2a.12) and letting lbm=lb mass and lbf=lb force

$$
\begin{aligned}
a_w &= e^{-\dfrac{(18\ \text{lb m/mol})\ (0.152\times10^{5}\ \text{lbf/ft}^2)}{(62.4\ \text{lbm/ft}^3)\ (1544\ \text{ft lbf/mol °R})\ (460+68\text{°R})}} \\
&= e^{-0.00538} \\
&= 0.995.
\end{aligned}
$$

Measurement of Water Activity

The following methods are used for measurement of water activity:

1 Decicators containing sulfuric acid or salt solutions.
2 Wink Method—sample suspended in a vessel with saturated salt solution and weighed continuously by a bottom weighing balance.
3 Dynamic Method—air of given relative humidity is passed through the sample in a sealed vessel.
4 Spring Sorption Balance—food sample is suspended from a calibrated coil spring whose extension or contraction is followed by a cathetometer or recorded electronically.

The above methods involve samples of known initial moisture content which go through the process of moisture gain or loss in a constant relative humidity atmosphere. A second approach is to place the food sample of known constant moisture content in a sealed enclosure with a head space. The atmosphere in the head space is then measured either for relative humidity changes or partial pressure until equilibrium is established.

There is still another method, which although not as precise, is less time consuming and usuable when it is not feasible to allow for samples to come to equilibrium. In this method weighed samples of foods at known moisture contents are exposed for a short period of time to various relative humidities. By interpolation, it is possible to determine the relative humidities at which samples neither gain nor lose weight. Such relative humidities, regarded as equilibrium relative humidities, along with samples at various moisture contents are then used to plot the sorption isotherm for the given food.

The details of these methods of water activity measurements are found in reviews by Gal (1967) and Toledo (1973). The methods of partial pressure measurement, using a simple manometric technique is given by Taylor (1961); while using a hygrometeric technique is given by Karel (1955).

Applications of Water Activity

As related to water sorption and desorption, water activity is important in the design of processing operations as well as in product development. Examples are:

1 Food dehydration
2 Moisture transfer between food components in processing of intermediate moisture foods
3 Moisture transfer during storage
4 Moisture transfer and equilibriation across plastic film barriers during storage
5 Control of chemical reactions, such as vitamin degradation, nonenzymatic browning, lipid oxidation, and other deteriorative reactions

To predict moisture equilibriation of a mixture of dehydrated foods, the equilibrium will be at equal a_w and not at equal moisture content. The following expression shows the importance of water activity and isotherms of a mixture of materials.

$$a_w\ (\text{equilibrium}) = \sum_{i=1}^{i=n} a_{wi} b_i W_i \Big/ \sum_{i=1}^{i=n} b_i W_i \qquad (2a.13)$$

where n = number of food components
a_{wi} = initial water activity of component i
b_i = slope of isotherm of component i
W_i = initial dry solids of component i.

Examples of influence of water activity for various foods are given by Karel (1974).

2a.3 CONTROL OF FOOD'S WATER CONTENT (PSYCHROMETRIC CONSIDERATION)

Partial pressure of water in solids increases with temperature at roughly the same rate as the increase in maximum partial pressure in gases (saturation water vapor pressure). Over narrow temperature spans, this follows closely the definition of relative humidity (Fig. 2a.9). The temperature difference between the material and surrounding air requires psychrometric consideration.

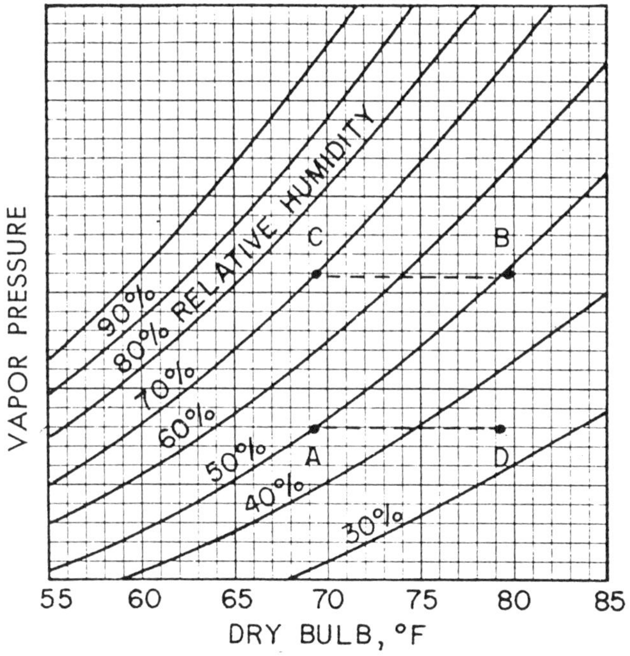

Figure 2a.9 The relationship between vapor pressure, dry bulb temperature, and equilibrium relative humidity (Quinn, 1967)

For example, a material with 50% E.R.H. at 70°F, shown as point A in Fig. 2a.9, is placed in a space at 50% R.H. and 80°F, shown as point B in Fig. 2a.9. If the temperature of the air and the material are assumed to remain stable at 80° and 70°F, respectively, under the 80° temperature condition, the material would have a higher vapor pressure and thus some moisture transfer must take place. If the mass of the air is considerably greater than that of the material, the air would remain essentially unchanged while the material would reach a balance at the higher vapor pressure similar to that of the air resulting in a condition of about 70% R.H. and 70°F, shown as point C in Fig. 2a.9. However, if the material is the dominant mass, the air could lose water to a balance condition with the lower vapor pressure of the material resulting in a condition of about 36% R.H. at the 80°F air temperature, shown as point D in Fig. 2a.9. The R.H. would be 35% which will result in a product with a moisture content of about 6.3 percent, or a loss of 2.2% from 8.5%. Finally, if the relative humidity were kept constant, heating of the product to point B, will result in a moisture content of about 8.3% or only 0.2% loss than the required moisture content. Thus it is seen how the psychrometric consideration can help to predict which method of control would be most effective and is to be used. To predict the end condition of the material, it is essential to establish the temperature versus E.R.H. (or a_w) for each specific material.

In another example, consider product A (Figure 2a.10, above) at 70°F and 8.5% M.C. under 50% E.R.H. condition. If the product and space are heated to 75°F without gain or loss of water by the material (point B), the balance condition is about 51% E.R.H. If these conditions were plotted on psychrometric chart (Figure 2a.10, below), when the air is heated to 75°F without gain or loss of water (no change in dew point), the resulting R.H. would be about 42%, shown as point E, instead of the required 51%. This results in a product with moisture content of about 7.3% or 1.2% less than the required 8.5%. If wet bulb temperature were to be kept constant, when temperature rises to 75° (point F).

Example

A rice mill in the humid Gulf Coast area at 80°F and 60% relative humidity wants to establish a market in the dry Arizona climate at 90°F and a relative humidity of 35%. Rice filled in packages at the mill would have 12.65% (w.b.) moisture. How much of the packages must be overfilled in order to maintain a 500 g weight in Arizona? Assume the equilibrium moisture content for the rice in Arizona 9.1% (w.b.).

Let X = sum of dry matter and moisture content at the mill
DM = dry matter
At the mill: $X = DM + 0.1265\,X$
In Arizona: $500 = DM + 0.091 \times 500$, or: $DM = 454.5$ g
At the mill: $X = 454.5 + 0.1265\,X$, or: $X = 520.32$ g
Percent overfill: $(520.32 - 500)/500 = 4\%$

Table 2a.4 Physical Properties of Water Under Standard Atmosphere (14.7 Psia or 10^5 Pa)

Property	Temperature		Units	
	°C	°F	SI System	English System
Boiling point			373°K	212°F
Sat. vapor pressure	20	68	2.33×10^3 Pa	0.36 Psi
Mass density	20	68	997 kg m^{-3}	1.936 lb S^2 ft^{-4}
Weight density	20	68	997 kg m^{-3}	62.3 lb m ft^{-3}
Viscosity (dynamic)	20	68	0.98×10^{-3} PaS	$2.05 \times 10-5$ lb S ft^{-2}
Viscosity (kinematic)	20	68	$1.02 \times 10^{-6} m^2 sec^{-1}$	1.06×10^{-5} ft^2 S^{-1}
Viscosity (dynamic vapor)	100	212		$2.6484 \times 10-7$ lb S^{-1} ft^2
Surface tension	20	68	7.23×10^{-2} Nm^{-1}	4.98×10^{-3} lb ft^{-1}
Latent heat vaporization	100	212	2.258×10^6 Jkg^{-1}	970 Btu lb^{-1}
Latent heat, fusion	0	32	334.9×10^3 Jkg^{-1}	144 Btu lb^{-1}
Specific heat (Cp)	20	68	4.18×10^3 Jkg^{-1} $°K^{-1}$	0.9987 Btu lb^{-1} $°F^{-1}$
Thermal conductivity	20	68	0.6 Wm^{-1} $°K^{-1}$	0.347 Btu ft^{-1} hr^{-1} °F
Thermal diffusivity	20	68	0.143×10^{-6} m^2 S^{-1}	$5.54 \times 10-3$ ft^2 hr^{-1}
Mass diffusivity (air)	20	68	24.6×10^{-6} m^{-2} S^{-1}	9.53×10^{-1} ft^2 hr^{-1}

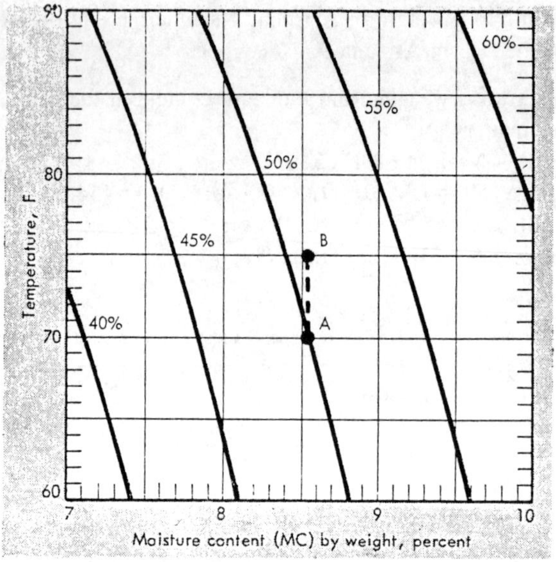

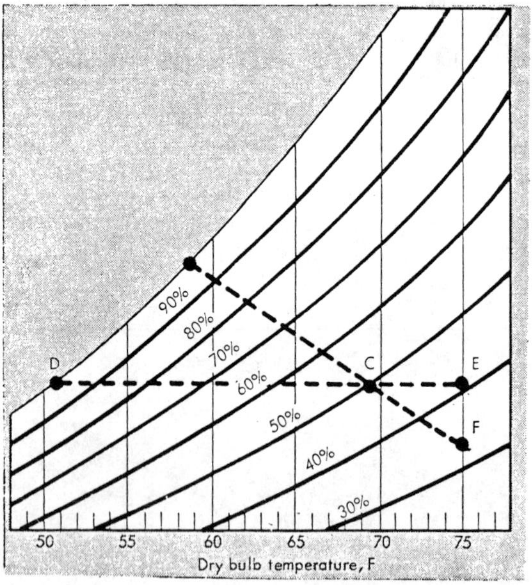

Figure 2a.10 (above) Moisture content versus temperature, relative humidity for a hygroscopic material, (below) conditions of (above) plotted on psychrometric chart (Quinn, 1966)

3

PHYSICAL CHARACTERISTICS

WHEN PHYSICAL PROPERTIES of grains, seeds, fruits and vegetables, eggs, forage, and fibers are studied by considering either bulk or individual units of the material, it is important to have an accurate estimate of shape, size, volume, specific gravity, surface area and other physical characteristics which may be considered as engineering parameters for that product. Table 3.1 is presented to show the application of several of the physical characteristics of potato and stones which have been utilized in design of the separating mechanisms. In this chapter methods for determination of such physical characteristics as shape, size, volume, specific gravity, bulk density and surface area are discussed.

3.1 SHAPE AND SIZE

Shape and size are inseparable in a physical object, and both are generally necessary if the object is to be satisfactorily described. Further, in defining the shape some dimensional parameters of the object must be measured.

In certain applications where both shape and size affect the process, the relationship can be shown by a single, two-dimensional system as follows:

$$I=f(sh,\ s) \tag{3.1}$$

where I is the index influenced by both shape, sh, and size, s. In other applications the index I may be a function of not only shape and size but also of such other parameters as orientation, o, packing index, p, firmness, f, etc.

$$I=f(sh,\ s,\ o,\ p,\ f,\ \ldots) \tag{3.2}$$

Table 3.1 Separation of potatoes and stones, and the properties which are involved[1]

Method of separation	Properties of stones and potatoes								
	Spec. grav.	Depending on spec. grav.			Resil-ience	Rolling resist.	Air resist.	Shape	Hard-ness
		Size	Weight	Mass					
Flotation in brine solution	*								
Flotation in sand	*								
Sieve apertures covered by weighted flaps	*	*	*						
Rubber-strip sieve	*	*	*					*	
Single and double coil-spring sieve	*	*	*					*	
Rubber fringed wheel	*	*	*					*	
Rotating brush on "Digger"	*	*		*				*	
Dropping on to rotating brush	*	*	*					*	
Dropping on to rotating wooden drum					*			*	
Dropping on to sloping and vibrating rubber-finger belt	*	*	*		*	*		*	
Placing on to sloping, undulated belt					*			*	
Placing on to a right-angled conveyor belt									
Rolling off wooden drum					*	*		*	
Rolling over sloping surface and wooden drums					*	*		*	
Separation on inclined conveyor belt					*	*		*	
Roller experiments					*	*		*	
Separation on a rotating disc						*		*	
Separation by means of a lath						*		*	
Separation of static objects by means of air blast from one side				*		*	*	*	
Dropping through horizontal air stream				*			*	*	
Separation in vertical air stream	*								
Horizontal movement through vertical air stream	*		*			*	*	*	
Separation by means of spikes									*

[1]Maak, 1957.

A good example is the problem of determining the number of a given fruit required to fill a container. If we substitute Y for I, x_1 for shape, x_2 for size, x_3 for orientation, x_4 for packing and x_5 for firmness, we may write the defining equation in multiple regression form as follows:

$$Y = b_1x_1 + b_2x_2 + b_3x_3 + b_4x_4 + b_5x_5 \tag{3.3}$$

This relationship can be evaluated by measuring a set of specimens and the magnitude of the contribution of each x to the variation in y can be estimated by means of analysis of variance and multiple correlation (Quenoville, 1952). This technique has been used in petrography to determine the relationship between such properties as permeability of an oil reservoir and the petrographic properties such as shape, size, mineral composition, arrangement and orientation, etc. of sediments (Griffith, 1958).

Seeds, grains, fruits and vegetables are irregular in shape and a complete specification of their form theoretically requires an infinite number of measurements. From practical point of view, measurement of several mutually perpendicular axes is, however, sufficient. The number of these measurements increases with increase in irregularity of the shape. It is, therefore, important to know what criterion should be used to decide when adequate number of measurements has been made to define the form of the object. In an attempt to establish such a criterion, Griffith (1964) related the volume of a set of specimens of pebbles to their axial dimensions using the following relationship:

$$V = a_1^{b_1}\, a_2^{b_2}\, a_3^{b_3} \dots a_n^{b_n} \tag{3.4}$$

where V is the volume of the specimen and $a_1, a_2, a_3, \dots a_n$ are diameters within the body considered as measures of size. Taking logarithm of both sides of the above equation yields the linear expression

$$\log V = b_1 \log a_1 + b_2 \log a_2 + b_3 \log a_3 + \cdots + b_n \log a_n \tag{3.5}$$

Now using multiple linear regression procedure (Goulden, 1952; Quenouille, 1952), volume was related to axial dimensions and the contribution of each axis to volume was determined using the analysis of variance technique (component analysis). It was found that a well-defined linear relationship existed between the log of axial dimensions and log volume of the pebbles. The three mutually perpendicular axes accounted for some

93 percent of variation in volume. Of this total percentage, the intermediate axis contributed only 4 percent to volume prediction. In other words, the measurement of only two axes, major and minor axes in this case, conveyed the bulk of the information on variation in log volume. As will be seen later, the third axis may be simultaneously measured, however, with little added inconvenience.

The above technique was used on 50 kernels of dry-shelled corn by measuring the major, minor and intermediate axes as well as weight and specific gravity of each kernel. The volume of the kernel was taken as one of the parameters defining the shape of the kernel and the three mutually perpendicular axes were taken as a measure of size of the kernels. The relationship between shape, size, weight, and specific gravity at different sections of the corn cobs of the same variety as well as a mixture of shelled corn consisting of several varieties are shown in Table 3.2. Note the increasing variance in axial dimensions for samples taken from the middle section of single ears of one variety as compared with those taken from the whole ear of one variety and the whole ears of several varieties. Also, note that correlation coefficients between volume and axial dimensions are higher for the single variety than the mixture of several varieties. From this experiment one may conclude that if estimates of volume are adequate criterion, then measurement of three axial dimensions (a, b, and c) should supply the bulk of the information on shape and size of such irregular objects as corn kernels. The methods for measuring these axial dimensions and their use in describing shape and size are discussed in the following sections.

3.2 CRITERIA FOR DESCRIBING SHAPE AND SIZE

Charted standards

In this method tracings of longitudinal and lateral cross sections of the material can be compared with the shapes listed on a charted standard. Figure 3.1 shows examples of charts prepared for apples, peaches, and potatoes. Using standard charts, the shape of the product can be defined either by a number on the chart or by descriptive terms such as the following prepared for fruits and vegetables (Mohsenin, ed., 1965):

Shape *Description*

Round—aproaching spheroid

Oblate—flattened at the stem end and apex

Table 3.2 Relationship between three axial dimensions, volume, and specific gravity in 50-kernel samples of shelled corn

Sample and statistics	Major diameter a mm	Intermediate diameter b mm	Minor diameter c mm	Weight W gm	Specific gravity SG	Volume W/density cc	Volume Reg. eq. cc
Single variety middle of ear							
Mean	10.78	7.58	4.57	0.27	1.22	0.222	0.221
Variance	0.475	0.214	0.151	0.001	0.002	0.001	—
Correlation with volume-r	0.66	0.59	0.27	—	—	1.00	—
Regression eq. $\ln V = -7.14 + 1.18 \ln a + 1.04 \ln b + 0.48 \ln c$							
Single variety whole ear							
Mean	11.15	7.30	4.86	0.27	1.24	0.222	0.221
Variance	1.071	0.670	0.586	0.003	0.001	0.002	—
Correlation with volume-r	0.75	0.78	0.30	—	—	1.00	—
Regression eq. $\ln V = -6.19 + 0.62 \ln a + 1.13 \ln b + 0.6 \ln c$							
Mixture of varieties							
Mean	12.66	8.5	5.24	0.35	1.24	0.28	0.29
Variance	1.72	0.77	0.5	0.004	0.004	0.002	—
Correlation with volume-r	0.35	0.41	0.32	—	—	1.00	—
Regression eq. $\ln V = -6.18 + 0.9 \ln a + 0.71 \ln b + 0.68 \ln c$							

Oblong—vertical diameter greater than the horizontal diameter
Conic—tapered toward the apex
Ovate—egg-shaped and broad at the stem end
Oblique (lopsided)—axis connecting stem and apex slanted
Obovate—inverted ovate
Elliptical—approaching ellipsoid
Truncate—having both ends squared or flattened
Unequal—one-half larger than the other
Ribbed—in cross section, sides are more or less angular
Regular—horizontal section approaches a circle
Irregular—horizontal cross section departs materially from a circle

Visual comparison of the shape of the object with charted standards is a very simple technique but is a psychophysical subjective assessment which suffers from personal prejudice so that different observers achieve different results. It requires elaborate precautions and an experienced observer to ensure reasonable reproductibility. The procedure has been used for estimate of color and various other physical attributes of various materials and shape and size study in geological work petrography (Curray, 1951).

Roundness

Roundness is a measure of the sharpness of the corners of the solid. Several methods have been proposed for estimating roundness. Those least objectionable are given below (Curray, 1951).

$$\text{Roundness} = \frac{A_p}{A_c} \tag{3.6}$$

where A_p = largest projected area of object in natural rest position
A_c = area of smallest circumscribing circle

The object area is obtained either by projection or tracing (Fig. 3.2).

$$\text{Roundness} = \frac{\sum r}{NR} \tag{3.7}$$

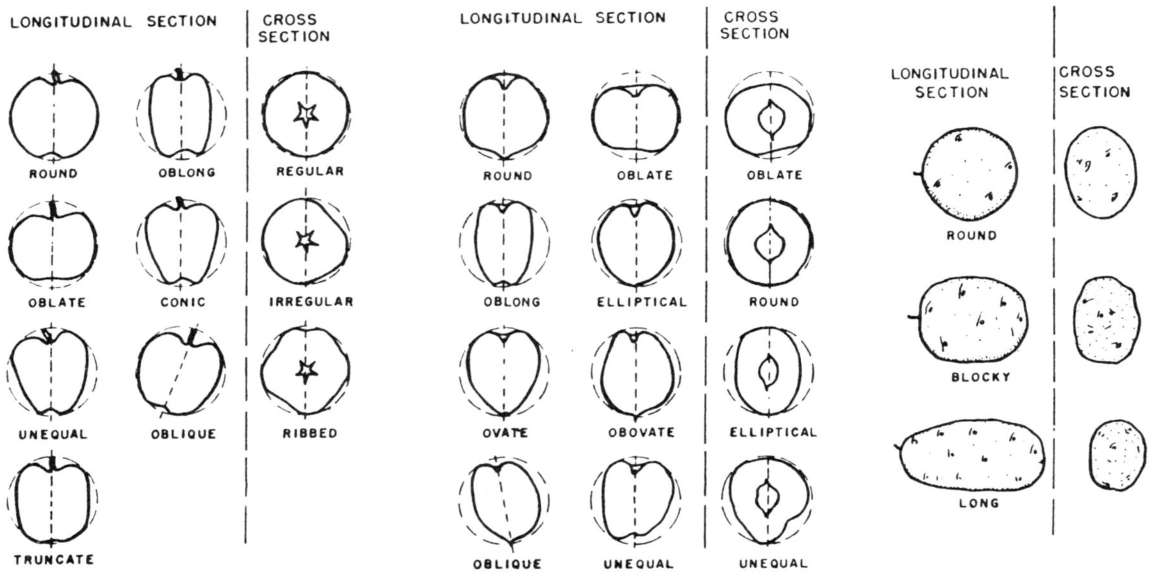

Figure 3.1 Example of charted standard for describing shape of fruits and vegetables. From left to right: apples, peaches, potatoes

where r = radius of curvature as defined in Fig. 3.2
R = radius of the maximum inscribed circle
N = total number of corners summed in numerator

$$\text{Roundness ratio} = \frac{r}{R} \tag{3.8}$$

where R in this case is the mean radius of the object and r is the radius of curvature of the sharpest corner.

The objection to this method is that the radius of curvature of a single corner determines the roundness or flatness (Fig. 3.2).

Sphericity

The geometric foundation of the concept of sphericity rests upon the isoperimetric property of a sphere. A practical three-dimensional expression

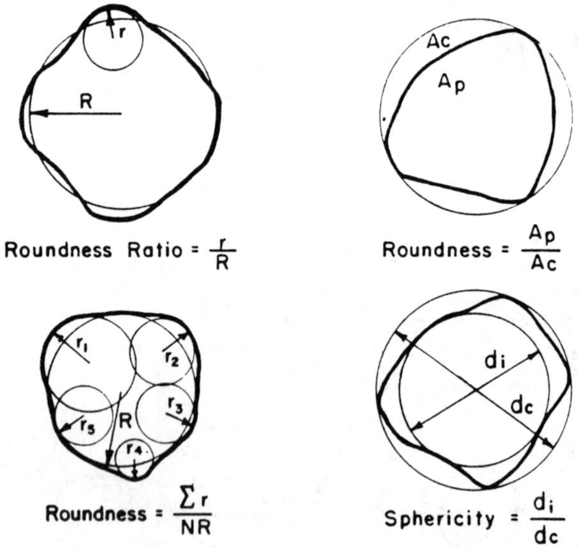

Figure 3.2 Roundness and sphericity as defined by geologists to describe shape of grains and pebbles (Curray, 1951)

can be stated for estimating the sphericity of an object using the following definition:

$$\text{Sphericity} = \frac{d_e}{d_c} \tag{3.9}$$

where d_e is the diameter of a sphere of the same volume as the object and d_c is the diameter of the smallest circumscribing sphere or usually the longest diameter of the object (Curray, 1951). This expression for sphericity expresses the shape character of the solid relative to that of a sphere of the same volume.

Assuming that the volume of the solid is equal to the volume of a triaxial ellipsoid with intercepts a, b, c, and that the diameter of the circumscribed sphere is the longest intercept a of the ellipsoid, the degree of sphericity can also be expressed as follows:

$$\text{Sphericity} = \left(\frac{\text{Volume of solid}}{\text{volume of circumscribed sphere}}\right)^{1/3}$$

$$= \left[\frac{(\pi/6)\,abc}{(\pi/6)\,a^3}\right]^{1/3} = \left(\frac{bc}{a^2}\right)^{1/3} \tag{3.10}$$

$$= \frac{\text{geometric mean diameter}}{\text{major diameter}} = \frac{(abc)^{1/3}}{a} \tag{3.11}$$

where a = longest intercept
b = longest intercept normal to a
c = longest intercept normal to a and b

The intercepts need not intersect each other at a common point.

Another definition of sphericity is given by

$$\text{Sphericity} = \frac{d_i}{d_c} \tag{3.12}$$

where d_i=diameter of largest inscribed circle and d_c=diameter of smallest circumscribed circle as shown in Fig. 3.2 (Curray, 1951). Sphericity of several fruits are given in Table 3.3.

Table 3.3 Percent sphericity of several fruits based on equation (3.10)

Product	Sphericity	Product	Sphericity
Apples		Blueberries	90
McIntosh	90	Cherries	95
Melba	92	Peaches	
Golden Delicious	92	Red Haven	93
Red Delicious	92	Elberta	97
Stayman	90	Pears	
Rome	89	Maxine	89

Measurement of axial dimensions

For small objects such as seeds, the outline of the projection of each sample can be traced using a photographic enlarger. The seed is placed on the plane where the negative is positioned, turned so that its shadow covers the largest area. Then the enlarger is focused to give a sharp boundary. A millimeter scale is also traced along with the seed image. The seed is then turned to show a minimum projection area whose long dimension is equal to the long dimension of the maximum projection area. After tracings of the projected maximum and minimum areas are obtained, the *a*, *b*, and *c* axes are measured from these drawings. The *a* and *b* axes are measured from the drawings of the maximum area. A set of parallel rules can be used to draw the tangents to the seed outline in the narrowest area of the tracing and the perpendicular lines to these tangents. The *a* axis is the longer of the rectangular sides, while *b* axis is the shorter. The same procedure is followed on the outline of the minimum projection to determine the *c* axis. Fig. 3.3 shows the shape and the three axes *a*, *b*, and *c* of seeds and grains obtained by the above method.

A faster and more accurate method is the use of a shadowgraph similar to that shown in Fig. 3.4. The use of two micrometers on this instrument permits direct measurement of two axial dimensions for the natural rest or other positions of the seed. A tracing of the seed outline can also be obtained from the shadow of the seed on the ground glass of the instrument.

Resemblance to geometric bodies

In some cases the shape can be approximated by one of the following geometric shapes:

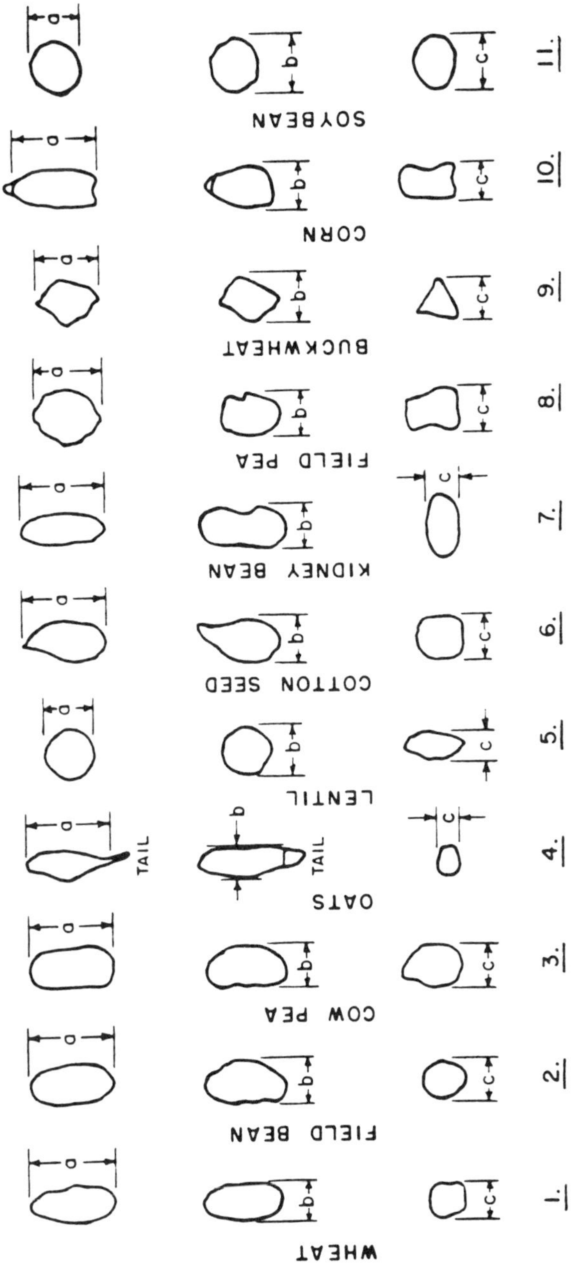

Figure 3.3 Tracings of shape and designation of the three intercepts for seeds and grains obtained by a photographic enlarger

Figure 3.4 Nikon shadowgraph for measurement of shape and size of seeds

1 Prolate spheroid which is formed when an ellipse rotates about its major axis like a lemon.
2 Oblate spheroid which is formed when an ellipse rotates about its minor axis like a grapefruit.
3 Right circular cone or cylinder like carrots and cucumbers.

After deciding on the shape of the body, its volume and surface area can be calculated using the appropriate equations. For some agricultural products, the following formulas may be applicable. Volume, V, and surface area, S, of a prolate spheroid are given by

$$V = 4/3\,(\pi ab^2); \quad S = 2\pi b^2 + 2\pi \frac{ab}{e} \sin^{-1} \mathrm{e} \tag{3.13}$$

where a and b are respectively major and minor semi-axes of the ellipse of rotation and e is eccentricity given by $e=\left[1-\left(\frac{b}{a}\right)^2\right]^{1/2}$. Volume, V, and surface area, S, of an oblate spheroid are given by

$$V = 4/3\ (\pi a^2 b);\quad S = 2\pi a^2 + \pi \frac{b^2}{e} \ln \frac{1+e}{1-e} \tag{3.14}$$

Volume, V, and surface area of the frustum of a right cone are given by

$$V = (\pi/3)\, h(r_1^2 + r_1 r_2 + r_2^2);$$
$$S = \pi(r_1 + r_2)\,[h^2 + (r_1 - r_2)^2]^{1/2} \tag{3.15}$$

where r_1 and r_2 are respectively the radii of base and top and h is the altitude.

Having volume and surface area estimated in this manner, the actual volume and surface area can then be determined experimentally and a correction factor can be established for the "typical" shape of each variety of the product.

Average projected area

In developing a sizing machine for lemons, Houston (1957) proposed a new criterion of size, defined as the average of the projected areas taken along three mutually perpendicular axes. This criterion, referred to as the criterion area A_c, is defined and illustrated in Fig. 3.5. To test the validity of this criterion, a relationship was developed between the true average projected area A (when all possible directions of projection are considered) and the volume of the object. The calculated volume based on A was then compared with the actual volume determined experimentally.

Based on the theories of convex bodies (Bannesen and Fenchel, 1948) it has been established that

$$\frac{V^2}{S^3} \geqq \frac{1}{36\pi} \tag{3.16}$$

where V is the volume and S is the surface area of the convex body. It has also been shown that the average projected area of a convex body is one-fourth the surface area (Polya and Szega, 1951). Substituting $S=4A$ in the above expression

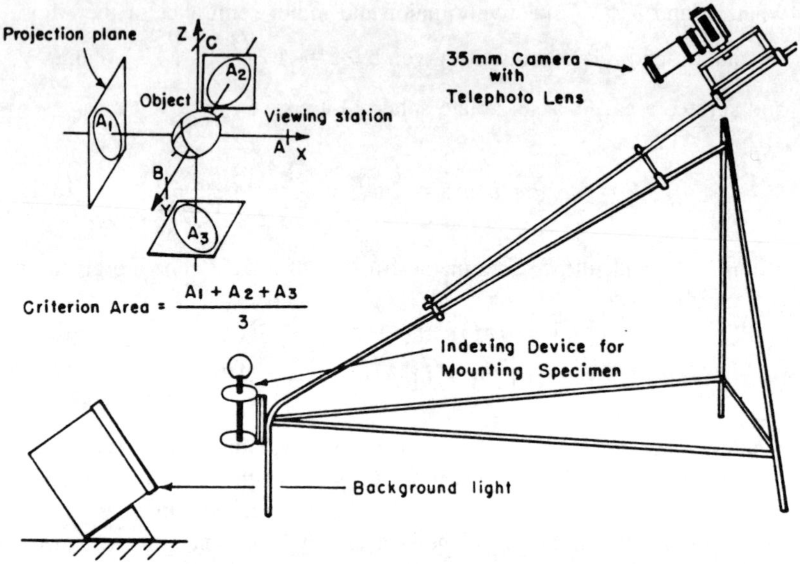

Figure 3.5 Camera set up for recording the criterion area (above left) of fruits and vegetables for several orientations (Redrawn from Houston, 1957a)

$$A \leqq \left(\frac{9\pi}{16}\right)^{1/3} V^{2/3}$$

or

$$A \leqq KV^{2/3} \tag{3.17}$$

where K is a constant. For a sphere where equality is achieved,

$$K = \left(\frac{9\pi}{16}\right)^{1/3} = 1.21$$

Note that this is another method for estimating the degree of sphericity of a convex body (Figs. 3.5 and 3.6).

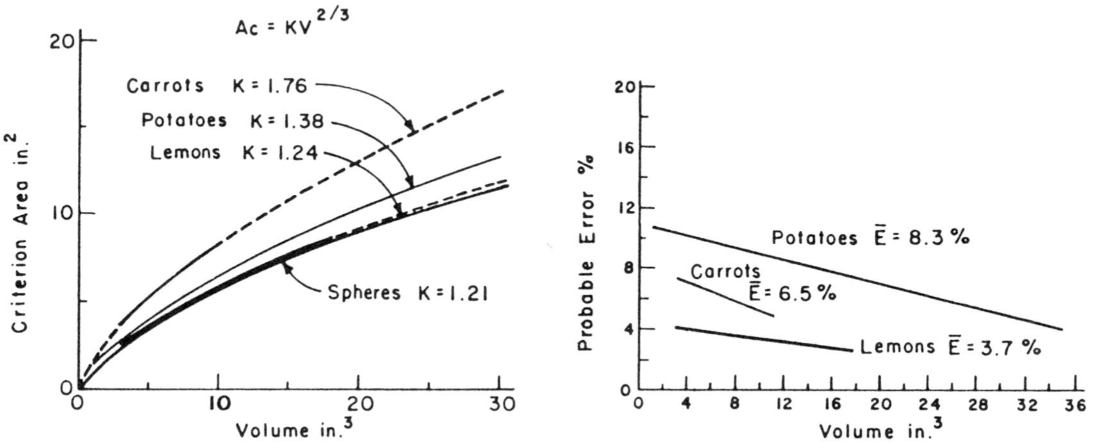

Figure 3.6 (left) Relationship between criterion area and volume for carrots, lemons, and potatoes as compared with the computed curve of spheres; (right) Probable error in volume measurement based on $A = KV^{2/3}$ and the true volume, $\bar{E}$ = mean probable error (Houston, 1957b)

In using Eq. (3.17) the following restrictions should be taken into consideration:

1 The equality is achieved only for spheres.
2 The dimensionless K constant varies with the characteristic dimensions of the object. To keep the K value reasonably constant, a "typical" shape must be defined for each product such that the shape factors of the individual samples fall within a specified range.
3 The variations of A_c, as defined in Fig. 3.5, for unrestricted orientation is too great for a satisfactory measure of volume. However, this variation can be reduced (Houston, 1957) to an acceptable level by restricting the bodies to a natural rest position and have the three axes of projection form equal angles with the horizontal plane. The unrestricted orientation would occur only if the object were to pass through the measuring unit of the sizing machine in free fall. Free fall excludes use of the machine for products susceptible to mechanical damage.

Houston (1957) obtained measurements on several products and determined the value of K for lemon, potato and carrot as shown together with that of a true sphere in Fig. 3.6. According to these curves, lemon represents a nearly spherical shape while carrot represents an elongated shape. Such products as potato, tomato, apples, pears, and plums lie between these two shapes. In other words, the computed curves for carrot and sphere bracket the range of shapes that a sizing machine would have to handle. The probable error in predicted volume as compared with the predicted true volume is shown in Fig. 3.6. Note that the large variation in characteristic dimensions of such products as potatoes results in large mean probable error. Also, note that as the size of the object decreases, the mean probable error increases.

3.3 VOLUME AND DENSITY

Density and specific gravity of food materials and agricultural products play an important role in many applications. Drying and storing of hay (Day and Panda, 1965), design of silos and storage bins (Otis, 1957), mechanical compressing of ensilages (Graham, 1965), stability of feed pellets and wafers (Gustafson and Kjelgaard, 1963), separation from undesirable materials (Maak, 1957), determining the purity of seeds (Stermer, 1965), separation and grading (Kunkel *et al.*, 1952), maturity evaluation

(Lee, 1948), texture and softness of fruits (LaBelle, 1964), estimation of air space in plant tissues (Davis, 1962), and quality evaluation of products such as peas, sweet corn, lima beans, and potatoes which increase in their density as they mature (Gould, 1957; Burton, 1938) are some of the examples where density (bulk or unit) or specific gravity of the material has found applications.

The irregular shape of most intact agricultural products, the small size of materials such as seeds and grains, and the porous nature of others such as feed pellets and wafers present certain problems in volume and density measurement. Because of the irregular shape of the product, volume is usually determined by water displacement.

Platform scale

A simple technique which applies to large objects such as fruits and vegetables is the platform scale illustrated in Fig. 3.7. The fruit is first weighed on the scale in air and then forced into the water by means of a sinker rod. The second reading of the scale with the fruit submerged minus the weight of the container and water is the weight of the displaced water which will be used in the following expression to calculate volume:

$$\text{Volume (in}^3) = \frac{\text{Weight of displaced water (lb)}}{\text{Weight density of water (lb/in}^3)} \tag{3.18}$$

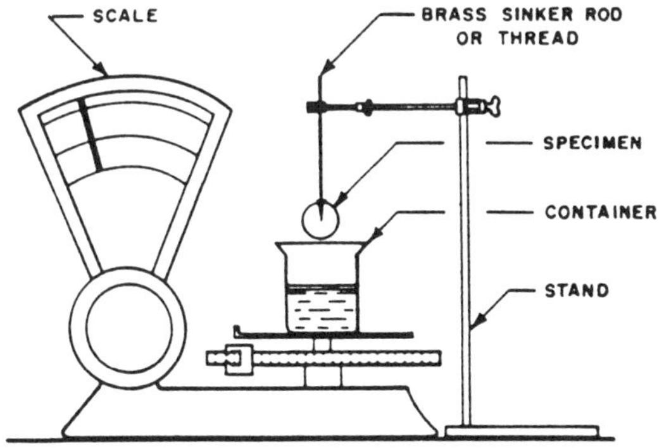

Figure 3.7 Platform scale for measurement of volume and density and specific gravity of large objects

Knowing the weight in air and the volume, weight density of the fruit is then obtained by the ratio of weight to volume. At the same time, specific gravity is calculated from

$$\text{Specific gravity} = \frac{\text{Weight in air} \times \text{specific gravity of water}}{\text{Weight of displaced water}} \tag{3.19}$$

Example

Assuming a specific gravity of 1.0 and a weight density of 62.4 lb/ft^3 for water, using the platform scale method, the volume and specific gravity of an apple was determined as follows:

weight of apple in air = 0.292 lb.

weight of container + water = 2.24 lb.

weight of container + water + apple submerged = 2.61 lb.

weight of displaced water = 2.61 − 2.24
= 0.37 lb.

$$\text{volume of apple} = \frac{0.37 \times 1728}{62.4}$$

$$= 10.16 \text{ in}^3$$

$$\text{specific gravity} = \frac{0.292 \times 1}{0.37}$$

$$= 0.787$$

Specific gravity balance

For smaller objects such as small fruits, peas and beans, kernels of corn, etc., an analytical balance or more conveniently a specific gravity balance (Fig. 3.8) can be used to determine volume, density, and specific gravity employing the following expressions.

If the solid is heavier than water,

$$\text{Volume} = \frac{\text{Weight in air} - \text{Weight in water}}{\text{Weight density of water}} \tag{3.20}$$

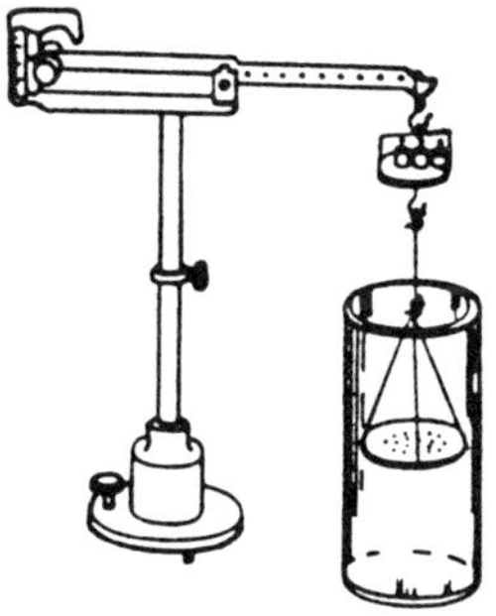

Figure 3.8 Specific gravity balance for measurement of volume, density and specific gravity of smaller objects (Courtesy Arthur H. Thomas Co.)

$$\text{Specific gravity} = \left[\frac{\text{Weight in air}}{\text{Weight in air} - \text{Weight in water}}\right][(SG)_L] \quad (3.21)$$

where $(SG)_L$ is the specific gravity of water.

If the solid is lighter than water, attach another solid, heavier than water to the object as a sinker and find specific gravity from the following expression:

$$\text{Specific gravity} = \left[\frac{(W_a)\text{ object}}{(W_a - W_w)\text{ both} - (W_a - W_w)\text{ sinker}}\right][(SG)_L] \quad (3.22)$$

where W_a=weight in air and W_w=weight in water.

A solution of 3 cc wetting agent in 500 cc distilled water will reduce errors due to surface tension and submergence in water.

Specific gravity gradient tube

A fast and accurate method for determining volume and weight density of agricultural products is the use of specific gravity gradient tube shown in Fig. 3.9. The technique is based on observing the level to which a test specimen sinks in a liquid column exhibiting a density gradient in comparison with standard glass floats. The only requirement is that the specimen must be impervious to the liquid in the column until the liquid and test specimen have reached equilibrium and a reading is obtained.

The tube illustrated in Fig. 3.9 is constructed basically according to the ASTM tentative method for density determination of plastics (ASTM Designation: D1505–63T). This particular apparatus utilizes a commercially

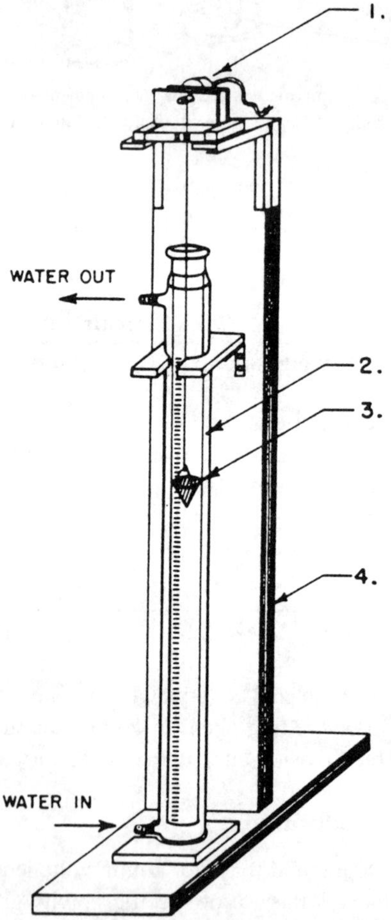

Figure 3.9 Specific gravity gradient column for small seeds, grains and other solids (1-clock motor, 2-water jacket, 3-elevating basket, 4-stand)

available water-jacketed tube and glass floats calibrated to four decimal places. However, any suitable graduated cylinder with ground-glass stopper can be used if temperature is kept constant-through a constant temperature bath or other means. Table 3.4 gives liquid systems for specific gravity gradient tube.

In conducting the specific gravity test, the specimen is placed gently in the tube allowing the liquid and specimen to reach equilibrium. At this time the height of the specimen in the tube is read and by reference to a calibration curve (Fig. 3.10), the specific gravity is determined. If equilibrium is not obtained, liquid may be penetrating the specimen. The calibration curve is merely a plot of float position versus float specific gravity. If calibration chart is not available, a method of interpolation can be used employing the following expression:

$$\text{Specific gravity at } x = a + \frac{(x - y)(b - a)}{(z - y)} \tag{3.23}$$

Table 3.4 Liquids suitable for preparation of a specific gravity gradient tube (ASTM Designation: D1505–63T)

Liquid	Specific gravity
Methanol – benzyl alcohol	0.80 to 0.92
Isopropanol – water	0.79 to 1.00
Isopropanol – diethylene glycol	0.79 to 1.11
Ethanol – carbon tetrachloride	0.79 to 1.59
Toluene – carbon tetrachloride	0.87 to 1.59
Water – sodium bromide	1.00 to 1.41
Water – calcium nitrate	1.00 to 1.60
Carbon tetrachloride – trimethylene dibromide	1.60 to 1.99
Trimethylene dibromide – ethylene bromide	1.99 to 2.18
Ethylene bromide – bromoform	2.18 to 2.89

where

a and b = specific gravities of two standard floats
y and z = distances of the two standard floats a and b, bracketing the unknown, respectively, measured from a reference level
x = distance of the unknown from the same reference level

A wire screen basket attached to a long wire and an electric clock motor forms an elevator for removing the old specimens from the tube at such a slow rate so that the gradient is not disturbed. The accuracy of the technique depends on gradient of specific gravity of the liquid per mm height of the column. Columns with sensitivities as high as 0.0001 per mm height can be constructed.

Air comparison pycnometer

This is a commercially available instrument (Beckman Instrument, Inc.) for volume measurement. The apparatus consists basically of two chambers and two pistons, a valve connecting the two chambers, a differential pressure indicator, and a digital counter calibrated for readings in cubic centimeters (Fig. 3.11a). With the connecting valve closed any change of the position of one piston must be duplicated by an identical stroke in the other in order to maintain the same pressure on each side of the differential pressure indicator. If the connecting valve is closed and both pistons are advanced the same amount to a given position, inserting a sample in the

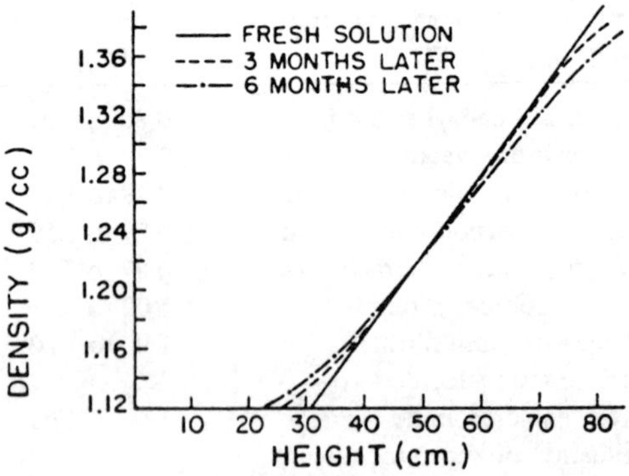

Figure 3.10 Effects of time and use of the density gradient column on the calibration curve (Shelef and Mohsenin, 1968)

measuring chamber would cause a pressure differential which is brought to zero by withdrawing the piston in this chamber. Under this condition, the distance that the measuring piston differs from its position before inserting the sample will be proportional to the volume being measured.

This instrument measures the true volume of a sample. For measurement of the apparent volume, i.e., the volume of a sample enclosed by its outer surface plus the volume of its open pores, the manufacturer recommends filling the pores first by immersing the sample in a molten wax bath. Knowing both the apparent and true volume of a sample, the open-pore volume can be calculated as an index of porosity. Figure 3.11b shows the use of this instrument in determining bulk density of grains.

Pycnometer method

For seeds and grains, the method of specific gravity bottle or pycnometer and toluene has been the practice for many years (Bailey, 1912). Toluene ($C_6H_5CH_3$) has the advantages of (1) little tendency to soak into the kernel; (2) a low surface tension, enabling it to flow smoothly over the kernel surface; (3) little solvent action on constituents of the kernel especially fats and oils; (4) a fairly high boiling point; (5) not changing its specific gravity and viscosity materially on exposure to the atmosphere; and (6) having a low specific gravity.

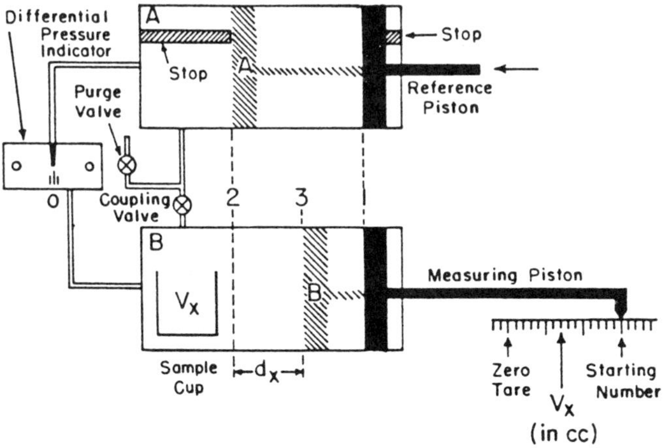

Figure 3.11a The air comparison pycnometer (Courtesy Beckman Instrument, Inc.)

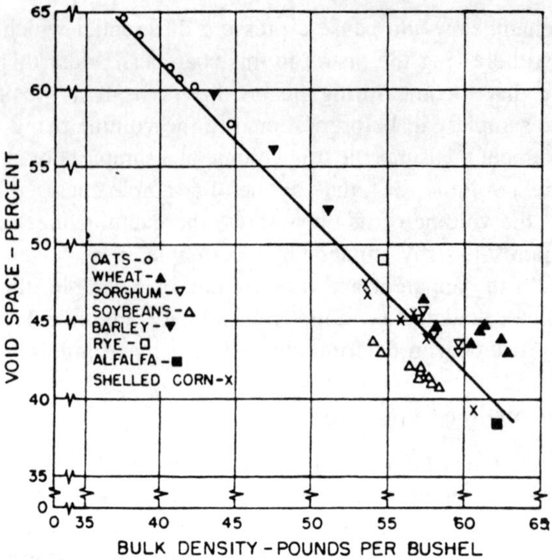

Figure 3.11b Correlation between void space and bulk density in different grains as determined by the air-comparison pycnometer shown in figure 3.11a (Thompson and Isaacs, 1967)

The procedure involves the following steps:

1 Determine the exact capacity of the pycnometer by weighing it when empty and likewise when filled with distilled water at 20°C.
2 Specific gravity of the batch of toluene is determined by comparing the weight of toluene which the bottle holds with the weight of the distilled water at the same temperature.

$$\text{Specific gravity of toluene} = \frac{\text{Weight of toluene}}{\text{Weight of water}}$$

3 Ten grams of the solid specimen are placed in the pycnometer with sufficient toluene to cover the sample.
4 Gradually exhaust the air from the bottle by a vacuum pump to promote the escape of the air trapped under the surface hairs and in the creases of the seeds or kernels.

5 When air bubbles cease to be given off after several cycles of vacuuming and releasing the vacuum, fill the bottle with toluene and allow the temperature to reach 20°C.
6 Weigh the bottle and calculate specific gravity of the seeds as follows:

$$\text{Specific gravity of grain or seed} = \frac{\text{Specific gravity of toluene} \times \text{Weight of the grain}}{\text{Weight of the toluene displaced by the grain}}$$

Weight of toluene displaced by the grain is found by subtracting the difference in bottle weights when filled with toluene and when containing the grain from the weight of the grain sample (10 grams in this case).

As an example, consider the volume measurement for a sample of 16 corn kernels coated with Pliabond.

weight of sample = 4.598 gm
weight of pyncnometer = 55.6468
weight of pyncnometer and toluene = 78.2399
weight of pyncnometer plus toluene and sample = 79.6226
weight of pyncnometer and water = 81.7709

$$\text{specific gravity of toluene} = \frac{78.2399 - 55.6468}{81.7709 - 55.6468} = 0.8648$$

$$\text{specific gravity of corn} = \frac{0.8648 \times 4.4598}{[4.4598 - (79.6226 - 78.2399)]}$$

$$= 1.256$$

volume of sample = 4.4598/1.256

= 3.558 cu.cm.

The same sample tested with an Air Comparison Pycnometer, previously mentioned, gave a volume of 3.49 cu. cm. or about 2% less than the conventional pycnometer method.

Unit bulk density of porous materials

For unit bulk density measurement of porous materials such as stones, collected along with potato in harvesting, feed pellets, hay wafers, and other products with surface openings several methods have been tried with different degrees of success.

For such porous materials as stones, the NE-44 Regional Research Project (Mohsenin, ed., 1965) has recommended the use of the platform scale technique to obtain the following data:

W_{ad} = dry weight of object in air
W_{w24} = weight of object in water after being submerged in water for 24 hours
W_{aw} = weight of object in air after being removed from water and free moisture removed
$(SG)_L$ = specific gravity of liquid

From these data the following can be determined:

$$\text{Specific gravity of dry object} = (SG)_D = \left[\frac{W_{ad}}{W_{aw} - W_{w24}}\right][(SG)_L] \tag{3.24}$$

$$\text{Specific gravity of wet object} = (SG)_W = \left[\frac{W_{aw}}{W_{aw} - W_{w24}}\right][(SG)_L] \tag{3.25}$$

$$\text{Specific gravity of material making up the solid} = (SG)_A = \left[\frac{W_{ad}}{W_{ad} - W_{w24}}\right][(SG)_L] \tag{3.26}$$

For measuring unit bulk weight density of feed pellets and wafers several methods have been tried to prevent the liquid from penetrating the packaged feed. Paraffin coating, plastic bag and even dipping in a fine granular material such as sand instead of a liquid, have been tried with little success. The present proposed ASAE standard (ASAE Yearbook, 1966) involves the placing of the specimen in a plastic bag and finding the volume by water displacement. Arching of plastic membrane over protruding surfaces and presence of air pockets in the package are among the objections to this method.

Radiation method

Measurement of silage density in an upright silo without removing samples by the use of a radioactive source has been reported (James *et al.*, 1962). The radioactive source (Cesium 137) is lowered in a pipe installed in the center of the silo before the silo is filled. A radiation detector outside the silo measures the amount of radiation coming through the silage. The transmitted radiation decreases with an increase in silage density as settling takes place. The method does not require making openings in the silo, which cause aeration and decomposition, and offers continuous determination of silage pressures. Excessive silage pressures cause silo rupture.

3.4 POROSITY

The percent voids of an unconsolidated mass of materials such as silage, hay, grain and other porous materials is often needed in air flow and heat flow studies as well as other applications. Day (1964) has given a simple method for determination of porosity which is illustrated in Fig. 3.12.

With the material in tank 2, valve 2 is closed and air is supplied to tank 1. When suitable manometer displacement is achieved, valve 1 is closed and after the manometer has come to equilibrium pressure p_1 is read. Under this condition according to the well known perfect gas law

$$p_1V_1 = MR_1T_1$$

where p_1 is absolute pressure, V_1 is volume in tank 1, M is mass of air, R_1 is gas constant for air, and T_1 is absolute temperature. Now, valve 3 is closed and valve 2 is opened and the pressure p_3 is read. Under this condition with valves 1 and 3 closed, the total mass of air, M, is divided into M_1 to fill tank 1 and M_2 to fill pore space V_2 in tank 2. Assuming $R_1T_1=R_2T_2=RT$, the following can be deducted:

$$M = M_1 + M_2$$

$$\frac{P_1V_1}{RT} = \frac{P_3V_1}{RT} + \frac{P_3V_2}{RT}$$

from which percent pore volume can be found to be

$$V_2/V_1 = \frac{P_1 - P_3}{P_3} \tag{3.27}$$

Example

To determine the porosity of dry shelled corn, tank 2 of the apparatus in Fig. 3.12 is filled with a sample of this corn to a bulk density of 47 lb/ft^3. The pressure readings were $P_1=15.2$ and $P_3=10.4$ inches of mercury.

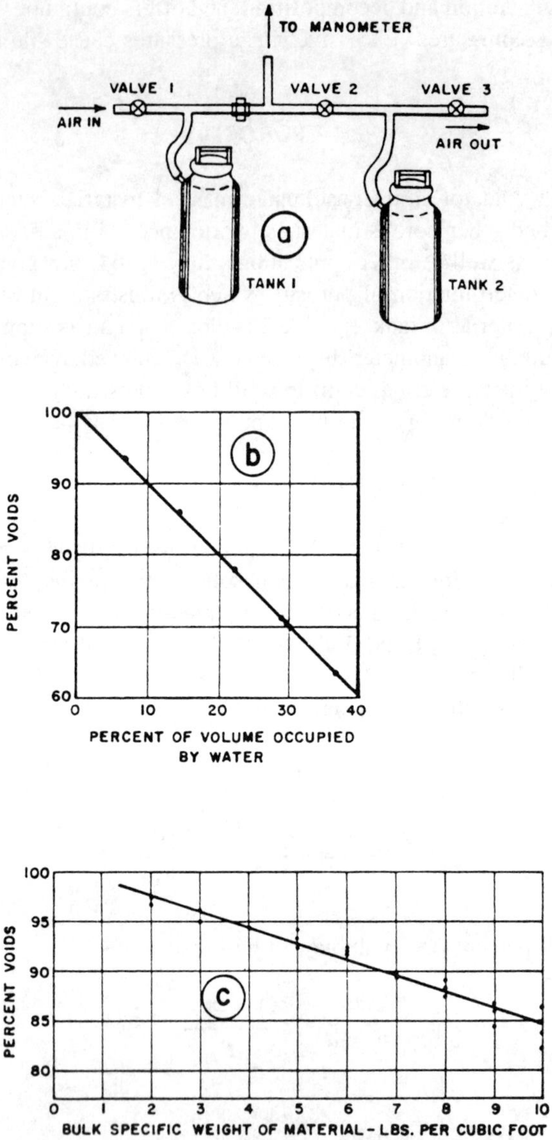

Figure 3.12 Porosity of granular and unconsolidated materials (Day, 1964), a) Apparatus, b) Calibration data with water, c) Bulk specific weight versus percent void in chopped alfalfa hay

Using Eq. (3.27), porosity of the corn at 47 lb/ft^3 density is found to be

$$V_2/V_1 = \frac{15.2 - 10.4}{10.4}$$

$$= 0.46$$

Figures 3.12b and 3.12c show calibration of the apparatus using water in tank 2 and percent voids in chopped alfalfa hay. Extrapolation of the straight line, fitted by the method of least squares, to zero porosity crosses the horizontal axis at a specific weight of 65 lb/ft^3. This value is in good agreement with specific weight of commercially available pressed wood.

Porosity, which is also referred to as packing factor, *PF*, may be calculated from the following relationship.

$$PF = \frac{\text{solid density of particles} - \text{density of packing}}{\text{solid density of particles}} \tag{3.28}$$

Solid density may be determined by one of the methods given for density measurement. Density of the packing may be found by weighing a given volume of the packed particles.

Example

A soil pycnometer (Pall and Mohsenin, 1978) is designed on the principle of porosity tanks of Figure 3.12. Using this apparatus, the field porosity of a soil was determined to be 16.78 percent. The soil moisture content was 2.65% (d.b.) and its dry bulk density was 1.225 gm/cc. If the volume of the measuring chamber were 728 cc. find the porosity of the dry soil (total porosity).

728 × 1.225 = 891.8 gm, dry soil
891.8 gm × 0.2965 gm water/gm soil = 264.42 gm water or cm^3 water
Total porosity = field porosity + porosity taken up by water
= 16.78 + 264.42/728
= 53.1%

3.5 SURFACE AREA

A knowledge of surface area of some parts of plant materials, such as leaf area and surface area of fruits, is important to plant scientists as well as engineers handling and processing the products. Leaf area is an indicator of photosynthetic capacity and growth rate of a plant and its measurement is of value in studies of plant competition for light and nutrients, plant-soil-water relations, determining application rates of insecticides and fungicides. In a crop like tobacco, where the leaf is the major commercial product, leaf area is a good indicator of yield potential. Likewise, surface areas of fruits are important in investigations related to spray coverage, removal of spray residues, respiration rate, light reflectance and color evaluation, and heat transfer studies in heating and cooling processes.

Leaf and stalk surface area

Some of the methods used for measuring leaf and stalk surface area are contact printing the surface on a light sensitive paper and measuring the area by a planimeter; tracing the area on a graph paper and counting the squares; use of a photographic projector similar to the method described for seeds and grains; light interception method and the use of photronic cell to measure the intercepted light (Miller *et al.*, 1959); use of an air flow planimeter which measures the area of a function of the surface obstructing the flow of air (Jenkins, 1956; Prince, 1966) and a measurement of length and width of a leaf and relating these measurements to the area of the leaf (Suggs *et al.*, 1960; McKee, 1964). Figure 3.13 shows an empirical relationship found between product of length and width of tobacco leaf and its area. The same type of relationship has been found for cabbage leaf, field corn and some other crops.

Jenkins (1956) compared three of the above mentioned methods and found that the method of air flow planimeter is a reliable and the fastest method as shown in Table 3.5.

Figure 3.14 shows the schematic diagram of an air flow planimeter constructed following the design suggested by Prince (1966). He employed the apparatus to measure the cross section area of corn stalks. The principal parts of the apparatus are a blower, a pitot tube, a horizontal and a vertical pipe, a 16-mesh screen attached to a wooden ring to support the specimen, and a micromanometer. Calibration curve of the apparatus is also shown in Fig. 3.14. The following example illustrates the use of the apparatus.

Example

To check the calibration curve for the air flow planimeter illustrated in Fig. 3.14 the averages of four readings on the micromanometer were as follows:

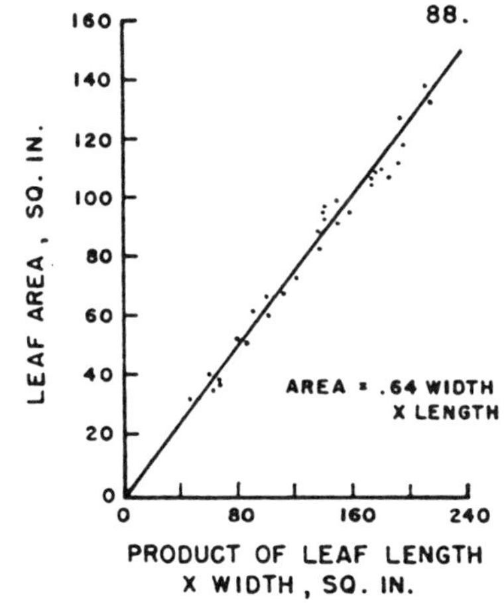

Figure 3.13 Prediction of surface area of tobacco leaf from the leaf length × width product (Suggs *et al.*, 1960)

Table 3.5 Comparison of leaf area measurements of rye grass by several methods (Jenkins, 1956)

No. of Leaves	Method of measurement		
	Length and width	Contact print	Air flow planimeter
		Area in sq.cm.	
5	35	37	36
5	34	37	36
10	72	72	74
10	67	71	71
15	113	105	103
15	125	126	127
20	143	141	142
20	153	146	148
		Time required in man-seconds per leaf	
	50	50	12

Reading with no air flow=0.049 inches of water

Reading with air flow and a 1.05 in^2 circular disk=0.309 inches of water

Corrected reading would be 0.309−0.049=0.260 inches of water.

Reference to the calibration curve in Fig. 3.14 gives an estimated area of 1.10 in^2 which is in error by about 5 percent.

Hall *et al.*, (1968) have reported the use of a Flying-Spot Particle Analyzer for measuring cross sectional area of alfalfa stems. Cross sectional areas of short sections of stems were first photographed on 35 mm film and the Flying Spot Particle Analyzer was used to scan the resulting image areas on the film. Analysis of 1000 specimens taken at one-millimeter intervals along the stem indicated considerable variation in cross-sectional area as a function of position along the stem. It was also pointed out that when hollowness of the stem is neglected, the area estimate based on outside diameter measurement only can become several times greater than that of the true stress-bearing portion of the stem.

For measurement of areas of tracings of leaf outlines a method which is simpler and less time consuming than the method of planimeter is cutting out the tracing for each leaf and weighing the cut out paper. For example, if the paper used for tracing weight 1 gram per 135 square centimeters, the surface area of each leaf in square centimeters can be obtained simply by multiplying the weight of the cut out paper for each specimen by 135.

Fruit surface area

For measurement of surface areas of fruits such as an apple, the fruit was

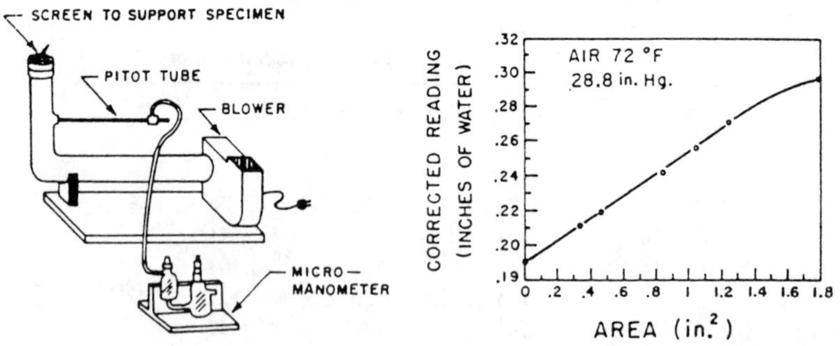

Figure 3.14 An airflow planimeter and its calibration curve

peeled in narrow strips and the planimeter sum of the areas of tracings of the strips was taken as the surface area of the apple (Baten and Marshal, 1943). This presumably actual surface area was then compared with the surface area estimated by each of the following methods:

1 A formula involving two elliptic integrals for surface area of an ellipsoid with axes $a>b>c$.
2 Areas of transverse cross sections.
3 Areas of axial or longitudinal cross sections.
4 Transverse diameters.
5 Axial or longitudinal diameters.
6 Weight of fruits.

It was found that the standard error of estimate obtained by predicting surface areas from areas of transverse cross sections, transverse diameter, and weight of fruits were smaller than those for other measurements tried and any one of these three methods may be used to predict approximate surface area. The most practical predicting equation for approximate surface areas of picked apples was found to be that based on weights of fruits and most suitable method for use on unpicked apples was the one based on measurement of transverse diameters. The latter can be done easily with a pair of calipers while the fruit is still attached to the tree.

Table 3.6 shows prediction equations for surface areas based on weights of several varieties of apples, pears, and plums. As seen from this table, the weights of plums and pears were also found to be the best indicator of their surface areas.

In a study of transient heat transfer in apples, the method of Baten and Marshal was employed to estimate the surface areas for McIntosh apples (Frechette and Zahradnik, 1965). A linear and a non-linear regression, using a least squares method and a second degree polynomial, respectively, indicated that the non-linear regression gave the best-fit curve for the data involving some 84 apples. For comparison, an equation for surface area of an equivalent sphere of the same weight as the apple was also derived. The prediction equations for the three methods and the corresponding curves are shown in Fig. 3.15. Because of the ease of computation, the high correlation coefficient (97.5%) and a maximum error of only 2.9% of the mean, and the fact that the non-linear fits cannot be used for extrapolation, the linear regression fit and the corresponding prediction equation were recommended for estimation of surface area.

Table 3.6 Prediction equations for estimating surface area from weights of fruits (Baten and Marshal, 1943)

Variety and year grown	Mean surface area (in^2)	Regression equation[1]	Std. dev. (in^2)	Coef. of var.[2]	Coef. of cor.
Apples:					
Jonathan, 1940	20.31	$S = 6.44 + 0.0136w$	0.48	2.31	0.97
Jonathan, 1941	18.22	$S = 5.31 + 0.146w$	0.52	2.85	0.96
Delicious, 1940	21.72	$S = 6.62 + 0.129w$	0.46	2.12	0.99
McIntosh, 1940	22.62	$S = 7.12 + 0.129w$	0.40	1.77	0.98
Stayman, 1940	24.35	$S = 8.42 + 0.11w$	0.62	2.55	0.99
Wagner, 1941	25.17	$S = 7.09 + 0.121w$	0.58	2.30	0.98
Gravenstein, 1941	25.49	$S = 8.24 + 0.116w$	0.52	2.04	0.96
Chenango, 1941	19.29	$S = 6.91 + 0.128w$	0.58	3.01	0.94
Grimes Golden, 1941	17.25	$S = 5.89 + 0.125w$	0.70	4.06	0.92
Average		$S = 6.72 + 0.129w$			
McIntosh, 1963[3]	24.32	$S = 7.82 + 0.11w$	0.70	2.9	0.98
Pears:					
Anjou, 1941	23.05	$S = 7.4 + 0.097w$	0.6	2.59	0.98
Bosc, 1941	22.18	$S = 7.49 + 0.101w$	0.8	3.58	0.96
Bartlett, 1941	22.54	$S = 7.58 + 0.101w$	0.69	3.07	0.97
Average		$S = 7.49 + 0.99w$			
Plums:					
Pond	7.01	$S = 2.49 + 0.138w$	0.22	3.14	0.95
Monarch	5.43	$S = 1.88 + 0.159w$	0.15	2.71	0.94
Average		$S = 2.18 + 0.149w$			

[1] S = fruit surface area (in^2), w = fruit weight (gm)
[2] Error in percent of the mean or 100 × std. dev./mean
[3] Frechette and Zahradnik (1965)

Egg surface area

The relationship between weight and surface area of products has also been studied for eggs. An empirical equation of the following type has been reported for fresh eggs (Besch *et al.*, 1968)

$$S = kW^m \tag{3.29}$$

where k is a constant with reported values varying from 4.56 to 5.07 and m=0.66. Units for S and W are cm^2 and grams, respectively.

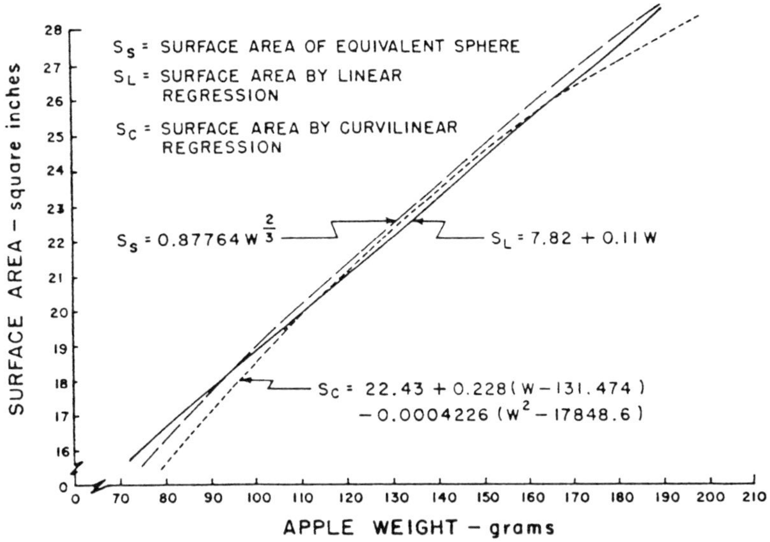

Figure 3.15 Prediction of surface area of apples (Western McIntosh) from their weights (Frechette and Zahradnik, 1965)

As in the case of apples, the actual surface area was measured by various techniques and then a correlation was established between weight of fresh eggs and their surface areas.

The simplest method of obtaining the surface area of a symmetrical convex body such as an egg is the projection method using a shadowgraph or a photographic enlarger. Having the profile of an egg, equally spaced, parallel, perpendicular lines can be drawn from the axis of symmetry to the intersection with the profile. Then using manual computation, integration or a computer program, the surface area S_s can be obtained by summing up the surfaces of revolution for all of the divided segments.

Example

To estimate the surface area of a fresh egg weighing 60 grams, the profile of the egg was first determined using a photographic enlarger. Next the profile was divided into the upper segment, the lower segment and the middle sections of Δy thickness. Figure 3.16 shows the profile and the measured dimensions. The surface area was estimated by the following methods.

1 By summation of surface areas. The two segments at the ends were assumed to be segments of spheres. The sections between the two ends were assumed to be cylinders with heights Δy and diameters d_i. Assuming that these sections were the frustoms of right circular cones did not change the surface area of this section of the egg significantly.

Surface area of a segment of sphere$=2\pi rh$ where r is the radius of the sphere found from

$$\frac{(AC)^2}{8BD} + \frac{BD}{2}$$

$$\text{Upper radius} = \frac{1}{8 \times 0.178} + \frac{0.178}{2} = 0.791 \text{ in.}$$

$$\text{Lower radius} = \frac{1}{8 \times 0.222} + \frac{0.222}{2} = 0.674 \text{ in.}$$

$$A_1 = \text{Surface area of upper segment} = 2\pi(0.791)(0.178) = 0.885 \text{ in}^2$$

$$A_2 = \text{Surface area of lower segment} = 2\pi(0.674)(0.222) = 0.904 \text{ in}^2$$

$$A_3 = \text{Surface area between the two segments} = \sum_{i=1}^{n} \pi \, di \, \Delta y$$

$$= \pi \, \Delta y \sum_{i=1}^{n} di$$

Measurement of di at Δy intervals and summation according to the above relationship yields

$$A_3 = 8.95$$

$$\begin{aligned}\text{Total area} &= A_1 + A_2 + A_3 \\ &= 0.885 + 0.904 + 8.95 \\ &= 10.74\end{aligned}$$

2 By resemblance to a prolate spheroid. From the profile of the egg the major and minor diameters were found to be 2.3 and 1.7 inches, respectively. Using Eq. (3.13) eccentricity

$$e = \left[1 - \left(\frac{0.85}{1.15}\right)^2\right]^{1/2} = 0.674$$

$$\text{and surface area} = 2\pi(0.85)^2 + 2\pi\frac{(0.85)(1.15)}{0.674}\sin^{-1} 0.674$$
$$= 11.29 \text{ in}^2$$

3 By direct measurement. Strips of narrow masking tape were used to cover the surface of the egg. This method gave a total area of 11.3 in^2.

Based on these results it appears that the approximation of the surface of the egg to that of a prolate spheroid is a simple and sufficiently accurate method for estimating the surface area. To check the constants K and m given in Eq. (3.29), substitution of 70.5 cm^2=11.3 in^2 for S and 0.66 for m yielded

$$K = \frac{S_s}{W^m} = \frac{70.5}{60^{0.66}} = 4.8$$

which is within the range reported by Besch *et al.* (1968).

Specific surface in a porous pack

Specific surface is defined as the surface of the pores in a porous media exposed to fluid flow either per unit volume or per unit weight of the solid particles. In studies of fluid flow and heat generation and transfer in stored products a knowledge of specific surface may be required.

For measurement of specific surface of non-uniform pore space, Carman-Kozeny equation may be employed (Collins, 1960; Scheidegger, 1957). Kozeny considered the structure of a porous medium as a bundle of capillary tubes which are not necessarily of circular cross section. Applying the classical hydrodynamic equations for slow, steady state flow through such a system, Kozeny showed that permeability, K, can be given by the following equation:

$$K = \frac{cP^3}{S^2} \tag{3.30}$$

where c is a dimensionless constant, P is porosity and S is specific surface in length squared divided by length cubed. Carman modified the above

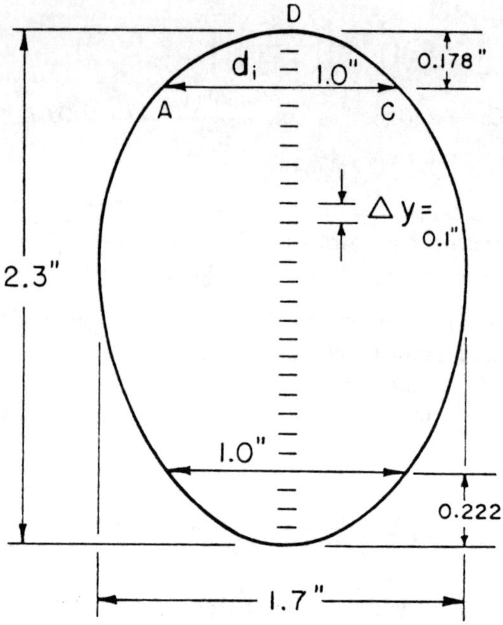

Figure 3.16 Dimensions of an egg profile for determination of the surface area

equation, considering $c = 1/5$ and proposed Carman-Kozeny equation given by

$$K = \frac{P^3}{5S^2(1-P)^2} \tag{3.31}$$

To calculate S, employing the above equation, it is necessary to determine first porosity and permeability of the porous material.

Porosity of the pack can be determined by one of the methods given in this chapter. Collins (1960) defines permeability as "that property of a porous material which characterizes the ease with which a fluid may be made to flow through the material by an applied pressure gradient." The equation which gives permeability in terms of measurable quantities is the Darcy's equation given as

$$K = \frac{q\eta}{A(\Delta p/L)} \tag{3.32}$$

where

K = permeability in length squared (ft^2)
q = flow rate in volume per unit time (ft^3/sec)
η = viscosity of fluid $\frac{lb - sec}{ft^2}$
Δp = pressure difference across the length of the pack (lb/ft^2)
L = length of porous pack in the direction of flow (ft)
A = cross section area of flow (ft^2)

The unit of permeability in length squared shows that this physical property is a measure of mean square pore diameter in the material. In an anistropic porous media, permeability has a directional quality and varies when measured with flow perpendicular to each face of a cube of the porous material. Permeability also varies with compaction and structural changes of the material due to mechanical forces.

For measuring permeability, steady state flow is established through a sample of the material in a flow apparatus and appropriate Darcy equation is applied to calculate K. The flow apparatus consists of a tube holding a cylindrical sample of the porous material having parallel ends with cross sectional area A and length L. The walls of the tube are supposed to be tightly bonded to the sample. A micromanometer is used to measure the pressure differential. To establish the validity of the method in any specific problem, a comparison can be made between the surface area as calculated from the permeability data and the actual surface area of packs of spheres of various sizes.

Specific Surface by the Coating Method

The specific surface area of a number of grains may be determined by coating the grains with a single layer of metal powder and measuring the change in weight. A control group, consisting of geometric shapes of known surface area and of a density close to the grain being tested, is run through the coating process with the grain. A factor representing the coating weight per unit surface area for the control group is used to calculate the surface area of the grain. The bulk volume of the grain sample is measured and entered into the calculation of specific surface along with the calculated surface area.

Example

To determine the specific surface of a sample of dry shelled corn (specific gravity=1.3), 10 grams of this corn was coated with nickel powder and varnish. Six nylon cylinders (specific gravity=1.13 g/cc) were also selected as control objects. The dimensions of the control objects were as in table below.

weight of control group before coating=5.45750 gm
weight of control group after coating=5.68122 gm
weight of grain before coating=10.45152 gm
weight of grain after coating=11.03364 gm
bulk volume of grain=16.0 cc

Using above data, calculate the specific surface of the grain in terms of ft^2/ft^3.

#	Diameter	Length
1	9.61 mm	12.30 mm
2	9.61 mm	11.30 mm
3	9.61 mm	9.60 mm
4	9.61 mm	16.00 mm
5	9.61 mm	10.10 mm
6	9.61 mm	7.40 mm

From the geometry, the surface areas of the six nylon cylinders are:

No.	Surface (mm^2)
1	371.35
2	341.15
3	289.83
4	483.05
5	304.93
6	223.41
	2013.72

Weight gain due to coating:

nylon cylinders: 5.68122 − 5.45750 = 0.22372
corn kernels: 11.03364 − 10.45152 = 0.58212

By proportionality, the surface area of the corn sample is

$$\frac{2013.72 \times 0.58212}{0.22372} = 5239.7 \text{ mm}^2$$

$$\text{Specific surface} = \frac{5239.7 \times 10^{-6} \text{ m}^2}{16 \times 10^{-6} \text{ m}^3}$$

$$= 327.48 \text{ m}^2/\text{m}^3$$

$$= 99.8 \text{ ft}^2/\text{ft}^3$$

Using the method of coating, Jindal *et al.* (1974) obtained surface area per unit volume of several agricultural grains and seeds. The data are reported in Table 3.7.

The coating method for several grains has also been used by Bakker-Arkema *et al.* (1971). These investigators resorted to this method of determination when such data were needed in the study of heat and mass transfer in fixed and moving beds of agricultural grains during heated air drying.

3.6 SURFACE AREA OF AGRICULTURAL GRAINS

Though it is possible to estimate the sum of the external surface areas of particles in a packed-bed, no available method can help to determine the non-contacting surface area in beds of agricultural materials. The Carman-Kozeny equation, in some modified form, has been suggested as a possible means by which the available surface area in a bed of granular materials can be measured. The major limitation of this method, which employs the principles of flow through porous media, has been its applicability to the fine powders consisting of particle sizes generally less than 50 microns.

To test the applicability of the Carman-Kozeny equation to agricultural grains, Jindal *et al.* (1974) employed a modified version of this equation to several agricultural grains and compared the results with those obtained by surface coating reported earlier in this chapter. The results are shown in Table 3.8.

Examination of these data shows that the specific surface values obtained by the air permeability method were consistently lower than those obtained by the coating method. This is expected because the coating method determines the maximum available external surface area whereas the air permea-

Table 3.7 Surface Area per Unit Volume of Various Agricultural Grains and Seeds

Product	Quantity tested, lb.	No. of test	Mean surface area per cu ft., sq. ft. per cu. ft.
Alfalfa, Saranac	0.5	1	774.5
Barley, Wong	0.6	1	425.6
Bean, large Lima	0.7	1	141.3
Bean, Marrow	0.9	1	114.9
Bean, Navy	1.0	2	145.0
Buckwheat	1.0	1	388.3
Corn, "Bonus Maker"	1.0	4	186.0
Corn, yellow Dented	1.0	3	222.4
Lentil	1.0	3	316.7
Oats	0.6	2	301.9
Pea, dried Whole green	1.0	1	160.0
Rye, Balboa	0.45	1	633.3
Sorghum	1.0	1	323.2
Soybean (Seed)	1.0	4	169.5
Wheat	1.0	2	400.8

bility method takes into account only the non-contacting surface area available in a packed bed. This difference should increase with the decrease in particle size and depends on particle shape and their arrangement in the packing. A log-log plot of the specific surface area by the coating method versus specific surface area as obtained by the air permeability method was approximated by a straight line. As the grain size gets smaller, the

Table 3.8 Specific Surface Area of Selected Agricultural Grains

Grain	"S" by air permeability, sq. ft. per cu. ft.	"S by coating method, sq. ft. per cu. ft.	"S" based on ASAE data* sq. ft. per cu. ft.
Soybeans	164.5	169.5	192.5
Corn	197.3	222.4	206.5
Oats	203.0	255.0	165.0
Buckwheats	280.2	388.0	—
Wheat	300.8	400.0	294.7

*Based on data given by Shedd (1953) extrapolated for air velocity range of 50 to 100 fpm and average porosities of grains in ASAE Yearbook.

difference between the surface area values obtained by the coating and the air permeability method increased. The difference, due to the areas taken up for particle to particle contact, varied from about 15 percent to 30 percent. For truly spherical particles generally the reduction in area is considered to be about 10 percent.

From the straight line relationship of the log-log plot the following equation was derived which can be used to estimate the specific surface area in a pack of grain from the specific surface area obtained by coating method.

$$S_{\text{coating}} = 0.2\, S_{\text{permeability}}^{1.333} \tag{3.32a}$$

It is interesting to note that the surface area values calculated by the air permeability method when compared with the ASAE data for most general conditions are not very far off from the experimentally obtained values. It may be concluded that the air permeability method can be applied for determining the surface area of granular materials provided reasonable estimates of porosity and resistance to air flow characteristics are available. It should be noted that in the air permeability method, the range of air velocity under consideration is very important. For different air velocity ranges, the nature of Eq. (3.32a) will be different. The above expression is valid for air velocities in the range of 50 to 100 feet/min.

3.7 GRAIN INTERPOROSITY

The interporosity of cereal grains can influence the variety differences as well as mechanical properties such as hardness, elasticity, plasticity, resistance to compression, and water movement in grain. When pore spaces in the grain kernel are in the range of macropores, surface tension forces are smaller than the gravitational forces. In characterizing the grain structure, free spaces with diameters below 10^3Å are classified as micropores, while pores larger than 10^3Å are classified as macropores.

Figure 3.17 shows the method which has been used to determine the pore distribution in several varieties of wheat. The method is based on the mercury penetration technique which has been employed for the study of pore size distribution in other materials (Ritter and Drake, 1945; Winslow and Shapiro, 1959). The method is based on the principle that the intrusion of mercury into a capillary is a function of its diameter as given by the following formula:

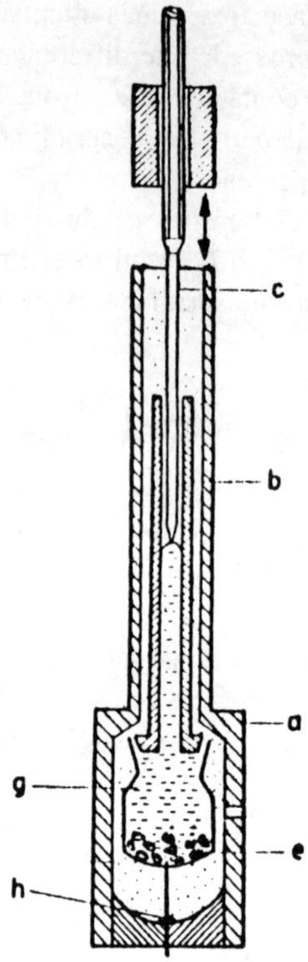

Figure 3.17 Apparatus for measuring interporosity of wheat grain. a—dilatometer chamber, b—alcohol, c—electronic sensing rod, e—grain sample, g—mercury, h—mercury contact (Stawinski and Szot, 1976)

$$r = \frac{2T_s \cos \delta}{p} \tag{3.32b}$$

where r is radius of the capillary in Å, T_s is surface tension in N/m, δ is the solid-liquid contact angle, and p is the external pressure in Pa.

The grain sample is placed in the dilatometer, and after its evacuation, it is filled with mercury. The dilatometer is then placed in chamber a, which in turn is filled with alcohol. The alcohol is used to transfer the high pressure, about 15 kg/cm^2, to the mercury, which in turn causes the intrusion of mercury into the pores of the grain sample. The pore volume is indicated by the decrease in the level of mercury in the dilatometer, which is sensed by the electrical contact and recorded along with the alcohol pressure. The recorder plots the pore volume as a function of the applied pressure.

Figure 3.18 shows the distinct difference in pore size distribution among the varieties of wheat investigated. As shown, pore size distributions for wheat have definite trends with distinct minimum and maximum values. The minimum was at a radius of about 10^3Å, while the maximum at about 10^5Å. The differences in varieties were largest at the values showing maximum pore size distributions.

3.8 SIZE REDUCTION AND ENERGY REQUIREMENTS

Milling and grinding, although it is one of the earliest food processing operations, is still based on empirical methods with very little, if any, of the mechanical properties of the material being used in the design and analysis. As a result, much of the energy used in the grinding operation is wasted as heat raising the temperature of the product and the equipment. In fact, only about 0.06 to 1% of the energy used in grinding is used to create breakage and new surfaces.

In the absence of any derived expressions for energy estimation, most of the engineers still use empirical equations given in terms of particle diameters and some constants. The following expression, seen frequently in textbooks, is based on the assumption that the energy necessary to cause a change in the particle size is directly proportional to the size of the particle raised to a power n (Bond, 1963) as shown below:

$$dE = -K(dB/D^n) \tag{3.33}$$

where E is the input energy, K and n are constants, and D is the characteristic diameter of the particle.

Rittinger (1867), assuming that particles, being infinitely brittle, do not break unless new surface are created, gave the following relationship in terms of the initial surface area S_1 and final surface area S_2 and a constant K_1:

$$E = K_1(S_2 - S_1) \tag{3.34}$$

Since surface area per unit mass is proportional to the reciprocal of particle

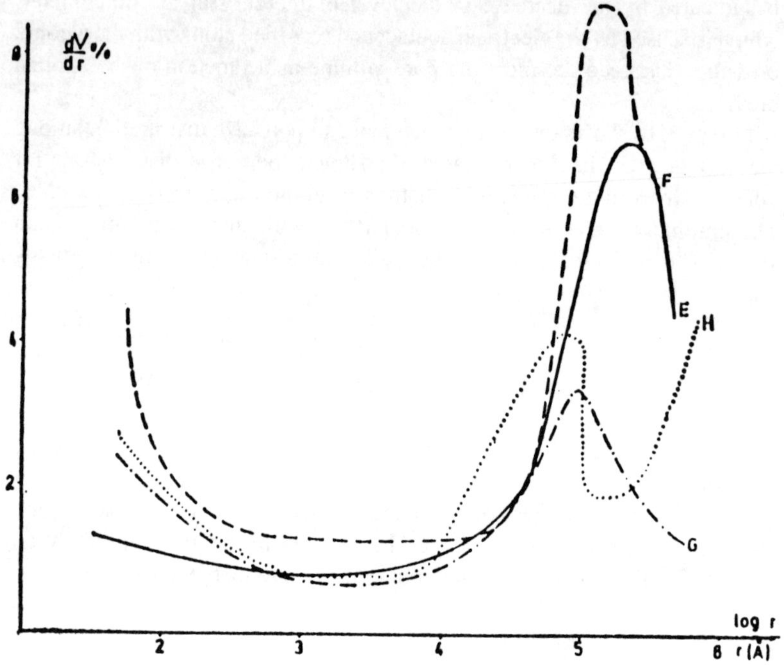

Figure 3.18 Pore size distribution as a function of pore diameter for four varieties of spring wheat (Stawinski and Szot, 1976)

diameter, the energy equation can be given in terms of particle diameters as follows:

$$E = K_2(1/D_2 - 1/D_1) \tag{3.35}$$

where K_2 is a new constant involving the relation between particle surface areas and diameters. The same relationship can be arrived at by integration of Eq. (3.33) assuming $n=2$. No attempts have been made to explore the relationship between the rheological properties, such as, hardness, fracture energy, etc. of grains, for example, to Rittinger's equation. Jindal (1975), who studied the mechanics of hammermill grinding of corn grains as related to its mechanical properties, did not consider these empirical relationships. It should be noted that Rittinger's equation has proven to be valid only for very brittle materials where no energy is spent for elastic or plastic deformations of the particle.

Another expression, introduced by Kick (1885), assumes that if one unit of energy reduces a certain mass of equal-sized particles to particles of half the size, one additional unit of energy must then reduce the half-size particles by half again to one-quarter of their original size. If in Eq. (3.33) $n=1$, the integration yields the following expression which is based on this postulation:

$$E = K_3 \log (D_1/D_2) \tag{3.36}$$

where $K_3=2.303$ K in Eq. (3.33). While Rittinger's equation holds best for large brittle particles, Kick's equation has shown to hold best for small particles.

Examples

Grain with an initial geometric mean diameter of 5 mm is ground to final diameter of 1 mm using a grinding energy of 3600 J/Kg. Determine how many times would the energy increase if the material is ground from 5 mm to 0.2 mm.

Assuming the grain being very brittle, from Eq. (3.35)

$$K_2 = 3600/(1/1 \times 10^{-3} - 1/5 \times 10^{-3})$$

$$= 4.5 \text{ Jm/Kg}$$

If K_2 remains constant for the finer grinding, substitution in Eq. (3.35) yields

$$E = 4.5(1/0.2 \times 10^{-3} - 1/5 \times 10^{-3})$$

$$= 21500 \text{ J/Kg or } 21600/3600 = 6 \text{ times}$$

If the grain is not that brittle and exhibits elasticity and plasticity, then from Eq. (3.36) the new constant

$$K_3 = 3600/\log (5 \times 10^{-3}\ 1 \times 10^{-3})$$

$$= 5.15 \times 10^{-3} \text{ J/Kg}$$

For size reduction to 0.2 mm

$$E = 5.15 \times 10^{-3} \log (5 \times 10^{-3}/0.2 \times 10^{-3})$$

$$= 7200 \text{ J/Kg or } 7200/3600 = 2 \text{ times}$$

Conclusion is that if the grain brittleness can be reduced by tempering of other means to show some elastic or plastic deformation in the milling process, with little change in quality of the finished product, there will be considerable saving in the grinding energy.

Taking the value of n in Eq. (3.33) midway between 1 and 2 or $n=1.5$, Bond (1954) introduced a third theory for estimating energy incorporating a new parameter called work index W_i. Using the value of $n=1.5$ in Eq. (3.33) and integrating yields the Bond equation

$$E = 2K(1/D_2)^{1/2} - 1/(D_1)^{1/2}) \qquad (3.37)$$

where $K=5\ W_i$.

Orr (1966) has given tabulated values for W_i for various minerals. The range of most values given is from 10 to 20 KWhr/ton or 40,000 to 80,000 J/Kg. For brittle food materials, such as grains, the W_i values are probably on the lower side of these figures.

Loncin and Merson (1979) have presented Eq. (3.37) in the following form:

$$E/W_i = \left(\frac{100}{D_2}\right)^{1/2} - \left(\frac{100}{D_1}\right)^{1/2} \qquad (3.38)$$

Using 100 μm as the reference particle size, D_2 is the diameter of sieve opening in μm which will pass 80% of the material after grinding, and D_1 is the opening which will pass 50% of the material before grinding. To determine the W_i for any combination of material and milling machinery, grinding experiments can be conducted to determine the energy requirement and particle sizes. It should be noted that the value of $n=1.5$ has apparently some theoretical significance related to the ratio of crack length to particle diameter. But again, no work has been reported to relate the rheological properties of the product to this empirical quantity.

3.9 · PARTICLE SIZE MEASUREMENT

Particle size specification of ground materials and powders is the basic data needed in many practical problems involving flow of powders and granular materials. Among the many techniques available for analysis of particle size distribution, the method of log normal distribution reportedly describes best the size distribution of dry powder food products. Likewise, among the numerous methods for describing size of particles, the size of the sieve opening through which 80% of the material by weight will pass is the simplest and sufficiently accurate enough to use. Expressing in equation form, Orr (1966) gives the following equation:

$$Pm = 80\ (d/d_{80})^a \tag{3.39}$$

where Pm is the weight percent of material of diameter less than d, d_{80} is the diameter below which 80% of the material by weight occurs, and, a is a constant determined from the slope of a log-log plot of the above equation, assuming it leads to a straight line. Deviation from linearity has been observed more for coarser materials. Reportedly there is little variation of, a, over a size range of a single grinding operation. In many cases it has a value of 0.5.

In recent years the method of image analysis, employing either scanning electron microscope (SEM) or optical microscope along with the supporting electronic and computer systems, has been developed which are most suitable for small particle size analysis. The system locates and calculates the diameters, classifies the particle parameters for distribution plots, and performs the statistical analysis.

4

SOME BASIC CONCEPTS OF RHEOLOGY

MECHANICAL PROPERTIES may be defined as those having to do with the behavior of the material under applied forces. Following this broad definition, such properties as stress-strain behavior of a material under static and dynamic loading as well as flow characteristics of the material in air or in water, can be classified as mechanical properties.

Rheology has been defined as "a science devoted to the study of deformation and flow." Therefore, when the action of forces result in deformation and flow *in the material,* the mechanical properties will be referred to as rheological properties. Moreover, rheology considers the time effect during the loading of a body. Rheologically then, mechanical behavior of a material is expressed in terms of the three parameters of force, deformation and time. Examples of rheological properties are time-dependent stress and strain behavior, creep, stress relaxation, and viscosity.

Mechanical properties other than rheological properties usually deal with the motion of the material under applied forces. Examples are drag coefficient and terminal velocity, rebound and coefficient of restitution in impact, and flow of the material in bulk.

4.1 BIOLOGICAL SYSTEMS AND MECHANICAL PROPERTIES

Each unit of the food and feed materials selected from agricultural products is in itself a biological system which differs from identical mass-produced products. These materials are alive, constantly undergoing changes in shape, size, respiration, and other aspects of life processes. During development and storage, the cells are sensitive to such external influences as humidity, temperature, oxygen, food supply, energy consumption, as well

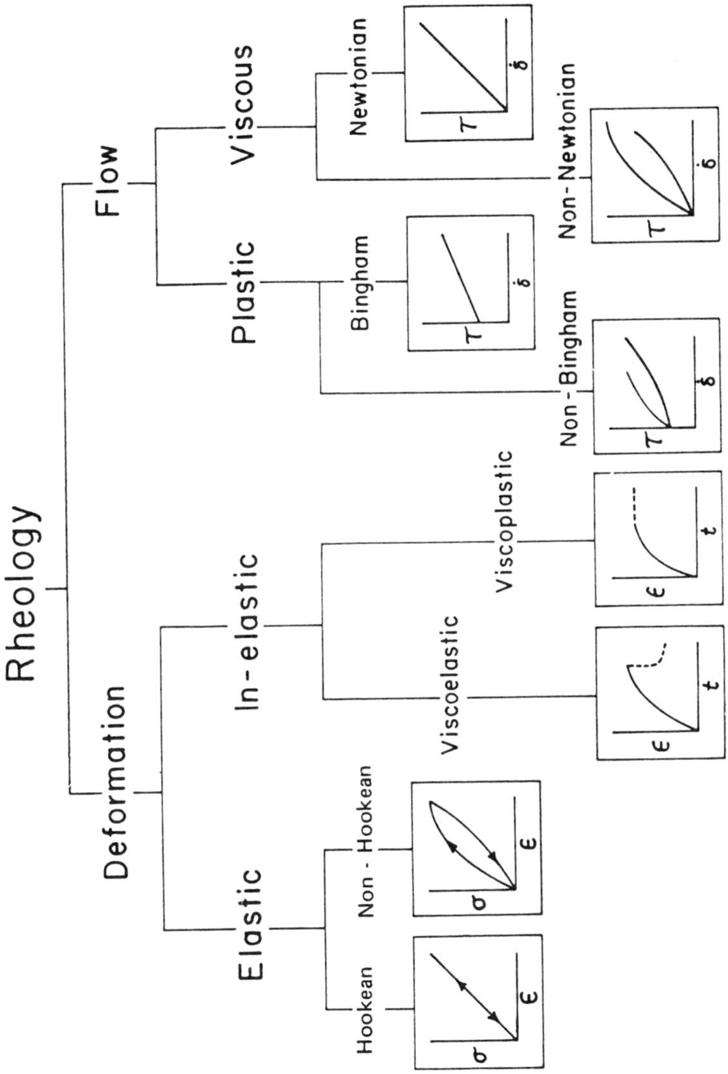
Rheology
Deformation
Flow
Elastic
In-elastic
Plastic
Viscous
Hookean
Non - Hookean
Viscoelastic
Viscoplastic
Non - Bingham
Bingham
Non - Newtonian
Newtonian

as the interplay of internal factors which are difficult to control. In biological solids elasticity varies with age and physiological conditions. Biological fluids are mostly non-Newtonian liquid which cause additional complications.

As a result of this complex situation, in studying the rheology of a biological system, only an empirical approach is possible. The treatment of the problem usually consists of either a simple description of observed facts or theoretical considerations which often lead to very complicated mathematical formulations containing many variables. Constants, as known in physical experiments, rarely exist.

Despite this apparently hopeless situation, the application of the fundamental principles of mechanics and rheology is a good start for the study of mechanical behavior in biological systems. Until specific laws and principles are derived and established, we can study the relative changes in one definite property by making a rather rough approximation of other variables influencing this one property.

4.2 ASTM STANDARD DEFINITION OF TERMS RELATED TO MECHANICAL PROPERTIES*

Strain

The unit change, due to force, in the size or shape of a body referred to its original size or shape. Strain is a nondimensional quantity, but it is frequently expressed in inches per inch, centimeters per centimeter, etc. (Fig. 4.2).

a *Linear (tensile or compressive) strain* The change per unit length due to force in an original linear dimension.
b *Axial strain* Linear strain in a plane parallel to the longitudinal axis of the specimen.
c *Transverse strain* Linear strain in a plane perpendicular to the axis of the specimen.
d *Shear strain (angular strain)* The tangent of the angular change, due to force, between two lines originally perpendicular to each other through a point in a body.

* Selected from ASTM Designation: E 6–65, last revised 1965.

e *True strain* In a body subjected to axial force, the natural logarithm of the ratio of the gauge length at the moment of observation to original gauge length.
f *Macrostrain* The mean strain over any finite gauge length of measurement large in comparison with interatomic distances.
g *Microstrain* The strain over a gauge length comparable to interatomic distances.

Stress

The intensity of a point in a body of the internal forces or components of force that act on a given plane through the point. Stress is expressed in force per unit of area (pounds-force per square inch, kilograms-force per square millimeter, etc., Figs. 4.1 and 4.2).

a *Nominal stress* The stress at a point calculated on the net cross section by simple elastic theory without taking into account the effect on the stress produced by geometric discontinuities such as holes, grooves, fillets, etc.
b *Normal stress* The stress component perpendicular to a plane on which the forces act. Normal stress may be either:

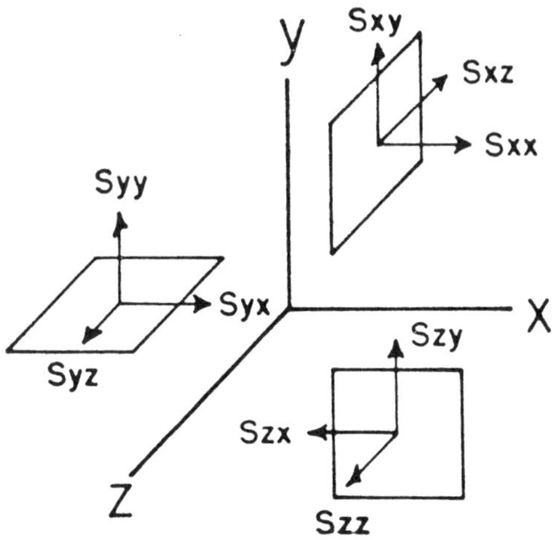

Figure 4.1 Components of force per unit area dA oriented normal to *x*, *y*, *z* directions

1 *Tensile stress* Normal stress due to forces directed away from the plane on which they act, or

2 *Compressive stress* Normal stress due to forces directed toward the plane on which they act.

c *Shear stress* The stress component tangential to the plane on which the forces act.

d *Torsional stress* The shear stress on a transverse cross section resulting from a twisting action.

e *True stress* The axial stress in a tension or compression test, calculated on the basis of the instantaneous cross sectional area instead of the original area.

f *Principal stress (normal)* The maximum or minimum value of the normal stress at a point in a plane considered with respect to all possible orientations of the considered plane. On such principal planes the shear stress is zero.

g *Fracture stress* The true normal stress on the minimum cross sectional area at the beginning of fracture.

Bearing load

A compressive load on an interface.

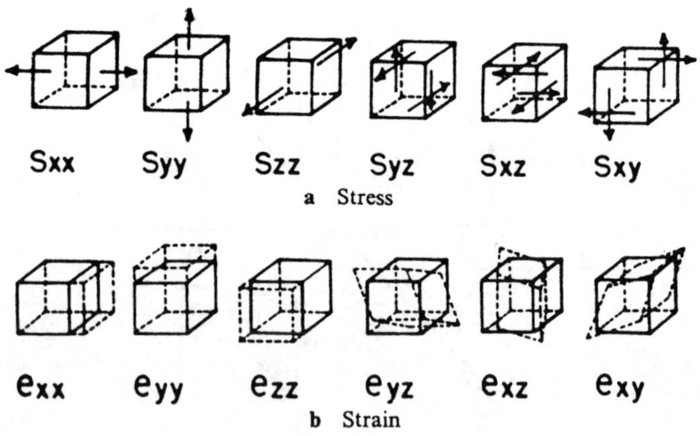

Figure 4.2 A volume element of a material showing the six components of stress and strain required for complete specification (Alfley, 1948)

Compressive strength

The maximum compressive stress which a material is capable of sustaining. Compressive strength is calculated from the maximum load during a compression test and the original cross sectional area of the specimen.

Elastic limit

The greatest stress which a material is capable of sustaining without any permanent strain remaining upon complete release of the stress.

Modulus of elasticity

The ratio of stress to corresponding strain below the proportional limit.

(*Note*) *For materials where the stress-strain relationship is curvilinear rather than linear, one of the four following terms may be used* (*Fig. 4.5*):

a *Initial tangent modulus* The slope of the stress-strain curve at the origin.
b *Tangent modulus* The slope of the stress-strain curve at any specified stress or strain.
c *Secant modulus* The slope of the secant drawn from the origin to any specified point on the stress-strain curve.
d *Chord modulus* The slope of the chord drawn between any two specified points on the stress-strain curve.

Poisson's ratio

The absolute value of the ratio of transverse strain to the corresponding axial strain resulting from uniformly distributed axial stress below the proportional limit of the material.

(*Note 1*) *For stresses beyond the proportional limit, the range of stress should be stated.*

(*Note 2*) *Poisson's ratio will have more than one value if the material is not isotropic.*

Proportional limit

The greatest stress which a material is capable of sustaining without any deviation from proportionality of stress to strain (Hooke's law).

Set

Strain remaining after complete release of the load producing the deformation

(Note 1) *Due to practical considerations, such as distortion in the specimen and slack in the strain indicating system, measurements of strain at a small load rather than zero load are often taken.*

(Note 2) *Set is often referred to as a permanent set if it shows no further change with time. Time elapsing between the removal of load and the final reading of a set should be stated.*

Shear strength

The maximum shear stress that a material is capable of sustaining. Shear strength is calculated from the maximum load during a shear or torsion test and is based on the original dimensions of the cross section of the specimen.

Tensile strength

The maximum tensile stress that a material is capable of sustaining. Tensile strength is calculated from the maximum load during a tension test carried to rupture and the original cross sectional area of the specimen.

Yield point

The first stress in a material, less than the maximum attainable stress, at which an increase in strain occurs without an increase in stress.

(Note) *It should be noted that only materials that exhibit the unique phenomenon of yielding have a "yield point."*

Yield strength

The stress at which a material exhibits a specified limiting deviation from the proportionality of stress to strain. The deviation is expressed in terms of strain.

(Note) *It is customary to determine yield strength by Offset Method (usually a strain of 0.2 percent is specified).*

4.3 OTHER DEFINITIONS RELATED TO MECHANICAL PROPERTIES

Pressure

A measure of the mean normal stress on a point of a body—(lb/in^2).

Deformation

Deformation or distortion is the relative displacement of points within a body. Deformation like stress is a vector quantity. In general, deformation is accompanied either by change of volume or by change of shape. The change of volume is caused by isotropic stress such as hydrostatic pressure. The change of shape is brought about by shear stresses—(inch). Material can be deformed by uniaxial compression, uniaxial tension, shear, and bulk compression.

Bioyield point

A point, such as *y* in Fig. 4.3 on the stress-strain or force-deformation curve at which there occurs an increase in deformation with a decrease or no change of force. In some agricultural products, the presence of this bioyield point is an indication of initial cell rupture in the cellular structure of the material. The term bioyield point is proposed for biological materials to differentiate this phenomenon from the yield point in engineering materials. The bioyield point may occur at any point beyond the point *LL,* where the curve deviates from the initial straight line portion.

Rupture point

A point on the stress-strain or force-deformation curve at which the axially loaded specimen ruptures under a load. In biological materials, rupture may cause puncture of shell or skin, cracking, or fracture planes. It may be stated that a "bioyield point" in these materials corresponds to a failure in the *microstructure* while a "rupture point" corresponds to a failure in the *macrostructure* of the specimen. In a force-deformation curve such as Fig. 4.3, the point of rupture of the specimen may occur at any point on the curve beyond the bioyield point. In a "brittle" material, rupture may occur in the early portion of the curve. In a "tough" material, rupture may take place after considerable plastic flow at such point as *R*.

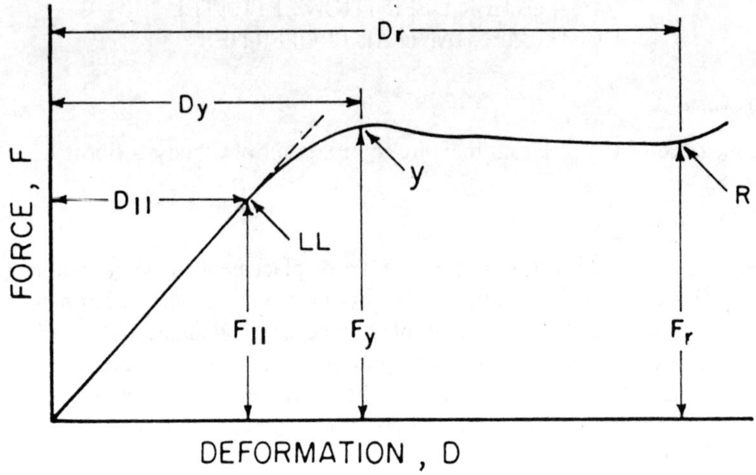

Figure 4.3 A possible force-deformation curve for an agricultural product. *LL*, linear limit; *y*, "bioyield point"; *R*, rupture point

Stiffness

Stiffness or rigidity is indicated by the slope of the initial straight line portion of the curve (Fig. 4.3). The ratio of stress to strain in this more or less elastic region of the curve may be referred to as the "modulus of elasticity" or "Young's modulus" (lb/in^2). In the case of nonlinear stress-strain behavior, stiffness or "apparent modulus" can be defined in terms of *initial tangent modulus, secant modulus,* or *tangent modulus,* as illustrated in Fig. 4.5. Initial tangent modulus is taken as the slope of the curve at the origin. Secant modulus is the slope of the line connecting the origin and a selected point *A* on the curve. Tangent modulus is the slope of the tangent line to a selected point *B* on the curve.

Elasticity

The capacity of a material for taking elastic or recoverable deformation. In those portions of the curve in Fig. 4.3 before the point *LL* is reached, elongations are, in large part at least, recoverable, and are a measure of elastic deformation.

Plasticity

The capacity of a material for taking plastic or permanent deformation. Since deformations from the bioyield point to the point of rupture are not all recoverable, the unrecoverable part can be taken as a measure of plastic deformation (Fig. 4.4).

Degree of elasticity

The ratio of elastic deformation to the sum of elastic and plastic deformation when a material is loaded to a certain load and then unloaded to zero load (Fig. 4.4).

Strength

The resistance to applied force (lb/in).

Ultimate strength

The stress corresponding to the rupture point (lb/in^2).

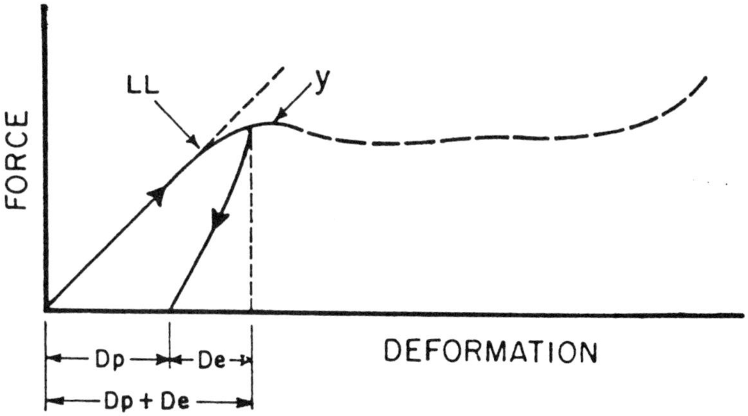

Figure 4.4 Degree of elasticity from a loading-unloading curve. D_e = elastic or recoverable deformation; D_p = plastic or residual deformation; $D_e/(D_p + D_e)$ = degree of elasticity

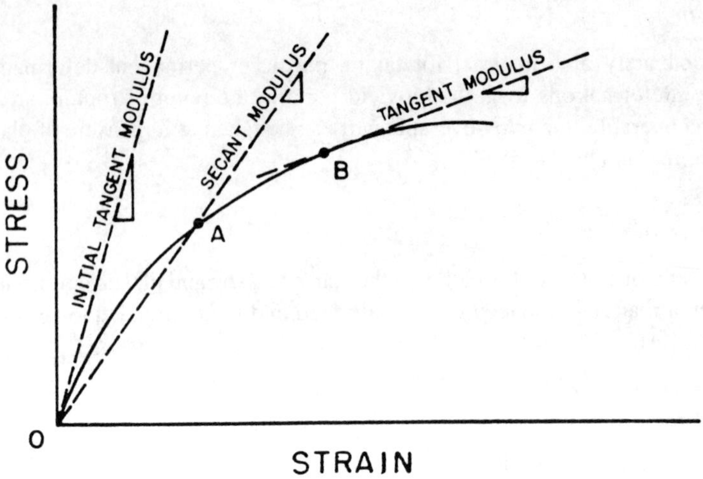

Figure 4.5 Methods for defining the modulus in non-linear stress–strain diagrams

Bioyield strength

The stress corresponding to the bioyield point. If the curve does not show a well-defined bioyield point, the stress corresponding to an arbitrary strain, similar to the *offset* strain in engineering materials, may be taken as the bioyield strength (lb/in^2).

Toughness

The work required to cause rupture in the material. This can be approximated by the area under the stress-strain or force-deformation curve up to the point selected as the rupture point ($in\text{-}lb/in^3$). If in estimating toughness, a force-deformation curve is used, the size of the specimen and the loading surface area should be specified.

Resilience

The capacity of a material for storage of strain energy in the elastic range ($in\text{-}lb/in^3$). Thus the area under the unloading curve in Fig. 4.4 is a measure of resilience of the material. As in the case of toughness, when a force-deformation curve is used, the size of the specimen and the loaded area should be specified.

Mechanical hysteresis

The energy absorbed by a material in a cycle of loading and unloading—evaluated as area between loading and unloading curve (Fig. 4.4). Mechanical hysteresis is a measure of the damping capacity or the ability of the material to dissipate the strain energy as heat (inch-pound per cycle).

Stressed object

An object under a "balanced" set of forces, tending to change its shape or size, or both.

Strained object

An object subjected to a change of shape or size, or both as a result of application of a "balanced" set of forces.

Elastic fore-effect and elastic after-effect

Delayed elastic deformations which take place upon loading and recovery upon unloading (Fig. 4.4).

Rigidity

Rigidity or shear modulus is the ratio between shearing stress and the elastic shearing strain (lb/in^2).

Deviatoric stress component

The stress which is responsible for change in shape.

Isotropic stress component

The stress which is responsible for change in volume.

Viscoelasticity

A combined solid-like and liquid-like behavior in which the stress-strain relationship is time dependent.

Linear viscoelasticity

A viscoelastic behavior in which the ratio of stress to strain is a function of time alone and not of the stress magnitude.

Stress relaxation

Decay of stress with time when the material is suddenly deformed to a given deformation—constant strain (Fig. 4.15).

Relaxation time

The rate of stress decay in a material subjected to a sudden strain. It is the time required for the stress in the Maxwell model, representing stress relaxation behavior, to decay to $1/e$ or approximately 37 percent of its original value.

Creep

Deformation with time when the material is suddenly subjected to a dead load—constant stress (Fig. 4.15).

Retardation time

The rate at which the retarded elastic deformation takes place in a material creeping under dead load. It is the time required for the Kelvin model, representing creep behavior, to deform to $(1-1/e)$ or about 63 percent of its total deformation.

Viscosity

Resistance to flow indicated by coefficient of viscosity.

Viscosity coefficient

The ratio of shearing stress to shearing rate in Newtoninan fluids $\left(\frac{\text{dynes} - \text{sec}}{\text{cm}^2}\right)$ = poise or $\left(\frac{\text{lb} - \text{sec}}{\text{ft}^2}\right)$ (Fig. 4.11).

Kinematic viscosity

Coefficient of viscosity divided by the mass density (ft^2/sec).

Newtonian liquid

An ideal fluid in which the relationship between shear stress and shear rate is a straight line passing through the origin (Fig. 4.12).

Non-Newtonian liquid

A liquid in which the relationship between shear stress and shear rate is non-linear. If the flow curve is concave to the shear stress axis, the flow is called *dilatant*. If the flow curve is convex to the shear stress axis, the flow is called *pseudoplastic* (Figs. 4.38 and 4.39).

Apparent viscosity

As applied to non-Newtonian liquid is the viscosity of a Newtonian liquid exhibiting the same resistance to flow at the chosen shearing stress or shear rate. It is determined from slope of a straight line connecting the chosen point on the non-linear curve to the origin (Fig. 4.39).

Thixotropic fluids

Those fluids which show a decrease in shear stress with time of shear at a given shear rate.

Rheopectic fluids

Those fluids which show an increase in shear stress with time of shear at a given shear rate.

4.4 PHYSICAL STATES OF A MATERIAL

The state of various materials depends at any instant on load and deformation history to which it is subjected to as well as environmental factors such as temperature. In the case of biological materials, the moisture content also affects the physical state of the material. The state of a material in terms of creep compliance function and relaxation modulus function plotted against time has been given by Sharma (1965), as given in Figs. 4.6 and 4.7. The creep compliance function $J(t)$ is the ratio of shear strain to shear stress at any given time t when the load or stress is kept constant. The relaxation modulus function $G(t)$ is the ratio of shear stress to shear strain at any time t when strain is kept constant.

All the materials having linear response will display behavior corresponding to one or more zones of these diagrams. Zone I corresponds to the behavior of perfectly elastic materials. The theory of elasticity (Timoshenko and Goodier, 1951), which explains the behavior of this class of materials,

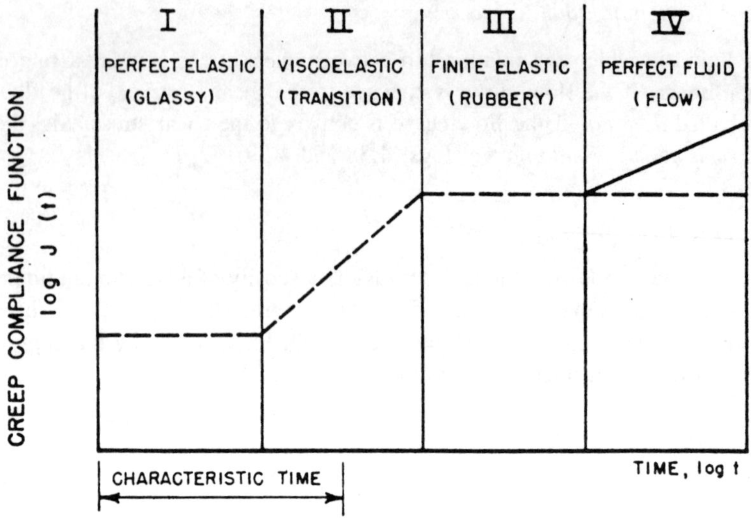

Figure 4.6 Creep compliance *vs*. time relations in shear. - - - - crosslinked polymers, —— linear polymers (Sharma, 1965)

is based on the assumption that the state of the material does not depend on time or the history by which the given deformation was reached. Zone II corresponds to the behavior of materials exhibiting both elastic and viscous effects. The behavior of these materials, classified as viscoelastic, is obviously time-dependent and is covered under the theory of linear viscoelasticity (Ferry, 1961). Zone III corresponds to the rubbery or soft materials exhibiting non-linear elastic behavior. The theory covering the behavior of these materials is called the finite elastic theory, dealing with large and finite deformations (Murnaghan, 1959). Zone IV shows the state of a material for which stress is proportional to the rate of strain or a perfect fluid. The classical hydrodynamics covers the theories concerned with these materials (Lamb, 1945). In real materials the linear response can be achieved only if a certain critical stress is not exceeded. If this critical stress is exceeded, plastic deformations are observed for which the theories of plasticity can be applied (Hill, 1950). The real fluids also can display non-linearity which can be studied by applying the theories of non-Newtonian fluids (Wilkinson, 1960).

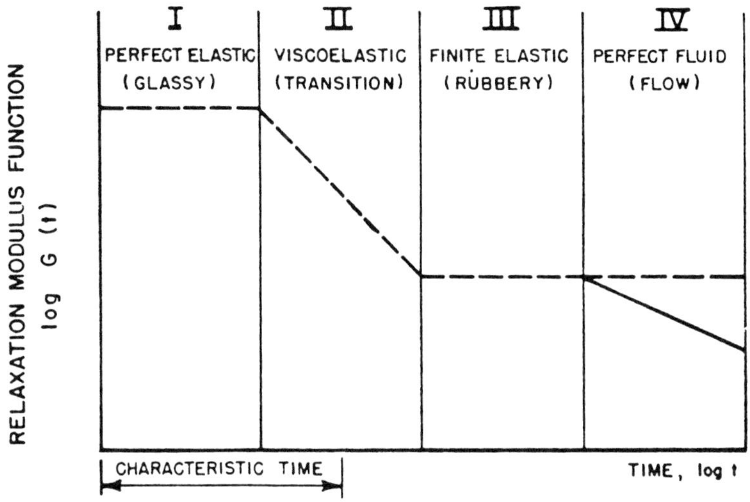

Figure 4.7 Relaxation modulus *vs.* time relations in shear, - - - - crosslinked polymers, —— linear polymers (Sharma, 1965)

No serious attempt has yet been made to characterize the biological materials and determine their characteristic times by which they can be positioned in the spectrum of the physical states of a material. Discussion of the theories dealing with the various states of a material is obviously beyond the scope of this book. The classical reference textbooks for each state of the materials are given in the above references. In the following sections we shall briefly present only those concepts which we need to apply in discussing the work reported for food and feed materials.

4.5 CLASSICAL IDEAL MATERIALS

As mentioned under definitions, there are two types of deformation, elastic deformation and flow. Flow is also divided into plastic flow and viscous flow. Therefore, elasticity, plasticity and viscosity are three fundamental properties by which the rheological behavior of a material can be described. The three classical ideal bodies representing these properties are Hookean body, St. Venant body, and Newtonian liquid. Since no real material behaves perfectly elastic or perfectly plastic, the three ideal bodies have been chosen to serve as standards of comparison in the analysis of the behavior of any real material.

Ideal elastic behavior (Hookean body)

In a Hooke body, stress is directly proportional to strain as illustrated in Fig. 4.8a. The relationship is known as the Hooke's law and the behavior is referred to as Hookean behavior. Although this behavior has been demonstrated for small strains (less than 0.1 percent) in certain solids, as stated by Reiner (1960), improvements in experimental techniques may prove that perfect elasticity may not exist in any real material. Furthermore, the definition of elasticity requires only a complete recovery of strain upon removal of stress. Therefore, we may differentiate between the Hookean or linear elasticity shown for steel and the non-linear elasticity shown for rubber (Fig. 4.8b).

Compression tests of a variety of food and feed materials such as fruits and vegetables, forage, cereal grains, and egg shell have indicated that Hookean elasticity, even for very small strains, apparently does not exist in these biological materials. Figure 4.8c shows the first cycle of loading and unloading for the horny endosperm of dry corn kernel. This type of curve which shows some residual deformation upon unloading is typical for most of the food and feed materials.

Based on Hooke's law and together with Poisson's ratio, the following relationships have been established for elastic, homogeneous and isotropic materials (See Fig. 4.9).

When the Hooke solid is under tensile or compressive stress, modulus of elasticity or Young's modulus is given by

$$E = \frac{\text{tensile or compressive stress } \sigma}{\text{tensile or compressive strain } \varepsilon} \text{ where } \varepsilon = \frac{\Delta l}{l} \tag{4.1}$$

When the Hooke solid is subjected to distortion by shear stresses, the shear modulus or modulus of rigidity is given by

$$G = \frac{\text{shear stress } \tau}{\text{shear strain } \gamma} \text{ where } \gamma = \frac{D}{l} = \tan\theta \tag{4.2}$$

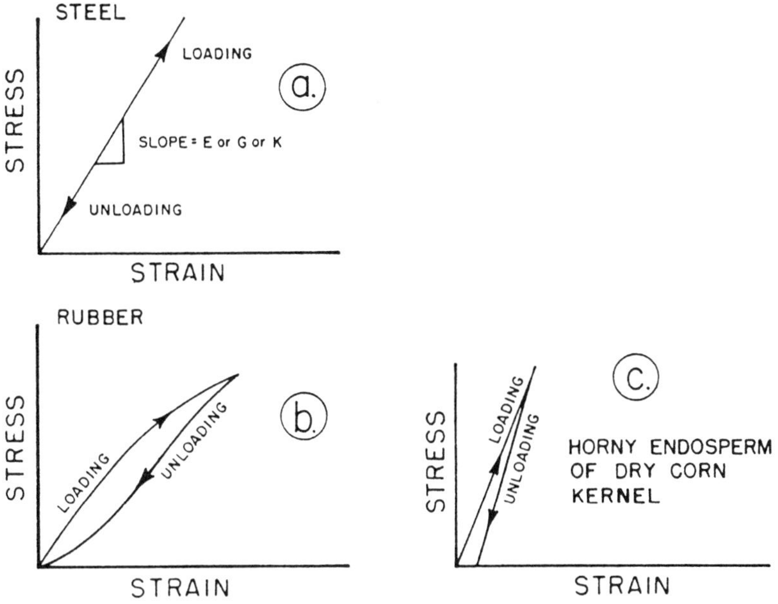

Figure 4.8 (a) Linear elasticity in steel, (b) non-linear elasticity in rubber, (c) inelasticity in corn

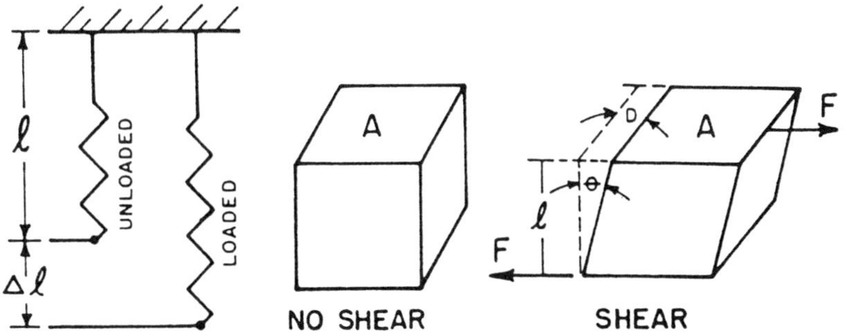

Figure 4.9 Hooke's elasticity in tension and shear

When the Hookean solid is under hydrostatic pressure the bulk modulus or incompressibility is given by

$$K = -\frac{\text{hydrostatic pressure } p}{\text{volume strain } \varepsilon_v} \text{ where } \varepsilon_v = \frac{\Delta v}{v} \tag{4.3}$$

The relationship between the above elastic constants and Poisson's ration μ is given by

$$\frac{1}{E} = \frac{1}{3G} + \frac{1}{9K} \tag{4.4}$$

$$E = 3K(1 - 2\mu) \tag{4.5}$$

$$E = 2G(1 + \mu) \tag{4.6}$$

$$\mu = \frac{(3K - E)}{6K} \tag{4.7}$$

$$\mu = \frac{(E - 2G)}{2G} \tag{4.8}$$

Table 4.1 shows the range of values for Poisson's ratio, E/G ratio, and K/E ratio calculated from the above equations.

For most materials Poisson's ratio is between 0.2 and 0.5. As seen from Table 4.1 the Poisson's ratio approaches 0.5 as the character of a material approaches that of rubber or liquids. This becomes evident when we recall that a liquid cannot support any shear stresses, resulting in a rigidity modulus of G equals zero. This will yield a zero value for E in Eq. (4.6). Since the incompressibility K of a liquid is a finite value, if E in Eq. (4.5) is zero the term $(1-2\mu)$ must also be zero, which yields a value of 0.5 for μ.

Some typical values for Poisson's ratio and bulk modulus for selected materials are given in the Appendix.

Example

Two block specimens of a soft solid food material are subjected to parallel direct shear as shown in Table 7.2 and Figure 4.9. The block dimensions are 1 in. wide, 1.25 in. long, and 0.5 in. thick. Block distortion under 2.2 lbs. is 0.007 in. Determine the shear modulus G of this material, with reference to Fig. 4.9.

Table 4.1 Elastic constants for rigid to liquid-like materials

Poisson's ratio	E/G	K/E
0	2.0	0.333
0.10	2.20	0.417
0.20	2.40	0.556
0.25	2.50	0.667
0.30	2.60	0.833
0.35	2.70	1.111
0.40	2.80	1.667
0.45	2.90	3.333
0.50	3.00	—

$$l = 0.5 \text{ in.}, D = 0.007 \text{ in.}, A = 1 \times 1.25 = 1.25 \text{ in.}^2$$

shearstress $\tau = F/2A$

$$= \frac{2.2}{2 \times 1.25} = 0.88 \text{ lb./in.}^2$$

From Eq. 4.2 shear strain $\gamma = D/l$

$$= \frac{0.007}{0.5} = 0.014 \text{ in./in.}$$

Shear modulus $G = \tau/\gamma$

$$= 0.88/0.014 = 62.8 \text{ lb./in.}^2$$

Ideal plastic behavior (St. Venant body)

A friction block such as shown in Fig. 4.10 can be used as the mechanical model for St. Venant Body representing ideal plasticity. In this model the solid friction between the block and the surface prevents any movement of the block. When the pull slightly exceeds the static friction, the block starts to move. Once the movement begins, the pull must overcome only the kinetic friction to keep the block moving. In this model the displacement gradient (in/in) stands for shearing strain γ in an ideal plastic material. The material does not flow until a limiting value of shearing stress τ_y, called the "yield stress," is reached. The material can sustain no stress

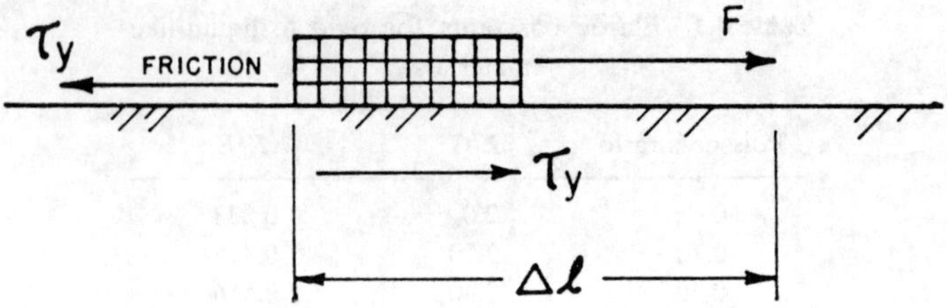

Figure 4.10 Friction block representing St. Venant Body for ideal plasticity

greater than this value and flows indefinitely under this stress unless distortion is restricted by some other factor.

Ideal viscous behavior (Newtonian liquid)

In an ideal plastic material we noted that a minimum stress, called the yield stress, is necessary before deformation and flow begins. In a liquid, deformation and flow begin as soon as a shear stress is applied, and when the stress is removed, like in the case of plastic flow, it will not return to its original state. In a liquid, strain is a function not only of stress but of time as well.

In a "simple" liquid with laminar flow the velocity gradient may be expressed as follows:

$$\frac{dv}{dy} = \phi\tau = \frac{1}{\eta}\tau \tag{4.9}$$

where dv is the velocity increment of one layer of the liquid passing over another layer a distance dy apart (Fig. 4.11). The movement is due to a shearing stress, τ, which is proportional to the velocity gradient with proportional factor ϕ, called fluidity, or $1/\phi=\eta$, called viscosity.

If we write the Eq. (4.9) in terms of shear strain rate γ or simply rate of shear, the Newton's law for ideal viscous liquids will result as given below

$$\tau/\dot{\gamma} = \eta \tag{4.10}$$

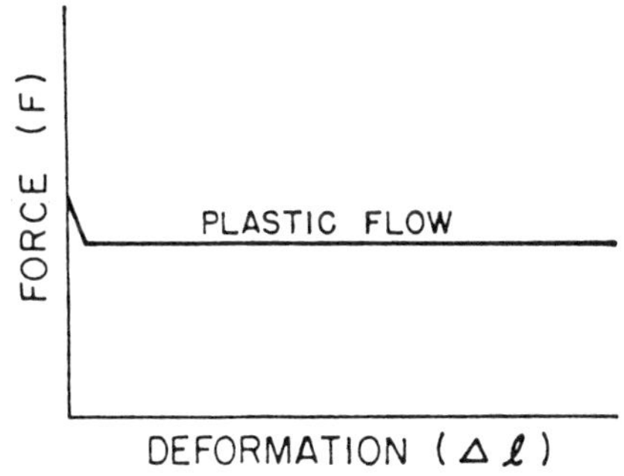

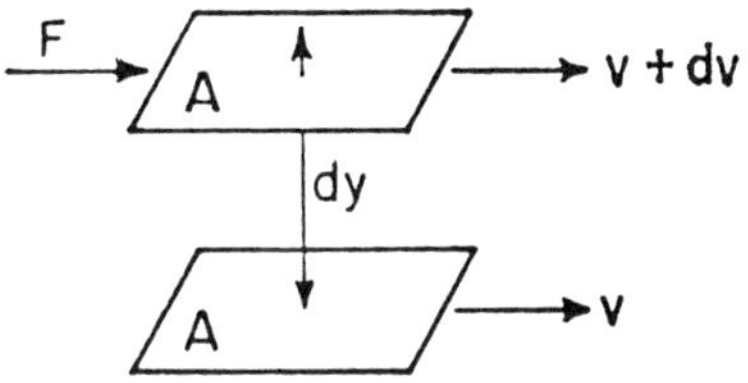

Figure 4.11 Laminar flow of a liquid under shear

Figure 4.12 shows the plot of Newton's law of viscosity and a simple dashpot representing the mechanical model for Newtonian liquids. The plot is a straight line passing through the origin. The slope of the line is by definition the viscosity. This shows that although the yield value in an ideal liquid is zero, there is an internal frictional resistance to flow called the viscosity. The mechanical model, called the viscous element, is supposed to be a weightless, loosely fitted piston moving in a medium filled with the liquid.

The viscosity of a liquid is measured by the tangential force and a unit area of either of two horizontal planes at unit distance apart, required to move one plane with unit velocity with reference to the other plane. When the force is one dyne over one square centimeter, the distance between planes is one centimeter, and the velocity one centimeter per second, the absolute viscosity is by definition one "poise." A centipoise, which is one $1/100$ of one poise, is usually taken as the unit of measurement. The kinematic viscosity or "stokes," used in engineering practice, is η/ρ, where ρ is the mass density of the liquid. The units of viscosity in the English

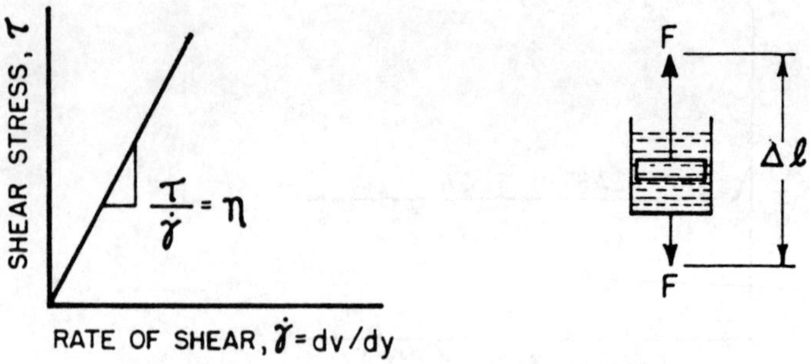

Figure 4.12 Newtonian liquid and the model

system are lb−sec/ft^2 for absolute viscosity and ft^2/sec for kinematic viscosity. Viscosity values for various materials are given in the Appendix.

4.6 TIME EFFECTS (VISCOELASTICITY)

Any real material shows deviation from the ideal materials discussed in the previous section. According to Ferry (1961), these deviations can be divided into two types. First, the relationship between stress and strain for solids and stress and rate of strain for liquids may be more complicated than the simple relationships given for Hookean solids or Newtonian liquids. Second, the stress-strain relationship may depend on the rate of strain as well as higher time derivatives of the strain. This time dependency results in a behavior called viscoelastic, which combines liquid-like and solid-like characteristics.

In a linear viscoelastic material the ratio of stress to strain is a function of time alone and not of the stress magnitude. For a number of viscoelastic materials linear viscoelastic response can be achieved experimentally if the deforming stresses are kept sufficiently small. If the magnitude of stress is such that the resulting strain is mostly nonrecoverable upon unloading, the ratio of stress to strain is a function of stress as well as time and the viscoelastic behavior is non-linear. This type of non-linear viscoelastic behavior is normally considered under viscoplasticity. Another type of non-linear viscoelasticity is the type observed in rubber and rubber-like materials, where non-linearity results due to large or finite strains, in contrast to infinitesimal strains considered in the theory of elasticity.

Based on experimental evidence agricultural products are viscoelastic. From the very limited data available in this area it appears, however, that the viscoelastic behavior is non-linear. Since no general theory for non-linear viscoelasticity is yet available, in an attempt to explain the rheological behavior of agricultural products, we are forced to make simplifying assumptions and apply the theories of linear viscoelasticity as in the case of some engineering materials.

4.7 RHEOLOGICAL MODELS

Mechanical models consisting of springs and dashpots are used to explain and interpret the rheological behavior of linear viscoelastic materials. The model is supposed to behave qualitatively, to some degree of approximation, in a manner similar to that of an actual material. If the mechanical behavior can be expressed in terms of force-deformation or stress-strain and time, the results can lead to a rheological equation. This equation can be used to explain and in certain cases predict the behavior of the material under various loading conditions.

The use of rheological models in studying the mechanical behavior of engineering materials has been questioned by some authorities, even though the model is to represent only the macroscopic behavior and not any insight into the molecular basis of the viscoelastic phenomena. In the case of biological materials, where liquids, simulating the viscous elements of the model, actually exist in the material, we have probably a better justification for the use of rheological models. It has been shown that despite the simplifying assumptions which have been made for applying the theories of viscoelasticity, the mechanical behavior of selected biological materials can be fitted to the physical behavior of rheological models.

The two basic mechanical elements used in mechanical models are a spring which obeys Hooke's law and a dashpot with a Newtonian liquid. These elements and their two basic combinations, Kelvin model and Maxwell model, are shown in Fig. 4.13. These curves are drawn for two strain rates $\dot{\varepsilon}_1$ and $\dot{\varepsilon}_2$. It is shown that for the spring the behavior is independent of time. For the dashpot and the other models containing a dashpot, stress-strain relationship is time-dependent.

In the Kelvin model, both the spring and dashpot are forced to move together at a constant rate. Therefore, the force due to the dashpot jumps to a constant value instantaneously and remains constant while the force on the spring starts at zero and gradually builds up. When the effects of

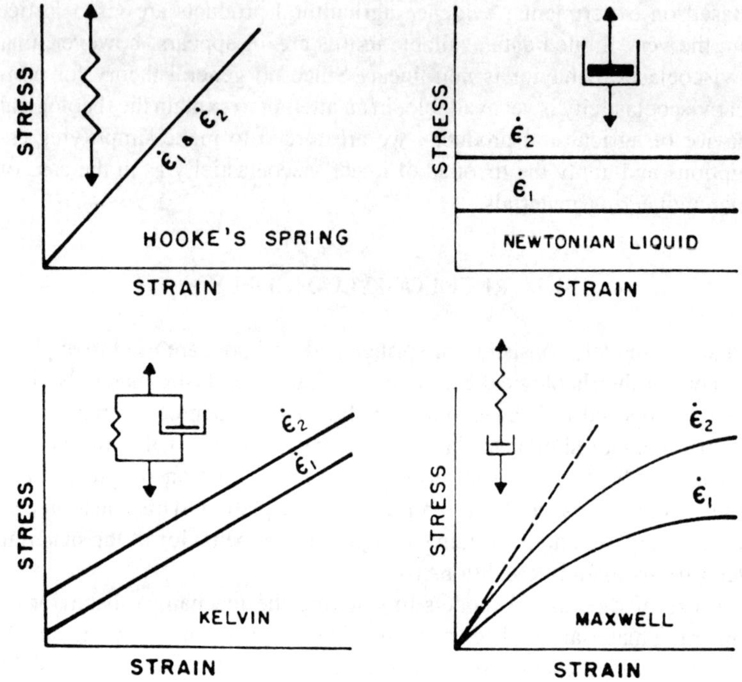

Figure 4.13 Effect of strain rate $\dot{\varepsilon}$ on stress–strain curves of four basic models $\dot{\varepsilon}_2 = 2\dot{\varepsilon}_1$ (Nielson, 1962)

the two elements are superimposed the net effect is the behavior of the Kelvin model.

In the Maxwell model, initially all the stress goes into stretching the spring, resulting in an initial straight line portion of the curves which obeys Hooke's law. As the spring changes, the dashpot carries more and more of the stress until the spring reaches the end of its elongation. At this point all the stress goes into flow in the dashpot and the stress-strain curve levels off.

Electrical equivalence of mechanical models

It has been said that if a physical phenomenon can be represented by a mechanical model, it can also be represented by an infinite number of other models. Since the differential equations governing the mechanical behavior of a material provides the most complete knowledge of its properties, and the most thoroughly studied examples of these equations are available for the linear electrical networks, the electrical analog of the mechanical models has been suggested for rheological studies.

In this analogy, the spring, representing elasticity, is replaced by a capacitance and the dashpot, representing viscosity, is replaced by a resistance. The tension and compression of the spring corresponds to charging and discharging of the capacitor. The work done on the dashpot is dissipated as heat just as the work done on the electrical resistance is converted to heat. Figure 4.14 shows the electrical analog of the basic mechanical element and the models. Note that when elements are coupled in parallel in the mechanical model, its equivalent in the electrical network is a series coupling. Similarly, a series coupling in the mechanical system has a parallel equivalence in the electrical system. In the electrical network, the stress is represented by the electric potential (voltage) and the strain by electric current.

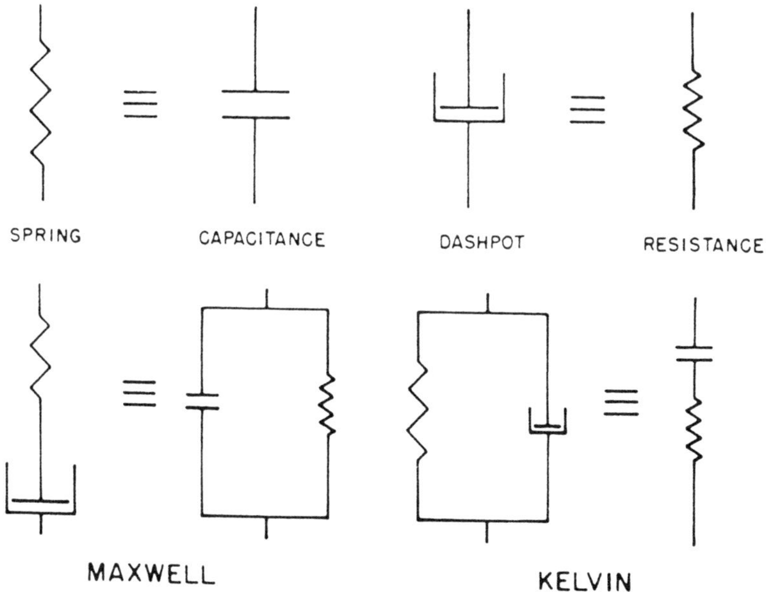

Figure 4.14 Electrical analogs of mechanical models

There are other mechanical-electrical analogies, proposed by various investigators (Stambaugh, 1952; Shoefield and Scott-Blair, 1933). The principle advantage of the electrical models is the simplicity and directness by which the model can be built and put into operation against the more complicated and elaborate mechanical system.

4.8 RHEOLOGICAL EQUATIONS

Total stress and total strain

When a material is subjected to tensile or compressive stresses resulting in both change in shape and change in volume, the deformation may be shown to be a combination of both shear and bulk deformation. Consequently, stress components at any point of this material can be decomposed into a deviatoric stress component, which is responsible for the change in shape, and an isotropic stress component which is responsible for the change in volume. Therefore, the total stress in the x direction, σ_x, can be written as follows:

$$\sigma_x = S_x + \bar{\sigma}$$

where S_x is the deviatoric stress and $\bar{\sigma}$ is the mean normal stress defined as

$$\bar{\sigma} = 1/3\ (\sigma_x + \sigma_y + \sigma_z)$$

Similarly, the total strain in the x direction, ε_x, can be resolved into a deviatoric strain e_x and *a* mean normal strain $\bar{\varepsilon}$ as follows:

$$\varepsilon_x = e_x + \bar{\varepsilon}$$

where

$$\bar{\varepsilon} = 1/3\ (\varepsilon_x + \varepsilon_y + \varepsilon_z)$$

It may be said that while the mean normal stress is a measure of pressure and the mean normal strain a measure of volumetric deformation, the deviatoric stresses isolate the shear stresses and strains. For example, the state of stress and strain in an isotropic elastic material would lead to the simple stress-strain relationships

$$\bar{\sigma}/\bar{\varepsilon} = K \text{ and } S_x/e_x = G \tag{4.11}$$

where K is the bulk modulus and G the shear modulus. (Compare Eqs. 4.2 and 4.3). Since in an isotropic elastic material K is much greater than G (the incompressible case), it is often justified to neglect volume changes in engineering materials and consider only the shear strain relationship.

In the above discussion only one of the six components of stress at a volume element of a material is given. However, the type of test and the frame of reference can be chosen such that σ_x and the corresponding ε_x become the principal stress and strain. The other components will either be smaller than the principal tensile or compressive stress, or become zero. An example for this situation is tensile test of a cylindrical specimen where the principal tensile stress in the direction parallel to the long axis is F/A, the principal shear stress is $\frac{1}{2}F/A$, and the other components of stress are zero. Therefore, the principal tensile stress, being twice that of the principal shear stress, should be the one to consider.

Since simple shear deformation can be easily produced on both solids and liquids and viscoelastic materials are intermediate between solids and liquids, it is customary to show the rheological equations in terms of shear stress and shear strain. In the following, however, the rheological equations have been developed for total axial stress σ and total axial strain ε in tension or compression. The equations are equally valid for shear parameters as well as bulk compression. Exceptions are indicated.

Maxwell model

To derive the rheological equations for the basic Maxwell and Kelvin models, we assume the Newtonian law of viscosity for the dashpots and the Hook's law for the springs. With reference to Fig. 4.15, if

σ = stress
ε = strain
t = time
E = stiffness or modulus of the spring representing the ideal elastic body
E_0 = instantaneous modulus or modulus at zero time
E_e = equilibrium modulus or modulus after infinite time
$E_d = E_0 - E_e$ = decay modulus
η = viscosity coefficient of the liquid in the dashpot
s, v = subscripts denoting respectively, spring and viscous element, in the Maxwell model for the spring:

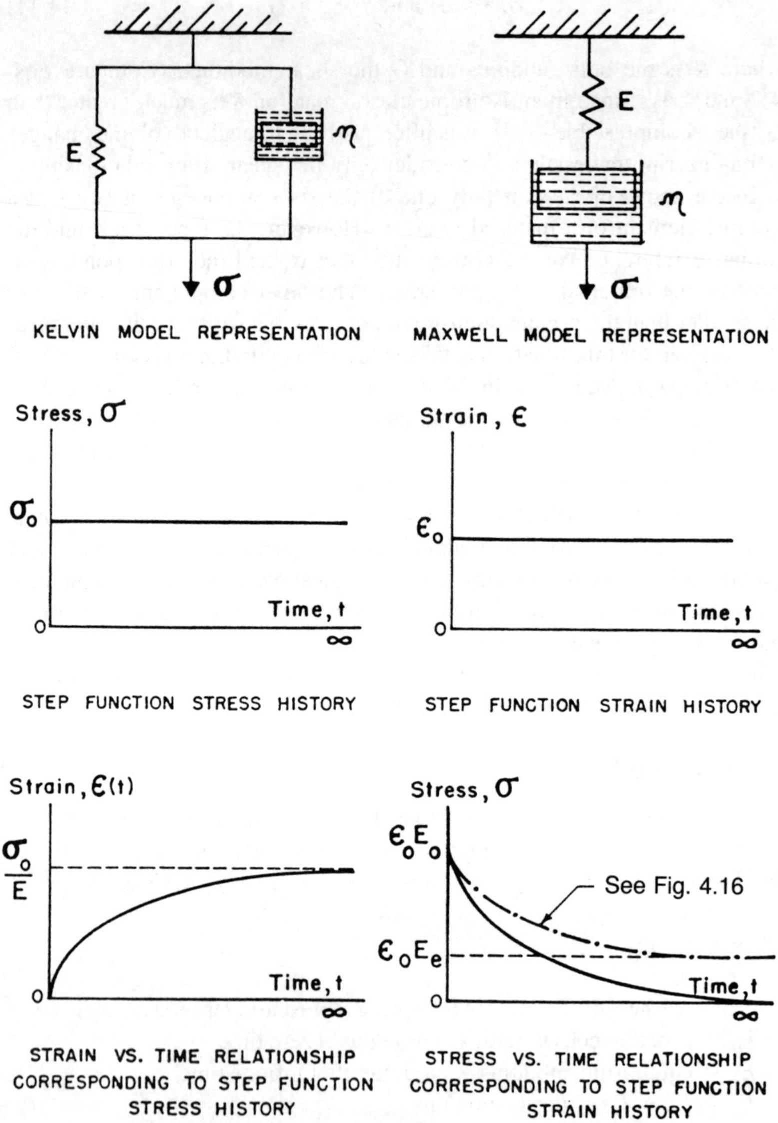

Figure 4.15 Kelvin and Maxwell models showing creep and stress relaxation characteristics (Sharma, 1964)

$$\sigma_s/\varepsilon_s = E \text{ and } \dot{\varepsilon}_s = \dot{\sigma}_s/E \qquad (4.12)$$

where dot on any symbol represents the time rate.

For the dashpot

$$\eta = \sigma_v/\dot{\varepsilon}_v \qquad (4.13)$$

the strains being additive in the Maxwell model, the total strain

$$\varepsilon = \varepsilon_s + \varepsilon_v \qquad (4.14)$$

differentiating (4.14) and substituting $\dot{\varepsilon}_s$ and $\dot{\varepsilon}_v$ from (4.12) and (4.13) yields

$$\dot{\varepsilon} = \dot{\varepsilon}_s + \dot{\varepsilon}_v$$

$$\dot{\varepsilon} = \dot{\sigma}_s/E + \sigma_v/\eta \qquad (4.15)$$

Since in a Maxwell model the same force is carried through the spring and the dashpot, substitution for $\sigma_s = \sigma_v = \sigma$ in Eq. (4.15) yields

$$\dot{\varepsilon} = \dot{\sigma}/E + \sigma/\eta \qquad (4.16)$$

If the model is subjected to a constant strain as in Fig. 4.15 ($\dot{\varepsilon} = 0$), and the term η/E is replaced by the symbol T_{rel}, called the time of relaxation, Eq. (4.16) can be reduced to

$$\dot{\sigma} + \sigma/T_{\text{rel}} = 0 \qquad (4.17)$$

The solution of the above differential equation is of the form

$$\sigma = A\, e^{-t/T_{\text{rel}}} + c \qquad (4.18)$$

where e is the base of Naperian logarithm ($e = 2.72$). Substituting the boundary conditions of

$$\sigma = \sigma_0 = E_0\varepsilon_0 \text{ at } t = 0$$

$$\sigma = \sigma_e = E_e\varepsilon_0 \text{ at } t = \infty$$

in (4.18), constants A and C are found to be

$$C = \varepsilon_0 E_e = \sigma_e$$

$$A = \varepsilon_0(E_0 - E_e) = \sigma_d$$

After substitution of the above constants, Eq. (4.18) can be written in terms of stress

$$\sigma(t) = \sigma_d \, e^{-t/T_{rel}} + \sigma_e \tag{4.19}$$

where $\sigma(t)$ denotes stress at any time "t", σ_d is the decay stress, and σ_e is the stress at equilibrium. In terms of time dependent modulus, $E(t)$, Eq. (4.18) can be written as

$$E(t) = E_d \, e^{-t/T_{rel}} + E_e \tag{4.20}$$

Note that after $t=\infty$ there is still some stress left in the spring E of the Maxwell model in Fig. 4.15. This can physically be visualized either by assuming the dashpot being limited and reaching the bottom while the spring is still strained, or by considering the dashpot being unlimited but adding another spring with modulus E_e in parallel with the Maxwell element. This latter situation can be seen in Fig. 4.16 where a generalized Maxwell model is illustrated. Sometimes the equation for a simple Maxwell model is shown as

$$\sigma(t) = \sigma_0 \, e^{-t/T_{rel}} \tag{4.21}$$

and

$$E(t) = E_0 \, e^{-T_{rel}} \tag{4.22}$$

In this case it is assumed that the dashpot is unlimited in flow resulting in zero stress in the spring after infinite time. In real materials, however, even after long periods of time, stresses do not vanish completely. Therefore, Eqs. (4.19) and (4.20) appear to represent the actual behavior more realistically.

Examination of Eqs. (4.19) and (4.20) shows that the Maxwell model is not sufficiently general to describe the behavior of a linear viscoelastic material. For example, if a constant stress is applied to a Maxwell model, the model exhibits only Newtonian flow and not a retarded elastic deformation which is experimentally observed in a creep or constant stress test.

To avoid this problem, an infinite number of Maxwell models are used in parallel and the resulting model is called a generalized Maxwell model.

Generalized Maxwell model

A generalized Maxwell model is composed of n Maxwell elements with a spring in parallel with the nth element, as illustrated in Fig. 4.16. The elastic modulus E_e of this last spring corresponds to the equilibrium modulus in a stress relaxation test.

The generalized Maxwell model is usually used to represent stress relaxation. If this model is subjected to constant strain ε_0 at time $t=0$, the total stress in the model

$$\sigma = \sigma_1 + \sigma_2 + \sigma_3 + \ldots + \sigma_n + \sigma_e$$

decreases with time and can be represented by

$$\sigma(t) = \varepsilon_0(E_{d1}\, e^{-t/T_1} + E_{d2}\, e^{-t/T_2} + \ldots + E_{dn}\, e^{-t/T_n} + E_e) \quad (4.23)$$

where T_1, T_2, T_3, . . . , T_n are the relaxation times, T_{rel}, corresponding to various elements in the model. The physical significance of T_{rel} can best be appreciated if we consider that the relaxation time, according to the simplified case given by Eq. (4.21), is the time required for the stress to decay to $1/e$ times its original value or to 36.8 percent of its original value. In other words, at constant strain, a Maxwell element relaxes its stress exponentially at a rate determined by T_{rel}.

Kelvin model

In the Kelvin model of Fig. 4.15, while the total stress is divided between the spring and the dashpot

$$\sigma = \sigma_s + \sigma_v \quad (4.24)$$

the dashpot is forced to take up the same deformation as the spring such that

$$\varepsilon_v = \varepsilon_s = \varepsilon \quad (4.25)$$

Substituting for σ_s and σ_v from (4.12) and (4.13) and considering (4.25), Eq. (4.14) can be written in the following form

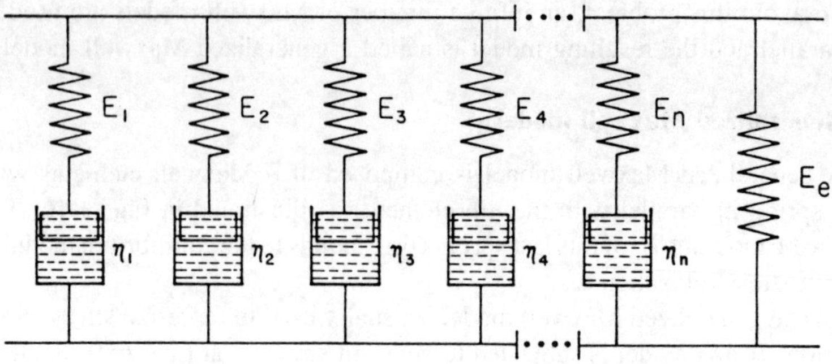

Figure 4.16 Generalized Maxwell model representation

$$\sigma = E\varepsilon + \eta\dot{\varepsilon} \tag{4.26}$$

If the ratio $\eta/E=T_{ret}$, called the time of retardation, Eq. (4.26) can be reduced to

$$\sigma/E = \varepsilon + T_{ret}\,\dot{\varepsilon} \tag{4.27}$$

After differention

$$\sigma/E = \dot{\varepsilon} + T_{ret}\,\ddot{\varepsilon} \tag{4.28}$$

Subjecting the model to a constant stress σ_0, as in Fig. 4.15, reduces Eq. (4.28) to

$$T_{ret}\,\ddot{\varepsilon} + \dot{\varepsilon} = 0 \tag{4.29}$$

By integration we obtain

$$\varepsilon = \sigma_0/E + (\varepsilon_0 - \sigma_0/E)\,e^{-t/T_{ret}} \tag{4.30}$$

Where σ_0 is the constant stress and ε_0 is the initial strain at $t=0$.

If the initial strain is zero, substituting $\varepsilon_0=0$ in Eq. (4.30) yields the following expression for creep under constant load:

$$\varepsilon = \sigma_0/E\,(1 - e^{-t/T_{ret}}) \tag{4.31}$$

For

$$t = 0, \qquad \varepsilon = \varepsilon_0 = 0$$

$$t = \infty, \qquad \varepsilon = \sigma_0/E$$

$$t = T_{ret}, \qquad \varepsilon = (\sigma_0/E)\,(1 - 1/e)$$

As shown above, the retarded elastic strain reaches the constant value of σ_0/E only after infinite time (leveling off of the curve). Note that the time of retardation, T_{ret}, in the case of creep is $(1-1/e)$ or about 63 percent of the time for total strain. This is the time required for the parallel combination to deform to this extent. (Compare with time of relaxation T_{rel}). In other words, at constant stress, a retarded elastic element relaxes exponentially into its equilibrium shape at a rate determined by its T_{ret} just as at constant strain a Maxwell element relaxes exponentially into its equilibrium stress at a rate determined by its T_{rel}.

If the constant load is removed from the strained body, it will not revert to the unstrained state until after infinite time. This can be seen from the following equation which is derived from Eq. (4.30) by letting

$$\sigma_0 = 0$$

$$\varepsilon = \varepsilon_0\, e^{-t/T_{ret}} \tag{4.32}$$

Equations (4.31) and (4.32) represent respectively the loading and unloading sections of the strain-time curve of a Kelvin model (See Fig. 4.24).

As in the case of Maxwell model, Eq. (4.31) is not sufficiently general to predict the behavior of a viscoelastic material under all conditions of loading. For example, if the model is subjected to a constant strain, the observed stress relaxation cannot be predicted from the Kelvin model, because σ_0 remains a constant stress and not a decreasing stress. To correct this situation, a more general model called a 4-element model has been proposed.

4-element model (Burgers model)

The 4-element model shown in Fig. 4.17 is one of the best known rheological models which has been used to predict the creep behavior in a number of materials. Application of this model to food materials is discussed in Chapter 6. The model is composed of a spring and dashpot in series with another spring and dashpot in parallel. This 4-element model is also known as Burgers model and has several other equivalent representations (Fig. 4.18).

Reference to Fig. 4.17 shows that in the 4-element model strains are additive while the stress is the same in all three units *A, B,* and *C*.

$$\varepsilon = \varepsilon_A + \varepsilon_B + \varepsilon_C$$
$$\sigma = \sigma_A = \sigma_B = \sigma_C$$

Substituting for σ_A, σ_B, and σ_C from Hookean elasticity, Eq. (4.26), and Newtonian viscosity, respectively

$\sigma_A = E_0\varepsilon_A$ (corresponding to instantaneous deformation)

$\sigma_B = E_r\varepsilon_B + \eta\dot{\varepsilon}_B$ (corresponding to retarded elastic deformation)

$\sigma_C = \eta_v\dot{\varepsilon}_C$ (corresponding to Newtonian flow)

These three equations can be combined to one equation relating stress and total strain (for derivation see Morrow, 1965).

$$\ddot{\varepsilon} + \dot{\varepsilon}/T_{ret} = 1/E_0$$
$$\left[\ddot{\sigma} + \left(\frac{E_0}{E_r T_{ret}} + E_0/\eta + 1/T_{ret}\right)\dot{\sigma} + \left(\frac{E_0}{T_{ret}\,\eta_v}\right)\sigma\right] \tag{4.33}$$

This differential equation is sufficiently general to describe the behavior of a linear viscoelastic material through the behavior of a 4-element model. Both constant stress (creep) and constant strain (stress relaxation) behavior can be predicted from this equation. However, it is much easier to use a generalized Maxwell model for stress relaxation. The equation is also valid for a material exhibiting elasticity, retarded elasticity, and flow.

For example, if a constant load (stress) is instantly applied and held constant, for the initial conditions of

$$\sigma = \sigma_0 = \text{constant at } t = 0$$

Equation (4.33) reduces to

$$\ddot{\varepsilon} + \dot{\varepsilon}/T_{ret} = \frac{\sigma_0}{T_{ret}\,\eta_v} \tag{4.34}$$

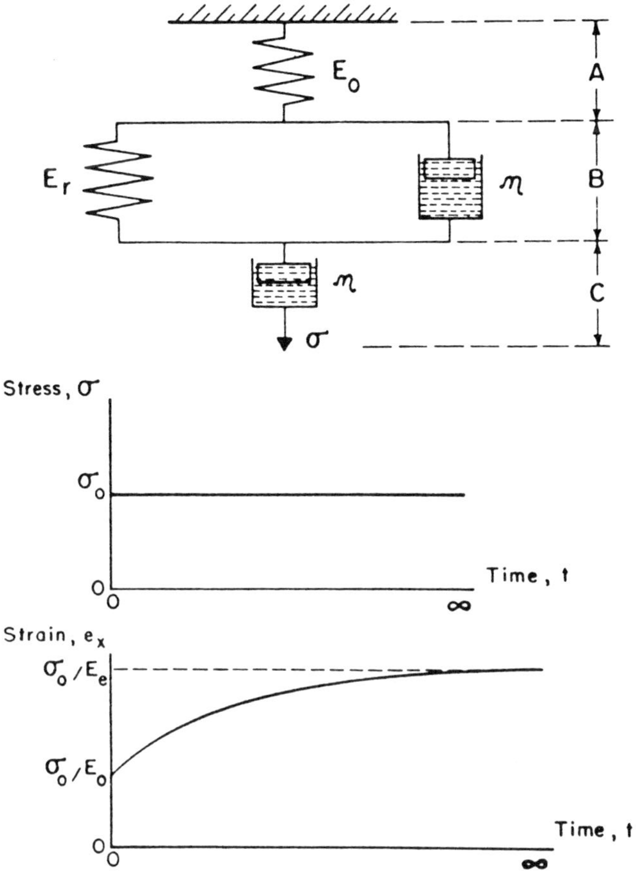

Figure 4.17 4-Element Burgers model characteristics

The solution of the above equation (as given by Morrow, 1965) is

$$\varepsilon(t) = \sigma_0/E_0 + \sigma_0/E_r\,(1 - e^{-t/T_{ret}}) + \frac{\sigma_0 t}{\eta_v} \qquad (4.35)$$

If we define a compliance function $D(t)$ as the reciprocal of a tensile or compressive modulus function $E(t)$ or the ratio of strain to stress, the above equation can be expressed in terms of compliance

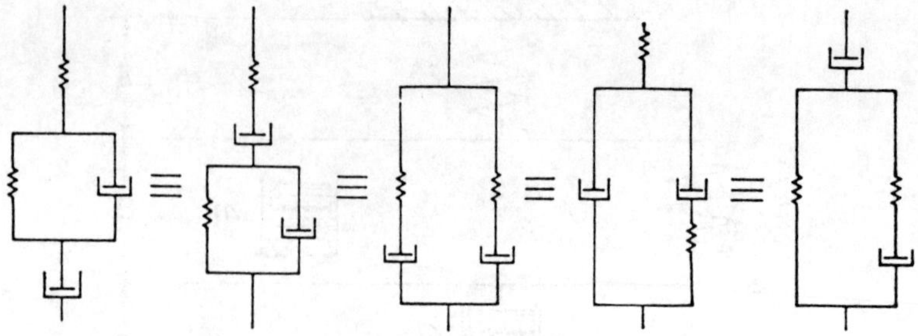

Figure 4.18 The four-element model and equivalent representations (Bland, 1960)

$$D(t) = D_0 + D_r(1 - e^{-t/T_{\text{ret}}}) + t/\eta_v \tag{4.36}$$

where

$D_0 = 1/E_0 =$ initial compliance

$D_r = 1/E_r =$ retarded compliance

The strain variation with time shown in Fig. 4.17 is actually the plot of Eq. (4.35). It resembles a typical creep curve for a linear viscoelastic material. Although Eq. (4.35) is derived for total stress and total strain in tension or compression, it is equally valid for shear. In the case of bulk compression, however, η_v, being the ratio of shear stress to the rate of shear strain in the Newtonian viscous element, will be infinitely large and the term $\sigma_0 t/\eta_v$ drops out because no shear strain may exist under hydrostatic compression. This results in the following equation for bulk creep compliance

$$B(t) = B_0 + B_r(1 - e^{-t/T_{\text{ret}}}) \tag{4.37}$$

where $B(t)$ represents compressibility, the inverse of bulk modulus K. This means that for bulk behavior interpretation, the viscous element with η_v is to be eliminated from the 4-element model.

Generalized Kelvin model

Experimental data on many viscoelastic materials, including biological materials, have shown more than one relaxation time or retardation time.

For these materials, the complete behavior cannot be represented by a single Maxwell model or a single Kelvin model or even the 4-element model. Each of these models has only one time constant. To present the viscoelastic behavior more realistically, like the case of generalized Maxwell model, a chain of Kelvin models, each with its own time of retardation, is assumed and the model is called a generalized Kelvin model. It consists of "n" Kelvin elements connected in series with an initial spring and a final viscous element, as shown in Fig. 4.19. The first spring with E_0 is to account for instantaneous elastic strain. The "n" Kelvin models are to account for retarded strain. And, the dashpot with η_v viscosity corresponds to the flow experienced in creep tests.

The equation for the generalized Kelvin model can be derived in the same manner shown for a 4-element model. This procedure would lead to the following equation which is the generalized form of Eq. (4.35).

$$\varepsilon(t) = \sigma_0 \, [1/E_0 + 1/E_{r1}\,(1 - e^{-t/T_1}) + 1/E_{r2}\,(1 - e^{-t/T_2}) + \cdots + 1/E_{rn}\,(1 - e^{-t/T_n}) + t/\eta_v] \tag{4.38}$$

where $T_1, T_2, T_3 \ldots T_n$ are different retardation times, T_{ret}, corresponding to various elements in the model.

4.9 VISCOELASTIC CHARACTERIZATION OF MATERIALS

There are a number of tests which may be used to study viscoelastic materials and determine the relations among stress, strain, and time for a given type of deformation and a given type of loading pattern. As mentioned before, the assumption of linear viscoelasticity can be valid if the stress is kept sufficiently small. The most important tests include stress-strain, creep, stress relaxation, and dynamic tests.

Stress-strain behavior

The force-deformation curve of Fig. 4.3 together with definitions of terms in section 4.2 give some of the mechanical properties related to the strength of a material. Figure 4.13 shows how the rate of straining affects the stress-strain behavior and how such behavior can be explained in terms of rheological models. Further illustrations of the time-effect, manifested through the rate of straining for selected food materials are given in Figs. 4.20 and 4.21. Note that in the case of fresh fruits the initial slope of the curve is independent of the speed of testing (Fig. 4.20). This phenomenon was also shown in the stress-strain curve of the Maxwell model in Fig. 4.13. After a given force (point E in Fig. 4.20), the reaction, as given by the Maxwell model, combines elastic and viscous types of behavior. This results in divergence of the curves according to the rates of loading. It is also seen that at large deformations, the force (stress) levels off to a limiting value. The greater the speed of testing, the greater is the limiting value of the force or stress.

Because of this time-effect phenomenon, in evaluating qualities of products such as tenderness of meat, hardness of grains, and firmness of fruits and vegetables, the compression apparatus must provide a constant rate of loading for the duration of the test. Otherwise, for any fixed displacement of the crosshead, the force readings may vary as the rate of loading varies. To illustrate, suppose the compression force exerted on samples of fruits at 100 units of displacement of the crosshead is to be taken as the firmness of the fruit (Fig. 4.20). If due to the change in oil temperature in a hydraulic type of machine, or other factors, the speed of the crosshead increases from 1 inch per minute to 13 inches per minute, the firmness reading may vary from 4 pounds to more than 6 pounds just because of the variation in the rate of loading.

The effect of rate of loading on stress-strain behavior can be predicted from the stress-strain equation of the Maxwell model which can be derived from the differential equation

$$\dot{\varepsilon} = \dot{\sigma}/E + \sigma/\eta$$

given previously. For a constant rate of strain $\dot{\varepsilon}=R$, the above equation can be rearranged to give

$$\dot{\sigma} + E\sigma/\eta = RE \tag{4.39}$$

The homogeneous solution of Eq. (4.39) for $RE=0$ is

$$\sigma = A\,e^{(-E/\eta)t}$$

The particular solution of Eq. (4.39) for the stress σ being some constant B is

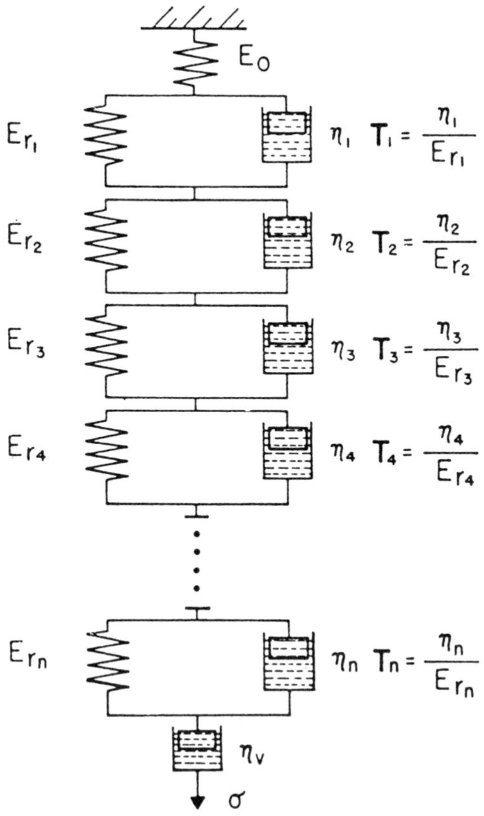

Figure 4.19 Generalized Kelvin model representation

$$EB/\eta = RE$$

or

$$B = R\eta$$

Now the complete solution of Eq. (4.39) can be written as

$$\sigma = A\ e^{(-E/\eta)t} + R\eta \tag{4.40}$$

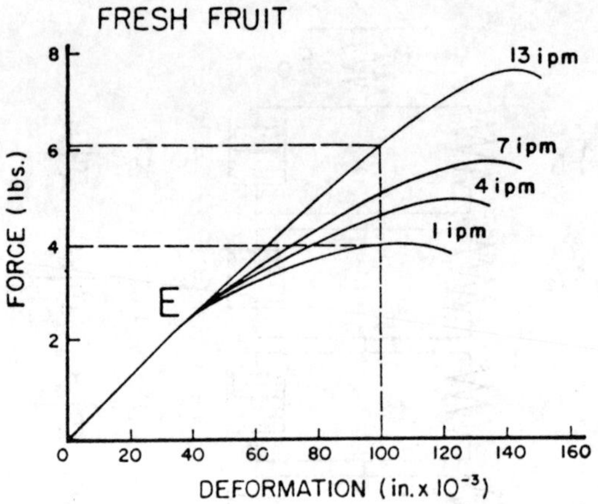

Figure 4.20 Firmness reading of a fruit can be influenced by the rate of loading

Constant A can be evaluated by considering the initial condition of $\sigma=0$ at $t=0$. This results in

$$A = -R\eta$$

Substituting for constant A and incorporating $\varepsilon=Rt$ in the exponent of Eq. (4.40), the stress-strain relationship for a Maxwell unit can be written as

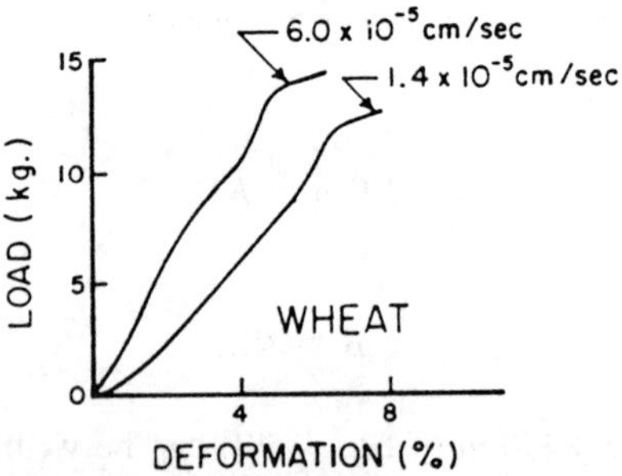

Figure 4.21 At high rates of loading grain is more brittle and produces coarsely ground wheat—at low rates of loading, flaky particles are produced (Sh. Polyanskaya, 1952)

$$\sigma = R\eta\,(1 - e^{-E\varepsilon/R\eta}) \tag{4.41}$$

In this equation, E is the slope of the initial portion of the stress-strain curve. For this portion of the curve, E is independent of R. At higher strains, however, the slope of the curve depends upon the rate of testing R. For large strains, the stress levels off to the limiting value $R\eta$, predicting the increase of the limiting stress and the modulus E with the increase of the rates of straining. The use of stress-strain curves along with creep and stress relaxation curves for viscoelastic characterization of a material and fitting the behavior to a 4-element model has been discussed by Frederick and Gaecia (1958). Also, six rules for analyzing stress-strain curves of non-linear viscoelastic materials have been presented by Halsey (1957). These rules suggest an orderly mode of experiment leading to the representation of the material by a hyperbolic sine law model.

There are a number of testing machines available for controlled stress-strain experiments. Zoerb 1967 has reviewed a number of these testing machines which have been used for testing of agricultural products. Figure 4.22 shows a testing machine which has been developed for investigation of rheological properties of food and feed materials. The details of construction and operation of the machine are given by Mohsenin (1963).

Figure 4.23 shows an Instron testing machine equipped with a test chamber for testing of materials under controlled temperatures ranging from −100°F to +600°F. If testing at various relative humidities is desired, the air in the test chamber can be conditioned and recirculated through an Aminco Aire unit and two flexible air ducts connected to the back of the test chamber. A specially designed rheometer, using the principle of extrusion, can also be used with this machine for direct recording of shear stress-shear strain characteristics of fluids at controlled temperatures. Another useful accessory for this testing machine is an integrator which integrates the area under stress-strain or force-deformation curves as the test is being conducted.

Creep

In this test the load (stress) is suddenly applied and held constant, and deformation (strain) is measured as a function of time. The results are expressed in terms of time-dependent parameters $E(t)$ or its compliance $D(t)$ in tension or compression creep, in terms of $G(t)$ or $J(t)$ in shear creep, and in terms of $K(t)$ or $B(t)$ in bulk creep.

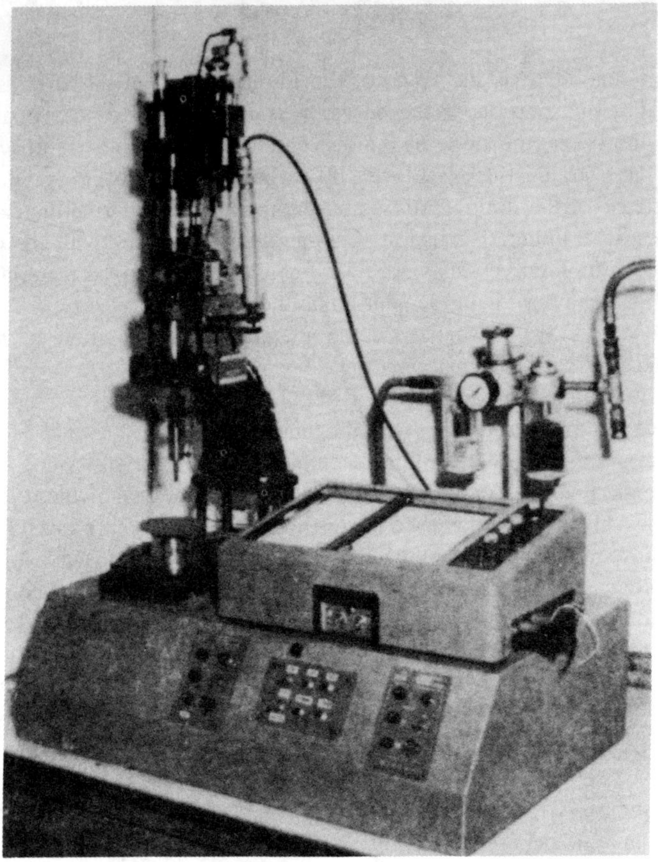

Figure 4.22 Compression testing machine developed for agricultural products

The rheological model to represent the creep behavior is the 4-element Burgers model illustrated in Fig. 4.17 or its equivalents given in Fig. 4.18. The rheological equation of the model is given by Eq. 4.35. If in a creep test of a viscoelastic material, results can be presented in terms of strain versus time, the complete curve for creep and recovery upon unloading will normally look like the curve shown in Fig. 4.24.

Interpretation of Fig. 4.24 in terms of a 4-element model is given in Fig. 4.25. It is seen that upon sudden loading spring E_0 is stretched by an

Figure 4.23 Instron testing machine equipped with test chamber and Integrator

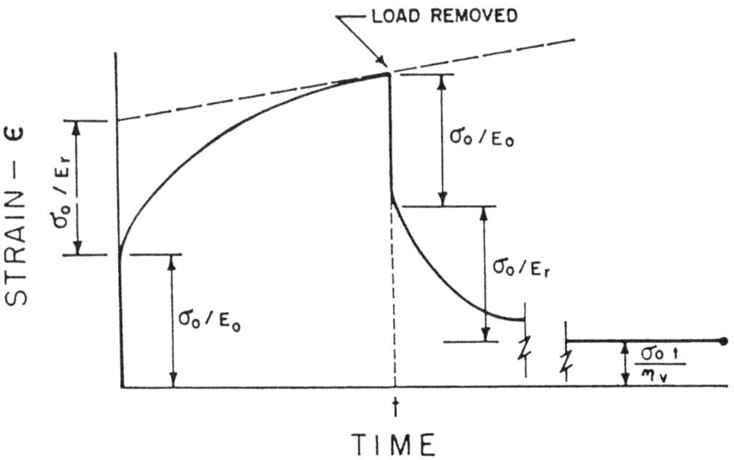

Figure 4.24 Typical creep and recovery curve in a viscoelastic material exhibiting instantaneous elasticity, retarded elasticity and viscous flow

amount σ_0/E_0. Knowing the constant stress σ_0, the value of E_0 can be calculated from the instantaneous strain. After this initial strain, creep starts at a high rate but gradually slows down due to dashpot η_v. At time "t" the load is removed and spring E_0 snaps back to its original state. The spring E_r, however, cannot contract to its original state instantaneously. Note that the instantaneously recovered elastic strain is equal to the initial instantaneous strain σ_0/E_0. During the recovery time, the spring E_r slowly forces the plunger of dashpot η back to its original position. Since after removal of the load, no force is acting on dashpot η_v, this element retains a nonrecoverable displacement representing permanent deformation in the material.

The retarded elasticity E_r can be evaluated from σ_0/E_r taken either from the loading portion or the unloading portion of the curve. This is shown graphically in Fig. 4.24. The graphical construction is evident if time "t" is let to be infinite in Eq. (4.35). The flow parameter $\sigma_0 t/\eta_v$ is found from the curve after the unloading curve has come to equilibrium. Knowing constant stress σ_0 and time of creep "t", viscosity η_v can be calculated from the experimental value of $\sigma_0 t/\eta_v$.

The time of retardation can be found from the retarded recovery part of the curve and its corresponding equation taken form Eq. (4.35)

$$\sigma_0/E_r(1-e^{\times t/T_{\text{ret}}}) = \varepsilon(t) - \sigma_0/E_0 - \sigma_0 t/\eta_v \tag{4.42}$$

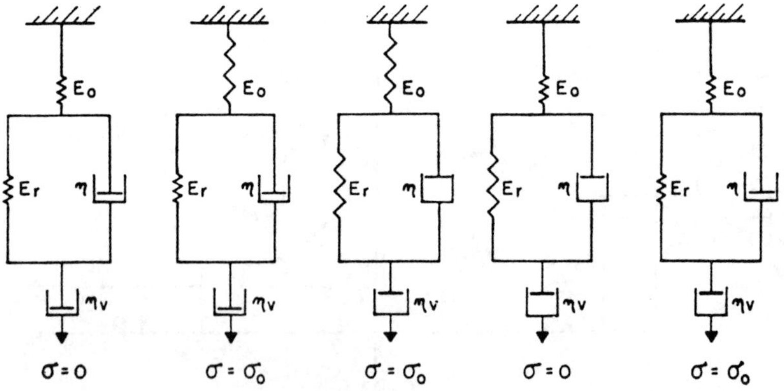

Figure 4.25 Creep and recovery behavior in terms of a four-element model (Alfrey, 1948)

If

$$\sigma_0/E_r(1 - e^{-t/T_{ret}}) = A$$
$$\sigma_0/E_r = B$$

the lefthand side of Eq. (4.42) (the recovery part) can be written as

$$A = B(1 - e^{-t/T_{ret}})$$

or

$$(A/B - 1) = -e^{t/T_{ret}}$$
$$\log(1 - A/B) = \left(\frac{-1}{2.3T_{ret}}\right)t \tag{4.43}$$

After selecting a few points for "A" at several time intervals "t," the results can be plotted on a semi-log paper. The slope of the resulting straight line is $\left(\frac{1}{2.3T_{ret}}\right)$, from which the time constant T_{ret} can be calculated. Since $T_{ret} = \eta/E_r$, knowing T_{ret} and E_r, η in the Kelvin unit of the 4-element model can be evaluated.

Reiner has given the time of retardation in a graphical method (Reiner, 1956, 1960). This method, with some modification, is shown in Fig. 4.26. The graphical presentation is self-explanatory and starts by drawing the flow line, $\sigma_0 t/\eta_v$, parallel to the straight line portion of the creep curve.

In real materials, the creep behavior is represented by a generalized Kelvin model containing several retardation times (Fig. 4.19). In that case, it is necessary to find the distribution of retardation times so that the most probable T_{ret} can be estimated. A simple method for obtaining the distribution of retardation times is to plot the instantaneous slope of the creep curve against logarithm of time. This method is also used in finding the best estimate for relaxation time when a generalized Maxwell model is considered for behavior of a material. The details of this technique are given by Andrews (1952) and will be discussed briefly in discussion of stress relaxation.

Figure 4.27 shows a creep and recovery test of a McIntosh apple. Using this curve and the procedure above, the various viscoelastic parameters can be obtained for this biological material. A creep test, in addition to

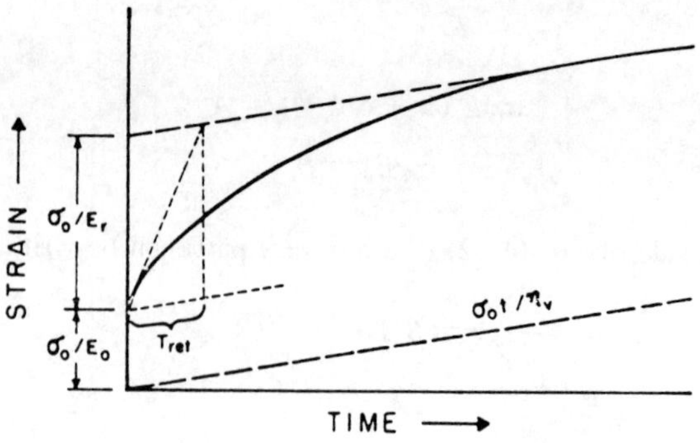

Figure 4.26 Graphical method for determination of retardation time in creep (Reiner, 1960)

providing information on viscoelastic characterization of the material, can be used to predict the deformation of agricultural products such as fruits, vegetables, silage, etc., under dead load.

Stress relaxation

In stress relaxation the test specimen is suddenly brought to a given deformation (strain), and the stress required to hold the deformation constant is measured as a function of time. The results are expressed in terms of time-dependent modulus $E(t)$ in tension or compression, $G(t)$ in shear, or $K(t)$ in bulk compression. The rheological model representing stress relaxation is the generalized Maxwell model given in Fig. (4.16). The corresponding rheological equation is given in Eq. (4.23).

One of the most important viscoelastic parameters which can be obtained from a stress relaxation test is the time of relaxation. As indicated before, relaxation time is the time at which the stress in a body resembling a simple Maxwell model decays to $1/e$ of initial stress. It is a measure of the rate at which a material dissipates stress after receiving a sudden force. There are a number of methods for treating experimental data on stress relaxation and finding the relaxation time. The following method permits the investigator to study the distribution characteristics of the "spectrum" of relaxation times which is usually found in biological materials.

The first step in analysis of stress relaxation data is to plot the logarithm of stress versus time. If the plot were linear, the behavior of the material is Maxwellian and the time of relaxation can be determined from the slope of the straight line.

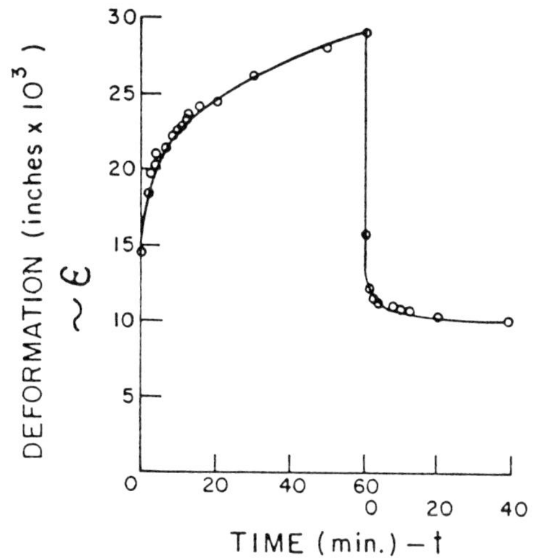

Figure 4.27 Distortion of McIntosh apple under dead load of 21 lbs determined by axial creep and recovery test with ¼-inch rigid plunger)

In most cases, however, the plot of logarithm of stress versus time is non-linear (see Fig. 4.28), indicating that the rheological behavior cannot be represented by a single Maxwell element but, as discussed previously, an array of Maxwell elements connected in parallel is required. Under such conditions, a "spectrum of relaxation times" exists from which we must select the mean relaxation time. It has been shown (Feltham, 1955) that this "spectrum of relaxation times" follows a Gaussian or normal distribution curve with a logarithmic time base.

To determine the closeness of distribution of relaxation times to a normal distribution, first the plot of stress versus logarithm of time is established (lower curve Fig. 4.29). When the negative of the instantaneous slopes from this curve are plotted against logarithm of time, the resulting curve usually approximates a normal curve if Maxwellian behavior exists. This technique has been applied to evaluation of stress relaxation data in wheat dough (Cunningham *et al.*, 1953). In this case, only the ordinate of the distribution function was considered in determining the effect of such factors as changing of the water content in dough. The ordinate of the approximately normal curve indicated the contribution to the dough modulus made by the Maxwell mechanisms in the material having that relaxation

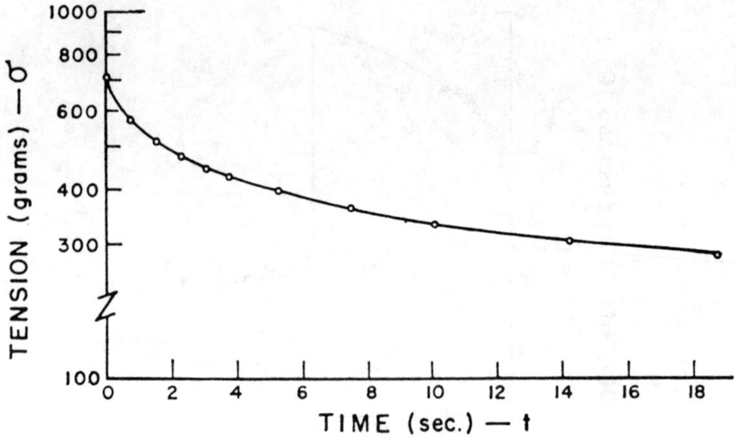

Figure 4.28 Relaxation curve for wheat dough (Cunningham *et al.*, 1953)

time. Figure 4.29 shows the presence of such a normal curve for data in Fig. 4.28.

To avoid the tediousness and inherent error of replotting the instantaneous slope, other investigators (Grogg and Helms, 1958) have made use of the "Central Limit Theorem" by which the mean relaxation time and the standard deviation of the mean can be estimated quickly and more accurately. In this method, the stress at various time intervals is calculated as percent of initial stress and plotted on probability coordinates against the logarithm of time. For this purpose, a log-normal probability curve takes on the shape of a straight line when plotted on a normal-probability paper. Therefore, if the plotted relaxation data result in an approximately straight line, it is reasonable to assume that the "spectrum of relaxation times" has a log-normal distribution. To obtain the best estimate of the relaxation time, we can approximately equate the half relaxation period to the geometric mean relaxation time. The half relaxation period can be determined by reading the time at the 50 percent stress level (see Fig. 4.30).

Similarly, the geometric standard deviation may be approximated by taking the ratio of the time at the 84 percent stress level to the time at the 50 percent stress level. If necessary, the time at the 84 percent stress level can be obtained by extrapolation. Note that the *ratio* of the two time values are taken because the time axis is expressed in logarithmic coordinates.

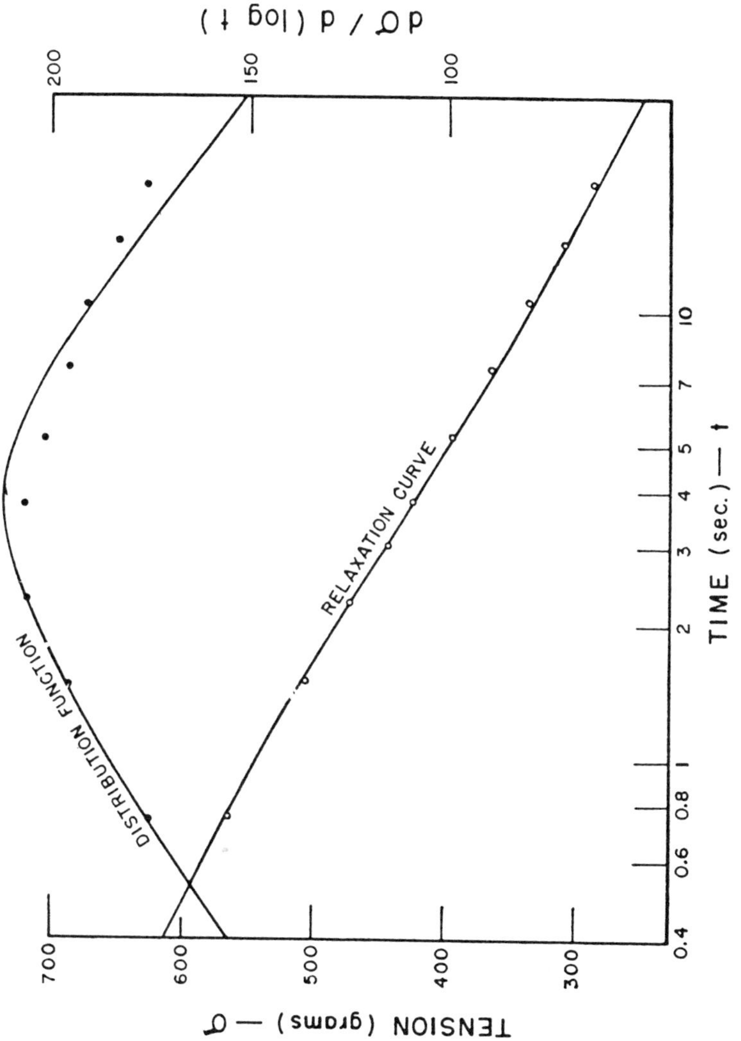

Figure 4.29 Distribution function obtained from from instantaneous slope of relaxation curve (data from Fig. 4.28)

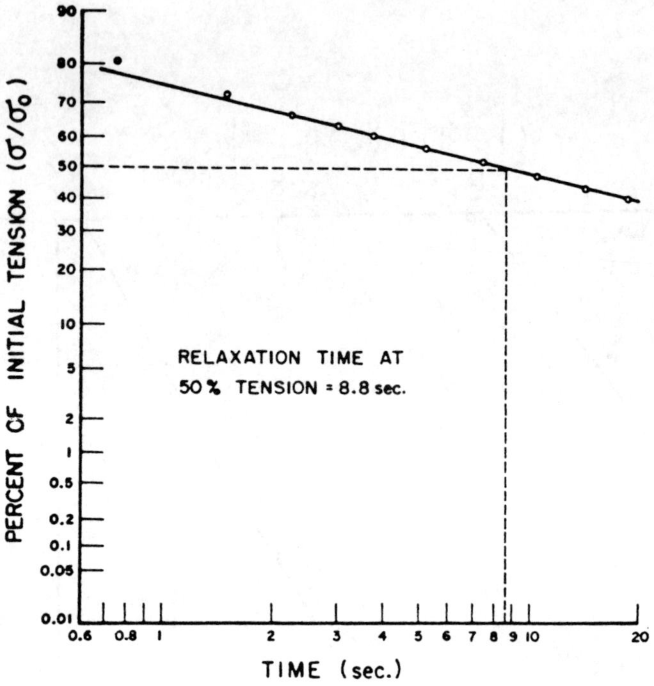

Figure 4.30 Relaxation data of Fig. 4.28 plotted on probability coordinate *vs*. logarithm of time

This method of the "Central Limit Theorem" can be better understood if we recall some of the characteristics of normal distribution and log-normal distribution curves and the definition of the geometric mean as follows:

1 A perfectly normal curve is symmetrical. This means that 50 percent of the cases, indicated by the area under the curve, lie below the mean and 50 percent above it.

2 The area under the normal curve between the mean and a point one standard deviation above the mean expressed as a proportion of the total area is approximately 0.34.

3 The proportion of cases in the first or the last part of the curve is (0.50−0.34) or 0.16.

4 The geometric mean, G, is defined as the Nth root of the product of N observations

$$G = \sqrt[N]{X_1 X_2 X_3 \ldots X_N}$$

or

$$G = \text{antilog} \frac{(\log X_1 + \log X_2 + \log X_3 + \cdots + \log X_N)}{N}$$

5 When a distribution is log-normal, the geometric mean is a more appropriate measure of central tendency than the arithmetic mean, because if the logarithms are normally distributed, the mean of the logarithms should be at the center of the distribution.

6 Accordingly,

$$\text{geometric standard deviation} = \log X_{0.84} - \log X_{0.50}$$

which in the case of stress relaxation is equivalent to the form

$$\begin{matrix}\text{geometric standard deviation} \\ \text{of relaxation time}\end{matrix} = \log \frac{\text{time at 84\% stress ratio}}{\text{time at 50\% stress ratio}}$$

The *method of Central Limit Theorem*, which is used when relaxation times have a normal distribution, is also referred to as the Alfrey Method (Sharma, 1965) in which the so called intensity function is plotted against log (t/T). This function specifies the intensity of the contribution of various regions of the distribution of relaxation times.

There are other methods for determination of relaxation time. The methods of successive residual and the point of inflection will be discussed here.

The *method of successive residuals* begins by plotting the logarithm of the stress against time. The straight part of this original curve for larger values of time gives the exponential term with the largest time of relaxation. The slope of this straight portion of the curve gives the time-constant T_{rel}, as illustrated below. If the straight line were extended to the ordinate, the intercept would give the coefficient of the first exponential term defining this segment of the curve (see Fig. 4.31). Next the difference between this straight line and the original curve is plotted on the same semi-log paper and the first residual curve is obtained. The straight line portion of this first residual would then be extended to intercept the ordinate and the second exponential term with the largest time-constant if found using again the slope and intercept method. By subtracting the corresponding ordinates of the second exponential line from the first residual curve, the second residual curve is obtained which yields the third exponential term. This procedure can be repeated several times until the true curve is represented by a sufficient number of exponential terms. The method of successive residuals is outlined by Lipka (1918) and the accuracy of the method is illustrated by Whitehead (1935). For any specific problem it may be desired to eliminate the long-time or the short-time variations and consider only the value of time-constant which describes the behavior over a range of time important to the problem.

Figure 4.31 shows the method of successive residuals for analysis of the same stress relaxation data used for illustration of the method of Central Limit Theorem. The straight line segment of the original curve of log-stress against time (the first exponential) is extended to the ordinate axis and its equation is obtained by slope and intercept method.

For calculation of time constant T_{rel} from the slope of the straight line segments, the two following equations can be used.

$$T_{rel} = \frac{t_2 - t_1}{\ln \sigma_1 - \ln \sigma_2} \tag{4.43a}$$

This equation is based on the fact that time constant T_{rel} in any of the exponential terms in Eq. (4.23) is the inverse of the slope of the straight line segment representing that particular term. When the scale factor in the semi-log paper is taken into consideration,

$$T_{rel} = \frac{\text{scale factor in terms of unit of time/cycle}}{2.3 \times \text{slope of straight line segment}} \tag{4.44}$$

Using Eq. (4.44), the longest time constant is found to be

$$T_1 = \frac{17 \text{ sec/cycle}}{2.3 \times 2.9/29.5} = 76 \text{ sec}$$

Since the ordinate intercept is at 372 gm., the first exponential term for the straight line segment of the original curve is

$$\sigma_1 = 372 \text{ e}^{-t/76}$$

Similarly the equation for the second exponential term is found by calculating T_2 and reading the intercept for the straight line segment of the second curve, which is obtained by difference.

$$T_2 = \frac{17 \text{ sec/cycle}}{2.3 \times 28.5/12.5} = 3.16 \text{ sec}$$

$$\sigma_2 = 230 \text{ e}^{-t/3.16}$$

Repeating this procedure once more yields

$$T_3 = \frac{17 \text{ sec/cycle}}{2.3 \times 21.5/1.5} = 0.5 \text{ sec}$$

$$\sigma_3 = 100 \text{ e}^{-t/0.5 \text{ sec}}$$

Thus the complete curve for stress relaxation can be represented by a three-term equation

$$\sigma = 372 \text{ e}^{-t/76} + 230 \text{ e}^{-t/3.2} + 100 \text{ e}^{-t/0.5}$$

If the straight line extends beyond a full decade of the log axis, it is simpler to use Eq. (4.43a) and take the time values t_2 and t_1 at two consecutive cycles. For example, the time constant for the second exponential term can be found as follows:

$$T_2 = \frac{10 - 2.5}{2.3 (\log 100 - \log 10)} = 3.26 \text{ sec.}$$

The *point of inflection method* for analysis of stress relaxation data makes use of the fact that if stress is an exponential function of time, for example $e^{-t/T_{rel}}$, then the second derivative of a function will be zero at $t=T_{rel}$. Since the second derivative of a function becomes zero at the point where the curve changes the direction of bending (point of inflection), each of such points on the plot of stress against log time must correspond to one time constant. This method can be used if the time constants are few and widely apart like the examples reported for textile fibers (Leaderman, 1943). However, if the curve of stress against logarithm of times does not indicate distinct points of inflection, the evaluation of time constants becomes difficult and very inaccurate. This method is illustrated in Fig. 4.32 for the data on wheat dough given previously. It is seen that only one inflection point, corresponding to only one time constant can be detected from this curve.

In another curve fitting attempt, using data on stress relaxation of bovine muscle, various combinations of stress and time functions were fitted to first, second, and third degree polynomials (Morrow, 1969). It was found that stress versus natural logarithm of time was the most suitable and simplest method. The plotting of the data gave an adequate fit for the first degree polynomials. For investigating the relaxation time distribution, the method of Central Limit Theorem was used. A plot of time-dependent stress divided by the instantaneous stress versus logarithm of time gave a closely fitting straight line indicating that the relaxation times in bovine

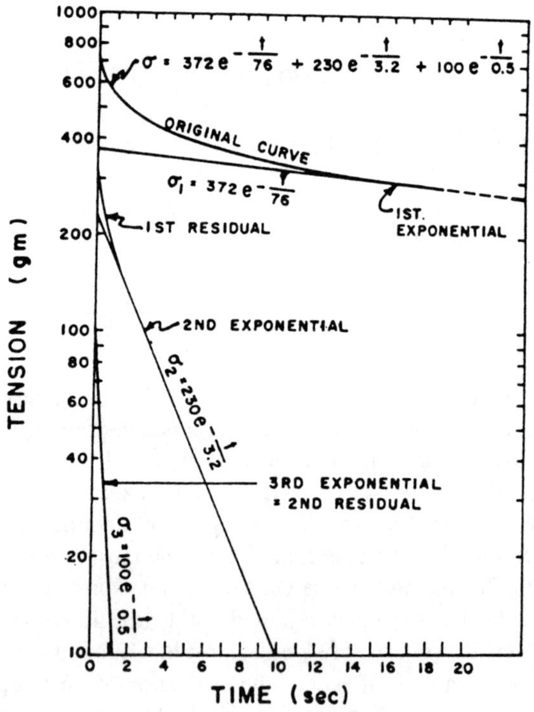

Figure 4.31 Analysis of stress relaxation of wheat dough of Fig. 4.28 by the method of successive residuals

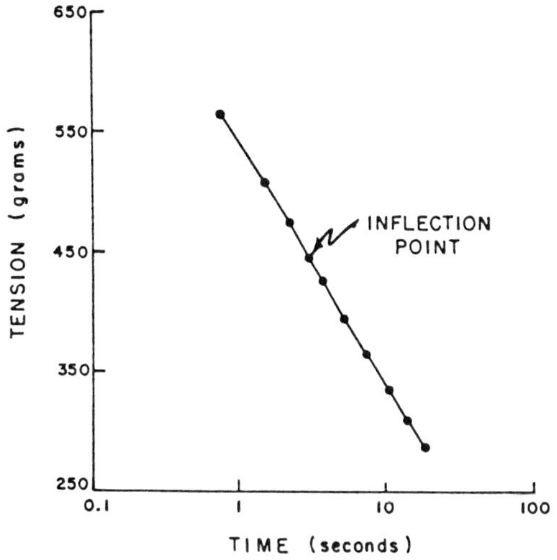

Figure 4.32 Point of inflection method for determination of relaxation time (same data as in Fig. 4.28)

muscle follows a log-normal distribution. Following the method of analysis for this technique, gave the mean relaxation time at a probability of 0.50 and the standard deviation time by taking the time at a probability of 0.84 divided by the time at a probability of 0.50.

Another method for curve fitting of rheological data is the use of an analog computer as employed by Clark *et al.* (1968) on force cyclic tests of the cottonseed. These investigators used an EAI TR-20 analog computer that gave the equation which best fitted the cyclic experimental data. The obvious advantage of the analog computer in such a case is the speed at which the experimental and theoretical curves can be plotted on the oscilloscope and compared until the best fit is obtained.

Due to the importance and usefulness of the analog computer in solving many of the problems dealing with time-dependent physical properties of biological materials, several examples are given in this book.

Example

A stress relaxation experiment was performed on a piece of bovine muscle and stress *vs.* time was found as given in the following table. It was found that the stress-relaxation data cannot be represented by a first order Maxwell model. It is desired, therefore, to use analog computer for curve fitting purpose and see whether or not the experimental data can be represented by a second or third order Maxwell model.

Time	Stress	Time	Stress	Time	Stress
0.00	0.529	0.40	0.325	2.00	0.226
0.01	0.485	0.50	0.308	3.00	0.220
0.02	0.457	0.60	0.303	4.00	0.209
0.04	0.430	0.70	0.286	5.00	0.198
0.06	0.413	0.80	0.281	6.00	0.187
0.08	0.402	0.90	0.275	7.00	0.182
0.10	0.391	1.00	0.275	8.00	0.176
0.15	0.374	1.20	0.270	9.00	0.171
0.20	0.352	1.40	0.264	10.00	0.165
0.25	0.347	1.60	0.248	11.00	0.165
0.30	0.336	2.00	0.277		

For the given stress-relaxation data which does not follow a first order Maxwell model, in general, n exponential terms are needed to represent the stress relaxation according to a Maxwell model, i.e. according to Eq. (4.23),

$$\sigma = \varepsilon_0 \left[E_{d1}\, e^{-\frac{t}{T_1}} + E_{d2}\, e^{-\frac{t}{T_2}} + \cdots + E_{dn}\, e^{-\frac{t}{T_n}} + E_e\right]$$

The equation above is the solution of the following differential equation:

$$\left(P + \frac{1}{T_1}\right)\left(P + \frac{1}{T_2}\right)\cdots\left(P + \frac{1}{T_n}\right)\sigma - (T_1 T_2 \ldots T_n)^{-1}\, \varepsilon_0 E_e = 0$$

where P denotes the differential operator d/dt.

The solution of the second equation on an analog computer requires that one must have the initial value of the σ and its first $n-1$ derivatives. These can be found from the experimental data. In the treatment to follow, n was set equal to 3 and attempt was made to match the experimental data with a third order Maxwell model.

Having set $n=3$ in the last equation, one gets:

$$\left(P + \frac{1}{T_1}\right)\left(P + \frac{1}{T_2}\right)\left(P + \frac{1}{T_3}\right)\sigma - \frac{\varepsilon_0 E_0}{T_1 T_2 T_3} = 0 \quad \text{or}$$

$$\left[P^3 + \left(\frac{1}{T_1} + \frac{1}{T_2} + \frac{1}{T_3}\right)P^2 + \left(\frac{1}{T_1 T_2} + \frac{1}{T_2 T_3} + \frac{1}{T_3 T_1}\right)P + \right.$$
$$\left. + \frac{1}{T_1 T_2 T_3}\right]\sigma - \frac{\varepsilon_0 E_0}{T_1 T_2 T_3} = 0$$

The block diagram of the solution of the above equation on TR-20 analog computer is shown in Fig. 4.32b. The explanation of the building blocks is given in Fig. 4.32a.

The solution given in Fig. 4.32b is one without any time or magnitude scaling. For the data given, it was necessary that a time scale change of $\tau = 20t$ and magnitude scale change of $\Sigma = 10\sigma$ to be made. Thus, based on the new variables from the data it was found that $\Sigma''(0) = 2.06$, $\Sigma'(0) = 1.7$, $\Sigma(0) = 5.3$ and $\Sigma(\text{steady state}) = 1.5$.

Using this information, it was found that if $\frac{1}{T_1} + \frac{1}{T_2} + \frac{1}{T_3} = 1.5$, $\frac{1}{T_1 T_2} + \frac{1}{T_2 T_3} + \frac{1}{T_3 T_1} = 0.204$, and $\frac{1}{T_1 T_2 T_3} = 0.208 \times 10^{-2}$, then one would have the closest match between the computer solution and that of the experimental data. The parameters $\frac{1}{T_1}$, $\frac{1}{T_2}$, and $\frac{1}{T_3}$ can be found by solving the following equation:

$$P^3 + 1.5P^1 + 0.204P + 0.208 \times 10^{-2} = 0$$

The three solutions of above are:

$$\frac{1}{T_1} = -1.35, \quad \frac{1}{T_2} = -0.14, \quad \text{and} \quad \frac{1}{T_3} = -1.1 \times 10^{-2}$$

The solution Σ thus becomes:

$$\Sigma = C_1\, e^{-1.35\tau} + C_2\, e^{-0.14\tau} + C_3\, e^{-1.1 \times 10^{-2}\tau} + 1.5$$

Using the initial values of Σ given before, one obtains a set of three equations and three unknown whose solutions C_1, C_2, and C_3 are:

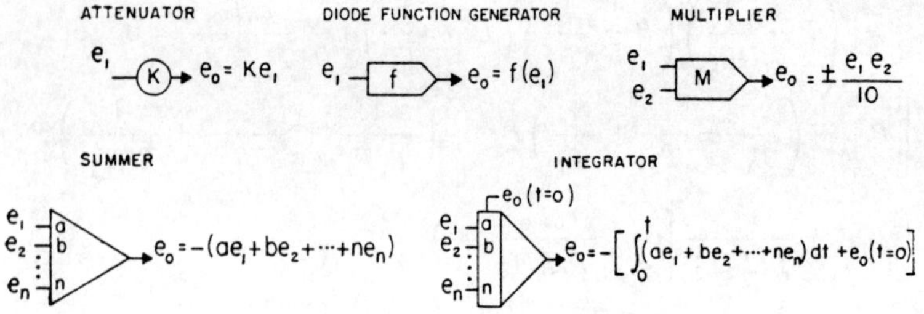

Figure 4.32a Building blocks in the analog computer program

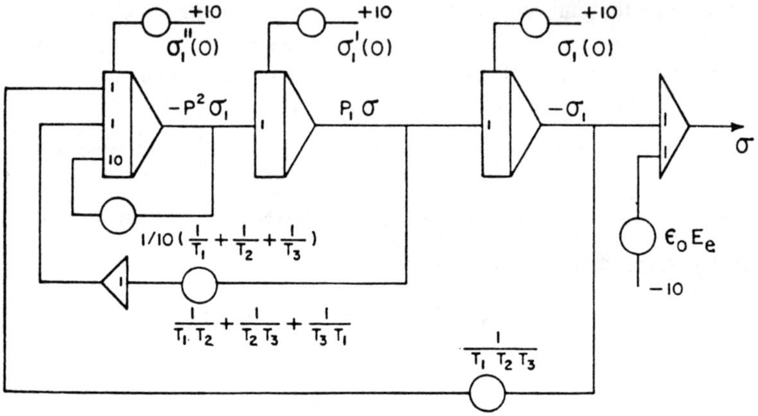

Figure 4.32b Computer program on TR-20 analog computer for solution of stress relaxation of bovine muscle

$$C_1 = 1.12, \quad C_2 = 1.28, \quad \text{and} \quad C_3 = 1.4$$

Thus, $\Sigma = 1.12\ e^{-1.35\tau} + 1.28\ e^{-0.14\tau} + 1.4\ e^{1.1 \times 10^{-2}\tau} + 1.5.$

Now switching back to the old variables σ and t, the final solution is found to be:

$$\sigma = 0.112\ e^{-0.675t} + 0.128\ e^{-7 \times 10^{-3}t} + 0.14\ e^{-5.5 \times 10^{-4}\tau} + 0.15$$

The Computer solution and the experimental data are shown in Fig. 4.32c.

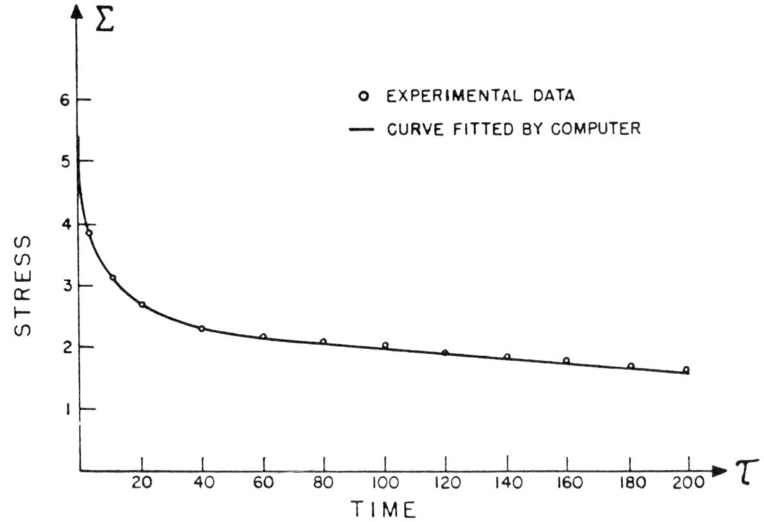

Figure 4.32c Experimental data and computer-fitted curve for stress relaxation of bovine muscle

In conducting a stress relaxation test, a step-function loading technique should be used so that the desired value of stress can be reached instantaneously (see Fig. 4.15). Such a loading technique would minimize the relaxation of stress during the loading cycle. Since it is difficult to simulate this instantaneous loading in a testing machine the fastest rate which is practical under the conditions of the experiment is usually selected.

Dynamic tests

Despite the simplicity of creep and stress relaxation experiments, there are two disadvantages in these tests. The first disadvantage is that in order to obtain complete information about viscoelastic behavior of a material, it is necessary to make measurements over many decades of time scales. This, in addition to prolonging the experiment, may cause, particularly in the case of biological materials, chemical and physiological changes which will affect the physical behavior of the material. The second disadvantage is the impossiblity of having a truly instantaneous application of load or deformation at the beginning of the experiment.

These disadvantages of long time creep and relaxation tests can be overcome by dynamic mechanical tests in which the specimen is deformed by a stress which varies sinusoidally with time. Since a periodic experiment at frequency ω is quantitatively equivalent to tests at time $= 1/\omega$, it is possible to provide considerable amount of information corresponding to very short times. The dynamic tests allow the calculation of elastic modulus and mechanical damping over a wide range of frequencies. Mechanical damping is associated with energy loss, heating, and toughness of the material. Dynamic tests have also shown their usefulness is studying the molecular structure and chemical composition of high polymers.

The most important vibrational methods for dynamic tests can be divided into the following types:

1 Direct measurement of sinusoidally-varying stress and strain (Fig. 4.33).

2 Resonant method in which the specimen with a known geometry is brought to its resonance (Fig. 4.34a) or allowed to oscillate freely (Fig. 4.34b).

3 Pulsed wave propagation in which high frequency waves, such as ultra sonic waves, are sent into the specimen and their times of travel through the specimen are measured (Fig. 4.35).

4 Transducer method where measured electrical impedance is converted to mechanical impedance from which dynamic properties are calculated (Fig. 4.36).

The mechanical properties which govern the ability of the material to transmit vibrations are basically its density, elasticity and internal friction. The basic principle involved in choosing one method in preference to another is the critical dimension of the specimen and the limitations of the experimental set up in relation to its useful range of available frequencies. The critical dimension is the dimension in the direction of the load for simple tension or compression and perpendicular to the plane of shear for shear. The ratio of the critical dimension, L, to the wave length, λ, of the elastic wave propagated at the frequency of measurement can be used as a criterion for selecting the most suitable method.

The wave length, λ, is given by

$$\lambda = \frac{\sqrt{E/\rho}}{f} = \frac{V}{f} \tag{4.45}$$

where E and ρ are respectively Young's modulus and mass density of the material, f is the frequency of measurement, and V is the wave velocity. When L/λ is greater than unity, elastic waves will propagate at the frequency of measurement. When L/λ is approximately equal to unity, resonance vibrations can be excited in the material. And finally when L/λ is less than unity, other methods such as direct measurement of stress and strain can be applied.

Sinusoidally-varying stress and strain method

When viscoelastic materials are deformed, part of the energy is stored as potential energy (in the spring element of its equivalent model), and part is dissipated as heat (in the viscous element of its equivalent model). Therefore, when a linear viscoelastic material is subjected to periodically varying stress, the strain will also vary periodically but out of phase with the stress (Fig. 4.33). This behavior results in a complex frequency-dependent modulus $E^*(\omega)$ which can be separated into an in-phase or real component, associated with the storage of energy, and an out-of-phase or imaginary component, associated with the loss of energy.

$$E^* = E' + iE'' \tag{4.46}$$

where

E' = real component or storage modulus

E'' = imaginary or loss modulus

The complex modulus E^* is experimentally given by the ratio of the peak stress to the peak strain in the sinusoidal stress-strain curve of the material

$$\sigma \max/\varepsilon \max = |E^*| = (E'^2 + E''^2)^{1/2} \tag{4.47}$$

The phase angle δ between stress and strain can also be obtained from the sinusoidally-varying curves

$$\delta = \omega \Delta t \tag{4.48}$$

where ω is the angular frequency (radians/sec) and Δt is the shift in time of the peak values of the stress and strain curves (Fig. 4.33). Other known relationships are as follows:

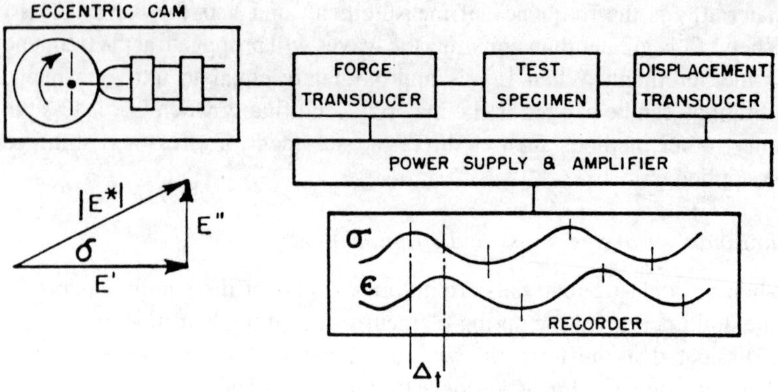

Figure 4.33 Block diagram for a low-frequency dynamic viscoelastic measurement (Vectorial resolution of complex modulus E^* is shown on the left)

$$E' = |E^*| \cos \delta \tag{4.48a}$$

$$E'' = |E^*| \sin \delta \tag{4.48b}$$

$$E''/E' = \tan \delta \tag{4.48c}$$

Sonic resonant method

For a given specimen which has been brought to resonance by an electromagnetic vibrator or other methods, the following basic relationship exists:

$$E = K_2 f^2 \tag{4.49}$$

where E is the Young's modulus of elasticity, f is the resonance frequency and K is a constant determined from the geometry and density of the materials, The exact form of the above equation for flextural vibration of a cantilever beam are given in various forms. The following is one form of resonance equation given by Ferry (1961) and Nielson (1962).

$$E = \frac{38.3 \rho L^4 f_r^2}{d^2} \tag{4.50}$$

where

E = Young's modulus of elasticity of dynes/cm^2
ρ = density in g/cm^3
L = free length of specimen in cm
d = thickness of specimen in cm
f_r = resonance frequency in cycles/sec.

The modulus calculated by this method is the storage modulus, E', of the complex modulus, E^*, given in Eq. (4.46). The loss modulus, E'', can be found from Eq. (4.48b) if damping factor tan δ is known. This parameter can be found from the resonance curve and the equations given in Fig. 4.34a. In these equations, 0.5 and 0.707 are subscripts denoting change in frequencies, $\Delta\omega$, at which the amplitude is either 0.5 or 0.707 its maximum value, respectively and ω_r is the resonance frequency (Ferry, 1961). It should be noted that the equations given for tan δ in Fig. 4.34a are valid only when tan $\delta \leqq 1$. For higher damping factors, which are not uncommon among soft biological tissues, a more complicated relationship has to be used. When free vibration technique is applied, as in a vibrating reed, the ratio of the loss modulus to storage modulus is given by the logarithmic decrement of the damped response curve as shown in Fig. 4.34b.

Ultrasonic pulsed technique

When elastic waves are generated in a solid which due to its characteristic dimensions and its inertia cannot move, the waves will propagate in the

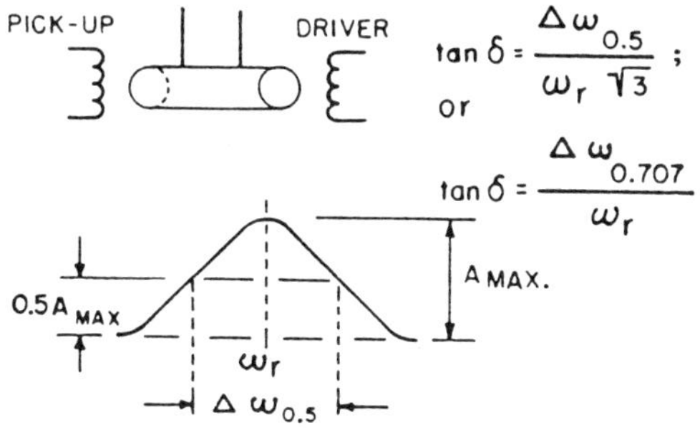

Figure 4.34a Resonant half amplitude method

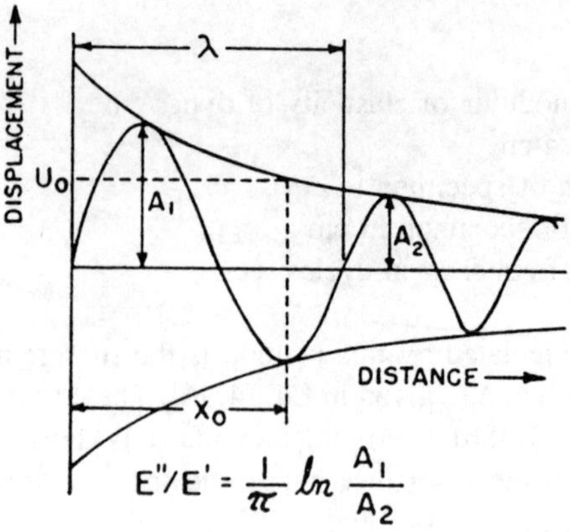

Figure 4.34b Free vibration and logarithmic decrement method

solid, resulting in both compression and shear of the solid. The types of waves causing compression or shear are longitudinal or compressional waves and transverse or shear waves. The rate of travel of these waves is related to modulus of elasticity, E, and mass density, ρ, of the solid. Compressional waves travel with the greatest velocity and thus arrive at the opposite end of the solid prior to the shear waves. The arrival time can be determined accurately using quartz transducers. If the length of path in the specimen is known, the wave velocity can be determined and from the velocity, the modulus can be calculated from the following expression (Leslie, 1951).

$$E = \rho V^2 \tag{4.50a}$$

The wave velocity, V, depends not only upon the elastic constants μ and E and mass density, ρ, but also upon the kinds of waves travelling through the specimen. For longitudinal or compressional waves

$$V = \left[\frac{E}{\rho}\frac{(1-\mu)}{(1+\mu)(1-2\mu)}\right]^{1/2}$$

For transverse or shear waves

$$V = \left[\frac{E}{\rho} \frac{1}{2(1 + \mu)}\right]^{1/2}$$

Theoretically in non-attenuating media the wave velocity can be determined from the simple relationship given in Eq. (4.50a). However in practice this equation also applies to attenuating media provided that the damping does not exceed 20 d.b./wave length. The attenuation of the material can be determined by measuring the amplitudes of two consecutive waves and logarithmic decrement. Attenuation is the inverse of damping factor, tan δ, shown in Fig. 4.34b, usually denoted by Q. The velocity measurement can be made by means of ultrasonic pulsed technique illustrated in Fig. 4.35.

Ultrasonic pulsed waves are characterized by pulsed width, t, and the pulse repetition period, T_p (see Fig. 4.35). The pulse repetition frequency, f_p, is given by the reciprocal of T_p. Note that the values of pulse period and pulse frequency are independent of the values of wave period, T, and its reciprocal wave frequency, f.

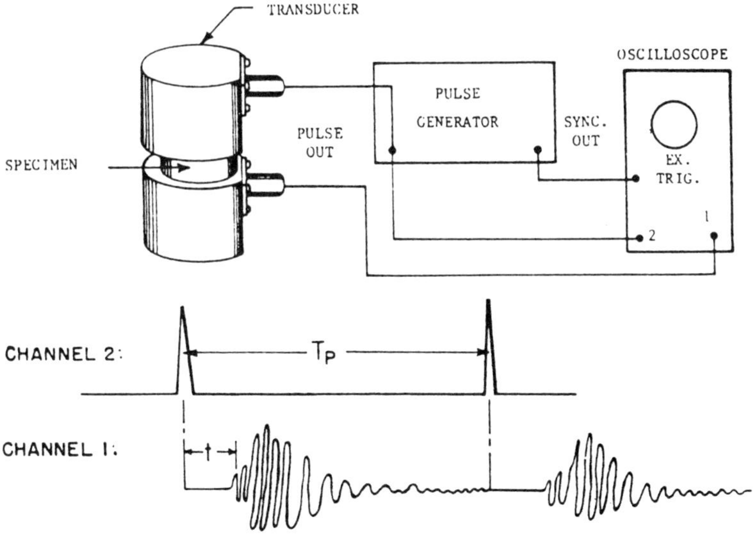

Figure 4.35 Ultrasonic pulsed technique (Wen and Mohsenin, 1970): t = Pulse width or time of travel; T_p = Pulse repetition period; $f_p = 1/T_p$ = Pulse repetition frequency

One major problem presented when this technique is to be applied to biomaterials is the condition and type of the contacting surface which makes acoustical coupling between the transducer and the specimen very difficult. A perfectly smooth contacting surface is essential if scattering of waves is to be minimized. When an ultrasonic transducer is placed on the surface of the material to be tested it will rest on a layer of air even though the surfaces of the transducer and the material may appear to be smooth and in good contact. Because the characteristic impedance of a solid medium is about 100,000 times that of air reflection occurs at the transducer air boundary and the waves do not enter the sample. Thus one normally uses a liquid to couple the transducer to the sample being tested. Water, which is present in soft, fleshy biomaterials is a good coupling fluid.

Electrical-mechanical impedance method

Another method for measurement of dynamic mechanical properties employs the measurement of electrical parameters from which the mechanical parameters can be calculated. Figure 4.36 shows the basic components of one such instrument (Smith *et al.*, 1949).

A contact pad is driven axially by a coil suspended in the air gap of a permanent magnet. The contact pad and coil are supported by a gas bearing. This provides a rigid and light moving system with practically no friction where the coil moves in the magnetic field. By applying a sinusoidally-varying current to the coil, the rod will be subjected to a sinusoidally-varying force which can in turn be used for deforming a viscoelastic solid.

The relation between electrical impedance, Z mot, due to motion and mechanical impedance, Z mech, is

$$Z \text{ mech} \times Z \text{ mot} = A \tag{4.51}$$

where

Z mech = force/velocity
Z mot = back emf/current
A = electromechanical constant

The mechanical impedance is related to storage and loss moduli through the following equation simplifed for a gas bearing transducer (Ferry, 1958).

$$Z \text{ mech (specimen} = (a/\omega)\, E'' + i(\omega M - (a/\omega)\, E')$$

where

a = shape factor of the specimen
ω = circular frequency
E' = storage tensile or compression modulus
E'' = loss tensile or compression modulus
M = total effective mass

The transducer instrument in Fig. 4.36 is suitable for a frequency range of about 30 to 200 cyles per second. The movement of the rod is less than 0.1 mm. An instrument built on the same principle by the Franklin Institute (Keiper, 1962) is being used for measurement of dynamic viscoelastic properties of leather and animal fat at the U.S. Department of Agriculture Eastern Utilization Laboratory. The same instrument was used on dry shelled corn. The results are discussed in Chapter 5. The usable range of frequency for this instrument is about 100 to 2000 cycles per second. A transducer instrument, which is available commercially, has been designed for a range of 25 to 5000 cycles per second (Fitzgerald and Ferry, 1953). At higher frequencies, dynamic mechanical properties can be measured by wave propagation methods (Ferry, 1961).

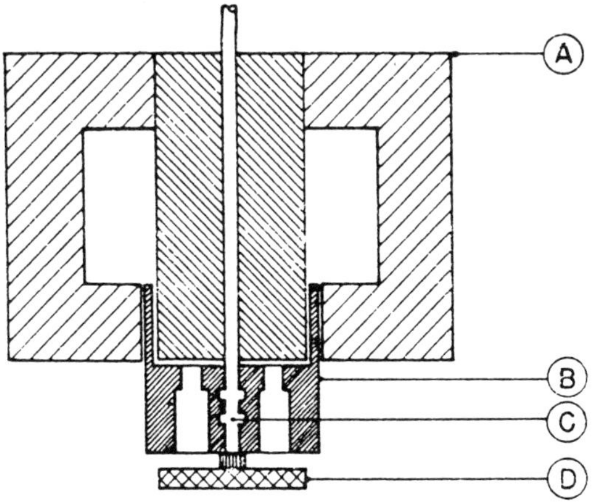

Figure 4.36 Schematic diagram of a transducer type apparatus for dynamic viscoelastic measurements in high frequency range. A = permanent magnet, B = moving coil, C = gas bearing supporting the moving coil, D = specimen contact pad

4.10 NON-NEWTONIAN FLUIDS AND VISCOMETRY

In section 4.4 we defined ideal Newtonian liquids and viscosity. The flow characteristics of many organic fluids are, however, non-Newtonian. In a Newtonian fluid, shear stress is proportional to shear rate and the straight line, representing the behavior, passes through the origin. In non-Newtonian fluids, however, the behavior is not that simple.

Types of flow curves

Figure 4.37 shows the behavior of a liquid which requires a minimum stress (yield value) before its "plastic flow" occurs. In this case, the velocity gradient dv/dy, can be modified to account for the yield stress τ_y

$$-dv/dy = (1/\eta')\,(\tau - \tau_y) \tag{4.52}$$

where τ_y is the yield value and η' is called plastic viscosity. The reciprocal of plastic viscosity is called mobility, which is the analog of liquid fluidity. When we compare Fig. 4.37 with Figs. 4.10 and 4.12, we find that plastic flow in Fig. 4.37 is nothing but the combination of plasticity and viscosity. This type of rheological behavior was first discovered by Bingham and is known as a Bingham-body. A classical example of a material which behaves in this manner is oil paint. If the curve of a plastic flow is not a straight line, the term "quasi-plastic" is applied, as shown in Fig. 4.38. Figure 4.39 illustrates the case where the flow curves do not obey the direct proportionality between shear stress and shear rate but they pass through the origin. This type of "quasi-viscous" flow is called non-Newtonian flow. Note that the difference between these curves and those in Fig. 4.38 is the absence of a yield stress. This presence or absence of a yield value has been considered by some rheologists as the distinction between solids and liquids.

The straight portion of curve 1 in Fig. 4.38 intercepts the τ axis where three types of yield values can be established. These yield points specify changes in the trend of the flow curve. For example, point *A* is the transition point at which flow changes from a true plastic to a quasi-plastic flow. Sometimes, when these changes are sufficiently abrupt, they may serve as useful concepts. At the lower yield value, the flow first becomes detectable. The upper yield value corresponds to the stress above which the plastic flow becomes linear. The Bingham yield stress τ_{yB} is an extrapolated constant corresponding to τ_y in Eq. (4.52).

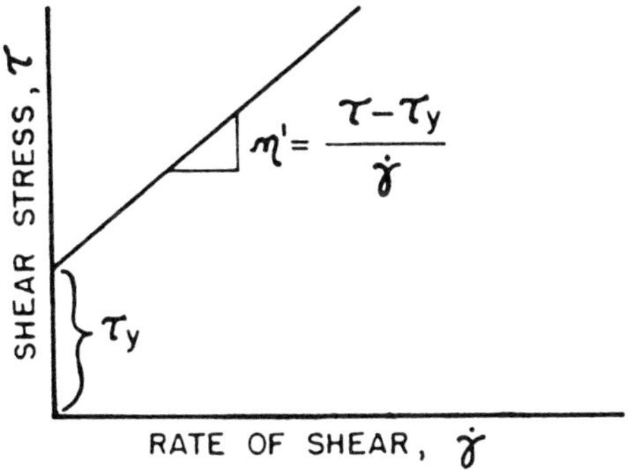

Figure 4.37 Ideal plastic flow (combination of plasticity and viscosity)

When the flow curve is non-linear, the term *apparent viscosity* is used to represent the viscosity of a Newtonian liquid exhibiting the same resistance to flow at the chosen shearing stress or shearing rate. The apparent viscosity is determined from the slope of the line connecting the chosen point on the non-linear curve to the origin (dotted line in Fig. 4.39).

Various formulas have been proposed for describing quasi-viscous and quasi-plastic flow. One form of these equations is presented by a power law equation given below.

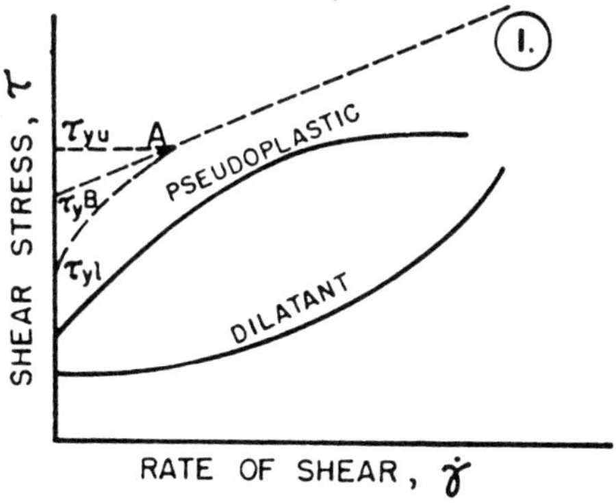

Figure 4.38 Quasi-plastic flow: τ_{yu} = upper yield stress, τ_{yB} = Bingham yield stress, τ_{yl} = lower yield stress

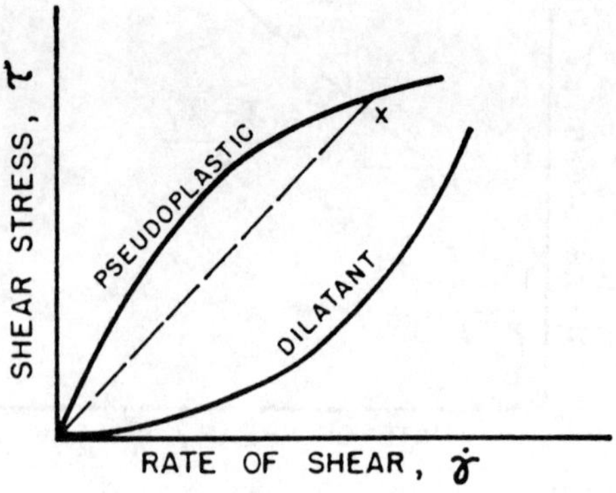

Figure 4.39 Quasi-viscous or non-Newtonian flow

$$\dot{\gamma} = 1/\eta'' \, (\tau - \tau_y)^n \qquad (4.53)$$

where n is the power law constant and η'' is apparent viscosity. In an ideal Newtonian liquid, where $\tau_y = 0$ and $n = 1$, $\eta'' = \eta$, the Newtonian viscosity. In Fig. 2.35, where the fluid shows no yield stress

$$\dot{\gamma} = 1/\eta'' \, (\tau^n) \qquad (4.54)$$

The flow curves with these power law equations should give straight lines when log rate of shear is plotted against log shearing stress.

The terms *pseudoplastic* and *dilatant* are frequently used for flow curves which are not straight lines. If the curve is concave towards the shear stress axis, the flow is called dilatant. If the curve is convex towards the shear stress axis, the flow is called pseudoplastic (Figs. 4.38, 4.39). In Fig. 4.40 the actual flow curves for some food materials are shown.

The types of flow described above are usually classified as time-independent non-Newtonian fluids. There are two other types of non-Newtonian behaviors which are referred to as time-dependent non-Newtonian fluids. The designation is meant for those materials in which the shear stress at a given shear rate and temperature is not constant and changes with duration of shear. If the shear stress increases with time of shear at a given shear rate, the material is called *rheopectic*. If the shear stress decreases with time of shear, the material is classified as *thixotropic*. An example of these

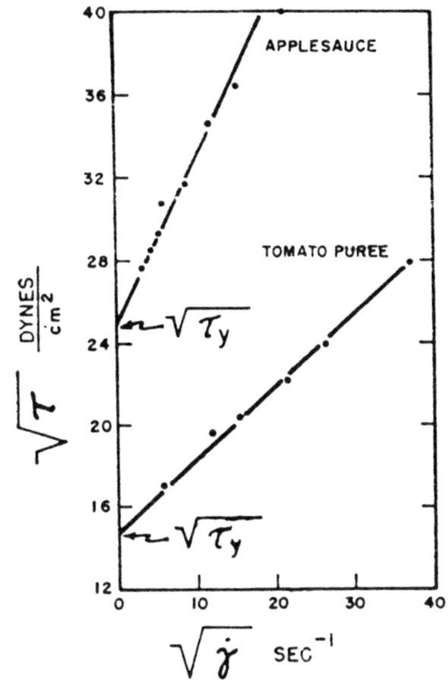

Plastic flow (Charm, 1960)

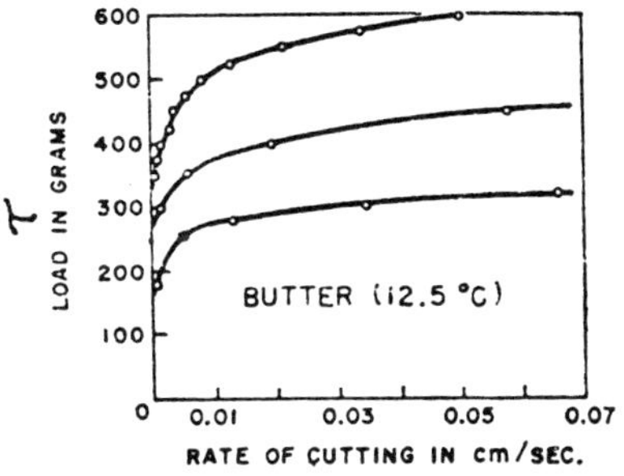

Quasi-plastic flow (Dolby, 1941)

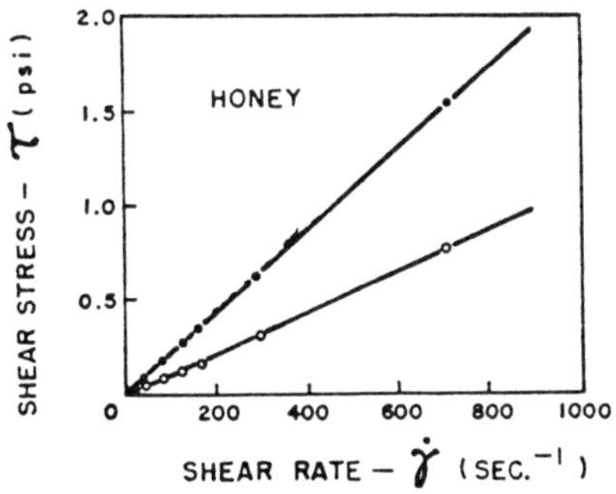

Viscous flow

Figure 4.40 Flow curves for food materials

rheological behaviors is the increasing or decreasing torque required in beating of viscous fluids in a mixer, running at constant speed. The former case is apparently very infrequent, but has been reported for some pigment and clay suspensions (Weltman, 1960). Thixotropic materials are, however, common to paints and such food materials as ketchup.

Later in this section, we will see that the mechanics of flow of fluids in tubes for all fluids can be represented with one general equation describing the behavior of non-Newtonian fluids [Eqs. (4.56) and (4.57)]. All the various types of flow described here are only special cases of this general case.

Viscometry

To understand the flow behavior of a fluid, it is required to establish the equation of motion of the fluid in a rheological instrument such as a viscometer. This equation, which is an expression for stress in terms of the velocity gradients, can then be solved for the given geometry and the appropriate boundary conditions. Usually a solution is obtained for an over-simplified geometry of the viscometer and then corrections are made for such factors as edge-effects, wall effects, etc. The measured quantity for the shear stress will be a form of torque for rotational viscometers and a pressure drop for tubes. The measured quantity for shear rate will be some function of the velocity of flow.

Based on these considerations, the apparent viscosity η'' in any type of viscometer can be given by

$$\eta'' = \mathrm{K}(F/v) \tag{4.55}$$

where F is the applied force to induce the flow velocity v, and K is a constant associated with the particular type of flow and the instrument. For example, in the case of capillary viscometers

$$\eta'' = (K_{\mathrm{cap}})\,(\Delta p/Q) \tag{4.56}$$

where Δp is the pressure drop to induce volume flow Q. And, in the case of rotational viscometers

$$\eta'' = (K_{\mathrm{rot}})\,(M/\Omega) \tag{4.57}$$

where M is the moment inducing the angular velocity Ω.

The K constant is found by solving the equation of motion of the fluid using the specified boundary conditions. In the following discussion the working equations, which include the form factor K, will be developed for the two basic types of capillary tubes and rotational viscometers. A mathematical treatment for the non-Newtonian fluids is available only for these types of viscometers. The geometry of other types of viscometers has been too complicated for the development of a mathematical treatment to be used for non-Newtonian fluids.

Mechanics of flow in capillary viscometers

In capillary viscometry, the liquid is forced through a tube and the viscosity is determined from the rate of flow, applied pressure, and the geometry of the tube. Consider the laminar flow of a Newtonian or non-Newtonian liquid in a tube of radius R and length L in Fig. 4.41. The dotted line shows an imaginary cylindrical column of the fluid which flows under pressure difference ΔP between the ends of the tube. If the flow is steady and wall-effects and end-effects are negligible, the viscous force tending to resist the flow is equal to the force tending to move the cylindrical column in the direction of flow. This simple relationship can be written in the form of the following equation which is the basic equation for capillary viscometers.

$$\tau\,(2\pi r L) = \Delta P\,(\pi r^2) \tag{4.58}$$

or

$$\tau = \Delta P\, r/2L \tag{4.59}$$

where τ is the shear stress at the surface of the moving column of the fluid, ΔP is the applied pressure, and r is the distance from the center. Since r varies from $r=0$ to $r=R$, the shear stress is zero in the middle of the tube and maximum at the wall.

Newtonian flow

For these types of fluids we presented Eq. (4.9) which, with reference to Fig. 4.41, can be written in the following form

$$-\mathrm{d}v/\mathrm{d}r = \tau/\eta \tag{4.60}$$

Substituting τ from Eq. (4.59) yields

$$-\mathrm{d}v/\mathrm{d}r = \frac{\Delta P\, r}{2L\eta} \tag{4.61}$$

Integration of Eq. (4.61) with respect to r for the boundary conditions $v=0$ where $r=R$, results in an expression for velocity distribution at any distance r from the center

$$-\int_{v}^{0} dv = \frac{\Delta P}{2L\eta}\int_{r}^{R} r\,dr \tag{4.62}$$

or

$$v(r) = (\Delta PR^2/4L\eta)\,[1-(r/R)^2] \tag{4.63}$$

The graphical presentation of this parabolic velocity distribution is shown in Fig. 4.42. The maximum velocity occurs at $r=0$, or at the center of the tube.

Integration of Eq. (4.63) with respect to r over the cross-sectional area yields the volumetric flow, q, known as the Poiseuille law for laminar flow in tubes.

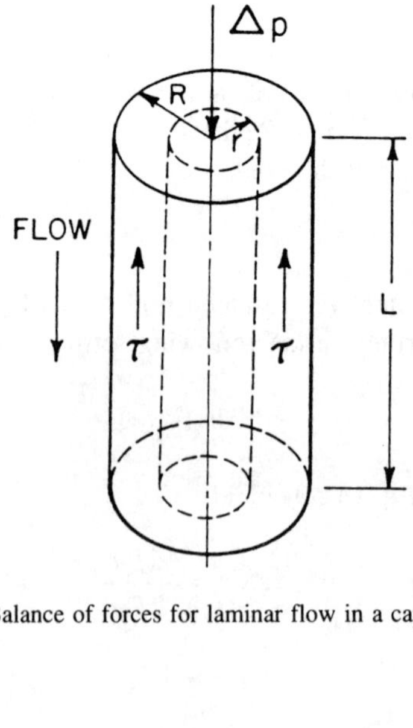

Figure 4.41 Balance of forces for laminar flow in a capillary tube

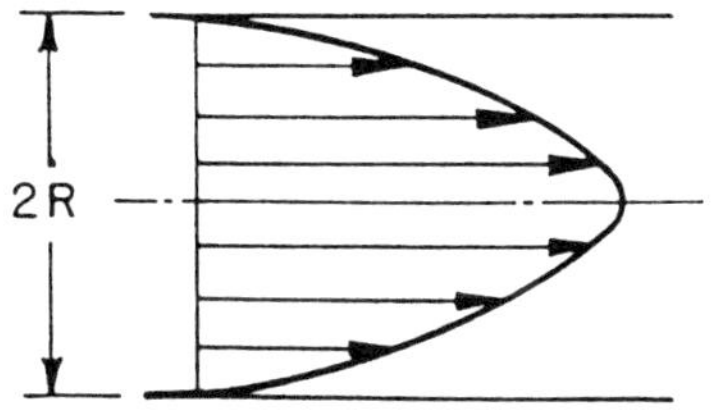

Figure 4.42 Parabolic velocity profile for Newtonian flow

$$q = \int_0^R v(r)\, 2\pi\, r\mathrm{d}r \qquad (4.64)$$

or

$$q = \pi R^4\, \Delta P/8\eta L \qquad (4.65)$$

Combining Eqs. (4.61) and (4.65), rate of shear at the wall can be expressed as

$$(-\mathrm{d}v/\mathrm{d}r)_w = 4q/\pi R^3 \qquad (4.66)$$

$$(-\mathrm{d}v/\mathrm{d}r)_w = 4V/R \qquad (4.67)$$

where V is the average velocity found by dividing the flow, q, by area πR^2.

A useful relationship between shear stress, τ, and average velocity to diameter ratio (V/D) can be found if we recall that in a Newtonian fluid, shear stress is simply the product of η and shear rate ($\mathrm{d}v/\mathrm{d}r$). Substituting for shear rate from Eq. (4.67) with $R=D/2$ yields

$$\tau = \eta\, (-\mathrm{d}v/\mathrm{d}r) \qquad (4.68)$$

$$\tau = \eta\, (8V/D)$$

Plastic flow

The equation for this type of flow was presented by Eq. (4.52), which can be repeated as

$$-(dv/dr) = (1/\eta')(\tau - \tau_y) \qquad (4.69)$$

Knowing the general equation for capillary viscometers [Eq. (4.59)], the expressions for yield stress, τ_y, and shear stress at the wall of the capillary, τ_w, would be

$$\tau_y = \Delta P\, r_y/2L \qquad (4.70)$$

$$\tau_w = \Delta P\, R/2L \qquad (4.71)$$

where r_y is the distance where the velocity is constant and consequently the rate of shear is zero, as illustrated in Fig. 4.43. Substituting in Eq. (4.69) for τ and τ_y from Eqs. (4.59) and (4.70), respectively and integrating the resulting equation with respect to r, using $v=0$ at $r=R$, yields the expression for velocity distribution for plastic flow in a capillary tube

$$v(r) = (\Delta PR^2/4L\eta')\,[1 - (r/R)^2 - 2(r_y/R)\,(1 - r/R)] \text{ for } r \geqq r_y \qquad (4.72)$$

and

$$v(r) = (\Delta PR^2/4L\eta')\,(1 - r_y/R)^2 \text{ for } r \leqq r_y \qquad (4.73)$$

As shown in Fig. 4.43, beyond the distance r_y, the velocity profile takes up the parabolic form given for Newtonian flow and the rate of shear increases proportional to distance r or the shear stress. The region of r_y is sometimes referred to as the region of plug flow. There is no flow until the pressure gradient, ΔP, exceeds the yield stress τ_y. At that point the flow starts in the capillary with the central portion moving as a solid cylindrical plug.

Following the same procedure as given for Newtonian flow, the equation for volumetric flow can be derived as follows:

$$q = \int_0^R v(r)\, 2\pi\, r\, dr$$

or

$$q = \pi R^4\, \Delta P/8\eta' L\, [1 - 4/3(r_y/R) + 1/3\,(r_y/R)^4] \qquad (4.74)$$

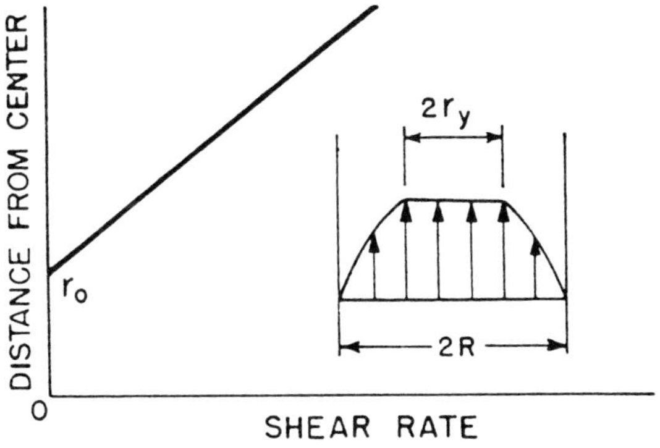

Figure 4.43 Velocity profile and ideal flow curve for plastic flow in a capillary

Since from Eqs. (4.70) and (4.71)

$$r_y/R = \tau_y/\tau_w$$

the flow expression Eq. (4.74) can also be written in terms of τ_y/τ_w instead of r_y/R

$$q = (\pi R^4 \; \Delta P 8\eta' L) \; [1 - 4/3(\tau_y/\tau_w) + 1/3(\tau_y/\tau_w)^4] \qquad (4.75)$$

In cases where the applied pressure and consequently the shear stress at the wall is much greater than the yield stress, the term $(\tau_y/\tau_w)^4$ in Eq. (4.75) can be neglected. With this consideration and substitution of τ_w for $\Delta PR/2L$ [Eq. (4.71)], Eq. (4.75) yields a working equation for plastic flow in capillary viscometers

$$\eta' = \left(\frac{\tau_w - (4/3)\,\tau_y}{4q/\pi R^3} \right) \qquad (4.76)$$

By rearrangement, Eq. (4.76) can be rewritten in the following form which is recognized as the slope-intercept form of the equation of a straight line.

$$\tau_w = (4/3)\tau_y + \eta'(4q/\pi R^3) \qquad (4.77)$$

If the flow rates, q, are obtained over a wide range of pressures and the results plotted to obtain a straight line represented by Eq. (4.77), the values of shear stress, τ_y, and plastic viscosity, η', can be obtained from the plotted data. Note that yield stress, τ_y, can be obtained either by direct recording or by extrapolating the straight line portion of the plotted data to shear stress axis and multiplying the resulting value by ¾ (see Fig. 4.44). If the term $(4q/\pi R^3)$ in Eq. (4.77) is set equal to $(8V/D)$, as given by Eq. (4.67), shear stress, τ_w, can be expressed in terms of V/D ratio as follows:

$$\tau_w = (4/3)\tau_y + \eta'(8V/D) \quad (4.78)$$

Non-Newtonian flow

As indicated before, the flow of non-Newtonian fluids can be represented by a power law equation. For the case of capillary tube, we may write an empirical power law equation as follows:

$$\tau = C(-dv/dr)^n \quad (4.79)$$

where C and n are constants. Substituting for τ from Eq. (4.59) in Eq. (4.79) yields

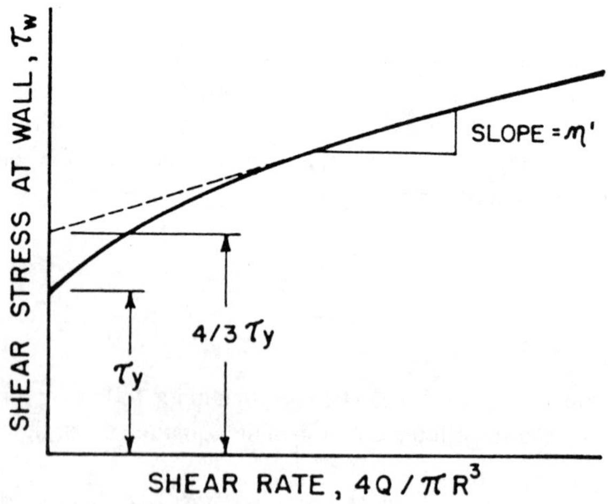

Figure 4.44 Shear diagram for plastic flow in a capillary

$$-\int_{v}^{0} \mathrm{d}v = (\Delta P/2CL)^{1/n} \int_{r}^{R} r^{1/n}\,\mathrm{d}r \tag{4.80}$$

Integration of the above equation with respect to r for the boundary conditions $v=0$ at $r=R$ yields the velocity distribution equation as follows:

$$v(r) = (\Delta P/2CL)1/n\ [n/(n+1)]\ R^{(n+1)/n}\ [1 - (r/R)^{(n+1)/n}] \tag{4.81}$$

To obtain volumetric flow, q, the above equation is integrated with respect to r over the cross-sectional area.

$$q = \int_{0}^{R} v(r)\, 2\pi r\,\mathrm{d}r = \pi\,(\Delta P/2CL)^{1/n}\ [n/(3n+1)]\ R^{(3n+1)/n} \tag{4.82}$$

For $n=1$, Eq. (4.82) reduces to

$$q = \pi R^4 \Delta P/8CL$$

which is the same as Eq. (4.65) or Poiseuille's equation with C being equal to η.

To obtain the constants C, n, and apparent viscosity, η'', take the logarithm of Eq. (4.82). The resulting expression should yield the slope-intercept form of the equation of a straight line. Therefore, when data in terms of log $\Delta P/L$ are plotted against log q, the slope of the straight line will be equal to n while the value of constant C can be obtained from the intercept. The apparent viscosity can be obtained from

$$\eta'' = \tau/(-\mathrm{d}v/\mathrm{d}r) = \tau/(\tau/C)^{1/n} \tag{4.83}$$

$$\eta'' = (C)^{1/n}\tau^{(1-1/n)}$$

Having n and C from the slope and intercept of the plotted data, η'', can be calculated from the above equation.

The mathematical relationship between shear stress at the wall, τ_w, and the term $(8V/D)$ is given as follows (Metzner, 1956)

$$D\Delta P/4l = \tau_w = K'(8V/D)^{n'} \tag{4.84}$$

In which n' is a constant equal to n if the shear diagram is linear on logarithmic coordinates. If the plot is not linear, Eq. (4.84) represents the tangent to the curve at any chosen value of τ_w or $(8V/D)$ and n' is the slope of the tangent line. The intercept at unit shear rate of tangent to shear diagram is taken as the value for K'.

The parameters n or n' indicate the extent of non-Newtonian behavior and can be referred to as flow behavior index (Metzner, 1956). Similarly, the factor K' can be called viscosity index. As shown in Fig. 4.46, if the flow behavior index is equal to unity, the fluid is Newtonian; values less than unity indicate pseudoplastic or ideal plastic (Bingham body), and values greater than unity characterize dilatant behavior. The shear rate, being proportional to the slope of the velocity profile, is zero for a flat velocity profile, and a constant for a triangular profile.

These velocity profiles and the accompanied shear diagrams show that capillary flow behavior of all the fluids discussed in this section can actually be presented by the general Eq. (4.79).

Capillary-tube viscometers

As seen in the preceding discussion for capillary-tube viscometry of any of the fluids discussed in this section, it is only required to provide a system for measurement of flow rate, q, at several different pressure drops ΔP. The complete system may consist of (1) a fluid reservior, (2) a capillary tube of known geometry, (3) a unit for applying a controlled pressure, (4) a unit for determining flow rate, and (5) a unit for controlling temperature (see Fig. 4.45). These units can be either built and assembled by the investigator or bought commercially. Commercially available capillary viscometers are divided into three main types: cylinder-piston type, glass capillary type, and orifice type. The details of construction and operation of a number of commercially available capillary viscometers are given by Van Wazer *et al.* (1963).

With the original data of flow rate and pressure drops available, the formulas and the methods of plotting data discussed in this section can be used to calculate shear stress at the wall, τ_w, shear rate, $(-dv/dr)$, and viscosity. These equations, however, do not account for various errors such as end effects, wall effects, time-dependency, energy losses, etc. The sources and methods for correcting 20 of the major errors are discussed by Van Wazer *et al.* (1963). This reference, Wilkinson (1960), and Metzner (1956) are recommended for further study of the subject of rheometry.

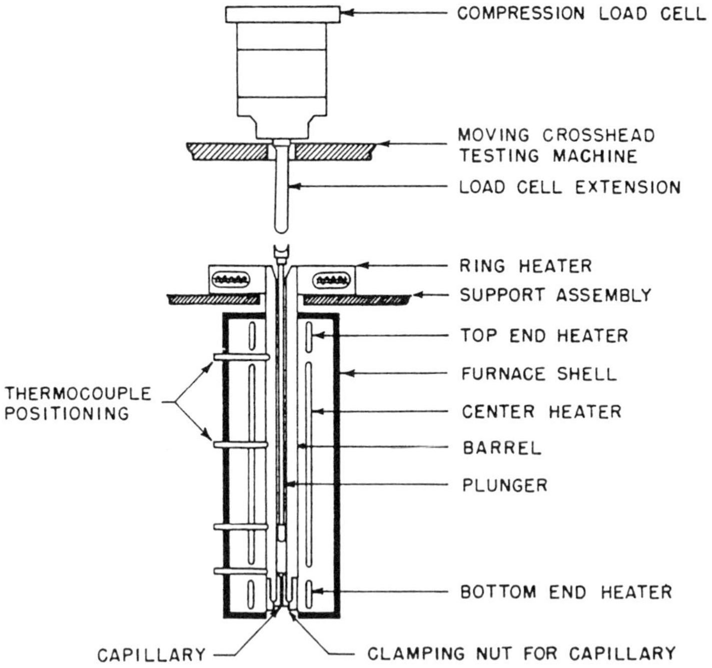

Figure 4.45 Instron capillary rheometer

Example

The following data were obtained in a tube viscometry (a capillary-type viscometer) of applesauce. If the diameter of the tube is 0.105 inch and its length is 35.8 inches, find the flow behavior index and viscosity index for this fluid food.

ΔP (lb/ft^2)	q (ft^3/sec)
2.722×10^3	3.21×10^{-3}
3.02	5.3
3.26	7.42
4.15	11.3
4.45	18
5.04	30
5.63	44.1

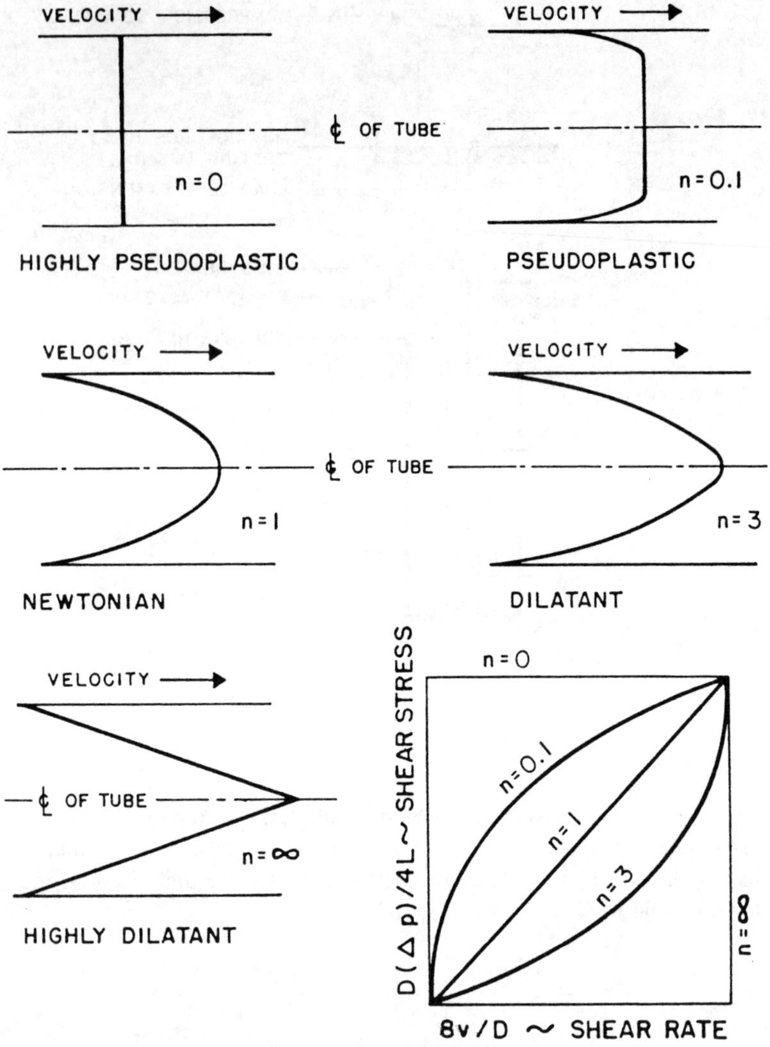

Figure 4.46 Velocity profile and shear diagrams for selected values of flow behavior index (Metzner, 1956)

By rewriting Eq. 4.82,

$$\log(\Delta P/2L) = [\log C - n\log\pi - n\log(n/3n+1) - (3n+1)\log R] + n\log q$$

which is in the form of: $y=a+bx$.
By plotting log $\Delta P/2L$ vs. log q, the slope is found to be $n=0.28$, meaning that the flow is pseudoplastic.
Having n, find C, the viscosity index, from the above equation. Susbstituting

$$\log q = -2.5 \text{ at } \log \Delta P/2L = 2.65.$$

$$2.65 = [\log \mathrm{C} - 0.28 \log \pi - 0.28 \log (0.25/3 \times 0.28 + 1) - (3 \times 0.28 + 1) \log 0.0525/12] + 0.28(-2.5)$$

or

$$C = 0.082 \text{ lb-sec}^n/\text{ft}^2$$

Mechanics of flow in rotational viscometers

The capillary viscometers are limited in applications where measurements are to be made at varying conditions in the same sample or time dependency of the fluid is to be studied. Because of these limitations, rotational viscometers are more widely used than the capillary viscometers. In these instruments, the viscous drag of a rotating body, immersed in a liquid, is measured as a function of speed. Thus from these two parameters, the shear stress versus shear rate diagram and the related relationships are established.

There are various types of rotational viscometers. A comprehensive review of the various types available commercially are given in reference Van Wazer *et al*. (1963). Since the most popular type is the coaxial-cylinder or concentric cylinder viscometer, we will discuss the derivation of the related equations for this type of rotational viscometer.

General expressions

Consider the concentric cylinder viscometer of Fig. 4.47 in which

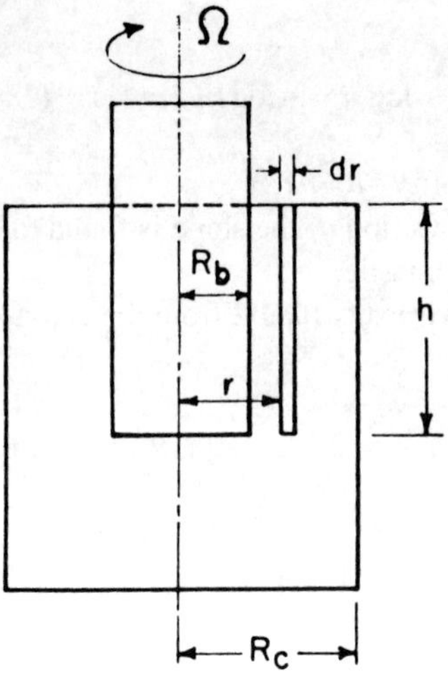

Figure 4.47 Rotational viscometer (concentric cylinder)

Ω = angular velocity of the inner cylinder
R_b = radius of the inner cylinder or the bob
R_c = radius of the outer cylinder or the cup
h = height of liquid in the inner cylinder

For an incompressible liquid with laminar motion and no slippage between the liquid and the cylinders, the following general expressions are valid.

In the steady state of flow, the external torque equals the opposing torque in the fluid such that

$$M = 2\pi rh\tau r \tag{4.85}$$

where M is the external torque, r, is any radius from the axis, and τ is the shear stress which tends to turn one cylinder surface with respect to another. Since under steady state conditions the moments in the fluid at any two radii in the gap are equal, for the two boundaries

$$M = 2\pi R_b^2 h\tau_b = 2\pi R_c^2 h\tau_c \tag{4.86}$$

As linear velocity r from the axis is $r\omega$, in passing from r to $(r+dr)$, the change in velocity is from v to $(v+dv)$. Therefore

$$(v + dv) = (r + dr)(\omega + d\omega)$$

Neglecting the second order terms

$$dv/dr = \omega + r\,d\omega/dr \qquad (4.87)$$

If no shearing takes place, the angular velocity of the viscometer is ω, therefore only $r\,d\omega/dr$ is the term contributing to the shear rate dv/dr. Consequently

$$dv/dr = r\,d\omega/dr \qquad (4.88)$$

Newtonian flow

With consideration of Eq. (4.88) and basic definitions of a Newtonian fluid

$$\tau = \eta(-dv/dr) = \eta(-r d\omega/dr) \qquad (4.89)$$

Substituting for shear stress, τ, in Eq. (4.85) and rearranging results

$$-d\omega = (m/2\pi h\text{yh})\,dr/r^3 \qquad (4.90)$$

Integrating Eq. (4.90) for the boundary conditions $\omega=0$ at the wall of the outer cylinder (with the assumption of no slippage) and $\omega=\Omega$ at the inner cylinder and r varying from R_b to R_c

$$\int_{\Omega}^{0} -d\omega = \int_{R_b}^{R_c} (M/2\pi h\eta)\,dr/r^3$$

or

$$\eta = (M/4\pi h\Omega)\,(1/R_b^3 - 1/T_c^2) \qquad (4.91)$$

With the use of this equation, known as Margules equation, one can calibrate a given concentric viscometer such that the viscosity can be obtained by measurement of the speed of rotation and the applied torque. The other parameters in the equation, as well as errors due to deviations from the basic assumptions, can be considered as the apparatus constant and determined by using a liquid of known viscosity.

To express the results in terms of shear stress

$$\tau = M/2\pi r^2 h \tag{4.92}$$

and for shear rate in terms of dv/dr, substitute for M from

$$M = 2\pi r^2 h\tau = 2\pi r^2 hy\eta\ (-dv/dr)$$

in Eq. (4.91) which yields

$$-dv/dr = (2\Omega/r^2)\ (R_b^2 R_c^2)/(R_c^2 - R_b^2) \tag{4.93}$$

Plastic flow

When a Bingham flow is used in a concentric cylinder viscometer, the shear stress at the outer cylinder will be less than the yield value while the shear stress at the inner cylinder may exceed the yield value. In the following derivation it is assumed that all the liquid in the gap between the two cylinders is undergoing shear.

The law of plastic flow given in Eq. (4.52) can be written as

$$-dv/dr = -r d\omega/dr = 1/\eta'\ (\tau - \tau_y) \tag{4.94}$$

Substituting for shear stress, τ, from Eq. (4.85) and rearranging Eq. (4.94) yields

$$-r d\omega/dr = (1/\eta')\ (M/2\pi r^2 h) - \tau_y/\eta'$$

$$\int_{\Omega}^{0} -d\omega = (1/\eta')\,(M/2\pi h) \int_{R_b}^{R_c} dr/r^3 - \tau_y/\eta' \int_{R_b}^{R_c} dr/r \tag{4.95}$$

or

$$\Omega = (1/\eta')\ (M/4\pi h)\ (1/R_b^2 - 1/R_c^2) - (\tau_y/\eta')\ \ln\ (R_c/R_b) \tag{4.96}$$

Considering the portion of the flow curve for which shear stress is greater than the yield stress (straight line portion of the curve), the shear stress at the wall of the inner cylinder, τ_w, is found from the torque and the geometry of the viscometer

$$\tau_w = M/2\pi R_b^2 h \tag{4.97}$$

Note that this stress τ_w is the most convenient shear stress in the gap to calculate in a rotational viscometer. It is analogous to τ_w, shear stress at the wall of the capillary tube in capillary viscometers.

Shear rate can be obtained by eliminating η' between Eqs. (4.94) and (4.96) and solving for dv/dr

$$-dv/dr = \frac{(\tau_w - \tau_y)\,\Omega}{(M/4\pi h)\,(1/R_b^2 - 1/R_c^2) - \tau_y \ln(R_c/R_b)} \tag{4.98}$$

where τ_w in Eq. (4.97) is equal to τ in Eq. (4.94). The yield value τ_y can be found experimentally, and the parameters related to the geometry of the instrument can be gathered as one or two constants for the viscometer.

Having the quantities τ_w and dv/dr calculated, the shear diagram can be established and the plastic viscosity determined from the slope of the straight line.

Non-Newtonian flow

As in the case of capillary viscometers, the non-Newtonian case is the general case where the rheological behavior can be shown with power law equations. The equations for each special case can be derived from the equations of the non-Newtonian case by using the correct value of exponent n. Following the same procedure as presented for capillary viscometry, Eq. (4.79) can be written as

$$\tau = C(-r\,d\omega/dr)^n \tag{4.99}$$

The moment equation will be

$$M = 2\pi r^2 h\tau = 2\pi r^2 hC(-r\,d\omega/dr)^n \tag{4.100}$$

Rearranging and integrating for the boundary values yields

$$-\int_{\Omega}^{0} d\omega = (M/2\pi hC)^{1/n} \int_{R_b}^{R_c} dr/r^{(n+2)/n} \tag{4.101}$$

or

$$\Omega = n/2(M/2\pi hC)^{1/n}\ (1/R_b^{2/n} - 1/R_c^{2/n}) \tag{4.102}$$

Substituting for M in terms of shear stress, τ_w, at the inner cylinder, Eq. (4.102) can be reduced to

$$\Omega = (n/2)\ (\tau_w/C)^{1/n}\ [1 - (R_b/R_c)^{2/n}] \tag{4.103}$$

where τ_w can be obtained from experimental data for M. Taking logarithm of both sides, Eq. (4.103) can be converted to the slope-intercept form of the equation of a straight line with log τ_w versus log Ω. The slope of this line is the exponent n.

With the value of the exponent known, constant C can be calculated from Eq. (4.103). Having C and n, shear rate $\left(-\frac{r d\omega}{dr}\right)$ can be computed from Eq. (4.99). Apparent viscosity may be obtained from Eq. (4.83).

Figure 4.48 shows an example of a simple rotational viscometer in which the shear stress is held constant rather than the rate of shear. The shear stress is applied by the free fall of a weight through a vertical distance. The moment, M, is given by

$$M\,(\text{dynes} - \text{cm}) = \frac{1.425 \times 980}{11} \times \text{weight (gm)}$$

where 1.425 is the radius of the winding drum, and 11:1 is the gear ratio in the instrument. The speed of rotation is obtained by determining the time for the number of revolutions indicated by the counter. Having the speed of rotation, moment and the instrument constants, shear stress and shear rate can be obtained by the use of Eqs. (4.92) and (4.93), respectively.

Example

The MacMichael viscometer (a rotational viscometer, Fig. 4.47), was used in determining the flow properties of catsup at 78 degrees F. Using the given information below, find (a) the shear diagram giving the flow equation with the appropriate constants, and (b) compare the flow constants with those given in Table 5.11.

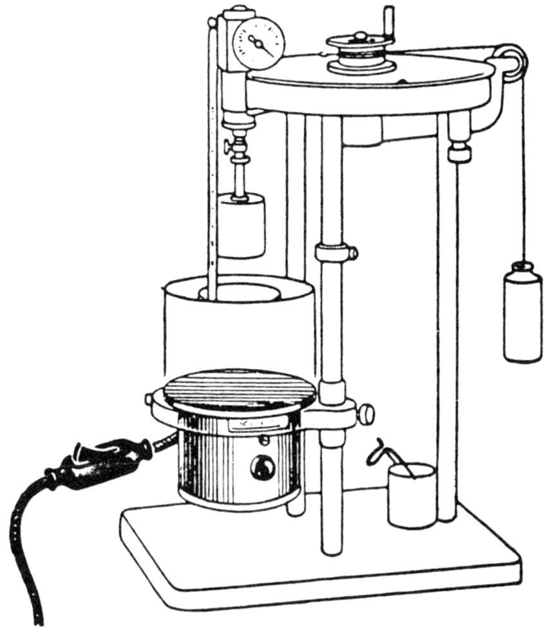

Figure 4.48 Stormer viscometer—Example of Rotational Viscometers (Courtesy of Arthur H. Thomas Company)

Inside Dia. of cup—6.98 cm
Outside Dia. of bob—4 cm
Ht. of fluid—3.9 cm
Wire size—No. 20

$$M_w = 1.36 \frac{\text{g} - \text{cm}}{°M}$$

Speed (rpm)	Deflection (°M)
6.3	27
10.5	29
23.5	33
34.9	35.5
47.6	38.0

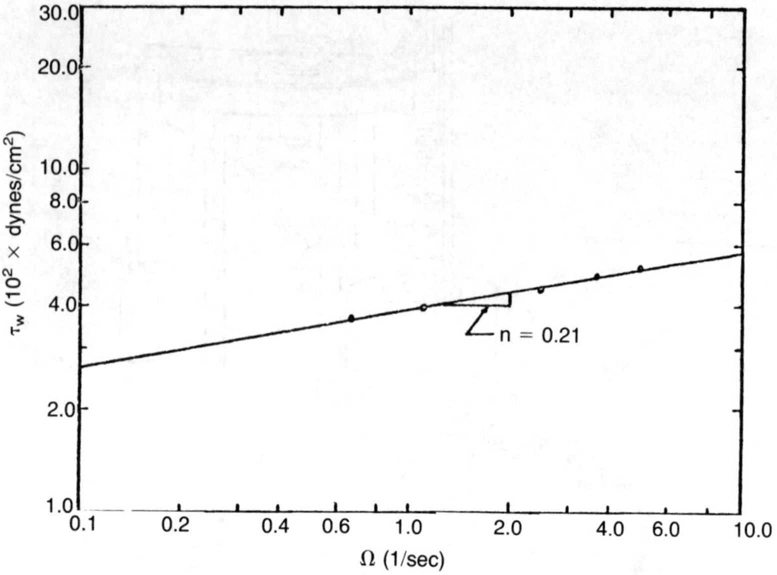

Figure 4.49 Plot of shear stress *vs*. angular velocity for catsup

Using the given data, the tabulated data and examples of the computations are shown below for run No. 1.

Run	Speed (rpm)	Ω (1/sec)	Deflection (°*M*)	Torque (dynes-cm)	Shear stress dynes/cm^2	Shear rate (1/sec)
1	6.3	0.660	27.0	36022.32	367.42	5.800
2	10.5	1.100	29.0	38690.64	394.64	9.664
3	23.5	2.461	33.0	44027.28	449.08	21.647
4	34.9	3.650	35.5	47362.68	483.10	34.
5	47.6	4.980	38.0	50698.08	517.12	43.796

$$\Omega = \text{rpm} \times \frac{2\pi}{60}$$

$$\Omega = 6.3\,(\text{rpm}) \times \frac{2\pi}{60} = 0.660\frac{1}{\text{sec}}$$

$$\text{Torque} = \text{Deflection} \times M_w \times g$$

$$\text{Torque} = 27 \times 1.3614 \times 980 = 36022.32 \text{ dynes-cm}$$

$$\text{Shear stress} = \text{Torque}/2\pi R_b^2 h$$

$$\text{Shear stress} = \frac{36022.32}{2\pi\left(\frac{4}{2}\right)^2 3.9} = 367.42 \text{ dynes/cm}$$

Plotting shear stress vs. angular velocity, results in Fig. 4.49 giving a flow behavior index of n=0.21 which means the flow is pseudoplastic. Using this information and the other computed data for run No. 4, from Eq. (4.103),

$$\Omega = \left(\frac{n}{2}\right)\left(\frac{\tau_w}{C}\right)^{\frac{1}{n}}\left[1 - \left(\frac{R_b}{R_c}\right)^{\frac{2}{n}}\right]$$

$$3.65 = \left(\frac{0.21}{2}\right)\left(\frac{483.1}{C}\right)^{\frac{1}{0.21}}\left[1 - \left(\frac{2}{3.49}\right)^{\frac{2}{0.21}}\right]$$

$$C = \left(\frac{0.21}{2 \times 3.65}\right)^{0.21}\left[1 - (0.573)^{\frac{2}{0.21}}\right]^{0.21} 483.1$$

$$= 230$$

From Eq. (4.99)

$$\text{Shear rate} = \left(-r\frac{d\omega}{dr}\right) = \left(\frac{\tau_w}{C}\right)^{\frac{1}{0.21}}$$

$$= \left(\frac{483.1}{230}\right)^{\frac{1}{0.21}}$$

$$= 34 \text{ 1/sec}$$

The values of the shear rate for each run are tabulated in the table. Now the plot of shear stress vs. shear rate can be established as shown in Fig. 4.50.

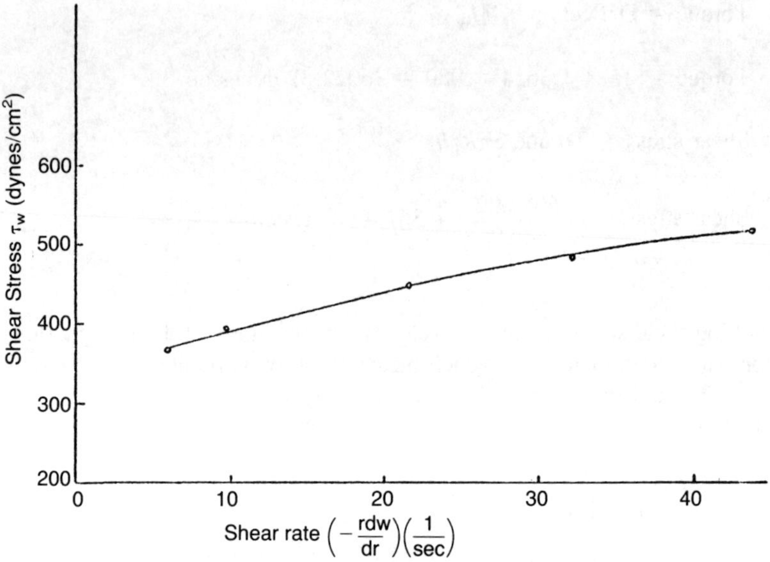

Figure 4.50 Plot of shear stress *vs*. shear rate for catsup

4.11 SELECTED BOOKS ON RHEOLOGY

Andrade, E. N. da C. *Viscosity and Plasticity*. Chemical Publishing Company, Brooklyn (1951).

Barkas, W. W., R. F. S. Hearman, and H. F. Rance. *Mechanical Properties of Wood and Paper*. New York: Interscience Publishers, Inc. (1953).

Bergen, J. T. *Viscoelasticity, Phenomenological Aspects*. Proc. of Armstrong Cork Co., Symposium Academic Press: New York (1960).

Biezeno, C. B. and R. Grammel. Theory of Elasticity; *Analytical and Experimental Methods*. 2nd ed., D. Van Nostrand Company, New York (1955).

Bingham, E. C. *Fluidity and Plasticity*. McGraw-Hill Book Company, New York (1922).

British Rheologists' Club. *Essays in Rheology* (Oxford Conference, 1944) Sir Isaac Pitman and Sons, Ltd., London (1947).

British Rheologists' Club. *The Principles of Rheological Measurement* (London Conference 1946) Thomas Nelson and Sons, Ltd., London (1949).

British Rheologists' Club. *Some Recent Developments in Rheology*. United Trade Press, Ltd., London (1950).

Burgers, J. M., J. J. Hermans, and G. W. Scott-Blair, Editors. *Deformation and*

Flow. Interscience Publishers, New York. Vols. I, II (1952); Vols. III-V (1953); Vol. VI (1955); Vol. VII (1956).

Frey-Wyssling, A. *Deformation and Flow in Biological Systems*. Interscience Publishers, Inc., New York (1952).

Godfrey, D. E. R. *Theoretical Elasticity and Plasticity for Engineers*. Macmillan Company, New York (1960).

Green, H. *Industrial Rheology and Rheological Structures*. John Wiley and Sons, Inc., New York; Chapman and Hall, Ltd., London (1949).

Gross, B. *Mathematical Structure of the Theories of Viscoelasticity*. Hermann, Paris (1953).

Harrison, V. G. W., Editor. *Proceedings of the Second International Congress on Rheology*. Academic Press, Inc., New York (1954).

Hermans, J. J., Editor. *Flow Properties of Disperse Systems*. North-Holland Publishing Company, Amsterdam (1953).

Johnson, W. and P. B. Mellor. *Plasticity for Mechanical Engineers*. D. Van Nostrand Company, New York (1961).

Leaderman, H. *Elastic and Creep Properties of Filamentous and Other High Polymers*. The Textile Foundation, Washington, D.C. (1943).

Lee, E. H. ed. Fourth International Congress on Rheology in four parts. Interscience Publishers, New York (1965).

Mason, P. and N. Wookey, Editors. *The Rheology of Elastomers*. Pergamon Press, New York (1958).

Meredith, R., Editor. *Mechanical Properties of Wood and Paper*. North-Holland Publishing Company, Amsterdam (1953).

Meredith, R., Editor. *Mechanical Properties of Textile Fibers*. North-Holland Publishing Company, Amsterdam (1956).

Merrington, A. C. *Viscometry*. Longmans, Green and Company, New York (1949).

Mill, C. C., Editor. *Rheology of Disperse Systems*. Pergamon Press, New York (1959).

Movozhilov, V. V. *Foundations of the Nonlinear Theory of Elasticity*. Graylock Press, Albany (1953).

Nadaj, Arpad. *Theory of Flow and Fracture of Solids*. Vol. I, 2nd ed. McGraw-Hill Book Company, New York (1950).

Planck, Max. *Mechanics of Deformable Bodies*. Macmillan Company, New York (1949).

Proceedings of the First International Congress on Rheology. Schweninger, Holland, North-Holland Publishing Company, Amsterdam (1948).

Reiner, Markus. *Lectures on Theoretical Rheology*. 3rd ed., Interscience Publishers, Inc., New York (1960).

Richardson, E. G. *Relaxation Spectrometry*. Interscience Publishers, Inc., New York (1957).

Roder, H. L. *Rheology of Suspension*. H. J. Paris, Amsterdam (1939).

Rosenthal, D. *Introduction to Properties of Materials*. D. Van Nostrand Company,

New York (1964).

Rotherham, L. A. *Creep of Metals*. Institute of Physics, London (1951).

Schmidt, A. K. and C. A. Marlies. *Principles of High Polymer Theory and Practice*. New York: McGraw Hill Book Co (1948).

Schmitz, J. V., ed. *Testing of Polymers*. Interscience Publishers, New York (1965).

Scott Blair, G. W. *Flow Properties of Blood and Other Biological Systems*. Pergamon Press, Oxford (1960).

Scott-Blair, G. W. *An Introduction to Industrial Rheology*. P. Blakiston's Son and Company, Philadelphia (1938).

Scott-Blair, G. W. *A Survey of General and Applied Rheology*. 2nd ed., Sir Isaac Pitman and Sons, Ltd., London (1949).

Sommerfeld, Arnold. *Mechanics of Deformable Bodies*. Academic Press, New York (1950).

Swainger, Keith. *Analysis of Deformation:* Vol. 1, *Mathematical Theory* (1954); Vol. 2, *Experimental and Applied Theory* (1955); Vol. 3, *Fluidity* (1957); Vol. 4, *Waves and Vibrations* (1960); Macmillan Company, New York.

Tobolsky, A. V. *Properties and Structure of Polymers*. John Wiley and Sons, Inc., New York (1960).

Treloar, L. R. G. *The Physics of Rubber Elasticity*. Clarendon Press, Oxford (1949).

4.11a SELECTED BOOKS AND BIBLIOGRAPHIES ON RHEOLOGY AND RHEOLOGY OF FOOD PRODUCTS

DeMan, J. M., P. W. Voisey, V. F. Rasper and D. W. Stanley (eds.), *Rheology and Texture in Food Quality*. AVI Publishing Company, Inc., Westport, Conn. (1966).

Mohsenin, N. and J. P. Mittal. Bibliography on Texture of Solid Foods. Dept. of Agr. Engng., Penn State University, University Park, Pennsylvania (1978).

S. C. I. Monograph No. 1. *Texture in Foods*. Soc. of Chem. Ind., London (1960).

S. C. I. Monograph No. 27. *Rheology and Texture of Foodstuffs*. Soc. Chem. Ind., London (1968).

Sherman, P. *Industrial Rheology, with Particular Reference to Foods, and Cosmetic Products*. Academic Press, New York (1970).

Sherman, P. The Textural Charactertistics of Dairy Products. In *Advances in Food Texture*. Eds. DeMan, Stanley, and Voisey. AVI Publishing Company, Inc., Westport, Conn. (1976).

Sherman, P. Evaluation of Textural Properties of Protein Foods in *Proteins in Human Nutrition*. Eds. J. W. G. Parker and B. A. Rolls. Academic Press, London (1973).

Sherman, P. Structure and Textural Properties of Foods in Texture *Measurement of Foods*. Eds. A. Kramer and A. S. Szczesniak. Reidel Publishing Company,

Dordrecht, Holland (1973).
Sherman, P. (ed.), *Food Texture and Rheology Academic Press,* New York (1977).
Sone, T. *Consistency of Foodstuffs*. Reidel Publishing Company, Dordrecht, Holland (1966).

4.12 LIST OF SYMBOLS FOR CHAPTER 4

a —semi-axis of ellipse or radius of the circle of contact
B —bulk compliance or compressibility; $B=1/K$
d —diameter
D —deformation or compression compliance, $D=1/E$
E —axial elastic modulus
E^* —complex modulus
E' —storage modulus
E'' —loss modulus
F —force
G —shear modulus
J —shear compliances; $J=1/G$
K —bulk modulus; $K=1/B$
n —power law constant in non-Newtonian flow
n' —flow behavior index
p —pressure
q —flow rate
r —radius from the axis of symmetry
R —radius of convex body or sphere
T_{rel} —relaxation time
T_{ret} —retardation time
(t) —time-dependent
γ —shear strain
$\dot{\gamma}$ —shear rate or velocity gradient
δ —phase angle between sinosoidally varying stress and strain; tan δ=damping factor
ε —tensile or compressive strain
Ω —angular velocity
η —Newtonian viscosity coefficient
η' —plastic viscosity
η'' —apparent viscosity
μ —Poisson's ratio

σ —tensile or compressive stress
τ —shear stress
τ_w —shear stress at the wall
τ_y —yield stress
τ_{yB} —Bingham yield stress

Subscripts

0 —instantaneous or after zero time
e —equilibrium
d —decay for stress relaxation
r —retarded for creep

5

RHEOLOGICAL PROPERTIES

In an attempt to obtain more meaningful and usable data on mechanical properties of agricultural products, many investigators have approached the problem by employing the testing procedures used for non-biological materials. In this chapter techniques for obtaining data on load-deformation; stress-strain data from compression, tensile, shear and bending tests; hydrostatic compression; and the basic tests for viscoelastic characterization of the materials are discussed. As the first approximation and in the absence of better techniques, this is a sound approach. Many examples are given where the use of this engineering approach has shown more clearly the changes in mechanical properties with variety, temperature, moisture content, maturity, size and other factors.

5.1 FORCE-DEFORMATION BEHAVIOR

Study of force-deformation behavior of agricultural products in their natural state has been an attempt to provide objective measurements resulting in more meaningful data usable in engineering analysis and design. Despite the complexity of the material and the test specimens and the lack of knowledge of stress-strain distribution within the usually convex body, this approach has been justified on the ground that the raw material is usually subjected to mechanical treatments in handling and processing in its natural form. Any modification of the intact material, in an attempt to prepare a standard test specimen, may not give the true behavior when it is subjected to mechanical treatments in practice. For example, to date there is no experimental or theoretical evidence to show that the resistance of a whole fruit or vegetable to mechanical damage can be predicted from tests of standard specimens prepared from the flesh of these materials.

Although, as will be discussed in Chapter 6, the force-deformation data can be used with Hertz and Boussinesq methods to calculate a better defined parameter such as an apparent elastic modulus, most of the reported data have not been obtained under conditions to make such analysis possible. As a result, most of the force-deformation data on mechanical properties of intact biological materials are empirical data with which the conditions of the specimen, testing procedure, probe size, and other details must be stated. Table 5.1 summarizes the work on force-deformation studies on selected food and feed materials.

Extensive tests have shown that if the initial part of the force-deformation curves of soft biological tissues are taken into consideration, the initial part of the curves are usually concaved towards the force axis. This is exactly opposite the force-deformation curves for polymeric materials which is usually convex towards the force axis (Fig. 5.1). The reason for this difference in the behavior is not known. It is possible, however, that the presence of moisture in biomaterials offers little resistance to shear stresses causing relatively large deformations in response to small initial stresses. Under larger strain, however, the force-deformation curves assume a sigmoidal shape with slope of the curve first increasing and then decreasing.

The presence of a sigmoidal shape force-deformation curve in biomaterials means that a modulus of elasticity calculated on the basis of the slope

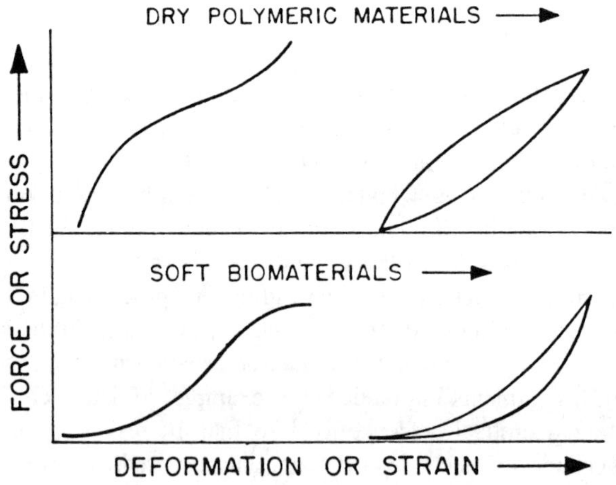

Figure 5.1 Basic difference in force-deformation and hysteresis curves of polymeric and biological materials in tension

of the force-deformation curve would always be, up to a point, greater for heavier loads or larger strains than for lighter loads and smaller strains. Fung (1967, 1968) has shown that because of this characteristic, the tangential modulus of soft biological tissues is almost zero at small strain but increases exponentially as the strain increases. Therefore, a statement of modulus of elasticity of a biomaterial must always be accompanied by the load or strain level at which the value of the modulus was calculated. It should be pointed out, however, that for closer approximation of rheological properties of a biomaterial, from theoretical point of view, one must choose data from the initial portion of the curve. At this stage, under small strains, the material's elasticity or linear viscoelasticity exists to a greater extent and thus the appropriate theories would be more applicable.

Experimenting with stems, roots and leaves of various plant materials, Treitel (1944) has shown that plants with greater number of air chambers show greater elasticity and thus are less stiff and have smaller modulus of elasticity. He also found that low hardness values (in terms of Brinell hardness number), as in the case of rubber and metals, are associated with low modulus of elasticities while high hardness values corresponded to high values of modulus of elasticity. As for the shape of stress-strain curves, Treitel (1944) found that both plant and animal materials exhibited curves which varied from sigmoidal shape, like rubber, to straight line, like metals. This experiment also showed that soft plant and animal tissues have usually flat curves (low modulus), while hard plant and animal tissues have steep curves (high modulus). From his experiments with many plant and animal tissues, Treitel concluded that growth is a biological phenomenon whose physical part may be best understood by stress-strain relationships and whose chemical part may be explained by using autocatalytic reactions.

5.2 STRESS-STRAIN BEHAVIOR

While some investigators have been concerned with mechanical properties of agricultural products in their natural form, others have prepared test specimens from the product and applied conventional engineering material testing techniques to express the data in terms of stress and strain.

Uniaxial compression

The compression test is merely the opposite of the tension test, at least insofar as the direction or sense of the applied stress is concerned. The

Table 5.1 Examples of force-deformation experiments with some agricultural products

Product	Type of test	Crosshead drive	Crosshead speed	Force measurement	Deformation measurement	Loading device	Specimen support	Information reported	Source*
Apple, Saltstock pickle (whole)	Magness–Taylor pressure test	Mechanical	6.48 in/min	Baldwin Lima load cell	Chart calibration	5/16-in tip of Magness–Taylor	Flat rigid plate	Force, displacement, energy to puncture	1
Apples, Berries, Peaches, Pears, Plums, Tomatoes (whole)	Die loading compression with skin intact	Pneumatic PSU testing machine	0.5 in/min	Daytronic force transducer	Daytronic displacement transducer	1/4-in rigid plunger	Rigid doughnut ring, cherries on rigid plate	Force–deformation, energy to bioyield (if any) and skin rupture	2
Apples, Apricots, Peaches, Pears, (spherical segments)	Die loading compression	Universal testing machines (Dillon, Riehle)	0.5 in/min	Mechanical force gage	Circular scale at end of worm gear	5/16-in rigid die	Rigid flat plate	Force, deformation data to bioyield and bruising	3
Apple with skin removed (whole)	Magness–Taylor pressure test	Instron testing machine	2.0 in/min	Instron load cell	Chart drive	5/16-in and 7/16-in tips of Magness–Taylor tester	Flat rigid plate	Force, displacement to puncture, effect of tip size, speed	4
Apricots, Grapes, Peaches, Pears,	Parallel plates compression	Dead weights	Static	Known weight	Dial gage	Flat rigid plate	Flat rigid plate	Resonant frequency from force–deformation and elastic	5

Table 5.1 (continued)

Product	Type of test	Crosshead drive	Crosshead speed	Force measurement	Deformation measurement	Loading device	Specimen support	Information reported	Source*
Tomatoes (whole)								modulus	
Barley, Corn, Oats, Soybeans, Wheat	Parallel plates compression	Mechanical	0.50 in/min	Load cell	Chart calibration	Rigid flat plates	Anvil of load cell	Breaking force and energy, position of grain moisture	6
Corn, Peabeans, Wheat	Parallel plates compression	Mechanical	0.078 0.267 0.467 in/min	Strain gage cantilever beams	Strain gage cantilever beams	Flat rigid plate	Flat rigid plate	Force–deforma–tion energy hysteresis, moisture	7
Corn (whole and endo-sperm), Wheat (kernel)	Compression rigid die, spherical indenter, flap plate	Instron testing machine	0.20 in/min	Instron load cell	Chart drive	0.016-in rigid die, 0.062-in steel ball, flat steel plate	Mounted on rigid flat plate	Linear limit force, deforma-tion, elastic modulus, effect of load, moisture	8
Corn silage	Compression with re-strained lat-eral expansion	Mechanical press	——	Dead weights applied manually	Calibrated scale	Flat plate	"Vessel" with rigid walls	Load–deforma-tion, bulk density compressibility % elasticity	9
Egg	Parallel plates compression	Mechanical	0.045 in/min	Shaevitz load cell	Displacement transducer	Cushioning materials backed by rigid plate	Cushioning materials backed by rigid plate	Force, deforma-tion and energy at failure	10

Table 5.1 (continued)

Product	Type of test	Crosshead drive	Crosshead speed	Force measurement	Deformation measurement	Loading device	Specimen support	Information reported	Source*
Egg	Compression rigid die and spherical indenter	Instron testing machine	0.020 in/min	Instron load cell	Chart drive	0.016-in rigid die, 0.062-in stell ball	Plaster support	Shell rupture force, deformation energy	11
Forage stalks (alfalfa, clover, timothy)	Bending of stalk as a single bean	Load applied manually with a depth micrometer	——	Strain gage cantilever beams	Depth micrometer	See Fig. 5.2	Anvil with plastic support	Ultimate bending force and displacement, effect of moisture	11
Sweet potato (half of 2-in middle portion)	Die loading compression	Instron testing machine	0.2 in/min	Instron load cell	Chart drive	3/16-in rigid die	Flat rigid plate	Effect of aging on force–deformation ratio, energy, elasticity up to proportional limit	13
Onion (whole)	Parallel plate compression	Hydraulic (Kramer shear press)	0.48 in/min	Daytronic transducer	Chart calibration	Flat rigid plate	Flat rigid plate	Firmness in terms of force to deformation ratio	14
Potato (whole)	Die loading with skin intact	Pneumatic similar to PSU machine	1.0 in/min	Load cell	Strain gage cantilever	1/4-in rigid die	Flat rigid plate	Rupture force–deformation, energy, age, variety	15

Table 5.1 (continued)

Product	Type of test	Crosshead drive	Crosshead speed	Force measurement	Deformation measurement	Loading device	Specimen suport	Information reported	Source*
Wheat (kernel)	Parallel plates compression	——	34×10^{-5} in/min	——	——	Flat rigid plate	Flat rigid plate	Effect of strain rate, moisture elastic–plastic properties	16

1 Pflug *et al.*, 1960
2 Mohsenin, 1962; Unpublished data, 1965
3 Fridley and Adrian, 1964
4 Bourne, 1965
5 O'Brien *et al.*, 1965
6 Billanski, 1965
7 Zoerb and Hall, 1960
8 Unpublished data, 1965
9 Yaremenko, 1956
10 Rehkugler, 1962
11 Unpublished data, 1965
12 Prince, 1961, 1965
13 Fluck *et al.*, 1964
14 Ang *et al.*, 1960
15 Finney *et al.*, 1964
16 Shpolyanskaya, 1952

* Numbers refer to sources given at the end of this table.

choice of one or the other test depends on such factors as the type of loading to which the material is to be subjected, differences in properties under compressive and tensile loading, and testing problems associated with each specific material. Problems of the preparation of test specimen and complications induced by the gripping of the specimen for tensile test on the one hand and the fact that mechanical damage to the product usually results from compressive loads have probably been the reasons for finding more data reported for compression tests than tensile tests.

A few of the basic requirements for uniaxial compression tests of engineering materials, which are applicable to biological material, are applying a truly concentric or axial load so that bending stresses are not set up as a result of irregularities in alignment, avoiding friction between the end surface of the specimen and the bearing plate of the testing machine due to expansion of the specimen, and selecting a specimen with such length to diameter ratio that a proper degree of stability is obtained and buckling action is avoided.

To evaluate an elastic modulus for the flesh of fruits and vegetables, the conventional test of cylindrical specimens of the material, subjected to uniaxial compression between two parallel rigid plates have been reported for grains (Zoerb, 1960; Shelef and Mohsenin, 1966), apple (Mohsenin, Cooper *et al.*, 1963, 1965), white potato (Finney, 1963), and cheese and butter (Davis, 1937). In the case of grains, core specimens were prepared from corn, bean and wheat by cutting off each end. The cross-sectional area was obtained by measurement with a micrometer or microscopically using a micrometer disc. The modulus of elasticity was found by taking the ratio of conventional stress to conventional strain as follows:

$$E = \frac{F/A}{\Delta L/L} \tag{5.1}$$

where

F = force
A = original cross-sectional area of the specimen
ΔL = deformation corresponding to force F
L = initial length of the specimen

Considering the fact that in biological materials, even at very small strains, part of the deformation is always nonrecoverable, the question which arises in using the above equation is what point of the stress-strain curve should bc considered for calculation of the above ratio? To answer this question some investigators have taken the slope of the stress-strain curve, if the slope were "approximately" linear (Zoerb, 1960). This at best can be referred to as modulus of deformability or apparent elastic modulus. Others have attempted to establish the elastic region of the curve by loading and unloading two or three times until no residual deformation was observed upon unloading (Mohsenin and Cooper, 1963). In the case of wheat, it was demonstrated, however, that the elastic deformation did not change upon further loading-unloading (Shelef and Mohsenin, 1966). In this case, the slope of the first unloading curve was used for calculation of the elastic modulus. If the unloading curve was not linear, the method of tangent modulus or secant modulus as illustrated in Fig. 4.5, was employed. For more information on this subject, see ASAE S 368.1 in the Appendix.

Uniaxial compression test has also been reported for forage and corn stalks (Prince, 1961, 1965, 1968). The ends of the specimen were cut parallel to each other and perpendicular to the plant axis using a high velocity fine-tooth power saw in a special fixture. Stalk specimens were then subjected to axial compression in a universal testing machine (Fig. 5.2). Because of the variation in cross-sectional area of the stalks, in the earlier work no attempt was made to express compressive force in terms of stress. The compressive forces were adjusted, however, to a common stalk diameter of 0.30 centimeter for comparison. The type of failure depended upon the straightness of the plant specimen and the length to diameter ratio. Specimens were 1.9 to 7.6 centimeters long. Apparently, for large length to diameter ratios, a column action existed because it was reported that the majority of the stalks failed in bending.

The relationship between depth of indentation of a spherical indenter and load on cheese has been reported (Scott Blair and Baron, 1949). In this case, the rate of increase of load was adjusted so that the rate of load increase just compensated the increase in contact area resulting in a virtually constant unit pressure. The following table shows the relationship between depth of indentation, D, time after application of load, t, and unit pressure, referred to as "stress," S. For a constant stress, S, was found that the relationship betwen indentation and time was exponential given by the following equation

$$\log D = a + b \log t$$

which is the equation of a straight line if D and t were plotted on log-log paper.

	Cheese I		Cheese II	
Time (Sec)	*D* (cm)	*S* (kg/cm²)	*D* (cm)	*S* (kg/cm²)
10	0.354	1.04	0.448	0.82
20	0.395	1.01	0.513	0.77
30	0.430	0.98	0.562	0.75
40	0.462	0.97	0.602	0.75
50	0.482	0.99	0.633	0.75
60	0.500	1.01	0.660	0.76

Uniaxial tension

The most difficult problem in tension test of biological materials is finding a gripping device which would hold the ends of the specimen tight without overstressing the living tissues. The shape, size, presence of water in the living tissues and the softness of biological specimens introduce extremely difficult problems in gripping, alignment, symmetry with respect to the longitudinal axis, stress concentration, avoiding the bending stresses during axial loading, and other requirements which are usually specified for tension tests of engineering materials.

Because of these problems, very few investigators have tried the axial tests for agricultural products. An attempt was made to measure the modulus of elasticity of the skin in sweet cherries and relate that to splitting and cracking of cherries due to water absorption (Levin *et al.*, 1959). Transverse and longitudinal sections of the skin (14/64 in.×12/64 in.×0.008) in.) were removed for axial tension tests. The modulus was determined by dividing the stress by the unit deformation of the skin. The details of the method of holding the specimen for tension tests were not reported.

In an attempt to explain the mechanism causing cracking of potatoes in handling, data were obtained on tensile stress-strain properties of potato skin, center flesh and halfway between skin and flesh (Huff, 1966). The tensile specimens consisted of 0.15-in. slices shaped as rectangular specimens with reduced section shown in Fig. 5.3. Stress was calculated on the basis of original cross-sectional area at the narrowest part of the specimen. Strain was expressed in terms of deformation per unit initial length of a given gage length on the specimen, measured with a specially designed extensometer (Fig. 5.3). Stress-strain curves obtained up to the point of

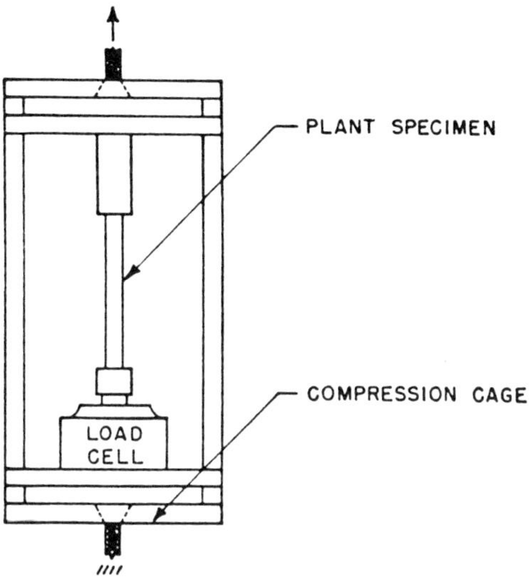

Figure 5.2 Compression cage and load cell for compression of forage stalks parallel to the plant axis (Prince, 1965)

failure proved to be non-linear. For this reason, a tangent modulus, referred to as "failure modulus," had to be reported for these data (Fig. 5.4).

To investigate the possibility of stress cracking in corn due to thermal stresses (Ekstrom *et al.*, 1965), tensile properties of individual corn kernels were determined by notching the kernel on each side with a file and grasping the kernel by means of a special tension jig (Fig. 5.5). Application of various types of cementing agents was unsuccessful due to the difficulties involved in producing a firm bond between the kernel and another material to be held by a tension jig. The distance between the two flat sides of the notch in the unstrained kernel was selected as the gage length (0.145 in., the width of the file). The change in the gage length was measured by a cantilever beam with strain gages at the root. Although the method of grasping the kernel must have introduced many problems such as insufficient alignment, stress concentration at the notches, and filing away a great portion of the test specimen, it is interesting to note that this was the first tensile test of cereal grains reported in literature. The results of these tension tests are compared with compression tests in Table 5.2. The higher value for elastic modulus reported in the last column is probably due to the fact

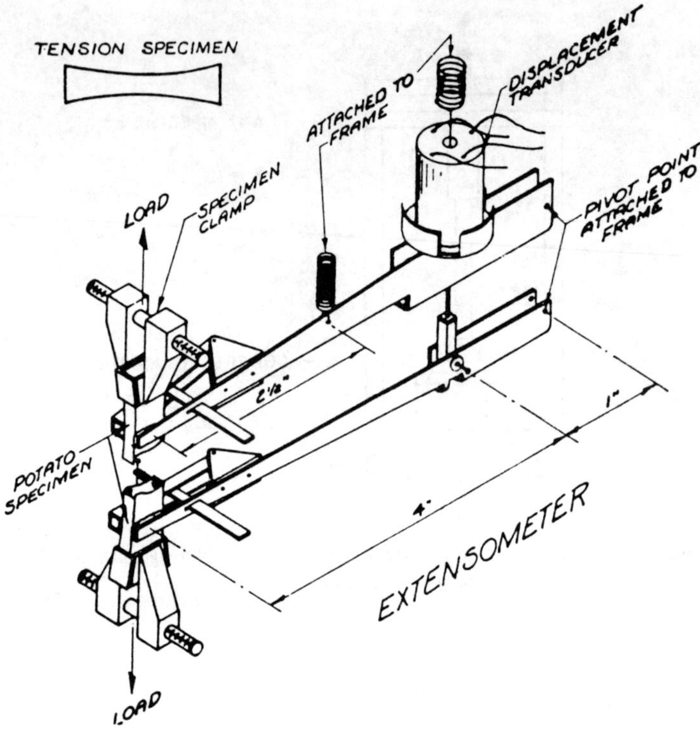

Figure 5.3 Extensometer and tensile specimen used in tensile test of potato slices (Huff, 1966)

that in this case the upper portion of the stress-strain curve was used for calculation of the modulus. Since this portion of the curve had a steeper slope than the initial part, this could have given a larger value for the calculated modulus. Another reported tensile test of corn is reported by Hammerle (1968), who isolated rectangular segments of the horny endosperm and subjected the small slabs at various moisture contents to tensile tests.

The ultimate tensile strength of alfalfa stems was measured using the fixture shown in Fig. 5.6 and a Scott Testing machine set for a loading rate of one inch per minute (Halyk, 1962). The tension jig, as reported, would firmly grip the ends of the stem without crushing or weakening them. The ultimate strength was computed on the basis of maximum load at the

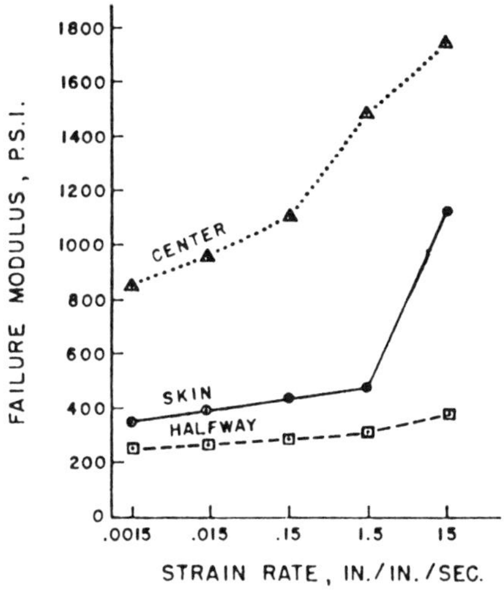

Figure 5.4 "Failure modulus" for Kennebec potato (Each point is a mean of 20 points) (Huff, 1966)

point of stem failure and the original cross-sectional area of the stem. Due to the non-uniformity of the cross-sectional area, an average cross sectional area had to be measured using a shadowgraphing technique. This technique, which is used in the textile industry, consisted of simply making a contact print of the stem on a high-contrast photographic paper. The average diameter of the stem was then determined, using a measuring microscope. It was assumed that the alfalfa stem was a solid right circular cylinder with a diameter equal to the average stem diameter, measured from the photographic print. The calculated ultimate strength was found to vary from 1260 to 5240 psi. Statistical analysis showed that about 80 percent of the variation in ultimate tensile strength could be attributed to the combined effects of moisture content and dry-matter bulk density. However, as shown by Hall *et al.* (1968) the variation in the actual stress-bearing cross-sectional areas of the stems could probably account for a greater portion of the variation in ultimate strength.

Shear

In contrast to compressive and tensile stresses which act normal to a plane, shear stresses act parallel to the shearing plane. If tensile or compressive

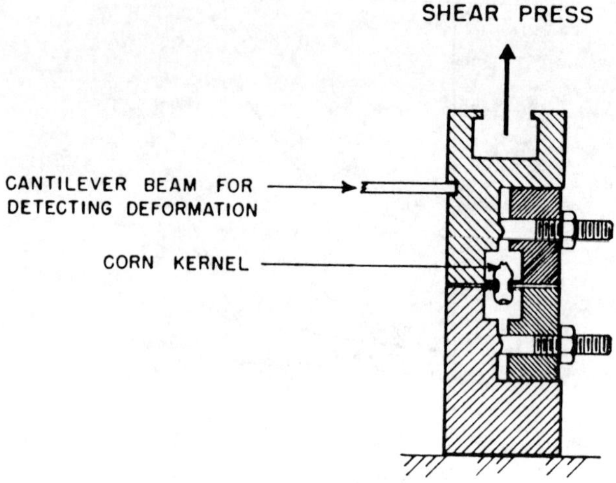

Figure 5.5 Tension jig for tensile test of corn kernel (Ekstrom, 1965)

stresses are acting only in one direction, the shear stresses are at 45 degrees thereto and one-half the magnitude of the applied direct stresses. Shear stresses are the resultants of parallel but opposed forces acting through the centroids of sections of the material that are spaced theoretically infinitesimal distances apart. In practical situations, however, the case of a rivet in shear is approximation of this condition. If the parallel and opposed shearing forces do not lie in a plane containing the longitudinal axis of the body, a couple is set up which produces a twisting action termed torsion.

In reporting the tests on mechanical properties of foods and agricultural materials, sometimes the term shearing has been used for a process which is actually a cutting process and not true shearing. Few tests have been reported which approximate the true shear tests. There are examples, however, of direct shear or transverse shear, torsion, and punching test, which is actually a form of the direct shear test.

To study the process of ripening in fruits, measurements were made of shear strength of the flesh of apples and pears (Schmidt, 1962). Shearing strength of the flesh was determined by shearing a plug from a slice of the flesh. Knowing shearing force, F, the diameter of the solid cylindrical die, d, and the thickness of the slice, t, shearing strength, S, was determined from

Table 5.2 Ultimate stress and modulus of elasticity of yellow dent corn obtained by tensile tests and compression tests

Property	Tension, vertical with notches on sides (Ekstrom, 1965)	Compression, vertical with ends cut off (Zoerb, 1960)	Compression, flat, germ side down, surfaced for full contact (Unpublished data)
Moisture content	10%, w.b.	15.4%, d.b.	12.2%, d.b.
Strain rate, in/min	0.125	0.08–0.46	0.020
Ultimate stress, psi	1270	4811	2390 (linear limit)
Elastic modulus psi	58300 (1% strain secant)	58600, from slope of "elastic region"	167600 (tan. at 40-lb load)

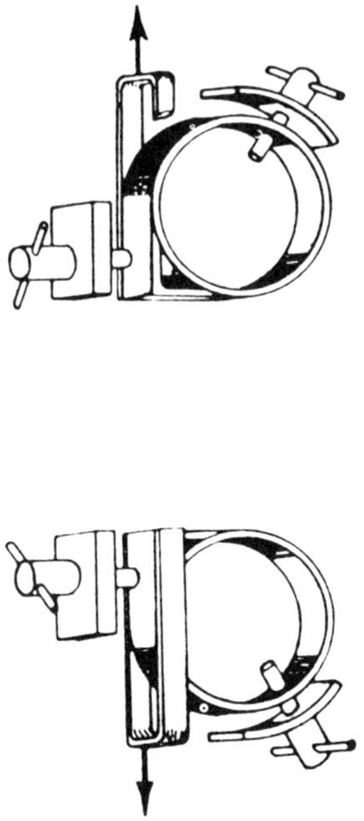

Figure 5.6 Jig for tensile test of alfalfa stem (Halyk, 1962)

$$S = \frac{F}{\pi \, dt}$$

Ripening of the fruit is closely related to the changes in pectic substance in the middle lamela of the cell walls, which act as a cementing agent holding the cells together (see Chapter 2), and these changes result in loosening of the cells as the fruit ripens. Therefore, the shearing strength of the flesh should determine the degree to which the cells are held together or the fruit has ripened. In an unripe fruit, the cells along the shearing surface are still tightly held together and will tear apart in response to the shearing stress. In a ripe fruit on the other hand, the cementing agent is soft and these cells glide alongside each other without being torn apart.

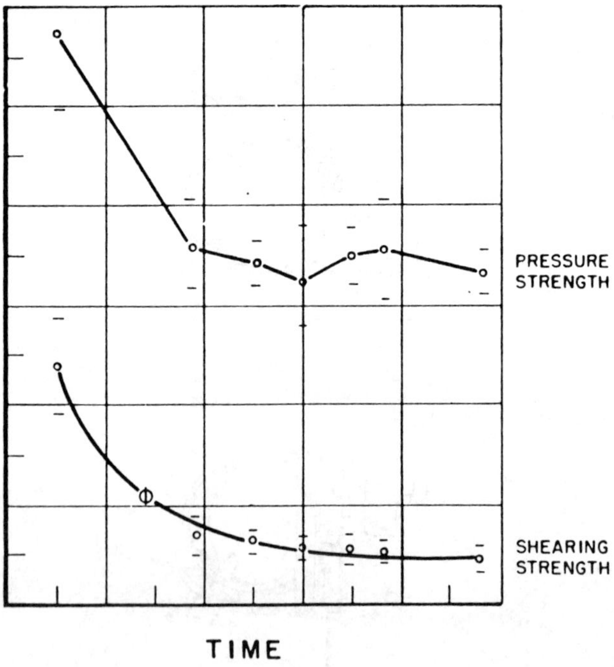

Figure 5.7 Shearing strength of apple tissues in the process of fruit ripening (Schmidt, 1962)

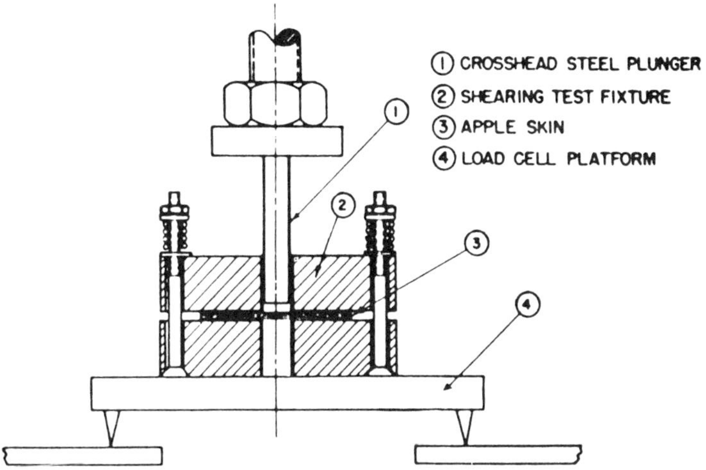

Figure 5.8 Apparatus for determining shearing strength of skin in apple fruit (Mohsenin and Goehlich, 1962)

Thus, as the ripening process continues and the pectic substances decompose, the connection between the cells weakens and the shearing strength decreases (See Fig. 5.7).

Shearing strength of the skin of apple was also measured using the fixture shown in Fig. 5.8. This was an attempt to determine to what extent the skin contributed to the resistance of fruits to shearing and puncturing forces (Mohsenin and Goehlich, 1962). Shearing strength was calculated from the shear force, the diameter of the steel solid plunger, and the thickness of the skin. The results produced values lower by about 42 percent than the compression rupture tests of the skin on the apple. In McIntosh apple, which has a tough skin, the shearing forces were only about 12 percent lower than the comparable compression tests of the skin supported by the flesh.

Static shear tests have also been reported for pea beans, wheat and corn using a steel punch to shear a plug from the specimen in flat position (Zoerb, 1960). In these tests shear stresses for pea beans were compared with those of yellow dent corn and red winter wheat. In the case of the latter two grains, however, even a punch of $1/16$-inch diameter was unsatisfactory because it tended to crush the entire grain rather than shearing a plug from the specimen. Shear stress values were obtained by dividing the maximum force required to shear the grain with the "shear bar" by the cross-sectional area parallel to the direction of applied load.

The ultimate shear strength of alfalfa stem internodes has been reported (Halyk, 1962) using a device illustrated in Fig. 5.9. This device, which applied shearing loads similar to the load exerted on a rivet in double-shear, permitted shear tests for the region extending from 3 to 83 percent of total shoot height containing an average of seven internodes. The ultimate shear strength, computed on the basis of maximum shear load and average stem diameter, varied from 58 to 2610 lb/in^2. The ultimate shear strength, like the ultimate tensile strength, was found to be inversely proportional to moisture content and directly proportional to dry-matter density. Again, the lack of accuracy in measuring the cross-sectional area could account for the major part of this variation.

Ultimate torsional strength of alfalfa stalks was also investigated measuring the angular displacement and the torque required for failure by torsion (Prince, 1965). Each end of the specimen was inserted into a roll pin and coated with tar in the molten state. The tar was quickly chilled to prevent

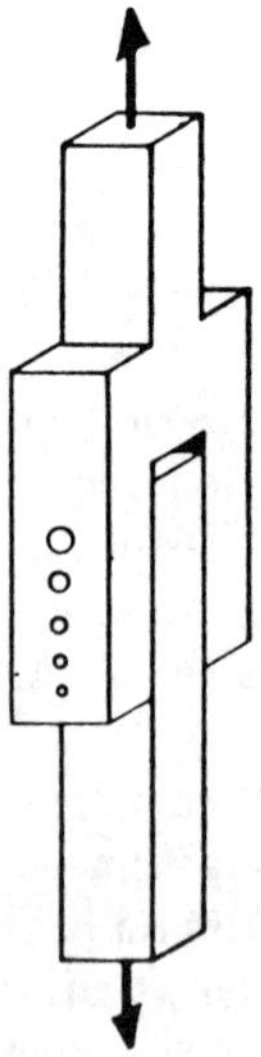

Figure 5.9 Jig for direct shear of alfalfa stems (Halyk, 1962)

damage to the living tissues. The influence of stalk diameter and moisture content on torsional strength was shown by regression equations derived from graphical illustrations such as Fig. 5.10.

Assuming the alfalfa stalk as a solid bar with circular cross section (neglecting the pith in the center), in a torsion test such as described above, the stress varies linearly from zero at the axis of twist to a maximum at the extreme fiber. By summing up the stresses over a cross section, the maximum shearing stress, S_s, on extreme fiber of the stalk in terms of the applied torque, T, and radius, r, is given by the torsion formula

$$S_s = \frac{2T}{\pi r^3} \tag{5.2}$$

The modulus of rigidity, G, is found from the ratio of stress, S_s, to torsional strain $r\theta/L$ to be

$$G = \frac{2TL}{\pi r^4 \theta} \tag{5.3}$$

where L is the length of the stalk, and θ is the angle of twist in degrees. If the stalk is to be considered as a tube and the inside and outside radii, r_1 and r, are known, the torsion formula for a tube given by

$$S_s = \frac{2Tr}{\pi (r^4 - r_1^4)} \tag{5.4}$$

may be applied.

For small angles of twist, the application of torsion formula appears to be a logical engineering approach as the first approximation of ultimate torsional strength and calculation of a modulus which may be referred to as apparent shear modulus. Even in the case of engineering materials, where the theory has not yet been developed for a particular configuration, the results of the simple torsion theory are often used in conjunction with appropriate correcting factors.

Table 5.3 shows the application of torsion formula to alfalfa stalks tested in the modified apparatus developed by Prince (1965). The modification included instrumentation for continuous recording of torque versus angle of twist on an x-y recorder (Prince, 1968).

Shear stress required for skinning of sweet potato was determined by an apparatus designed to skin a circular area under an applied torsional force (Fluck *et al.*, 1968). the torsional force was applied by the crosshead of an Instron testing machine moving at the rate of one inch per minute (Fig. 5.11). Crosshead movement affected an angular motion of a shaft upon which a rubber-backed circular piece of abrasive paper gripped a constant area of skin and sheared it as it turned. Normal force was applied through a gear rack and a dial indicator. A circular impression was cut through the skin on 1″×¾″ plugs of sweet potato to define the area to be skinned by shearing. This enabled a measurement of shear stress between the periderm and cortex without a confounding periderm to periderm shear stress.

For calculation of skin shear stress it was assumed that the applied stress increased linearly from center to the boundary of the circular section and thus cell distortion increased from center to outer edge. Equating the torque applied by the pulley with the torque due to shearing force over the circular area yields the torque equation

$$Fa = \int \tau r \, dA$$

where F is the applied force, a is the pulley radius, r is the radius from the center, and τ the shear stress. If the ring area, dA, is given by

$$dA = 2\pi r \, dr$$

and the maximum shear stress by

$$\tau_{max} = kr$$

then the above torque equation for maximum shear stress can be written as

$$Fa = 2\pi k \int_0^r r^3 \, dr$$

Integrating the above equation yields the expression for constant k which, by substitution results in maximum shearing stress given by

$$\tau_{max} = \frac{2Fa}{\pi r^3}$$

Figure 5.12 shows the change of skin shear stress for three varieties of sweet potato with time in storage, as determined by this method.

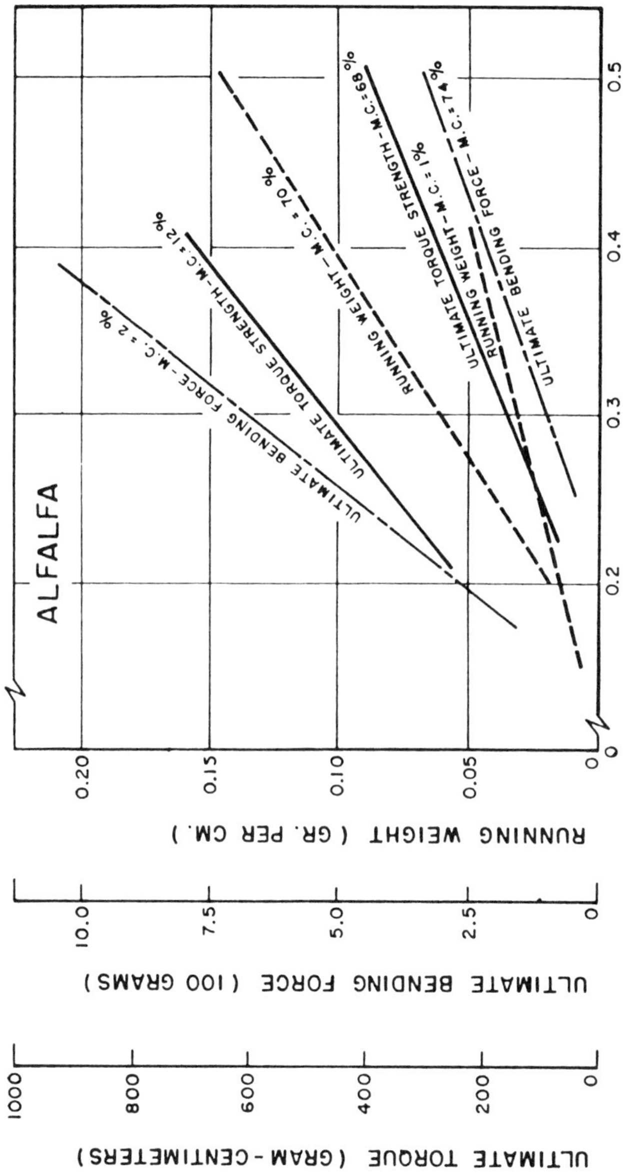

Figure 5.10 Ultimate bending strength and torque strength of alfalfa stem as affected by moisture and stalk diameter (Prince, 1965)

Table 5.3 Torsional rigidty of green and dry alfalfa stems[1]

Radius (outer) (in)	Radius (inner) (in)	Twist angle (radians)	Torque (lb-in)	Rigidity modulus[2] tube 10^5 psi	solid 10^5 psi	Section of specimen
A. Green: Tangent modulus from linear torque-twist relation at different sections of stem but same twist angle.						
0.066	0.042	0.174	0.036	0.338	0.283	3.1—Bottom
0.065	0.043	0.174	0.036	0.360	0.290	3.2
0.062	0.042	0.174	0.036	0.448	0.354	3.3
0.055	0.042	0.174	0.025	0.594	0.392	3.4
0.046	0.034	0.174	0.006	0.303	0.212	3.5—Top
B. Green: Secant modulus at same section of stem but different twist angles.						
0.066	0.042	0.174	0.036	0.338	0.283	3.1
0.066	0.042	0.349	0.068	0.311	0.262	3.1
0.066	0.042	0.524	0.096	0.296	0.247	3.1
0.066	0.042	0.698	0.120	0.277	0.233	3.1
0.066	0.042	0.873	0.141	0.260	0.219	3.1
C. Dry: Tangent modulus at different sections and different twist angles.[3]						
0.057	0.035	0.463	0.226	1.271	1.170	1.1—Bottom
0.043	0.028	0.437	0.151	3.292	2.484	1.2
0.054	0.044	0.433	0.148	1.635	1.176	1.3
0.047	0.037	0.850	0.109	1.057	0.638	1.4—Top

[1] Data were obtained by C. T. Morrow in cooperation with R. P. Prince at Connecticut Agr. Exp. Sta., 1966. Stem (10 cm long) was assumed as a right circular cylinder with diameter equal to that of the smallest end. See also Prince (1968).

[2] Compare modulus of rigidity of typical cellulose fibers 65% r.h., cotton, 3.6×10^5 psi; ramie, 2.5×10^5 psi; flax, 2×10^5 psi (Meredith, 1956).

[3] Torque-twist lineraity was much better than green stems.

A factor called "brittleness" was evaluated for several varieties of wheat grain using a technique which placed the grain under "shear" (Naumov, 1957). The grain was held between two knife edges with the upper edge cutting slowly into the grain (Fig. 5.13). The load on the upper knife edge and the cutting depth necessary to rupture the grain were determined. It was assumed that the stress was distributed uniformly along the cross section of the grain, and a shear stress was calculated by dividing the rupture load by the cross section area at the point of loading. Whole grains as well as the endosperm, after removing the hull, were tested, taking the shear

Figure 5.11 Torsional force applied by the crosshead of a universal testing machine to measure shear stress required for skinning of sweet potato (Fluck, 1964)

stress value as the brittleness of the grain. Differences in varieties and moisture content were detected using this technique.

Resistance of silage material (Sudan grass) to shear was determined using a direct shear unit (Yaremenko, 1956). The silage material was first subjected to a normal load, F_n, allowing sufficient time for the initial deformation to fade out. A steadily increasing load was then applied to a point where shearing of the silage was affected. Resistance to shear force, F_s, was equated to cohesion force, C, and friction force, $F_n \tan \alpha$, where α is the internal frictional angle resulting from motion between silage particles.

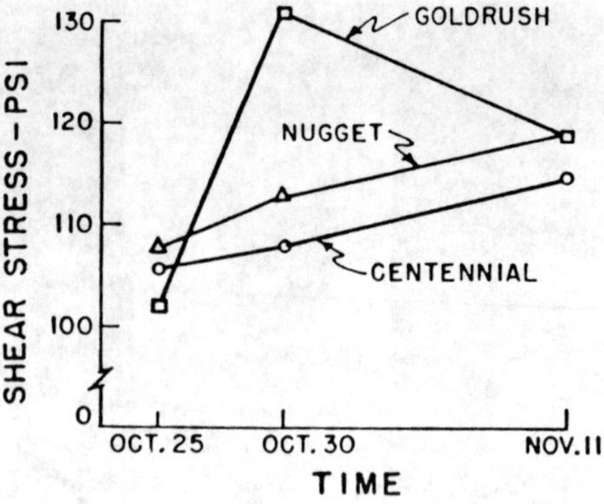

Figure 5.12 Shear stress between periderm and cortex in skinning test of sweet potato during curing–storage sequence (Fluck, 1964)

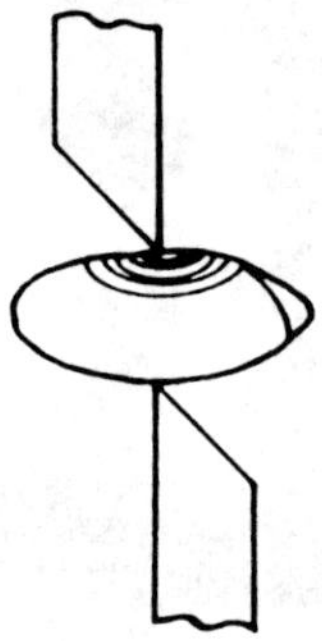

Figure 5.13 Shearing of single grain of wheat to evaluate "brittleness" (Naumov, 1957)

$$F_s = C + F_n \tan \alpha \tag{5.5}$$

Using this equation the shear coefficient, F_s/F_n, the internal frictional angle, α, and cohesion force, C, were determined for grass silage chopped to different lengths of cut. The important conclusion from this study was that the silage material is not a "friable" substance and its mechanical handling requires working elements with a positive action.

This same principle has been used in establishing design criteria for flow of bulk solids in bins. This will be discussed later. Equation 5.5 is also employed to predict theoretical soil thrust on traction members (Bekker, 1960).

In evaluating the physical properties of haylage (chopped forage), important in design of handling and storing equipment, such as the digging and cutting reel in a silo unloader, mass shear strength of cut and chopped forage after several months of storage in silos were determined using a punch-and-die method of shearing (Bright and Kleis, 1964). Two ¼-inch steel plates with four holes on the circumference of a circle on each plate, together with 0.476-inch diameter punch and inserts, were employed to measure the shear stress and shear energy required to shear a sample with a given volume and a known bulk density compressed between the two plates. Shear stress was calculated from the shearing force and shearing area. Shearing energy was obtained from the area under the force-deformation curves expressed in inch-pound per unit shear area. Figure 5.14 shows the effect of haylage density on shear stress and energy for alfalfa with 57 percent (w.b.) moisture content and after three months in silo. Similar experiments on orchard grass, timothy, and alfalfa at other moisture contents showed that moisture content and the type of forage had significant effects on shear stress and shear energy while the effect of storage was not significant.

In the search for design and development of more economical mechanical handling systems to handle frozen high-moisture grain from storage during cold weather, it was realized that basic information on mechanical properties of such a material is lacking. Kranzler and Witz (1967) conducted some compression and shear tests on $3'' \times 2''$ plugs of frozen high-moisture barley prepared by freezing the grain in standard soil cans with the above dimensions. The grain plugs were subjected to uniaxial compression and single shear. In shear tests, a special jig restrained one half of the grain plug specimen while the other half was forced downward, shearing the specimen along a vertical plane passing through the center line of the plug. Results of these tests are given in Table 5.3a. It was noted that the compression and shear properties of the grain in the frozen condition undergoes a temperature-related transition from elastic to viscoelastic. Elastic behavior and sudden failure from yield, as seen from Table 5.3a, was exhibited at subfreezing temperatures. As temperature increased, the heat energy generated as the result of loading was sufficient to initiate a gradual breakdown of kernel-to-kernel ice bonding mechanism.

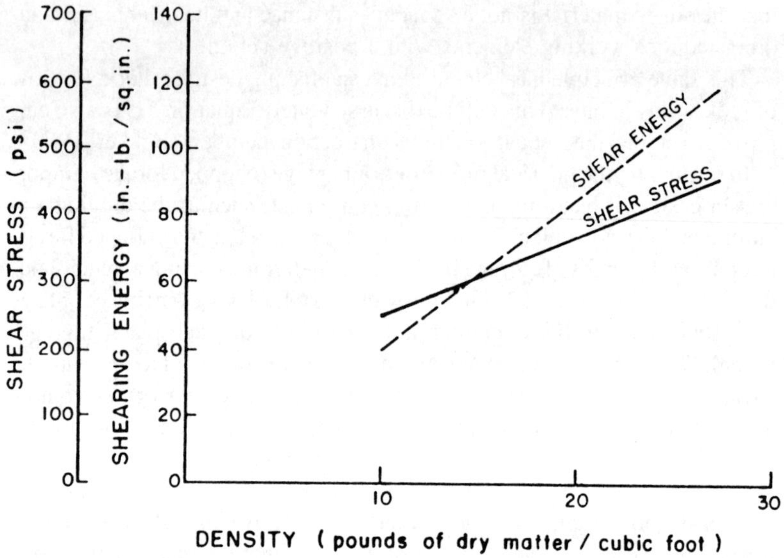

Figure 5.14 Effect of alfalfa haylage density on shearing stress and shearing energy after three months storage silo (Bright and Kleis, 1964)

Bending

Certain agricultural materials can be tested as simple beams as done in flexture tests of engineering materials. Forage and some vegetable stalks in their natural state may be tested as beams. Rectangular bars of more homogenous materials such as cheese and butter can also be tested as beams sagging either due to their own weight or addition of small concentrated loads. Knowing the load and deflection, the following formula for a simple beam may be used to calculate an apparent stiffness or modulus of elasticity:

$$D = \frac{FL^3}{48EI} \tag{5.6}$$

where D is deflection at mid span, F is the concentrated load at mid span, L is the effective length, I is moment of inertia about the neutral axis, and E is the modulus of elasticity. Indicating depth or diameter of the beam section by d and breadth by b, the moment of inertia, I, is given by

Table 5.3a Mechanical properties of frozen high-moisture barley (Kranzler and Witz, 1967)[1]

Temperature (°F)	Bulk density (lb/ft^3)	Compression test: Stress (lb/in^2) at yield	Compression test: Stress (lb/in^2) at failure	Compression test: Energy ($in.lb/in^2$) at yield	Compression test: Energy ($in.lb/in^2$) at failure	Modulus of elasticity (lb/in^2)	Shear test: Stress (lb/in^2) at yield	Shear test: Stress (lb/in^2) at failure	Shear test: Energy ($in.lb/in^2$) at yield	Shear test: Energy ($in.lb/in^2$) at failure
−20°	43.1	54.5	53.6	1.3	1.5	3025	29.8	25.0	1.3	1.8
	46.2	62.7	58.4	2.0	2.1	2405	40.8	34.0	2.4	3.1
	49.6	97.6	95.9	3.6	4.1	3600	66.8	58.7	6.2	8.2
0°	43.1	41.7	40.7	1.3	1.5	1580	31.8		1.6	
	46.2	54.9	52.1	2.1	2.4	1920	36.8		1.8	
	49.6	63.8	56.6	2.7	4.0	2130	45.0		2.4	
28°	43.1	14.6		0.7			7.7		0.4	
	46.2	19.2		0.8			12.2		0.8	
	49.6	25.5		1.7			17.7		1.1	

[1] Grain was soaked in water for 24 hours and after removing "loose" water was frozen at moisture content of 47% to 49% (w.b.)

$$I = \frac{bd^3}{12} \tag{5.7}$$

for a beam with rectangular section and

$$I = \frac{\pi d^4}{64} \tag{5.8}$$

for a beam with circular section. In cases where the material in its natural state is of non-uniform cross section, the formulas for beams of variable cross sections may be used.

The stiffness of individual forage stalks, for example, appear to be an important relevant property of the material in design of the cutting mechanism in the mowing machines. The individual stalks can be tested either as a cantilever beam or a beam supported at the ends to evaluate the elastic modulus which should be an indication of stiffness of the material. This can be done by assuming the stalk is solid and using the above beam formulas or estimating the mean outside and inside diameters of the stalk and using the formulas for hollow beams.

Figures 5.16 and 5.17 show apparatus used for determining the bending strength of forage stalks. The data, however, were reported (Prince, 1965) in terms of force and deflection at various moisture contents (Fig. 5.10) and linear densities of forage stalks. Modulus of elasticity of the forage material in flexture was not calculated. With some modification in the design of the apparatus and selection of the size of the forage stalk, it should be possible to test the material for relatively pure bending so that a flexture modulus can be calculated.

To study the rigidity and strength of tissues in potato tubers as affected by various chemicals, slices of potato were mounted as cantilever beams (Somers, 1965) and deflection of the beams under their own weight were observed. The slices were 1 mm thick and were trimmed to 13 mm by 28 mm beams. The downward deflection, D, of the free end of the beam, under its own weight, w, was related to elasticity using the following equation for uniformly loaded elastic cantilever beams with L as the projected beam length

$$D = \frac{wL^4}{8EI} \tag{5.9}$$

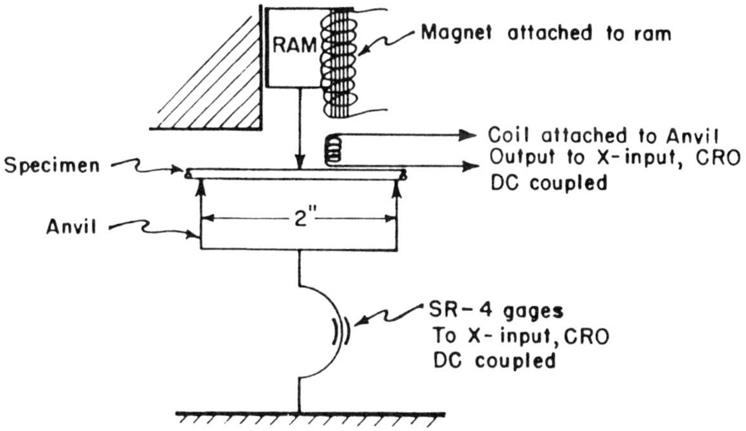

Figure 5.15 Apparatus for measurement of ultimate bending strength of common pasture plants (McClelland and Spielrein, 1958)

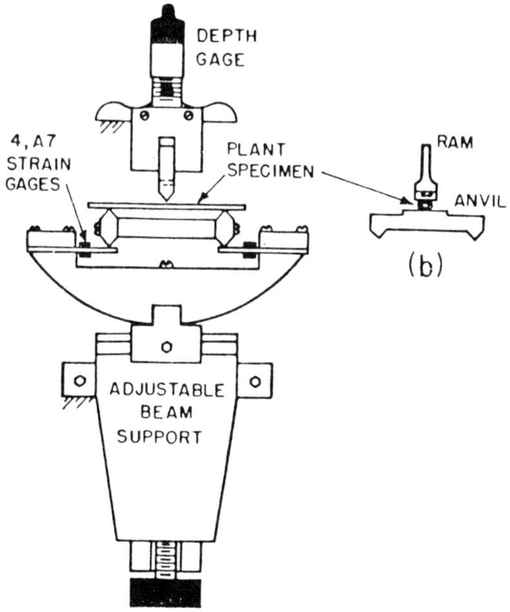

Figure 5.16 Apparatus for measurement of ultimate bending force and deflection of forage stalks—"b" shows the stalk under compression (Prince, 1965)

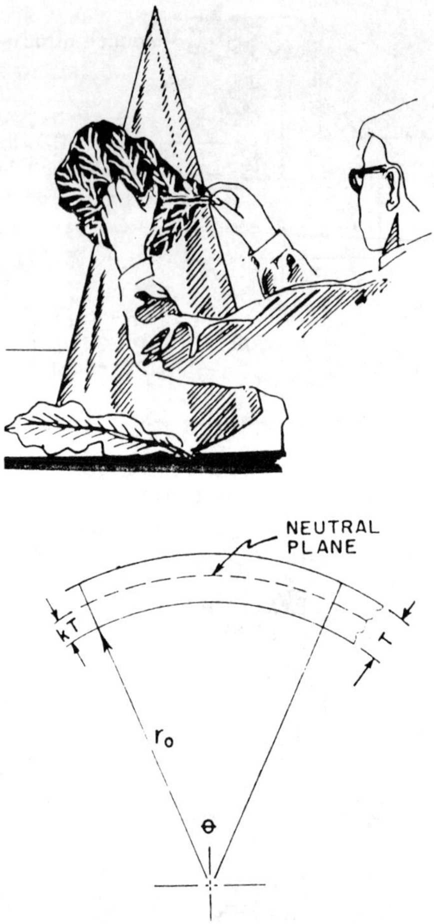

Figure 5.17 Measuring critical radius of curvature of tobacco leaf midribs in bending (Suggs *et al.*, 1962)

The reciprocal of the product of E, elastic modulus, and I, cross-section moment of inertia, known as flextural rigidity in mechanics, was then taken as the "moment of flaccidity." This parameter was used to study the effects of various chemicals or the loss of rigidity and fresh weight of the tissue. We will discuss this subject again under mechanics of texture in another section.

The critical radius of curvature of tobacco leaf midribs subjected to bending was investigated as an engineering parameter of the product important in design of machines handling tobacco leaves (Suggs *et al.*, 1962). Since the leaf midrib fractures by bending, a bending strain at the critical radius was calculated and the computed values were compared with experimental data. If (See Fig. 5.17)

T = thickness of the beam or midrib
kt = distance from the inside surface of the curved beam to the neutral plane (kt=0.5 means neutral plane lies in the center line of the beam)
S_0 = original length of the beam
S = length after deformation
r = radius of bending to a given fiber
r_0 = radius of curvature of the compression side of the beam (critical radius of curvature)
θ = bending angle

the strain Y is given by

$$Y = \frac{S - S_0}{S_0}$$

Since $S=r\theta$ and $S_0=(r_0+kt)\theta$, the above strain equation can be written in terms of radius r as follows:

$$Y = \frac{r - (r_0 + kt)}{(r_0 + kt)}$$

where (r_0+kt)=radius from the center of curvature to the neutral plane.

If the beam is homogeneous, the outer fibers of the beam, either the ones in tension or the ones in compression, will fail first. A tobacco leaf midrib, being a biological material, cannot be assumed having a neutral plane lying in its center line. Substituting the mean values in the above equation of the experimental data for r_0, T and Y resulted in a value of k=0.84. Values for individual midribs varied from 0.65 for the smallest critical radius to 0.93 for the largest. Whether this non-central location of the neutral plane is to the difference in modulus of elasticity in tension and in compression due to the structural mechanics of the midrib was not established.

Figure 5.17 shows the method for measuring critical radius of curvature by bending the leaf around the outside surface of a right circular cone and gradually sliding it toward the cone apex. The cone radius at the point where midrib fracture occurred was taken as the critical radius. Variations of critical radius with leaf level on the stalk, and variety is also shown in this work. Other factors affecting the critical radius were soil fertility, leaf ripeness, and midrib thickness. Based on statistical analysis of the data a critical radius of curvature of 3 inches was found reasonable for design of leaf handling machines with mechanisms such as belts and rollers to move leaves around small radii.

The lodging resistance of cornstalks was studied by Pickett *et al.* (1968) who made field measurements of both stalk flextural stiffness and resistance to penetration of the rind at the middle of the first internode approximately 6.5 inches above the ground. Stiffness was defined as the ratio of force to the deflection. Resistance to penetration of the cornstalk rind was defined as the ratio of the force applied by means of a conical indenter to the displacement of the indenter. The following relationship between stiffness, S, and resistance to penetration, RP, was established through a fourth-order polynominal regression analysis.

$$S = -247.88 + 7.7656RP - 0.0259RP^2$$

where both S and RP have units of lb/in. It was concluded that both stalk stiffness and resistance to penetration were affected by diameter of the stalk. Therefore, stiffness adjusted for internode diameter is probably the best method for predicting lodging resistance.

In the study of the deflection of tobacco stalks Suggs *et al.* (1962) and Casada *et al.* (1968) have treated the material as a hollow tapered cylinder with the xylem layer providing the structural strength. The former applied the equation for deflection of a cantilever beam to the intact stalk while the latter employed the equation for a simple beam loaded at mid point, [Eq. (5.6)] to the xylem tissue of the stalk. To determine the applicability of the calculated modulus from the deflection tests of the xylem tissue specimens in predicting flextural behavior of the tobacco stalk, Casada *et al.* (1968) suggested the use of the following integral equation.

$$D = \int_0^L \frac{Mm}{EI}\,dx$$

In this equation D is deflection at the center of the beam, L is the length of the beam, M is the bending moment at any point when one pound load is applied at the center of the beam, and E and I are modulus of elasticity and moment of inertia at any point. Using the modulus of elasticity of the xylem and solving the above equation by approximation, these investigators found that the predicted deflections of intact specimens compared closely with the measured deflections (correlation coefficient of 0.84).

Rehkugler and Buchele (1967) have reported on rheological properties of forage as related to wafering. Considering the various parameters which influence the density and expansion of forages compressed in a closed-end die (wafer), they proposed the following equation

$$\varepsilon = \frac{\Delta}{(\sigma/\alpha) + 1} \quad \frac{\sigma - \sigma_b}{K_1} \quad \frac{\sigma}{K_2}$$

In this equation ε indicates the unit strain of wafer expansion, Δ is permanent strain due to consolidation and reorientation of the forage particles, α is the energy capacity of the material, σ is axial pressure, σ_b, cohesive pressure, and K_1 and K_2 are elastic moduli representing elastic properties of the forage. Since they could not directly obtain the coefficients appearing in this equation, a simple bending test and a multiple regression analysis were used to derive an empirical equation to predict forage wafer densities from given physical and mechanical properties of the material.

Bending tests have also been reported for cotton stalks used as hollow cantilever or simple beams supported at two ends (Curtis and Hendrick, 1968). The maximum bending force, F, for cantilever beam stalks for one variety of cotton was found to be a linear function of the mean stalk radius, R, cubed as given below.

$$F = 395.1R^3 + 0.1037 \text{ (Cotton variety Auburn 56)}$$

The values of modulus of elasticity for stalk internodes, found using Eq. (5.6), varied considerably.

5.3 ELASTIC-PLASTIC BEHAVIOR

In Chapter 4 elasticity was defined as the capacity of a material for taking elastic or recoverable deformation. The ratio of the recovered deformation, as soon as the load is brought to zero, to total deformation is referred to as the degree of elasticity. To determine the degree of elasticity in agricultural products, force-deformation curves for loading and unloading cycles have been obtained for cereal grains, fruits and vegetables, eggs, cheese and butter, and some other materials.

As it has been indicated before, none of the biological materials tested so far show perfect elasticity. Regardless of the level of load there always seems to be some residual deformation remaining after the first loading and unloading cycle.

The major part of the residual deformation is due to initial setting which may be caused by the presence of pores or air spaces, weak ruptured cells on the surface, microscopic cracks in brittle materials such as grains, and other discontinuities which may exist in the structure of the material. This can be viewed as an analog to the phenomenon of slip and dislocation in metals due to imperfection in their crystal structure. These defects in the crystal structure are believed to be responsible for plastic or permanent deformation which results from slip, or glide, of one part of the body over the other. Since each additional amount of plastic deformation requires an additional increment of stress, the process of loading and unloading in metals results in a behavior called strain hardening. The ability of metals to strain harden, therefore, depends on their plastic properties which in turn depends on the crystal structure.

Loading and unloading of biological materials for several cycles has also shown the reduction of plastic and residual deformation and strain hardening phenomenon (Fig. 5.18). When the slope of the unloading curves (elastic recovery curves) for the first and the subsequent cycles in Fig. 5.18 are compared, no change in slope can be detected. This means that for calculation of elastic modulus, force and the corresponding elastic deformation, F/D, can be obtained from the first loading and unloading cycle. The F/D values obtained from the second and the subsequent cycles, for that same level of loading, will not be different than that obtained from the first cycle. This observation shows that the modulus is not affected by strain hardening, which is a variable property depending on the factors which will influence the plasticity of the material.

If in the process of loading and unloading, there is a complete cycle resulting in a closed loop, like in the case of rubber, the behavior is called elastic hysteresis. If there are any residual deformations, the behavior is called elasto-plastic hysteresis. In either case, some energy is lost in the process of loading and unloading. This energy loss, referred as hysteresis

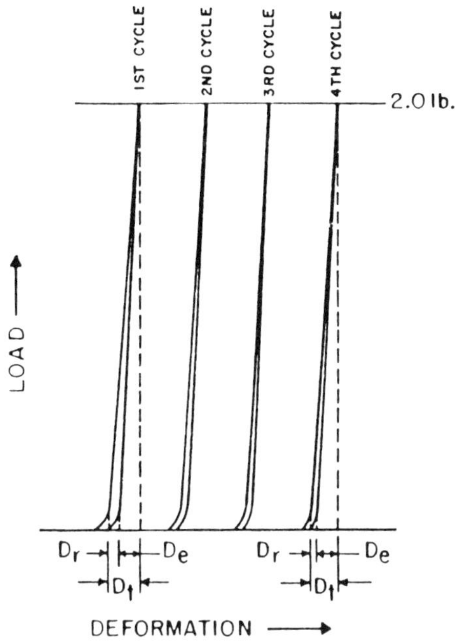

Figure 5.18 Loading-unloading curves for Seneca wheat at 10% m.c. showing equal elastic deformation, D_e, but decreasing residual deformation, D_r, in subsequent cycles (Shelef and Mohsenin, 1966)

loss, is obtained by taking the difference between the work of loading and the work of unloading. The relative amount of hysteresis loss is a measure of elasticity. The closer the material is to being perfectly elastic, the smaller is the hysteresis loss. Since this energy loss is converted mostly to heat, hysteresis loss may also be taken as a measure of temperature rise in the material. In engineering materials, hysteresis loss is also referred to as specific damping capacity, useful in evaluating the material for its ability to damp vibrations.

In the case of corn, Fig. 5.19 shows that the higher the moisture content, the greater was the hysteresis loss. This would be expected because the addition of water increases the plasticity of the grain which in turn will increase the hysteresis loss. Figure 5.20 shows a typical picture of load deformation curves and hysteresis loss for fruits and vegetables.

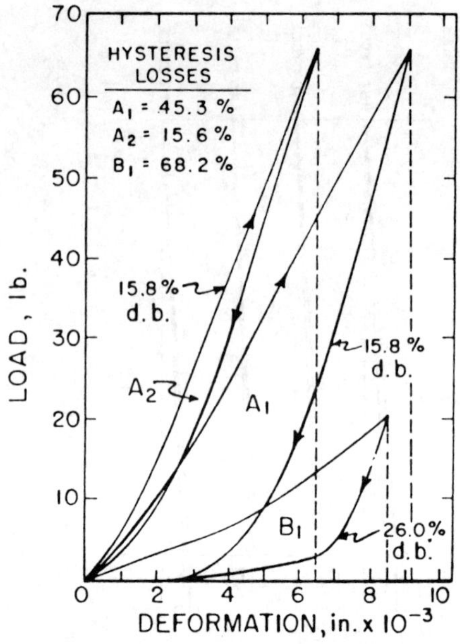

Figure 5.19 Hysteresis loops for yellow dent corn showing strain hardening and the effect of moisture content on hysteresis loss (A_1 and B_1, maiden loading cycles, A_2, second loading cycles) (Zoerb, 1960)

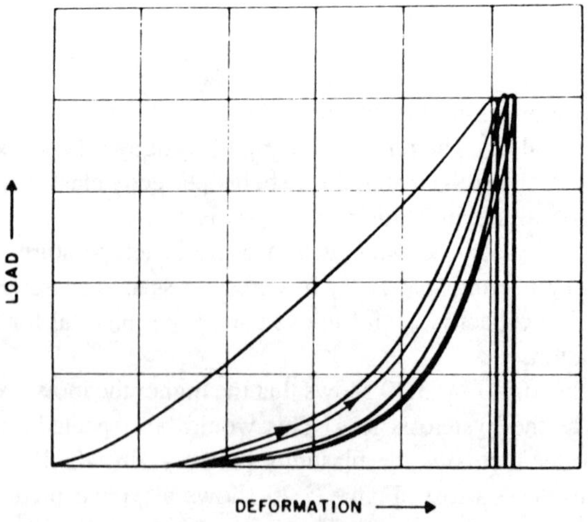

Figure 5.20 Hysteresis curves typical for fruits and vegetables

5.4 EVALUATION OF POISSON'S RATIO

Poisson's ratio as defined in Chapter 4 can be determined experimentally but it is usually calculated from such known relationships as Eq. (4.5) given below

$$E = 3K(1 - 2\mu) \tag{4.5}$$

Using the above relationship, Poisson's ratio, μ, for potato tuber was calculated to be about 0.492 (Finney, 1963). This calculation was based on a mean bulk modulus K=11,300 psi for the whole tuber and a mean uniaxial elastic compression modulus E=543 psi for cylindrical plugs of potato flesh.

The Poisson's ratio for McIntosh apple was calculated by solving Eq. (4.5) above and the Boussinesq equation $E=F/D\dfrac{(1 - \mu 2)}{2a}$ (see Chapter 6) simultaneously for μ. In this case both bulk modulus, K, and elastic modulus, E, were determined for the whole fruit with skin intact. For K, the elastic bulk moduli for each of the specimens subjected to hydrostatic pressure were evaluated by considering that the volumetric strain at a time of 0.08 min after the beginning of application of pressure was entirely elastic. This procedure gave an average elastic bulk modulus of K_0=524 psi. For E_0, elastic moduli for specimens subjected to die loading were calculated in terms of μ by the Boussinesq equation mentioned above, considering the strain at a time of 0.01 min after load application was entirely elastic. This procedure gave an average uniaxial elastic compression modulus of E_0=464 $(1-\mu^2)$. Substituting the values of K_0 and E_0 in for K and E in Eq. (4.5) above, a mean value of μ=0.37 was found for Poisson's ratio. Considering the fact that about 23 percent of volume in an apple is air, it seems logical that the value of μ for apple to be lower than potato. Direct measurements of Poisson's ratio in apples using cylindrical specimens under compression and a measuring microscope (cathetometer) as well as the Nikon Shadowgraph of Fig. 3.4 showed a value of about 0.34 for Poisson's ratio of McIntosh apples.

Chappell and Hamann (1967) have also made direct measurement of Poisson's ratio for apples. They found that Poisson's ratio can be represented by a general form $\mu=at^b$, where t is time and a and b are stress dependent coefficients. This time and stress dependency of Poisson's ratio revealed the non-linear viscoelastic nature of the apple flesh.

5.5 BULK STRESS-STRAIN BEHAVIOR

When a material is subjected to equal compressive stress of magnitude p, bulk deformation is produced which results in change of volume Δv without change in shape. The application of hydrostatic pressure, using water or other fluids, for determination of bulk stress-strain behavior of agricultural products appears quite attractive in that the product can be tested in its natural state. To avoid penetration of the fluid into the material, sometimes minor modifications which will not affect the mechanical properties of the product are needed. For example, in hydrostatic compression of corn kernels illustrated below, no sealing agent could be found to obtain a firm bond with the cutinized outer hull at the tip of the kernel where the fluid penetrated. However, by removing the hull over the germ, tip and adjacent areas, a firm bond was obtained with duPont's Duco Cement.

In additional to volumetric stress (hydrostatic pressure) versus volumetric strain (unit change of volume), sometimes called cubical dilatation, creep and relaxation tests can also be conducted if either pressure or change in volume is kept constant. The slope of volumetric stress versus volumetric strain curve is the bulk modulus, K, which is a measure of incompressibility of the material. Bulk modulus can be determined from equation

$$K = \frac{\Delta p}{\Delta v / v} \tag{5.10}$$

This equation is analogous to the equation for uniaxial compression relating stress to strain. In this a three-dimensional state of stress and strain is considered. The inverse of K is called bulk compliance, B, or compressibility.

$$B = 1/K = \frac{\Delta v}{\Delta p v} \tag{5.11}$$

Bulk creep compliance is presented in Eq. (4.37).

Figure 5.21 shows an apparatus for simultaneous measurement and recording of volume stress versus volume strain behavior of agricultural products. A pressure chamber provided hydrostatic pressure up to 4000 psi. The pressure is obtained by the intrusion of a plunger into a confined fluid. The change in volume is determined by multiplying the plunger cross-sectional area by the displacement of the plunger measured with a linear variable displacement transducer (LVDT). Both the change in volume and the change in pressure, detected with a pressure transducer, are recorded electronically. The loading force on the plunger is from an air motor with speed and stop controls on the stroke. With this apparatus bulk modulus and compressibility, along with creep and recovery parameters of shelled corn have been determined. A silicone fluid (Dow Corning 560 Fluid) was used in the pressure chamber. This fluid was not absorbed by the corn and its wetting ability eliminated the problems of surface tension and air pockets introduced by a fluid like mercury. Stress relaxation parameters can also be determined.

In using the compression chamber along with Eq. (5.10), the change in volume can be considered to be made up of three components, the specimen, the confining fluid, and the expansion of the chamber. To incorporate all these components in the calculation, Eq. (5.10) is modified as follows:

$$\frac{\Delta v}{\Delta p} = B_{sp}V_{sp} + B_fV_f + B_aV_a \tag{5.12}$$

where the subscripts *sp*, *f* and *a* refer to the specimen, fluid and apparatus, respectively. Expansion factor of the apparatus, B_aV_a, can be found by running a test with no specimen in the chamber and using a fluid of known compressibility, B_f. Knowing B_aV_a and the fluid compressibility factor, B_fV_f, compressibility of the specimen can be found from Eq. (5.12) for each change in pressure and volume.

$$B_{sp} = \frac{1}{V_{sp}}\left(\frac{\Delta v}{\Delta p} - B_fV_f - B_aV_a\right) \tag{5.13}$$

The inverse of the instantaneous values of B_{sp} will give the bulk modulus K.

Bulk compression tests have also been used to obtain an estimate of bulk modulus of potato tuber (Finney and Hall, 1967) and to evaluate bulk creep viscoelastic parameters for apples (Morrow and Mohsenin, 1966). Both Finney and Morrow used water as the compression fluid and a bulk compression chamber along with a graduated transparent tubing, similar to the apparatus shown in Fig. 5.22, to measure changes in volume of the specimen under various hydrostatic pressures. Compressed air supplied the pressure for the hydrostatic fluid.

Finney obtained data to plot volumetric stress-strain curve for potato tuber, shown in Fig. 5.23. The tendency of the curve toward the stress axis at higher pressures indicated that potato tuber becomes more incom-

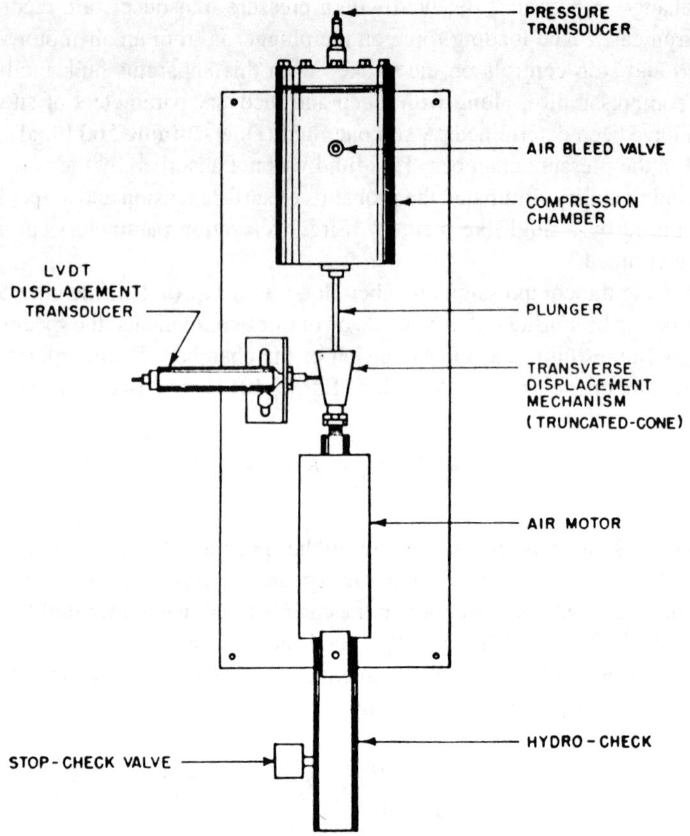

Figure 5.21 Bulk compression apparatus for simultaneous measurement and recording of bulk stress–bulk strain behavior of agricultural products (White and Mohsenin, 1967)

pressible under hydrostatic pressure. Considering the fact that about 85 percent of potato (by weight) is water and incompressibility of water (bulk modulus) is about 300,000 psi, such result is to be expected. Based on his experimental results, Finney reported an average bulk modulus of 11,300 psi, with a coefficient of variation of 15 percent, for mature Kennebec potato tubers (Finney and Hall, 1961).

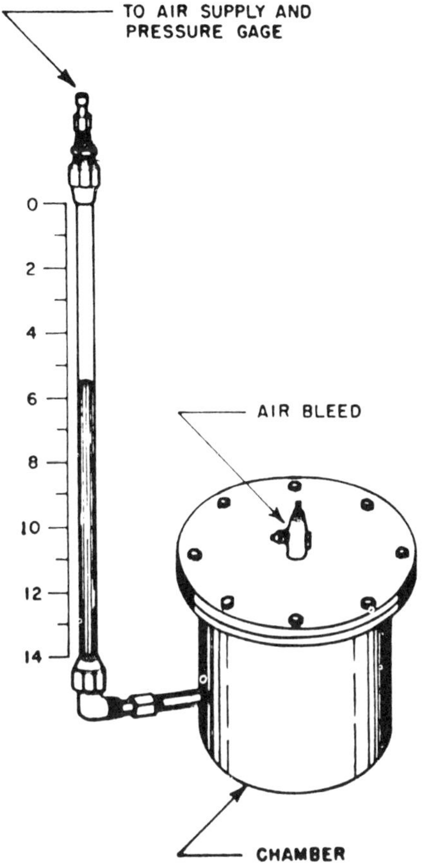

Figure 5.22 Low-pressure bulk compression apparatus (Morrow and Mohsenin, 1966)

In the case of tests on McIntosh apples, Morrow found that at levels of 40 psi and above damage to the fruit tissue resulted. Furthermore, the variation of bulk modulus as hydrostatic pressure increased was insignificant. He held the pressure for a period of 60 minutes and obtained data for calculation of bulk creep compliance, *B*, by observing the change in specimen's volume, at 5-minute time intervals. As will be shown later,

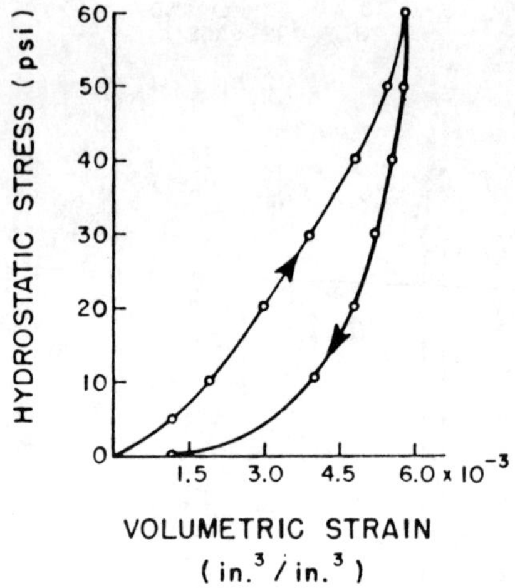

Figure 5.23 Volumetric stress–strain curve for a potato fiber (Kennebec) under hydrostatic pressure (Finney, 1967)

Morrow used his bulk creep compression data to estimate the various viscoelastic parameters needed in deriving the constitutive equation representing the bulk creep behavior of the apple.

The average value of bulk modulus found for McIntosh apple was about 524 psi. This was considerably lower than the value found for potato. This low value of K for apple is believed to be attributed to the presence of about 23 percent of air (by volume) in apple. The amount of intercellular air in potato (about 2%) and from the denser cellular structure and higher specific gravity in potato (1.1 compared to 0.8 for apple), one would expect that the bulk modulus of apple to be much lower than that of potato.

Data on hydrostatic compression of materials have been used as a nondestructive technique to estimate voids and determine such thermal properties as specific heat and bulk coefficient of thermal expansion in materials (Milloway *et al.*, 1961; Surland, 1960). Both of these applications are useful techniques for determination of these physical properties in intact solid food materials.

Internal hydrostatic pressure has been used to determine strength of egg shell (Sluka *et al.*, 1965). The egg was mounted between a rubber seat and rubber cap while water was injected into the egg by a small syringe (Fig. 5.24a). The pressure at which shell failure occurred was recorded and compared with drop test. In this manner pressure was applied equally to all parts of the shell. Hammerle and Mohsenin (1967) who used air for internal hydrostatic pressure in a study of egg shell strength, found that rate of loading influences the results. Accordingly, they suggested the use of a universal testing machine or other means for pressing the syringe at a constant rate of displacement. This technique, where applicable, is preferred to uniaxial die loading or plate loading. In uniaxial testing, variations due to composition, thickness and other physical characteristics of the shell at the loaded point may cause difference in strength values which may not be representative of the whole shell. A high correlation between the hydrostatic test and drop tests reported by the authors supports this idea. Despite this good correlation between impact and hydrostatic pressure test, no significant correlation was found between either of these tests and shell thickness. And yet shell thickness is the measurement used commonly for studying the effccts of temperature, bird's ration and other factors contributing to shell strength.

The authors mention the following formulas (Flugge, 1962) for shell strength in the latitudinal, S_t, and meridional, S_a, directions, respectively (Fig. 5.24b)

$$S_t = \frac{pr_2}{2t}\left(2 - \frac{r_2}{r_1}\right) \tag{5.14}$$

$$S_a = \frac{pr_2}{2t} \tag{5.15}$$

where r_1 is the meridional radius of curvature at the point of maximum latitudinal radius, r_2 is the maximum latitudinal radius, p is the internal hydrostatic pressure, and t is the shell thickness. For a spherical shell $r_1 = r_2 = r$ and

$$S_t = S_a = \frac{pr}{2t} \tag{5.16}$$

For a cylindrical shell r_1 is very large compared to r_2 and $\frac{r_2}{r_2}$ approaches zero. Therefore

$$S_t = \frac{pr_2}{t} \quad \text{and} \quad S_a = \frac{pr_2}{2t}$$

Figure 5.24a Apparatus for application of internal hydrostatic pressure to determine strength of egg shell (Sluka *et al.*, 1965)

For an egg, the values of the stresses lie between S_t for sphere and S_t for cylinder. The exact formula has not been found.

Although the authors mention the shell formula, no attempt was made to express the shell strength in terms of S_t and S_a. Since the relationship for S_t includes the thickness and size, perhaps a significant correlation could be established between shell strength S_t, drop test, and shell thickness.

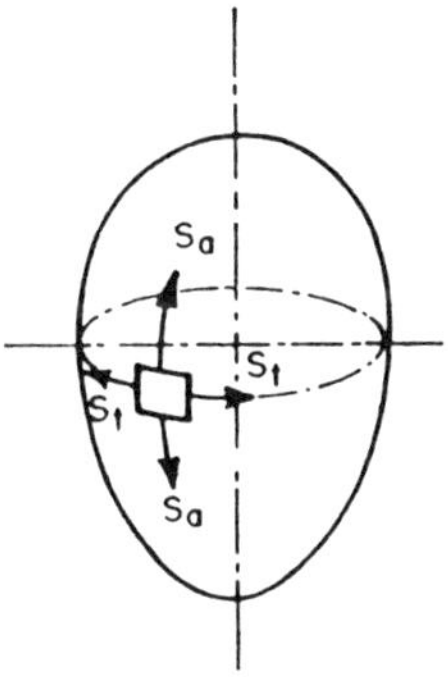

Figure 5.24b Meridional and latitudinal stresses in egg shell (Redrawn from Sluka *et al.*, 1965)

5.6 VISCOELASTIC BEHAVIOR

Mechanical behavior of agricultural products, being time dependent, must logically be studied by applying the principles of rheology and viscoelasticity in which both viscous and elastic responses are taken into consideration. However, this approach has been employed only recently and by a few investigators.

Stress relaxation behavior

Stress relaxation experiments have been conducted on several agricultural products. As discussed before in this rheological test, the material is subjected to a constant strain and the decay of force or stress as a function of time is recorded. Since the deformation of the product under load is held constant, it is usually assumed that the loaded area of contact remains constant during the relaxation test and the recorded force-time curve is representative of the stress-time curve.

Once stress relaxation tests are completed, the data must be analyzed by one or more of the several methods described in Chapter 4.

Figure 5.25 shows the application of the method of successive residuals for analysis of stress relaxation in alfalfa at 40 percent (w.b.) moisture content compressed in a closed end cylinder to 37 psi. The mechanical behavior is approximated by the four-term equation involving four times of relaxation. In this case, if alfalfa is compressed for the purpose of wafering and the plunger displacement is to be held for a short period of time to improve wafer density and stability, we will be interested in the behavior of the material during the hold-times of very short magnitude. Hold-times greater than probably one second will not be of interest if the wafering machine is to be efficient and economical. In this case, the shortest relaxation time (T_4=0.47 second) will be chosen as the rheological parameter to compare to effects of such factors as moisture content, length of cut, cultural factors, variety and other variables which may influence compressibility and expansion characteristics of the forage.

The method of successive residuals has also been used to analyze force relaxation in pea beans at various moisture contents (Zoerb and Hall, 1960), and to derive a rheological model for stress relaxation of potato tuber (Timbers, 1964) and wheat straw (Mustafa *et al.*, 1966). The pea beans were compressed between parallel plates in their natural state in the flat position. Figure 5.26 shows one of the force relaxation curves of pea beans represented by a two-term exponential equation. It is seen that after a time of 1.4 seconds the curve becomes a straight line and a two-term exponential equation is sufficient to represent the decay of stress as a function of time. From a short-time relaxation test of Fig. 5.26 it is not apparent, however, whether or not the stress within the pea bean will approach zero as the time will approach infinity. If the long-time test showed that stress did not level off and approached zero with time, the rheological model representing behavior would consist of only two Maxwell models in parallel corresponding to the two-term exponential equation. No elastic element is needed to go in parallel with the Maxwell units to represent equilibrium stress, as explained under the discussion of generalized Maxwell model.

Stress relaxation of potato tuber by Timbers showed that force relaxation curve can be represented by a three-term exponential equation up to a period of ten minutes. The relaxation curves matched closely those obtained by another investigator (Finney, 1964) up to a ten-minute period. Beyond this time, a fourth-term exponential, as proposed by Finney, was needed to explain the experimental curve. The generalized Maxwell model proposed by Timbers to represent qualitatively the relaxation behavior of potato contained a dashpot in parallel and a spring in series with Maxwell units. No attempt was made to fit this proposed model to the relaxation curves obtained experimentally. On the other hand, the rheological model proposed by Finney for stress relaxation in potato consisted of an undetermined num-

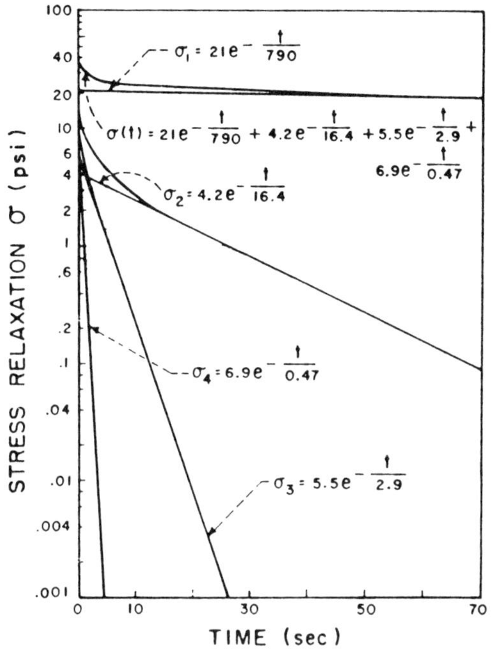

Figure 5.25 Stress relaxation in compressed alfalfa at 40% moisture content (w.b.) represented by a four-term exponential function

ber of Maxwell models in parallel. This agreed with the experimental curves obtained by this investigator. The stress in these curves did not level off but continued to decrease with time.

The stress relaxation of wheat straw showed the same type of behavior as illustrated in Fig. 5.26 but the authors propose a generalized Maxwell model, consisting of two simple Maxwell units and a single dashpot in parallel to represent the behavior. The addition of the dashpot to the Maxwell units was not explained.

In the work on pea beans, the importance of the rate of deformation before relaxation begins was well demonstrated. It was found that the initial rate of deformation had more effect on the rate of stress relaxation than moisture content or the initial amount of deformation. Although the effect of rate of deformation on stress relaxation was not investigated for the same initial stress, the relaxation times were reported to be essentially

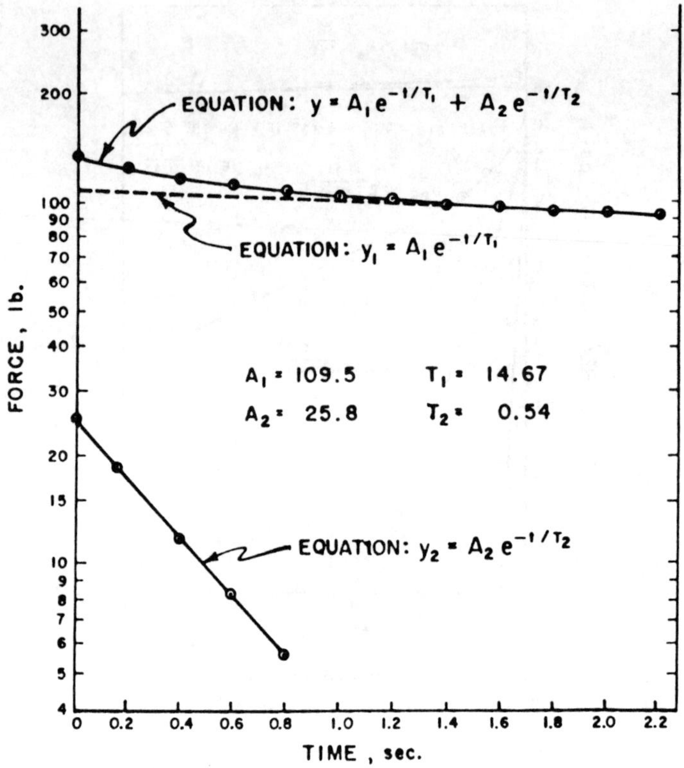

Figure 5.26 Force-relaxation for pea beans at 18.5% moisture content and 45.5% deformation represented by a two-term exponential equation (Zoerb and Hall, 1961)

constant with various amounts of deformation. Some of these relationships can be seen in the following table.

For analysis of stress relaxation data the sum of exponential terms can be considered to be a linear equation, defining force or stress at any given time. Having obtained the stress relaxation data, the best fit can then be found following a procedure for finding the linear equation of best fit to a number of points. This method has been employed to evaluate viscoelastic parameters for McIntosh apples under axial compression relaxation test (Morrow and Mohsenin, 1966). Whereas other investigators had so far presented the results of their relaxation tests in terms of force relaxation against

Table 5.4 Effect of rate of deformation on relaxation times of pea beans at 32.8% moisture (Zoerb and Hall, 1960)

Speed	Per cent average deformation	Time of deformation (sec)	Relaxation time (sec)	
			T_1	T_2
Slow	44.2	71.0	17.5	0.55
Medium	43.7	20.7	15.3	0.43
Medium	36.0	16.6	16.7	0.44
Fast	44.6	11.8	10.4	0.33

time, Morrow calculated $E(t)$, the relaxation modulus, as a function of time, using the Boussinesq method for die loading and the Hertz method for plate loading (See Chapter 6). For deriving the equation representing the relaxation behavior of the specimens the following procedure was followed:

1 Relaxation moduli $E(t)$ were calculated from experimental force-relaxation data at selected intervals of time.
2 That value of $E(t)$ which did not change with longer time was taken as the equilibrium modulus E_e.
3 A simple Maxwell model with a spring in parallel (to represent the equilibrium modulus) represented by the following equation was selected as the mechanical analog for the relaxation behavior.

$$E(t) = E_d e^{-t/T_{\text{rel}}} + E_e$$

4 The quantity ln $[E(t)-E_e]$ was plotted against time.
5 Best-fitting straight line for the above semi-log relationship was found using the method of least square.
6 Time of relaxation, T_{rel}, was found from slope of the straight line and the decay modulus, E_d, was determined from the ordinate intercept.
7 Substituting the known quantities in the equation of the assumed mechanical model resulted in

$$E(t) = 104 e^{-t/56.6} + 134$$

for die loading and

$$E(t) = 213e^{-t/52.5} + 200$$

for plate loading.

It should be noted that the above equations were obtained from the mean values of each of the parameters and may not be representative of any particular specimen. As it has been mentioned before, there should have been no differences between the viscoelastic behavior of specimens subjected to plate loading and die loading. The differences noted in the two above equations are partly due to variations in the methods of calculating relaxation modulus, $E(t)$. The method of die loading and the application of Boussinesq equation always gives lower values than the method of point loading and the application of Hertz equation (See discussion in Chapter 6).

Having the equation representing the relaxation behavior, values of relaxation modulus at various time increments were calculated. When these calculated values were compared with the experimental values, the agreement, as seen from Table 5.5, was rather poor for the initial portion of the curve, but improved considerably as time increased. Data in Table 5.5 are for one single specimen. Note that in this specimen $E(t)$ approached equilibrium modulus, E_e, after about 90 min and remained at that level for the remainder of the time.

The above illustrated example shows that the method of best-fitting straight line is not suitable if we are interested in short-time relaxations, which concern the initial portion of the relaxation curve. The use of this method for analysis of relaxation data, particularly if the rheological model contains only one Maxwell unit, may result in considerable difference between experimental and theoretical values. This difference may be reduced by using a generalized Maxwell model containing a larger number of Maxwell units. In that case, the mathematical manipulation becomes so involved that the choice of this method for analysis of data becomes questionable. However, if viscoelastic parameters at long times are to be considered, this method is quite simple and useful.

The use of stress relaxation technique is also reported for investigation of the viscoelastic properties of storage tissues in potato, apple and pear (Somers, 1965). In this work, 1-mm thick slices of potato tubers were sized to 1 cm by 4 cm specimens and held in pneumatic jaws of an Instron testing machine for tension and then stress relaxation. Cubes of ripe apple and pear tissues (0.5 cm on each side) were also subjected to compression

Table 5.5 Compression relaxation modulus, $E(t)$, versus time for McIntosh apples subjected to constant deformation and loaded by a $^1/_4$-inch rigid cylindrical die (Morrow, 1965)

Time	Relaxation modulus $E(t)$ psi	
(min)	Experimental	Theoretical
0	122.2	111.8
1	101.3	107.0
2	96.6	102.5
3	92.6	98.3
4	90.4	94.4
5	—	90.8
10	76.7	76.1
15	68.5	65.9
20	62.2	58.6
30	51.6	50.1
40	45.0	45.9
50	41.8	43.8
60	41.8	42.8
70	41.8	42.3
80	41.8	42.0

and stress relaxation. When the specimen relaxed to 20 percent of its initial stress, the load was removed and then reloaded to a second relaxation. This cyclic loading-unloading was repeated several times. The results were reported as time required for 20 percent relaxation and apparent elastic modulus, taken from various parts of the stress-strain curves. It was found that treatments which wilted the potato slices and cubes resulted in shorter time for 20 percent relaxation of stress while treatments which produced turgid specimens resulted in longer time for the specified relaxation. In potato short relaxation time was associated with wilting, dehydration and freezing. With fully ripe pears and apples, short relaxation times were associated with a tendency to disintegrate on soaking for three hours in water.

As to the phenomenon of stress relaxation in biological materials, Somers states that in a heterogeneous system such as plant cell wall stress, relaxation "may result from a variety of factors such as slippage of cellulose microfibrils through the amorphous matrix of the cell wall, from a flow of the matrix, from molecular rearrangements of various polymeric cell wall constituents, especially cellulose, or various combinations of such factors." He further states that, since the elastic part of the cell wall does not show relaxation, the differences in relaxation time with various treatments may reflect properties of the amorphous matrix of the cell wall. These observations, discussed later, should prove to be valuable in understanding such textural factors as turgidity in fresh produce and other food materials.

Creep behavior

Creep experiments have been reported for several agricultural products by having the material subjected to dead loads and observing deformation as a function of time. Plugs of cheese 3 cm by 1.7 c in diameter were placed under dead loads for 60 seconds to obtain creep and recovery data (Davis, 1937). Deformation values plotted against time gave creep and recovery curves similar to Fig. 5.27. The mechanical behavior was represented by a model consisting of two coil springs, imbedded in vaseline and joined together with screw clips (Fig. 5.28). In this model, extension of the springs due to external force results in recoverable deformation of the springs plus the slight "give" at the screw clips, which represents the unrecoverable deformation. The vaseline introduces a viscous effect resulting in elastic aftereffect, which we have called delayed or retarded elasticity, E_d. In order to obtain a hysteresis effect, the analogy of having sand or grit in the vaseline was given. By getting between the coils of the spring, the grit prevents the attainment of a true equilibrium just as a rusty spring fails to return to its initial position after an extension.

From his creep and recovery curves, Davis obtains a number of rheological properties including shear stress, viscosity, time of relaxation, etc. Since the calculation of some of these parameters are based on definitions not acceptable in mechanics, they will not be discussed here. However, he attempted to use some of the elastic and plastic properties obtained from his creep experiments to explain such terms as "springiness," "hardness," "body" and other textural attributes used by graders in evaluating the cheese quality. This will be discussed later under texture evaluation.

In a study of rheological properties of wheat grain, continuous deformation of single grains of wheat with time has been investigated (Shpolyanskaya, 1952). Taking the straight portion of the deformation-time curves, a coefficient of viscosity was calculated for the wheat grain. This portion of the creep curve can be represented by the following equation which can

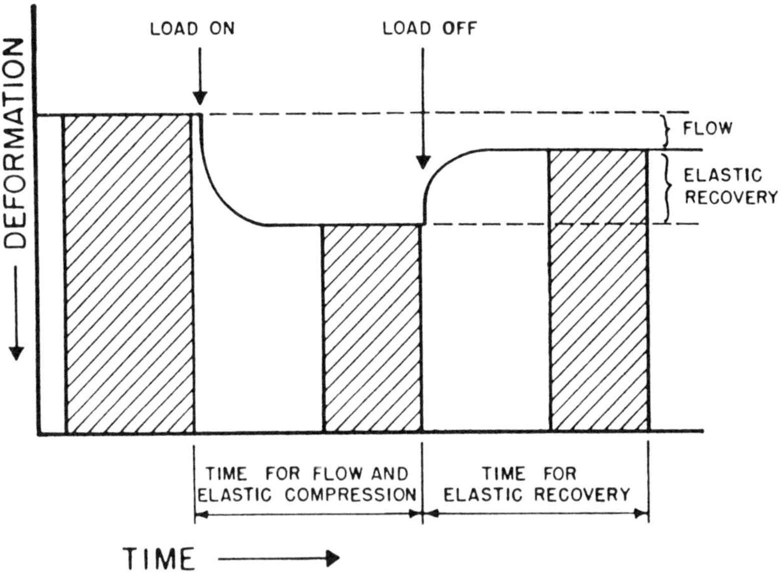

Figure 5.27 Creep and recovery in plugs of cheese (Davis, 1937)

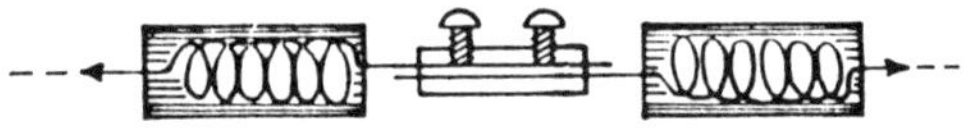

Figure 5.28 Mechanical analog of creep and recovery of cheese illustrated in Fig. 4.30. Coiled springs imbedded in vaseline and joined by screw clips (Davis, 1937)

be written from consideration of Figs. 4.24, 4.26 and the related discussion in Chapter 4.

$$D = a + \frac{F_0}{\eta} t$$

Where D=total deformation, a is the ordinate intercept, F_0 is the dead load, and t is the time. Solving the above equation for η

$$\eta = \frac{F_0 t}{D - a} \tag{5.17}$$

The units of η from the above equation are (lb-sec)/in instead of (lb-sec)/in^2, for Newtonian viscosity. This is because the deformation was expressed in terms of in and not in terms of strain in in/in. For this reason, the viscosity calculated by this equation was referred to as "specific viscosity." Table 5.6 shows the specific viscosity and deformation of the grain under applied load for Lyutestsens 62 wheat.

The variation among the individual grains was reported as the reason for variation in calculated viscosity. The increase of deformation was also reported being basically due to increase of residual deformation with load. Shpolyanskaya also proposed a mechanical model for the wheat grain shown in Fig. 5.29. This is basically the 4-element model with a telescopic mechanism added to represent the plastic element producing residual or unrecoverable deformation. In this model, combination of elastic and plastic elements in series represents the total deformation characterized by deformability modulus discussed in Chapter 6.

Attempts to find a mechanical model and the corresponding equation to fit the experimental uniaxial creep data of apples under dead load resulted in a 4-element Burgers model (see Fig. 4.18) and the following rheological equation (Mohsenin *et al.*, 1963).

$$F + (1/T_{rel}) \int F dt = K(L + T_{ret} \dot{L}) \tag{5.18}$$

where F is the applied dead load on the fruit, L is the deformation, and K is the constant representing the elasticity in the mechanical model. The creep and recovery curve was similar to that shown in Fig. 4.27. This general trend in creep and recovery was also observed in potato tubers subjected to dead loads and then removing the load (Timbers, 1964).

The creep work on apples was carried out using bulk creep compression and uniaxial creep compression with a rigid die or parallel plates (Figs. 5.22 and 5.30). From the bulk creep tests a bulk modulus was calculated for the apples and the inverse of this modulus, bulk compliance, was used in Eq. (4.37) to find if a 4-element rheological model can be used to represent the bulk creep behavior of the fruit. This equation

$$B(t) = B_0 + B_r (1 - e^{-t/T_{ret}})$$

is written in logarithmic form,

Table 5.6 "Specific viscosity" and deformation of Lyutestsens 62 wheat at 11.7 percent moisture calculated from creep curves (Shpolyanskaya, 1952)

Experiment No.	Load (kg)	Deformation (% per kg)	Viscosity $\left(\frac{\text{kg — min}}{\text{mm}}\right)$
1	1.0	0.75	25000
2	3.0	0.75	21400
3	4.2	0.81	26300
4	5.2	1.00	32500
5	6.3	1.24	22500
6	7.0	1.42	18400
Mean	—	—	24400

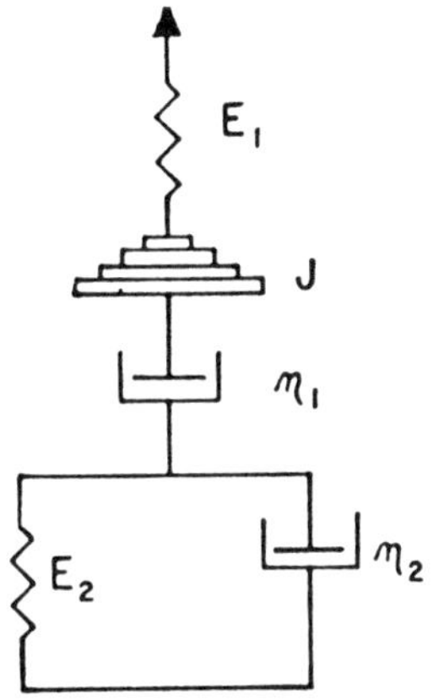

Figure 5.29 One mechanical analog for creep behavior proposed for wheat grain under dead load (Shpolyanskaya, 1952)

$$\ln [B_0 + B_r - B(t)] = \ln B_r - t/T_{\text{ret}} \qquad (5.19)$$

This is the equation of a straight line which can be established from the experimental data. The quantity B_0 is the elastic bulk compliance taken as the value at a time of 0.08 min. At equilibrium $B(t)=B_r+B_0$, from which it was possible to calculate B_r. Knowing B_0, B_r and $B(t)$ at various times,

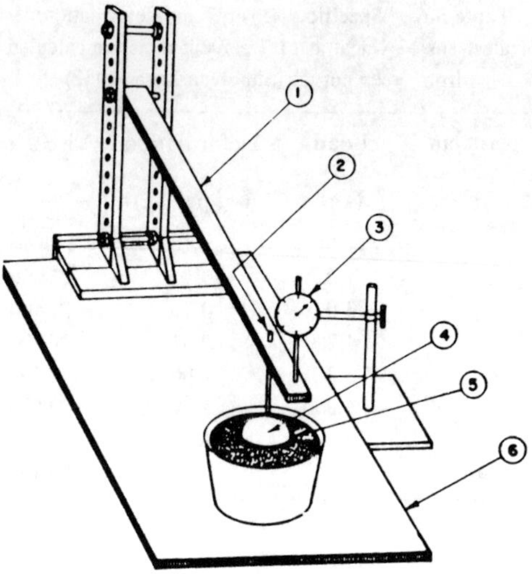

Figure 5.30 Apparatus for uniaxial creep test of fruits supported by quick-setting plaster mold (Morrow and Mohsenin, 1966)

other quantities in Eq. (5.19) were calculated. Then the method of best fitting straight line was employed to determine and compare the various bulk creep viscoelastic parameters as discussed for the case of stress relaxation. The mean values of these parameters for 20 bulk compression tests, computed assuming a 4-element Burgers model represented by Eq. (4.37), are given below:

P_0 psi	K_0 psi	K_r psi	$B_0 \times 10^{-3}$ 1/psi	$B_r \times 10^{-3}$ 1/psi	T_{ret} min	$\eta \times 10^{-3}$ psi-min	η_v psi-min
25	510	3225	1.96	0.31	30.8	99.6	infinity

The hydrostatic pressure P_0 is the mean of 20 and 30 psi. No significant difference in the viscoelastic parameters was observed for these two levels of pressure. The value of η, viscosity in the Kelvin element of the Burgers model, was calculated from the relationship $\eta/K_r = T_{ret}$.

Axial compression creep tests on these apples were analyzed using the same technique described for bulk creep compression and stress relaxation. In this case, Eq. (4.36)

$$D(t) = D_0 + D_r\,(1 - e^{-t/T_{\text{ret}}}) + t/\eta_v$$

was taken as the representative of deformation-time behavior following the 4-element Burgers rheological model. Utilizing the Boussinesq and Hertz methods and taking the calculated value of 0.37 for μ, it was possible to determine the various values of axial creep compliance, $D(t)$, by considering deformations for various time intervals. In Eq. (4.36) the quantity D_0 is the elastic compliance and was taken at a time of 0.01 min. Examination of the plots of $D(t)$ versus time showed a linear relation for times greater than thirty minutes. By considering the portion of the curve between thirty and sixty minutes, it was possible to evaluate η_v. This was done by using Eq. (4.36) for thirty minutes and sixty minutes and assuming that after thirty minutes $e^{-t/T_{\text{ret}}}$ had become approximately equal to zero. This yields

$$D(60) - D(30) = 30/\eta_v$$

Knowing η_v, the retarded compliance, D_r, was found from Eq. (4.36) for time of 60 min where the quantity $e^{-t/T_{\text{ret}}}$ approaches zero. This yields

$$D(60) = D_0 + D_r + 60/\eta_v$$

In the case where recovery data are available and compliance approaches an equilibrium value t/η_v according to Eq. (4.36), η_v can be calculated from the permanent deformation (see Fig. 4.24). Knowing $D(t)$ *and* η_v for the straight portion of the curve, the retarded compliance was calculated again using Eq. (4.36). These viscoelastic parameters were then used to derive the best fitting straight line to fit the 4-element behavior. The viscoelastic parameters computed in this manner for the uniaxial tests using a rigid die or parallel plates are shown below. The values are the mean of 20 tests using applied loads of 1, 2, 3, and 4 pounds. Like the case of hydrostatic pressures, no significant difference in viscoelastic parameters was observed for these loading levels accompanied by small deformations.

	F_0 lb	E_0 psi	E_r psi	$D_0 \times 10^{-3}$ 1/psi	$D_r \times 10^{-3}$ 1/psi	T_{ret} min	$\eta_v \times 10^{-3}$ psi-min
Rigid die	2.5	394*	599*	2.54	1.67	6.61	98
Flat die	2.5	781*	1298*	1.28	0.77	6.57	492

Having the various viscoelastic parameters, equations 4.36 and 4.37 were used to predict bulk and uniaxial compliances as given by the 4-element Burgers model. Table 5.7 shows the comparison of predicted and experimental compliances at various time intervals. This comparison shows large difference during the first portion of the curve. This difference, as mentioned in the case of stress relaxation, could be reduced by using a generalized Kelvin model containing a larger number of elements. If recovery data are available, the retardation time, T_{ret}, may also be obtained from the slope of logarithm of retarded recovery part, $\log D_r(1-e^{-t/T_{ret}})$ plotted against time.

Clevenger and Hamann (1968) have reported on stress relaxation and creep of apple skins under tensile loading. Rectangular sections of skins 0.098-in. × 1.060-in in size with flesh scraped off were placed under tension until failure. From the stress-strain curves a secant modulus was determined at selected strain intervals. It was found that the skin of Golden Delicious was much weaker in tension than that of the Red Delicious even though both of these varieties have about the same initial elastic moduli. The explanation offered for this difference was the presence of large amounts of breaks and cracks in the cuticle which could cause isolated weaknesses. The skin of Winesap variety was slightly stronger than the Golden Delicious in tensile strength but had a lower initial elastic modulus. All the skins tested were anisotropic with the tensile strength greater parallel to the core. Stress relaxation and creep tests confirmed the viscoelastic nature of the apple skin with the four-element model representing the creep behavior and the Burgers model with the series dashpot removed representing the relaxation behavior. The range of values for tensile strength and elastic modulus of the varieties tested are given in the Appendix.

Dynamic viscoelastic behavior

Applications of resonance frequency to biological materials are reported by plant physiologists for determination of rigidity in plant tissues and es-

* For explanation of differences between moduli values by rigid die and flat plate see Fig. 6.14 in Chapter 6.

Table 5.7 Comparison of experimental and theoretical values of mean creep compliance *versus* time relations for McIntosh apples as predicted by a 4-element Burgers model (From Morrow, 1965)

Time (min)	Bulk deformation (at 20 psi)		Uniaxial deformation			
			Rigid die		Flat plate	
	$B_1(t)$ $\left(\times 10^3 \frac{1}{\text{psi}}\right)$		$D_1(t)$ $\left(\times 10^3 \frac{1}{\text{psi}}\right)$		$D_1(t)$ $\left(\times 10^3 \frac{1}{\text{psi}}\right)$	
	Pred.	Exp.	Pred.	Exp.	Pred.	Exp.
0	1.96	1.96	2.54	2.54	1.28	1.23
1	1.97	2.11	2.78	3.50	1.39	1.74
2	1.99	2.16	3.00	3.69	1.49	1.81
3	2.00	2.17	3.18	3.80	1.57	1.85
4	2.01	2.18	3.34	3.88	1.64	1.88
5	2.02	—	3.48	—	1.70	—
10	2.08	2.22	3.94	4.17	1.90	2.00
15	2.12	2.25	4.19	4.31	2.00	2.05
20	2.16	2.27	4.33	4.42	2.05	2.09
25	2.19	2.29	4.43	—	2.08	—
30	2.22	2.30	4.50	4.62	2.10	2.14
35	2.24	—	4.56	—	2.12	—
40	2.26	2.32	4.61	4.80	2.13	2.19
45	2.28	—	4.67	—	2.14	—
50	2.30	2.34	4.72	4.95	2.15	2.22
55	2.31	—	4.77	—	2.16	—
60	2.32	2.45	4.82	5.05	2.17	2.24

timation of turgor pressure. Figure 5.31 shows a simple apparatus for observing the vibration amplitude of a specially mounted specimen of plant tissue brought to resonance by an electromagnet and an oscillator. A small piece of de-tempered steel is mounted at the free end of the specimen for attraction by the electromagnet. As rigidity of plant tissues is altered by changes in water content and turgor pressure, these changes can be measured by determining the resonance frequency.

From the resonance frequency obtained by the above method, attempts were made (Falk *et al.*, 1958) to determine modulus of elasticity of the tissue using the formula for a rod clamped at one end and vibrating in vacuum. For any given specimen of a given material, the basic Eq. (4.49) given in Chapter 4 may be used. It should be noted that the addition of an additional

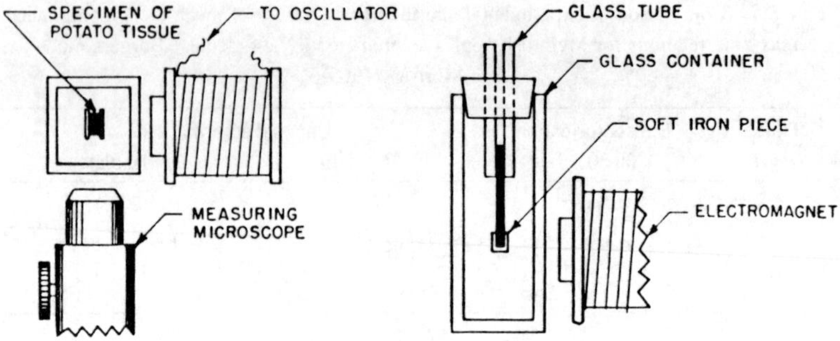

Figure 5.31 Apparatus for measurement of resonance frequency of potato tissue (Virgin, 1955)

mass to the free end of the specimen decreases the resonance frequency by increasing the effective mass, and Eq. (4.49) cannot be used for exact determination of E from the resonance frequency. However, it is possible to obtain an approximation of the modulus if the additional mass is small compared with the mass of the specimen.

This technique was later modified and used to obtain resonance curves of specially prepared (6×12×50 mm) specimens of food materials (Drake, 1962). These cantilever beams of food material, fixed at one end, were subjected to forced vibrations transverse to the longitudinal axis by two electromagnets acting on a small piece of iron inserted into the specimen. The amplitude of vibration was determined with a lamp-phototube-amplifier arrangement. From the response curves recorded by an x-y recorder (Fig. 5.32), damping factor and modulus of elasticity were calculated using the following equation also given in Fig. 4.34a.

$$\tan \delta = \frac{\Delta \omega_{0.5}}{\omega_r \sqrt{3}} \tag{5.20}$$

or

$$\tan \delta = \frac{\Delta \omega_{0.707}}{\omega_r} \tag{5.21}$$

where δ is the phase angle discussed in Chapter 4, tan δ is the damping factor, 0.5 and 0.707 are subscripts denoting change in frequencies at which the amplitude is either 0.5 or 0.707 its maximum value, respectively, and ω_r is the resonance frequency (Ferry, 1961). The resonant curves recorded for several food materials are shown in Fig. 5.32. As shown here, various food materials exhibit different resonant curves and damping characteristics under forced vibration.

Having the resonance frequency from resonant curves, modulus of elasticity of the material was determined from Eq. (4.50). In measuring the resonance frequency of several food materials, Drake found that the frequency of each material changed with time elasped from the moment of cutting. Since this change for some food materials is apparently considerable, the change in calculated modulus will be even greater.

The modulus calculated by this method of resonance vibrations is the real part, E', of the complex modulus, E^*, discussed in Chapter 4. The imaginary part, E'', of the complex modulus can be obtained from

$$E'' = E' \tan \delta$$

where $\tan \delta$ is calculated by either of the two equations given above. It should be noted that the equations given for $\tan \delta$ are valid only when $\tan \delta \leqq 1$. For higher damping factors, which are not uncommon among soft biological tissues, a more complicated relationship has been used.

Finney (1966) employed the resonance technique to evaluate a dynamic elastic modulus for tissues of potato tubers and apples. Following the procedure described by Spinner and Tefft (1961), a speaker driver was used as the exciter at one end of a cylindrical specimen of the tissue and a phonograph pickup transducer at the other end. Having the resonant curve,

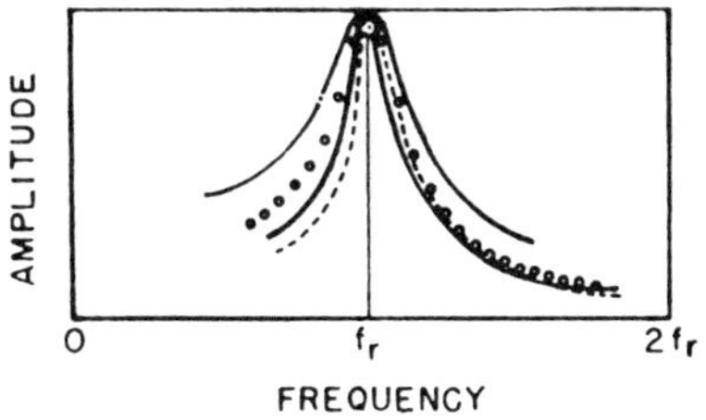

Figure 5.32 Resonant curves recorded for food materials. From the right: cheese, theoretical values for δ=0.2, potato, pear (Drake, 1962)

Eq. (5.20) was used to calculate damping factor. The following equation, which is another form of Eq. (4.50), was used for calculating the modulus of elasticity of the cylindrical specimen under longitudinal vibration.

$$E = 4(\rho f_r^2 L^2) \tag{5.22}$$

The units are the same as those in Eq. (4.50). In the above equation it is assumed that the wave length of sound, λ, is long compared to the cylinder diameter. If the wave length is not long compared to the diameter, Spinner and Tefft (1961) give a correction factor in terms of Poisson's ratio and the diameter to wave length ratio. In longitudinal vibration the axis of the cylinder remains stationary while the end faces of the specimen vibrate in the direction parallel to the axis of the cylinder, resulting in alternating compression and tension. Having the resonance frequency the damping factor can also be determined using the method of half-amplitude width band illustrated in Fig. 4.34a. As suggested by Finney, the dynamic modulus E can be taken as an indication of tissue rigidity in food materials while damping factor is an indication of internal friction attributed to fibrousness or mealiness. If the technique could be used on the whole fruit without cutting it, it would provide a very useful technique for evaluating the internal friction nondestructively.

The relationship between shear modulus, G, and the resonance frequency f_r of the cylindrical specimen resonating in torsion is also given by 5.22 except that G will be substituted for E. For torsional experiment Pickett (1949) has found that if a radius-to-length ratio of 0.10 is maintained, the experimental difficulty of exciting torsional resonance within the specimen will be minimized.

For longitudinal resonance, Finney and Norris (1968) had the cylindrical specimen suspended horizontally near its center within two thread loops with one end excited with a speaker driver. The transmitted energy was then picked up at the other end with a phonographic pickup cartridge. For torsional resonance these investigators introduced the vibration energy to the suspended cylindrical specimen through one suspension thread using a record-cutting head. The transmitted vibrational energy to the other thread was then picked up with a phonographic detector. A mechanical coupling between the frequency generator and a graphic-level recorder provided a record of frequency versus response which identified the resonance frequencies (Fig. 5.34).

In an attempt to apply the resonant method to the intact fruit, Nametre Company made use of sonic energy to measure the inner texture of apples

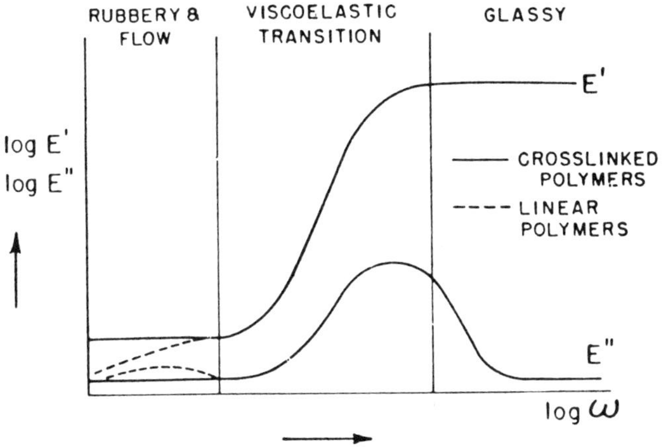

Figure 5.33 Variation of storage and loss moduli with frequency (Redrawn from Sharma, 1965)

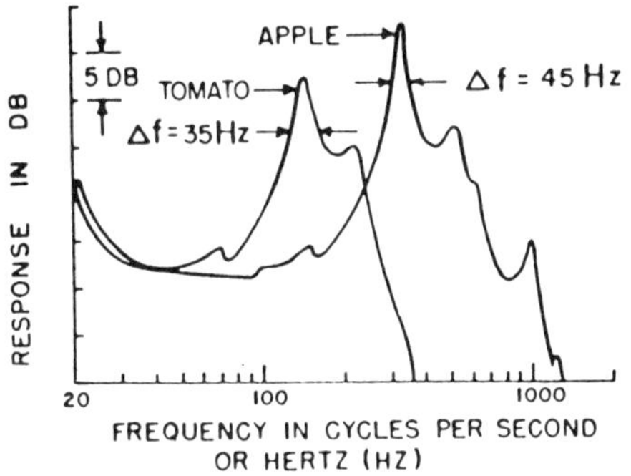

Figure 5.34 Resonant curves for a 185-gm apple and a 200-gm tomato (Finney and Norris, 1968)

and other fruits and vegetables (Abbott *et al.*, 1967). This work, which was done under contract with the U.S. Department of Agriculture, employed resonant technique both to cylindrical sections cut from apples using flextural vibration as well as to whole apples. Figure 5.35 shows the schematic diagram for both applications. A series of resonant curves was obtained for each test specimen to determine which mode of vibration could be correlated to given qualities of the fruit. The square of frequency of these resonances multiplied by the mass of the apple, designated by f^2m, was termed the "stiffness coefficient" of the apple. This is expected because as shown by Eq. (4.49), the modulus of E is directly proportional to f^2 and a constant K_2 which is dependent on the specimen's geometry and density. Due to the complexity of the problem it is difficult to determine the exact form of the relationship. However, some relative measurements can be made using the quantity f^2m as the evaluating parameter. With this in mind, it was reported that the position of the second mode was influenced by change in certain characteristics of the apple. This made it possible to correlate the "stiffness modulus" with the size, Young's modulus, harvest dates, Magness-Taylor pressure test, and soluble solids of the fruit. Also, it was reported that the frequencies of the various modes did not change with temperature in the range of 0° to 40°C.

Example

Cylindrical specimens of Red Delicious apples 5 cm long by 1 cm diameter were subjected to flextural vibration using the experimental arrangement illustrated in Fig. 5.35. The recorded resonance curves showed a resonance frequency f_r=29.5 cps. The frequencies read at half-amplitude band width were 31.5 and 27.5 cps. If density of the apple flesh is 0.8 g/cm³ find the storage and loss moduli and the damping factor for this specimen.

From Eq. (4.50)

$$E' = \frac{38.3\,(0.8)\,(5)^4\,(29.5)^2}{(1)^2}$$

$$= 16.6 \times 10^6 \text{ dynes/cm}^2 \text{ or } \quad 244 \text{ psi}$$

$$\tan\delta = \frac{\Delta f_{0.5}}{f_r\sqrt{3}} = \frac{31.5 - 27.5}{29.5\,(\sqrt{3})}$$

$$= 0.078$$

$$E'' = E' \tan\delta = 16.6 \times 10^6 \times 0.078$$

$$= 1.29 \times 10^6 \quad \text{or} \quad 18.7 \text{ psi}$$

Finney and Norris (1968) have continued the research on the application of resonant technique to intact fruits and vegetables. Their method of trans-

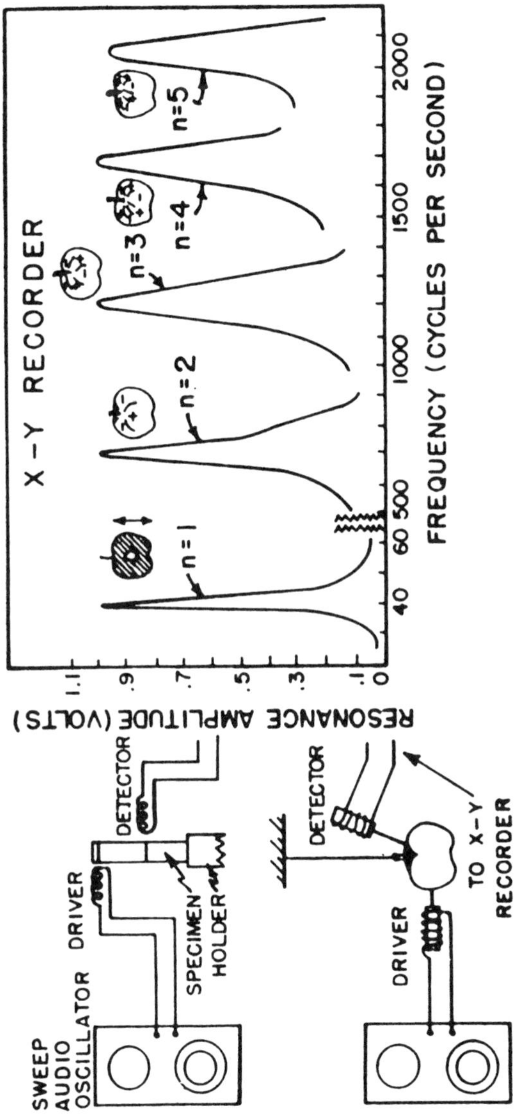

Figure 5.35 Application of resonance vibration to a whole fruit and cylindrical specimen of the fruit flesh (Abbott *et al.*, 1967)

mitting and receiving sound energy through the whole intact fruit is illustrated in Fig. 5.36. More research is needed to clarify the significance of the resonance curves as related to fruit size, orientation, shape as well as internal qualities. The relationship between the dynamic mechanical properties of fruits as found by this method and subjective evaluation of texture is discussed in Chapter 7.

An application of resonance frequency, elastic modulus, and damping characteristics of fresh fruits, for example, was reported in a study of vibration bruising of fruits in transit (O'Brien, 1965). To investigate the reasons for in-transit injury of various fruits due to vibration attempts were made to determine the natural frequencies of the fruits and relate them to the vibration characteristics of the transport vehicles. The natural frequencies of the fruits were determined using the equations for longitudinal vibration of a bar with one end fixed.

$$f_n = \frac{1}{4\lambda}\left(\frac{Eg}{\rho}\right)^{1/2} \tag{5.23}$$

where f_n is the natural frequency or resonant frequency of the fruit in cps, λ is the depth of fruit column in inches, ρ is density in lb/in^3, E is elastic modulus in lb/in^2, and g is 386 in/sec^2. Using this equation, natural frequencies of various fruits were obtained for various depth in the shipping container as illustrated in Fig. 5.37. As seen, the natural frequencies of the fruits tested vary from 8 to 52 cps. depending on the kind of fruit and fruit depth. The vibration frequencies of fruit transport trucks and freight cars

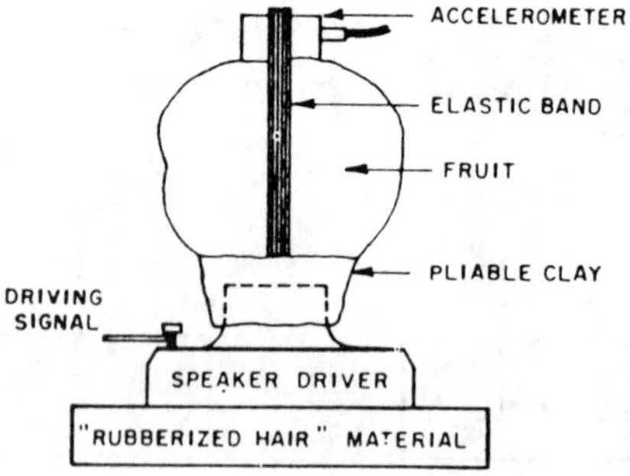

Figure 5.36 Experimental arrangement for exciting and measuring resonance within a whole intact fruit (Finney and Norris, 1968)

ranged from 2.5 to 20 cps with sufficient amplitudes to cause continuous accelerations above 0.2 g with peaks over 1.0 g. If the natural frequency for a given kind of fruit at a given depth falls in the middle of the range of that of the transport vehicle, resonant vibration is expected to occur. It was noted that this results in maximum amplitudes of vibration in the top layer of fruit causing more in-transit injury than if the fruit's natural frequency were out of the range of that of the transport vehicle. It was also noted that in tight-filled containers the elastic modulus of the pack will approach the modulus of the individual fruits to which the curves presented in Fig. 5.37 apply. Since the increase of pressure on the fruit increases the natural frequency of the fruit, tight-filling of containers by vibration settling may move the natural frequency of the fruit out of the range of that of the transport vehicle resulting in reduced resonant vibration and vibration bruising.

Another interesting observation in this work was that if the internal damping capacity of a certain fruit is high, even if its natural frequency may fall within the range of truck-bed vibrations, it may not receive as much vibration injury as a fruit with a low damping capacity. The low injury to apricots, for example, was attributed to its high damping capacity.

Another reported example of application of damping characteristics of agricultural products is the use of damping characteristics of an egg mounted on a torsion pendulum as a measure of its viscosity and internal quality. Recognizing the desirability of a non-destructive technique and the fact that the internal quality of an egg is directly related to variations in the viscosity of the contents of the egg, Wilke (1936) used the principle of the torsion pendulum to measure an index representing the combined viscosity of the entire contents of the egg in terms of the number of swings of the pendulum. The basic parts of the apparatus are shown in Fig. 5.38. Damping occurs when the direction of rotation of the egg holding a liquid is suddenly reversed. Using glycerol of known viscosity in the eggshell (contents removed), a calibration curve was obtained in terms of viscosity versus number of swings. A refined version of the same technique was employed later by Rowan *et al.* (1958) who concluded that the technique was usable for separating low-quality eggs from those of high quality, but not sufficiently accurate for purposes of egg grading. Figures 5.39, 5.40, and 5.41 show the relationship of damping factor and weight, egg number produced by the same hen, and albumen quality. Damping factor was found by determining the number of oscillations occurring during a rotational angle change from approximately 20 to 10 degrees. Factors which

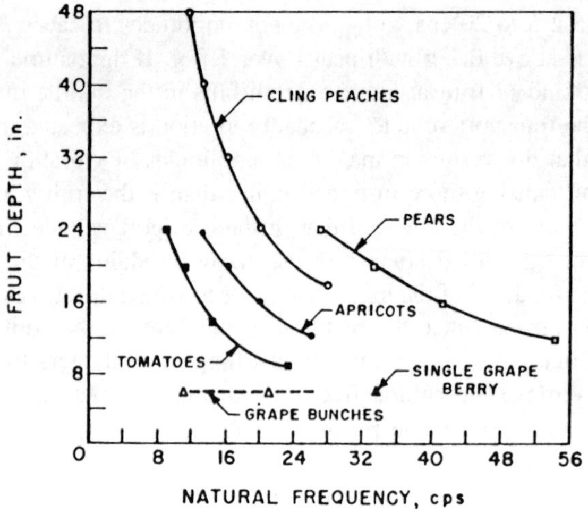

Figure 5.37 Natural frequency for several fruits at various depths in shipping containers (O'Brien, 1965)

were considered to contribute to variation of damping factor for the same albumen quality (Fig. 5.41) were temperature, air cell, size, shape, yolk size, yolk location, and yolk sticking. Evaluation of these factors, however, failed to explain this variation. Likewise, damping factor was found to vary for eggs produced by the same hen (Fig. 5.40). These variations, however, could be due to variation in shape and internal characteristics of the eggs from the same hen, which were assumed to be similar. Considering the biological system involved in producing an egg, such an assumption is rather doubtful.

Some dynamic mechanical properties of apple flesh in terms of complex axial compression moduli have been determined by Hamann (1968). Cylindrical flesh specimens were subjected to sinusoidal forcing at cyclic rates of 50 to 365 cycles per second using an MB Electronics Model SD electromagnetic shaker as the force generator. Shear complex moduli were also obtained using the same dynamic apparatus. A knurled surface was used to provide a sufficiently high coefficient of friction between the specimen and the surfaces (Fig. 5.42).

Plot of log $G(t)$ versus time, t, resulted in relaxation moduli expressions that were equivalent to a simple Maxwell model with element constants

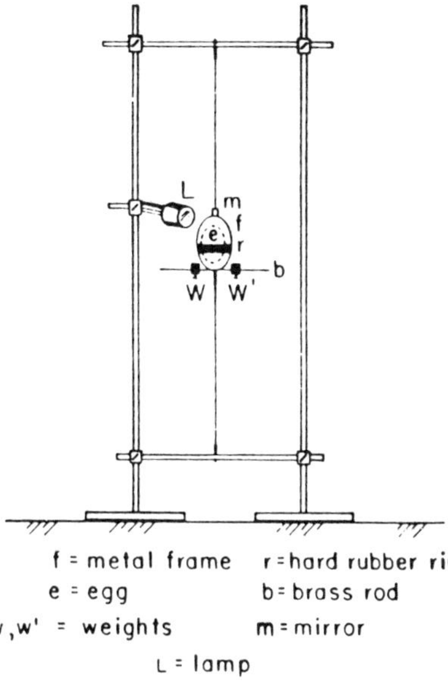

Figure 5.38 Torsion pendulum for determining damping resistance and interior quality of eggs (Wilke, 1936)

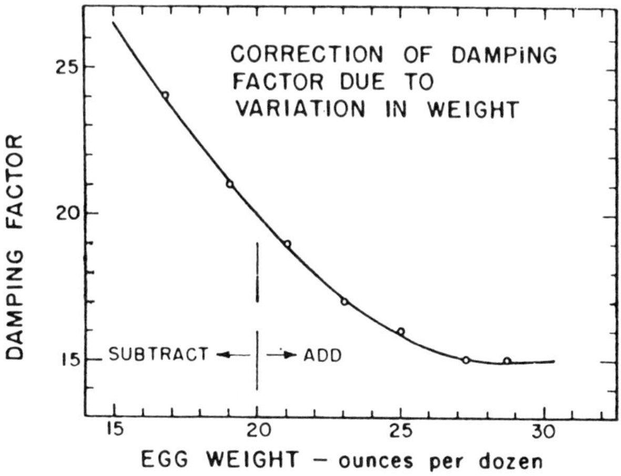

Figure 5.39 Variation of dampling factor with egg weight (Rowan *et al.*, 1958)

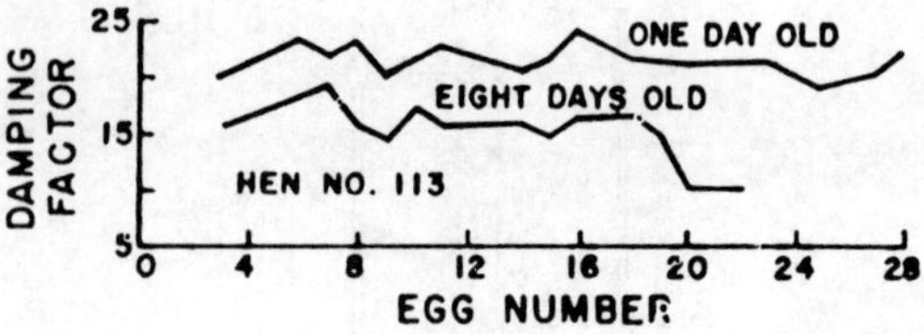

Figure 5.40 Variation of damping factor for eggs from the same hen (Rowan *et al.*, 1958)

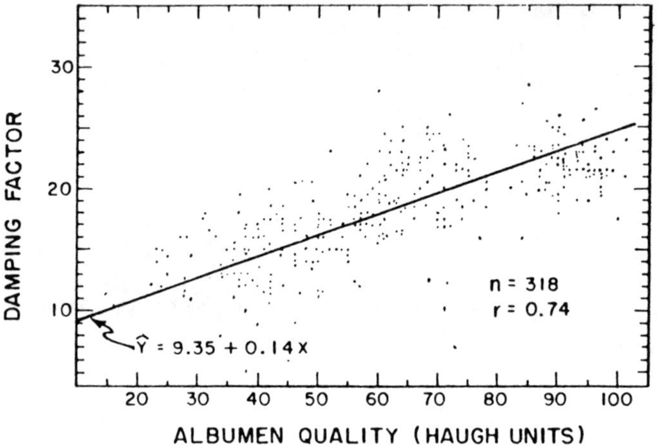

Figure 5.41 Variation of damping factor with albumen quality (Rowan *et al.*, 1958)

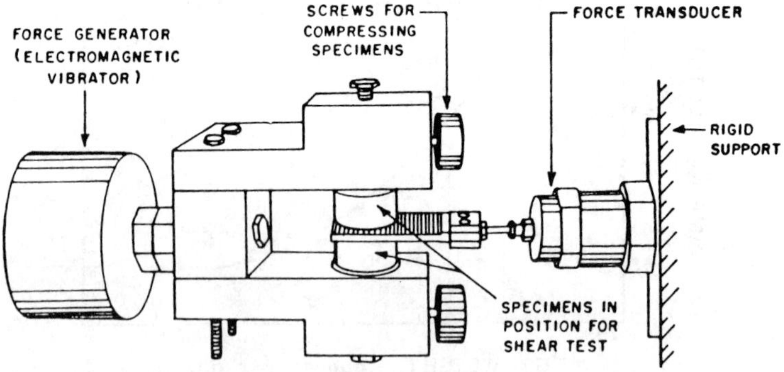

Figure 5.42 A sinusoidal forcing appartus using an electromagnetic vibrator (Hammann, 1968)

slightly larger than those obtained by direct computation using the spring and viscous element constants from the complex moduli. The expression for the relaxation modulus, $G(t)$, was given as

$$G(t) = G_0 \, e^{-t/T}$$

with $G_0=E$, the spring constant, and $T=\eta/E$, the time constant. The following relationship given by Bland (1960) were then used for the direct computation.

$$E = \frac{(E')^2 + (E'')^2}{E'} \quad \text{and} \quad \eta = \frac{(E')^2 + (E'')^2}{\omega E''}$$

where E' and E'' are storage and loss moduli respectively and ω is the angular frequency. The constants found for elastic elements varied from about 3000 psi for the hardest apples at harvest to approximately 750 psi for the softest apples from storage. Similarly viscous element constants varied from about 20 psi/sec to about 4.5 psi/sec depending on the variety and storage condition. The complex shear moduli obtained were approximately the magnitude expected based on the knowledge of the compression moduli and consideration of the fact that imperfect pure shear state existed during the test.

An application of the sinusoidally-varying force and deformation method to corn grain is given by Wen and Mohsenin (1968). Figure 5.43 shows the apparatus which employs a variable speed electronic motor and an eccentric instead of the electromagnetic vibrator used by Hamann. In the latter method the outputs from the transducer and the accelerometer were fed into an oscilloscope which traced a Lissajous figure in the form of an ellipse which gave sufficient data to calculate stress, strain, and the phase angle. This is a simple method for producing sinusoidally-varying direct stress and direct strain and would enable gathering data for a wide range of frequencies. However, with the use of the electromagnetic vibrator it is not possible to maintain a constant amplitude of vibration as the frequency changes. Unless it is assumed that the material under investigation is a linear viscoelastic material, the technique employed for dynamic viscoelastic testing must incorporate the concept of constant amplitude for any given frequency. The apparatus illustrated in Fig. 5.43 is mechanical in nature and thus is limited to work in a frequency range not exceeding 10 cps but it should fulfill the requirement of constant amplitude at all frequencies.

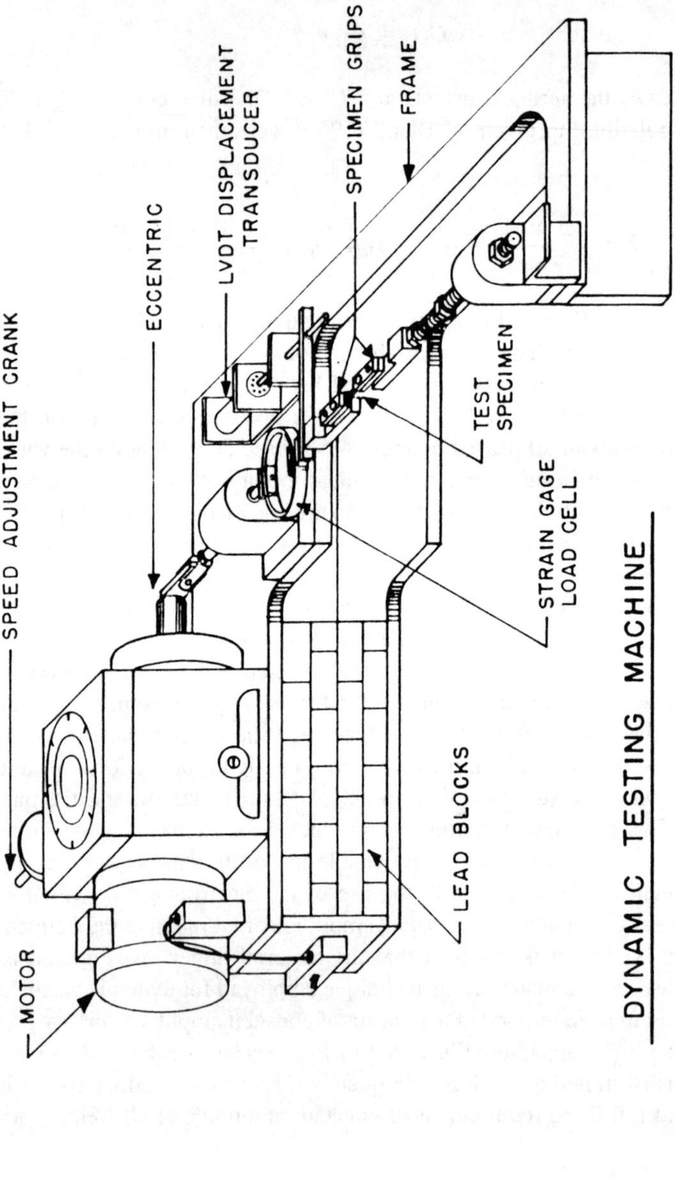

Figure 5.43 A sinusoidal forcing apparatus using a variable speed motor and an eccentric (Wen and Mohsenin, 1968)

Test specimens used in the testing machine of Fig. 5.43 were slabs of horny endosperm of corn kernel finished to a thickness of about 0.040 in. with a rectangular cross sectional area to fit the notch of the spacers in the specimen grips. The slab specimens conditioned to a given moisture content were then subjected to sinusoidally-varying tensile stress-strain for a frequency range of 0 to 10 cps at a preset amplitude fixed by the eccentric. When the two-channel recorder showed three or more consecutive undistorted sinusoidal waves for recording of maximum amplitudes, the chart speed was increased for a better estimate of the phase angle. The graphical technique for determination of the phase angle is shown in Fig. 5.44. From the maximum amplitudes of the stress and strain curves the complex modulus E^* was directly calculated using Eq. (4.47). Having the time shift Δt and the angular frequency ω, the phase angle δ was calculated using Eq. (4.48).

Table 5.8 shows the change of tensile complex modulus E^* with moisture content. One significant finding in this experiment was the sudden decrease of complex modulus as moisture content of horny endosperm increases from 16% d.b. to about 18% d.b. No explanation has yet been found for this phenomenon which has been observed also in both tensile tests and compression tests of horny endosperm employing quasi static methods of loading.

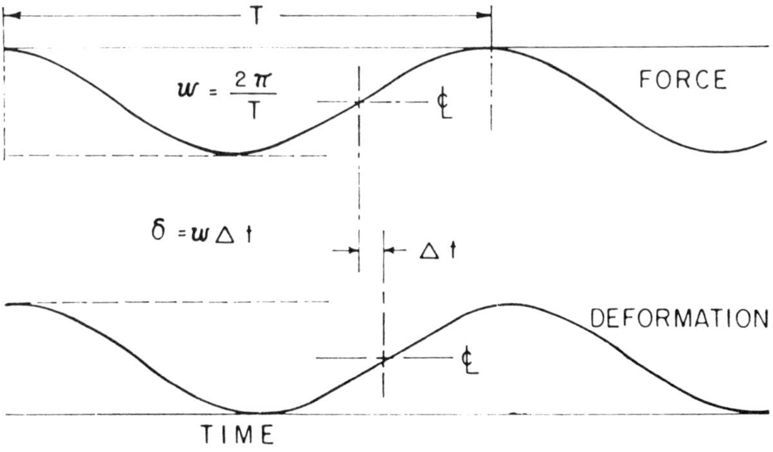

Figure 5.44 Sampling of sinusoidally-varying force and displacement curves enlarged to show the graphical method for estimation of time shift Δt and phase angle δ (Wen and Mohsenin, 1968)

Table 5.8 Complex tensile modulus of elasticity for slab specimens of corn horny endosperm at various moisture contents (Wen and Mohsenin, 1968)

Moisture content (% d.b.)	Number of tests completed	Complex modulus E^* mean (psi)	coefficient of variation (%)
24.75	2	31,100	—
19.90	2	36,350	—
17.70	2	42,100	—
16.00	6	83,633	11.8
14.50	9	84,567	20.5
12.15	7	88,328	14.9
7.15	7	97,914	26.0

Due to crudeness of the method for determination of phase angle, no specific trend in change of phase angle with change of moisture content could be detected. Since the tangent of the phase angle represents the damping factor, one would think that such a factor should increase with the increase of moisture content in the material. The average value of δ for 26 tests was found to be about 9° for a range of 7% d.b. to 17% d.b. moisture contents. Using this value for δ and the mean complex modulus of 83,633 psi at 16% d.b. for E^* in Eq. (4.48a) and (4.48b) resulted in values of 82,500 psi and 13,000 psi for storage and loss moduli of the corn horny endosperm, respectively.

Example

A slab of corn horny endosperm with 0.0344 in$\times$0.134 in. cross-sectional area at 5% (d.b.) moisture content is subjected to sinusoidally-varying tensile test using the apparatus of Fig. 5.43. The recorded sinusoidal curves show a peak deformation of 1.18×10^{-3} in. and a peak force of 2.73 lb. The rate of chart travel is 25 mm/sec with 15 mm taken for a complete cycle and 0.7 mm for time shift between the sinusoidal curves of force and deformation. If the gage length of the slab specimen is 0.23 in., determine the complex modulus E^* and the phase angle δ for the specimen.

From the maximum deformation and the gage length of the slab, peak strain is

$$\varepsilon_{max} = \frac{\Delta l}{l} = \frac{1.18 \times 10^{-3}}{0.23}$$

$$= 5.12 \times 10^{-3} \text{ in/in}$$

From the maximum force and the cross-sectional area of the slab

$$\sigma_{max} = \frac{F}{A} = \frac{2.73}{0.0344 \times 0.134}$$

$$= 592 \text{ lb/in}^2$$

Therefore,

$$|E^*| = \frac{\sigma_{max}}{\varepsilon_{max}} \frac{592}{5.12 \times 10^{-3}}$$

$$= 115{,}500 \text{ lb/in}^2$$

From the information on chart speed and displacements, the phase angle can be calculated as follows

$$\omega = \frac{25 \text{ mm/sec}}{15 \text{ mm/cycle}}$$

$$= 1.67 \text{ cps}$$

$$= 3.34\,\pi \text{ rad/sec}$$

$$\Delta t = \frac{0.7 \text{ mm}}{25 \text{ mm/sec}}$$

$$= 0.028 \text{ sec}$$

$$\delta = \omega \Delta t$$

$$= 3.34\pi \times 0.028$$

$$= 0.292 \text{ rad} \quad \text{or} \quad 16.6^0$$

The transducer-type instrument illustrated in Fig. 4.36 and discussed in Chapter 4 was used on dry shelled corn to estimate the storage and loss moduli E' and E''. The phase angle δ was calculated from the relationship given by Eq. (4.25). It has been shown that variation of storage and loss moduli with frequency are similar to the curves given in Fig. 5.33. At very high frequencies the storage modulus has a constant maximum value. The material behaves like a perfect elastic solid. As frequency is reduced, the storage modulus gradually decreases until it reaches a constant low value, representing the equilibrium modulus. The transition region which extends over several decades of frequency is associated with viscoelastic behavior. At lower frequencies, the material behaves like a rubbery material and shows low modulus and high elasticity. Finally a stage is reached where a decrease in storage modulus is observed with further decrease of frequency for linear polymers and the material behaves like a viscous liquid. The corresponding variation of loss modulus is shown below the storage modulus.

Examples of wave propagation technique for determination of rheological properties of biomaterials are also found in recent publications. Nybom (1962) placed soft fruits such as raspberries between the metallic diaphragms of earphones and measured the intensity of vibrations transmitted through the fruit (Fig. 5.45). An alternating current of 50 cps from a 6-volt electric bell transformer supplied power to one of the earphones causing its diaphragm to vibrate. The fruit, depending on its rigidity, transmitted part of the vibrations to the diaphragm of the other earphone which generated a secondary alternating current. This current was then amplified with a small battery-operated transistor amplifier, built into the drawer of the cabinet, and measured by means of a sensitive meter. To provide a constant force on the fruit specimens, the weight of the upper earphone was first counterbalanced by lead shots on the left scale and then a 20-gram weight was placed on the earphone to compress the fruit presumably to the same pressure in all tests. Using this technique, the course of fruit maturation was followed by studying the relationship between soluble solids and anthoyanin content of raspberries and their firmness expressed in microamperes. Figure 5.46 shows varietal differences in firmness for 24 raspberry varieties determined by measuring ten samples of each variety.

The dynamic complex shear modulus of goby muscle tissues and tissues of various salt-water fish were determined by Lebedeva (1965) employing a wave propagation technique. In this method a plane shear wave was excited in a flat tissue specimen and the vibrational phase and amplitude were measured. Figure 5.47 shows the specimen holder with transmitting and receiving transducers and the associated instrumentation. A stack of Rochelle salt wafers cemented between two brass bars form the transducers. The sinusoidal voltage to the piezoelectric stack of the transmitter causes

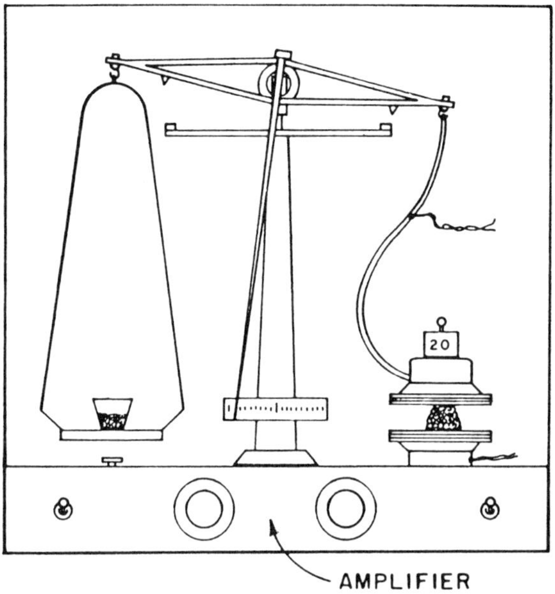

Figure 5.45 One wave propagation methods used for measuring firmness of soft fruits (Nybom, 1962)

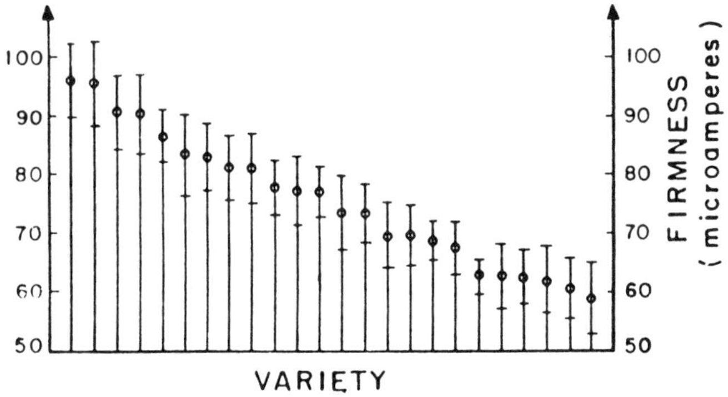

Figure 5.46 Varietal differences in firmness of raspberries detected by the use of the technique of Fig. 5.45 (Nybom, 1962)

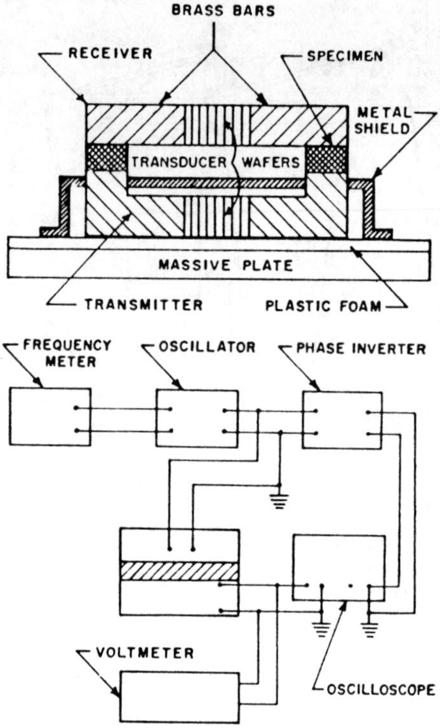

Figure 5.47 Shear vibration technique applied to texture evaluation of fish muscle (Lebeveda, 1965)

longitudinal vibrations of the bars of this transducer which are transformed into shear vibrations in the specimen. These shear vibrations are then transformed back to longitudinal vibration in the receiving transducer and then to electrical signals to be read on the oscilloscope. For computing the complex shear modulus, G^*, the storage shear modulus, G', and the loss factor in shear, tan δ, were determined using the following expressions.

$$G' = \frac{\omega^2 (t_2 - t_1)^2 \varrho}{(\alpha^2 - \beta^2)(1 + \tan^2 \delta)}$$

$$\tan \delta = \frac{2\alpha\beta}{\alpha^2 - \beta^2}$$

where ω is in radians per second, t_2 and t_1 are thickness of two pairs of specimens, ρ is density, and β and α are defined by

$$\beta = \text{arc sinh}\sqrt{\frac{F^2 + M^2 - 1}{2} + \frac{\sqrt{(F^2 + M^2 - 1)^2}}{4} + M^2}$$

$$\alpha = \text{arc cos}\left(\pm\sqrt{F^2 + M^2 - \sinh^2\beta}\right)$$

For convenience of calculations three pairs of specimens with three different thicknesses t_1, t_2 and t_3 were prepared such that $2(t_2-t_1)=(t_3-t_1)$. The values of F and M were then obtained from directly measurable voltage ratios $V_1/2V_2$ and $V_3/2V_2$ and phase differences $(\phi_1-\phi_2)$ and $(\phi_3-\phi_2)$ using the following relationships

$$F = V_1/2V_2 \cos(\phi_1 - \phi_2) + V_3/2V_2 \cos(\phi_3 - \phi_2)$$

$$M = V_1/2V_2 \sin(\phi_1 - \phi_2) + V_3/2V_2 \sin(\phi_3 - \phi_2)$$

where subscripts 1, 2, 3 refer to the corresponding values for specimens t_1, t_2, t_3, respectively.

Using the above technique the effect of fiber orientation and frequency on loss factor and shear modulus were determined (Figure 5.48). It was found that the shear modulus was twice as large for specimens in which the fibers were oriented perpendicular to the direction of shear wave propagation as for those with their fibers coinciding with the direction of wave propagation. Furthermore, as seen from Fig. 5.48, both the shear modulus and loss factor were highly frequency dependent.

The ultrasonic pulsed technique of Fig. 4.35 was used on one inch cube specimens of yellow poplar to determine the effect of moisture on the modulus of elasticity of the wood (Wen and Mohsenin, 1968). To insure the acoustical contact, a coating of high vacuum silicon lubricant was applied on the surfaces of contact between the specimen and the transducers. In Fig. 4.35, a train of short pulse, produced by means of a pulse generator, is converted into mechanical vibration in one transducer and sent through the specimen. The second transducer receives the mechanical vibrations and converts them back to electrical signals which are exhibited on the oscilloscope. The pulse width or time of wave travel, t, was measured and substituted in Eq. (4.50a) to determine the modulus of elasticity. The approximate time required for a single pulse to travel through the wood speci-

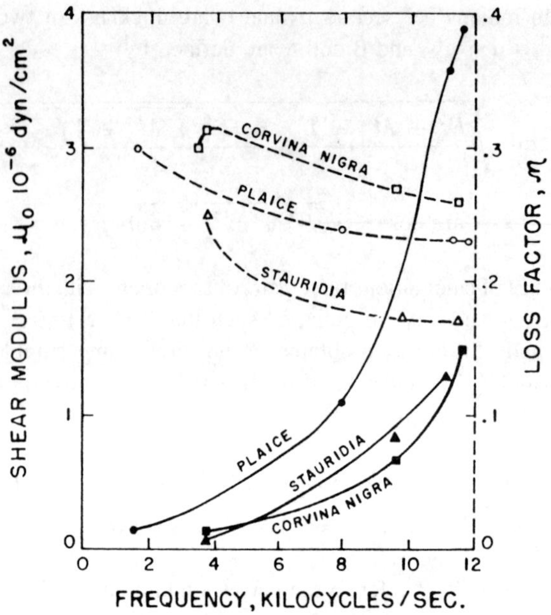

Figure 5.48 The effect of fiber orientation and frequency on loss factor and shear modulus of fish muscles (Lebedeva, 1965)

men was about 5×10^{-6} seconds. The pulse period or the time required for a transmitted pulse to fade out was about 25×10^{-6} seconds. This gave a pulse frequency of $\frac{1}{25 \times 10^{-6}}$ or 40,000 cps. The specimens were tested in the direction parallel to the grain at moisture contents varying from 4% to 20% (Fig. 5.49).

To determine the validity of the experimental data the values of the modulus were compared with published data obtained by conventional quasi static compression tests. For yellow poplar at 12% (w.b.) moisture content Drow and McBurney (1946) and Wood Handbook (USDA Handbook No. 72) have reported a modulus of elasticity of about 1.58×10^6 psi. This was about 10% lower than that shown in Fig. 5.49. In light of the fact that dynamic tests at such high frequencies should yield modulus values higher than quasi static tests, which are always subject to creep and relaxation, the difference was considered reasonable. To further verify the results, data were substituted in the well known exponential relationship between Young's modulus of elasticity and moisture content of wood as given in Fig. 5.49. Once the validity of the experimental technique was established, the instrumentation and technique was applied on soft and hard biomaterials selected from a number of food products.

Example

In using the pulse technique illustrated in Fig. 4.35 cube specimens of yellow poplar with an average moisture content of 19% (w.b.) and a density of 0.4 gm/cm^3 were tested for determination of the Young's modulus of elasticity. For a pulse at a frequency of 38,500 cps, the average of the pulse duration (t in Fig. 4.35) for five specimens with an average thickness of 1.012 in. was about 4.98×10^{-6} seconds. Assuming $\mu=0$, find the wave velocity through the specimen and the Young's modulus of elasticity.

$$V = \frac{d}{t} = \frac{1.012 \text{ inch}}{4.98 \times 10^{-6} \text{ sec}}$$

$$= 2.032 \times 10^5 \text{ in/sec}$$

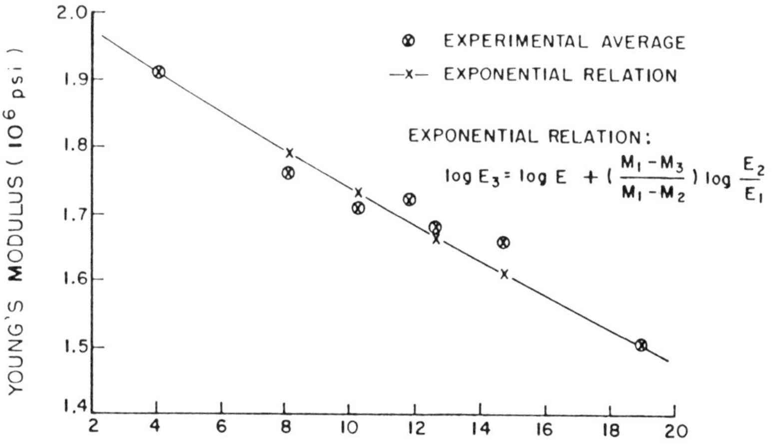

Figure 5.49 Variation of modulus of elasticity of yellow poplar with moisture content as determined by the ultrasonic pulsed technique (Wen and Mohsenin, 1970)

From Eq. (4.50a)

$$E = \rho V^2 = \left(\frac{0.4 \text{ gm}}{\text{cm}^3}\right)(2.032 \times 10^5 \text{ in/sec})^2 \left(\frac{2.54 \text{ cm}}{1 \text{ inch}}\right)^2$$

$$\times \left(1.45 \times 10^{-5} \frac{\text{psi}}{\text{dynes/cm}^2}\right) = 1.545 \times 10^6 \text{ psi}$$

During working, softening of the butter occurs which is apparently duc to breakdown within the product structure. However, when the product is allowed to set, the structure reforms or rebuilds with time. These observations indicate the importance of rheological properties of butter in design of handling and processing of this material. Both quasi static and dynamic tests were conducted on freshly churned samples of butter to determine if rheological properties of butter based on its structure can be explained (Diener and Heldman, 1968). Quasi static tests consisted of loading specimens of butter at 62°F and 48°F temperatures with a flat circular die at several rates of loading. For dynamic tests, cantilever beam specimens of butter 1/4 by 1 by 8 in. were supported vertically with a small metal shim imbedded in the end of the beam. A Wayne-Kerr B731A vibration meter was used to measure the oscillations at the end of the free end. In the static tests, the Boussinesq solution for die loading, as will be explained in Chapter 6, was used to derive the expression $3F(t)/8a\dot{D}$, where $F(t)$ is force on the die at any time t, a is radius of the die, and $\dot{D}$ is the rate of displacement. Using the experimental force-time data, the above expression was plotted against experimental time as shown in Fig. 5.50. The superposition for various rates was said to indicate the linear viscoelastic behavior over the ranges used. The departure from the master curve was associated with yield stress. At lower temperature, butter was found to have a higher elastic modulus and smaller relaxation time which means that less energy is lost in the material during the loading cycle.

On the basis of static response, the elements for a rheological model to represent the behavior of butter was developed. Figure 5.51 shows the rheological model designated as VMB (viscous, Maxwell, Bingham). The y element in this model was added to represent yield. The physical relationship of the mechanical elements of the model to the structure of butter was designated based on the description given by King (1964). According to this description, under certain conditions, the fat globules are capable of large elongations without rupture. This indicated that the membrane sur-

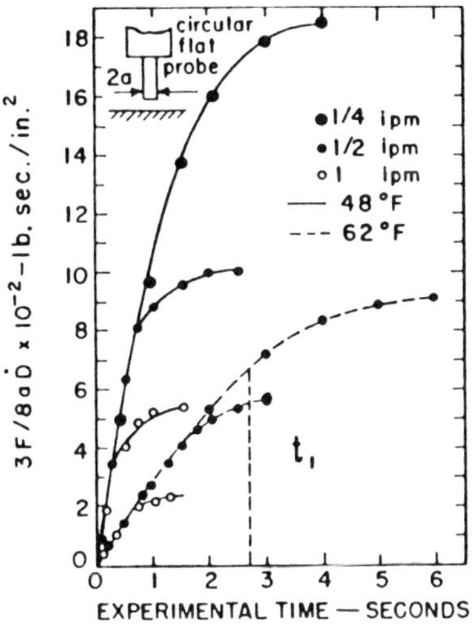

Figure 5.50 Treatment of static response of butter to show a "yield stress" (Diener and Heldman, 1968)

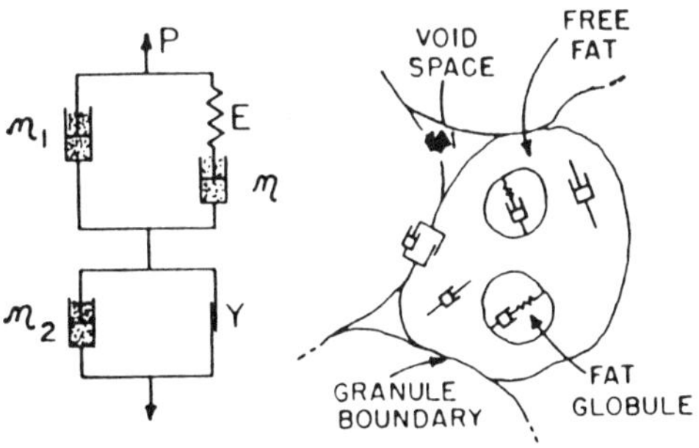

Figure 5.51 Rheological response and physical structure of butter (Diener and Heldman, 1968)

rounding the fat globule may be elastic in nature. Accordingly, a fat globule was approximated by a purely viscous fat surrounded by an elastic material. This combination was designated as a Maxwell element. Likewise, free fat in the granule was associated with free viscous elements, while the yield element was theorized to represent the behavior of the granule boundary. Failure would be likely to occur in the granule because of the presence of void spaces. It was further postulated that failure and yield may occur as a result of shear failure of the free fat or even rupture of the fat globules.

Using the stress relaxation versus experimental time curves, Diener and Heldman (1968) also determined the specific values for their proposed model for an experimental time of one second.

For the case of dynamic tests, the logarithmic decrement and damped natural frequency from the oscillation curves similar to Fig. 4.34b, for the free end of the cantilever beam specimens of butter were used to calculate storage modulus and loss tangent properties. It was found that over a frequency range of 1⁄10 to 10 cps the storage modulus, E', for butter increased from 360 to 1300 psi at 62°F and from 1300 to 4600 psi at 48°F. Also within the same range of frequency, the loss tangent, tan δ, varied from about 0.05 to 0.25, for both temperatures.

Viscoelastic characterization by compliance and strain energy methods

The solution of a spherical indentation in viscoelastic solid food materials was considered by Mohsenin *et al*. (1968) in terms of a numerical approach similar to that of Nolt and Meier (1965). This was based on the extension of the Hertz problem in elasticity to viscoelastic bodies as given by Lee and Radok (1960). They give the following relationship for the case of spherical indentation which is the basis for Eq. (6.16) and (6.17) in Chapter 6.

$$\int_0^t {}^1/_2\, J(t - t')\left\{\frac{[dF(t')]}{dt'}\right\} dt' = \frac{8}{3}\sqrt{R}[D(t)]^{3/2} \qquad (5.24)$$

where $J(t)$ is shear creep compliance, $F(t)$ is total force as a function of time or the loading history, $D(t)$ is central indentation at time t, and $R=$ radius of indenter. The solution is valid provided that the material is assumed to be incompressible and the central indentation is a nondecreasing function of time. If both $F(t)$ and $D(t)$ are known from $t=0$ to $t=t_{n+1}$, a numerical method of integrating Eq. (5.24) can be used to obtain shear creep compliance, $J(t)$ by measuring force on the sphere, $F(t)$, and central indentation of the material, $D(t)$. By dividing the time scale into intervals, t_i, where $i=1, 2, \ldots (n+1)$, with $t_i=0$ and $t_{n+1}=t$, the following solution is found for Eq. (5.24).

$$J\,(t_{n+1/2}) = \frac{-Q(t_{n+1}) - \sum_{i=1}^{n-1} J(t_{i+1/2})\,[F(t_{n+1} - t_{i+1}) - F(t_{n+1} - t_i)]}{F(0) - F(t_{n+1} - t_n)} \tag{5.25}$$

where Q is a function of the central indentation $Q(t)=\frac{8}{3}\sqrt{F}$ is a function of the load, t is the time and J is the shear compliance. A computer program was adapted to the above equation to analyze force-deformation-time data.

This method was used in the analysis of the problem of a spherical indenter compressing specially prepared 0.10-inch thick slabs of horny endosperm of shelled corn (Fig. 6.11) at different equilibrium moisture contents. The indenter was a $\frac{1}{16}$-inch smooth steel ball pressed onto the specimen at the rate of 0.020 in/min. Having the shear creep compliance function $J(t)$, the Laplace transform was used to convert $J(t)$ to relaxation modulus function $G(t)$. From $G(t)$, the modulus of elasticity in compression $E(t)$ was found assuming incompressibility, i.e. $\mu=0.5$. Since many biomaterials do not lend themselves to the conventional shear tests, this technique would be a suitable method for evaluation of time-dependent shear modulus using a simple indentation test. The shear modulus can be taken as a criterion of hardness of the grain or be used as a parameter needed for studying viscoelastic properties of the material (see Fig. 4.6).

The results of this experiment for 10 slab specimens of corn horny endosperm at one level of moisture content and one level of load are shown in Table 5.9.

The application of Eq. (5.24) to this problem is based on the assumptions that (1) the material is incompressible, (2) that it is adequately represented by the theory of linear viscoelasticity, and (3) that the material occupies infinite halfspace. The first two assumptions, based on previous work, were believed to be sufficiently valid for practical purposes. As for the third assumption, according to the spatial strain field given by Timoshenko and Goodier (1951), the minimum sample thickness, in relation to both indenter radius and the central indentation, should be about ten times the radius of the circle of contact in order that the compressive stresses along the axis of penetration reduce to 1% of their maximum value [see Eq. (6.23)]. As seen from Table 5.9, even for the largest radius of the circle of contact, the specimens with 0.10-inch thickness were large enough to satisfy the condition of vanishing compressive stresses some distance below the point of loading. The radius of the circle of contact was calculated both from

Table 5.9 Some viscoelastic functions of slabs of corn horny endosperm at 8% (d.b.) moisture content obtained by means of a 1/16-inch spherical indenter (Mohsenin *et al.*, 1968)

Sample	$J(t)$ (1/psi)	$G(t)$ (psi)	F (lb)	t (sec)	G (10^5 psi)	$E(\mu = 0.5)$ (10^5 psi)	D (10^{-4} psi)	a_1 (10^{-3} in)	a_2 (10^{-3} in)
1	$1.7592 \times 10^{-5} t^{0.2836}$	$0.4962 \times 10^5 t^{-0.2836}$	0.8	1.632	0.4318	1.2954	5.44	5.81	4.77
2	$2.4489 \times 10^{-5} t^{0.2972}$	$0.3516 \times 10^5 t^{-0.2972}$	0.8	2.376	0.2718	0.8155	7.92	6.99	5.57
3	$2.0147 \times 10^{-5} t^{0.3117}$	$0.4208 \times 10^5 t^{-0.3117}$	0.8	1.896	0.3447	1.0341	6.32	6.25	5.14
4	$1.7148 \times 10^{-5} t^{0.3483}$	$0.4736 \times 10^5 t^{-0.3483}$	0.8	1.536	0.4078	1.2234	5.12	5.63	4.86
5	$1.2286 \times 10^{-5} t^{0.2254}$	$0.7476 \times 10^5 t^{-0.2254}$	0.8	1.296	0.7052	2.1156	4.32	5.18	4.05
6	$1.3101 \times 10^{-5} t^{0.3000}$	$0.6552 \times 10^5 t^{-0.3000}$	0.8	1.32	0.6028	1.8085	4.40	5.23	4.27
7	$1.6338 \times 10^{-5} t^{0.2057}$	$0.5701 \times 10^5 t^{-0.2057}$	0.8	1.584	0.5186	1.5558	5.28	5.72	4.49
8	$2.1087 \times 10^{-5} t^{0.3267}$	$0.3952 \times 10^5 t^{-0.3267}$	0.8	1.908	0.3200	0.9601	6.36	6.27	5.27
9	$2.7978 \times 10^{-5} t^{0.2802}$	$0.3130 \times 10^5 t^{-0.2802}$	0.8	2.4	0.2449	0.7347	8.00	7.03	5.76
10	$1.6271 \times 10^{-5} t^{0.3162}$	$0.5184 \times 10^5 t^{-0.3162}$	0.8	1.524	0.4537	1.3612	5.08	5.61	4.69
			mean	1.747	0.430	1.290	5.82	5.98	4.88
			s	0.392	0.1465	0.4395	1.31	0.654	0.5797
			cv	22.4%	34.1%	34.1%	22.5%	10.9%	11.9%

a_1 = radius of the circle of contact from the geometry of the indentation
a_2 = radius of the circle of contact from $a_2 = 0.721\ (FAD)^{1/3}$ based on the Hertz contact problem (Chapter 6)
s = standard deviation
cv = coefficient of variability

the geometry of the indentation and the indenter as well as the use of the Hertz equation for contact of elastic bodies (Chapter 6). The reason for the smaller radius by the Hertz contact theory is the "sinking-in" effect generally observed in hardness tests of materials by spherical indenters (Tabor 1951).

To check the validity of the results, the solution given for sample 1 in Table 5.9 was substituted into Eq. (5.24) for a constant rate of loading equal to 0.568 lb/min. This substitution resulted in 13% difference as shown below

$$k \int_0^t J(\tau)\, d\tau = 0.568 \int_0^{1.632} 1.7592 \times 10^{-5}\, t^{0.2836}\, dt$$

$$= 1.38 \times 10^{-5}$$

$$\frac{16}{3} \sqrt{R}\, [\alpha(t)^{3/2}] = \frac{16}{3} \sqrt{0.03125}\, [5.44]^{3/2}\, 10^{-6}$$

$$= 1.2 \times 10^{-5}$$

$$\text{difference} = \frac{0.18}{1.38} = 13\%$$

The k factor being the constant rate of loading in the above equation was estimated from the straight line portion of the force-time curve of the indentation tests. Nolt and Meier (1965), who employed the same technique but maintained a constant rate of loading by having the load suspended from a spring, found a maximum difference of only 5.4 percent for the equality of Eq. (5.24). One of the major factors responsible for greater error in establishing equality for both sides of Eq. (5.24) was felt to be the lack of maintaining a constant rate of loading.

Strain energy relations quite often provide a convenient method of characterizing viscoelastic behavior of materials. Morrow (1968) developed energy functions for bovine muscle subjected to transient loading (stress relaxation). He then used these relations to predict the storage and loss energies in the same muscle subjected to cyclic ramp loading. This predicted energy was then compared with the experimental energies obtained for the same type of loading.

This study indicated that the bovine muscle which was studied was definitely non-linear viscoelastic. Response was found to be isotropic under static conditions and partially anisotropic under dynamic conditions. Comparison between theoretical and experimental dynamic energies revealed only a partial correlation indicating that additional studies were needed.

Similar technique for analysis of cyclic compressive force application on cottonseed has been reported by Clark *et al.* (1968).

Example

Bovine muscle test specimens with a gage length of 1.0 in. and a cross section of 0.2 in by 0.2 in were subjected to tensile cyclic loading and unloading under 5% maximum strain. An automatic integrator gives the loading and unloading energy for a sawtooth pattern recorded on the chart for a total of 5 cycles. Using the following data determine (1) secant modulus for unloading curves and (2) express the energy difference and energy ratio for the loading-unloading cycles in the form of continuous functions.

	Force (10^3 dyne)		Energy (inch–gram)	
Cycle	Max.	Min.	Loading	Unloading
1	25.2	20.8	1.21	1.41
2	24.5	20.6	2.34	2.27
3	24.3	20.5	3.46	3.39
4	24.1	20.5	4.58	4.50
5	24.0	20.5	5.68	5.62

Using the information on maximum strain, cross sectional area and the above data, the following calculated values are obtained.

Cycle	Stress (10^3 dyne/cm^2)		Secant modulus (10^3 dyne/cm^2)	Energy (inch–gram)	
	Max.	Min.		Difference	Ratio
1	97.7	80.5	364	0.07	1.06
2	95.0	79.9	310	0.07	1.03
3	94.1	79.5	292	0.07	1.02
4	93.5	79.5	280	0.08	1.015
5	93.0	79.5	270	0.09	1.01

When calculating secant modulus for the unloading curves, for simplicity the minimum stress 79.5 was assumed as zero reference as shown below for the first cycle

$$\text{Secant modulus} = \frac{\Delta\sigma}{\Delta\varepsilon} = \frac{97.7 - 79.9}{0.05} \times 10^3 = 364 \times 10^3 \frac{\text{dyne}}{\text{cm}^2}$$

A continuous function for the difference in energy appears to be approximately the simple function $\Delta E = 0.07$ inch-gram. The ratio of loading energy to unloading energy, however, turns out to be a logarithmic function of the following form

$$\text{Ratio} - 1 = 0.062\, e^{\frac{-t}{2.74}} + 0.9\, e^{\frac{-t}{0.25}}$$

which, with the exception of the first cycle, is in good agreement with the experimental data. This function was obtained by plotting first the quantity (Ratio−1) versus cycles on a semi-log paper and then determining the best fitting equation of the curve by the method of successive residuals.

5.7 FLOW OF ORGANIC FLUIDS

Most organic fluids are non-Newtonian in nature but research employing the principles of non-Newtonian fluids on these materials is quite recent and very little basic data are available on their flow characteristics. Considerable research is currently in progress on human blood. Disposal and utilization of agricultural wastes such as farm manure in the form of liquid and liquid feeding of livestock for higher efficiency and easier handling of the feed are among the problems on the farm where basic data on flow properties of the material are needed (Van Gilst *et al.*, 1966). Viscosity and consistency of liquid food products are considered important parameters for design of process equipment as well as quality evaluation and yet research employing the non-Newtonian approach is only a few years old even in this area.

Herum *et al.* (1966) developed a rheometer to study the flow properties of a swine ration consisting of ground feed and water at various feedwater ratios and a finely ground meat mixture consisting of one part each of pork, beef and water. The rheometer consisted of two pistons activated with hydraulic cylinders to force the organic fluid back and forth several passes at pressures of about 300 psi through the test section of tubing. Pressure inside the tubing was detected with strain gages bonded to the outside of the tube. In this manner shear stress-shear rate data were obtained for each test sample at various temperatures controlled with an oil bath surrounding the cylinder and tubing assembly. Figure 5.52 shows the shear diagram and effects of temperature for a swine ration. The data are plotted in terms of power-law parameters on log-log coordinates using Eq. (4.84) in Chapter 4. The slope of each straight line in Fig. 5.52 is equal to the flow behavior index, n', and the intercept at unit shear rate is equal to the viscosity index k'. These parameters of flow for several feed slurries and meat pastes are given in Table 5.10.

Time-dependency of the fluids are shown in Fig. (5.53) and (5.54). As seen, gradual absorption of water by the feed caused a consistent increase in shear stress with increase of holding time after mixing both for low feed-water and high feed-water ratios. The high feed-water ratio, however, showed a rheopectic behavior (Fig. 5.54) while a low feed-water ratio exhibited a thixotropic behavior (Fig. 5.53). As expected, less water in the liquid feed and longer holding time resulted in greater shear stress.

With consideration of Eq. (4.84) and the experimental values of n' and k' as well as knowledge of the flow behavior, such design parameters as diameter of the pipe, pressure losses, flow rate, type of pump and power

Table 5.10 Flow behavior index (n') and viscosity index (k') for organic slurries and pastes (from Herum *et al.*, 1966)

Feed-Water Ratio	Nominal Temperature °F	Nominal Tube Diameter 1/2-Inch k'	1/2-Inch n'	1-Inch k'	1-Inch n'	1−1/2-Inch k'	1−1/2-Inch n'
		Ground Corn Slurries					
1:1.10	Room			0.00308	1.56	0.00201	1.84
1:1.25	Room	0.306	0.936	0.00994	1.23		
1:1.25	106	0.0855	0.686				
1:1.25	150			28.6	0.478		
1:1.40	144			0.0128	1.41		
		Mixed Swine Ration Slurries					
1:1.50	Room	5.76	0.449	15.7	0.279	4.89	0.393
1:1.50	110			9.97	0.282		
1:1.50	143			10.6	0.282		
1:2.00	Room	0.0579	0.806	1.03	0.406	0.786	0.449
1:2.50	Room	0.00369	1.03				
		Ground Meat					
—	60	0.0173	1.03	1.70	0.419		
—	110	0.0432	0.727				
—	140	2.20	0.0676				

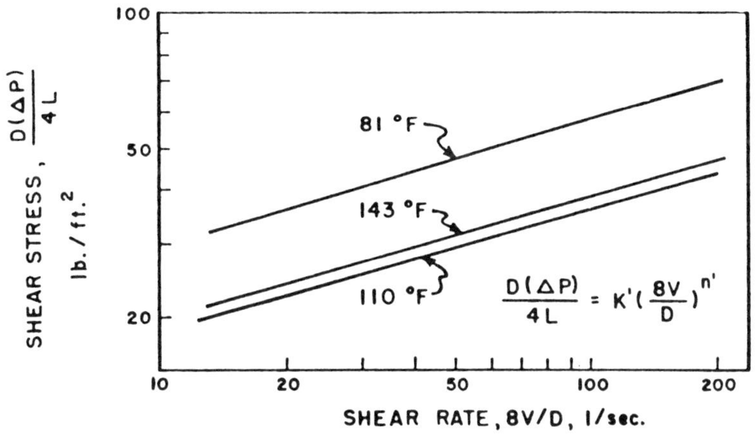

Figure 5.52 Shear diagram for mixed swine ration showing effects of temperature (Redrawn from Herum *et al.*, 1964)

requirements of fluid transfer can be determined on a more rational basis than the laborious and costly trial-and-error method. For example, Fig. 5.53 shows that if the liquid feed of the medium feed-water ratio has been left to stand in the pipe or in the mixing tank, the pump must be sufficiently powerful to start the flow. Once flow has started, the fluid, according to Fig. 5.53, will break down under shear and the load on the pump will be

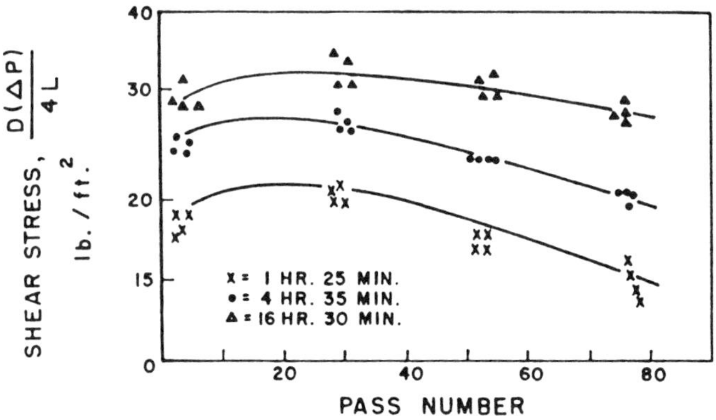

Figure 5.53 Shear stress in a ground corn slurry with a *medium* feed–water ratio showing thixotropic behavior (Redrawn from Herum *et al.*, 1964)

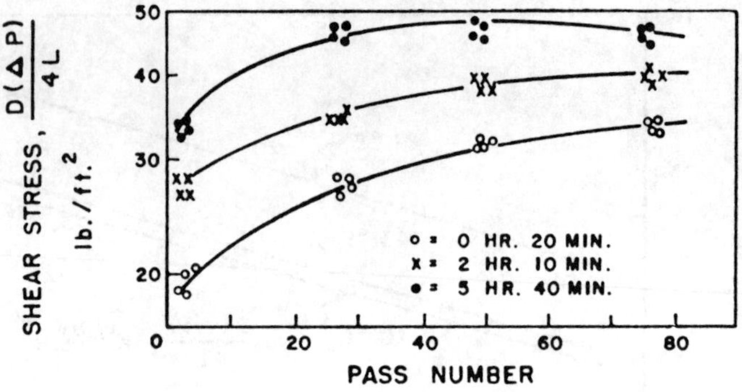

Figure 5.54 Shear stress in a ground corn slurry with a *high* feed–water ratio showing rheopectic behavior (Redrawn from Herum *et al.*, 1964)

considerably reduced. In the case of a high feed-water ratio liquid, however, according to Fig. 5.54, the critical load on the pump would be developed some time after the flow has started. Therefore, while in the case of a thixotropic fluid the pump is to be selected for the start-up condition, a rheopectic fluid requires a pump which would handle the fluid after the final steady state condition has been reached.

Mitschell and Peart (1968) later provided more basic data on flow properties of several slurries consisting of ground corn and water with moisture content ranging from 63.6% to 75% (w.b.). A tube-type rheometer similar to that used by Herum *et al.* (1966) was used for the experiments which also allowed recirculation of the slurry. It was noted that a minimum velocity of 40 ft/min was necessary to prevent settling. This was accomplished by incorporating also a circulating pump in the rheometer. Before taking the pressure readings, the slurry was allowed to circulate for 30 min after which the hand valves were closed to shut off the pump section of the rheometer (Fig. 5.55). Tests under two different velocities provided data for plotting of the parameters $8V/D$ versus $D(\Delta P)/4L$ on logarithmic coordinates. The values of the exponents k' and n' showed that slurries with 66.7% moisture or less were pseudoplastic and the slurries with 69.3% or more were dilatent. The time dependency showed that practically all the slurries seemed to reach a maximum pressure drop, ΔP, showing rheopectic effect after being worked in the rheometer, for about 20 to 30 min. Slurries with low moisture contents continued to be rheopectic after 30 min, while the high moisture slurries showed a decrease in ΔP reading (thixotropic) or reached a steady state.

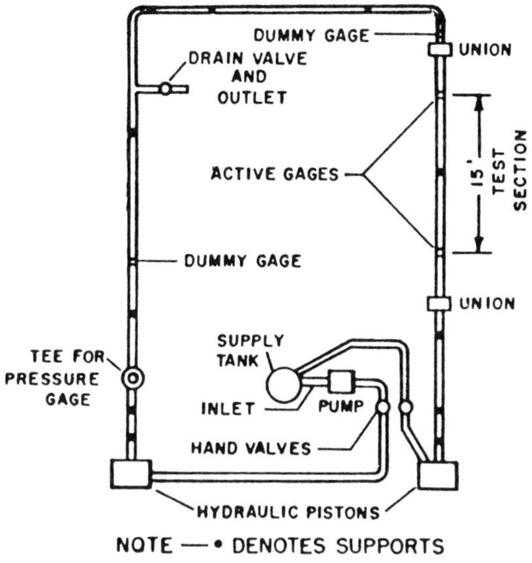

Figure 5.55 Rheometer used for study of flow properties of organic slurries (Mitchell and Peart, 1968)

Example

It is desired to find the horsepower necessary to convey a hog slurry composed of corn ground through a 1/16-inch screen and water with 1:2.500 feed-water ratio (71.5% moisture) at 70°F with a velocity of 360 ft/min around a 30×100-ft long hog house through a ½-inch pipe. Assume a 50% efficiency for the pump.

From Table 5.10 for a 1:2.5 feed-water ratio in a 1/2-inch pipe at room temperature $k' \times 0.00369$ and $n' = 1.03$. Using these constants in Eq. (4.84), the required horsepower can be calculated as follows:

$$\frac{D(\Delta P)}{4L} = k'\left(\frac{8V}{D}\right)^{n'}$$

$$D = 1/2 \text{ inch} = 0.0416 \text{ ft}$$

$$L = 260 \text{ ft}$$

$$V = 360 \text{ ft/min} = 6 \text{ ft/sec}$$

$$\frac{0.0416(\Delta P)}{4 \times 260} = 0.0037\left(\frac{8 \times 6}{0.0416}\right)^{1.03}$$

$$\Delta P = \frac{5.22 \times 4 \times 260}{0.0416} = 130{,}000 \text{ lb/ft}^2$$

$$Q = AV = \frac{\pi}{4}(0.0416)^2(6)$$

$$= 0.00816 \text{ ft}^3/\text{sec}$$

$$Hp = \frac{130{,}000 \times 0.00816}{550} = 1.93$$

For 50 percent efficiency $Hp = 2 \times 1.98 = 3.86$ or about 4 horsepower.

Investigation of rheological properties of fluid foods, particularly viscosity and consistency, has been in progress for many years. For example, relationships have been established between the viscosity of milk and fat content, protein content, *PH*, total solids, seasonal variations, etc. Effects of chemical composition and additives as reflected by viscosity and consistency are among other examples (see Scott-Blair, 1953). The measurements reported are, however, usually on empirical basis which are quite inadequate for design of handling and processing equipment. Harper and Kamel (1966) studied the flow properties of a large number of tomato and fruit puree samples with several types of viscometers and concluded that there is a good possibility that useful correlations can be developed among the results on a viscosity parameter basis. The measurements of viscosity as reported are commonly in terms of single-point measurements using a given technique rather than reporting the curves of shear stress–shear rate. These curves not only characterize the fluid but, if properly reported, would yield many of the flow parameters usable in the design equations as well as characterizing the product quality on a more fundamental basis. Following are a few examples along this line of approach which have been reported in literature.

One of the classical studies of flow of fluid foods is rheological investigations on honey. Viscosity of honey is very markedly influenced by the water content. Since there has been no simple method of water content determination for honey by evaporation or chemical analyses, the rheological approach, using a simple viscometer, appears to be quite attractive. Among the various types of information a shear stress–shear rate diagram has also been reported for honey. Figure 5.56 shows the shear diagram for three types of honey as reported by Pryce-Jones (1953). The clover honey according to this diagram is a true Newtonian liquid as shown also in Fig. 4.40 in Chapter 4 and Table 5.11. Heather honey is apparently more pseudoplastic than clover honey if we assume that extrapolation of curves *A* and *B* will intercept the shear stress axis at the origin. If not, there must be a yield stress value for the two types of honey which apparently increases with time of rest.

Charm (1969, 1963) and Harper (1964, 1965) have reported some data on flow constants for several organic fluids and food materials as shown in Table 5.11. Two important conclusions from these studies were apparent: (1) most of the organic fluids tested were power-law fluids following flow Eq. (4.79) and (4.83); and (2) the flow constants vary considerably for

Table 5.10 Flow behavior index (n') and viscosity index (k') for organic slurries and pastes (from Herum *et al.*, 1966)

Feed-Water Ratio	Nominal Temperature °F	Nominal Tube Diameter					
		1/2-Inch		1-Inch		1-1/2-Inch	
		k'	n'	k'	n'	k'	n'
		Ground Corn Slurries					
1:1.10	Room			0.00308	1.56	0.00201	1.84
1:1.25	Room	0.306	0.936	0.00994	1.23		
1:1.25	106	0.0855	0.686				
1:1.25	150			28.6	0.478		
1:1.40	144			0.0128	1.41		
		Mixed Swine Ration Slurries					
1:1.50	Room	5.76	0.449	15.7	0.279	4.89	0.393
1:1.50	110			9.97	0.282		
1:1.50	143			10.6	0.282		
1:2.00	Room	0.0579	0.806	1.03	0.406	0.786	0.449
1:2.50	Room	0.00369	1.03				
		Ground Meat					
—	60	0.0173	1.03	1.70	0.419		
—	110	0.0432	0.727				
—	140	2.20	0.0676				

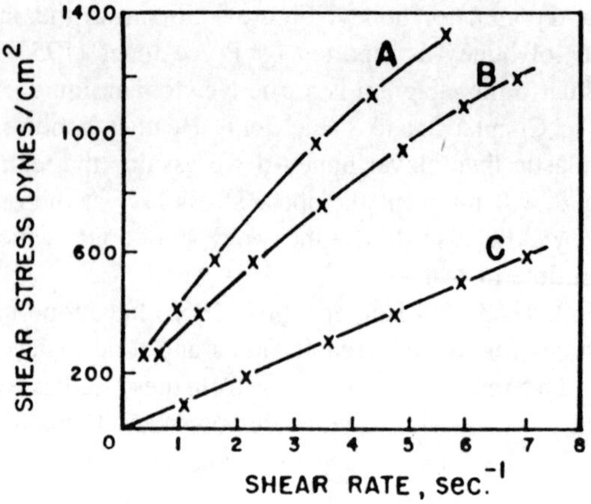

Figure 5.56 Shear stress–shear rate for honey. A—Heather honey after 12-hours rest; B—Heather honey after stirring; C—Clover honey, 80 poises (Pryce-Jones, 1953)

the same food material. Since more than one brand of the food material and several types of viscometers were used in the experiments, it is not evident which contributes more to the variation in the flow constants, the type of viscometer or the brand of the food material.

Charm (1963) employed a concentric cylinder (Merrill) and a single cylinder (Brookfield) viscometer to determine shear stress–shear rate of tomato puree and applesauce as shown in Fig. 4.40 in Chapter 4. To evaluate the flow constants three methods were proposed.

1 Assume that the material is a power-law fluid and follows an empirical power law equation such as Eq. (4.79) with a yield stress, τ_y, added to it as follows:

$$\tau = C\dot{\gamma}^n + \tau_y \tag{5.26}$$

where the notations are same as in Chapter 4, with the exception of $\dot{\gamma} = -dv/dr$=velocity gradient or shear rate. Based on this assumption, Eqs. (4.79) and (4.102) can be employed to calculate yield stress, τ_y, and flow constants C and n. Shear stress and shear rate can then be found from the above power-law equation.

2 Through the use of a narrow gap concentric viscometer, shear rate can be solved for directly from

$$\dot{\gamma} = -\frac{d\nu}{dr} = \frac{V}{G} = \frac{2\pi N R_b}{G} \tag{5.27}$$

where

V = linear velocity of moving cylinder
G = gap between concentric cylinders
N = speed of cylinder in r.p.s.
R_b = radius of the cylinder

If G/R_b is less than 0.05, for a Newtonian fluid, the variation across the gap in both shear stress and shear rate will be about 10% (Wilkinson, 1960). And shear stress can be calculated from torque exerted using Eq. (4.86).

3 Use of Casson's equation (Casson, 1959)

$$\sqrt{\tau} = K\sqrt{\dot{\gamma}} + \sqrt{\tau_y} \tag{5.28}$$

where K is a constant. For fluids obeying this equation a plot of $\sqrt{\tau}$ versus $\sqrt{\dot{\gamma}}$ should yield a straight line whose intercept is $\sqrt{\tau_y}$ as shown in Fig. 4.40. Having shear stress and shear rate, the power-law equation can again be used to evaluate the flow constants.

Harper (1964, 1965) who employed a specially designed concentric cylinder viscometer for obtaining shear stress-shear rate data on pear puree (1964) and tomato concentrates (1965), states that the Casson's equation is more difficult to use and does not give as good agreement with the experimental data as a power law equation. Harper's power-law equation in terms of apparent viscosity, η'', was similar to equation 4.83 which can be written in the following form:

$$\eta'' = C\dot{\gamma}^{(n-1)} \tag{5.29}$$

assuming that the power-law equation is of forms given by

$$\tau = C\dot{\gamma}^n$$

According to Eq. (5.29) a plot of log apparent viscosity versus log shear rate for a power-law fluid should yield a straight line. Harper found that for tomato concentrate (Harper, 1965) and pear puree (Harper, 1964), con-

Table 5.11 Examples of flow constants for some organic fluids (see equations 4.79 and 4.83)

Fluid	Temp. °F	Shear rate "$\dot{\gamma}$" sec^{-1}	Power law Const. "n"	Proportionality Const. "C" $\frac{dynes-sec^n}{cm^2}$	Apparent viscosity poise	Yield stress τ_y dynes/cm^2	Viscometer	Source
Applesauce A	75	—	0.645	5.0	—	0	Coaxial[1]	Charm, 1960
Applesauce B	75	—	0.408	6.6	—	0	Tube[2]	Charm, 1960
Banana puree A	75	—	0.458	65.0	—	0	Coaxial[1]	Charm, 1960
Banana puree B	75	—	0.333	107	—	0	Tube[2]	Charm, 1960
Honey A	75	—	1.00	56	56	0	Brookfield[3]	Charm, 1960
Honey B	75	—	1.00	61.8	61.8	0	Tube[2]	Charm, 1960
Human blood	75	—	0.89	3.84×10^{-2}	—	0	Tube[2]	Charm, 1960
Tomato puree	—	51–1370	0.554	9.2	—	213	Hauke[4]	Charm, 1963
Applesauce	—	8.4–457	0.470	56.3	—	595	Hauke[4]	Charm, 1963
Pear puree								
18.3% T.S.	90	2600[6]	0.486	22.5	—	35	Coaxial[5]	Harper and
45.6% T.S.	90	2600[6]	0.479	355	—	339	Coaxial[5]	Lebermann, 1964
Tomato juice								
5.8% T.S.	90	500–800	0.59	2.23	0.175[7]	—	Coaxial[5]	Harper and
30% T.S.	90	500–800	0.40	187	4.5[7]	—	Coaxial[5]	El Sahrigi, 1965
Tomato juice reconstituted	90	500–800	0.435	50.0	1.48[7]	—	Coaxial[5]	Harper and El Sahrigi, 1965

[1] Narrow-gapped, radius of outer cylinder 2.6 cm, gap 0.01 cm, outer cylinder remained stationary while the inner cylinder rotates (see Merrill, 1956).

[2] Compressed air was employed to force fluid through tube viscometer, gage measured pressure (see Charm, 1959).

[3] Single cylindrical or disk-type spindle.

[4] Coaxial with a spindle radius 4.08 cm and gap 0.0954 cm.

[5] Coaxial cylinders about 6 in. long and 3 in. diameter, gap 0.08 in. for tomato and 0.03 and 0.08 in. for pear puree. Torque and speed of stationary inner cylinder were recorded by an x–y recorder (see Harper, 1961).

[6] The high shear–rate ends of the logarithmic plots around which flow constants were obtained.

[7] At $\dot{\gamma} = 500\ sec^{-1}$.

siderable portions of the curves plotted were straight lines (Fig. 5.57). The slope of the approximated straight line was taken as $(n-1)$ and C was obtained from the y intercept. From Eq. (5.29) it is evident that at shear rate equal to unity (1 sec^{-1}), C is equal to apparent viscosity. To avoid the errors due to extrapolation of the lines to intercept the y axis, Harper reported the apparent viscosity for tomato concentrate at selected shear rates as shown in Table 5.11. Later, by taking examples from actual cannery operation in production of tomato catsup, he shows that the range of shear rate selected for reporting the apparent viscosity was appropriate for application to typical flow condition. An interesting observation by Harper in his work on tomato concentrate was that the apparent viscosities of the reconstituted samples were only about one-third those of the original concentrates (Table 5.11), indicating that this method of producing concentrates does not appear attractive. Figures 5.59 and 5.58 show the flow behavior of pear puree fitted approximately to power-law Eq. (4.79). No general equation was found to represent the entire range of the shear rates. Watson (1967) also has reported in rheological properties of fruit purees. He reported that a consistent measurable yield stress could not be found for apricot and apple purees and thus equations such as 5.26 are actually simplified to 4.79. He also found that the values of C in the power law equation were significantly correlated with solids and pectin content of apricot purees. An analysis of multiple regression showed that the soluble solids, insoluble solids, total solids, total pectin, and water soluble pectin together accounted for as much as 95% of the variation in the value of C. Particle size, temperature, and the degree of pectin degradation also affected the parameter C.

In reviewing the literature occasionally one finds some basic data on flow properties of fluid foods in connection with heat transfer studies of these materials. For example, Saravacos and Moyer (1967) have found values for power law constants C and n in studying the heating rates of liquid fruit products.

Frictional properties of fluid milk products have been investigated to determine the deviations from that of water. Betscher *et al.* (1966) determined viscosity and frictional characteristics of several fluid milk products by means of viscometers and measurement of pressure drops in a 1-inch sanitary pipe for a range of Reynold's number and several temperatures. The reliability of the apparatus and technique was established by using water as the reference liquid and the following relationship given by Moody (1947).

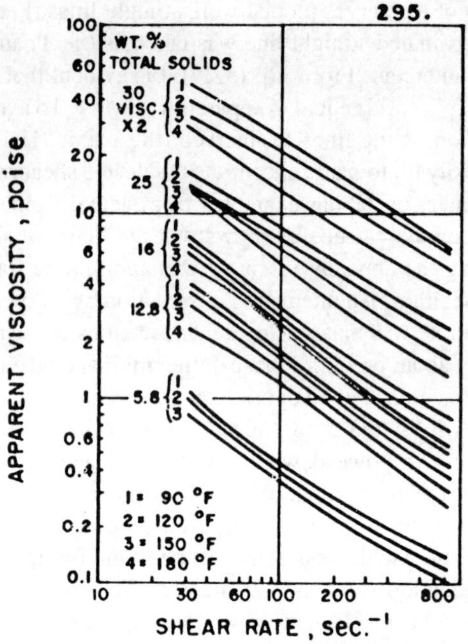

Figure 5.57 Apparent viscosity *vs.* shear rate of tomato concentrate as affected by temperature and % total solids (Harper and El Sahrigi, 1965)

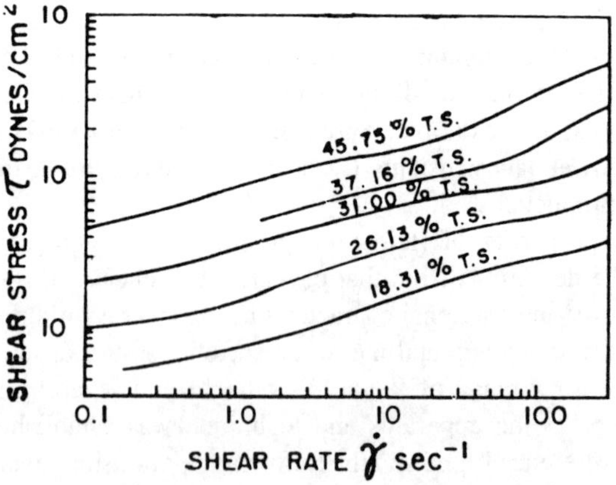

Figure 5.58 Shear stress *vs.* shear rate for pear puree at 90°F as affected by % total solids—log-log scale (Harper and Lebermann, 1964)

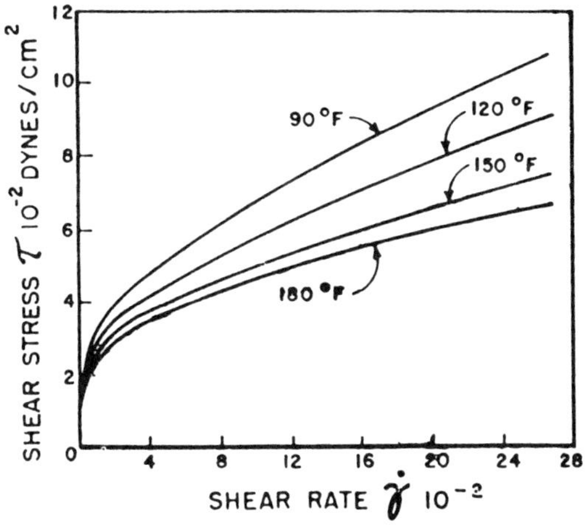

Figure 5.59 Shear stress *vs*. shear rate for pear puree at different temperatures—linear scale (Harper and Lebermann, 1964)

$$f = 0.0055\left\{1 + \left[20{,}000\left(\frac{e}{D}\right) + \left(\frac{10^6}{Re}\right)\right]^{1/3}\right\}$$

$$= 18180\,\Delta Pw/M^2 \qquad (5.30)$$

where f is Darcy friction factor, ΔP is pressure drop in lb/ft^2-ft, w is weight density in lb/ft^3, M is flow rate in lb/hr, e is pipe roughness in ft., D is pipe diameter in ft., and R_e is Reynold's number. The equation should be valid for Reynold's number in the range of 4×10^3 to 1×10^7 and roughness ratio $\frac{e}{D}<0.01$.

Results of this investigation (Figs. 5.60, 5.61) revealed that friction factors for the 9% and 20% skim milks, the regular milks, and the creams generally fell within about 5% of those for water. However, the friction factors for reconstituted skim milk (solids-not-fat, SNF, of 12% and 20%), butteroil (98%) and fortified milk (10% fat, 12% SNF) deviated appreciably from water. The skim milks and the fortified milk exhibited less friction than water, whereas butteroil exhibited more. In general, increases in fat content tended to increase the friction factor but decreased SNF had the opposite effect. The reason for these deviations was postulated to be the non-Newtonian behavior, surface tension at the particle-solvent and colloid-wall interfaces, as well as sedimentation drag coefficients.

An interesting example of viscometry of animal materials is the rheological studies on bull semen. Glover and Scott Blair (1966) investigated the effect of viscosity of bull semen on sperm mobility and viability which are assessed by primarily subjective means for artificial insemination. An instrument similar in principle to MacMichael viscometer was used to record continuous changes in viscosity at known rate of shear and temperature. Raising the temperature gradually increased the semen viscosity which may well be due to an activation of the sperm. As the sperm die, the viscosity falls, passes through a minimum and then rises steeply. The reason for the second rise is thought to be the aglomeration and clustering of dead sperms.

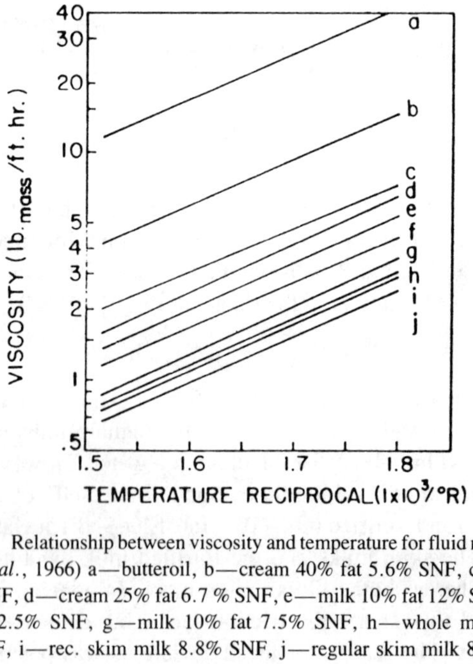

Figure 5.60 Relationship between viscosity and temperature for fluid milk products (Betscher *et al.*, 1966) a—butteroil, b—cream 40% fat 5.6% SNF, c—rec. skim milk 20% SNF, d—cream 25% fat 6.7 % SNF, e—milk 10% fat 12% SNF, f—rec. skim milk 12.5% SNF, g—milk 10% fat 7.5% SNF, h—whole milk 3.8% fat 8.7% NSF, i—rec. skim milk 8.8% SNF, j—regular skim milk 8.8% SNF

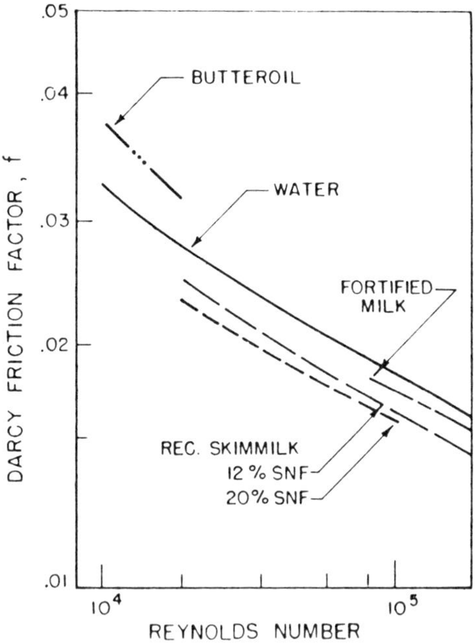

Figure 5.61 Comparison of friction factor of selected fluid milk products with that of water (Betscher *et al.*, 1966)

Example

Tomato juice at 90°F. and 5.8% and 30% total solid is to be transported 200 ft. in a pipe with 2 in. inside diameter at a rate of 20 tons per hour. Determine horse power and the resulting pressure drop for each case. Assume bulk densities of 68 lb./ft.^{3} and 70 lb./ft.^{3}, respectively, for each case.

From $Q=AV$, flow velocity

$$V = Q/A = 9.8/\pi(1/12)^2\ (60) = 7.49 \text{ ft./sec.}$$

$$Q = 20 \times 2000/68 \times 60 = 9.8 \text{ ft.}^3\text{/min.}$$

From Eq. (4.84), pressure drop

$$\Delta P = K'\left(\frac{8V}{D}\right)^{n'} (4L/D)$$

For 5.8% total solid,

$$\Delta P = 2.23\left(\frac{8 \times 7.49}{2/12}\right)^{0.59}\left(\frac{4 \times 200}{2/12}\right)$$

$$= 34469 \text{ Pa}$$

$$= 0.5 \text{ psi}$$

$$\text{HP} = \frac{(5 \times 144 \text{ lb/ft}^2)(9.8 \text{ ft}^3\text{/min})}{33000} = 0.214 \text{ hp}$$

For 30% total solid

$$Q = 20 \times 2000/70 \times 60 = 9.52$$

$$\Delta P = 187\left(\frac{8 \times 7.49}{2/12}\right)^{0.4}\left(\frac{4 \times 200}{2/12}\right)$$

$$= 944870 \text{ Pa}$$

$$= 137 \text{ psi}$$

$$\text{HP} = \frac{(137 \times 144 \text{ lb/ft}^2)(9.52 \text{ ft}^3\text{/min})}{33000}$$

$$= 5.69 \text{ HP}$$

Example

The relationship between the shear stress, τ, and the shear rate, γ, of a fruit juice concentrate is

γ, sec^{-1}	τ, lb/m^2
0.2	7.62×10^{-4}
0.5	1.10×10^{-3}
1.0	1.45×10^{-3}
2.0	1.91×10^{-3}
4.0	2.52×10^{-3}
6.0	2.97×10^{-3}
10.0	3.64×10^{-3}

Find the pressure drop required to pump this fruit juice concentrate through a circular pipe of 1″ diameter and 100″ long at a flowrate of 5 in^3 per sec.

Plot γ vs. τ on a piece of log-log paper. The linear relationship on the plot indicates that the fluid can be represented by the power law equation

$$\tau = C\gamma^n$$

From the graph we obtain

$$C = 1.45 \times 10^{-3} \text{ lb/m}^2$$

$$n = 0.4$$

The flowrate, q, through a cylindrical pipe can be expressed by Eq. 4.82

$$q = \pi\left(\frac{\Delta P}{2CL}\right)^{1/n}\left(\frac{n}{3n + 1}\right) \mathrm{R}^{(3n + 1)/n}$$

$$\Delta P = 2CL\left[\frac{q}{\pi}\,\frac{3n + 1}{n}\right]^n \frac{1}{R^{(3n + 1)}}$$

$$= 2(1.45 \times 10^{-3})(100)\left[\frac{5}{\pi}\,\frac{1.2 + 1}{0.4}\right]^{0.4} \frac{1}{(.5)^{2.2}}$$

$$\Delta P = 3.17 \text{ psi}$$

If fluid foods follow the power-law relations, the friction in pipes can be estimated from charts designed for pipe friction versus a generalized Reynolds number, given by Heldman (1975), as follows:

$$GN_R = \frac{\rho V^{-(2 - n)} d^n}{2^{n - 3} c\left(\dfrac{3n + 1}{n}\right)^n} \tag{5.31}$$

where ρ and V are mass density and velocity of the fluid, n and c are flow behavior index and viscosity coefficient of the fluid, and d is the pipe diameter. Note that GN_R varies with the same parameters as N_R except it includes the flow behavior index of the power-law fluid.

Example

Applesauce B with the characteristics shown in Table 5.11 is being pumped through a 2-inch diameter pipe at a velocity of 10 ft/sec. Determine the pipe friction using the chart given in Fig. 5.61a.

Using the data in Table 5.11 for applesauce B and the given data, GN_R can be computed as follows:

$$n = 0.408,$$

$$c = 6.6\,(\text{dynes-sec}^n/\text{cm}^2)(1.45 \times 10^{-5} \times 144) = 1.38 \times 10^{-2}\,\text{lb-sec}^n/\text{ft}^2$$

Assume weight density of applesauce is 70 lb/ft^3

$$GN_R = \frac{(70/32.2)(10)^{-(2\,-\,0.408)}(2/12)^{0.408}}{2^{(0.408\,-\,3)} \times 1.38 \times 10^{-2}[(3 \times 0.408 + 1)/0.408]^{0.408}} = 8960$$

Reference to Fig. 5.61a gives a friction factor of f=0.0042 for n=0.4.

Once the friction factor is known, the pumping problem of any non-Newtonian food is not any different than the pumping of any other liquid in that the general balance or Bernoulli equation is used to compute the work required to transport the fluid food. The contributing terms of this equation, static pressure, location above a reference line, and W will remain the same as in transport of other fluids. The kinetic energy term, however, is calculated from the following expression to account for the flow behavior index.

$$KE = V^2/\alpha \times 2$$

where α=(4n+2) 5n+3)/3 (3n+1)^2.

Furthermore, the friction term, which appears on the right side of the equation with a positive sign will be the sum of all the friction factors, found from charts such as Fig. 5.61a, converted to a quantity with units consistant with the other terms of the equation using the expression

$$f(V^2L/gR)$$

where V is flow velocity, and L and R are length and radius of the pipe section.

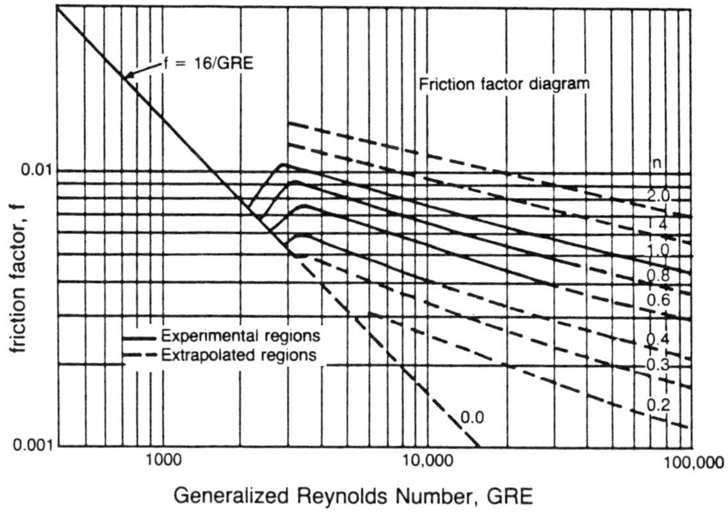

Figure 5.61a Generalized Reynolds number versus friction factor for fluids with different n factor (Dodge and Metzner, 1959)

The mechanics of coating of non-Newtonian fluids has been studied extensively. The flow characteristic of such fluids in the coating process is given by Linke (1969). The entrainment of non-Newtonian fluids on a flat plate being withdrawn from the fluid has also been studied by several investigators. The thickness of the film of the liquid, t, was converted to a non-dimensional thickness given in terms of density ρ, acceleration of gravity, g, and surface tension T_s, as a lump parameter $5(\rho g/T_s)^{0.5}$, and was related to the rheological parameters of the material (Tallmadge, 1970). Groenreld (1971) has also published several papers on this subject and suggested a non-dimensional thickness which included the rheological parameters of the fluid. For fluids, Newtonian or non-Newtonian with no yield stress, when vertical velocity of the plate approaches zero, the film thickness also approaches zero. For fluids with a yield stress, when vertical velocity of the plate is zero, an equilibrium is reached when the weight of the film, Atw, equals the force necessary to overcome the yield stress of the film, $A\,T_y$. From such a relationship Loncin and Merson (1979) give the following expression for the minimum thickness of the film:

$$Atw = A\,T_y$$

or

$$t = T_y/w \tag{5.32}$$

where A is the plate area, w is weight density of the fluid, t is the minimum thickness of the film and T_y is the yield stress of the fluid.

Example

A chocolate paste has a yield stress of 0.418 lb f/ft^2 (20N/m^2), a weight density of 68.66 lb/ft^3 (ρg=1000 Kg/m^3), a temperature of 122°F (50°C). If a flat plate is submerged in this paste and withdrawn, what is the film thickness entrained on the plate after the drainage is complete?

Using eq. 5.31,

$$t = 0.418 \text{ lb/ft}^2/68.66 \text{ lb/ft}^3$$
$$= 0.0061 \text{ ft}$$
$$= 1.86 \text{ mm}$$

The application of the flow behavior index n and a consistency constant to the design of food extruders has been shown by Levine (1982). Through the use of a simplified extruder model, dimensionless correlations of extruder output and power to pump the product was developed. Given rheological constants, these correlations predicted extruder performance independent of scale, thus providing a technique for scale-up.

For example, Fig. 5.61b shows an idealized conceptual model of an extruder for which the power to pump the dough for a certain finished product is

$$\text{Power} = QnP \tag{5.33}$$

Where Qn, as shown in Fig. 5.61b, is the extruder output and P is the pressure drop through the slot of the extruder. This pressure drop was estimated by the following relationship:

$$P = m\left(\frac{2(2n+1)}{nt_e h^2}\right)^n\left(\frac{2L}{h}\right)e^{A/T}e^{BM}q^n \tag{5.34}$$

In this expression, t_e, h, L are dimensional parameters of the ideal model, A and B are constants, T and M are temperature and moisture content of

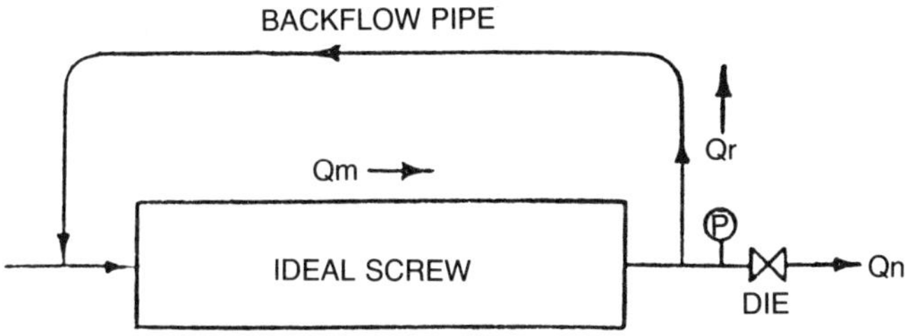

Figure 5.61b An idealized conceptual model in an extruder (Levine, 1982)

the dough, q is the flow rate through the die, and m and n are consistency and flow behavior index, respectively.

In recent years, much interest has developed in liquid handling of animal waste slurries and considerable research is being conducted on physical charactertistics and flow properties of these materials. Kumar (1969) has applied the principles of rotational viscometry to study the flow behavior of liquid manure. A coaxial-cylinder viscometer designed, constructed and used by him for this purpose is shown in Fig. 5.62. The inside cylinder (bob) of the viscometer was open at the bottom and had four vent holes at the top similar to the bob of a Stormer viscometer. The outer cylinder (cup) was coupled through a train of pulleys and belts to a variable speed drive. The gap between the cup and the bob was made adjustable by inserting in the gap sleeves of different diameters. Torque was measured by mounting a torque meter on the stem of the bob and a counter measured the RPM. Shear rate was represented as $\Omega=2\pi N$ and shear stress was evaluated from $\tau_w=\dfrac{T}{2\pi R_b^2 h}$, where T was the torque meter's reading. With this apparatus it was possible to take the readings on torque versus speed of rotation of the cup for liquid wastes of different physical make-up and properties. Equations (4.83) and (4.103) were used for calculating the viscosity index, flow behavior index, and apparent viscosity.

Using this technique Kumar (1969) studied the effects of dilution, addition of sawdust, and temperature on the shear characteristics of liquid manure (Fig. 5.63). In general, manure with approximately 5.3 percent and below total solids content showed a Newtonian behavior and above approximately 6 percent total solids showed a pseudoplastic (non-Newtonian) flow characteristic. The apparent viscosity of manure slurry decreased with an increase in dilution (reduction in total solids content). The decrease in

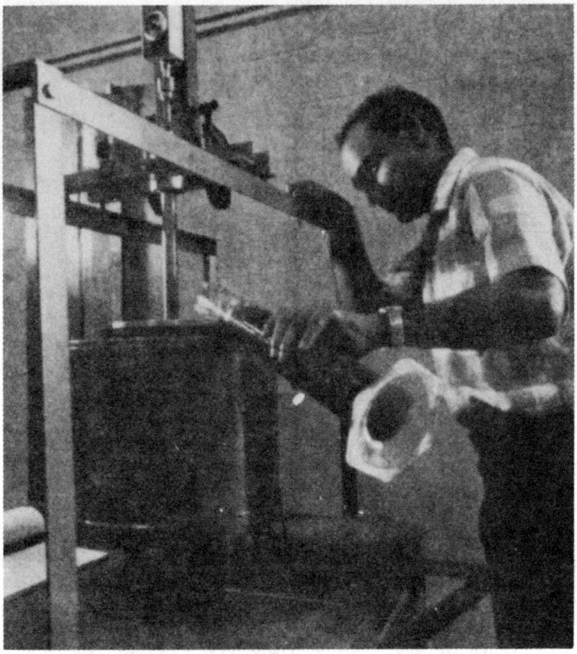

Figure 5.62 Rotational viscometer for flow behavior study of liquid manure (Kumar, 1969)

temperature increased the apparent viscosity of the manure slurry. The addition of sawdust up to 10 percent of the mixture decreased the apparent viscosity of the slurry up to approximately 8.7 percent total solids content. The conclusion from this study was that a waste slurry with a total solids content of 4 to 6 percent is a good compromise between excessive volume of handling and power requirement.

Example

The following data were obtained, using the viscometer in Fig. 5.62, for a manure slurry at 9.28 percent total solids content and 76°F.

Inner diameter of outer cylinder = 26.0 cm
Outer diameter of inner cylinder = 20.4 cm
Height of fluid on the bob = 20.0 cm
Apparatus factor coefficient = 0.838

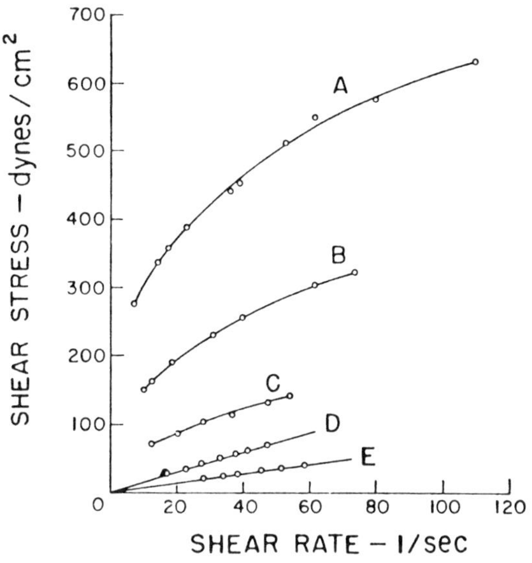

Figure 5.63 Shear diagrams for liquid manure showing the effects of temperature, addition of sawdust, and dilution (Kumar, 1969) A—7% T.S.C., 46°F; B—7.7% T.S.C, 76°F; C—7.8% T.S.C with 5% sawdust, 76°F; D—4.1% T.S.C., 46°F; E—5.3% T.S.C with 5% sawdust, 76°F

Rotating speed N rpm	Angular velocity Ω $^1/_{sec}$	Torque T gm-cm	Shear stress τ_w dynes/cm^2
9.8	1.026	4032.60	302.33
19.8	2.072	4867.20	364.95
28.2	2.948	5593.25	419.44
36.4	3.086	6011.20	451.27
52.6	5.508	6684.60	501.27
80.6	8.440	7590.70	569.28

Determine: (*a*) flow behavior index, (*b*) viscosity index, (*c*) type of flow, and (*d*) apparent viscosity.

(*a*) The rpm and torque were converted to angular velocity Ω, and shear stress, τ_w, as shown below for the first run

$$\Omega = 2\pi N = 2 \times 3.14 \times 9.8/60$$

$$= 1.026 \text{ l/sec}$$

$$\tau_w = \frac{T}{2\pi R_b^2 h} = \frac{4032.60}{2\pi (20.4)^2 (20)}$$

$$= 0.3086 \text{ gm/cm}^2 \quad \text{or } 302.33 \text{ dynes/m}^2$$

Data on angular velocity Ω and shear stress were then plotted on log-log paper. The resulting graph was a straight line. The slope of this line gave the value of flow behavior index which was found to be n=0.2980.

(b) From Eq. (4.103)

$$\Omega = \frac{n}{2}\left(\frac{\tau_w}{C}\right)^{1/n}\left[1 - \left(\frac{R_b}{R_c}\right)^{2/n}\right]$$

The straight line of the log-log graph was used to select any point for values of angular velocity and the corresponding shear stress. The point selected gave Ω=5.1 l/sec and τ_w=490 dynes/cm^2. Also

$$\frac{R_b}{R_c} = \frac{\text{outer radius of inner cylinder}}{\text{inner radius of outer cylinder}} = \frac{10.2}{13.0}$$

$$= 0.7848$$

Substituting the value of n, Ω, τ_w and $\frac{R_b}{R_c}$ in Eq. (4.103), viscosity index C was found to be

$$C = 60{\cdot}25 \frac{\text{dynes} - \text{sec}^n}{\text{cm}^2}$$

(*c*) Type of flow is pseudoplastic (non-Newtonian) because the value of n is less than 1.

(*d*) From Eq. (4.83)

$$\eta'' = C^{1/n}\tau_w^{(1 - 1/n)}$$

Substituting numerical values of n, C, and τ_w in above equation

$$\eta'' = 1150.30 \text{ centipoise}$$

$$\begin{aligned}\text{True apparent viscosity} &= \eta'' \times \text{apparatus constant} \\ &= 1150.30 \times 0.8380 \\ &= 963.93 \text{ centipoise}\end{aligned}$$

Substituting the values of C, and n for any selected shear stress, the corresponding shear rate can be computed using Eq. (4.99). For example:

$$\begin{aligned}(-r\, d\omega/dr) &= \left(\frac{\tau_w}{C}\right)^{1/n} \\ &= \left(\frac{302.33}{160.25}\right)^{1/0.298} \\ &= 8.44 \text{ 1/sec}\end{aligned}$$

In addition to the work reviewed in this section there are several books and numerous technical papers on the subject of emulsions and suspensions and handling of slurries, mostly of inorganic nature (Bowen, 1961; Caldwell, 1941; Hermans, 1953; Mill, 1959; Steiner, 1959). Many of the rheological principles developed for these fluids can be applied to organic fluids of interest to food and agricultural industries.

5.8 COMPACTION OF AGRICULTURAL MATERIALS

Compaction of agricultural materials, such as forages, grains and numerous food products, enables more efficient mechanical handling and requires less space for storage and transportation.

The economics of compacting milled wheat has been investigated by Hassan *et al*. (1973) at hydrostatic pressures up to 4.1 MPa. They concluded that substantial savings could be made in the storage and handling costs of compacted agricultural granular materials. Optimum low cost storage density for wheat was reported as 1080–1100 kg/m^3.

Kumar (1973) investigated the axial compaction behavior of ground corn at pressures up to 74 MPa and a punch velocity of 0.024 mm/s. He reported that his data were adequately described by an exponential relationship between the density and the pressure. Loewer et al. (1977) have studied the low pressure compaction of ground corn (7–70 kPa) as it relates to forces in bulk storage structures.

Mohsenin and Zaske (1976) investigated the axial compaction behavior of forage and wood by-products at pressures up to 70 MPa and a piston velocity of 330 mm/s. They found that a definite relationship existed between the stress relaxation and the expansion, final density, and handling durability of their test materials being compressed to a pellet or wafer. The compaction behavior of hay has been investigated by Mewes (1958, 1959) and Busse (1964) and that of various food items by Rahman et al. (1969, 1971, 1976). An instrumented compression chamber developed by Menzies (1976) which satisfactorily measured wall forces, friction forces on walls, applied load and piston displacements when compacting hay to pressures of up to 11.2 MPa. More recently, stable wheat straw briquettes have been produced in which the total energy input was only about 6 percent of the calorific value of the material (Smith *et al*. 1977).

McNulty and Mohsenin (1979) investigated the compaction behavior of bulk whole kernel at pressures leading to sample failure as manifested by oil expression or sample extrusion to determine if high pressure compaction would result in a temperature rise in the material of sufficient magnitude to alter its chemical, microbial or physical properties.

In a related paper, McNulty and Mohsenin (1977) have reported on the effect of moisture content and internal tissue gas on compaction behavior of whole corn kernel under pressures which did not result in either oil expression or sample extrusion. It was also investigated to determine the effect of pressure on the microbial activity of compacted whole corn kernels (McNulty *et al*., 1977).

Compaction Models

Bockstiegel (1966) and Orr (1966) have reviewed the various theories which have been applied to describe the compaction of both metallic and nonmetallic powders. These theories have also been applied to granular materials.

An early empirical relationship suggested that porosity, U, was an exponential function of pressure, P, as follows:

$$U = U_0 \exp(-kP) \tag{5.30}$$

where U_0=porosity at zero pressure=initial fractional void volume: k= constant.

Comparing the densification of metallic powders to a first order chemical reaction, Heckel (1961) developed the following relationship:

$$\ln [I/(I - D)] = kP + a \tag{5.31}$$

where D=relative apparent density=ρ/ρ_t=$I-U$; = bulk density at pressure, P; ρ_t=density of the compact or non-porous material; a, k=constants.

Using a thermodynamic approach, Kumar (1973) proposed the following relationship for compaction of ground whole corn:

$$\ln = Kp + a \tag{5.32}$$

Equations given above do not adequately describe compaction behavior especially at low pressures. Cooper and Eaton (1962) have attempted to resolve this problem by deriving a two-part expression for the fraction of attained compaction as follows:

$$(V_0 - V)/(V_0 - V_t) = a_l \exp(-k_l/P) + a_s \exp(-k_s/P) \tag{5.33}$$

where V_0=initial sample volume at zero pressure; V=volume at pressure P; V_t=compact volume; a_l, k_l and a_s, k_s=constants relating to the filling of large and small pores respectively.

Bockstiegel (1966) notes that: None of the formulae proposed so far seem to be generally applicable as most of them deliver an acceptably accurate description of the experimental data only over a rather limited range of pressures.

The work by McNulty and Mohsenin (1979) involved high pressure compaction of whole corn kernels, shown in Figure 5.64, using a cylindrical punch and die set. Compression loading rates of 0.028, 0.28, and 2.8 mm/s were applied at moisture contents of 6.6 to 24.2% (W.G.). In general, the structure of low moisture corn (6.6 percent w.b.) failed as manifested by oil expression which was first observed at pressures in the range of 140–310 MPa, whereas high-moisture corn (24.2 percent w.b.) failed as manifested by sample extrusion at the lower pressure range of 40–70 MPa. Oil expression or sample extrusion occurred only after the unit or kernel density

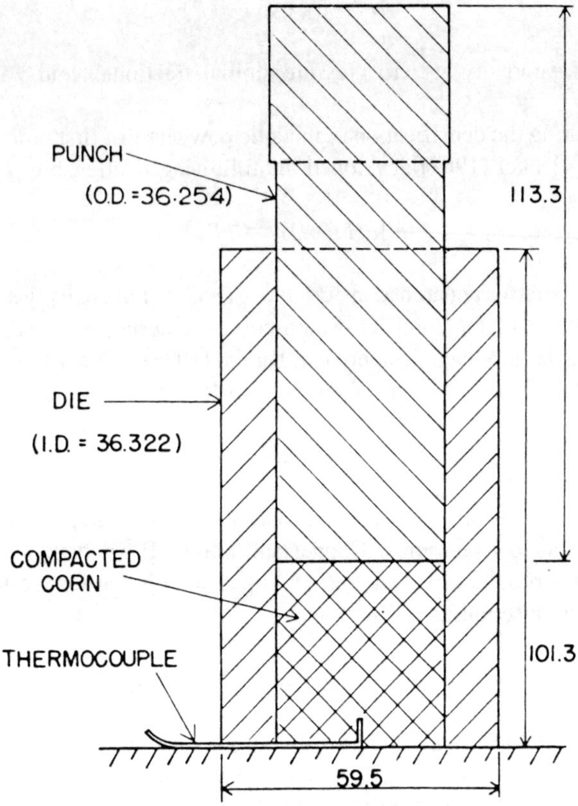

Figure 5.64 Punch and die set for compacting whole corn kernels under high pressures (McNulty and Mohsenin, 1979)

had been exceeded (i.e., after all gas external to the corn kernels had been expelled) and after internal tissue gas had been compressed. The presence of internal tissue gas greatly complicated the analysis of experimental data. All compaction profiles revealed curvature whether plotted in the form of log density versus pressure or log porosity versus pressure. This behavior supported Bockstiegel's (1966) conclusion that existing models do not adequately represent compaction behavior of particulate matter. After load removal, all samples expanded to a zero pressure final density which was less than the kernel density. Final compression ratios at zero pressure ranged from 1.78 to 2.18 corresponding to reduction in corn volume of 44 to 57 percent.

After ejection from the die, samples failed to hold their cylindrical shape. Thus, commercial application of this technique to corn compaction would necessitate bonding and/or packaging of the compacted sample to facilitate mechanical handling.

Table 5.12 shows a summary of compression tests resulting in various densities and temperature rises at various loading rates. Figure 5.65 gives the notation and the density terminology used in this study. The effect of loading rate on the relationship between pressure and density as well as pressure and porosity is shown in Figures 5.66 and 5.67.

In another study, McNulty and Mohsenin (1977) employed an 18.5 mm diameter die and pressures up to 34.6 MN/m^2 to determine the effect of moisture content on porosity and density and stress relaxation of whole corn kernels. Figures 5.68 and 5.69 show the results. In the stress relaxation studies, two moduli were defined as follows:

$$E_t(t) = \sigma(t)/\varepsilon_{fp} = Ph_0/(h_0-h_{fp}) \qquad (5.34)$$

$$E_t(t) = \sigma(t)/U_{fp} = P/U_{fp} \qquad (5.35)$$

where $\sigma(t)$=stress at time, t; ε_{fp}="at pressure" final strain which varied from a low of 42.3% at 6.6% moisture to a high of 53.8% at 22.0% moisture. In general, the strain moduli decreased systematically as moisture content increased. It should be noted that the resistance of the material to compaction increases as strain increases, i.e., as porosity decreases. Thus, the 22.0% moisture sample which was compacted to the highest strain recorded of 53.8% would be expected to exhibit relatively high relaxation moduli (Fig. 5.70). The fastest rate of relaxation was at 15.9% moisture level which also produced the most stable compacted corn samples.

The work on compaction of bulk whole corn kernels was extended into the effect of high pressure compaction on microbial growth (McNulty *et al.*, 1977). The effect of compaction resulted in a drastic reduction of viable cell count of bacteria and mold in corn samples during the first 24 hours of storage. Thereafter, up to 21 days, the viable cell counts increased gradually. The "compression shocks" observed in this study was similar to injury of micro-organisms by other physical and chemical treatments of food materials.

a	constant	
A	area	m^2
CR	compression ratio	
d	diameter	m
D	relative density $=\rho/\rho_t=1-U$	
E	relaxation modulus	N/m^2
h	sample height	m
ID	internal diameter	m
k	constant	
m	mass	kg
OD	outer diameter	m
P	pressure	N/m^2
t	time	s
T	temperature	°C
U	porosity	
V	volume	m^3
ρ	density	kg/m^3
ε	strain	
σ	stress	N/m^2

SUBSCRIPTS

B	bulk	max	maximum
e	external	o	initial
ext	extrusion	oil	oil expression
f	final	P	at pressure
i	internal	P_o	at zero pressure
k	kernel	s	small
l	large	t	compact

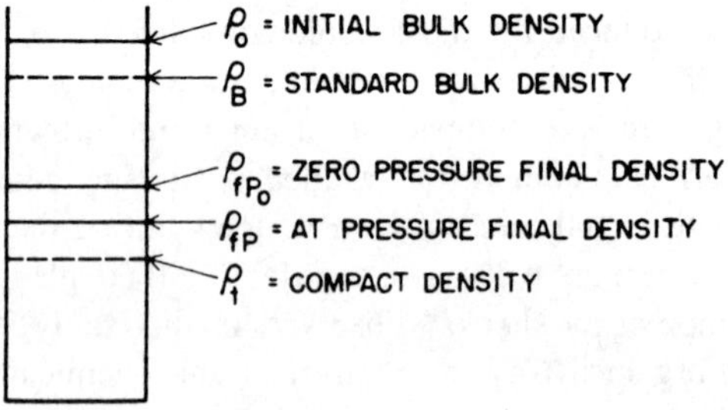

Figure 5.65 Density terminology and notation used in studying compaction of whole corn kernels (McNulty and Mohsenin, 1979)

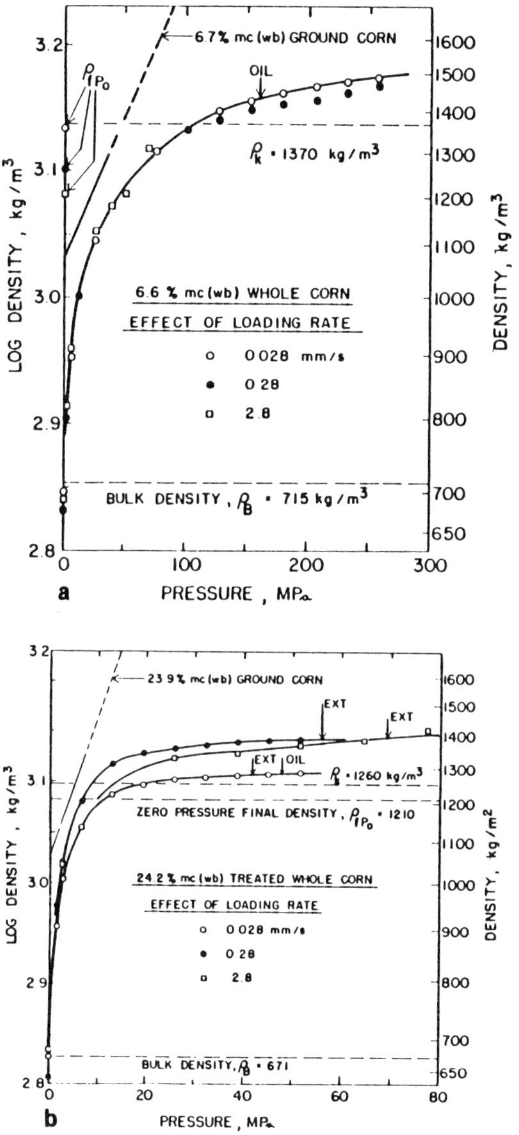

Figure 5.66 Effect of corn kernels and ground corn (data from Kumar, 1973) showing the pressures at which sample extrusion (EXT) and oil expression (OIL) occurred (McNulty and Mohsenin, 1979).

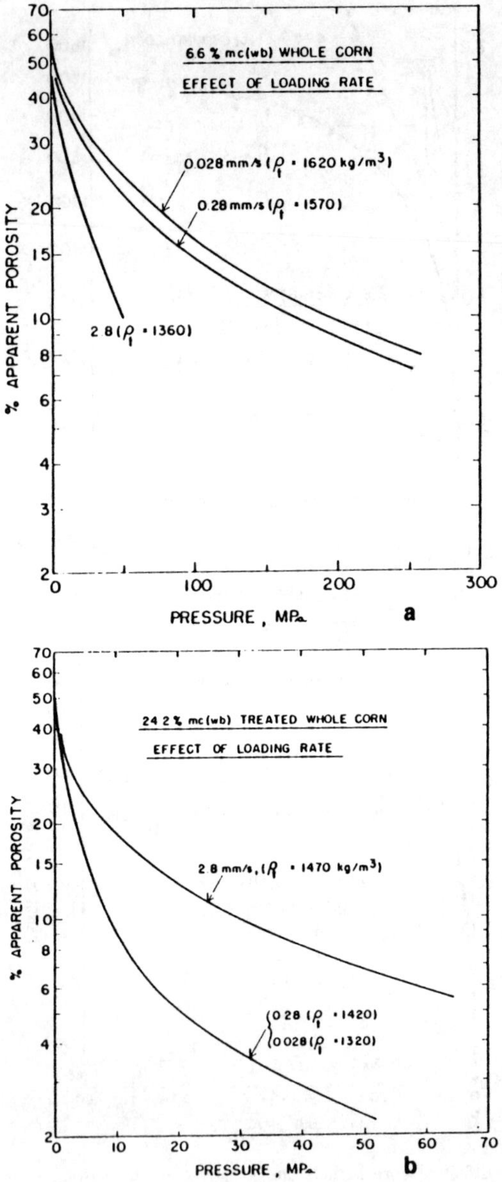

Figure 5.67 Effect of loading rate on apparent porosity of low-moisture and high-moisture corn kernels (McNulty and Mohsenin, 1979)

Table 5.12 Summary of compaction tests on whole corn kernel at 28°F (McNulty and Mohsenin, 1979)

Percent moisture, w.b.	Loading rate, mm/s	Density, kg/m^3				CR_{P0}	Maximum pressure, MPa	ΔT, °C
		ρ_0	ρfP	ρfP_0				
6.6	0.028	703	1490	1360	1620	1.93	258	2
6.6	0.28	677	1460	1260	1570	1.86	252	8
6.6	2.8	694	1310	1210	1360	1.74	69.6	4
24.2	0.028	673	1280	NM	1320	NM	84.1	1
24.2	0.28	641	1390	1210	1420	1.89	85.5	3
24.2	2.8	681	1390	1210	1470	1.78	83.3	3

CR_{P0} = compression ratio at zero pressure = h_0/h_{fP0} = $\rho FP_0/\rho_0$; NM = not measured.

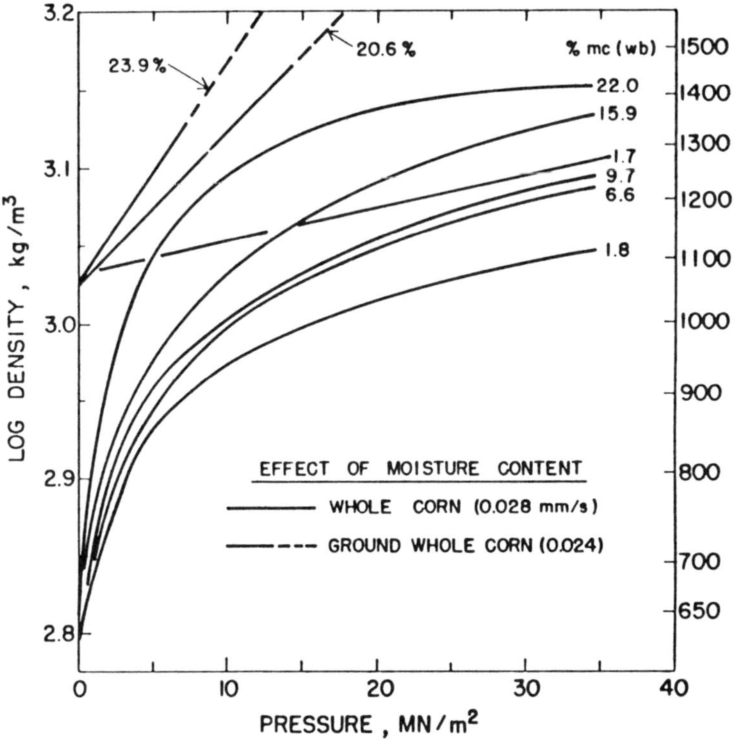

Figure 5.68 Effect of moisture content on density–pressure relationship of corn kernels and ground corn (data from Kumar, 1973) at 0.028 mm/s loading rate in an 18.5 mm diameter die (McNulty and Mohsenin, 1977)

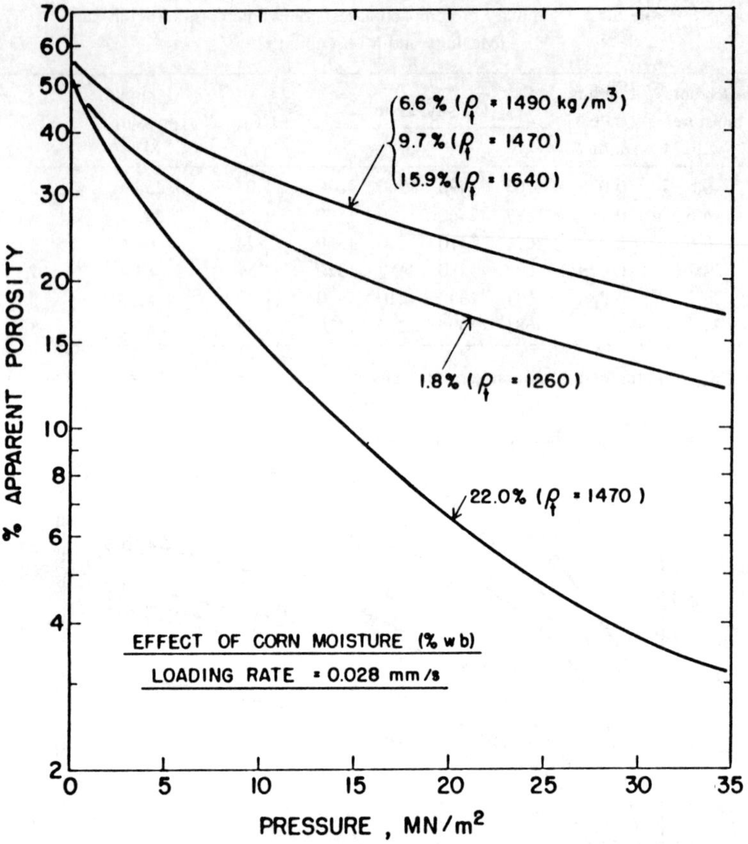

Figure 5.69 Effect of moisture content on porosity–pressure relationship of corn kernels at 0.028 mm/sec loading rate in an 18.5 mm diameter die (McNulty and Mohsenin, 1977)

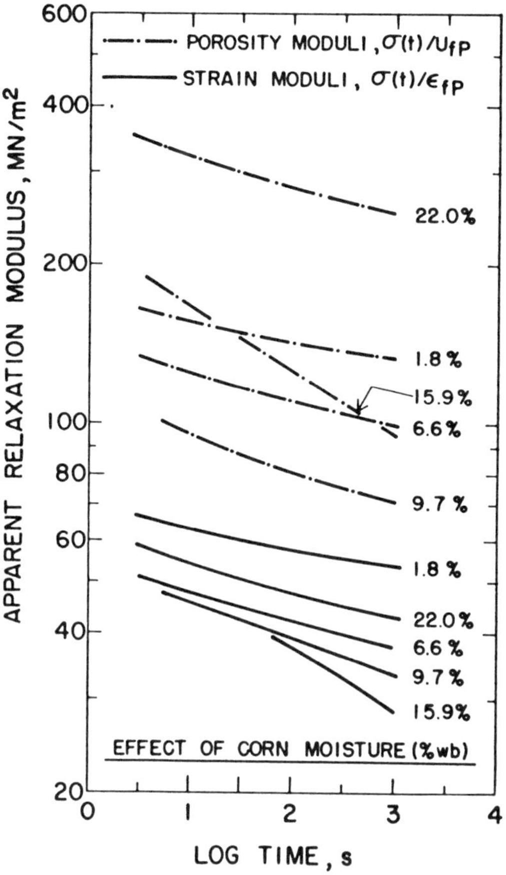

Figure 5.70 Relaxation moduli of compacted whole corn kernels subjected to a maximum pressure of 39.6 MN/m^2 (McNulty and Mohsenin, 1977)

6

CONTACT STRESSES BETWEEN BODIES IN COMPRESSION

AMONG THE VARIOUS mechanical tests available for solid biomaterials the simplest and most common is compression tests. Solids in their intact forms or as prepared specimens are subjected to simple axial compression tests employing a rigid cylindrical die, a steel ball, or rigid flat plates as the loading devices. In the case of compression tests of intact materials, such as seeds and grains, an egg, fruits and vegetables, the loading device is compressed against a convex body which results in a complex stress distribution particularly if the skin of such materials as fruits is left intact.

This inability to express the test results in terms of stress and strain, has forced the researchers to report their data in terms of the observed force and deformation which are not readily usable in the constitutive equations* available from solid mechanics. The problem is, therefore, to express these force–deformation data in terms of stress–strain or some other parameters, such as a modulus which can be used in the mathematical treatments discussed in Chapter 4. In this Chapter, several methods which have been employed by various investigators will be discussed.

6.1 HERTZ PROBLEM OF CONTACT STRESSES

In 1896 Heinrich Hertz (Hertz, 1896) proposed a solution for contact stresses in two elastic isotropic bodies, such as the case of two spheres of the same material touching each other. In this problem, Hertz attempted to find answers to such questions as the form of the surface of pressure, the magnitude

* Constitutive equation, which is a term used in continuum mechanics, simply refers to an equation which expresses the relationship between certain physical properties of a material. Hooke's law, for example, is a constitutive equation.

of the curve of pressure, normal pressure distribution on the surface of pressure, the magnitude of the maximum pressure, and the approach of the centers of the bodies under the pressure.

Before discussing the Hertz problem it is necessary to examine some of the fundamental assumptions which were originally made by Hertz in developing his theory of contact stresses. These assumptions, given by Kosma and Cunningham (1962) are the following:

1 *The material of the contacting bodies is homogeneous.* Obviously this restriction is not as easily achieved for biological materials as for any engineering material. However, if we assume random in-homogeneity for the body under load, the mechanical behavior may be considered as being equivalent to that of a homogeneous body.
2 *The loads applied are static.*
3 *Hooke's law holds.* At very low levels of load this condition can be approximated. This has been demonstrated in the case of several products.
4 *Contacting stresses vanish at the opposite ends of the body (semi-infinite body).* As will be discussed, it is possible to test the validity of this condition if certain parameters are known.
5 *The radii of curvature of the contacting solid are very large when compared with the radius of the surface of contact.* In an illustrative example, where the pressure of steel spheres against a rigid horizontal plane was to be determined, Hertz considered the ratio of 1/10 between the radius of the circle of contact and the radius of the sphere too large for his equations to be applicable.
6 *The surface of the contacting bodies are sufficiently smooth so that tangential forces are eliminated.*

Formulas based on the Hertz theory give the contact area, maximum compressive surface stresses, and the combined deformation of the contacting bodies at the point of contact. Figure 6.1 shows the general case where the two bodies, 1 and 2, with minimum and maximum radii of curvature R_1 and R_1' for body 1 and R_2 and R_2' for body 2 are pressed together with load F. Hertz has proved that the intensity of the surface stresses is presented by the ordinates of a semi-ellipsoid constructed on the surface of contact. The maximum contact stress, being at the center of the surface of contact, is denoted by $S_{\max}$ and is given by

$$S_{\max} = 3/2\left(\frac{F}{\pi ab}\right) \tag{6.1}$$

where a and b are the major and minor semiaxes of the elliptic contact area. Note that the maximum contact stress is 1½ times the average pressure on the surface of contact. It can be shown that the values of a and b in Eq. (6.1) may be calculated from the following equations:

$$a = m\left[\frac{3FA}{2\,(1/R_1 + 1/R_1' + 1/R_2 + 1/R_2')}\right]^{1/3} \tag{6.2}$$

$$b = n\left[\frac{3FA}{2\,(1/R_1 + 1/R_1' + 1/R_2 + 1/R_2')}\right]^{1/3} \tag{6.3}$$

$$A = \frac{1-\mu_1^2}{E_1} + \frac{1-\mu_2^2}{E_2} \tag{6.4}$$

where m and n are constants given in Table 6.1, the quantity A is defined by Eq. (6.4), radii of curvatures R are defined in Fig. 6.1, μ and E are Poisson's ratio and modulus of elasticity of the contacting bodies, and subscripts 1 and 2 refer to bodies 1 and 2. The combined deformation of the two bodies along the axis of load at the point of contact (approach of the centers of the two bodies) is given by

$$D = k/2\left[\frac{9F^2A^2}{\pi^2}(1/R_1 + 1/R_1' + 1/R_2 + 1/R_2')\right]^{1/3} \tag{6.5}$$

The values of k, m and n depend on the principle curvatures of the bodies at the point of contact and the angle ϕ between the normal planes containing the principle curvatures. These values are obtained from Table 6.1 by first calculating $\cos T$ from

$$\cos T = \frac{[(1/R_1 - 1/R_1')^2 + (1/R_2 - 1/R_2')^2 + 2(1/R_1 - 1/R_1')(1/R_2 - 1/R_2')\cos 2\phi]^{1/2}}{(1/R_1 + 1/R_1' + 1/R_2 + 1/R_2')}$$

Derivation of the above equations is given by Kosma and Cunningham (1962), Seely and Smith (1961), and Timoshenko and Goodier (1951). The use of Table 6.1, which is a condensed form of a more extensive table given by Kosma and Cunningham (1962), is the most convenient way of solving the above equations when the principle planes of curvature of the two contacting bodies do not coincide.

Table 6.1 Values of m, n, and k corresponding to Cos T (After Kosma and Cummingham, 1962)

Cos T	m	n	k	Cos T	m	n	k
0	1.000	1.000	1.3514	0.760	2.111	0.571	1.111
0.020	1.013	0.987	1.3512	0.780	2.195	0.558	1.091
0.040	1.027	0.974	1.3507	0.800	2.292	0.545	1.070
0.060	1.041	0.961	1.3502	0.820	2.401	0.530	1.047
0.080	1.056	0.958	1.3494	0.835	2.494	0.518	1.027
0.100	1.070	0.936	1.3484	0.845	2.564	0.511	1.013
0.120	1.085	0.924	1.3469	0.855	2.638	0.502	0.998
0.140	1.101	0.912	1.3453	0.865	2.722	0.494	0.983
0.160	1.117	0.901	1.343	0.875	2.813	0.485	0.966
0.180	1.133	0.889	1.341	0.885	2.915	0.476	0.948
0.200	1.150	0.878	1.339	0.895	3.029	0.466	0.929
0.220	1.167	0.866	1.336	0.905	3.160	0.455	0.908
0.240	1.185	0.855	1.334	0.912	3.262	0.448	0.892
0.260	1.203	0.844	1.330	0.916	3.326	0.443	0.882
0.280	1.222	0.833	1.327	0.920	3.395	0.438	0.872
0.300	1.242	0.822	1.323	0.924	3.468	0.433	0.862
0.320	1.262	0.812	1.319	0.928	3.547	0.428	0.851
0.340	1.283	0.801	1.315	0.932	3.631	0.423	0.839
0.360	1.305	0.790	1.310	0.936	3.723	0.418	0.828
0.380	1.327	0.780	1.305	0.940	3.825	0.412	0.815
0.400	1.351	0.769	1.300	0.944	3.935	0.406	0.801
0.420	1.375	0.759	1.294	0.948	4.053	0.399	0.787
0.440	1.401	0.748	1.288	0.952	4.187	0.393	0.772
0.460	1.428	0.738	1.281	0.956	4.339	0.386	0.756
0.480	1.456	0.728	1.274	0.960	4.509	0.378	0.739
0.500	1.484	0.717	1.267	0.964	4.700	0.370	0.719
0.520	1.515	0.707	1.259	0.968	4.94	0.361	0.699
0.540	1.548	0.696	1.251	0.972	5.20	0.351	0.676
0.560	1.583	0.685	1.242	0.976	5.52	0.341	0.650
0.580	1.620	0.675	1.232	0.980	5.94	0.328	0.621
0.600	1.660	0.664	1.222	0.984	6.47	0.314	0.586
0.620	1.702	0.653	1.211	0.988	7.25	0.298	0.545
0.640	1.748	0.642	1.200	0.991	8.10	0.281	0.504
0.660	1.796	0.631	1.187	0.993	8.90	0.268	0.472
0.680	1.848	0.619	1.174	*0.995	10.14	0.251	0.432
0.700	1.905	0.608	1.160	*0.997	12.26	0.228	0.376
0.720	1.966	0.596	1.145	*0.999	18.49	0.185	0.278
0.740	2.035	0.584	1.129				

* These intervals cannot be interpolated.

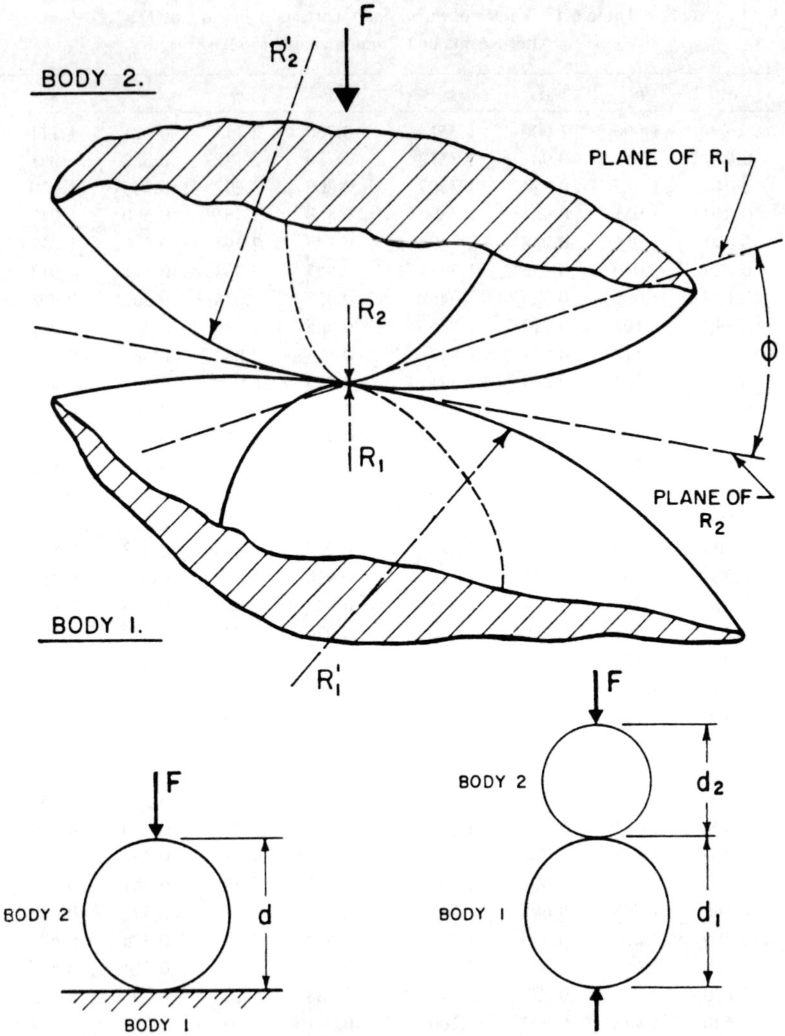

Figure 6.1 Hertz problem (above) for two convex bodies in contact, (below left) sphere on a flat plate, (below right) sphere on sphere

The greatest stress is the maximum compressive stress which is located at the center of the surface of contact (Figs. 6.2 and 6.8). The point of maximum shear stress is on the z axis at a depth equal to about one-half of the radius of the surface of contact (Timoshenko and Goodier, 1951). The maximum shear stress is approximately equal to one-half the difference between the stresses normal to and parallel to the flat plate pressing the sphere (Fig. 6.2).

Methods for calculating compressive, shear and tensile stresses at any point other than the center, on the elliptical surface of contact, are given by Seely and Smith (1961), Timoshenko and Goodier (1951) and Thomas and Hoersch (1930).

Two special cases of the Hertz problem which have applications in mechanical treatment of agricultural products are illustrated in the lower

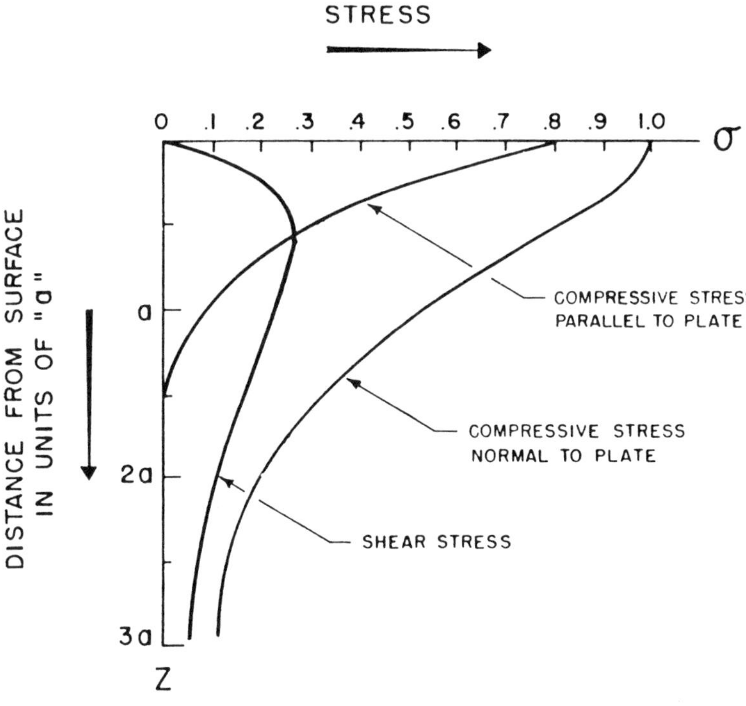

Figure 6.2 Stress distribution within an elastic sphere with Poisson's ratio = 0.3 compressed with a flat plate

part of Fig. 6.1. The equations for the case of a spherical body on a flat plate can be obtained by assuming

$R_1 = R_1' = \infty$ for the plate

$R_2 = R_2' = d/2$ for the spherical body

where d is the diameter of the sphere. Substituting these values in Eq. (6.6), (6.2), (6.1) and (6.5), yields the radius of the circular contact area, a, between the spherical body and the plate

$$a = 0.721\,(Fad)^{1/3} \tag{6.7}$$

the maximum contact stress

$$S_{max} = 0.918\left(\frac{F}{A^2d^2}\right)^{1/3} \tag{6.8}$$

and the combined deformation at the point of contact

$$D = 1.04\left(\frac{F^2A^2}{d}\right)^{1/3} \tag{6.9}$$

Similarly, the equations for the special case of two spherical bodies in contact can be obtained by substituting $R_1 = R_1'$ and $R_2 = R_2'$ in Eq. (6.6), (6.2), (6.1), and (6.5). This substitution yields the corresponding equations for the case of two contacting spheres as follows:

$$a = 0.721\left[\frac{FA}{1/d_1 + 1/d_2}\right]^{1/3} \tag{6.10}$$

$$S_{max} = 0.918\left[\frac{[F(1/d_1 + 1/d_2)^2}{A^2}\right]^{1/3} \tag{6.11}$$

$$D = 1.04\,[F^2A^2\,(1/d_1 + 1/d_2)]^{1/3} \tag{6.12}$$

When stacks or piles of agricultural products in the form of spherical bodies are considered, the Hertz method can be used to determine the contact forces and displacements on individual units. This has been done in studying the force-displacement relation for particles in a medium composed of a hexagonal close-packed array of elastic spheres (Duffy and Mindlin, 1957; Duffy, 1959).

If the convex body under load is not spherical, the radii of curvature R_1 and R_1' are to be measured and considered in the calculations. Fig. 6.3 shows a radius of curvature meter designed for products relatively large in size, such as fruits and vegetables. For smaller bodies such as a kernel of wheat, an approximation of R_1 and R_1' can be obtained as illustrated in Fig. 6.4. For the case where the product is loaded with a steel flat plate or a steel spherical indenter, the elastic modulus, E, of the material being tested is found from Eqs. (6.5) and (6.4) to be

$$E = \frac{0.338K^{3/2}F(1-\mu^2)}{D^{3/2}}\left(\frac{1}{R_1}+\frac{1}{R_1'}\right)^{1/2} \tag{6.13}$$

for a convex body under a steel flat plate and

$$E = \frac{0.338K^{3/2}F(1-\mu^2)}{D^{3/2}}\left(\frac{1}{R_1}+\frac{1}{R_1'}+\frac{4}{d_2}\right)^{1/2} \tag{6.14}$$

for a convex body under a steel spherical indenter with a diameter d. Note that for K=1.3514 from Table 6.1, the quantity 0.338K3/2=0.531. Also, deformations at the bottom of convex bodies are assumed negligible.

Example

Hydrostatic bulk compression of a sample of a specific fruit shows an average bulk modulus of k= 500 psi. Uniaxial compression tests of cylindrical specimens of tissues from the same batch of fruits give a compression modulus of E=600 psi. Samples of the whole fruits are then compressed between parallel plates. Assume a spherical shape for the fruit with a diameter of 2.4 in. and an axial deformation of 0.038 in. for a force of 5 lbs. Estimate a modulus of elasticity for the intact whole fruit using the Hertz formula for contact stresses.

From Eq. 4.5

$$\begin{aligned}\mu &= 1/2\left(1-\frac{E}{3K}\right)\\ &= 1/2\left(1-\frac{600}{3\times 500}\right)\\ &= 0.3\end{aligned}$$

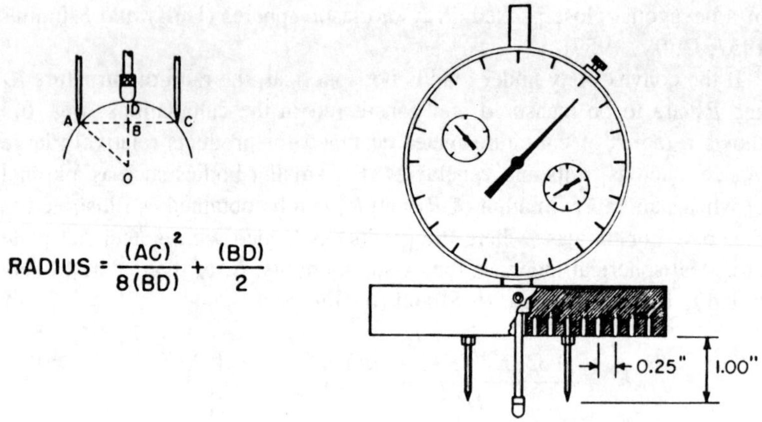

Figure 6.3 Apparatus for measuring the radii of curvature of convex bodies

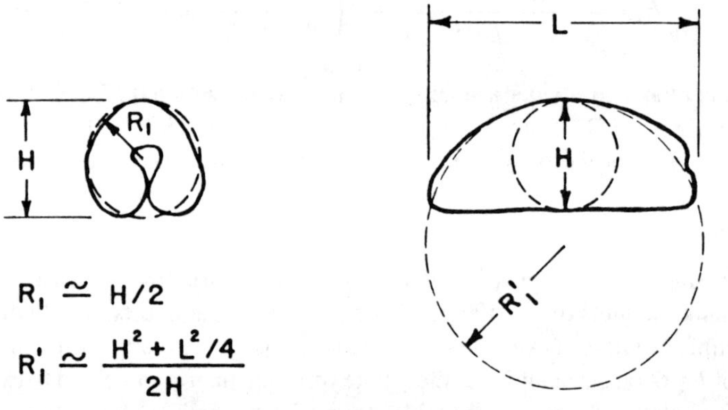

Figure 6.4 Approximation of R_1 and R_1' for convex bodies such as a kernel of wheat

since

$$R_1 = R_1'$$

Eq. 6.6 yields cos T=0. From Table 6.1

$$K = 1.3514 \text{ for } \cos T = 0$$

Then from Eq. 6.13

$$E = \frac{0.338(1.3514)^{3/2}(5)(1 - 0.09)}{(0.038)^{3/2}}\left(\frac{2}{1.2}\right)^{1/2}$$

$$E = 420 \text{ psi}$$

6.2 CONTACT PROBLEM FOR VISCOELASTIC BODIES

Solution of the viscoelastic counterpart of the Hertz problem in elasticity can be deduced from the elastic solution (Lee and Radok, 1960a). Since in a viscoelastic body the contact region varies with time (Fig. 6.5), for a smooth rigid sphere of radius R pressed against the plane surface of a viscoelastic half space, the normal contact pressure p is dependent on time and any radius r from the axis of symmetry. Lee and Radok (1960a) have given such a relationship by simply writing the Hertz solution in terms of t and r.

$$p(r, t) = \frac{4}{\pi R}\left(\frac{G}{1 - \mu}\right) \mathrm{Re}\left[\frac{2}{a(t)} - r^2\right]^{1/2} \tag{6.15}$$

where G, shear modulus, is substituted for E, a (t) is the radius of the surface of contact at time t, and Re indicates the real part of the square root expression. From this basic relationship, expressions for contact pressure distribution, for the case when deformation is continuous and the material behaves as a Maxwell body, can be derived (Lee and Radok, 1960a). Graphical presentation of this case is given in Fig. 6.6. It is seen that at times shorter than the relaxation time, T_0, the behavior is essentially elastic, following the Hertz elastic solution for contact stresses. As time increases, a gradual departure of the pressure distribution from the elastic case is observed. At time equal to the relaxation time, the pressure distribution is flattened and at longer times a central dip occurs. The extension of this problem to the more general case of the contact of two viscoelastic bodies is also discussed by Lee and Radok (1960a).

When a smooth spherical rigid body such as a steel ball is pressed against a linear isotropic viscoelastic half-space with a constant load, F, deformation as a function of time, $D(t)$, is given by the following expression presented by Lee and Radok (1960b).

$$[D(t)]^{3/2} = [3/(16\sqrt{R})]\ J(t)\ FH(t) \tag{6.16}$$

where $J(t)$ is the creep compliance and $FH(t)$ is the constant load applied according to Heaviside step function. For comparatively long times the difference between a step function loading and a ramp function loading, which is the actual loading in a creep test, becomes negligible. On the basis of this assumption, Eq. (6.16), written in the following form

$$J(t) = [(16\sqrt{R})/3]\ [1/F]\ [D(t)]^{3/2} \tag{6.17}$$

has been used to compare creep compliance in shear, $J(t)$, for low modulus viscoelastic materials obtained by a simple creep test, with creep compliance values obtained from torsion tests (Hadley, 1966). Results showed good agreement between the two methods. Figure 6.7 shows comparison of the two methods of tests for a filled epoxy resin.

The close agreement between the two methods of testing is not surprising when one considers the fact that Eq. (6.17) is actually the Hertz solution for a steel ball on a half space which can be obtained from Eq. (6.14). If we assume a Poisson's ratio of $\mu=0.5$ for the material under the steel ball, from Eq. (4.6).

$$E = 2G\ (1 + 0.5) = 3G$$

And, for half space, $R_1=R_1'=\infty$. Substituting these values in Eq. (6.14) and letting $1/G(t)=J(t)$, Eq. (6.17) will result.

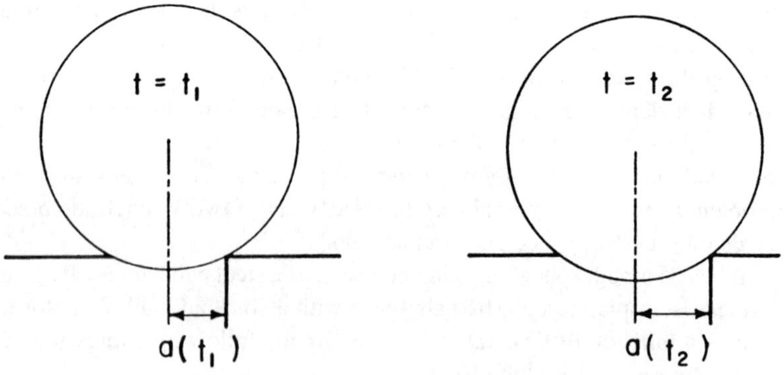

Figure 6.5 Change of contact region for a smooth rigid sphere on a plane surface of a viscoelastic half space

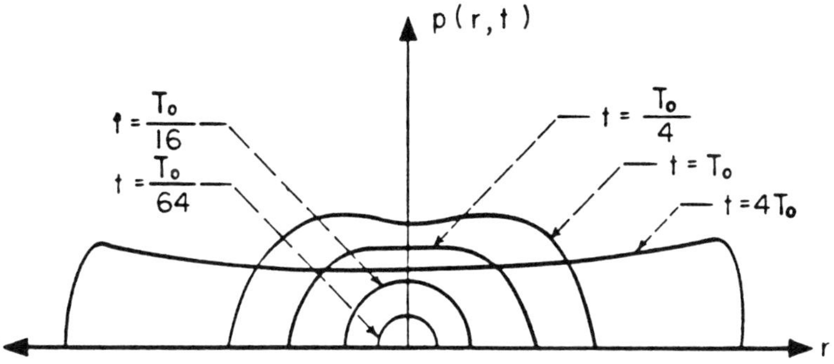

Figure 6.6 Pressure distribution as a function of time if deformation is continuous and the material behaves as a Maxwell body (Lee and Radok, 1960a)

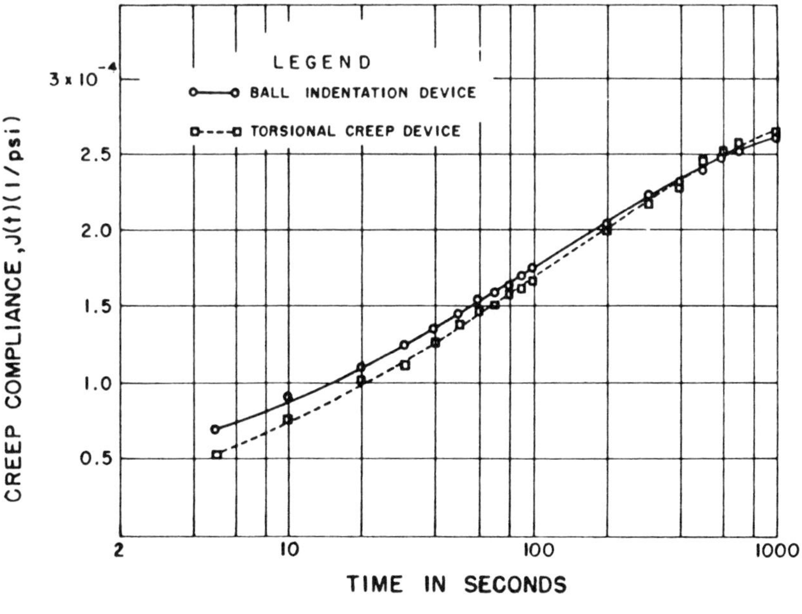

Figure 6.7 Creep compliance versus time for a filled epoxy resin obtained by two methods of testing (Hadley, 1966)

This shows that the elastic solution for the Hertz problem is valid for linear isotropic viscoelastic materials if we are interested in response for times comparatively longer than the duration of the ramp loading function.

6.3 BOUSSINESQ PROBLEM OF DIE LOADING

For evaluating mechanical properties of food materials such as fruits and vegetables and cereal grains several investigators have used a rigid cylindrical die to study load-deformation behavior of the material. This method of testing, as suggested by Finney (1963), approximates the Boussinesq problem in which a rigid die, in the form of a circular cylinder, is pressed against the plane boundary of a semi-infinite elastic solid. The original solution of this problem by Boussinesq (1885) has been expanded by Timoshenko and Goodier (1951) to include three dimensional considerations. While the Hertz method of analysis may be suggested for determining the maximum surface pressure and elastic modulus when one convex body is pressed against another convex body or against a flat plate, the Boussinesq method may be applied for the study of pressure distribution under a rigid die as well as the evaluation of elastic modulus.

In developing the equations for Boussinesq problem, the basic assumptions are the same as those given for the Hertz problem. Assuming that displacement is constant over the circular base of the die, Timoshenko and Goodier (1951) have given several equations for the problem of loading a semi-infinite homogeneous elastic body with a rigid die. As shown in Fig. 6.8 distribution of pressure, p, at the surface of the body is not uniform and varies according to the Boussinesq equation given by

$$p = \frac{F}{2\pi a \sqrt{a^2 - r^2}} \tag{6.18}$$

where

F = total load on the die
a = radius of the die
r = distance from the center of the area over which the die is acting

This equation shows that the maximum pressure exists along the boundary of the die where $r=a$, while minimum pressure exists at the center of the die where $r=0$. This minimum pressure is equal to half the average pressure on the circular area of the die

$$p_{\min} = 1/2\left(\frac{F}{\pi a^2}\right) \tag{6.19}$$

The relationship between deformation, D, under the die and the applied force, F, has been shown by Timoshenko and Goodier to be

$$D = \frac{F(1 - \mu^2)}{2aE} \tag{6.20}$$

where μ is the Poisson's ratio and E is the elastic modulus. When Eq. (6.20) is solved for E, the modulus can be calculated from the slope of the force-deformation curve of the material as shown below

$$E = F/D\,\frac{(1 - \mu^2)}{2a} \tag{6.21}$$

This equation gives an estimate of the elastic modulus in terms of the slope of the force-deformation curve, F/D, Poisson's ratio, μ, and the diameter of the die, $2a$. Since the extent of variation of μ is from zero to 0.5, the extent of modulus variation will be from $\left(\frac{F}{2aD}\right)$ to $0.75\left(\frac{F}{2aD}\right)$.

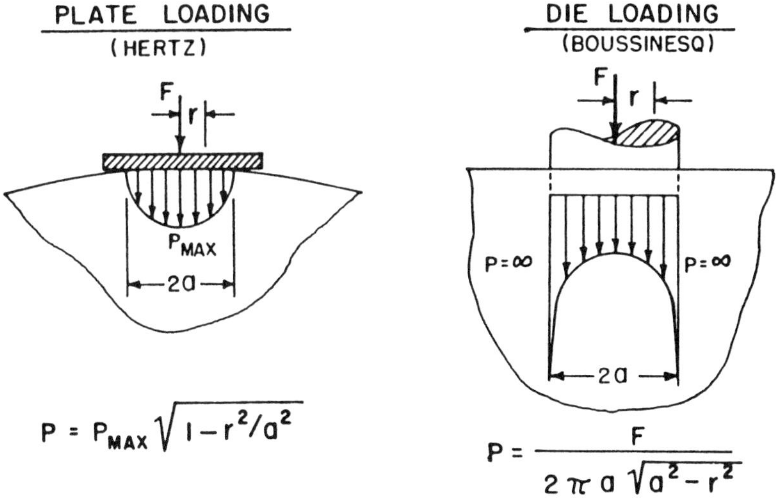

Figure 6.8 Pressure distribution in plate loading and die loading of a semiinifinite body according to Hertz and Boussinesq solutions

6.4 APPLICATIONS OF HERTZ AND BOUSSINESQ TECHNIQUES

The first known application of the Hertz solution for contact stresses in agricultural products is reported by Shpolyanskaya (1952) for determination of modulus of deformability of the wheat grain compressed between two parallel plates (Fig. 6.9). In this work, the total deformation of the grain, D_t, was considered to be

$$D_t = D_u + D_1 + D_0$$

where

D_u = deformation at the point of contact with the upper plate
D_1 = deformation at the point of contact with the lower plate
D_0 = deformation distributed over the body of the grain

Neglecting D_1 and D_0, the total deformation was assumed to be D_u, which was equalled to the combined elastic and residual deformations obtained from the first loading-unloading cycle. This value of D_u was then substituted in an equation similar to Eq. (6.9) and a modulus called the modulus of deformability was obtained for the wheat grain. If the parallel plates are steel with $E_2 = 30 \times 10^6$ psi, a simplified equation for the modulus, obtained by combining Eq. (6.9) and (6.4), would be (6.22)

$$E = \left[1.13\,(1 - \mu^2)^2 \frac{F^2}{D^3\, d} \right]^{1/2} \tag{6.22}$$

Although an apparent modulus of elasticity can be obtained from subsequent loading cycles, Shpolyanskaya considered modulus of deformability of greater practical significance because in the milling operation "it reflects the mechanical properties of the grain in the form in which it arrives at the rollers of the first milling system."

In calculating the modulus of deformability, Shpolyanskaya assumed a spherical shape for the grain of wheat (Fig. 6.9) and a value of 0.3 for the Poisson's ratio.

To obtain the modulus of elasticity reported in these data, the grain was first loaded and unloaded through several cycles until the residual deformation, which was referred to as plastic deformation, remained constant and the material behaved like an elastic body. The elastic deformation at this stage was then used in Eq. (6.22) to calculate the elastic modulus. This

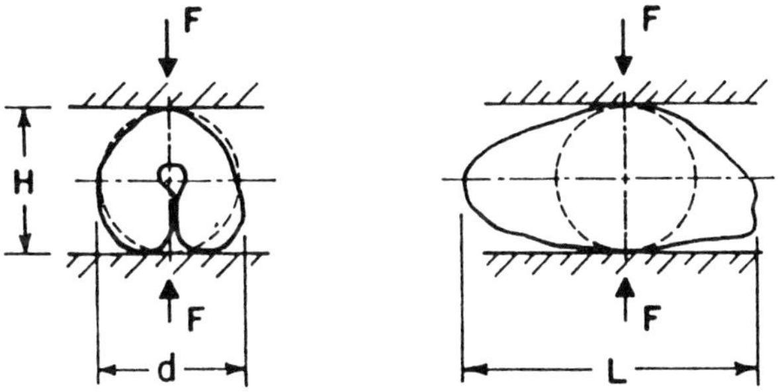

Figure 6.9 Grain of wheat loaded between two parallel rigid plates (Shpolyanskaya 1952)

behavior, which is similar to strain hardening in metal, and has been observed by the author in several other organic solid, is apparently due to initial set of air spaces, soft regions, and other internal irregularities in the material.

It can be shown from Eq. (6.22) that the ratio

$$\left(\frac{\text{Modulus of elasticity}}{\text{Modulus of deformability}}\right) = \left(\frac{\text{Total deformation}}{\text{Elastic deformation}}\right)^{3/2}$$

Using the above ratio for data reported by Shpolyanskaya the values of modulus of elasticity found to be several times greater than those reported by other investigators (see Table 6.2). Besides this apparent error in calculation of the modulus, the assumptions of $\mu=0.3$, spherical shape for the grain, and zero deformation for contact with the lower plate are also questionable. Although the term "modulus of deformability" appears to be a practical term for this application, the Hertz method is based on the assumption that deformation in the material is completely elastic and that the deformation to be used in Eq. (6.22) should be the deformation within the elastic range of the material and not a sum of elastic and residual deformations.

The method and the assumptions of Shpolyanskaya were used to determine the two moduli for Seneca wheat. The results of the tests, using the Instron testing machine illustrated in Fig. 4.23, showed moduli of elasticity three to four times larger than the moduli of deformability. This work was continued to include loading whole grains by means of parallel plates, a smooth spherical indenter, and a cylindrical indenter (Shelef and Mohsenin,

Table 6.2 Modulus of elasticity for wheat grain under uniaxial compression

Wheat variety	Loading rate, in/min	Moisture content (%)	Testing method	Modulus of elasticty E, 10^5 psi	Standard error, 10^5 psi	Source
Seneca	0.02	10 (d.b.)	Parallel plate—whole grains	4.12[a]	0.25	Shelef and Mohsenin
			Spherical indenter—whole grains	8.30[a]	0.25	
			Cylindrical indenter—whole grains	1.57[a]	0.03	
			Parallel plate—core specimens	2.30	0.13	
Lyntestens 62		11–12	Parallel plate—whole grains	21.30		Shpolyanskaya (1952)
Gordeiforme 10		11–12	Parallel plate—whole grains	28.40		
Soft red Winter	0.08–0.5	15.7 (d.b.)	Parallel plates—core specimens	0.46		Zoerb (1960
Seven Austrialian variety	0.26	11.5–13 (w.b.)	Parallel plate—whole grains	17.4–5.95[b]		Arnold and Roberts (1967)
			Parallel plate—core specimens	20.5–4.11		

(a) The value is for kE where $k = 1/1 - \mu^2$ and $1 < k < 1.33$.
(b) The value is for k times modulus of deformability.

1967). Also, core specimens, prepared by cutting off both ends of the grains, were loaded by means of parallel plates. Under a constant rate of deformation of 0.020 in/min, the load-deformation relation in all four compression tests was linear up to a certain load, nonlinear beyond it. Cyclical loading-unloading to a low constant load within the linear portion of the load-deformation curve showed that the deformation was partly recovered and partly residual. The residual deformations gradually decreased to small constant values in the third cycle, whereas the elastic deformations remained constant. Under these conditions the behavior of the wheat grains was considered approximately Hookean. Hertz's solution for convex bodies was used for compression of the whole grain by means of parallel plates and by the spherical indenter, and Boussinesq's solution for semi-infinite bodies subjected to concentrated compressive loads was used for the cylindrical indenter. For core specimens, Eq. (5.1) with ΔL being the elastic deformation, was used to calculate the modulus. Four values for the apparent modulus of elasticity were thus obtained for Seneca wheat grains of 10% moisture content, ranging from 1.6×10^5 to 8.3×10^5 psi.

Further work on compressive stresses in wheat grain is reported by Arnold and Roberts (1966, 1967). In the work reported by Shelef and Mohsenin (1967), the wheat grains, crease side down, were glued to the base plate which prevented them from movement during loading and provided a relatively large base contact area which caused the deformation at the bottom surface to be negligible. The work of Arnold and Roberts (1967) has confirmed this assumption. Grains glued to the bottom plate and assuming negligible deformation at the bottom end produced a mean value of modulus of elasticity equal to $2.10\,(1-\mu^2)\times 10^5$ psi with a standard deviation of 23%. Grains left loose on the bottom plate and allowing for deformation at top and bottom of the specimens produced the value $2.31(1-\mu^2)\times 10^5$ psi with a standard deviation of 32%.

Table 6.2 shows a comparison of elastic modulus in wheat reported by several sources. The methods of tests are indicated as: (1) loading the grain in flat position between parallel plates, (2) loading in flat position by a spherical indenter, (3) loading in flat position by a rigid cylindrical die, and (4) by cutting the ends of the grain off and loading the core specimen as a "cylinder."

When the four testing methods of loading are compared, the die loading method gives the lowest value while the parallel plates method gives the highest values. The reason for this lower value by cylindrical indenter is explained later in this Chapter.

In studying the force causing distortion of onion bulbs of different varieties and various sizes, the use of an equation from an engineering handbook has been reported (Ang *et al.*, 1960). This equation given as

$$D = KF^{2/3}$$

where

D = total lineal distortion
F = vertical compressive force
K = a "constant"

is obviously a simplified form of Hertz Eq. (6.9). In this work, sponge rubber balls were compressed between parallel plates; "constant" K was determined; and a good correlation was found between observed D and D calculated from the above equation. The use of the equation for predicting distortion of onion bulbs, however, was not reported.

The Boussinesq solution for concentrated force acting through a rigid die has been used for potato, apple, portions of corn kernels at various moisture contents, kernels of wheat, eggs, and peaches and pears. Eq. (6.20) was used by Finney (1963) to predict rupture force and rupture energy when potato tubers were loaded with a rigid die and rupture deformation was known. To accomplish this, Finney assumed a linear relationship between force and deformation so that the energy of rupture, E_r, could be presented by ½ of the product of rupture force and rupture deformation.

$$E_r = \tfrac{1}{2} (F) (D)$$

Substituting for F from Eq. (6.21)

$$E_r = \frac{aED^2}{(1 - \mu^2)}$$

The elastic modulus E was taken as 543 psi, which was determined by compressing cylindrical plugs of potato flesh between two parallel plates. The Poisson's ratio μ was taken to be 0.50, based on previous calculations. Comparison of calculated and experimental values of the mean rupture force and mean rupture energy showed good agreement for contact areas of rigid dies from 0.01 to 0.05 sq. in. The compared values tended to diverge, however, for loading areas approaching 0.50 sq. in. and greater (Fig. 6.10).

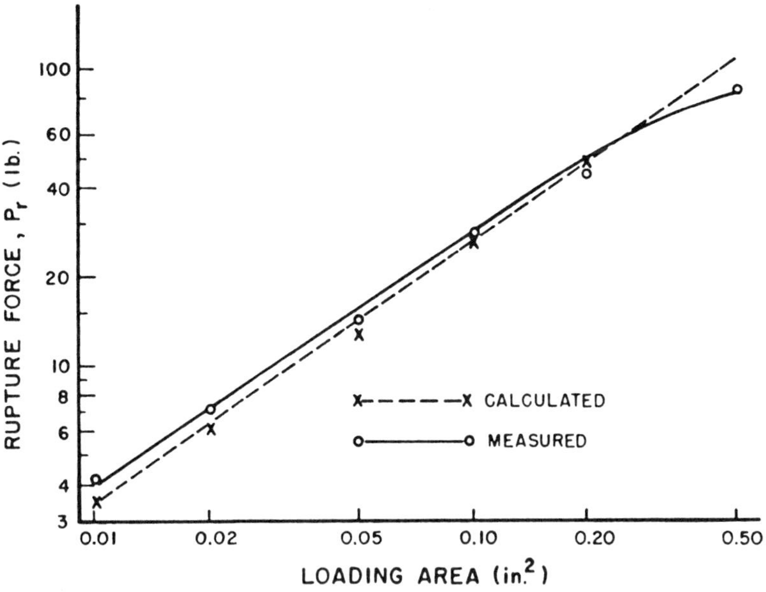

Figure 6.10 Comparison of measured rupture force with that calculated by Boussinesq method when potato tubers were loaded with a rigid die (Finney, 1963)

This was expected because as the surface of contact approached the same order of magnitude as the projected area of the tuber itself, there was a greater deviation from the assumption of the load acting on a semi-infinite body and the Boussinesq method was no longer valid.

The Boussinesq method has also been used to estimate an apparent modulus of elasticity for 1-inch cylindrical cores of potato tuber, loaded with rigid dies of various diameters (Timbers *et al.*, 1965). According to the results of this study, shown in Table 6.3, as the diameter of the rigid die approached the diameter of the core specimen, the value of the modulus approached the value for parallel plates.

Although in applying the Boussinesq formula in this case the assumption that the diameter of the die must be small compared with the loaded body has been violated and the force-deformation curves obtained for potato exhibited considerable residual deformation and thereby the requirement of elastic body was not met, it is interesting to note that the mean value of the modulus, 518 psi, is relatively close to the mean value of the modulus

Table 6.3 Apparent elastic modulus of l-inch cylindrical cores of potato determined by the Boussinesq die-loading method (Timbets *et al.*, 1965)

Plunger diameter in.	Poisson's ratio	Slope of *F-D* curve	Apparent elastic modulus lb/in^2
1/4	0.492	160	481
1/4	0.492	161	483
1/4	0.492	152	457
1/4	0.492	171	527
3/8	0.492	243	493
3/8	0.492	266	533
3/8	0.492	256	512
3/8	0.492	283	567
1/2	0.492	477	538
1	0.492	770	579
1	0.492	710	533

Mean by Boussinesq method: 518
Mean by parallel plates tests: 556

using the conventional method of parallel plates testing. This close agreement can be attributed to the fact that in the case of this particular product, as the diameter of the loading surface, $2a$, increased, the slope of the force-deformation curve, F/D, also increased in the same proportion. This resulted in relatively little change in the value of the modulus as calculated by Eq. (6.21). In Table 6.3 the slopes F/D were calculated from the data given by the authors.

If such behavior could be established for other food products, it appears that having the data on F/D and knowing the size of the die, Eq. (6.21) can be used to calculate a modulus (not necessarily elastic modulus) in terms of the expression $(1-\mu^2)$. However, tests of similar 1-inch core specimens of apple flesh with parallel plates and various size plungers resulted in considerable difference between the values obtained by the Boussinesq method and those by the parallel plate method.

The Hertz and Boussinesq techniques have also been applied to apples (Morrow and Mohsenin, 1966). McIntosh apples, supported by a quicksetting plaster mold so as to minimize support stress affects, were subjected to constant load (creep test) or constant deformation (stress relaxation test). Loading was accomplished by either a flat rigid plate or a ¼-inch rigid die. Axial constant loads for creep tests were 1 lb., 2 lb., 3 lb. and 4 lb. Axial constant deformations for stress relaxation tests were 1%, 2%, 3% and 4% of the fruit diameter. These small loads and deformations were well below the bio-yield point of the fruits. Since the data for relaxation and creep experiments were in terms of force and deformation, to avoid the complexity of converting these data to stress and strain, relaxation moduli and creep compliances were calculated which could directly be used in the appropriate rheological equations. For this purpose, the data for plate loading were analyzed by the Hertz method using Eq. (6.22) to calculate time-dependent relaxation modulus, $E(t)$, or the inverse of that equation to calculate creep compliance, $D(t)$. The data for die loading were analyzed by the Boussinesq technique using Eq. (6.21) for relaxation modulus, $E(t)$, and the inverse of that equation for creep compliance, $D(t)$.

Results of the uniaxial compression creep and relaxation tests, using both plate and die loading, were fitted to the appropriate rheological models. There was close agreement between the retardation and the relaxation times for the two types of loadings indicating that the viscoelastic decay function was of the same form. There were differences, however, between other terms in the prediction equations which were considered being probably due to the difference in Hertz and Boussinesq methods for calculating the moduli.

As indicated before, one of the basic considerations in using the Hertz or the Boussinesq equations for calculating the modulus or compliance is the assumption that contact stresses vanish at the opposite ends of the body under load. In the case of McIntosh apples loaded with a ¼-inch cylindrical rigid die, the normal stresses in the z direction (direction parallel to the longitudinal axis of the cylindrical die) at various points below the surface of the fruit was calculated using the following expression from Timoshenko and Goodier (1951).

$$\sigma_z = F\left(\frac{z^3}{(a^2 + z^2)^{3/2}} - 1\right) \tag{6.23}$$

For the die radius, a, equal to 0.125 in., the above expression showed that stress σ_z was approximately zero at a depth of z equal to two inches below the surface of the fruit. The reverse pressure effect caused by the fruit's support was neglected because of the large contact area in the plaster mold.

Using the same procedure, it was possible to show that the condition of vanishing stresses also existed for radial and tangential stresses. It was noted that in this case the diameter of the fruit was greater than ten times the diameter of the cylindrical die.

To determine the effect of moisture content on mechanical properties of the horny and floury sections of dry shelled corn, Shelef and Mohsenin (1969) continued the applications of various loading devices on surface-ground corn kernels 0.14 inch thick mounted on a rigid plate (Fig. 6.11). The germ-side surface of each kernel was lightly sanded and the kernels were glued, germ-side down, to a flat metal plate. The plates were then placed in dessicators with saturated salt solutions for kernels to be equilibrated to the desired moisture content. For higher relative humidities the Thiokol resin (Chemical Corporation, Trenton, New Jersey) was used for gluing the kernels because this material will dry at high humidity atmosphere.

Prior to testing, kernels were equilibriated at each of several relative humidity atmospheres to give a range in moisture contents of 6.53% to 28.00%. In addition, one sample was dried in a vacuum oven to a residual moisture level of 0.74%. The Instron testing machine was used and loading was either by cylindrical indenter (0.016-inch diameter), parallel plates, or spherical indenter (0.065-inch diameter). Three moisture-dependent parameters were evaluated from load-deformation relations obtained at different loading levels: (a) the linear limit load, up to which the load-deformation relation was linear, (b) the apparent modulus of elasticity, and (c) the modulus of deformability. Each of these parameters decreased with increase in moisture content of the kernel. Under the testing conditions employed, the major contributor to the mechanical properties observed was the horny endosperm. At low moisture levels the horny endosperm was very stiff. Increasing moisture content reduced the friction coefficient of the system, and caused deformations to increase and moduli to increase as pressure was applied. Results from tests conducted on the floury endosperm indicated considerably lower values for the three parameters evaluated but were not consistent.

Tables 6.4 and 6.5 show the results of these tests. As in the case of wheat reported by these investigators, modulus of deformability is based on the total deformation whereas modulus of elasticity is based on the elastic deformation only. Equation (6.14) was used for the case of spherical indenter. For cylindrical indenter Eq. (6.21) was employed. For slabs of surface-ground kernels, modulus was simply the ratio of stress to elastic strain. Two significant results were revealed from this work. One was that

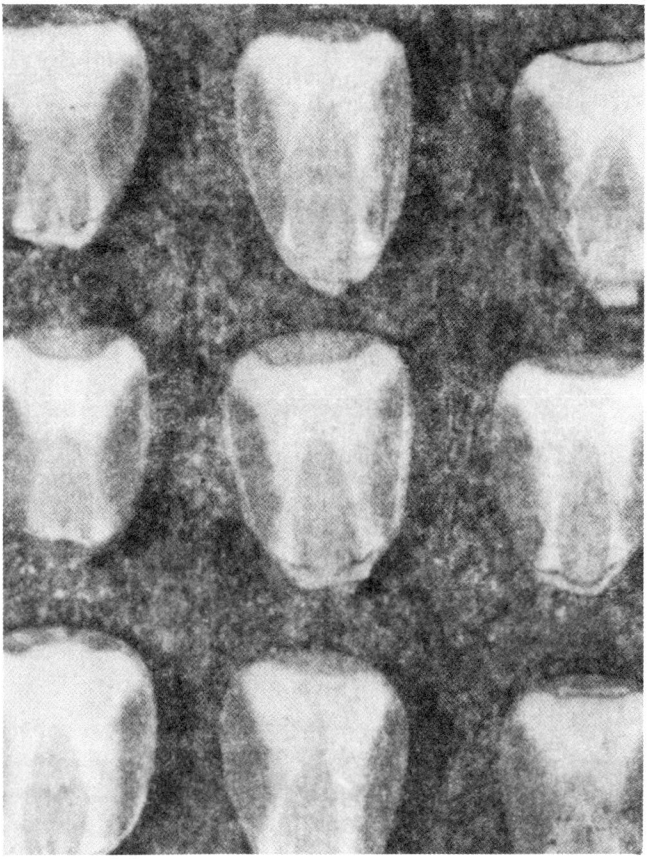

Figure 6.11 Surface-ground corn kernels mounted on a rigid plate for uniaxial compression tests.

the use of spherical indenter produced much less variability that the other methods (compare coefficients of variation in Table 6.4). The other point was the larger values of both types of moduli resulting from the application of the spherical indenter. Reference to Table 6.5 shows that at any given moisture level the values of E_a and E_d are relatively in the same order of magnitude for the cylindrical indenter and the parallel plates. These values, particularly for the lower moisture levels, are considerably higher for the case of the spherical indenter. Knowing the depth of indentation and the radius of the indenter the surface unit pressures under the spherical indenter were estimated and compared with those under the two other loading devices. As suspected, the values of unit pressures for the case of spherical indenter were much greater than those for the other two cases. Reference to Fig. 6.12 shows that because of the sigmoidal shape of the curves, the use of lower levels of unit surface pressure can result in smaller values of modulus (initial part of the curves) whereas, the use of higher values of unit pressure, as it has apparently been the case with the spherical indenter, can give higher values of modulus. Furthermore, Fig. 6.12 shows that such a trend would be less pronounced for higher moisture levels. The reported modulus values for spherical indenter support this observation. In other words if the force and deformation values used in Eq. (6.14) were taken from a line tangent to the initial portion of the force-deformation curves, the values obtained by the spherical indenters would have been closer to those by the other types of loading devices. Later indentation experiments with the same size spherical indenter, as well as tensile and

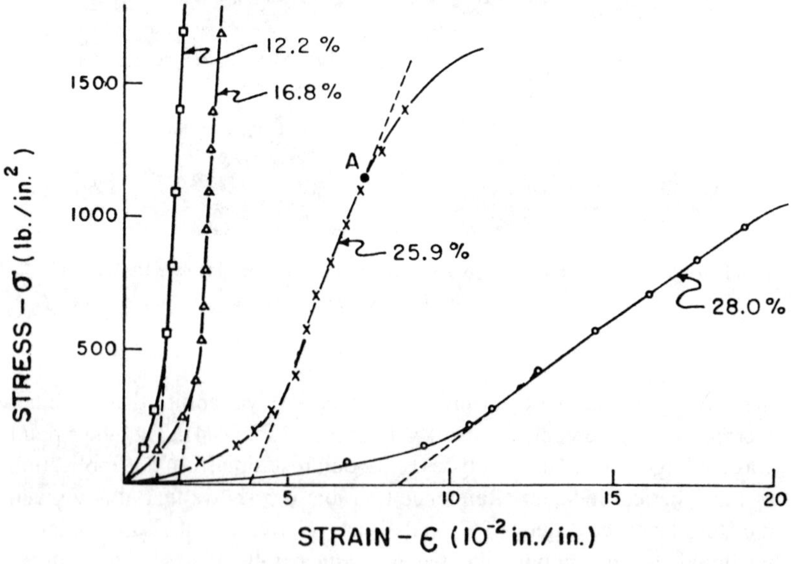

Figure 6.12 Stress-strain curves for parallel-plate testing of surface-ground corn kernels (Shelef and Mohsenin, 1968)

Table 6.4 Experimental results for uniaxial compression of corn kernels at different moisture levels (Shelef and Mohsenin, 1968)

Moisture content % (d.b.)	0.74	6.53	9.75	12.24	16.80	25.90	28.00
			Cylindrical Indenter				
Elastic Deformation[1]							
Mean (10^{-3} in.)	0.21	0.23	0.19	0.32	0.25	0.35	0.56
Std. Deviation		0.20	0.12	0.14	0.24	0.15	0.30
Coeff. of Var. %		43.1	31.9	21.1	47.0	20.9	26.6
Residual Deformation[1]							
Mean (10^{-3} in.)	0.03	0.12	0.08	0.34	0.19	0.19	0.77
Std. Deviation		0.09	0.07	0.16	0.16	0.06	0.38
Coeff. of Var. %		78.6	87.8	46.8	81.8	33.1	49.0
			Parallel Plates				
Elastic Deformation[2]							
Mean (10^{-3} in.)	0.41	0.46	0.37	0.65	0.92	1.23	1.99
Std. Deviation		0.169	0.217	0.14	0.26	0.25	0.40
Coeff. of Var. %		54.5	47.2	21.3	28.1	20.3	20.2
Residual Deformation[2]							
Mean (10^{-3} in.)	0.37	0.42	0.74	1.08	0.83	3.74	6.40
Std. Deviation		0.14	0.18	0.60	0.32	1.10	1.15
Coeff. of Var. %		33.6	22.2	55.7	39.1	32.6	18.0
			Spherical Indenter				
Elastic Deformation[3]							
Mean (10^{-3} in.)	0.55	0.66	0.67	0.70	0.92	1.62	3.18
Std. Deviation		0.07	0.05	0.09	0.08	0.27	0.48
Coeff. of Var. %		11.6	8.1	13.1	8.3	16.6	15.1
Residual Deformation[3]							
Mean (10^{-3} in.)	0.21	0.29	0.40	0.67	0.77	3.67	10.14
Std. Deviation		0.139	0.106	0.208	0.135	0.755	2.28
Coeff. of Var. %		47.7	26.2	31.0	17.4	22.4	2,5

[1] values determined for 0.5 lb. load.
[2] values determined for 50 lb. load.
[3] valued determined for 5 lb. load.
Each value is a mean of 50 tests.

dynamic tests of rectangular segments of horny endosperm subjected to very low levels of stress have given modulus values which are in the range reported for cylindrical indenter and parallel plates in this paper.

The problem of contact stress and the application of Hertz and Boussinesq theories to agricultural products has also been investigated by Fridley, *et al.*, (1968). These investigators obtained force-deformation curves for peaches and pears to determine to what degree these biomaterials follow the theory of elasticity and to evaluate the relative merits of die-loading and plate loading. Using a universal testing machine, tests were conducted on the cheek of whole fruits and fruit spherical segments without removing the skin. For plate tests the contact area was determined by measuring final area on the plate printed during loading. To study the strain without fruits, the fruit was cut in two pieces and a grid marked off on the exposed surface. A picture of the fruit was taken prior to the application of force and subsequent pictures were taken during the test to determine the deformation of cross-hatched lines. The shadowgraph apparatus shown in Fig. 3.4 when equipped with the straining mechanism and load measuring transducer can be used for this purpose more conveniently. This instrument would also give greater accuracy of observed results than those obtained photographically by these investigators.

Figure 6.13 shows the typical force-deformation points obtained in this work. The calculated theoretical curves were obtained by using the Hertz equation for plate against sphere [Eq. (6.9) and Boussinesq equation for die loading [Eq. (6.20)]. A value of 0.49 was used for Poisson's ratio and modulus of elasticities were calculated for each case from Eq. (6.13) and (6.21) using, in the case of plate loading, the data for a deformation of 0.006 in. and in the case of die loading, the slope of the straight line portion of the curve. Such calculations gave $E=770$ psi for pears and $E=150$ psi for peaches. Arnold and Roberts (1967), who also compared theoretical and experimental force-deformation curves in their grain studies, assumed a range of modulus of elasticity for the wheat and substituted the modulus directly in Eq. (6.9) and (6.20). The reason for experimental values in plate tests of pears were less than the theoretical values was stated to be stress relaxation. A faster rate of loading should have improved this situation but the authors reported there was no change when faster rate of loading was used. The reason for experimental points in plate loading of peaches, particularly at larger deformation, being higher than the theoretical values was believed to be in part due to the nonuniform curvature of the fruit which would influence the calculated deformation. As a whole, the agreement between the actual force-deformation curve and that calculated on the basis on the theories of elasticity appears to be reasonable.

When force-deformation data were plotted on log-log paper (Fig. 6.14), the slope of the straight lines, gave the degree to which the Hertz and

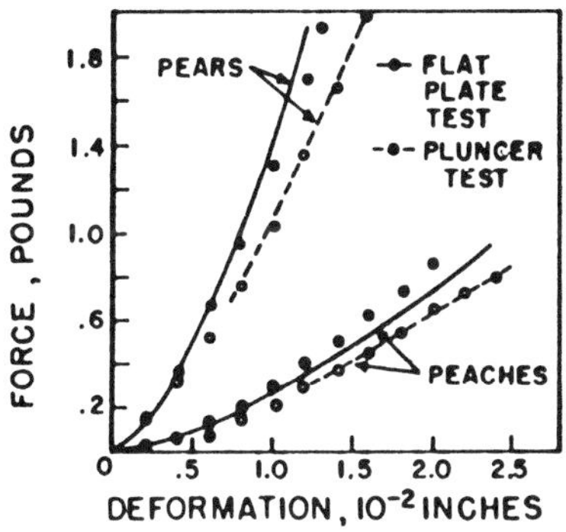

Figure 6.13 Force-deformation relationship for peaches, pears, and elastic spheres subjected to tests with a plate and ¼-in plunger. Points represent data from fruit. Lines are calculated for elastic spheres of comparable size and elasticity (Fridley *et al.*, 1968)

Boussinesq theories are applicable in such materials as fresh fruits. From the Hertz's theory, force is proportional to deformation to the power three halves. In other words, the plot of the force-deformation data on log-log paper should yield a straight line with a slope of 1.5. Similarly for the case of die loading and Boussinesq theory, force is proportional to deformation to the first power. Plotting of the data on log-log paper should yield a straight line with a slope of 1.0. Reference to Fig. 6.14 for die loading

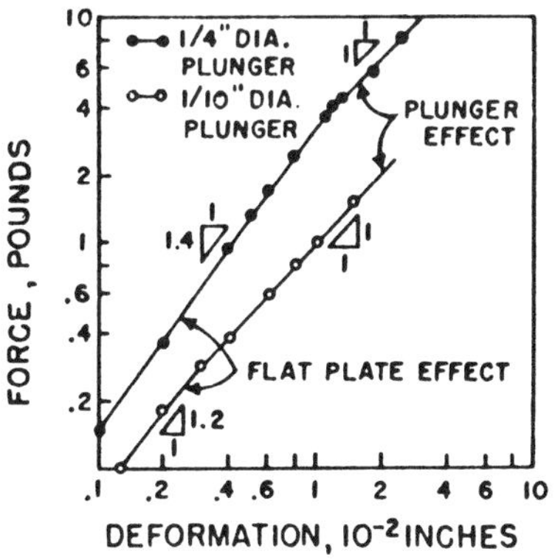

Figure 6.14 Force-deformation data for plunger tests on curved surface of a segment of pear showing flat plate effect before making full contact (Fridley, *et al.*, 1968)

and other data reported by the authors for plate loading shows that although the force-deformation relationship does not precisely follow the theories, considering many other factors, it is reasonably close.

In die loading of a curved surface such as that of a fruit, the circular die first tends to act as a flat plate until the entire end of the die is contacting the fruit surface. At that point the die effect would show up by giving a straight line of 1:1 slope. Note that the 1/10-in diameter plunger shows this effect much sooner than the 1/4-inch plunger (Fig. 6.14). By this technique, the experimenter can tell what formula should be used for any specific portion of the curve. As seen in Fig. 6.14, for small deformations the flat plate effect requires the use of Eq. (6.13) while for larger deformations, when the entire end of the plunger comes into contact with the fruit and the curve diverges to a slope of 1:1, Eq. (6.21) would be applicable for modulus calculations. This change of slope of the straight line from a maximum value of 1.5:1 to 1:1 is probably the reason for lower values which one obtains for modulus of elasticity E by means of a cylindrical indenter compared to that for spherical indenter. The latter makes use of the Hertz theory which is associated with the steeper slope and thus the higher modulus. Figure 6.14 also explains the effect of plunger size in firmness testing of fruits and vegetables where firmness is to be expressed in terms of some well-defined physical parameter such as modulus of elasticity. Based on these observations, Fridley *et al.* (1968) concluded that plate test is preferable to plunger test for fruits because the test procedure is less critical and results compare more favorable with those predicted from theory. Furthermore, this type of loading is more representative of the type of load application found in practice.

In their work with peaches and pears, Fridley *et al.* (1968) also evaluated the bruises occurring within the fruit due to various types of loading by comparing the location of the bruise with that predicted from theory. As seen from Fig. 6.2, the maximum compressive stress occurs at the surface of an elastic sphere subjected to compression by a flat plate while maximum shear stress occurs beneath the surface, a distance of about one-half the radius of the circle of contact. Experiments with pears and peaches by Fridley *et al.* (1968) and with apples by others have shown that in plate tests bruise usually occurs under the center of the area of contact at a small distance beneath the surface of the fruit. This indicates that bruising probably is caused by shearing.

To further evaluate the validity of the Hertz and Boussinesq equations, which are proposed to be used for expressing force and deformation in terms of a modulus, maximum stress or other engineering parameters, two types of cast resins, namely Hysol Resin Ru-2085 (a photoelastic material) and Dow Corning Silastic 588RTV (an opaque material) were used as standard materials. Two ½-inch thick disks with 1⅜-inch and 2-inch diameters, two spheres of 1-in. and 2-in. diameters, as well as an ASTM cylindrical specimen for compression and an ASTM tensile specimen were cast from these plastic resins. The Young's modulus of elasticity of the materials were determined from compression and tensile tests of the ASTM specimens. The results of these tests for Hysol Resin showed an average compression modulus of 582 psi and tensile modulus of 478 psi. The average of the values of compression tests of the cast materials was assumed to be the true value of the modulus for these materials. The results of indentation tests on the disks and spheres using cylindrical dies and spherical indenters of different sizes were then analyzed to determine how shape and size of the loading device would affect the calculated values of the modulus. The Hertz equations were used for spherical indenters. The Boussinesq equation was used for cylindrical indenters. Results showed no significant difference between the sizes of the indenters or the shapes and sizes of the specimens. Furthermore, the values of the modulus determined by the use of the spherical indenters on spherical or plate-like specimens were close to that found from compression test of the ASTM cylindrical specimens. The degree of difference noted was considered acceptable as far as standardization of testing methods for biological materials is concerned. The cylindrical indenters and the values of moduli obtained by the Boussinesq equation showed values much less than those of the ASTM specimens and involved considerable variation. These tests further confirmed the merit of the Hertz theory for use with intact agricultural products.

6.5 IMPORTANCE OF SMALL STRAINS IN COMPRESSION TESTS OF FOOD MATERIALS

The ASAE Standard S 368-1 was developed for compression tests of food materials of convex shape. According to this standard, which is given in the Appendix, first a point of inflection is found on the force-deformation curve, then the force and deformation corresponding to the point on the curve at one-half of the total deformation up to the point of inflection are chosen for use in the computing equations. By using such a technique, force and deformation corresponding to as small a strain as practical are chosen

Table 6.5 Apparent modulus of elasticity, modulus of deformability and linear limit load for corn kernels at different moisture levels (Shelef and Mohsenin, 1968)

	Cylindrical Indenter			*Parallel Plates*			*Spherical Indenter*		
Moisture Content (% d.b.)	$E_a^{\,1}$ (10^5 psi)	$E_d^{\,1}$ (10^5 psi)	*LL* Load (lb)	$E_a^{\,2}$ (10^5 psi)	$E_d^{\,2}$ (10^5 psi)	*LL* Load (lb)	$E_a^{\,3}$ (10^5 psi)	$E_d^{\,3}$ (10^5 psi)	*LL* Load (lb)
0.74	1.48	1.29	13.80	2.37	1.24	—	15.95	9.80	38.2
6.35	1.36	0.90	7.20	2.11	1.10	307.0	12.10	7.02	18.5
9.75	1.64	1.16	5.99	2.63	0.88	—	12.00	5.98	13.0
12.24	0.96	0.47	4.05	1.50	0.57	223.0	11.10	4.15	10.0
16.80	1.25	0.71	3.48	1.07	0.56	154.0	7.45	2.94	8.8
25.90	0.89	0.58	1.97	0.80	0.20	66.6	3.07	0.55	3.9
28.00	0.56	0.24	1.90	0.49	0.12	47.0	1.16	0.14	3.9

[1] values determined for 0.5 lb. load.
[2] values determined for 50 lb. load.
[3] values determined for 5 lb. load.

which make the application of the theories in solid mechanics less objectionable. Tables 6.6 and 6.7 show the use of this technique in computing the modulus of deformability for apple flesh by two methods based on the theory of elasticity. Correlations of these moduli with the bioyield point of the flesh tissue are also given. The bioyield point was taken as the point of failure initiation when the initial cell rupture and tissue bruising take place.

In the die-loading tests shown in Table 6.6 the skin of the whole fruit was sliced off and the flat surface of the flesh was loaded with a cylindrical die having a rounded end with a curvature of 1 in. radius. To minimize deformation of the whole fruit, the specimen-holding fixture and the preloading arrangement specified in the ASAE Standard S 368-1 was used. The following equation, based on the Hertz contact stress in elasticity for spherical indenter on a flat surface was used to calculate the modulus of deformability:

$$E_d = \frac{0.531\,F}{D^{3/2}}\left(\frac{4}{d}\right)^{1/2}(1-u^2)$$

where F and D are force and deformations corresponding to the point on the loading curve at one-half of the total deformation up to the point of inflection, d is the diameter of curvature of the spherical end of the die, and μ is Poisson's ratio which has been found to be between 0.24 and 0.34 for a few varieties of apples. For these tests, $\mu=0.3$ was assumed. In the parallel plate loading tests shown in Table 6.7 a cylindrical specimen of the flesh, taken from the opposite side of the same fruit, was loaded between two parallel plates. The following equation was used to compute the modulus of deformability:

$$E_d = (\mathrm{F/A})/(\Delta L/L)$$

where F and D are force and deformation as explained above, A is the surface area of the specimen, L is the height of the specimen and ΔL is the deformation on loading.

Examination of the data in Tables 6.6 and 6.7 supports the following observations:

1 Food materials with soft tissues, even under small loads, exhibit percent strains which are rather large compared with engineering materials. It is believed, however, that the range of values shown in Table 6.7 (about

Table 6.6 Modulus of deformability of the flesh of whole apple by the method of rigid die and the Hertz contact theory in elasticity

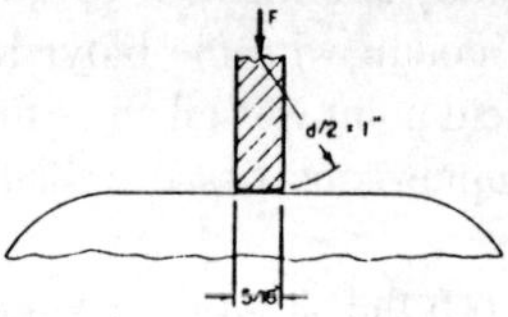

$$E_d = \frac{0.531\,F\,(1-\mu^2)}{D^{3/2}}\left(\frac{4}{d}\right)^{1/2}$$

Sample Number	Force, Fy, at bioyield (lb)	Deformation, Dy, at bioyield (10^{-2} in.)	Force, F, at ½ of the deformation at point of inflection (lb)	Deformation, D, at ½ of the point of inflection (10^{-2} in.)	$Ed/1-\mu^2$ (psi)	Ed @ μ=0.3 (psi)
1	3.8	5.5	1.35	2.0	358.4	326.2
2	4.3	6.0	1.60	2.25	356.00	323.96
3	3.7	5.0	1.20	2.25	267.00	242.97
4	5.1	8.0	1.50	2.6	264.85	241.01
5	4.3	5.0	1.36	1.85	406.82	370.21
6	4.1	5.5	1.25	2.0	331.88	302.01
7	4.1	5.0	1.80	2.5	341.96	311.18
8	4.6	6.5	1.20	2.5	227.97	207.45
9	3.6	5.5	1.10	1.85	328.28	298.73
10	3.6	4.0	1.60	2.0	424.80	386.57
11	1.7	4.5	0.87	2.0	232.26	211.36
12	2.1	4.5	0.60	1.35	287.25	261.40
13	2.5	4.0	1.25	2.15	303.03	275.76
14	3.6	5.0	1.15	2.25	255.88	232.85
15	2.9	5.0	0.94	1.65	331.77	301.91
16	2.6	4.0	0.94	1.65	331.77	301.91
17	2.4	5.5	0.90	2.15	218.18	198.54
18	1.8	4.5	0.90	2.25	200.25	182.23
19	2.4	5.0	1.15	2.50	218.47	198.81
20	3.0	7.0	0.95	2.25	211.38	192.36
Average	3.3	5.25	1.18	2.10	294.91	268.37

1.5%–3% strain) can be tolerated if one is going to use, as the first step, the concepts valid for small deformations in solid mechanics.

2 The difference between the modulus of deformability of the flesh of the whole fruit found by the use of the Hertz theory in elasticity (18.48×10^5 pa or 268 psi) and by the conventional method of compressing a cylindrical specimen of the flesh between two parallel plates (17×10^5 pa or 246.5 psi) is relatively small. Statistical analysis, using a t-test,

Table 6.7 Modulus of deformability and percent strain of cylindrical speciments of apple flesh

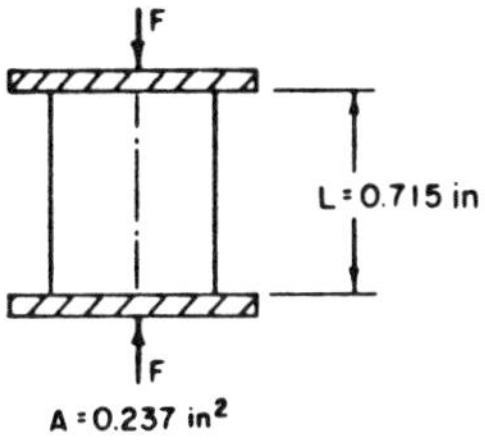

$$E_d = (F/A)/(\Delta L/L)$$

Sample Number	Force, F (lb)	Deformation, ΔL (10^{-2} in.)	Ed (psi)	% Strain
1	1.00	1.06	283.9	1.48
2	1.28	1.31	292.9	1.83
3	1.07	1.12	268.7	1.57
4	1.35	1.75	232.2	2.45
5	1.60	1.5	321.0	2.10
6	1.30	1.19	328.8	1.66
7	1.10	1.50	220.7	2.10
8	1.38	1.38	299.9	1.93
9	1.30	2.00	195.6	2.80
10	1.13	1.38	245.3	1.93
11	0.80	2.00	120.4	2.80
12	1.15	1.25	276.9	1.75
13	1.35	1.50	270.9	2.10
14	1.55	2.00	233.2	2.80
15	1.40	1.75	240.8	2.45
16	1.45	2.13	204.9	2.98
17	1.35	2.00	203.1	2.80
18	1.30	1.50	260.8	2.10
19	1.25	1.75	215.0	2.45
20	1.25	1.75	215.0	2.45
Average	1.26	1.59	246.5	2.23

showed that the two values were not significantly different at the 5% level. This observation confirms the fact that if rheological terms and methods of measurements and analysis are employed correctly, some of the concepts can be used in food research for practical purposes, even though the conditions upon which the theory is based are not fully met.

6.6 LIST OF SYMBOLS FOR CHAPTER 6

a —major semi-axis of elliptic contact area, or radius of the circle of contact, or radius of the rigid die
a(t) —time-dependent radius of the circle of contact
A —given by equation 6.4
b —minor semi-axis of elliptic contact area
Cos T —given by equation 6.6
d —diameter of the sphere in contact
d_1, d_2 —diameters of the lower and the upper spheres, respectively
D —combined deformation at the point of contact
D(t) —time-dependent deformation
E —Young's modulus of elasticity
F —normal load
G —shear modulus
$J(t)$ —creep compliance
k, m, n—constants given in Table 6.1
P —normal contact pressure
r —distance from the center of the area over which the die is acting or radius from axis of symmetry (Fig. 6.8)
R —radius of sphere
R_1, R_1' —radii of curvature for body 1 (lower)
R_2, R_2' —radii of curvature for body 2 (upper)
S_{max} —maximum contact stress
μ —Poisson's ratio

7

RHEOLOGY AND TEXTURE OF FOOD MATERIALS

IN CHAPTER 5 certain concepts of rheology as applied to plant and animal materials in general were discussed. In this chapter evaluation of several physical characteristics, usually known as textural qualities of raw and processed food materials, will be considered.

7.1 DESCRIPTIVE TERMS

It has been estimated that out of 350 descriptive terms dealing with food qualities, about 25 percent are concerned with those qualities known as texture. It is not difficult to visualize the numerousness of these terms if one considers such examples as hardness, softness, brittleness, firmness, ripeness, toughness, tenderness, crustiness, stickiness, gumminess, fibrousness, mealiness, blandness, smoothness, chewiness, juiciness, crispness, flakiness, fleshiness, flabbiness, lumpiness, oiliness, grittiness, springiness, shortness, etc. Texture in foods includes these descriptive terms as well as other terms, such as elasticity, plasticity, viscosity, etc., which are well-defined physical concepts with absolute units. Because of this vast coverage of complex physical qualities and characteristics, no satisfactory definition of the word "texture," as related to foods, has yet been offered. Perception in the mouth, properties which deal with a sense of touch, kinesthetic properties, manifestation of the rheological properties, are some of the definitions given by several workers. (See S.C.I. Monograph No. 7, 1960; Matz, 1962; Kramer and Twigg, 1966; Szczesniak, 1966.)

7.2 CLASSIFICATION OF METHODS OF TEXTURE EVALUATION

Texture characteristics manifest themselves either during food manufacturing and processing or during food consumption. Texture, together with appearance and flavor, are three sets of qualities which govern the acceptability of a food by the consumer. The present methods for evaluation of textural characteristics are classified into (1) subjective or sensory evaluation and (2) objective or instrumental measurements. In the subjective methods, sensory impressions of test panels to assessment of a quality are considered using correlation or factor analysis or other statistical techniques. In the objective methods those qualities which lend themselves to measurement by physical instruments are evaluated. Szczesniak (1966) divided the instrumental measurements into (1) fundamental or rheological, (2) empirical, and (3) imitative measurements. The objective or instrumental methods are particularly useful in evaluating those texture characteristics which are important in evaluating raw materials or foods in the initial stages of processing. Often, changes in a property due to a given process could be followed more closely and specifications for design of equipment and process control could be given with a greater reliability if instrumental measurements were available.

7.3 SUBJECTIVE OR SENSORY MEASUREMENTS

Those characteristics which are significant in acceptability of the finished product to the consumer must be evaluated either directly by sensory methods or be correlated with human senses, the ultimate "judge" of quality.

Sensory evaluation of quality is a field in itself, and no attempts will be made to describe the specialized techniques in this field involving both psychological and biological factors. Since the food industry relies upon consumer's judgement for texture preference, no texture test is considered to be of much value unless it can be correlated with human evaluation. The method employs trained test panels and is conducted on a statistical and psychometric basis, trying to minimize the subjective nature of human element involved (See Matz, 1962; Kramer and Twigg, 1966).

Despite all this, one cannot expect too much from these psychophysical tests where the impressions and preferences of individuals for judging items of foods are sought. According to Szczesniak (1966), "This approach has much merit, but it also has many pitfalls. Numerous promising instruments have fallen into disrepute because they failed to show statistically significant correlations with organoleptic panels of questionable reliability." In certain sensory evaluations only the fingers come into direct contact with the material. The feel of "firmness" in fruits and "body" in cheese and flour dough are examples of this type. Different groups of muscles and sense organs in different combinations are used on various occasions. In other sensory evaluations, the muscles and sense organs of the mouth and their associated anatomy and physiology come into play. Quantitive studies, reported by Scott Blair and Coppen (1940) on the judgment of relative "firmness," lead to the statement that the hand is comparatively sensitive to changes in elastic modulus but insensitive to changes in viscosity.

Obviously the cultural tastes and food habits as well as appearance, which may form impressions prior to consumption, influence the subsequent judgment. These factors, however, can usually be controlled by careful selection and blindfolding of the panel.

Recognizing the need for more information, the experts, in general, are in agreement that in order to make the use of test panels more effective, more must be learned about the physiological, psychological and mechanical processes involved in sensory evaluation.

Physiological aspects

These aspects involve the organs of the body which are responsible for control and performance of such processes as evaluating "firmness" by fingers or "chewiness" by mastication. To gain some insight into the various mechanisms involved, the physiologist suggests classifying the constituents of the sensory functions and various aspects involved in the perceptual processes. To accomplish this task, Oldfield (1960) recommended the use of synthetic foods, made of materials of known properties, and modern anaesthetic techniques to mask out different sensory channels selectively in order to evaluate their respective roles and contributions.

Psychological aspects

The relationship between what is perceived and what is measured is dependent on the psychological aspects. Psychologists have attempted to extract sensory qualities from psychological experiments and to correlate them with physical measurements. It has been shown, for example, that the relationship between the intensity of sensation and the intensity of stimuli (light, sound, force, etc.) can be expressed by a power law relationship as follows:

$$Y = mX^n \tag{7.1}$$

where Y is the sensation or sensory response, X is the stimulus, and m and n are constants depending on the type of stimulus. This relationship, which was originally proposed by Plateau (1872), has been reassessed by Stevens (1957). In analyzing the data, the sensory evaluation is taken as sensation Y and a physically measured parameter as stimulus X. Having Y and X plotted on a log-log paper, a straight line defined by

$$\log Y = \log m + n \log X$$

can be established using the method of least squares. From the slope and intercept of the straight line, the values of n and m, respectively, are obtained.

In quality evalution of materials, it has often been found that a logarithmic relationship exists when subjective assessment is compared with physically-measured stimuli. However, the choice of the physically-measured stimulus is not an easy task because single rheological parameters, such as shear stress, elastic modulus, etc., do not easily correlate with sensory measurements. Therefore, the decision on which parameters to measure is usually based on experience, intuition or pure chance. Examples of these measurements are given for testing viscosities of silicone fluids by Stevens and Guirao (1964) and for consistency of liquid foods by Wood (1968). Figure 7.1 illustrates this principle.

Mechanical aspects

These aspects include the mechanics of the organs such as the chewing apparatus involved in texture evaluation. To learn more about the mechanical processes involved when the organs of the human mouth evaluate food texture, Drake (1963) recorded the vibrations produced in the audible region when chewing various hard and soft food materials. It was found that there are differences in chewing sound for different foods and different people. Following this line of research, which was called biorheology of mastication, Drake concluded that our knowledge of the mastication process is highly empirical and suggested an integration of more extensive biological considerations with both theoretical principles and practical knowledge of

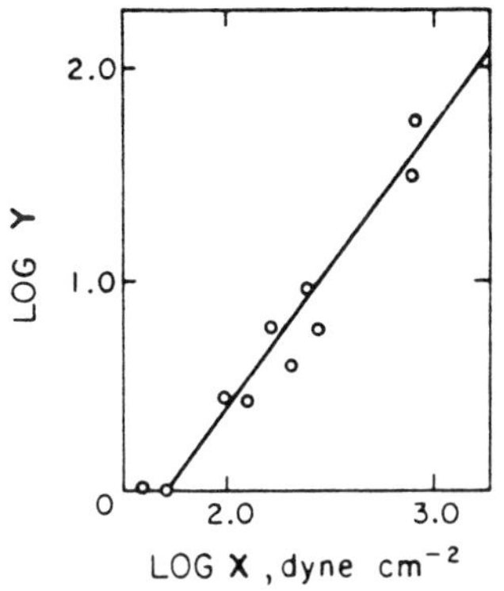

Figure 7.1 Relationship between subjectively assessed texture of cream soup and the measured rheological parameter. *Y* is estimated thickness determined subjectively. *S* is shearing stress at 50 sec^{-1} (Wood, 1968)

food rheology. A rather comprehensive review of the work on the chewing apparatus and the act of chewing, involving the teeth, jaws, muscles, motions, surface contacts, and forces, were given by Drake (1968). (See Figs. 7.2 and 7.3).

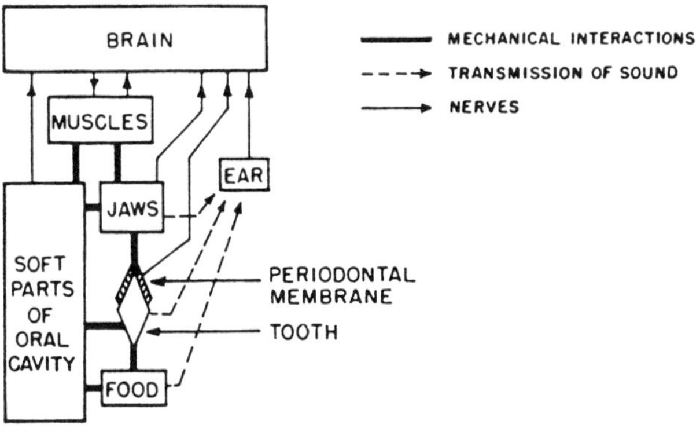

Figure 7.2 Biorheology of mastication (Drake, 1968)

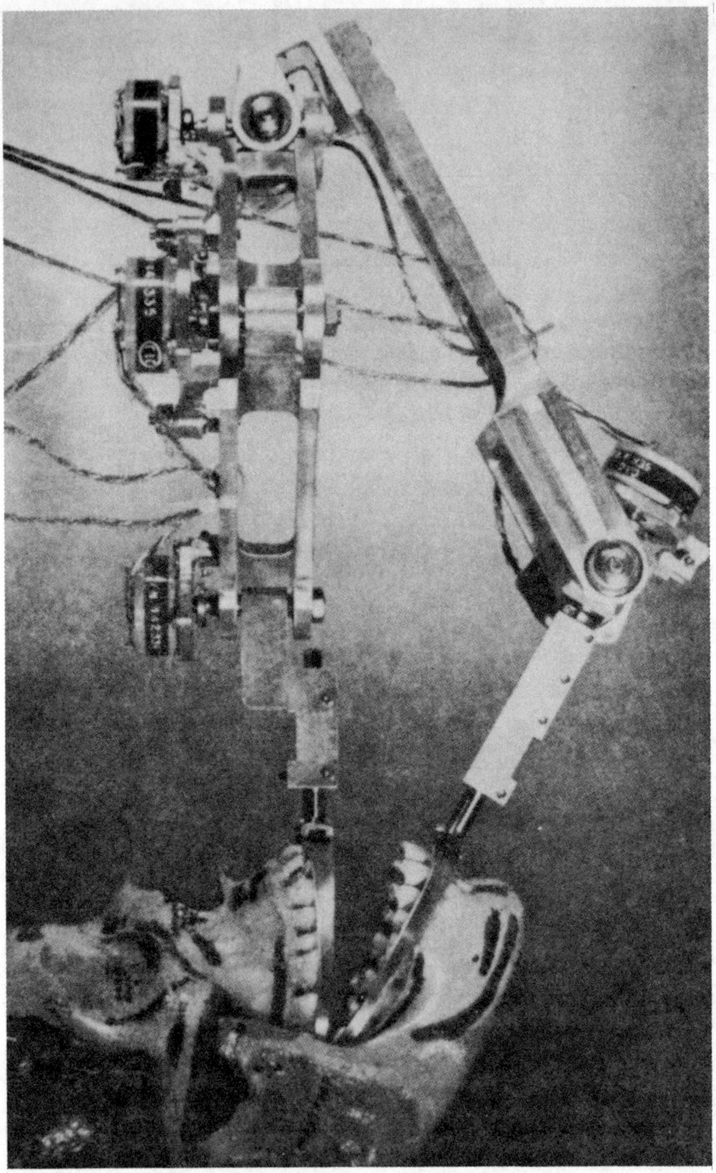

Figure 7.3 Developed for dental research, this apparatus weighs 20 ounces and measures six angles of jaw movement in inches and degrees of rotation. Such a device may be adapted to measurement of the nature and magnitude of forces involved in chewing of foods (Courtesy General Electric Research and Development Center)

7.4 OBJECTIVE OR INSTRUMENTAL MEASUREMENTS

Objective measurements are intended to eliminate personal factor and human element from measurements which are supposed to give the mechanical behavior of the material itself. Not all instrumental measurements are, however, free from personal factors. For example, the manually operated instruments are subject to variation in operating speed unless they are operated by a constant-speed drive mechanism. In this section, each of the three methods of instrumental measurements will be discussed and, through selected examples, the merit and limitations in each group will be pointed out.

Fundamental rheological tests

The following statement by Lord Kelvin is frequently seen in discussions dealing with measurement: "I often say that when you can measure *what you are speaking about* and express it in numbers you know something about it; but when you cannot measure it, when you cannot express it in numbers, your knowledge is of a meager and unsatisfactory kind; it may be the beginning of knowledge, but you have scarcely, in your thoughts, advanced to the stage of science, whatever the matter may be."

The above quotation seems to be most appropriate when one is dealing with the subject of food texture. Obviously we cannot measure *what we are speaking about* unless we define it. Therefore, the first step in fundamental mechanical tests in food texture is to define what specific mechanical property is to be measured. Then, we must obtain numerical data in such form which will enable the use of fundamental equations in mechanics and rheology. In this respect, it is essential that the measurement be unaffected by extraneous physical properties and complexities in geometry and that results be expressible in absolute units independent of the design and dimensions of the testing equipment. Examples of such measurements for fruits and vegetables, cereals and fluid foods were given in Chapters 5 and 6.

Rheological approaches with the idea of correlating results to sensory evaluation have been employed in some food materials for many years. Some dairy products, flour dough, honey and some other fluid foods are among the most frequently seen examples in the literature. Due to the complex nature of some materials, many of the original problems are still unsolved. However, one may also find examples where the fundamental approach has proven useful and of practical value.

Spreadability of butter, as judged by test panels, has been correlated with the force needed to extrude the butter through an orifice at a standard rate (Scott Blair, 1957). This measurement is clearly related to capillary rheometry discussed in Chapter 4.

Cheese making consists essentially of clotting the milk and separating the resulting curd. Differences in various types of cheese are due to the methods of milk clotting, and separating and subsequent handling. It has been shown that in ripening the cheese, the physical properties of the curd will control the resultant chemical and bacteriological changes leading to production of a particular type of cheese. "Body" in cheese is a textural characteristic which relates to specific physical characteristics. It includes several factors such as "firmness," "springiness" and "smoothness." Davis (1937) and Scott Blair (1938) have suggested that these characteristics can be expressed in terms of rheological parameters. For example, "firmness" is associated with relaxation time or the ratio of viscosity to elasticity. Likewise, "springiness" with elasticity and "smoothness" with homogeneity.

For evaluation of "body" in cheese, Scott Blair (1957) developed a constant stress creep test in which cylinders of cheese were loaded with a device that increased the load in proportion to the increasing cross section of the specimen. It was found that the relationship between strain, ε, and time, t, could be expressed by

$$\varepsilon = \psi t^k \tag{7.2}$$

where ψ and k are parameters which were constant for each type of cheese. Despite the claimed high correlation between this technique and measurement of "body" in cheese, a non-destructive apparatus had to be developed to resemble the process of judging "body" by pressing the thumb into the rind. In this apparatus, a steel ball was pressed into the cheese by means of a weight operating on the handle and the resulting penetration was read on the dial gauge after half a minute. The load was then removed, and after another half minute, a second reading of deformation was taken. The difference between the two readings (the elastic deformation) was related to the "body" of the cheese.

"Strength" of flour dough is highly related to the quality of the baked bread. For optimum strength, the weak and strong flours are usually blended by millers or bakers. While the rheology of wheat dough has been studied quite extensively and the answers to many questions still remain unrevealed, the fact remains that many of the physical characteristics involved in the "strength" quality of dough have direct relationship with the term strength when used in connection with engineering materials. This relationship was recognized as early as 1907 when Kosutany (1907) developed a device for extending pieces of dough in rectangular cross section at a constant rate and comparing the shapes of stress-strain curves for different flours. Several years later Hankoczy made one machine for blowing a bubble in a thin sheet of dough and another machine for measuring the work required to knead a dough (Scott Blair, 1957). The volume and internal pressure of the largest bubble which the dough would tolerate without breaking was taken as a measure of dough quality. The basis of this test lies in foam mechanics which is a rather well-developed area of engineering mechanics (Handleman *et al.*, 1961; Gent and Thomas, 1963). The blowing of the bubble actually imitated the swelling of dough under the influence of carbon dioxide in the baking process. The work required to knead the dough was found to increase first until the dough was formed and then decrease with further kneading at a rate depending on the strength of the flour.

Relationship between fundamental tests and sensory evaluations

If sensory evaluation of foods involves not only the basic mechanical properties of the food itself but also unknown physiological and psychological factors which vary with individuals in the panel, it seems unfair to underestimate the use of fundamental approach in mechanical tests because of the lack of correlation with results of sensory evaluation. There seems to be two distinctly different systems involved in this evaluation. One is the "mechanical system" with known mechanism, controls and idiosyncracies. The other is the "human system" with unknown mechanism, controls and idiosyncracies. Take, for example, evaluation of "toughness" in meat. We will first define toughness as work per sample or the area under the stress-strain curve when a sample with known geometry is brought to failure by a compression testing machine under a given strain rate and other specified conditions. Next, samples from the same meat are submitted to taste panels and the average reactions of this "human system," whose operating conditions are unknown, is compared with the results of the "mechanical system." Obviously, we cannot expect to get consistent correlation between the two methods of evaluation if we do not understand and thus cannot control the operating conditions of the "human system."

In light of this argument, while more is being learned about all aspects of human response prior and during eating, to all kinds of foods, fundamental mechanical tests should be of much value in research and development work. The following are examples for justifying this approach insofar as food texture is concerned.

1 Techniques and methods of measurements, capable of producing clearly-defined rheological parameters, are needed for a better understanding of the basic nature and properties of the biological material itself. In this respect, some of the problems involved were discussed in Chapters 5 and 6.
2 Once the material is characterized, it should be possible to see how it ties in with known systems and what are the critical factors. When these critical factors have been isolated, attempts can be made to alter the product rheologically and make it conform to a desired pattern of flow which may lead to improvements in the quality of the product.
3 Basic information on mechanical properties of foods is essential in developing theories to explain the principles of those instrumental measurements which have given high correlations with subjective tests. Once the principles involved in these usually complex measurements are understood, we may be able to explain the reasons for the high correlations.
4 The availability of basic data may contribute in connecting the rheological behavior and molecular and gross structure of food materials so that the mechanical behavior can be explained in light of structural mechanics of the material rather than the current practice of phenomenological model approach. In development of synthetic foods, this approach may prove to be quite valuable.

Elasticity and modulus of elasticity

Often there is a misunderstanding in using the terms elasticity and modulus of elasticity. The term "modulus of elasticity" is often associated with the degree of elasticity of a food material. Considering the fact that by definition modulus of elasticity is the ratio of stress to strain within the elastic range of the material, it is actually an indication of rigidity and stiffness of the material and not its degree of elasticity. Two materials can have the same degree of elasticity but different values of the modulus of elasticity. For example, the degree of elasticity of soft rubber and spring steel are about 100%, but their moduli of elasticity are about 0.0005×10^6 psi and 30×10^6 psi, respectively. In other words, steel is about 60,000 times more rigid than rubber. Therefore, in sensory correlations of food materials, it is incorrect to compare the elasticity of materials as judged by the panel (sensory elasticity) with the modulus of elasticity. This comparison shows an inverse relationship (Szczesniak 1975). One should compare sensory elasticity with the degree of elasticity of the material found from elastic and plastic deformations, if the viscoelastic portion of the total deformation is to be neglected. The latter requires more elaborate experiments (Mittal *et al.* 1977).

There seems to be more misunderstanding of the term modulus of elasticity—its usefulness and significance, and its correlation with food's textural parameters—than of any other rheological term. Even for engineers, this term is somewhat of a misnomer since it refers, as mentioned earlier, to stiffness in the elastic range rather than to elasticity. Since the modulus of elasticity is determined from the ratio of stress to the corresponding strain within the elastic range, one must know the upper limit of the elastic range. Since most of the theories in solid mechanics applicable to food materials are valid only for small deformations, one must also know how much deformation is too much. Therefore, in conducting well-defined mechanical tests on food materials, besides selecting the correct sample size and geometry, it is essential to interpret the force-deformation curve in a manner conforming to an accepted or standard procedure. (See ASAE S 368.1 in Appendix.)

Compression tests of a variety of soft and hard foods have indicated that elasticity as defined by Hooke's law does not really exist in food materials. For the first loading-unloading cycle, even such hard materials as the dry, horny endosperm of corn kernels show some unrecoverable deformation upon unloading (see Fig. 4.8C). However, this should not discourage and prevent the food researcher from using the known concepts in solid mechanics, since the engineers have been using them for structural materials which also are not perfectly elastic. They have defined arbitrarily a limiting value beyond which strains are too great for Hooke's law to be applicable. One such arbitrary but accepted procedure is the "offset" method for the determination of yield strength in engineering materials. Another one is the Johnson's apparent elastic limit (see discussion of Fig. 7.24). Similar arbitrary procedures can be applied to food materials to determine the limit beyond which strains are too great or an arbitrary elastic limit has been exceeded. To maintain the purity of the term "modulus of elasticity," the term 'modulus of deformability' should be substituted in research with food materials. In this case, the total elastic and plastic deformation at some selected point on the force deformation curve is taken for computing the strain.

Besides using the correct definitions of rheological terms, if rheological parameters are to be correlated with instrumental and sensory measurements, it is essential to examine such a correlation between measurements which are compatible and related to the same zone of the stress-strain curve of the material (Fig. 7.3a). A sensory test, which is a failure type of test, should not logically be expected to correlate significantly with small deformation instrumental tests which do not subject the material to failure. For example, some researchers have sought correlations between the modulus of elasticity of a solid food and sensory evaluation of the same food when subjected to biting and chewing. The former being a small strain test should not logically correlate significantly with a failure type of test which the food experiences during biting and chewing. This lack of compatibility between measurements to be correlated together with loose definitions and misuse of well-defined rheological concepts are believed to be responsible for the confusion and reluctance of some researchers to use more of the well-defined rheological terms and measurements.

In an attempt to correlate the information on the mechanics of solid foods to sensory evaluation, it is important to understand the boundary conditions, the assumptions, and the geometry of the measurement needed to have a well-defined mechanical property. For example, consider measur-

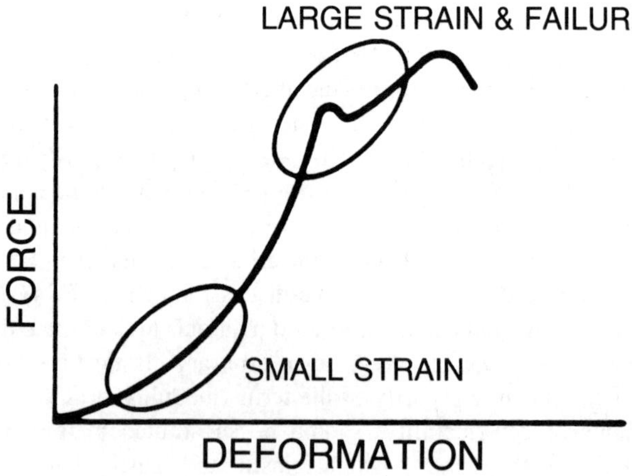

Figure 7.3a The two zones in the force-deformation curve related to small strain and large strain and failure

ing the firmness of an apple by the feel of the fingers. If we assume that the human feel measures the stiffness of the fruit without exceeding its elastic limit, then it is logical to expect the modulus of elasticity of the fruit (if determined correctly by consideration of comparable strain rate, proper geometry, etc.) to correlate significantly with the consumer firmness test. However, if the consumer is to test the apple firmness by biting into the fruit, then a failure property (such as a well-defined punch shear test) is probably more logical to correlate with the sensory evaluation. Likewise, in sensory evaluation of tenderness of meat, if we assume that the first sensation produced by the cooked meat on the teeth is a measure of elasticity, then elastic properties, defined and measured properly, may be expected to be one of the mechanical attributes to be correlated with tenderness. However, if the human test involves biting and tearing the meat apart, then we are dealing with a different zone of the stress-strain relationship of the material, namely yielding and fracturing. In this case, a well-defined failure test is more suitable.

7.5 IMITATIVE AND EMPIRICAL TESTS

An imitative test in foods is similar to the so-called "service" test in engineering, where some actual service condition is imitated as closely as possible. Like many engineering service problems, often the mechanical aspects of texture evaluation are so complex in geometrical form and the action of forces that it seems extremely difficult if not impossible to analyze the problem theoretically and predict how and to what extent the various physical properties affect the results of the imitative test. Under such conditions an imitative test which can lead to a final decision is fully justified.

The empirical test stands somewhere between the fundamental rheological test and the service or imitative test. Its usefulness depends either on the accuracy with which it predicts the behavior of the material under complex service conditions or on how accurately it permits the use of fundamental equations for calculations and analysis. If neither of the two objectives can be accomplished, it would be difficult to justify this type of test.

Most of the work reported as "objective" evaluation of foods fall in this category. Sometimes it is difficult to understand the significance of many of these tests. In some cases, this is the fault of the test method itself. Often the test methods are not conducive to measuring scientifically meaningful quantities.

Numerous examples of and references to work on imitative and empirical tests for texture in foods are found in Scott Blair (1953), S.C.I. Monograph (1960), Matz (1962), Szczesniak (1963), and Kramer and Twigg (1966). The following instruments, applied to various types of texture on foods, are given as examples to illustrate the nature of these tests and the physical parameters involved in the measurements.

Tenderometer consists of a grid assembly which simulates jaw action in the eating of peas. The upper and lower sets of grids are hinged together and the sample is first compressed and then sheared and extruded. It is operated hydraulically and force is shown by an indicating hand. This instrument has given high correlation between quality of raw pea and alcohol-insoluble solids of the processed product (Walls and Kemp, 1940). *Puncture tester* uses a needle (about 1⁄16 inch in diameter) to penetrate into products such as sweet corn or pea (Caldwell *et al.*, 1939). *Texturemeter* consists of a group of 25 rods which travel through the mass of sample until they pass through matching holes in the bottom of a cylindrical cup. Power is provided by a crank and force is measured by a pressure gage. *Maturometer* has each of its 143 rods "shearing" through a single pea. Total force for all the peas is measured on a single gage (Lynch and Mitchell, 1950). *Fibrometer* uses a channel shaped to contain an asparagus stalk and a three-pound weight which holds a wire. A stalk of canned asparagus is placed in the channel and the wire is placed through the slits in the channel until it rests on the stalk. Any segment of the stalk which the wire does not cut through is considered fibrous (Wilder, 1948). *Succolumeter* measures the volume of extractable juice under controlled conditions of time and pressure. Its measurements have been correlated with the maturity of sweet corn, the storage quality of apple, and the oil and water content of products.

L.E.E.-Kramer shear press is an attempt to combine all textural measurements in one testing machine. Different test cell assemblies are provided for different tests with the power unit. The press consists of a hydraulic drive moving the crosshead of the machine at a rate of travel, adjustable from 15 to 100 seconds for full stroke. Force is measured by the compression of a proving ring dynamometer varying in range from 100 to 6000 pounds. The unit can be obtained with a force transducer and a recorder to obtain a force-time curve as the product is being tested (Kramer and Twigg, 1966). The most common type of test cell used in this machine is shown in Fig. 7.4. The shear bars are attached to the crosshead of the shear press or any compression testing machine with adequate load capacity. In low-

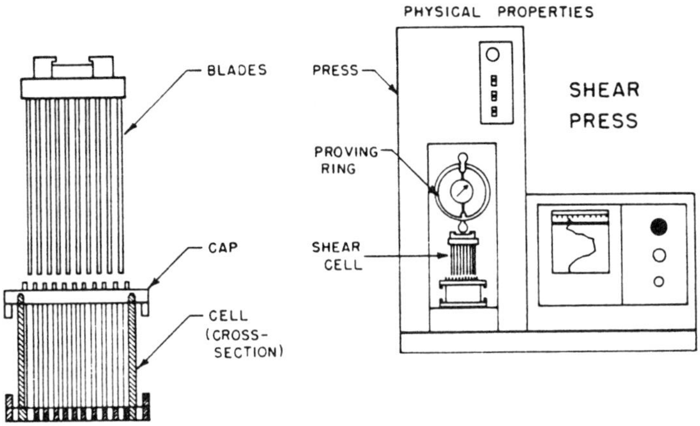

Figure 7.4 Right: L.E.E.-Kramer shear press. Left: Shear cell cross section

ering of the crosshead, the shear bars pass through the sample placed in the cell. The sample is first compressed and then extruded through the slots at the bottom of the cell.

Denture tenderometer was designed to simulate the denture surfaces and motions of mastication in the mouth (Proctor *et al.*, 1956). A complete set of human dentures is used for mechanical chewing. The forces are measured by strain gages fitted on the arm connecting to the upper jaw. An electronic modification of this device was adapted to tenderness measurement of individual peas recorded in digital form (Davidson *et al.*, 1959). The *General Food Texturometer* is a modification of denture tenderometer in which a plunger replaces the dentures (Friedman *et al.*, 1963).

Bloom gelometer is a standard instrument which measures the rigidity of edible gelatin. It employs the weight of lead shot required to push a standard plunger 4 millimeters deep into the gel (Bloom, 1925). *Baker Compression meter* utilizes a plunger for measuring the firmness of bread in terms of force and deformation (American Association of Cereal Chemists, 1947). Platt and Powers (1940) describe a number of other devices for measuring bread softness. One method uses a dead load for a fixed time. A modified *Bloom Consistometer* (Clardy *et al.*, 1952) consists basically of a plunger and a barrel and gives relative pressure necessary to force a semi-solid through an opening. *Bostwick Consistometer* measures the maximum distance a given amount of catsup travels down a trough (Townsend, 1956).

Warner-Bratzler shear device consists of a blade and two shear bars (Fig. 7.5). A core of meat, placed in a hole in the blade, is double sheared as the shear bars are lowered. The maximum force is related to toughness of the meat (Bratzler, 1932). Recently this shear device was mounted on the L.E.E.-Kramer shear press to compare sensory and instrumental measurements of beef tenderness (Sharrah *et al.,* 1965). Voisey and Hansen (1967) described an apparatus designed to give a record of Warner-Bratzler shear force on a strip chart. (See Fig. 7.6.)

Brabender farinograph uses a dynamometer to measure the torque of a given quantity of flour dough under continued kneading and stretching. The measurement has been related to the handling and gluten properties, stability and fermentation tolerance of the dough (Greup and Hintzer, 1953).

A meat *slice tenderness evaluation* apparatus illustrated in Fig. 7.7 was developed to be used with universal testing machines (Kulwich *et al.,* 1965). The penetrator first punctures the slice of the meat and then shears off a circular portion of the slice. The correlation coefficient with sensory evaluation was found to be very close to that obtained by using a Warner-Bratzler shear device.

Curd-o-meter was developed for measuring the so-called "curd tension" in dairy processing. With this device, the force needed to draw a standard

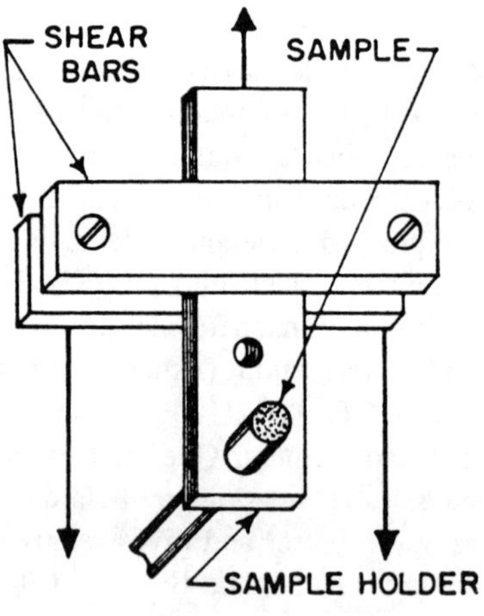

Figure 7.5 The principle of Warner-Bratzler shear device (Redrawn from Sale, S.C.I. Monograph, 1960)

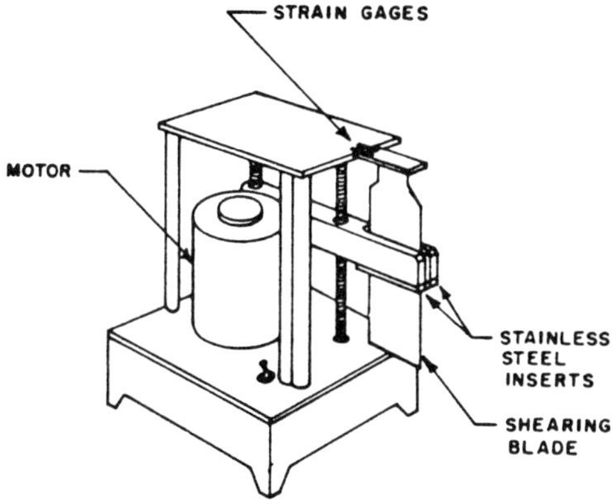

Figure 7.6 An automated version of Warner-Bratzler shear device (Voisey and Hansen, 1967)

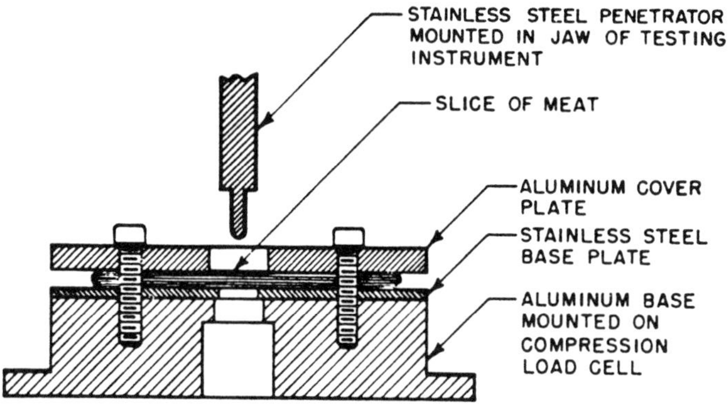

Figure 7.7 Device for measurement of meat slice tenderness (Kulwitch *et al.*, 1963)

knife through the curd is measured. Attempts to relate this measurement with digestability of milk have failed (Scott Blair, 1957).

Firmness meter shown in Fig. 7.8 was designed to prestress the specimen, apply a load, and measure deformation of the specimen during a given time. The right weight prestresses the sample and insures good contact. The left weight deforms the sample for a given time. After this time, a solenoid sets the brake and stops the deforming force. The force

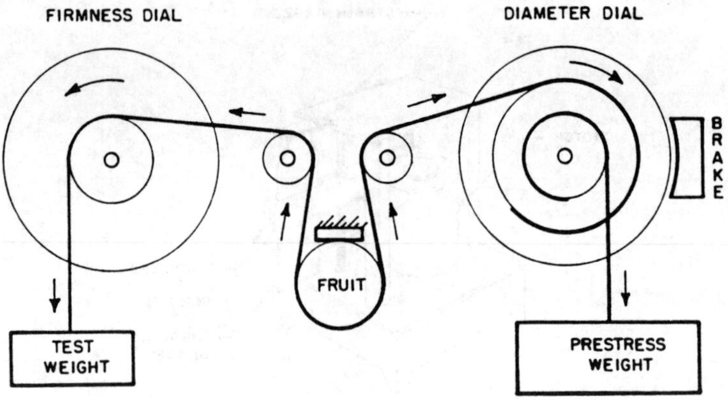

Figure 7.8 Firmness meter (Courtesy Agricultural Specialty Company)

is applied through a selection of roller chains, flat surfaces, and fiber glass tape to products such as apples, onions, tomatoes, etc. The force is applied uniformly around the circumference of the object simulating firmness measurement of fruits and vegetables when held in hand.

7.6 TEXTURE PROFILE METHOD

Since what is perceived by human senses is apparently not what is measured by instruments, some experts have come to the conclusion that perhaps texture is a combination of several rheological properties and no single measurement can be expected to predict human response. There are numerous examples reported in literature to support this idea.

For example, based on a psychophysical experiment on subjective judgment of "firmness" of vulcanized rubber, Harper (Scott Blair, 1953; Harper, 1968) concluded that more than one form of physical behavior may be involved in the experience called "firmness." Halton and Scott Blair (1937) have indicated that the baker's impression of "springiness" in dough is not related to its elasticity alone but a combined effect closely related to its relaxation time, i.e., to the ratio of dough viscosity to its elasticity.

To define texture in terms of more than one rheological parameter, Brandt *et al.* (1963) introduced the concept of "texture profile" method. The basis of this method was a new classification of textural characteristics proposed by Szczesniak (1963) and the development of standard rating scales for mechanical parameters of food texture (Szczesniak *et al.*, 1963).

The classification of textural parameters included three main types: (1) mechanical properties, such as hardness, cohesiveness, viscosity, elasticity, adhesiveness, gumminess, etc. dealing with strength and flow characteristics; (2) geometric properties, such as graininess, flakiness, lumpiness, etc. derived from shape, size, orientation, arrangement of the constituents, and appearance of the food; and (3) other characteristics referring mainly to moisture and fat content. The standard rating scales covered the intensity range of hardness, brittleness, chewiness, gumminess, viscosity and adhesiveness found in food products. Each point on the scale was represented by a food product selected on the basis of availability, familiarity, and constancy of textural characteristics.

Table 7.1 shows the standard rating scales and the intensity range of mechanical parameters given by food products. Other information given with these scales are the brand or type of the food, the manufacturer, the sample size, and the temperature. Foods selected to represent the individual points on the sensory rating scales were also evaluated by instrumental methods and the resulting correlation curves were established as shown in Fig. 7.1. These standard curves make it possible to assign numerical ratings for the unknown texture of a food product. Using the flavor profile method as a model, it was suggested that the entire texture characteristics of a given food can be recorded systematically in the order of their appearance or perception from the first bite through complete mastication (Brandt *et al.*, 1963).

Table 7.1 Standard rating scales used in texture profile method (Condensed from Szczesniak *et al.*, 1963)

Standard scale for mechanical parameter	Intensity range given by food products
Hardness	Cream cheese (1)*—Rock candy (9)
Brittleness	Corn muffin (1)—Peanut brittle (7)
Chewiness	Rye bread (1)—Tootsie rolls (7)
Gumminess	40% flour paste (1)—60% flour paste (5)
Adhesiveness	Hydrogenated vegetable oil (1)—Peanut butter (5)
Viscosity	Water (1)—Condensed milk (8)

*Number in parenthesis shows the rating by the sensory evaluation panel.

The claimed advantages for texture profile method are (1) its flexibility of application to any food product, and (2) its objectivity through "rigidly defined points of reference and nomenclature."

Although several of the definitions and the nomenclature regarding the mechanical parameters in texture profile method have been questioned, as stated by Szczesniak (1966), it is an attempt to form a bridge over fundamental rheological properties and popular nomenclature regarding food texture.

7.7 NEED FOR STANDARDIZATION OF NOMENCLATURE AND MEASURING TECHNIQUES

In reviewing the ASTM standards for measurements of physical properties of engineering materials, one would find a considerable number of techniques which can be classified as imitative and empirical. Yet these techniques, which are based on terms, definitions, procedures, and test apparatus which have been agreed upon, are meeting the needs of the industry.

For example, brittleness of asphalt plank is measured by taking a standard specimen supported with timber and driving a standard nail vertically through the face of the plank at each of the two diagonally opposite corners between 1½ to 2 inches apart. The driving force is applied by a 10-lb weight falling freely from a constant height of 20 inches above the surface of the plank. Brittleness of a specimen is compared by the appearance of crackings when the nail is driven at least one inch into the plank (ASTM D517-50). There are also brittleness tests conducted as flexture test using the material as a beam loaded in the middle.

Consistency of lubricating grease is measured by a penetration test using a cone penetrometer. The penetration is the depth that a standard cone penetrates the sample under prescribed conditions of weight, time, and temperature.

Hardness of rubber and plastics in a wide range of hardness or softness is measured by means of a durometer. The method permits measurements of either initial indentations or of indentations after a specified period of time or both (ASTM D2240-64T). Hardness of plywood is measured by a ¼-inch hardened steel ball penetrator with a minimum load of 10 kg and a major load of 100 kg.

Likewise, many of the techniques classified as empirical and imitative in food texture evaluation have their merits and usefulness. What is needed is refinement, standardization, and cataloging of those techniques which have shown repeatedly high correlation with sensory evaluation.

7.8 TEXTURE AND STRUCTURAL MECHANICS OF FOOD MATERIALS

A glance at the list of descriptive terms mentioned earlier in this chapter shows that texture in food must be closely associated with the structural makeup of the material. Despite this rather obvious relationship, very little work is found in literature to describe the observed mechanical responses of foods under examination in terms of structural mechanics of the material. Wood is a biological material which has been studied rather extensively both rheologically and structurally. Shafizadeh and Nearn (1966) have given an excellent review of structure in wood, tracing the origin of the anisotropic physical and mechanical properties of wood or its cellular structure and chemical composition. It has been found that differences in the orientation and continuity of the basic structural units of wood in longitudinal or transverse directions influence the physical and rheological behavior of the material. The degree of nonhomogeneity and anisotropy in many food materials is probably not any greater than in wood. A recent review of literature on mechanical behavior of biological tissues indicated that no investigation has yet been conducted to determine the viscoelastic response functions treating these materials as anisotropic and nonhomogeneous. Such a study may lead to the development of a macroscopic theory of mechanical behavior of food materials based on their structural factors.

It is known that the biological tissues in solid food materials are composite in nature and are made up of liquid and solid substances that are combined in a random fashion. Some of the structural aspects of biological tissues which are significant in their response to mechanical forces were discussed in Chapter 2. These included the structural aspects of the cell wall, cell contents, cementing agents, such as pectic substances, and the relationship of structure to turgor pressure and tissue rigidity. Perhaps such descriptive terms as "firmness," "mealiness" and "crispness" in fruits and vegetables can better be defined in terms of known chemical and structural changes associated with turgor pressure and pectic substances responsible for tissue rigidity. It has been shown, for example, that when turgor in apple and pear tissues is eliminated by placing the tissue under the influence of chloroform for 12 to 24 hours, its shearing strength is considerably reduced (Schmidt, 1962).

It was noted in Chapter 2 that mechanical properties of plant and animal tissues in their gross structure generally reflect the properties of the walls of the cells (viscoelastic solid) and the cell contents (viscous liquid). The cellular structure is admittedly too complex to analyze in terms of properties of the individual components. But, isn't it possible to treat the network of the interacting cellular units in selected solid food materials, such as fruits and vegetables, as viscoelastic closed-cell foams and apply what is known in the area of foam mechanics to this problem? For example, theoretical treatments have been developed to relate the properties of the foamed elastic materials (rubber foam) to those of the constituent materials (Gent and Thomas, 1963). In the area of food texture, it is known that the cells in a cake originate as bubbles in the batter. Hence physical characteristics and mechanical properties of these bubbles determine the strength and other textural characteristics of a cake. This being generally accepted, the known mechanics of foam formation were successfully employed to explain some of the observations made in practical cake-batter systems (Handleman *et al.,* 1961).

If certain food materials can be shown to be continuous in their structure, i.e., the smallest element cut from the body possesses the same specific physical properties as the body, it may be that the knowledge of the properties of the viscous fluid extracted from the material and the elastic or viscoelastic properties of the solid constituents would be sufficient to characterize the material and predict its response to mechanical actions. There are some evidences that this technique, which has been employed for engineering materials (Oldroyd, 1956), might have applications in food materials. For example, changes in textural characteristics such as firmness of apples during maturation and storage were found closely associated with the alcohol-insoluble solids abbreviated as AIS (Kertesz *et al.,* 1959; Wiley and Stembridge, 1961). These solids, which are considered the basic structural materials in apples, are composed mainly of starches, pectic acids, hemicelluloses, and cellulose.

Some food materials are composed of biological tissues and fluids which are capable of sustaining and recovering from large deformations. For solids in this class of materials, stress-strain relationship is non-linear and classical elasticity theory, which is restricted to linear stress-strain behavior and infinitesimal strains, is not applicable. For liquids, the flow behavior markedly deviates from Newtonian flow behavior and some elasticity is exhibited upon removal of the load. For description and understanding of such behavior in terms of structure, one should look into the fields of elastomers and elastic fluids. In recent years considerable progress has been made in this area. Saunders (1964) has reviewed some of the important and well-established theories applied to natural and synthetic high polymers. Although chemical structure of these materials is much simpler than the corresponding food materials, nevertheless, it is important to review and understand what has been achieved for these relatively simple materials before studying the more complex relationship between structure and mechanical properties of biological materials.

In reviewing literature, one would find considerable information linking chemical and mechanical structure of foods to certain textural characteristics. The proteins in wheat combine together to form a strong rubbery network known as gluten which is believed to be responsible for many of the textural characteristics of flour dough. Some rheological properties of noodles, which are essentially boiled dough strings made from wheat flour and salted water, have been linked with crude protein content (Shimizu, *et al.,* 1958). Studying the extension of the noodles under dead load (creep test), it was found that the gluten network plays an important role in a gelatinized and denatured dough. Viscoelastic properties and chemical nature of wheat gluten were also studied by Barney *et al.* (1965), employing stress relaxation tests on isolated gluten specimens (Fig. 7.9). To achieve instantaneous straining, the cord was cut, a 100-gram brass weight fell to the machine crosshead in less than 0.5 second, producing a 300% elongation of the specimen.

The theory of elasticity and simple concepts of viscoelasticity have been applied to plant and animal tissues to find explanations for the loss of tissue rigidity (Virgin, 1955; Falk *et al.*, 1958; Nilsson *et al.,* 1958; Drake, 1962; Somers, 1965 a, b). Chemicals which will influence either the mechanical properties of the cell walls or the hydrostatic pressure of the cell contents have been isolated. Citric acid, for example, was found to induce loss of rigidity in potato tissues possibly because of its effect in extracting calcium from the cell walls (Somers, 1965 a). Calcium apparently plays an essential role in the structure of plant cell walls.

Aging of certain fruits and vegetables in air or in water at various temperatures before canning was found to give a "firmer" texture and a greater bulk volume (Whittenberger and Hiels, 1953; Van Buren *et al.,* 1960). The reason for the increased rigidity was traced to a greater rigidity in the cell walls of the product aged before canning (see Fig. 7.10), but the chemical changes responsible for the greater rigidity of the cell walls were not fully explained.

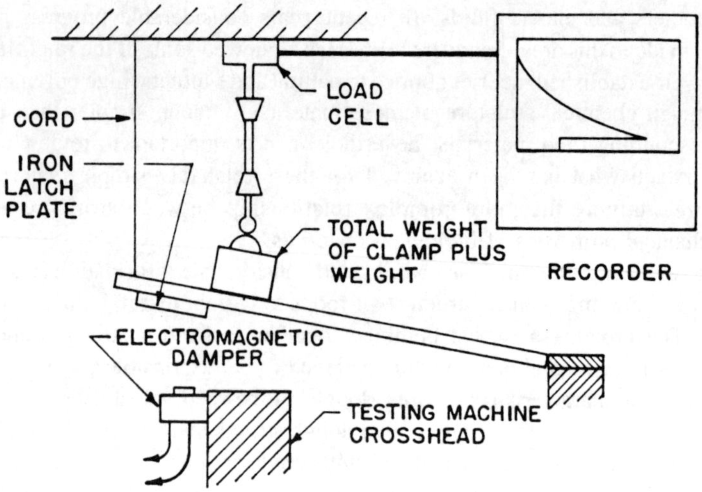

Figure 7.9 Experimental apparatus for stress relaxation of wheat gluten (Barney *et al.*, 1965)

Since, like many other food materials, an absolute definition of texture for fruits and vegetables is very difficult, some authors have defined textural changes in terms of chemical and structural changes in the plant tissue (Griffin and Kertesz, 1946; Reeve, 1953, 1954; Kertesz, 1959). Meyer (1960) and Matz (1962) list several plant factors which influence the texture of fruits and vegetables. These factors are turgidity of the cells, occurrence of supporting tissues, cohesiveness of cells, and assimilation such as conversion of starch to sugar, etc. Isherwood (1960) discusses the relationship of texture in fruits and vegetables to some of these factors as well as chemical composition of the various constituent parts.

It has been suggested that a toughening reaction in beef muscle fibers to heat could be due to tightening of the network of protein structure during denaturation (Hamm and Dentherage, 1960). Tenderness studies in meats at the cellular and molecular level have just begun. Herring *et al.* (1965) have reported studies on bovine muscle tenderness as influenced by sarcomer length and fiber diameter. It has been shown that toughness in fish increases during storage in frozen state (Love, 1960). This has been related to the changes in the structural proteins of the fish muscle giving increased "toughness" and mechanical rigidity to myofibrils (Connel, 1960). Sink

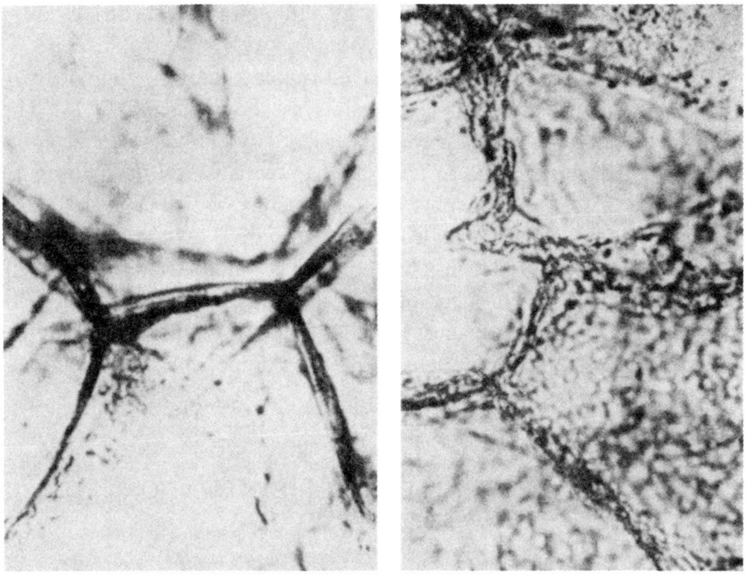

Figure 7.10 Effect of 24-hour aging on turgidity of cell walls in red tart cherries. Left: control; Right: firmed (Buch *et al.*, 1961)

has given a literature review on the biophysical and viscoelastic changes that occur in postmortem in procine muscle. The significance of these changes in the transformation of muscle to meat and their influence on the quality of the meat as food have been pointed out (Sink, 1965).

Textural characteristics of flour dough such as "strength," "body," "shortness," etc., are quite important in predicting the quality of the bread and process control at the bakery plant. Dough rheology has been studied for the last sixty years, employing both phenomenological model approach as well as empirical and imitative instrumental techniques. It is claimed that such intensive studies on other materials such as natural and synthetic fibers, paper, plastics, and rubber have proved to be more fruitful because of their relatively simpler structures. In the case of rubber, for example, it has been possible to connect rheological behavior and structure quantitatively (Treloar, 1949). This has not been possible for flour dough apparently because of its complex structure of which very little is known. In this regard, Muller (1967) made the following observation: "It is impossible as yet to correlate structure with rheology of dough since too little is known of the structure."

Shama and Sherman (1960) connected rheological properties of frozen ice cream with its structural components as shown in Fig. 7.11. They proposed that frozen ice cream consists essentially of an aerated ice crystal structure which is modified in textural characteristics by a superimposed fat network. Those components of the structure which appear to exert major effects on the six rheological parameters are shown in Fig. 7.11 (see Generalized Kelvin Model in Chapter 4). The authors claim that if this rheological model-structure component representation is substantially correct, "creep tests on ice cream of 'good' and 'poor' texture should indicate which parameters, and their related structural components, contribute most to texture."

7.9 DYNAMIC TESTS FOR EVALUATION OF FOOD TEXTURE

One aspect of sensory texture evaluation, which apparently has not yet been investigated, is the correlation between instrumental measurements employing dynamic techniques and the panel rating of the same food material. It is known that the light-touch organs in the human body cease to respond to a constant stimulus after about half a second and that continued response is only obtainable by movement of the stimulus over the sensory surfaces (Oldfield, 1960). In hard-touch organs, such as the sense organs in the teeth, the adaptation time is longer but it may still be shorter than the time required for a compression testing machine to record the maximum force in compressing or shearing a specimen of meat. This requirement for short response time and the deteriorating nature of food materials with time are two important factors suggesting the need for investigation of the correlation of instrumental and sensory measurements employing dynamic tests. Furthermore, the short duration dynamic tests give more information at the short end of the time scale, and it is possible to apply low level of stress in such a way that linear viscoelastic behavior may prevail.

Examples of application of dynamic testing methods for evaluation of mechanical attributes of texture in foods are given in Chapter 5. The basic principles involved in dynamic tests applied to solid food materials for their viscoelastic characterization are also reviewed by Morrow and Mohsenin (1968). An attempt to relate the parameters obtained by the sonic

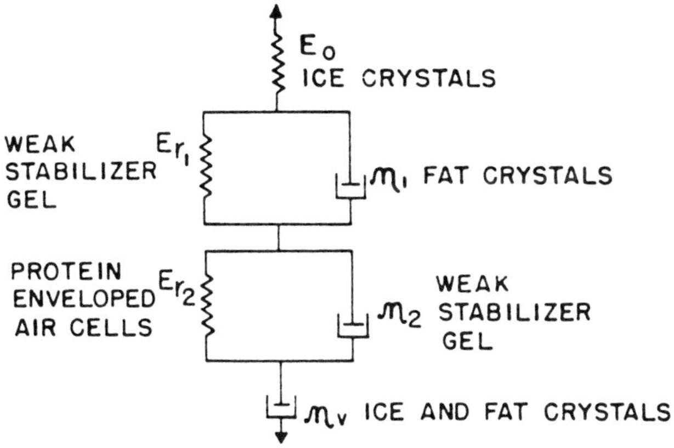

Figure 7.11 Relation of rheological behavior with structure in frozen ice cream (Shama and Sherman, 1960)

resonant method to subjective evaluation of texture in Red Delicious apples is reported by Finney (1968). The squared of resonant frequency, f, multiplied by the mass of apples, yielding the parameter f^2m, was found to be highly correlated with the modulus of elasticity of the fruit flesh but did not correlate with taste-panel scores evaluating the so-called "texture" of the fruit. The resonance parameter did correlate, however, with juiciness and crispiness which are characteristics of high-quality apples. On the other hand, the mealy, low-quality fruits resonated at the lower frequencies, giving lower values for f^2m parameter. In other words, by means of this nondestructive dynamic test conducted on the whole intact fruit, according to the author, one can predict changes in the elastic properties of the apple flesh as well as the consumer reaction of the eating quality of the fruit.

7.10 DIMENSIONAL ANALYSIS OF FOOD TEXTURE

Many complicated problems in engineering have been solved by a combination of incomplete theory, physical testing, and dimensional analysis. Charm (1963) proposed the use of dimensional analysis in food texture studies considering the variables affecting the chewing texture of a food in the form of cubes. In his analysis, it was assumed that the power expended by the mouth was a measure of texture. Using the system of mass-length-time of dimensional analysis with M, L and T denoting mass, length, and time, respectively, the combinations of units involved would be as follows:

P = power = $(ML/T^2)(L/T) = ML^2/T^3$
E = Young's modulus = $(ML/T^2)(1/L^2) = M/T^2L$
G = shear modulus = M/T^2L
μ = Poisson's ratio = dimensionless
S_s = shear stress = M/T^2L
S_t = tensile stress = M/T^2L
L = length of cube = L
N = number of chews per minute = $1/T$

Considering the parameters involved in chewing a food in the form of a cube, the following expression can be written for dimensional analysis.*

$$P/L^3NE = f(G/E,\ S_s/E,\ S_t/E,\ \mu) \qquad (7.3)$$

Such an analysis should yield the exact form of the equation containing one or several constants. The values of these constants may be determined either from physical analysis of the problem or from experimental measurements.

7.11 MECHANICAL TESTS APPLICABLE TO FOOD MATERIALS

In Table 7.2 is given an illustrated summary of mechanical tests applied to food materials. Many of these tests have already been applied in evaluation of mechanical properties of agricultural materials including food products (Mohsenin, 1970). Others are suggested to be considered for future investigations. Details of these tests and assumptions for validity of the computing formulas are given in other parts of this book. The shape, size and geometry of the test specimen must conform either to the available standards, such as those of the American Society for Testing and Materials (ASTM) or be compatible with the basic assumptions used in the derivation of the computing formulas.

*This is a dimensionally corrected form of the expression given by Charm (1963).

Table 7.2 Summary of mechanical tests applicable to food materials

Type of test	Parameter and calculating formula	Page
ASTM COMPRESSION OR TENSION 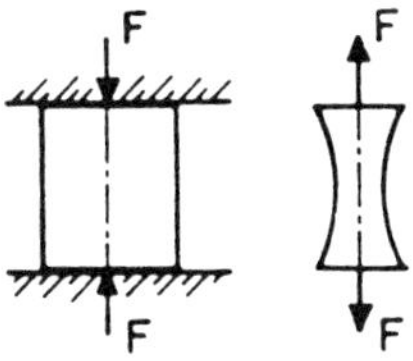	$\sigma = \dfrac{F}{A}$; $\varepsilon = \dfrac{\Delta L}{L}$; $E = \dfrac{F/A}{\Delta L/L}$	181
	$E = 3K(1 - 2\mu)$; $E = 2G(1 + \mu)$	
	$1/E = 1/3G + 1/9K$	105
POINT LOADING	$a = 0.721m(FAd)^{1/3}$	284
PLATE ON SPHERE 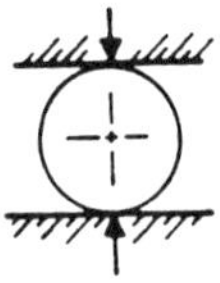	$S_{max} = \dfrac{0.918}{mn}\left(\dfrac{F}{A^2d^2}\right)^{1/3}$	284
	$D = 0.769k\left(\dfrac{F^2A^2}{d}\right)^{1/3}$	284
	$E = \dfrac{0.338k^{3/2}F(1-\mu^2)}{D^{3/2}}\left(\dfrac{4}{d}\right)^{1/2}$	286
	$R_2 = R_2' = \infty$; $R_1 = R_1'$; $\cos T = 0$;	
	$m = n = 1$; $k = 1.3514$	
PLATE ON CONVEX BODY 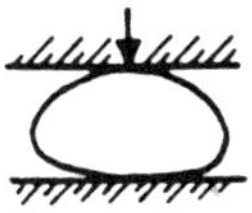	$a = m \times 1.145\left[\dfrac{FA}{(1/R_1 + 1/R_1')}\right]^{1/3}$	
	$S_{max} = \dfrac{0.365}{mn}\left[\dfrac{F}{A^2}(1/R_1 + 1/R_1')^2\right]^{1/3}$	
	$D = 0.485k[F^2A^2(1/R_1 + 1/R_1')]^{1/3}$	
Deformation at bottom assumed negligible	$E = \dfrac{0.338k^{3/2}F(1-\mu^2)}{D^{3/2}}(1/R_1 + 1/R_1')^{1/2}$	286
	$R_2 = R_2' = \infty$; $R_1 \neq R_1'$	
	Find m, n, k from eqn. 6.6	

Table 7.2 (continued)

Type of test	Parameter and calculating formula	Page
SPHERE ON SPHERE 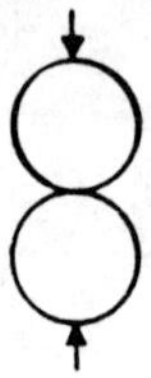	$a = 0.721m\left[\frac{FA}{1/d_1 + 1/d_2}\right]^{1/3}$ $S_{\max} = \frac{0.918}{mn}\left[\frac{F(1/d_1 + 1/d_2)^2}{A^2}\right]^{1/3}$ $D = 0.769k[F^2A^2(1/d_1 + 1/d_2)]^{1/3}$ $E = \frac{0.338k^{3/2}F(1-\mu^2)}{D^{3/2}}(4/d_1 + 4/d_2)^{1/2}$ $R_2 = R_2'; \quad R_1 = R_1';$ $m = n = 1; \quad k = 1.3514$	284
SPHERE ON CONVEX BODY  Deformation at bottom assumed negligible	$a = 1.145m\left[\frac{FA}{1/R_1 + 1/R_1' + 4/d_2}\right]^{1/3}$ $S_{\max} = \frac{0.365}{mn}\left[\frac{F}{A^2}(1/R_1 + 1/R_1' + 4/d_2)^2\right]^{1/3}$ $D = 0.485k[F^2A^2(1/R_1 + 1/R_1' + 4/d_2)]^{1/3}$ $E = \frac{0.338k^{3/2}F(1-\mu^2)}{D^{3/2}}(1/R_1 + 1/R_1' + 4/d_2)^{1/2}$ $R_2 = R_2'; \quad R_1 \neq R_1';$ Find $\cos T$, m, n, k	284
DIE LOADING  Deformation at bottom assumed negligible	$p = \frac{F}{2\pi a\sqrt{a^2 - r^2}}$	289
	$D = \frac{F(1-\mu^2)}{2aE}$	290
	$E = F/D\,\frac{(1-\mu^2)}{2a}$	290
SHEAR PUNCH SHEAR 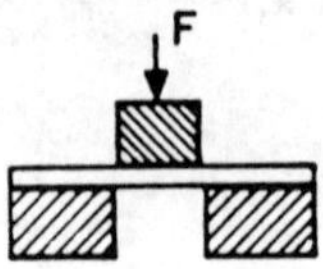	$S = \frac{F}{\pi dt}$	188

Table 7.2 (continued)

Type of test	Parameter and calculating formula	Page
DOUBLE SHEAR 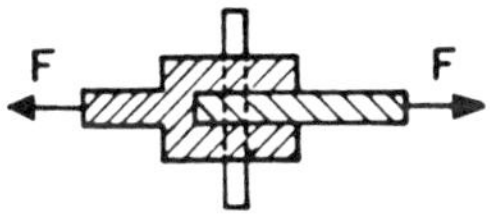	$S = \dfrac{F}{2A}$	191
PARALLEL SHEAR 	$S_s = F/2A$; $\varepsilon_s = \tan\theta$; $G = S_s/\varepsilon_s$; $E = 2G(1+\mu)$; $1/E = 1/3G + 1/9K$	104
PLATE SHEAR 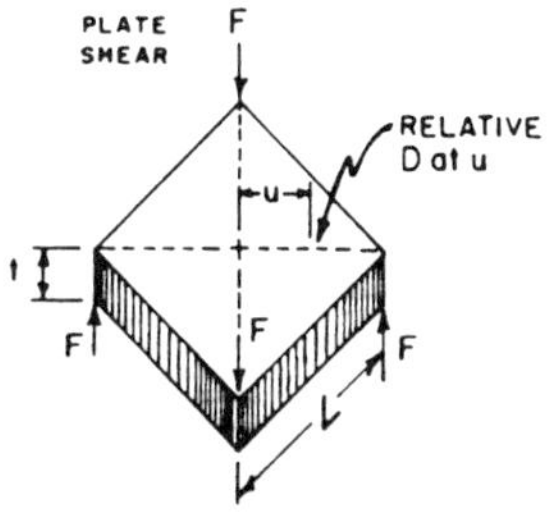	$S_{s_{max}} = \pm \dfrac{3F}{t^2}$; $\varepsilon_s = \pm \dfrac{3F}{Gt^2}$ $G = \dfrac{3Fu^2}{2Dt^3}$; $50 > L/R > 25$ R = radius of curvature at u	
TORSION TORSION BAR 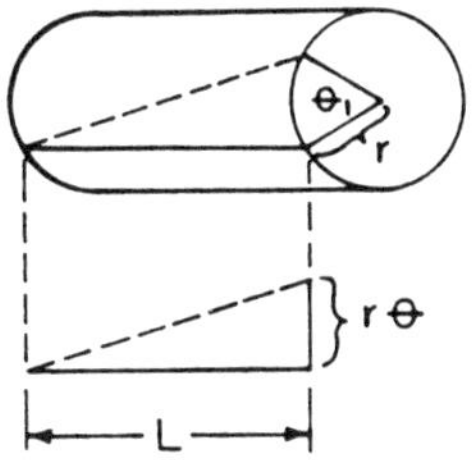	$S_s = \dfrac{2T}{\pi r^3}$; $\varepsilon_s = \dfrac{r\theta}{L}$; $G = \dfrac{2TL}{\pi r^4\theta}$ $\theta_1 = \theta$	193
TORSION PENDULUM 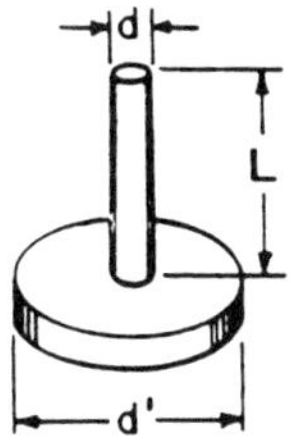	$G = \dfrac{16\pi f^2 W(d')^2 L}{gd^4}$ W = weight of disk	

Table 7.2 (continued)

Type of test	Parameter and calculating formula	Page
BENDING		
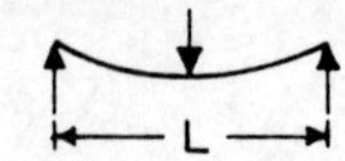	$E = \dfrac{FL^3}{48DI}$	200
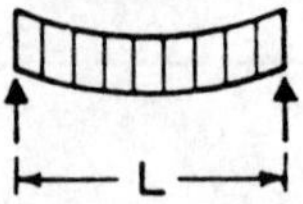	$E = \dfrac{5wL^4}{384DI}$	200
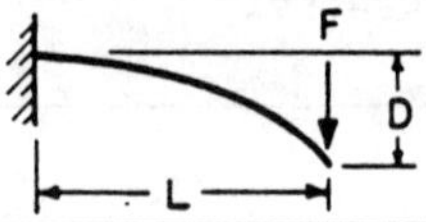	$E = \dfrac{FL^3}{3DI}$	200
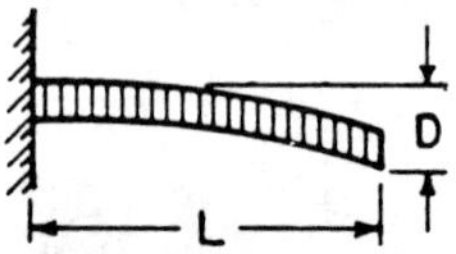	$E = \dfrac{wL^4}{8DI}$	202
BULK COMPRESSION		
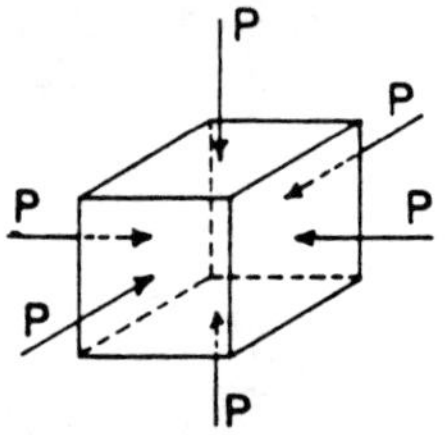	$K = -\dfrac{\Delta p}{\Delta v/v}$; $B = 1/K$ $1/E = 1/3G + 1/9K$	210 105
DYNAMIC LOADING		
DIRECT STRESS AND STRAIN 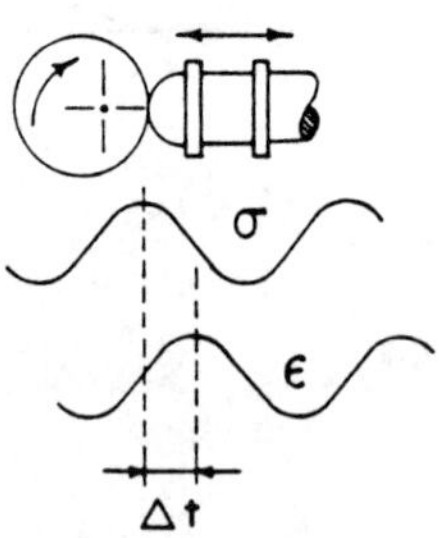	$\lvert E^*\rvert = \sigma_{max}/\varepsilon_{max}$; $\delta = \omega\Delta t$ $E' = \lvert E^*\rvert \cos\delta$; $E'' = \lvert E^*\rvert \sin\delta$	146

Table 7.2 (continued)

<table>
<tr><th>Type of test</th><th>Parameter and calculating formula</th><th>Page</th></tr>
<tr><td>RESONANCE
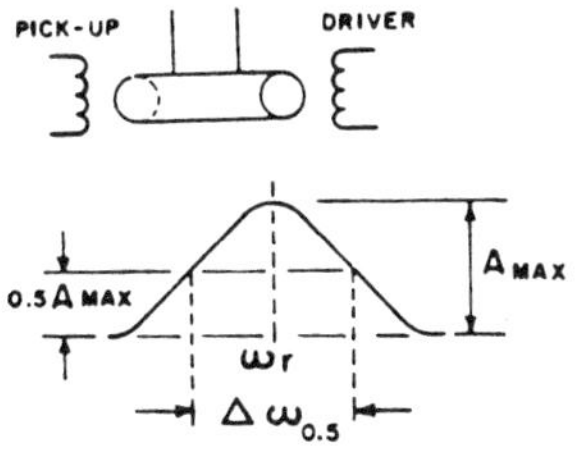</td><td>$\tan\delta = \dfrac{\Delta\omega_{0.5}}{\omega_r\sqrt{3}}$; $E = \dfrac{38.3\rho L^4 f_r^2}{d^2}$</td><td>147</td></tr>
<tr><td>PULSE TECHNIQUE
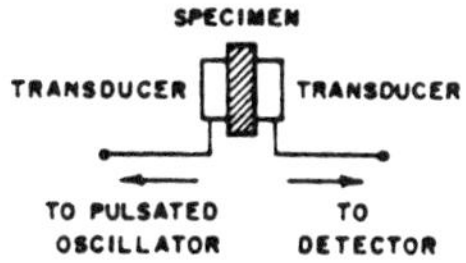</td><td>$E = K\rho V^2$</td><td>148</td></tr>
<tr><td>TRANSIENT LOADING
STRESS RELAXATION
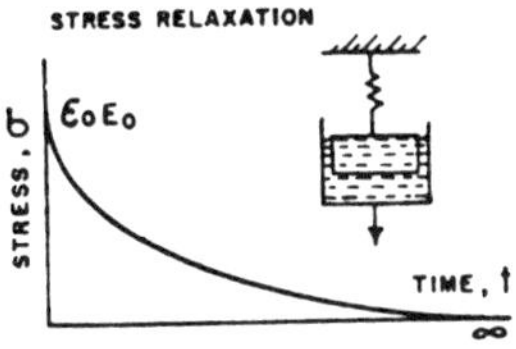</td><td>$\sigma(t) = \varepsilon_0 (E_{d1} e^{-t/T_1} + E_{d2} e^{-t/T_2} + \cdots + E_{dn} e^{-t/T_n} + E_e)$</td><td>123</td></tr>
<tr><td>CREEP
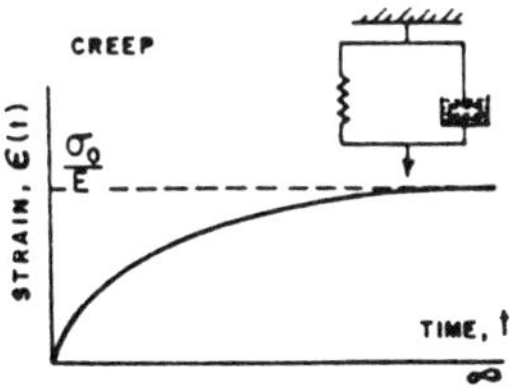</td><td>$\varepsilon(t) = \sigma_0(1/E_0 + 1/E_{r1}(1 - e^{-t/T_1}) + 1/E_{r2}(1 - e^{-t/T_2}) + \cdots + 1/E_{rn}(1 - e^{-t/T_n}) + t/\eta_v)$</td><td>123</td></tr>
</table>

Table 7.2 (continued)

Type of test	Parameter and calculating formula	Page
PLATE ON SEMI-CIRCULAR RING 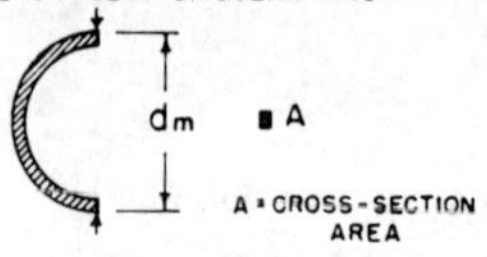	$D = \dfrac{0.0981 F d^3 m}{EI}$ $\sigma_{\text{tension}} = \dfrac{FdmC}{2I} - \dfrac{F}{A}$ $\sigma_{\text{compression}} = \dfrac{FdmC}{2I} + \dfrac{F}{A}$	
PLATE ON FULL RING 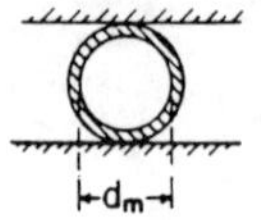	$D = \dfrac{0.0187 F d^3 m}{EI}$ $\sigma_{\text{tension}} = \dfrac{0.159 FdmC}{I}$ $\sigma_{\text{compression}} = \dfrac{0.159 FdmC}{I}$	
D = diametral deformation		

LIST OF SYMBOLS IN TABLE 7.2

A	= cross-section area (in^2)
A	$= \dfrac{1-\mu_1^2}{E_1} + \dfrac{1-u_2^2}{E_2}$ in point loading (l/psi)
a	= radius of the circle of contact or radius of the die (in)
B	= compressibility (l/psi)
b, h	= sides of a rectangular bar (in)
C	= distance from the neutral axis of the cross-section to the extreme fiber, in.
D	= Deformation at single contact (in)
d	= diameter of spherical indenter or rod (in)
d_1, d_2	= diameters of sphere 1 and sphere 2 (in)
e	= base of Naperian log (e = 2.72)
E	= Young's modulus (psi)
E^*	= complex dynamic modulus (psi)
E'	= storage modulus (psi)
E''	= loss modulus (psi)
E_d	= decay modulus
E_e	= equilibrium modulus
t	= time
E_r	= retarded modulus
F	= force
f	= frequency (cps)
G	= shear modulus (psi)
I	= moment of inertia (in^4)
I	$= \dfrac{bh^3}{12}$ rectangular bar
I	$= \dfrac{\pi r^4}{4}$ round bar
K	= bulk modulus (psi)
K	= constant in $K\varrho V^2$
L	= length (in)
m, n, k	= constants for analysis of convex bodies (see Table 6.1, p. 281)
R_1, R_1'	= radii of convex body (in)
R_2, R_2'	= radii of rigid loading devices
r	= distance from center of the area over which the die is acting or radius of solid bar (in)
S	= shear stress (psi)
S_{max}	= maximum normal stress (psi)
T	= torque (lb-in)
T	= time constant = η/E
t	= thickness of plate (in)
μ	= Poisson's ratio

v = volumn (in^3)
V = compressional pulse velocity (in/sec)
w = weight per unit length (lb/in)
ΔL = change in length (in)
Δp = change in pressure (psi)
Δv = change in volume
δ = phase angle (radians)
ε = strain (in/in)
ε_{max} = peak strain (in/in)
ε_0 = strain at time 0
$\varepsilon(t)$ = strain at any time (t)
η_v = viscosity
θ = angle of twist (radians)
$\tan \theta$ = shear strain
ρ = mass density $\left(\frac{\text{lb-sec}^2}{\text{in}^4}\right)$
σ = stress (psi)
σ_{max} = peak stress (psi)
σ_0 = stress at time 0
$\sigma(t)$ = stress at any time (t)
$\sigma_{tension}$ = maximum tensile stress in the extreme fiber of the ring or semi-circular ring at failure, psi
$\sigma_{compression}$ = maximum compressive stress in the extreme fiber of the ring or semi-circular ring at failure, psi
ω = angular frequency (radians/sec)
ω_r = resonance frequency (radians/sec)

7.12 FIRMNESS AND HARDNESS

Despite the importance of firmness and hardness in marketing and quality evaluation of raw and processed food and feed materials, today there is no general agreement as to the definition or measurement of these mechanical properties. Numerous devices have been proposed and used for measuring firmness and hardness of food materials. Results obtained are, however, usually expressed in arbitrary units which make it practically impossible to compare samples tested by means of different and sometimes even the same instrument. A good example is the difficulty involved in trying to convert the readings from a Magness-Taylor pressure tester (Fig. 7.12) using a 7/16-inch tip to those using a 5/16-inch tip (see Bourne, 1965).

Firmness is an important textural attribute in fruits and vegetables in connection with readiness of the crop for harvest, quality evaluation during storage for fresh market, as well as prior to processing, and its influence on the correlation between the quality of the raw material and that of the processed or manufactured product. Such correlations are due to the fact that many changes in physical, chemical and structural properties of fruits and vegetables are reflected in changes in firmness of the material (Meyer, 1960). In addition to textural factors, firmness of fruits has been correlated to such biological and cultural factors as respiratory rate and soil fertilization (Haller, 1941).

Hardness of grains is important in evaluating their feeding value as well as their size reduction and milling characteristics. For example, corn quality, in terms of higher percentage of alcohol soluble nitrogen or zein, is associated with the "horny" endosperm content of corn. Also, corn grits could be best made from hard corn with an average of 70% "horny" and 30% "floury" endosperm. It has been shown that horny endosperm content of corn as determined by the percent of non-floaters in a 1.275 specific gravity solution is an indirect measure of hardness (Wichser, 1961). Hardness of several agronomic crops as well as some manufactured food products is also important in ascertaining other physical and chemical properties during and after processing.

Firmness

Firmness of fruits and vegetables

To imitate the pressing of fruit with the ball of the thumb to determine its ripeness, the pounds of force required to press a marble into the side of an apple was noted on the dial of a spring scale (Morris, 1925). This development was followed by the Magness-Taylor fruit pressure tester (Magness-Taylor, 1925). This instrument, which is still widely used, consists of a plunger either 7/16-inch or 5/16-inch in diameter attached to a calibrated spring which is graduated in pounds (Fig. 7.12). The round tip of the plunger is pressed into the fruit to a depth of 5/16-inch, marked on the plunger, and the penetrating force is read on the scale. Haller (1941) describes other types of fruit pressure testers, and discusses the relationship of such factors as maturity, temperature, size, water content, and color of

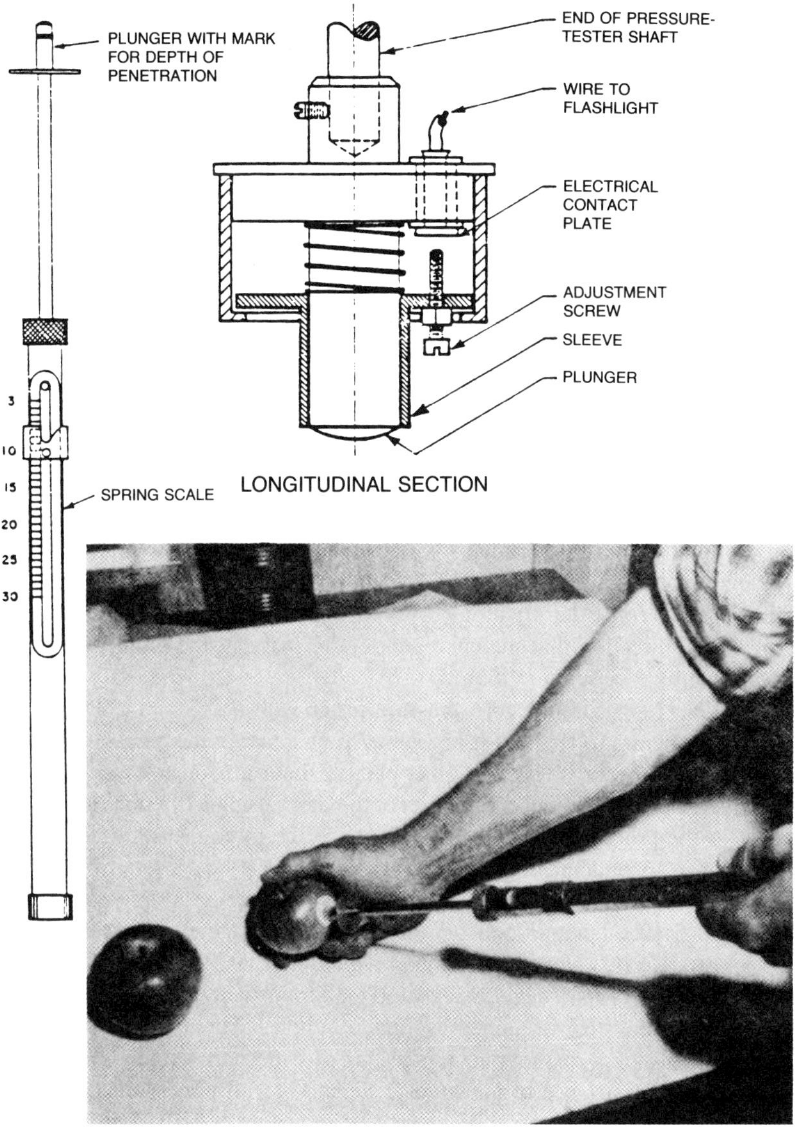

Figure 7.12 Clockwise: Magness-Taylor pressure tester; "Mechanical thumb" (Schomer *et al.*); Pressure tester in use.

the fruit on its firmness. Ross (1949) described a "pressure-hardness" tester for peas which used the gas pressure necessary to force a given size plunger a short distance into the fruit as the criterion of firmness.

The Magness-Taylor pressure tester was recently modified (Schomer and Olsen, 1962) to reduce the depth of penetration of the tip to 0.055-inch indicated by means of a small light bulb connected to a battery (Fig. 7.12). The extent of tissue damage by using this device is, however, of about the same magnitude as that induced by normal thumb tests. For some fruits, such as apples, the skin must be removed to obtain a satisfactory correlation with ripeness of fruits using the Magness-Taylor pressure tester while for others the skin can be left intact. Also, the relation between readings obtained with tips of 5⁄16-inch and 7⁄16-inch diameter is not understood. Being manually operated, the instrument reading is obviously subjective and depends on the operator. To eliminate the "human element" from the reading of this instrument, Pflug *et al.* (1960) employed a mechanical press and automatic recording. Recently universal testing machines have been used to press the Magness-Taylor pressure tester tip into fruits (Bourne, 1965). To determine what is being measured, Bourne attempted to analyze the shape of force-deformation curves qualitatively. He divided the observed force-deformation curves for apples into three classes according to the direction that the curves took after passing the bioyield point, proposed by Mohsenin (1962b).

Frazier (1934) used a corn pressure tester with a 0.5 mm plunger on skin of tomatoes to find out if resistance to puncture in the creases of fruit where cracks occur is less than other parts of the fruit. Johannessen (1949) conducted similar skin puncture tests of tomatoes using a 0.5 mm platinum wire as the probe and an ordinary balance as the force measuring device. Resistance to puncture of several varieties and various locations on the fruit was recorded in terms of pressure. Most of these tests showed significantly less resistance to puncture at the stem end of the fruit. A number of reversals were, however, included which indicated significant interactions. Voisey (1964) devised a motorized strain gage puncture tester to accelerate the testing of numerous tomato samples. The use of this instrument was also reported for puncture testing of sweet corn kernels, cabbage, and apples. Resistance to puncture of tomato has also been related to its size and infection by microorganisms (Rosenbaum, 1920). Firmness of tomatoes has also been expressed in terms of the force required to puncture a ½-inch slice of the center flesh by a given size plunger (Garrett *et al.*, 1960). Kattan (1957) devised an apparatus called "Firm-o-meter" which measured firmness in a manner similar to that determined by compressing the fruit in the band. A commercial version of the same apparatus is shown in Fig. 7.8.

Ang *et al.* (1960) used a flat plate at the end of the crosshead of a LEE-Kramer shear press to evaluate firmness of onion bulbs. In this work, the ratio of force to deformation (F/D) required for breaking of the onions under pressure was taken as the criterion of firmness. Note that in the Boussinesq solution of die loading [Eq. (6.21)], the stiffness or apparent modulus of elasticity is also in terms of F/D if the size of the plunger were constant and variations in Poisson's ratio were negligible. It was found that onion varieties could be separated on the basis of their F/D firmness. Also, it was found that pressure did not impair the seed producing ability of the onions unless breaking of the bulbs occurred.

Firmness of cherries has been a subject of considerable interest. Firm cherries would remain plump during baking and processing and would result in higher yield and usable solid portion of the processed product. Soft cherries may lose excessive juice during pitting or become excessively pitter-torn and flabby. One of the most satisfactory methods reported for firmness measurement of red tart cherries has been the measuring of the deflection of whole cherries subjected to a dead load applied for a given time over a constant area (Parker *et al.*, 1966). The cherry was placed under the free-traveling rod of a simple dial micrometer with the stem end down. The weight of the rod provided the dead load. The deformation after two minutes in percent of total diameter along the stem was taken as the criterion of cherry firmness. The satisfactory results reported for the use of this simple technique are due to the fact that the measurement is a better defined measurement. This is actually a familiar rheological test, namely a creep test. One of the objections noted for the test was the slow rate of measurement. To increase the rate of testing, one would first obtain sufficient data to construct creep curves for the total period of two minutes. Analysis of these curves might show that the initial part of each curve is all that is necessary and would give the same information about "firmness" as the total curve lasting for two minutes. In that case a shorter time test could be selected which would increase the capacity of each testing unit.

Bulk density, defined as the weight per unit volume of material in the aggregate, has been proposed as the measure of firmness for several food materials. LaBelle (1964) showed that where resistance to compression and flexure is involved, bulk density, measured by consideration of several factors, can be related to food texture. These factors were product characteristics such as unit density, plasticity and irregularity of shape, duration of disturbing force, container geometry, and the method employed to reorient the aggregate into "limiting close-pack" condition. The method was employed for measurement of firmness of blanched apple slices, degree of bruising in red tart cherries, and the extent of texture changes in precooked beans. The basis for the use of bulk density parameter was that softened material, as a result of special processing or mechanical damage, deforms more readily under its own weight reducing interstitial space and increasing bulk density.

Schmidt (1962) gives a comprehensive review of the method of firmness measurements for apples and discusses the relationship of firmness to the state of pectic substances and other chemical constituents, variety, stage of maturity, climate, location and several other factors. This investigator is the first who used a conventional punch shear test on the slices of the flesh and expressed firmness in terms of shear strength of the flesh (see discussion of "shear" in Chapter 5 and Fig. 5.7). Data in Table 7.3 shows the results of similar shear tests for one season's crop of apples. Note the high correlation coefficient between the flesh shear strength and both the bioyield pressure and the rupture pressure. This relationship suggests that the mechanics involved in the sudden rupture of the tissue structure in apples under the force of a rigid cylindrical die is probably a shear process which allows relative displacement of cells along the shearing surface.

In the development of selectors for mechanical harvesting of head lettuce, it has become evident that the available marketing standards, referring to head solidity and firmness, lack much precision and objectivity if a mechanical or manual selector is to be incorporated in the field machine. To determine those physical characteristics which correlate best with acceptability of head lettuce as judged by an experienced man, the measurement of weight, volume, density, diameter, and firmness were considered (Garrett *et al.*, 1968). Firmness was evaluated by recording vertical force versus head height on an *y-x* recorder. It was found that density and volume were better correlated with acceptability than any other factors. In squeezing a lettuce head for judgment of firmness with fingers or palm of the hand apparently one "feels" the density and at the same time estimates the volume visually. For a more objective measurement, the following relationship was found to give correlation with lettuce head acceptability.

$$\rho > 0.385 - (0.475V)\,10^{-4}$$

Table 7.3 Correlation coefficients between shear strength of apple flesh and some physico-chemical parameters of the fruit (1964, University Park Orchards)[1]

Physico-chemical parameter	Melba		McIntosh		Golden delicious		Red delicious			
							Before harvest		After harvest	
	w	*wo*	*w*	*wo*	*w*	*wo*	*w*	*wo*	*w*	*wo*
Bioyield pressure	0.49	0.93**	0.69*	×	0.99**	×	0.81**	×	0.81**	0.48
Rupture pressure	0.89**	0.97**	0.74**	0.75**	0.98**	0.97**	0.64	0.53	0.73*	0.70*
Days after full bloom	−0.80*		−0.65*		−0.86		−0.53		−0.50	
Size	−0.46		−0.57		−0.37		−0.21		×	
Color	×		−0.55		−0.91*		−0.58		×	
Sugar	0.82*		0.26		−0.69		−0.62		0.36	
Specific gravity	−0.48		−0.11		0.63		−0.09		×	
Correlation coeff. at										
5% level	0.69		0.63		0.88		0.67		0.67	
1% level	0.80		0.73		0.95		0.78		0.78	

[1] Continuation of work reported by Mohsenin *et al.*, (1965a). Each value is based on 20 apples per test day for a period extending to 168 days (for Red delicious) after full bloom.

w = with skin

wo = without skin

* = significant at the 5% level

** = significant at the 1% level

× = no test was conducted

where ρ is density and V is volume. Apparently, firmness is used primarily as an indicator for density. On the basis of this finding, a standard line for evaluating a mechanical or manual selector was proposed on a density-volume plot which separates acceptable and unacceptable heads (Fig. 7.12a).

Firmness of jellies

The firmness or rigidity of jellies, such as those of gelatin and pectin, is an important textural quality and a truly objective method for its evaluation is important to the industry. To make possible a comparison of results obtained by various workers, Campbell (1938) proposed a method of calibration based on measurement of torsional rigidity of the sample. He devised a "jelly tester" which utilized the principle of a rotational viscometer with readings which could be taken to a calibration chart and converted to modulus of rigidity in grams per square centimeter. The calibration device was simply a rotational viscometer shown in Fig. 7.13 and the equation for firmness in terms of rigidity was derived using the conventional rotational viscometer theory (see Chapter 4). The space between the inner and outer cylinders was filled with a gel solution. Upon setting, the gel adhered to both the other wall of the inner cylinder (bob) and the inner wall of the outer cylinder (cup). When the inner cylinder was rotated, the layer of gel adhering to it rotated with it but the outermost layer adhering

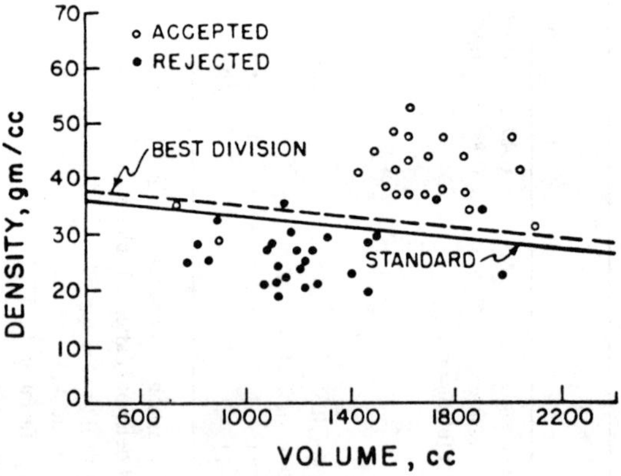

Figure 7.12a A proposed standard line for evaluating "firmness" of crisp, head lettuce (Garrett *et al.*, 1968)

to the inner wall of the outer cylinder remained stationary. The modulus of rigidity G was defined as the ratio of shear stress τ on the wall of the inner cylinder to the torsional strain ε_s and could be determined from the angle of twist undergone by the inner cylinder on applying a known rotational force. From an analysis similar to that given in rheometry (Chapter 4), the rigidity modulus can be shown to be equal to

$$G = M/4\pi\omega h(1/R_b^2 - 1/R_c^2) \tag{7.4}$$

where M is the torque or the weight of water necessary to produce a given angle of twist, ω, multiplied by the radius of the pulley, h, is the height of gel in the cup, and R_b and R_c are radii of the bob and cup, respectively.

Example

In a rotational apparatus similar to Fig. 7.13, the diameter of the pulley is 2.16 cm, the weight of water required to produce an angle of rotation of 10° is 23 gm, height of gel in the cup is 14.10 cm, and R_b and R_c are 0.774 cm and 4.79 cm, respectively. Calculate the modulus of rigidity for the gel.

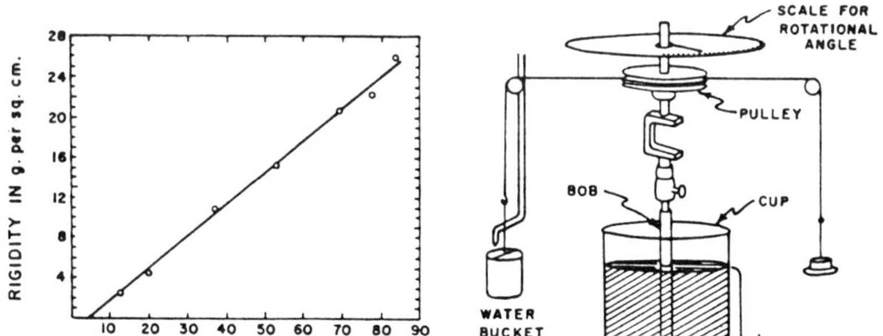

Figure 7.13 Right: The principle of a rotational viscometer used as a jelly tester. Left: calibration curve for the commercial jelly tester (Campbell, 1938)

Solution

$$G = \frac{23 \times 1.08}{4 \times 3.14 \times 0.1745 \times 14.10}(1/0.774^2 - 1/4.79^2) = 1.3 \text{ gm/cm}^2$$

Note that the commercially available Mac Michael viscometer can also be used for measuring rigidity in gels based on Eq. (4.92).

For testing a number of gels each day, gel specimens were molded in separate containers and then tested by an apparatus constructed on the above principle (Campbell, 1934). To express firmness of these specimens in terms of rigidity modulus, a calibration curve shown in Fig. 7.13 was also prepared. This is an example of a well-defined rheological measurement which was adapted to commercial use for quality evaluation. Recently, Jones (1968) reported the use of a modified version (commercially available as "Jelly Tester") of this apparatus which plots load versus angle of rotation or stress versus strain.

Bioyield pressure

In an attempt to determine the maximum allowable static and dynamic loads that apples can withstand in mechanical handling and storage, Mohsenin and co-workers introduced the concept of bioyield pressure (Fig. 7.14), showed the importance of rate of loading in compressive force readings (Fig. 4.20), applied the bioyield point concept for non-destructive firmness evaluation before and after harvest (Fig. 7.15), developed a testing machine for rheological properties of agricultural products (Fig. 4.22), and suggested a technique for predicting "readiness" for harvest of apples based on a combination of physico-chemical properties of the fruit (Mohsenin *et al.*, 1962a, b; 1963a, b; 1965a, b). In the course of experiments with apples, certain force-deformation-time relationships were observed which were thought to be of significance in texture evaluation of solid food materials (Mohsenin *et al.*, 1963).

A number of tests were conducted under known compression and rate of loading. A ¼-inch steel plunger was pressed into the fruit at the rate of 0.150 inches per minute and the resulting force-deformation curve was plotted by the *x-y* recorder of the testing machine shown in Fig. 4.22. A series of such curves, which were obtained for specimens of apple with the skin intact, is shown in Fig. 7.14. The first break in curve (a) is an indication of the initial fracture in the cell structure of the fruit which initiates browning and discoloration. This break in the curve was called the bioyield point. The surface pressure (force per unit area) corresponding

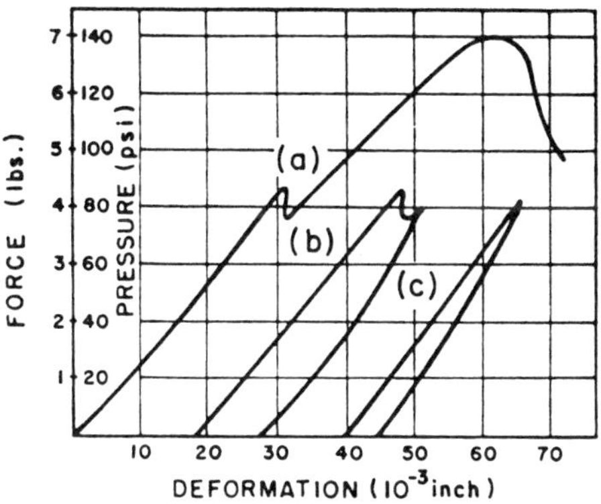

Figure 7.14 Loading and unloading curves of apple fruit showing a bioyield point

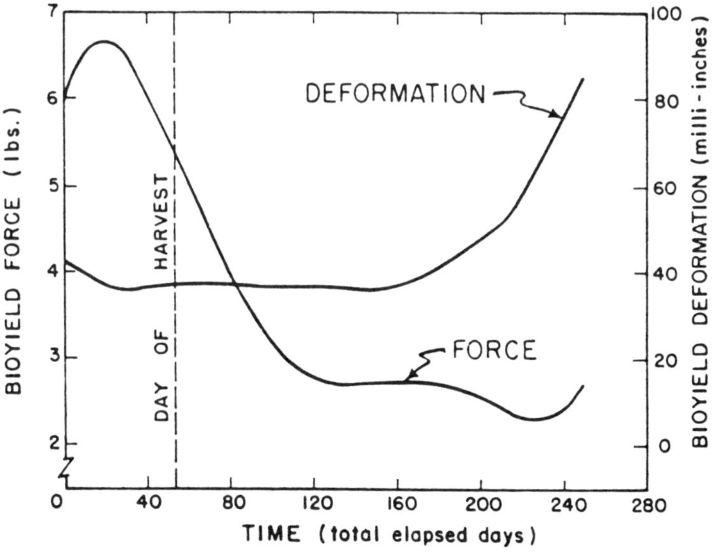

Figure 7.15 Changes of bioyield parameters of same golden delicious apples with time (Mohsenin *et al.*, 1965b)

to the bioyield point can be referred to as bioyield pressure for that fruit at that stage of development. Until a bioyield point is observed in the force-deformation curve of the fruit, no browning of the apple tissues can generally be detected upon sectioning of the test specimen.

The browning of tissues in fruits such as apples, pears, peaches, apricots, cherries, grapes, and bananas is the result of oxidation of cellular contents by enzymes. It is believed that the rupture of plant cells exposes the cell contents to the intercellular air and results in enzymatic oxidation and discolorations of plant tissues (Meyer, 1960). When loading is continued beyond the bioyield point, the puncture point is reached at which the plunger breaks the skin and a sudden decrease in force is observed.

If the terminology used in testing of engineering materials were employed, the slope of the straight portion of curve (a) would be an indication of stiffness. The area under the curve up to the bioyield point would be the work required to cause yield or initial cell rupture.

If a constant contact area were maintained during the test (plunger loading), the bioyield unit work can be expressed in terms of inch-pounds per square inch of contact area. The area under the complete curve up to the puncture point can be taken as an indication of toughness of the skin and the supporting tissues. Numerically, toughness would be indicated by inch-pounds per square inch of plunger contact area.

If the rate of unloading in the testing machine can be adjusted to equal the rate of loading, some additional information can be obtained which is important in evaluation of textural factors as well as understanding the elastic behavior and characterization of the plant material. Curves (b) and (c) are two such curves. Comparison of permanent deformation upon unloading for these two curves shows the increase of "plastic" deformation upon yielding. The ratio of deformation recovered upon unloading to the total deformation at any given load can be referred to as elastic recovery or percent elasticity.

The difference between the work of compression and the work of retraction is represented by a loop which can be referred to as the hysteresis loop. The relative size of the hysteresis loop is a measure of elasticity and resilience of the plant material. The smaller the hysteresis loop, the more elastic the product. The hysteresis loop is also a measure of the ability of the plant material to absorb and store energy without being deformed beyond its bioyield point. Accordingly, the area under the unloading curve below the bioyield point can be taken as the elastic resilience or energy capacity of the plant material.

Limited experiments with other products, such as peaches, pears, plums, cherries, tomatoes and potatoes, have resulted in force-deformation curves showing the same trend as the apple characteristic curve. The bioyield point, however, has been found to be less abrupt in products other than apples and pears or not shown at all.

Because of the lack of visible damage on the surface of fruits exhibiting a bioyield point and the negligible browning of the fruit tissues under the skin, the bioyield technique was suggested as a possible "non-destructive" method for studying the changes in fruit firmness prior to harvest and during storage (Mohsenin *et al.*, 1965b). To explore the possibilities of this technique, a study was conducted on 50 Golden Delicious apples tagged on the tree. The same apples were tested every 7 to 18 days from 53 days prior to commercial harvest until 249 days after harvest (Fig. 7.15). Bioyield tests of the specimens, while still attached to the tree, were made with a portable version of the testing machine shown in Fig. 4.22 (see Fig. 7.16). To insure that each successive test would be conducted at a different point on the fruit, each test location was marked with a felt marker as soon as bioyield point was recorded. After the specimens were harvested, periodic tests were continued with the laboratory testing machine. Experimental and predicted values of bioyield force and deformations for the post-harvest period of the fruits were compared. The predicted values were obtained from the best fitting curves which were derived using a polynomial derivation by orthogonal least squares. The best fitting polynomials for bioyield force, F_y, and bioyield deformation, D_y, with respect to time, t, after harvest were

$$F_y = 5.941 - 1.640(10)^{-1}t + 3.466(10)^{-3}t^2 - 3.386(10)^{-3}t^3 + 1.497(10)^{-7}t^4 - 2.428(10)^{-10}t^5 \qquad (7.5)$$

and

$$D_y = 4.431(10)^{-2} - 7.004(10)^{-4}t + 1.814(10)^{-5}t^2 - 2.111(10)^{-7}t^3 + 1.127(10)^{-9}t^4 - 2.053(10)^{-12}t^5 \qquad (7.6)$$

In the above expressions, F_y, is in pounds and D_y in inches.

After consideration of seasonal and environmental variations, such relationships may be established for any apple variety. Using these relations,

Figure 7.16 Bioyield test of apples while on the tree. From top clockwise: field test; modified version of the testing machine in Fig. 4.22; location of bioyield tests on one apple during preharvest and post-harvest

firmness of the fruit in terms of bioyield parameters may then be predicted at any time and storage life and related functions estimated (Fig. 7.15).

A thorough analysis of the mechanics of pressing a ¼-inch rigid plunger into a fruit with the skin intact is not a simple problem. Currently various methods, including photoelastic stress analysis techniques, are being employed to obtain a better understanding of the stresses in convex biological materials when loaded by rigid bodies of various shapes and sizes. Until a theory is developed to explain the experimental observations, it is suggested to use the Boussinesq and Hertz solutions, as discussed in Chapters 5 and 6, and express such textural factors as firmness of fruits in terms of an apparent elastic modulus. Reference to Eqs. (6.14) and (6.21) shows that the apparent modulus obtained by this method is not only a function of resistance of the fruit to force and deformation, but also considers the geometry of the loading device and the loaded specimen (Mohsenin, 1970).

Often a simple function such as F/D, slope of the force-deformation curve, in an apparently complex formula can be used as a useful parameter in texture evaluation or determining the effect of certain treatments. For example, the effect of size of papaya fruit on the slope of its force-deformation curve, obtained by plate loading, has been reported to have high sensitivity to fruit maturity (Burkner and Kinch, 1967). The section of the hollow geometrically ellipsoid fruit was considered to be a ring with rectangular cross section. A theoretical analysis considering the deflection of a quadrant of the ring, assumed to a curved beam, lead to the conclusion that the ratio of force to deformation (F/D) had greater sensitivity to papaya maturation than two other indices of maturity. The range of F/D was reported to be 20 times greater than either of the two other indices.

If force-deformation curves such as those shown in Fig. 7.14 are to be used as better defined textural parameters for such fruits as apples, it is essential to establish satisfactory correlation between certain parameters of these curves and several of the chemical and physical properties of the fruit which have proved to be good indices of maturity or ripeness. No single index of readiness of apples for harvest has yet been found. Elapsed time from full bloom, percent soluble solids, pectic substances, acid and starch content, and "firmness," as measured by Magness-Taylor pressure tester, are some of the more reliable indices. The "readiness for harvest" condition for fruits and vegetables is dependent upon the final use intended for the product, the treatment received after harvest, and the elapsed time between harvest and use. The optimal period for mechanical harvesting, for example, is the time at which both the quality and the resistance to mechanical damage are high. These plus other factors such as the perishable nature of the material, scheduling of machines and labor forces, and the various processes involved in preparation of the product for fresh market or processed foods, demand such exacting requirements which cannot be met by the present subjective means of determining when to harvest.

Mohsenin *et al.,* (1965a) have shown that several of the mechanical properties obtained from force-deformation curves of six varieties of apples give high correlation with the number of days between harvest and full bloom as well as with some other indices of maturity. The properties evaluated were "firmness," specific gravity, color, fruit size, fruit separation force, percent sugar, yield force, yield deformation, yield energy, rupture force, rupture deformation, rupture energy, modulus of elasticity of flesh, Magness-Taylor pressure firmness reading, and shear strength of flesh (Tables 7.4, 7.5, 7.6 and 7.7).

Readiness for harvest

Based on the weighted percentages of the properties affecting maturation, a method for determining the optimum harvest data for apples was suggested. This method included a composite curve, depicting the trends of the combined physical properties versus stage of the fruit development in terms of days after full bloom. Figure 7.17 shows such trends for several varieties of apples. Letter "H" indicates the time of commercial harvest for that particular variety. It should be pointed out that the time of commercial harvest may be influenced also by the availability of labor, capacity of the packing and processing plant, etc.

In an attempt to develop and test several other possible methods of predicting the optimal harvest date from physical properties of the fruit, Lindenmuth (1966) took the same data used by Mohsenin *et al.,* (1965a) and investigated the application of two operations research optimization techniques, namely simulation and linear programming. Linear programming was found to be inapplicable to the non-linear relationships in apple properties data. The simulaton method, which was the summation of derivatives or slopes of changes of the combined properties with time, produced a parabolic form which contained a minimum point. Using the minimum point as a criterion for harvest, compared with maximum point in Fig. 7.17, and applying the same weight factors as those used by Mohsenin *et al.,* (1965a), the predicted optimal harvest dates were found to be within three days of the commercial harvest dates for the seasons tested.

Table 7.4 Correlations among various properties of McIntosh apples and the number of days after full bloom (Mohsenin *et al.*, 1965)

	Days	F_y	D_y	E_y	F_r	D_r	E_r	E	F_d	D	Color	Sugar	Firm
F_y	−0.81**												
D_y	−0.74**	0.79**											
E_y	−0.79**	0.94**	0.94**										
F_r	−0.94**	0.85**	0.72**	0.81**									
D_r	−0.92**	0.78**	0.83**	0.82**	0.88**								
E_r	−0.93**	0.78**	0.81**	0.82**	0.93**	0.96**							
E	−0.86**	0.74**	0.62*	0.71**	0.87**	0.83**	0.81**						
F_d	−0.82**	0.70**	0.44	0.53*	0.81**	0.71**	0.71**	0.70**					
d	0.81**	−0.69**	−0.67**	−0.76**	−0.76**	−0.77**	−0.79**	−0.73**	−0.38				
Color	0.77**	−0.86**	−0.80**	−0.87**	−0.81**	−0.81**	−0.81**	−0.80**	−0.57*	0.73**			
Sugar	0.87**	−0.69**	−0.69**	−0.69**	−0.80**	−0.85**	−0.86**	−0.66**	−0.76**	0.65**	0.69**		
Firm	−0.93**	0.62*	0.62*	0.72**	0.93**	0.83**	0.89**	0.80**	0.82**	−0.76**	−0.66**	−0.76**	
Sp. Gr.	−0.16	−0.18	−0.19	−0.05	0.35	0.14	0.16	0.33	0.53	0.12	−0.15	−0.10	0.28

* Significant at the 5% level, $r = 0.51$.
** Significant at the 1% level, $r = 0.64$.

Days — Days after full bloom
F_y — Yield force in pounds
D_y — Yield deformation in milli-inches
E_y — Yield energy in inch-pounds
F_r — Rupture force in pounds
D_r — Rupture deformation in milli-inches
E_r — Rupture energy in inch-pounds
E — Modulus of elasticity in pounds per square inch
F_d — Detachment or separation force in pounds
D — Size of fruit in inches
Color — Color meter readings in per cent
Sugar — Per cent sugars (sugar content)
Firm — Magness-Taylor pressure tester readings in pounds
Sp. Gr. — Specific Gravity

Table 7.5 Mechanical and physical properties and percent sugars of apple in relation to the number of days after full bloom and commercial harvest, 1962 (Mohsenin *et al.*, 1965)

Cultivar and harvest date		F_y (lb)	D_y (milli-in)	E_y (in-lb)	F_r (lb)	D_r (milli-in)	E_r (in-lb)	E (psi)	F_d (lb)	D (in)	Color	Sugar (%)	Firm (lb)	Sp. Gr.
Melba	*A**	3.1	37	0.057	7.6	124	0.47	613	3.0	2.54	39	13.1	8	0.701
(*FB*+95)	*B*	2.9	33	0.054	8.0	121	0.48	743	3.3	2.34	33	12.8	9	0.721
	C	0.4	4	0.012	1.3	10	0.12	207	0.8	0.18	6	0.8	2	0.034
McIntosh	*A*	3.0	36	0.052	9.4	127	0.63	693	3.5	2.59	48	13.9	9	0.740
(*FB*+132)	*B*	3.5	39	0.078	10.0	129	0.66	698	3.9	2.45	40	13.5	9	0.728
	C	0.4	5	0.023	1.7	18	0.21	111	1.3	0.15	12	0.5	1	0.015
Delicious	*A*	4.9	36	0.089	7.2	72	0.32	486	—	3.30	77	14.3	9	0.827
(*FB*+149)	*B*	5.7	44	0.132	8.4	84	0.46	724	—	2.99	478	12.4	10	0.808
	C	1.1	7	0.043	0.9	9	0.09	155	—	0.18	191	2.1	1	0.021
Golden	*A*	5.3	40	0.107	8.2	84	0.41	549	5.0	2.60	29	14.3	10	0.750
Delicious	*B*	5.6	48	0.143	8.8	96	0.52	695	4.7	2.56	28	13.2	10	0.761
(*FB*+149)	*C*	1.0	10	0.055	1.5	17	0.17	178	0.8	0.16	6	1.9	2	0.018
Stayman	*A*		7.3	46	0.176	11.0	87	0.52	895	3.6	2.70	62	15.1	13
0.822														
(*FB*+156)	*B*	6.9	50	0.182	12.1	108	0.69	916	6.2	2.44	41	14.0	14	0.828
	C	0.4	6	0.038	1.5	16	0.18	144	2.0	0.21	14	1.6	2	0.036
Rome Beauty	*A*	8.5	47	0.201	10.8	79	0.57	799	3.2	2.95	67	12.4	11	0.792
(*FB*+162)	*B*	10.4	55	0.303	13.2	87	0.78	1031	4.9	2.85	48	12.0	13	0.792
	C	2.0	13	0.129	2.7	14	0.27	257	1.5	0.24	20	1.3	3	0.011

* *A*—Data taken on day of harvest only. Harvest date is reported as number of days following full bloom. Melba, for example, was harvested 95 days following full bloom. *FB* is abbreviation for full bloom.

B—Mean of all data taken.

C—Standard deviation of all data taken.

Table 7.6 Correlation coefficients between bioyield pressure and some physico-chemical parameters of apple (Mohsenin *et al.*, 1965a)

Physico-chemical parameter	Red Delicious			Golden Delicious			Stayman		Rome Beauty	
	1961	1962	1964	1961	1962	1964	1961	1962	1961	1962
Days after full bloom	−0.76**	−0.76**	−0.63*	−0.93**	−0.79**	−0.81*	−0.82*	−0.39	−0.95**	−0.94**
Skin rupture pressure	0.94**	0.84**	0.67*	0.98**	0.68**	0.98**	0.89**	0.57*	0.98**	0.97**
Modulus of elasticity of flesh	0.65*	0.84**	0.14	0.91**	0.84**	0.61	0.85**	0.28	0.81*	0.79**
Shear strength of flesh	×	×	0.81**	×	×	0.99**	×	×	×	×
Magness-Taylor pressure test	0.57	0.69**	×	0.95**	0.80**	×	0.90**	0.44	0.75*	0.89**
Color	−0.78**	−0.74**	−0.85**	0.32	−0.53*	−0.89*	−0.51	−0.27	−0.77*	−0.90**
Sugar	−0.83**	−0.72**	−0.56	−0.86**	−0.67**	−0.66	−0.94**	−0.50	−0.88**	−0.88**
Size	−0.52	−0.65*	−0.29	−0.92**	−0.74**	−0.31	−0.74	−0.59*	−0.95**	−0.73**
Specific gravity	0.59	−0.81**	−0.22	0.87**	0.14	0.68	0.84*	−0.25	0.56	0.15
Correlation coeff.										
at the 5% level	0.63	0.55	0.58	0.63	0.47	0.81	0.75	0.53	0.75	0.50
at the 1% level	0.75	0.68	0.71	0.75	0.59	0.92	0.87	0.66	0.87	0.62

× = No test was conducted.
* = Significant at the 5% level.
** = Significaant at the 1% level.

Table 7.7 Correlation coefficients between skin rupture and some physico-chemical parameters of apple (Mohsenin *et al.*, 1965)

Physico-chemical parameter	Red Delicious			Golden Delicious			Stayman		Rome Beauty	
	1961	1962	1964	1961	1962	1964	1961	1962	1961	1962
Days after full bloom	−0.90**	−0.94**	−0.64*	−0.92**	−0.92**	−0.79	−0.85**	−0.94**	−0.98**	−0.97**
Bioyield pressure	0.94**	0.84**	0.67*	0.98**	0.68**	0.98**	0.89**	0.57*	0.98**	0.97**
Bioyield deformation	0.66*	0.82**	0.65*	0.57	0.75**	0.76	0.10	0.74**	0.05	0.93**
Bioyield energy	0.79**	0.91**	×	0.95**	0.76**	×	0.78*	0.75**	0.58	0.96**
Modulus of elasticity of flesh	0.63*	0.88**	×	0.84**	0.81**	0.65	0.72**	0.64*	0.86*	0.80**
Shear strength of flesh	×	×	0.64*	×	×	0.98**	×	×	×	×
Size	−0.69*	−0.82**	−0.19	−0.91**	−0.83**	−0.38	−0.83*	−0.98**	−0.94**	−0.76**
Color	−0.86**	−0.85**	−0.65*	0.28	−0.56*	−0.92**	−0.72	−0.79**	−0.78*	−0.89**
Sugar	−0.90**	−0.89**	−0.64*	−0.86**	−0.92**	−0.62	−0.82*	−0.88**	−0.81*	−0.90**
Magness-Taylor test	0.74*	0.87**	×	0.93**	0.90**	×	0.97**	0.88**	0.86*	0.94**
Specific gravity	0.67 *	−0.85**	−0.30	0.83**	0.07	0.56	0.81*	−0.61*	0.65	0.20
Correlation coeff.										
at the 5% level	0.63	0.55	0.58	0.63	0.47	0.81	0.75	0.53	0.75	0.50
at the 1% level	0.75	0.68	0.71	0.75	0.59	0.92	0.87	0.66	0.87	0.62

× = No test was conducted.
* = Significant at the 5% level.
** = Significaant at the 1% level.

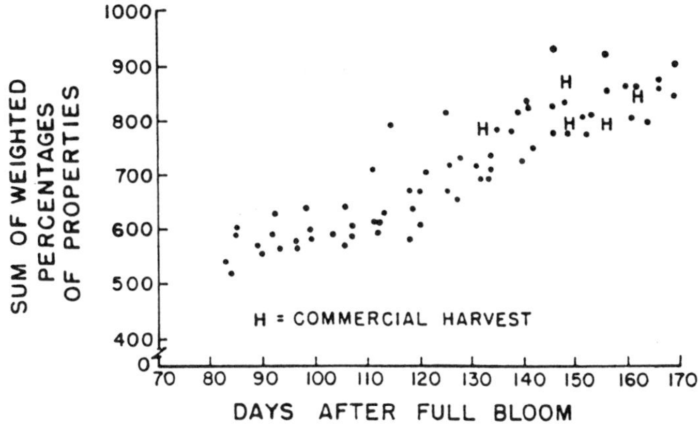

Figure 7.17 Predicting readiness for harvest of 5 varieties of apples from changes of their physical properties with time (Mohsenin *et al.*, 1965a)

This experiment showed that the possibilities of analog and digital computers for multidimensional analysis of the complex relationship which exists in texture of foods are yet to be explored. For example, Lindenmuth (1966) in using the digital computer to predict the optimum time for harvest of apples found it quite time consuming when any of the weight factors for various properties of the fruit was changed to obtain a better matching of the predicted harvest versus the actual harvest time. The use of an analog computer, however, proved to simplify this problem to a great extent.

It was noted that if one multiplies the equations of two straight lines, one with positive slope (ascending) and one with negative slope (descending), the resulting equation will be a quadratic equation which will yield a maximum point at the intersection of these two lines. If one were to raise the equation of these lines to a power and then multiply them together, then it is possible to vary the position of the resultant peak by varying the exponents of the lines. With these observations in mind, four properties of the data on Delicious Apples for one season were chosen. The choices were made on the basis of their importance in determining the final quality of the apples. The four properties chosen were: *S.C.*, sugar content; E, modulus of elasticity of flesh; E_r, rupture energy of whole fruit and F_y, yield force of whole fruit. Then, the following equation was programmed on a TR-48 analog computer (Fig. 7.17a)

$$W_1 = (S.C.)^{m_1} (E)^{m_2} (E_r)^{m_2} (F_y)^{m_4}$$

By varying one of the m exponents, while keeping the other three constant, a particular curve for W_1 was plotted. If the fruit was harvested that year at time H and that this time was the optimum time for harvest of the apple in that year, then the exponents m_1, m_2, m_3 and m_4 were chosen by trial and error on the computer in such a manner that the resulting W_1 peaked at H. Having done this, four other properties of Delicious apples for the same season were chosen. These were: C, color; d, diameter; $Sp.Gr.$, specific gravity and E_y, yield energy. Now, W_2 was set to be

$$W_2 = (C)^{m_5} (d)^{m_6} (Sp.Gr.)^{m_7} (E_y)^{m_8}$$

and m_5 to m_8 were chosen so that W_2 would give a maximum value at H. It was not decided to find out whether a combination of the properties from the two groups of 4 properties would yield a peak at H or not. This was done by finding,

$$W_3 = (S.C.)^{m_1} (E_r)^{m_3} (d)^{m_6} (E_y)^{m_8}$$

and noticing where the curve peaked. It was found that the peak was very near the H. When the four parameters of sugar ($S.C.$), color (C), flesh modulus (E), and rupture force of fruit with skin (F_r) were considered for the Red Delicious variety, it was found that assignment of equal weights to each one of these parameters yielded a curve which peaked at harvest time. The same results were obtained for data on McIntosh apples (Fig. 7.18).

The above technique is useful in determining the optimum harvest time of crops at any locality providing one has access to the physical and chemical properties of the crop throughout the growth period and after harvest for several years. In the absence of such data, one must find reaction equations and other mathematical expressions for each of the properties and use them instead of limited experimental data. Obviously, some experimental work and measurements are necessary in order to test the accuracy of the mathematical model. Thus, the next step in pursuing the requirements of such study is to find the mathematical model of those properties which represent certain physical and chemical changes in the fruit, in this example, as a function of time. One usable parameter isolated so far, is the reaction equation of pectic substances which, as a cementing agent, is assumed to affect the structural rigidity and firmness of the fruit.

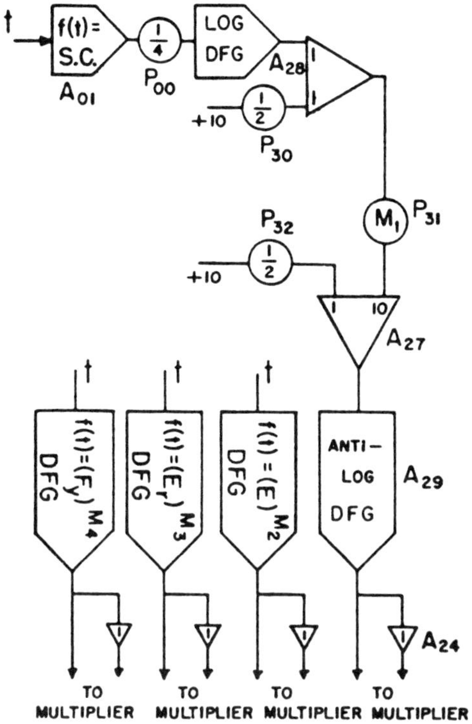

Figure 7.17a The analog computer program to evaluate maturity index *W*

Another example of the use of an analog computer in determining the effect of a certain treatment on several properties of a fruit is illustrated in Fig. 7.19. In water handling of cherries from the field to the processing plant it is desired to know the optimum soak time in order to affect the optimum quality of the harvested product and the highest efficiency of the process. As illustrated, up to a point firmness and soluble solids increased with soak time while at the same time pitter yield and pitter efficiency decreased with soak time. The use of an analog computer, searching for a curve showing the combined effect of all four parameters, resulted in curve *C* which peaked at about 4 hour soak time. The following example shows how the analog computer was used in obtaining the optimum time.

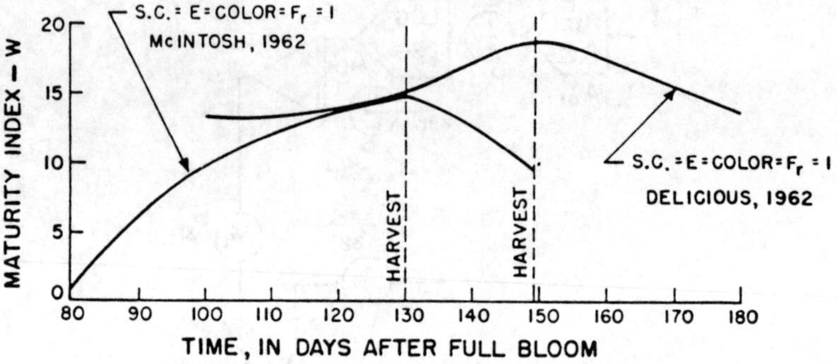

Figure 7.18 Comparison of analog computer prediction of optimum maturity and commercial harvest for two varieties of apples

Example

Cherries were first bruised by dropping them from a height and then soaking them in water (Tennes *et al.*, 1967). Four properties of the soaked cherries namely cherry product yield, cherry pitter efficiency, cherry firmness, and cherry soluble solids were measured as a function of soak time. The four variables were found to be given by the following polynomial equations.

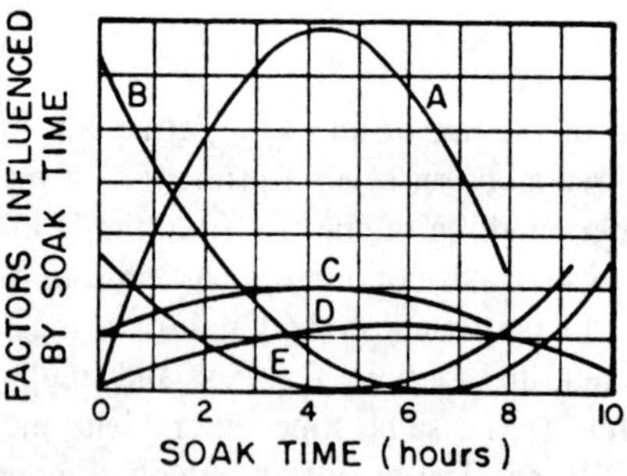

Figure 7.19 Analog computer optimization of soak time for bruised cherries (Original data from Tennes *et al.*, 1967). A—Firmness; B—Pitter yield; E—Pitter efficiency; D—Soluble solids; and C—Combined curve peaking at optimum soak time

pitted yield $= -0.138t^2 + 1.145t + 81$
pitter efficiency $= -0.177t^2 + 2.1t + 80.87$
firmness $= 0.379t^2 - 3.23t + 64$
soluble solids $= 0.039t^2 - 0.448t + 14.07$

It is desirable to find a soak time such that each one of the above parameters is as close to its maximum value as possible.

This is a problem in optimization. Analog computers can be used to determine the required optimum value. Assume that the four variables are represented by $f_1(t), f_2(t), f_3(t)$, and $f_4(t)$ respectively and that the variables have a maximum value of $f_{M_1}, f_{M_2}, f_{M_3}$, and f_{M_4} respectively, in the range $0 \leqq t < a$. Then, the optimum soak-time value (t_0) is a time such that

$$[f_{M_1}(t_1) - f_I(t)],\ [f_{M_2}(t_2) - f_2(t)],$$

$$[f_{M_3}(t_3) - f_3(t)],\ [f_{M_4}(t_4) - f_4(t)]$$

are as close to zero as possible. This is a rather hard program to do on the computer, however, if we add $[f_{M_1}(t_0 - f_I(t)]$ and other three similar expressions together and find the minimum of this new expression, then the time at which this minimum occurs is the desired optimum time.

A more realistic approach is to normalize each expression of the form $[f_{M_1}(t_1) - f_1(t)]$ by its maximum value and then add them together. Hence, the sum S is equal to:

$$S = \frac{f_{M_1}(t_1) - f_1(t)}{f_{M_1}} + \frac{f_{M_2}(t_2) - f_2(t)}{f_{M_2}} + \text{etc.}$$

In our example, $t_1 = 4.15$ hour and $f_{M_1}(t) = 83.4$; $t_2 = 5.93$ hour, and $f_{M_2}(t_2) = 87.25$; $t_3 = 0$, and $f_{M_3}(0) = 64$; $t_4 = 0$, and $f_{M_4}(0) = 14.07$. The block diagram given in (A) Fig. 7.19a is a sample of how we can obtain each of the $\dfrac{f_{M_1}(t_1) - f_1(t)}{f_{M_1}}$ terms.

$$f_1(t) = 10.138t^2 + 1.145t + 81$$

$$f_{M_1} - f_1 = 0.138t^2 - 1.145t + 2.4$$

Similarly other three expressions can be programmed on the computer (Fig. 7.19a).

A second method to solve this problem is to multiply all four $(f_{M_1} - f_1)$ terms together and find the maximum point of the resultant expression, that is, let

$$P = (f_{M_1} - f_1)(f_{M_2} - f_2)(f_{M_3})(f_{M_4} - f_4)$$

and find its maximum value. The computer block diagram for this method is shown in (B) Fig. 7.19a.

Hardness

For years engineers and metallurgists have been making hardness tests as a means of assessing the mechanical properties of metals and other engineering materials. Despite the fact that the term is widely used, measurements are being made and materials are being specified by their hardness. No general agreement as yet exists as to the definition of hardness in terms of fundamental physical units. Most experts are in agreement, however, that hardness as used in metals is resistant to permanent deformation associated primarily with their plastic properties and only to a secondary

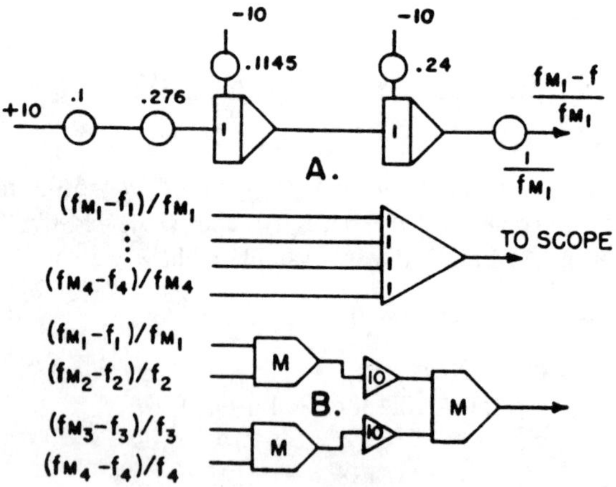

Figure 7.19a Analog computer program for optimizing soaking time for tart cherries

extent with their elastic properties (Tabor, 1951). The same concept can be applied to food materials as there are few firmness and hardness tests which do not result in permanent indentation or deformation of the raw or processed material. Even at the bioyield point of products exhibiting this parameter, deformation is largely permanent. The deformation at this point is, however, so small that it is not considered objectionable and, therefore, the test can be referred to as "nondestructive."

A number of attempts have been made to provide a quantitative measure of hardness of individual seeds and grains or of the average hardness of bulk samples. Determination of percent of "floury" portion in kernels of corn (Robinson, 1939), crushing the grain between the jaws of a force measuring device (Culbertson *et al.*, 1940), the torque required for crushing a sample of grain between two crushing wheels (Bennett, 1950), and application of soft metal hardness testers (Katz *et al.*, 1959) are among the methods reported. Figure 7.20 shows the schematic of an imitative method employed to measure the hardness of bulk samples of corn in terms of the average frame torque required for crushing the grain. The average torque of the cradled frame was taken as proportional to the readings of a mechanical integrator incorporating the hydraulic pressure and the rotation of the outer crusher wheel. This wheel was propelled by the pinning action of the crushing grain. When the results of this aggregate test were compared with the hardness test of individual kernels, considerable reduction in standard error was observed (Bennett, 1950, I). The validity of either of the two tests was based on the assumption that such physical characteristics as the small size of the kernels and higher proportions of the horny endosperm are usually associated with higher values of hardness of corn. Later microscopic examination of structural characteristics of the hardest and the softest corn samples showed that mature kernels of hard type of dent corn have smaller starch granules and more dense appearing protein matrix than softer types (Bennett, 1950, II).

Katz *et al.*, (1959) modified a soft metal hardness tester for hardness test of a number of points on kernel slabs of wheat grain (Fig. 7.20, 7.21). To provide a means for mounting and positioning of wheat kernel under the microscope and the tester, slabs of transverse sections of the kernel prepared by means of a freezing microtome were cemented on glass slides and positioned on the microscope stage mounted under the hardness tester. The index of hardness was taken as the distance that the spring loaded stylus of the tester was displaced into its case when the tester was pressed against a point on the wheat cross section. This displacement distance was

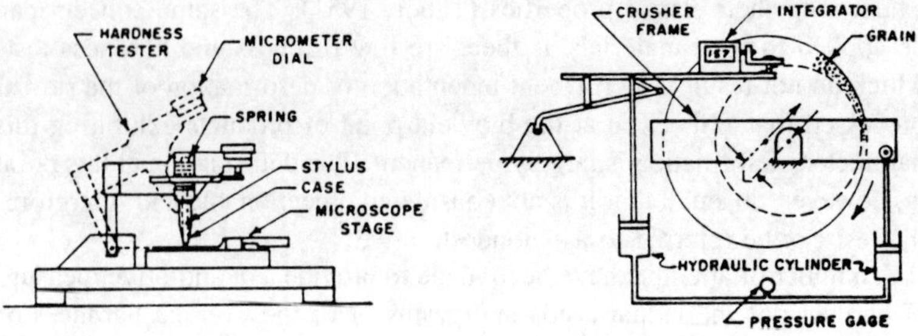

Figure 7.20 Right: An imitative method for testing "hardness" of a sample of corn kernels (Bennet, 1950). Left: A Barber-Colman (Barcol) soft metal hardness tester adapted to testing of hardness of kernel slabs of wheat (Katz, 1959)

read on the micrometer dial of the tester as hardness. Figure 7.21 shows the values of individual readings obtained on sections of the grain taken from various parts of the kernel.

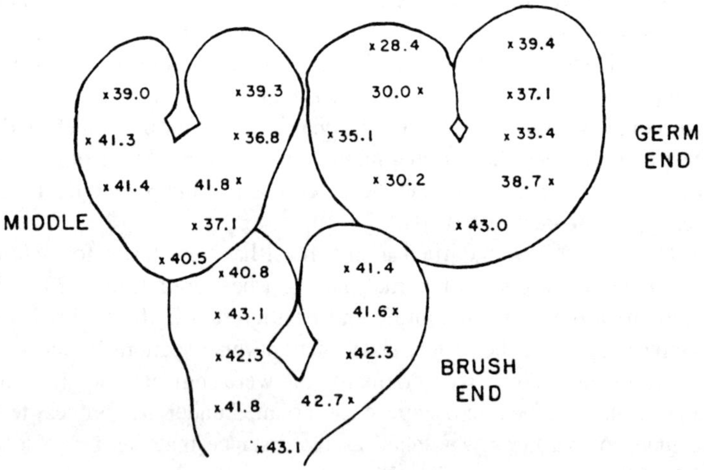

Figure 7.21 Variation of hardness of transverse sections of ponca wheat (Katz *et al.*, 1959)

Hardness of food materials is also of interest for control of the processing or manufacturing operations. In judging that textural characteristic of cheese known as "body," the thumb is pressed into the cheese and held momentarily and then removed. In this test several factors are sensed. One is "hardness" which has been associated with elastic modulus. Next is "springiness" which has been associated with elasticity and time of relaxation, $T=\eta/E$ (Davis, 1937). A material of low modulus but high viscosity would be considered less elastic than a material of equally low modulus but with lower viscosity. And finally, the last sensation in judging the body of cheese is the speed of recovery. In a series of tests conducted by Davis (1937), it appeared that both modulus and viscosity values are needed in order to place the cheese in the same order as subjective judgment of their "body" classification.

Scott Blair (1938) proposed an apparatus for measuring elastic and plastic deformations of cheese curd under a compressive load at the time when it is ready for cutting. He proposed that the elastic moduli of the curd must be the most important criterion of its firmness or hardness. His apparatus measured instantaneous and delayed elasticity (30 seconds both for compression and recovery). By subtraction plastic deformation was obtained. Since loads and times were always the same, he proposed that if the area under compression was small compared with the size of the vat, the elastic modulus, as a criterion of hardness, would be inversely proportional to the total elastic deformation [Boussinesq solution, Eq. (6.21)]. Firmness or hardness of cheese has also been determined by pressing a spherical indenter into the cheese under a constant stress and recording time and deformation. A sample of data reported by Scott Blair and Baron (1949) is found on page 183.

To determine the effect of such mechanical treatments as work softening by extrusion, the hardness of butter was measured with a cone penetrometer (Vasic and de Man, 1968). In this work, hardness was expressed as the ratio of the load to the area of the impression. The hardness number was calculated from the weight of the cone assembly, depth of penetration, and the radius of the flat surface of the cone tip. Penetrometer tests of butter have been related to its hardness as affected by variation in cattle feed and whether the cows were stall or meadow fed (Hill and Palmer, 1938). Hardness of peanut butter was also measured with a penetrometer in an attempt to relate hardness to spreadability (Vincent, 1947). Firmness of bread and baked products has been measured by determining the force required to cut slices of bread (Combs, 1944).

Mechanics of hardness measurement

If we accept the proposition that hardness in metals and other engineering materials is analogous to firmness and hardness in food materials, the available theories should be studied to investigate their possible applications. For example, spherical indenters have been used on food materials but analysis of the mechanics of the system, as known for Brinell and Meyer hardness tests in metals, has not, apparently, been applied to such materials as fruits, seeds and grains.

Brinell hardness number, *B.H.N.*, is expressed as the ratio of load, F, to the *curved area of the indentation* with a chordal diameter d_2 produced by a steel ball of diameter d_1.

$$B.H.N. = \frac{2F}{\pi d_1^2 \left[1 - \sqrt{1 - (d_2/d_1)^2}\,\right]} \text{ kg/mm}^2 \tag{7.7}$$

In testing, a hard steel ball is pressed on the surface of the metal under examination employing a constant load for a time long enough to reach equilibrium indentation. After the load is removed, the chordal diameter of the indentation, d_2, is measured and *B.H.N.* is calculated using the above expression. It can be shown that for geometrically similar indentations (Fig. 7.22), *B.H.N.* is independent of load and diameter of the indenter. This concept is widely used in practical hardness measurement for engineering materials. For example, a 3000 kg load and a 10 mm ball would give essentially the same *B.H.N.* as a load of 30 kg and a 1 mm ball, if the indentations obtained are geometrically similar.

If we take the ratio of the load to the *projected area of the indentation*, the results are known as Meyer hardness number (Tabor, 1951). This value, which is actually the mean pressure, P_m, can be given by

$$\text{Meyer hardness} = P_m = \frac{F}{(\pi/4)\, d_2^2} = \text{kg/mm}^2 \tag{7.8}$$

The relationship between load and size of indentation is an exponential function known as Meyer's law.

$$F = k d_2^n \tag{7.9}$$

According to the above expression, plotting of load, F, and diameter of

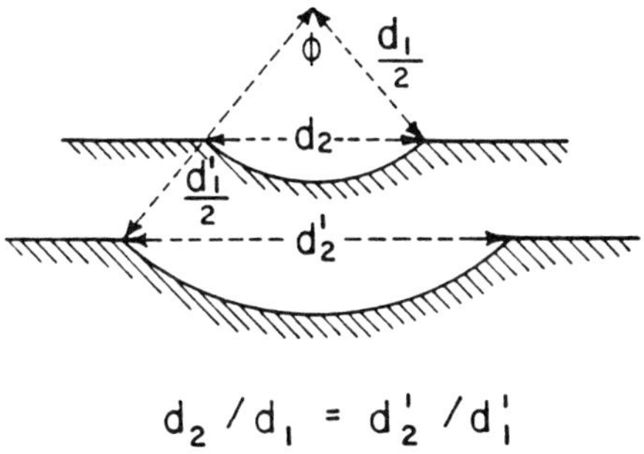

Figure 7.22 Geometrically similar indentations

indentation, d_2, on log-log paper should yield a straight line with k and n as the y-intercept at $d_2=1$ and slope of the line, respectively.

In metals, for the lower limit of $(d_2/d_1)=0.1$, the exponent $n=2$. Under this condition the Meyer hardness number is independent of load and diameter of spherical indenter. For $n=2$, the behavior is fully plastic. For $n=3$, the behavior is fully elastic. For $3>n>2$, the behavior is in the transition range. The Meyer's law is valid for the plastic range where exponent $n=2$ or $(d_2/d_1)=0.1$. For the elastic range, the Hertz formulas (Table 7.2) are valid. The diameter of the circle of contact can be calculated using the expression for the radius of the circle of contact given in Table 7.2 and Eqs. (6.7) and (6.10). Reference to Eq. (6.7) shows that the projected area of contact is proportional to $(F)^{2/3}$ and the mean pressure P_m over the region of contact is proportional to $(F)^{1/3}$. [See Eq. (6.8)]. This yields a pressure distribution shown in Fig. 6.8 (left). Table 7.8 illustrates the use of Brinell and Meyer hardness measurements in evaluating hardness of corn kernels. The data show that for the same size indenter the total deformations are of about the same order of magnitude for both horny and floury endosperms. As expected, however, the floury endosperm shows more residual deformation than the horny endosperm. Note that by this technique, in the case of floury endosperm, the average values of hardness parameters were little affected by magnitude of loads or the size of the indenters.

The two sets of expressions (Meyer's law and Hertz equations), valid for plastic and elastic ranges of the material, respectively, and a simple

Table 7.8 Brinell hardness and Meyer hardness measurements applied to corn kernels

	Horny endosperm (1/16-inch indenter)					Floury endosperm (1/16-inch indenter)					Floury endosperm (3/16-inch indenter)				
Kernel No.	Load F (lb)	Residual d_2 measured (in)	Total d_2 calculated (in)	*BHN* (kg/mm^2)	*MHN* (kg/mm^2)	Load F (lb)	Residual d_2 measired (in)	Total d_2 calculated (in)	*BHM* (kg/mm^2)	*MHN* (kg/mm^2)	Load F (lb)	Residual d_2 measured (in)	Total d_2 calculated (in)	*BHN* (kg/mm^2)	*MHN* (kg/mm^2)
1	9.95	0.020	0.041	21.7	22.4	11.15	0.037	0.044	5.8	7.3	29.5	0.057	0.085	8.0	8.1
2	12.10	0.023	0.036	19.8	20.6	11.19	0.035	0.044	7.5	8.4	33.0	0.065	0.099	6.8	7.0
3	9.10	0.014	0.036	44.8	41.7	12.0	0.034	0.046	8.6	9.3	30.1	0.059	0.095	7.6	7.8
Ave.	10.6	0.019	0.037	28.7	28.2	11.4	0.035	0.045	7.3	8.3	30.8	0.063	0.093	7.5	7.6

ball test might enable one to locate the points at which the biological material is fully elastic, fully plastic, or in the transition range. As we have seen, certain food materials do not exhibit a bioyield point or any other distinct mark on their load-deformation curves to indicate the initiation of tissue injury. Such a distinct mark as a break in the curve, particularly if it occurs at low levels of load with minimum tissue injury, would be a helpful tool in quality and texture evaluation. A simple ball test which yields a load versus deformation curve or a pressure versus load curve through the plastic range of the material might lead to a useful criterion which could be used in firmness and hardness measurements of agricultural commoditites. For the sake of illustration, a pressure-load curve for pressing a hard steel ball into the flat surface of a metal is shown in Fig. 7.23. The initial part *OA* represents the elastic region where equations in elasticity are applicable. The transition region is shown by the dotted line. The condition of full plasticity is shown by *MN*. Note that pressures are given in terms of *Y* which denoted the yield strength of the metal. It has been shown that the point at which the composite curve begins to deviate from the elastic curve (Point *A*, Fig. 7.23) corresponds to a mean pressure $P_m \approx 1.1Y$ while the full plasticity begins at $P_m \approx 2.8Y$ (Tabor, 1951; Bowden and Tabor, 1964).

It might be possible to establish similar pressure-load curves for food materials. The yield strength criterion can be taken either as the pressure at the bioyield point, or the linear limit *LL* (Fig. 4.3) on a pressure-deformation

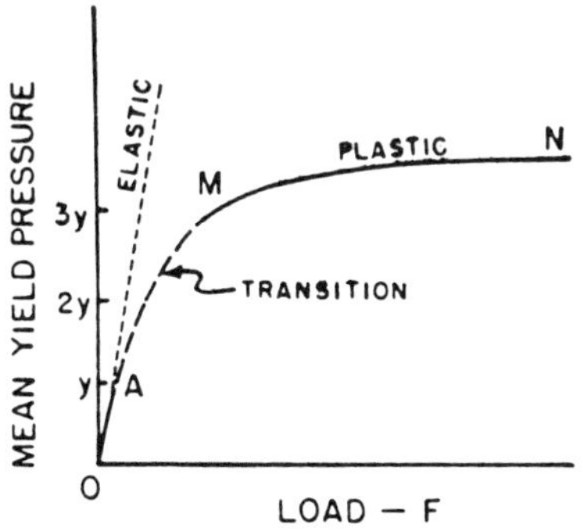

Figure 7.23 Theoretical pressure-load characteristics of a metal deformed by a spherical indenter (Tabor, 1951)

curve, or an arbitrary value such as the stress at the offset strain of 0.002 in. per in. recommended by the ASTM for metals. For food materials, however, a technique similar to Johnson's apparent elastic limit for engineering materials would seem more appropriate. In this method, the yield stress is taken as the stress where the slope of the stress-strain curve is 50 percent less than the initial slope. To determine this stress graphically, first a horizontal line *AB* is drawn (Fig. 7.24). This line is then extended to point *C* such that $BC=0.5AB$. Next, line *OC* is drawn and the line tangent to the curve and parallel to *OC* is established. The stress corresponding to the point of tangency is Johnson's apparent elastic limit or the arbitrary chosen yield stress. Note that the *Y* and *X* coordinates of the diagram in Fig. 7.24 are given in terms of pressure and deformation which are more applicable to food materials than stress and strain coordinates.

Example

To evaluate the hardness of horny endosperm of a sample corn at 12 percent moisture content, the kernels mounted on a rigid plate, were first surface ground and then subjected to indentation tests using 1/16 in. and 3/16 in. steel balls and the testing machine of Fig. 4.22. After each test, the chordal diameter of the impression was measured under a microscope.

(a) If the total indentation for the 1/16-in. ball is 0.1835 mm under a load of 4.51 kg and for the 3/16-in. ball is 0.2385 under a load of 14.68 kg, show to what degree a geometric similarity exists in these indentation tests.

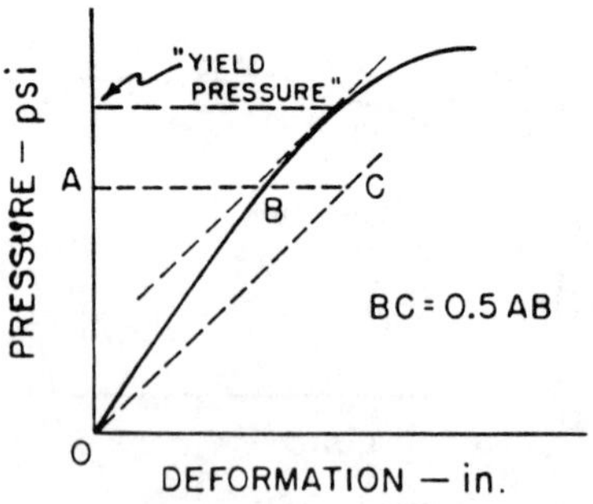

Figure 7.24 Graphical construction for an arbitrarily defined "yield pressure"

For geometric similarity of indentations the following relationship must hold:

$$d_2/d_1 = d_2'/d_1' \qquad (7.10)$$

which means that the chordal diameter of the indentation by means of the 3⁄16-in. indenter would have to be three times that of the 1⁄16-in. indenter. From the central indentations given in the problem, the chordal diameters can be calculated from the geometry of the indentation shown in Fig. 7.22.

$$(0.03125)^2 = (d_2/2)^2 + [0.03125 - 0.1835/25.4]^2$$
$$d_2 = 0.0405 \text{ in.}$$

Similar calculations for the 3⁄16-in. indenter gives d_2=0.0825 in. For geometric similarity d_2' should be 3×0.0405=0.1215 in. Therefore, the two indentations are not geometrically similar. To approach the condition of geometric similarity, the load should have been increased on the 3⁄16-in. indenter.

(b) Determine the degree of elasticity of the horny endosperm for the 1⁄16 in. ball test if the measured chordal diameter of the identation is 0.504 mm.

Degree of elasticity can be determined by comparing the total central indentation, which corresponds to the sum of elastic and plastic deformations, with the plastic central indentation based on measured chordal diameter. Again, from the geometry of indentation

$$(0.03125)^2 = \left(\frac{0.504}{25.4}\Big/2\right)^2 + \left[0.03125 - \frac{C.I.}{25.4}\right]^2$$
$$C.I. = 0.0318 \text{ in.}$$

where $C.I.$ is the central indentation corresponding to the measured or plastic chordal diameter.

$$\% \text{ elasticity} = 100 \times \frac{\text{Total indentation} - \text{Plastic indentation}}{\text{Total indentation}}$$
$$= 100 \times \frac{0.1835 - 0.0318}{0.1835} = 82.5$$

(c) Using data given in (a) calculate the $B.H.N.$ for both the 1⁄16-in. and the 3⁄16 in. indenters. $B.H.N.$ for the 1⁄16-in. indenter and 4.51 kg load is according to Eq. (7.7)

$$BHN = \frac{2 \times 4.51}{3.14\,(1/16 \times 25.4)^2\left(1 - \sqrt{1 - \left(\dfrac{0.0405}{0.0625}\right)^2}\right)}$$

$$= 4.79 \text{ kg/mm}^2$$

Similar calculations for the 3⁄16-in. indenter shows $B.H.N.$=3.98 kg/mm^2. Theoretically, if the two indentations were geometrically similar, there would have been a closer agreement between the two values of $B.H.N.$

(d) Given the following data from the load-indentation curve for the 1⁄16-in. indenter, show that the relationship between load, F, and the residual chordal diameter of indentation, d_2, is of the form $F=kd_2^n$. The residual chordal diameter was calculated from residual deformation after unloading.

Load F (kg)	d_2 (mm) measd.	d_2 (mm) calcd.	Load F (kg)	d_2 (mm) measd.	d_2 (mm) calcd.
1.5	0.33	0.38	3.99	0.48	0.51
2.00	0.36	0.41	5.10	0.58	0.61
2.27	0.41	0.43	6.99	0.74	0.79

Plotting the data on log-log coordinates yields the two straight lines shown in Fig. 7.25. Note that the slope of the line changes when the load exceeds a certain value. The same phenomenon has been observed in metals when loads versus permanent indentations are plotted on log-log coordinates. The inflection point is the beginning of transition from elastic to plastic behavior. The larger values of n and k correspond to stiffer and harder material.

When small differences in hardness are to be detected, it is possible that the parameters n and k may prove to be a better criterion of hardness than such pressure parameters as Brinell hardness number. For example, tests of both horny and floury endosperms of a sample of corn kernels by the method discussed in the above illustration gave the following average values for horny and floury endosperm:

	Horny endosperm	Floury endosperm
B.H.N.	6.26	4.53
n	5.94	2.67
k	6.27	3.42

As seen from the above data, there is a larger difference in hardness between floury and horny endosperms by considering *n* and *k* values than *B.H.N.* values.

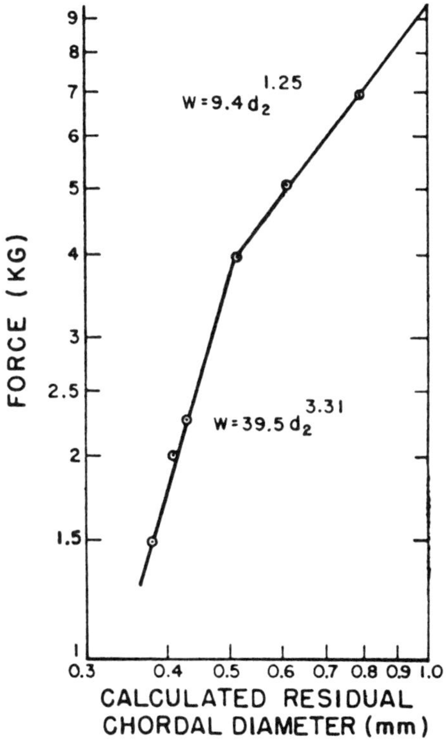

Figure 7.25 Plot of load versus residual chordal diameter for a 1/16 in. steel ball on the horny endosperm of a surface-ground corn kernel

Example

Grain hardness is given in terms of force readings of a steel ball deforming the grain and the chordal diameter of indentation as shown below. Express the results in the form of a power law equation $H = Ad^B$, where H is hardness, d is total chordal diameter of indentation, and A and B are constants for the specific grain. Find the values of A and B.

Force H	Diameter of Indentation d
2	0.17
3	0.23
4	0.27
5	0.35

Plot the data on a log-log paper as in Fig. 7.25

$$H = Ad^B$$
$$\log H = \log A + B \log d$$
$$A = y \text{ intercept at } d = 1, B = \text{slope of the straight line.}$$

Dynamic hardness

If in a hardness test the indenter is allowed to fall onto the specimen surface, it rebounds to a certain height and leaves an indentation in the specimen. An energy balance equation can be written for the system as follows

$$W_1 = W_2 + W_3$$

where W_1 is the energy of impact, W_2 is the rebound energy, and W_3 is the work done as plastic energy. The analytical expression for each of the above energy terms will be given in the next chapter when we discuss mechanical damage due to shock and impact. At this time it is sufficient to state that the above energy balance leads to the conclusion that the volume of the indentation is directly proportional to the difference between the energy of impact and the energy of rebound. This implies that the softer material offers an average pressure of resistance to indentation. This resistance, which can be referred to as dynamic hardness, is denoted by P_d and is given by the following expression (Tabor, 1951)

$$P_d = \frac{\text{mg}(h_1 - (3/8)h_2)}{Va} \tag{7.11}$$

where m is the mass of the indenter, g is the acceleration due to gravity, Va is the apparent volume of the indentation which would be obtained if the indentation were considered to have the same radius of curvature as the indenter, and h_1 and h_2 are, respectively, height of drop and height of rebound. Further analysis by Tabor (1951) has shown that Va can be eliminated from Eq. (7.11) and dynamic hardness P_d can be calculated from

$$P_d^5 = 6.3 \times 10^{-3}\,(\text{mg}/R^3)\left(\frac{1}{h_1 - (3/8)\,h_2}\right)^3 (h_2/A)^4 \tag{7.12}$$

where mg is the weight of the pendulum, R is the radius of the indenter and A is a function of the moduli of elasticity and Poisson's ratios of the two impacting bodies given by Eq. (6.4). It may be pointed out that the dynamic hardness, P_d, is also referred to as dynamic yield pressure in metals.

Some preliminary work has shown that the dynamic yield pressure can be taken as a criterion for resistance of fruits to mechanical injury under dynamic conditions. Nelson (1967) has shown that under constant impact energy, the greater dynamic yield pressure corresponds to greater resistance to bruising in McIntosh apples.

If A in Eq. (7.12) is not known, the dynamic yield pressure can be determined from Eq. (7.11) by measuring h_1 and h_2 and the chordal diameter of the dynamic indentation. One can also simultaneously conduct quasistatic tests to determine the bioyield pressure which would produce the same diameter indentation as the corresponding dynamic yield pressure. This technique might develop into a useful method for correlating the resistance of the material to deformation under static and dynamic conditions. This has been done for metals (Bowden and Tabor, 1950). The results showed that dynamic yield pressure is always greater than the static yield pressure. This effect, which suggests the presence of viscous forces, should be even more pronounced for viscoelastic biological materials. (See Chapter 8.)

Dynamic hardness test, as described above, is applicable to many agricultural commodities where hardness and firmness evaluation, particularly under dynamic loading is desired. For this purpose, an apparatus employing the principle of a simple pendulum (Fig. 7.26) can be set up to measure

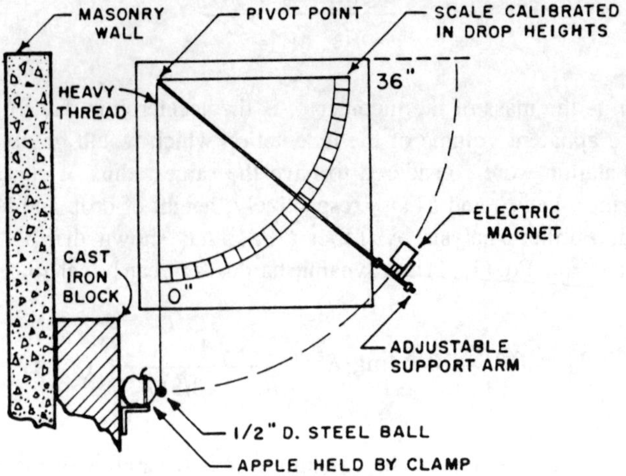

Figure 7.26 Pendulum apparatus for evaluation of dynamic firmness and hardness of agricultural commodities

the height of drop h_1, and the height of rebound, h_2. The pendulum would be a steel ball, used as the indenter, impacting the fixed test specimen, or the specimen itself hung by a string striking a rigid surface. In the latter case the dynamic hardness, P_d, can be expressed as a function of the height of rebound h_2. In Fig. 7.26, the pendulum is a ½-in. steel ball striking the approximately flat surface of apple specimens fixed on a rigid wall.

Example

Five apple specimens selected from two samples of apples at two different stages of maturity were subjected to dynamic firmness tests using the pendulum apparatus shown in Fig. 7.26. Prior to the impact by the steel ball, one force-deformation curve was obtained for each apple using the same steel ball as the dynamic indenter. The following table shows the original data for static and dynamic tests.

Sample (*x*)	*Quasi-static test* deformation *D* at $F = 1$ lb. (in)	*Dynamic test* rebound height h_2 at $h_1 \doteq 12$ in. (in)
1	0.044	1.5
2	0.044	1.0
3	0.045	1.5
4	0.040	1.5
5	0.072	1.5
Mean	0.049	1.4
Sample (*y*)		
1	0.024	2.75
2	0.033	2.50
3	0.027	2.75
4	0.025	2.75
5	0.025	2.75
Mean	0.027	2.70

If the indenter is a ½-in. steel ball weighing 0.019 lb., find the dynamic firmness values for the two samples of apples.

Solution

All parameters in Eq. (7.12) are known except the parameter A. For computing the parameter A, omit the quantity $\left(\dfrac{1-\mu_1^2}{E_1}\right)$ between Eq. (6.14)

$$E_1 = \frac{0.531F(1-\mu_1^2)}{D^{3/2}}\left(\frac{1}{R_1} + \frac{1}{R_1'} + \frac{4}{d_2}\right)$$

and Eq. (6.4)

$$A = \frac{(1-\mu_1^2)}{E_1} + \left[\frac{(1-\mu_2^2)}{E_2} \approx 0 \text{ for steel ball}\right]$$

which yields

$$1/A = \left(\frac{0.531\,F}{D^{3/2}}\right)\left(\frac{1}{R_1} + \frac{1}{R_1'} + \frac{4}{d_2}\right)$$

where d_2 is the diameter of the steel ball. Since the points on the fruit chosen for dynamic firmness tests were the points with least radii of curvature ($R_1 \approx R_1' \approx \infty$), the parameters $1/R_1$ and $1/R_1'$ were considered negligible. For more accurate measurement, these radii could be determined with a radius meter shown in Fig. 6.3 and considered in the computations.

Substituting the given values in the expression for A and Eq. (7.12), the values of P_d were found as given below.

Apple No.	Dynamic firmness P_d (psi)	Bruising Depth (in)	width (in)	Volume in^3 × 10^4
1	7.02	0.10	0.28	36
2	4.99	0.08	0.22	18
3	6.79	0.10	0.16	15
4	7.84	0.10	0.28	36
5	3.89	0.10	0.20	21
Mean	6.10	Ave. 0.096	0.23	25
1	25.5	0.10	0.20	21
2	16.5	0.05	0.20	8
3	21.2	0.05	0.25	13
4	23.9	0.05	0.25	13
5	23.9	0.05	0.20	8
Mean	22.2	Ave. 0.06	0.22	12

Note that, based on this set of data, the dynamic firmness parameter shows a greater difference between the two samples of apples than either the static indentation by the steel ball or the dynamic rebound height given previously. The usefulness of this technique, particularly for the cases where resistance of the fruit to shock and impact loading is of interest, is yet to be explored.

A further simplification of dynamic firmness measurement is to define, from Eq. (7.12), dynamic firmness, F_d, given by the following expression

$$F_d = P_d^5 A^4 = 6.3 \times 10^{-3} \left(\frac{mg}{R^3}\right) \left[\frac{1}{h_1 - (3/8)\, h_2}\right]^3 h_2^4 \qquad (7.13)$$

In this manner, factor A, which is a function of the modulus of elasticity of the deformable material, need not be known because it is incorporated in the dynamic firmness term F_d. Any consistent units can be used in Eqs. (7.11), (7.12), and (7.13).

7.13 EFFECT OF AGE, WATER CONTENT AND TEMPERATURE ON TEXTURE OF FOODS

Age, water content and temperature influence the textural characteristics of such food materials as fresh produce. Figures 5.7 and 7.15 illustrate, respectively, the effect of time and maturity or aging on certain mechanical properties of apples. Similar findings were reported by Finney (1964) for potatoes (Fig. 8.54) and several other commodities (Fig. 7.27, 7.28). The biochemical process involved in softening of fruits and vegetables upon aging has already been discussed in connection with Fig. 5.7.

The water content of fruits influence their firmness. Haller *et al.*, (1933) found an indirect relationship between the moisture content of strawberries and their resistance to flattening. The same trend has been reported for apples and pears (Haller, 1941). Bartlett pears and Gravenstein apples grown under high temperatures and low humidities were found distinctly firmer than those under lower temperatures and higher humidities. This was attributed to the difference in moisture content of the fruits. In grains, effect of moisture content is important in hardness measurement. In Fig. 5.19, examples of the effect of moisture on mechanical properties of agricultural grains are given. Figures 8.42 and 8.43 are further evidence of the influence of moisture in hardness and strength of several grains. Figure 8.42 shows that greater energy is required to break by impact the grains having a higher moisture content than those having a lower moisture content. In these tests, 20 grains were tested at each level of impact energy. Figure 8.43 shows the amount of deformation and load that grains with higher and lower moisture contents can withstand. A high-moisture grain which takes a greater amount of energy to break (Fig. 8.42) is not as resistant to deformation as a low-moisture grain (Fig. 8.43). If hardness of grain is defined as resistance to deformation, addition of moisture reduces hardness. On the other hand, if breaking energy is to be taken as a criterion

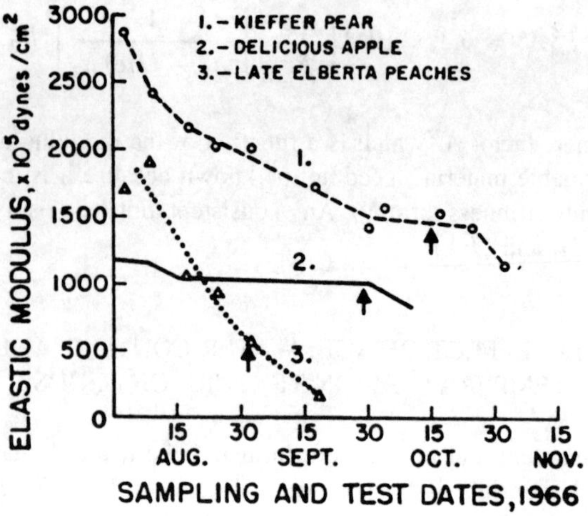

Figure 7.27 Changes in dynamic elastic modulus of fruits during growth and development (Finney, 1967)

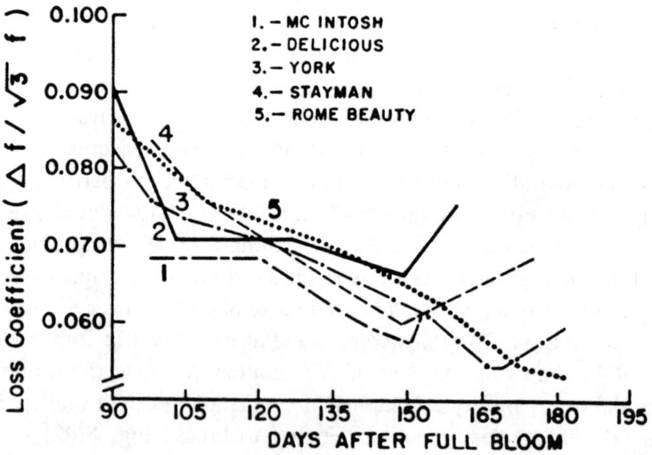

Figure 7.28 Changes in loss coefficient (internal friction) of apples during growth and development. Δf is half-amplitude bandwidth (Finney, 1967)

for strength and hardness, addition of moisture increases resistance to breakage. Such evidence was reported by both Zoerb (1960) and Bilansky (1966).

Matthews and Hall (1967) reported a decrease in modulus of elasticity for the flesh of potato tuber with increase in exposure to heat. They considered this behavior in agreement with increasing "toughness" of meat due to an increase in heat treatment as reported by Tuomy *et al.*, (1963).

Conflicting results have been reported for firming of cherries due to aging by holding either in water or in air at low or high temperatures. LaBelle (1960) stated that although the physiological mechanism involved might be the increase in turgor which enables easier cutting of the pitting needles through the skin of the fruit in the pitting operation, it has little influence on the firmness of the final product. He reported no difference in firmness and drained weight (a measure of usable solid portion of the pack) when cherries were held for several hours in water or in air at 40°F or 80°F temperatures. The higher temperature merely affected color by increasing scald damage. This finding seems to be conflicting with the previous studies on cherries and strawberries reporting that chilling promoted firming and increased resistance to mechanical damage (Whittenberger and Hills, 1953; Rose *et al.*, 1934).

Later studies by LaBelle *et al.*, (1964) on Montmorency cherries showed that firming is actually a matter of aging and takes place more rapidly at elevated than at lower temperatures. An interesting finding in this work was that bruised cherries gained tissue strength during repeated aging and bruising. The firmness in this case was determined by measuring the resistance of pitting of individual cherries using the pitting needle on the crosshead of a universal testing machine. The physiological process involved in this strength recovery of the tissue, which is similar to work hardening in metals, is not understood but its practical implications in repeated bruising and holding of the product in the processor's tank is significant. LaBelle *et al.*, (1964) have shown how the successive experiences of initial bruise, aging by holding in water at various temperatures for various times, rebruising, and reaging affect the firmness of the cherry at each stage.

Further discussion on effect of temperature, water content, variety and maturity is given in discussion of mechanical damage.

7.14 FAILURE STRENGTH AND CHARACTERIZATION OF SOLID FOODS

Although the history of rheology of solids foods goes back several decades and considerable basic work has been done in the past on the mechanical properties of these biomaterials, our knowledge is still in infancy with few well-defined terms, definitions and concepts, very few reference books, and a literature abounding with misconceptions, terms and definitions inconsistent with those known in solid mechanics. As far as the application of solid foods rheology to texture evaluation is concerned, it is important to note that in most cases the sensory evaluation of food texture involves forces at strain levels beyond the ultimate strength of the food, thus causing yielding and fracturing defined as failure in solid mechanics. In sensory evaluation of solid foods, a mechanical breakdown in the form of yielding, tearing or fracturing usually takes place. Knowledge of the characterization and failure phenomena is essential to understanding the role of the mechanical properties in food texture evaluation.

Soft and Hard Solid Foods

A solid food material is defined to be a material with high enough rigidity under normal handling conditions to support its own weight under gravitational forces. Considering an arbitrary dividing line, soft solids can be defined (Ferry, 1961) as those with an elastic modulus less than 10^7 N/m^2 (1450 psi). Hard solid foods are those with an elastic modulus greater than this value. Whereas, in engineering materials temperature and frequency range determine whether or not a material is soft or hard, in food materials "softness" or "hardness" depend not only on the temperature and frequency range but also on the moisture content. Under normal handling conditions, for most food materials, the moisture content is the most prominent factor determining softness or hardness. For example, the apparent modulus of elasticity of a corn kernel changes from about 2.4×10^5 psi to 0.5×10^5 psi as its moisture content on the dry basis increases from 0.75% to 28% (Shelef and Mohsenin, 1968). Table 7.9 shows the modulus of elasticity values for some solid foods.

Besides knowing the degree of rigidity, it is important to know the structure of solid foods influencing the strength of the material. To show the complexity of the structure which can affect the stress distribution within the material when subjected to mechanical forces, consider the tissue structure of the apple flesh, a kernel of corn and a beef muscle.

Table 7.9 Order of magnitudes of elastic modulus and Poisson's ratio of selected solid foods

Material	Type of Test	Apparent Elastic Modulus $\left(10^7 \frac{N}{m^2}\right)^*$	Poisson's Ratio	Source
Apple				
whole	plate compression	0.35–0.7	—	Arnold and Mohsenin (1971)
flesh	axial compression	0.33–1.0	0.21–0.34	Chappel and Hamann (1968)
skin	axial tensile	0.95–1.4	0.37	Clevenger and Hamann (1968)
Banana	dynamic	0.08–0.12	—	Finney and Norris (1968)
Beef	axial compession	0.0234–0.0237	—	Segars *et al*. (1974)
Carrot	bending	0.3–2.9	—	Kapsalis and Segars (1972)
Celery	bending	0.99–3.15	—	Kapsalis and Segars (1972)
Cheese	axial compression	0.016	0.5	Mohsenin and Morrow (1968)
Corn grain	axial compression	40–115	0.32	Mohsenin (1970)
Gels (protein)	axial tensile	0.00016–0.00103	—	Nakayama *et al*. (1971)
Noodles (cooked)	axial tensile	0.006	—	Shimizu *et al*. (1958)
Pear	plate compression	0.37–1.0	—	Fridley *et al*. (1968)
Peach	plate compression	0.05–1.0	—	Fridley *et al*. (1968)
Potato (white)	axial compression	0.32–0.43	0.49	Finney (1967)
Wheat grain	axial compression	120–570	0.3	Mohsenin (1970)

*Divide by 6900 to convert to psi

Figure 2.1 (Chapter 2) shows the transition of cellular structure for an apple from skin to parenchyma. The cuticle and epidermis tissue consist of tough, closely pressed cells to guard the plant material from mechanical injury and attack by fungi and microorganisms. The parenchyma, or edible tissue, lies under the epidermis and consists of thin-walled cells containing water, sugar, acids, and some solids. These cells are separated by intercellular spaces filled with water and air (as much as 25% by volume). A cementing agent, such as pectic substances, holds these cells together to provide the rigidity in the structure. Dimensional data on cell number, size, volume and wall surface area in different regions as well as on intercellular spaces are available (Reeve, 1953).

The main structural components of the corn kernel are the hull, the endosperm, and the germ as shown in Fig. 2.15. The hull consists of the pericarp and the seed coat. The seed coat is a nonporous membrane and completely encloses the germ and the endosperm. The endosperm consists of a thin layer of aleurone cells containing oil and protein, and a large inner portion of storage tissues containing starch and protein. The endosperm comprises over 80% of the corn kernel weight and is divided into horny and floury endosperm. The horny endosperm is hard and translucent, whereas the floury endosperm is soft, mealy, and opaque. The germ is the seed embryo and comprises over 10% of the kernel weight. Data on details of the structure and composition of corn kernels are available (McMasters, 1965; Anderson *et al.,* 1962).

Natural muscle tissue (Fig. 2.16) consists of fiber bundles lying parallel to one another in a matrix of connective tissues. The fibers are long, narrow, multinucleated cells. An individual fiber is composed of numerous smaller units called myofibrils, consisting of alternating layers of actin and myosin filaments which slide past each other as the muscle fibers stretch or contract. A thin elastic sheath, the sarcolemma, surrounds each fiber and retains sarcoplasm, a fluid phase around each fiber. Outside the sarcolemma is the connective tissue known as the endomysium. The perimysium, an associated connective tissue containing blood vessels and nerves, separates the fiber bundles. Encompassing the entire muscle is the epimysium or muscle sheath. Synthetic systems utilize two-phase systems of fiber and binder in producing meat analogs. Soy fiber and edible binder simulate the muscle fiber and the connective tissue of natural meat, respectively.

The relationship between structure and texture of foods has also been discussed by Jowitt (1977), Diehl *et al.* (1979), and Petrell and Mohsenin (1980).

Defining Failure in Solid Foods

One of the first problems is to define failure in a solid food material. Is failure, as in the case of engineering materials, due to excessive elastic deformation or inelastic deformation and yielding, or fracturing in the form of tearing and cutting the material? To what extent, therefore, does the structure of the material affect the strength and failure? With this simplified engineering approach given for the structure of the three food materials, let us proceed by defining a threshold of failure which is of some practical significance.

In the case of apple tissue, we consider the threshold of failure being the initial cell rupture in the parenchyma tissue resulting from excessive elastic or inelastic deformation, exhibited by a discontinuity in the force-deformation curve called a bioyield point (see Fig. 7.14). At this point, a fracture plane appears to develop along an arc within the tissue structure. The contents of the ruptured cells are then released into the intercellular air spaces causing darkening or browning, believed to be an enzymatic process in the fresh fruits taking place a few hours after pressure bruising (Tuckey and Mohsenin, 1966) (Fig. 8.1). The mechanics of the tissue failure, particularly in relation to the cell wall failure, is not yet fully understood.

In the case of corn kernel a break in the endosperm may develop from exceeding the elastic or inelastic deformation of the aleurone layer causing visible or invisible cracks. Initiation of these cracks can be taken as the threshold of failure in such foods as cereal grains. The stress and strain conditions at which these cracks begin to develop are not, however, as easily detectable as in the case of cell rupture in the apple tissue.

Finally, in the case of meats, the question of tissue failure is even more complicated, partially because of the many biochemical and structural changes occurring in the structural proteins and mechanical rigidity of the myofibrils (Connel, 1960). Due to the fact that the amount of connective tissue may determine the "toughness" of meat, one may consider the failure threshold as being the stress level at which tearing and separation of connective tissue take place. This has been done by measuring the force required to pull the fiber bundles in strips of cooked meat placed transversely to the applied strain (Bouton and Harris, 1972). Also, the extensibility and breaking strength of the fibers can be taken as the point of failure. The breaking strength of fiber bundles in uncooked pork muscle parallel to the direction of force has been shown to account for 60–90% of variations in tenderness (Stanley *et al.*, 1972).

Of course, the above definitions of failure are arbitrary and are given as a starting point. But even if we can agree on a definition of failure for a selected food material, considering the fact that the food materials are generally nonlinear, nonhomogeneous, anisotropic, changing their mechanical properties with age, moisture content, temperature, and other natural and environment conditions, is there any hope for a rational solution of practical value? It is difficult to answer this question since relatively little work of truly basic nature has been done on the theoretical aspects of rheology of solid foods. Surely, the complexity of the problem is not any greater than the biomaterials being investigated with applications to biomedical engineering.

Much of the work during the past two decades has been on fruits and vegetables because of the interest in mechanized harvesting and attempts to reduce mechanical damage. Compared with the other two examples of food materials given, fruits and vegetables are also simpler in structure, easier to prepare for testing and, if we accept the initial cell rupture as the threshold of failure, the question of when failure occurs is less complex. A selected example of work on the mechanics of quasi-static and dynamic loading, as related to fruits and vegetables, is given in Fig. 7.29. Its examination reveals some basic facts on the characterization and failure phenomenon of this class of solid foods. These will be discussed below.

Shear Failure of Tissue

A simple punch shear test of fruits, such as apples, which shears a plug out of a slice of the fruit held between two rigid plates, must be some indication of the strength of the cementing agent (pectic substances) in the middle lamella holding together the cells in the parenchyma tissue. Data given by Schmidt (1962) show the general trend of decreasing shearing strength as the fruit ripens (Fig. 7.29). This observation is in agreement with the biochemical theory of the pectic substances known today. The quantitative values of the shearing strength have been reported by Mohsenin (Table 7.3), Miles and Rehkugler (1973), and Yang (1973). Table 7.3 shows a highly significant correlation between the shearing strength of the apple tissue and the bioyield pressure, rupture pressure, and sugar content of Melba apples. Miles and Rehkugler (1973) did not determine the shear strength directly but calculated it by dividing the axial compression strength by two. This is true when the material follows the ideal Tresca failure criterion and the plane of failure is 45° with respect to the loading axis.

Figure 7.29 Mechanical characterization and failure of fruits and vegetables under quasi-static and dynamic loading

	Assumptions and Testing Method	Characterization	Failure-Related Findings	Reference
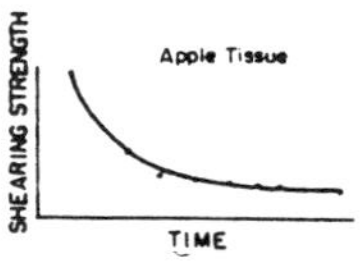	Punch shear of slice of apple tissue	—	Shear strength was related to pectic substances in the middle lamela cementing cells together	Schmidt (1962)
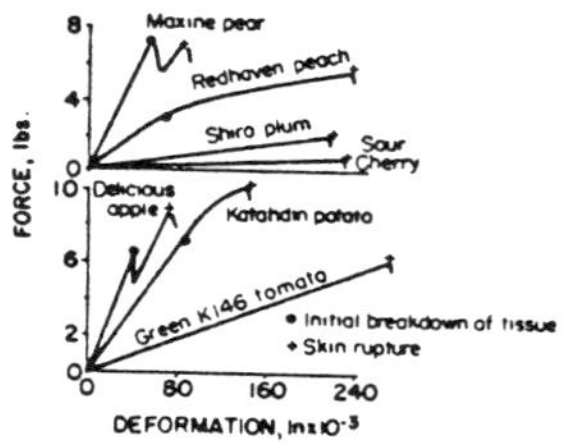	Whole apple rigid die loading	Elastic and plastic response on unloading	Initial cell rupture at "bioyield" point of curve	Mohsenin *et al.* (1962)
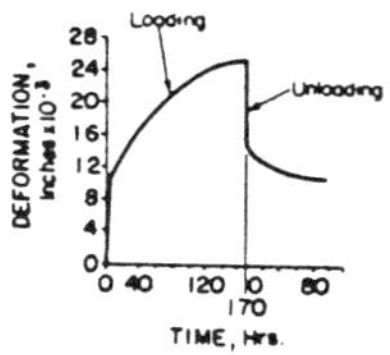	Whole apple rigid die and plate loading	Viscoelastic, 4-element rheological model	Distortion under dead load in storage	Mohsenin (1962)

Figure 7.29 continued

	Assumptions and Testing Method	Characterization	Failure-Related Findings	Reference
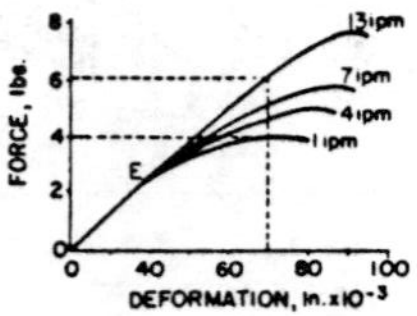	Whole apple rigid die loading	Deformation rate dependent	Effect of deformation rate in determining strength of tissue	Mohsenin (1963)
	Elasticity Tensile test of skin specimens of tomato	—	Tensile strength of tomato skin not an indication of cracking resistance	Voisey (1965)
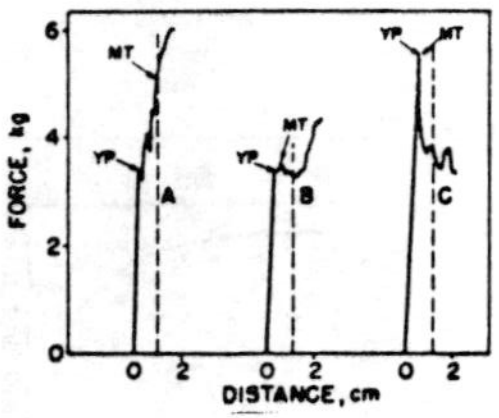	Rigid die with spherical end on whole apple	—	Fruits and vegetables exhibit different types of failure curve depending on stage of maturity and storage condition	Bourne (1965)

Figure 7.29 continued

	Assumptions and Testing Method	Characterization	Failure-Related Findings	Reference
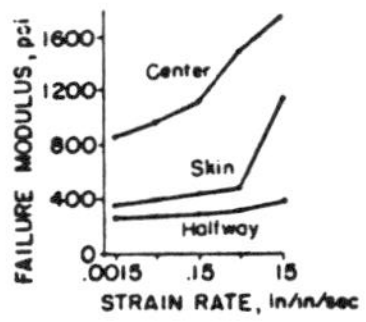	Elasticity. Tensile test of potato flesh	Strain rate dependent. Failure modulus varies from surface to center	Tensile strength and strain at failure. Modulus increased with increased strain rate	Huff (1967)
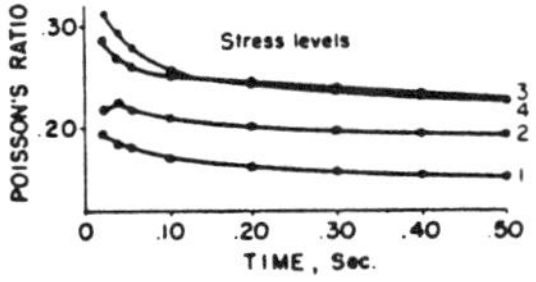	Cylindrical specimen axial loading of apple flesh	Non-linear viscoelastic	Dependency of Poisson's ratio on time and stress (1968)	Chappel and Hamann
	Tensile test, relaxation, creep of apple skin	Viscoelastic, anisotropic. Strength greater parallel to core	Tensile strength of skin and secant modulus	Clevenger and Hamann (1968)
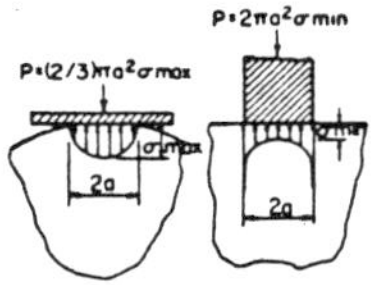	Elasticity. Hertz and Boussinesq contact problems. Flat plate and rigid die loading on whole and segment of apples, pears, and potatoes	Elastic below cell rupture	Pressure distribution under plate loading and plunger loading	Fridley *et al.*

Figure 7.29 continued

	Assumptions and Testing Method	Characterization	Failure-Related Findings	Reference
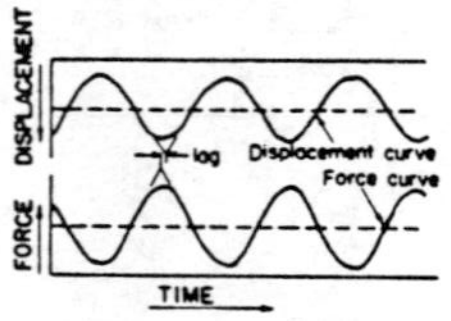	Viscoelasticity, whole papaya fruit loaded sinusoidally	Viscoelastic	Energy absorption before damage marked by formation of watery spots in the flesh along the axis of loading	Kumar and Wang (1969)
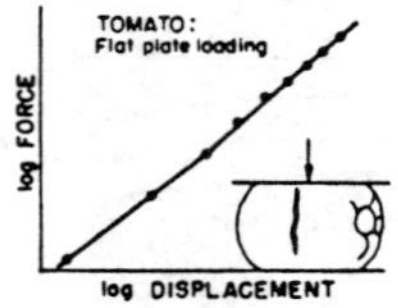	Whole tomato fruit, flat plate loading	Viscoelastic	Failure caused by tensile hoop stresses expanding the fruit diameter. Skin provides the strength	Miles *et al.* (1969)
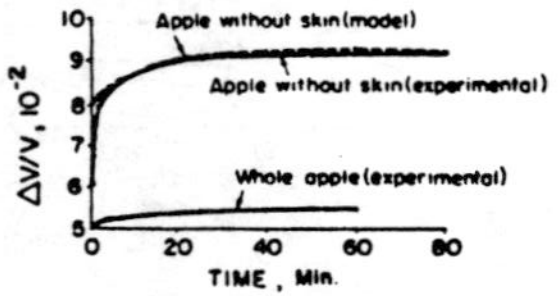	Elastic skin, linear viscoelastic flesh. Hydrostatic compression of whole and cylindrical specimens of apples	Linear viscoelastic at one stress level	Hydrostatic stresses resisted by skin and by flesh as a function of deformation	Sharma and Mohsenin (1970)

Figure 7.29 continued

	Assumptions and Testing Method	Characterization	Failure-Related Findings	Reference
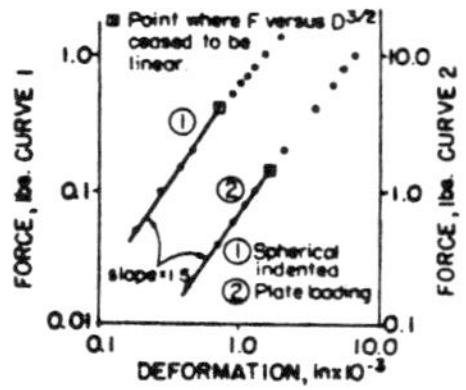	Elastic below initial cell rupture. Plate loading and rigid die loading of whole apple	Elasticity theory and Hertz contact stress valid for small deformations	Stress index and modulus of elasticity at initial cell rupture	Arnold and Mohsenin (1970)
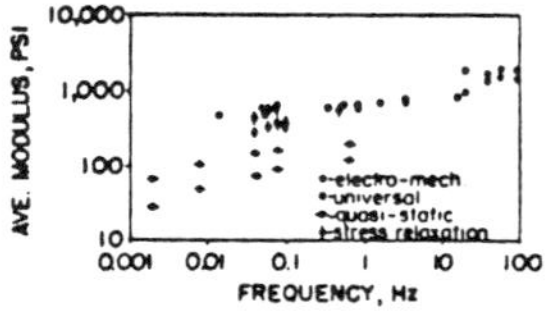	Hertz contact stress for quasi-static loading valid. Cylindrical flesh, whole fruit with and with skin—apple	Non-linear coelastic. Large strains even at small levels of stress	Spectrum of modulus vs. frequency. Stress-strain data through failure	Morrow *et al.* (1971)
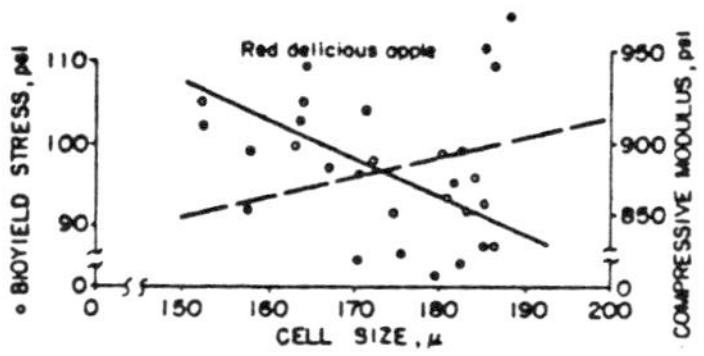	Cylindrical specimen of apple parallel plate loading	—	Relation between structural and biochemical factors to bioyield stress and to compressive modulus	Lozada, (1974)

Figure 7.29 continued

	Assumptions and Testing Method	Characterization	Failure-Related Findings	Reference
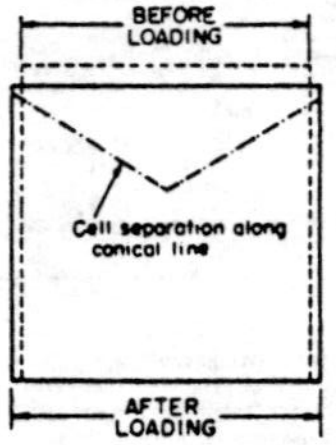	Elasticity theory. Triaxial test of cylindrical specimens of apple	Non-linear visco-elasticity seen from strain rate data (not mentioned by the authors)	Limiting shear stress cause failure failure	Miles and Rehkugler (1973)

Figure 7.29 continued

	Assumptions and Testing Method	Characterization	Failure-Related Findings	Reference
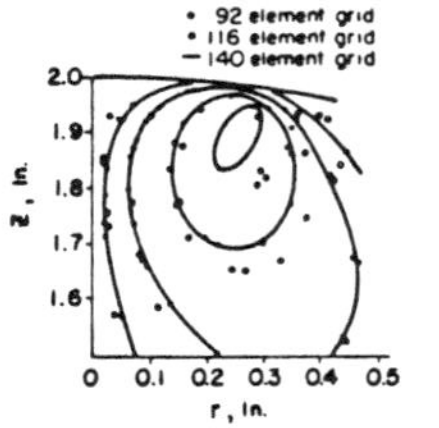	Spherical, linear viscoelastic shear and an elastic bulk response, cylindrical specimens of apple flesh. Final element formulation	Linear visco-elastic	Shear stress and compressive stress distribution within the fruit	Rumsey (1974)
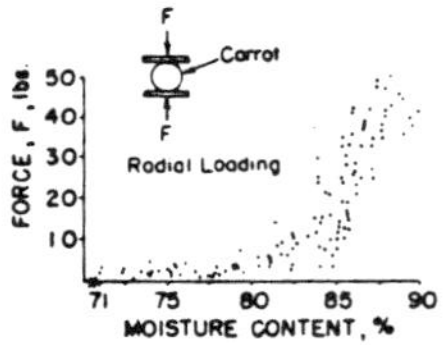	Contact problem of cylinder and flat plate in elasticity. Diametrical loading of carrots	—	Axial loading when center part is more rigid not related to rigidity of outer part	Segerlind *et al.* (1977)

Figure 7.29 continued

	Assumptions and Testing Method	Characterization	Failure-Related Findings	Reference
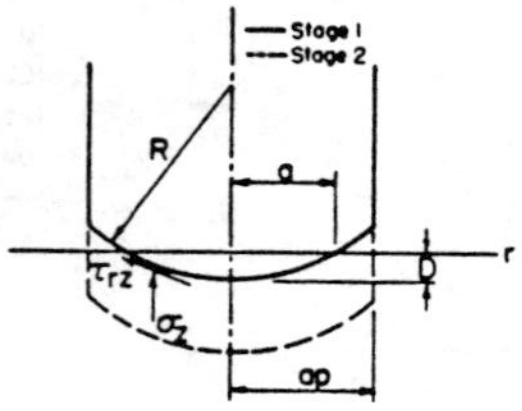	Inelasticity theory for 1st stage, piercing theory for 2nd stage. Whole fruit rigid die with spherical end. Cylindrical specimen of apples	Constitutive equation in terms of force and deformation	Mechanics of failure of flesh under probe of different sizes	Yang and Mohsenin (1974)
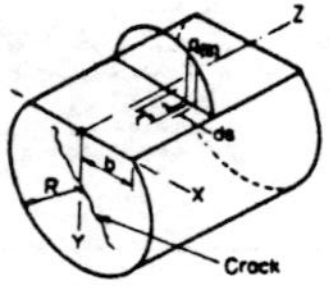	Axisymmetric, elastic, isotropic, homogeneous large deformation. Finite element formulation flat plate loading on potato, apple flesh, peaches	Elastic, nearly incompressible	In potato and peach, failure due to large deformation. In potato, tensile and shear stresses in center cause crack. In apple, bruising in the vicinity of the contact surface. In peaches, some distance below the contact surface due to maximum shear stresses	Sherif (1976)

As seen in Fig. 7.29, such a phenomenon has been reported by these investigators.

Another technique to determine the relationship between the strength of the tissue and the ripening process is a dynamic test, stressing the tissue sinusoidally to determine the storage and loss moduli. These mechanical parameters may then be related to turgidity and mealiness of the fruit as the ripening progresses (Hamann, 1969; Finney, 1971; Patrell *et al.*, 1980).

The initial cell rupture in the tissue causing a discontinuity in the force-deformation curve (see Fig. 7.29) (Mohsenin, 1962; Bourne, 1965) may also be associated with the shearing strength of the tissue. The mechanics of tissue failure at the bioyield point, as discussed earlier, was observed microscopically and recorded by motion picture (Tuckey and Mohsenin, 1966) as the fruit tissue was loaded uniaxially with the skin sliced off. Miles and Rehkugler (1973) also took motion pictures of the apple tissue failure but loaded the specimen triaxially. As the load was applied, a critical loading was reached and the sample failed. At this point a large group of cells, forming a conical pattern, separated from the test of the sample as diagramed in Fig. 7.29. Because of this pattern of failure, Miles and Rehkugler (1973) considered failure as being by shear. It was noted that most of the cells outside and inside the line of fracture appeared undamaged. The load measuring device showed a substantial decrease in the amount of force transmitted through the sample when failure occurred. Holt and Schoorl (1977) stated that the boundary of the bruise, as they observed, was curved rather than conical as reported by Miles and Rehkugler (1973). The curved pattern of tissue failure is what should be expected under uniaxial loading as the author also observed using a microscope and a movie camera (Mohsenin, 1962).

Some of the information in Fig. 7.29 may appear contradictory at the first glance. For example, Voisey and Lyell (1965) reported that the tensile strength of tomato skin is not an indication of its cracking resistance, while Miles *et al.* (1969) pointed out that the skin provides the strength in tomatoes since its failure was caused by tensile hoop stresses expanding the fruit's diameter. The paper by Miles *et al.*, however, does not show how the hoop stresses were determined and whether or not the tensile strength of the tomato skin was actually used in the calculations.

It is interesting to note that failure in fruits may not be associated with cell fracture. In the case of papaya, failure is shown to be manifested through the formation of watery spots in the flesh along the axis of loading at the bottom of the fruit prior to cell fracture (Wang and Chang, 1970).

In the case of apple, distortion of the fruit under dead load in cold storage was found to be rather extensive, but examination of the flesh showed no sign of cell fracture and browning (Mohsenin, 1962).

Characterization of Fruits and Vegetables

Figure 7.29 shows how the investigators have made assumptions as to the nature of the material at the beginning of the tests, and how their findings confirmed or rejected the original assumptions. Based on the information given in Fig. 7.29, the viscoelastic nature of fruits and vegetables has been shown repeatedly. Creep behavior under dead load, stress relaxation behavior under constant strain, and strain rate dependency have been shown by most of the researchers leading to the conclusion that the material is viscoelastic.

On the other hand, in cases where researchers have gone a few steps further and tried different stress levels, or tested specimens taken from different parts of the same fruit, or analyzed strains carefully, they have found their fruits, such as apples, are nonlinear viscoelastic, anisotropic, and show large strains even at small levels of stress. Chappel and Hamann (1968) have shown the dependency of Poisson's ratio on time and stress. A series of intensive tests on one variety of apples over a two-week period at the Pennsylvania State University reported by Morrow *et al.* (1971) included uniaxial quasi-static and dynamic tests, as well as stress relaxation tests. Stress relaxation at various levels of stress indicated the nonlinearity of the viscoelastic behavior. Composite results from four different tests gave a modulus change from about 300 to 2000 psi over a frequency range of 0.5 to 100 Hz (Fig. 7.29). Strain analysis in terms of longrangian, natural, Eularian, finite, and Eurlarian strain tensor showed significant differences between the five methods with strains ranging from 1.6% to 5.3%. This indicated the presence of large strains even for small levels of stress. Sherif (1976) has reported the presence of large strains for potatoes and peaches at the point of failure.

Although fruit materials have been assumed to be homogeneous and isotropic, experiments by Hugg (1967) on potatoes showed change in the failure modulus from the skin to the center (Fig. 7.29). Clevenger and Hamann (1968) demonstrated the anisotropic nature of apple skin by observing that the skin exhibits greater strength in the direction parallel to the core. Currently, experiments are underway at the Pennsylvania State University to test the isotropic or anisotropic nature of the apple flesh in a study involving the mechanics of ripening of the apple fruit.

Methods of Loading

One of the most controversial problems in the study of mechanics of failure of fruits and vegetables has been the method of loading the specimen. Researchers have often taken a cylindrical specimen and subjected it to uniaxial loading. This method, besides the problem of friction at the load bearing surfaces, has been found to be completely unsuitable for testing materials in which the interior structure is stronger than the outer portion. This problem was demonstrated by Segerlind *et al.* (1977) in attempting to develop an appropriate test for the rigidity and hardness of carrots as tested by the consumer. In deforming the sample axially, the center core dominates the reaction to the applied force at a given deformation because it is stiffer and has a higher moisture content than the outer ring. Being less rigid, the outer ring follows the behavior of the center core under the applied load.

One solution to the problem of axial loading is radial loading tried by Segerlind *et al.* (1977) and Sherif (1976) as shown in Fig. 7.29. In the case of the carrots, the compressive force for a given deformation obtained in this manner followed the moisture content much more accurately than that derived from axial loading. Sherif *et al.* (1976) developed the appropriate equation for relating the modulus of elasticity and the total deformation of a food cylinder pressed between two flat plates, consistent with the assumptions in the Hertz contact theory.

The Hertz and Boussinesq contact theories have been used by several investigators in relation to plate and die loading. The Hertz contact problem is used when the convex body of a food is pressed between two parallel plates, or the surface of a spherical fruit is subjected to loading by a rigid die with a spherical end. Loading of a convex body by a rigid flat-ended die is analogous to the Boussinesq problem in mechanics (Finney, 1963). The pressure distribution for these two types of problems is shown in Fig. 7.29. The justification, assumptions and limitations of these theories for application to food materials are given in Chapter 6. It is true that these theories are basically applicable to materials that are ideally elastic. It is also true that food materials are apparently nonlinear viscoelastic and in some cases exhibit large strains. It has been shown, however, by the above investigators and several others that from the practical engineering point of view and for small levels of stress, the Hertz contact solution is a useful technique. It enables expressing the force-deformation relationship for convex bodies in terms of a failure parameter, such as maximum stress to yield or fracture. The extension of this theory to the domains of viscoelasticity, plasticity, and large strains appear to be too complex at this time for applications to food materials (Yang and Mohsenin, 1974).

Stress Distribution and Finite Element Formulation

Even after characterizing the material and finding a well accepted and experimentally proven failure criterion, the problem still remains of how to determine or estimate the stress levels within whole intact fruit of different shapes when it is subjected to static and dynamic loading. This information yields important engineering data which must be obtained before systems for minimizing damage can be designed.

The use of the finite element formulation has shown considerable promise in answering questions of stress distribution within an intact fruit and in eliminating the nature and magnitude of these stresses. Rumsey (1974) and Sherif (1976) applied the finite element technique to the characterization of stress distribution within the intact product using experimental values for cylindrical specimens (Fig. 7.29).

Rumsey (1974) considered an apple as an axisymmetric, spherical shaped body under quasi-static conditions. They assumed that the material exhibits a linear viscoelastic shear and an elastic bulk response. Impact tests were used to compare the elastic solutions with the viscoelastic finite element solutions. It was concluded that, in the dripping of the fruit, the elastic theory in the Hertz problem would be used to calculate the maximum stresses if the impact time were less than ¼ of the material's relaxation time. However, the authors recommended the use of the finite deformation theory in cases where large deformations are expected.

Sherif (1976) used the finite element method to calculate the stress components within cylindrical and semi-spherical specimens of apples, peaches, and potatoes. Two dimensional and axisymmetric computer programs valid for different values or Poisson's ratio and for small and large deformations were developed. Sherif considered that potatoes and peaches do not fail until large strains are attained. In flat plate loading of semi-spherical specimens of apples and potatoes, a tensile stress existed at the circular boundary of the contact region as well as on the axis of symmetry. He made the observation that the reason for the white potato splitting at or near the center when subjected to quasi-static loading might be the existence of maximum tensile and shear stresses in that region. In peaches, Sherif's conclusion was that the maximum shear stress occurs in the axis of symmetry at a point less than a quarter of the contact width below the surface. These findings seem to confirm the mechanical damage occurring in these two products under practical conditions. The analytical approach used by these investigators, strengthened by the application of the finite element formulation, appears to be a basic, sound and promising approach to be followed in developing failure phenomena for solid food materials.

Other Failure Theories

Hamann, in the chapter "Structural Failure in Solid Foods" (Peleg and Bagley, eds., 1983), has taken several rather homogeneous and isotropic foods as a starting point in quantifying and understanding conditions at failure. He has used fairly simple stress and strain theories in solid mechanics and more recent information to determine failure conditions. The foods used in the experiments were protein gel systems, raw fruits and vegetables, butter, margarine, and cheese. The torsion test and analysis cited allowed for a curvilinear moment-displacement relationship, while true stress and true strain in aniaxial tests were employed to compensate for dimensional changes during deformation.

The maximum shear strain theory seemed to apply to the protein gels tested, while maximum normal strain theory appeared more appropriate for raw fruits and vegetables. Turgor pressure in tissue cells was found to be an important factor affecting tissue strength. Torsion tests seemed more sensitive to turgor pressure changes and reduced pressure resulted in an increase in failure strain level. Increased turgor increased the tensile strength of parenchyma tissue, while the opposite effect was observed for axial compression strength. More complex foods, such as red meats, were found to require more intensive studies applying more complex theories and analyses.

Figure 7.29 shows only a sample of the research done in this area attempting to find a criterion applicable to failure of raw fruits and vegetables. Any specific failure criterion may only have application to a specific class of foods. It is interesting to note that the methods and theories shown in Fig. 7.29, those investigated by Hamann, discussed earlier, and others, such as the fracture element models proposed by Peleg (1977) and Drake (1979) have not really considered to structural and biochemical changes which take place upon loading up to the point of failure. It is generally accepted that little progress can be made in this area without consideration of the histological structure and biochemical changes occurring in the sample upon loading. A theoretical structure analysis, involving constitutive equations, was attempted by Akyurt *et al.* (1972). However, it appeared to be too complex and to require much data not presently available and difficult to collect experimentally. The application of the mixture theory or multi-phase interacting media has also been tried. Brusewitz and Stout (1970) employed a reduced case of the mixture theory where the flow of fluids through a porous deformable solid is considered. Working with the potato, considered to be a mixture of solid, liquid, and gas, these investigators attempted to model the plant material as a mixture of deformable solid (basically the cell wall) and incompressible fluid (cellular liquid). Although some experiments were conducted, the reported work does not show the relationship between the experimental data and the work structural analysis proposed by the authors. The conclusions indicated only the potential for consideration of stress-strain behavior according to the interacting media and the mixture theory. The author's work also considered using several of the theories of heterogeneous materials, but it was concluded that the development of the theory, even for engineering materials, has not yet reached a point where it can be applied to the more complex biomaterials. The only engineering approach which was taken to relate the structural and biochemical factors to the mechanical behavior of the fruit is that reported by Lazada (1974). In that work, the bioyield stress and the apparent compressive modulus of apple cortex were related to the cell size, intercellular space size and volume, and alcohol insoluble solids.

8

MECHANICAL DAMAGE

8.1 ECONOMIC IMPORTANCE

One by-product of mechanization in production and handling of agricultural products has been mechanical damage to the crop during harvesting and subsequent handling as well as on-the-farm and off-the-farm processing. It has been found that some varieties of crops are more susceptible to mechanical damage than others. This offers an opportunity for plant breeders to develop new varieties of products which would withstand the mechanical forces imposed during mechanical handling. Such new varieties have been developed for some seeds and such vegetables as tomatoes.

Seeds and grains

In mechanical handling of seeds and grains, most of the damage occurs in the threshing as well as mechanical conveying by screw conveyors and other equipment. Damage is the failure of the product under either excessive deformation when it is forced through fixed clearances or excessive force when it is subjected to impact. Damage to the kernels of seeds and grains could affect their milling quality, result in greater losses in sifting, and would lower the germination capacity and seedling development. Mechanical damage to the hull of some products increases the tendency toward mold development in storage. One example is peanuts.

The damage resulting from the impact forces is usually referred to as cracking which may vary from complete splitting of the kernels to small hairline cracks invisible to the naked eye. Even small cracks in the seed coat may allow soil bacteria to enter the seed and destroy the food supply before a plant is established. In cottonseed, rupture of the seed coat permits a rise in free fatty acid content which results in low germination and reduced high-quality oil refined during processing.

The factors affecting the extent of grain damage are history of the crop involving variety, stage of maturity, previous storage and drying conditions, and moisture content of the seed. In addition, design and operating characteristics of the machine which subjects the kernels to certain impact forces at certain rates for a certain duration also influence grain damage. (See also "Symposium on Grain Damage, 1968.)

In single-stage threshing of wheat four types of damage to the embryo were detected. These types of damage together with their germinative energy are given in Table 8.1. As shown in this table, even minor damage sharply depresses the seed's germinative energy. More serious damage leads to a loss in vitality of growth and development of the seedling (see Fig. 1.2). In certain crops, such as cotton and sugar beet, this can cause quite a problem in mechanized operations if fast and uniform emergence of planted seeds cannot be assured.

Mechanical damage to pea beans with the outer cover cracked is a problem faced by the pea bean processing industry. This damage causes the skins to separate from the beans and lower the quality of the product. The extent of this damage has been estimated to be between 2 and 10 percent. Its cause is claimed to be mechanical handling. Susceptibility of beans to this type of damage, like many other seeds and grains, depends primarily upon the moisture content of the crop during harvest and subsequent handling.

Table 8.1 Types of damage to embryo of wheat grain and the resulting depression in germinating energy (Summarized from Kolganov, 1956)

Nature of damage to embryo	Per cent of damaged seed	Per cent emergence	Weight of 100 plants (grams)	
			Top growth	Root system
Whole grain (control)	0	98	6.3	13.3
Embryo knocked out	7	0	0	0
Embryo damaged	31	43	3.5	7.0
Skin over embryo torn off	42	56	3.7	8.5
Skin cracks in vicinity of embryo	20	69	4.3	9.1

Mechanical damage to some oil-bearing seeds such as castor seed during harvesting, hulling, handling and cleaning tend to cause a buildup of oil bearing fragments in critical areas in the mechanical equipment. This may cause eventual clogging together with further increase in seed damage.

Fruits and vegetables

Contrary to the general belief, the processors are just as much concerned about mechanical injury to fruits which they receive as the retail stores selling the fresh fruits. Direct costs to the processor include loss of fruit flesh being trimmed and increased labor required for removing the bruised sections. Indirect costs include a general slowing of the entire processing line as well as lowering the quality of finished products.

Field studies conducted in Pennsylvania (Wennergren and Lee, 1961) to determine economic aspects of bruising in the apple processing industry revealed that (1) bruised flesh represented 2.8 percent of the total weight of the apples, (2) removal of objectionable bruise required five trimmers for each peeler in the processing line, (3) trimmers removed 69 percent of the total bruises in the production of apples slices and 21 percent in the production of applesauce, and (4) in slice production labor represented 85 percent of the material and labor cost resulting from bruising as compared to 67 percent for sauce production.

Transit losses in fruits due to vibration have been reported to be up to 10 percent (O'Brien *et al.*, 1960). This extensive injury was not recognizable, however, until the 1000-pound bulk bins were introduced and the resulting injuries were compared with those in lug boxes.

Brown spot in pears is an internal bruising of tissues due to impact which does not extent to the surface. Like apple, after a period of storage, the bruised flesh begins to dry out and becomes corky. Not being visible on the surface, the damaged pears cannot be sorted out before processing. Removing of discolored flesh during processing, in addition to being an expensive process, yields lower canning grades of the trimmed sections. Tests with Bartlett pears (Mattus *et al.*, 1960) have shown that drop heights exceeding six inches on a hard surface produced internal bruises which developed into brown spots (see Fig. 8.1).

With the advent of mechanical harvesting of red tart cherries, like many other tender fruits and vegetables, considerable attention is being focused on bruise damage associated with mechanical handling. Bruised cherries result in cherry scald which is the movement of red pigment from the skin to the flesh. Whittenberger (1958) measured the distances that cherries

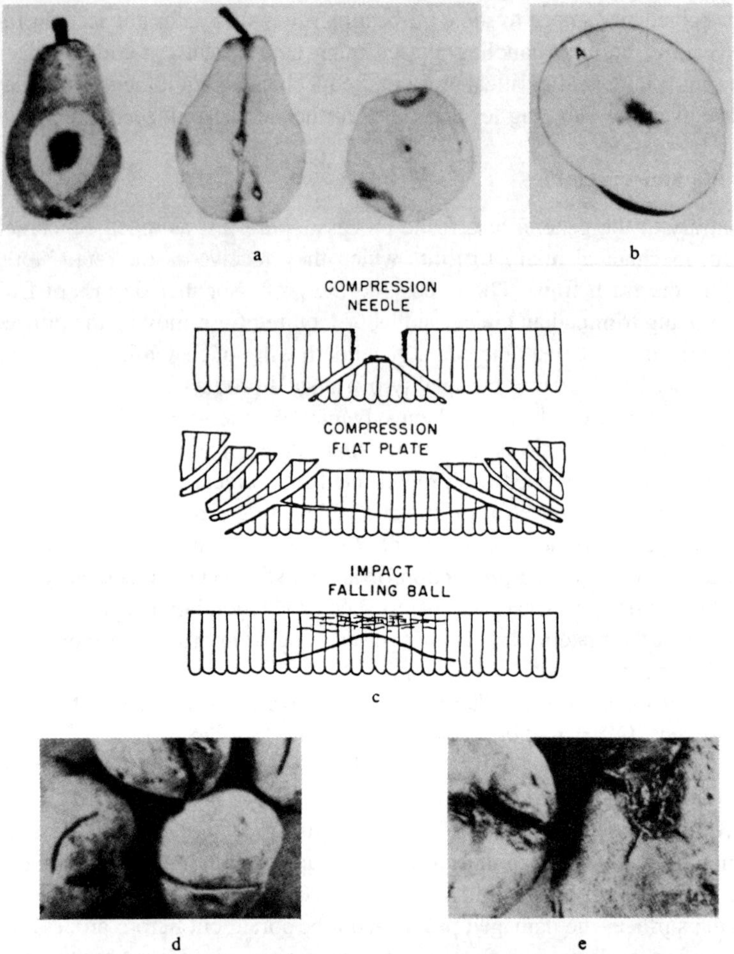

Figure 8.1 Forms of mechanical damage: a—brown spot in pear (Mattus *et al.*, 1960); b—black spot in potato (Finney, 1964); c—damage to eggshell shown in radial section of crystal structure (Tyler and Moore, 1965); d—air cracks in potato (Donaldson and McFeeley, 1951); e—cuts and bruises in potato (Donaldson and McFeeley, 1951)

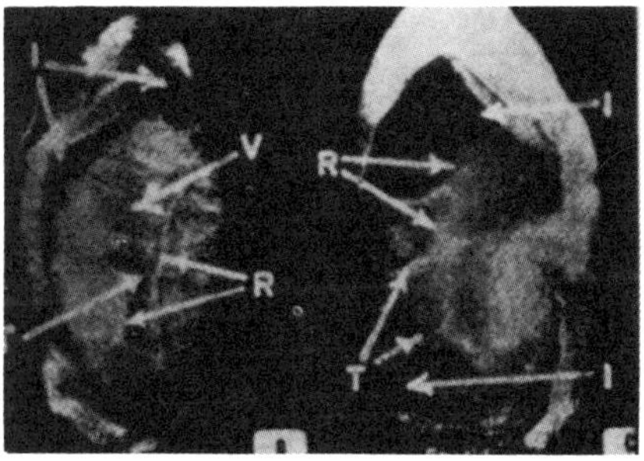

f

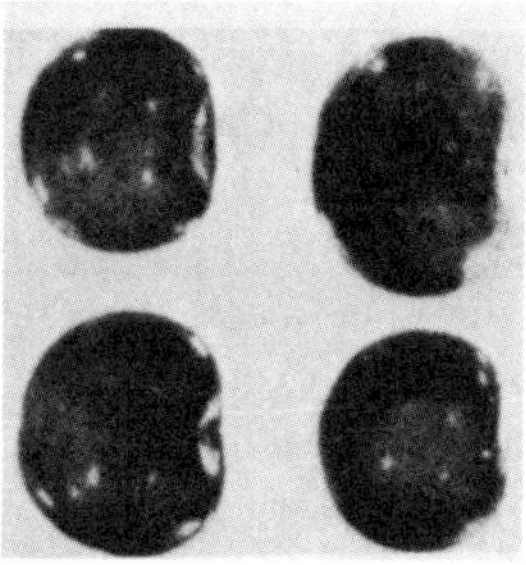

g

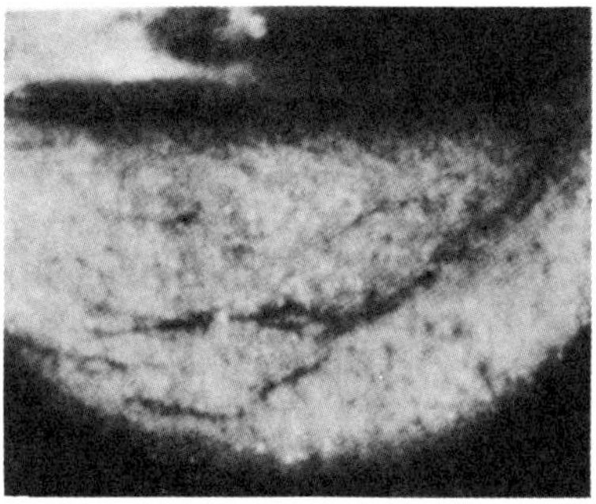

h

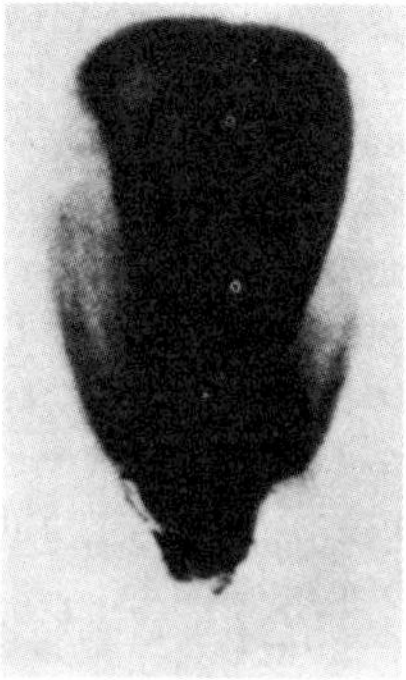

i

Figure 8.1 (continued) Forms of mechanical damage: f—stress crack in wheat (Grosh and Milner, 1959); g—scald in cherries resulted from bruising (Dekazos, 1966); h—initial cell rupture in apple tissue at bioyield point; i—stress crack in corn (Thompson and Foster, 1963) fall, drop, or tumble on the processing line and found that cherries were dropped an average cumulative total of 22.8 feet from the delivery truck to the final container. The pitter efficiency, product yield, and the quality of processed cherries are affected by the amount of bruising the fruit receives during handling from the orchard to the processing line. Product yields is defined as the ratio of the weight of cherries before and after pitting also expressed as a percent. Water handling of cherries in recent years and soaking in chilled water at the processing plant has produced firmer fruits, as discussed in Chapter 7, resulting in higher pitter efficiency and product yield. Loss or gain of weight when cherries are soaked in water, the effect of bruising on cherry scald, and the yield of pitted cherries are illustrated in Table 8.2. As seen, the effect of bruising is cumulative. As the number of recurrent bruising increased, the cherries decreased in fresh weight and in yield of pitted product. Maximum yield of pitted cherries was obtained from unbruised cherries in water at 50°F. Therefore, loss of weight due to bruising becomes of economic importance to both the grower and the processor.

Tennes *et al.*, (1967) have reported on the effect of bruise level on product yield and pitter efficiency of red tart cherries as related to fruit firmness, soluble solids, soak time, and temperature of soak water, and temperature of the cherry at harvest. Bruise levels consisted of four levels of 0-*x*, 1-*x*, 2-*x* and 3-*x*, where *x* referred to the number of 30-inch free fall drops of a cherry on a hard surface with 0-*x* being the check sample or unbruised cherry. The results showed that product yield is related to a combination of bruise level and soak time (Fig. 7.19). The 1-*x* bruise level

Table 8.2 Effect of recurrent bruising on scald, change in weight, and product yield of red cherries held for 24 hours in 50°F water before processing (Wittenberger *et al.*, 1964)

Treatment (Time and number of bruises)	Scald (%)	Change in weight of whole cherries (%)	Yield of pitted cherries (%)
Control, not bruised	0	+4.2	88.7
Bruised at 0 hr. (harvest)	0	−0.5	85.6
Bruised at 0 and 3 hrs.	34	−5.3	80.2
Bruised at 0, 3, and 6 hrs.	76	−7.9	77.2
Bruised at 0, 3, 6, and 24 hrs.	74	−8.6	74.0

and 5-hours soak time at 54°F resulted in optimum values of product yield of tart cherries. It was stated that apparently a small amount of bruising is necessary to give the best pitting results as excessive firmness may be as undesirable as excessive bruising.

Onions, which are generally piled in storage from wall to wall to a depth of 8 to 12 feet, resting on slatted wooden floors, are subjected to distortion in the bottom two feet of the pile. According to some growers "hard" onions store better than "soft" ones. Ang *et al.,* (1960) have used the concept of force to deformation ratio (F/D) as a criterion for selection of hard onion bulbs. Since planting experiments showed that compression tests of the bulbs, unless they broke, did not impair their seed producing ability, the method was recommended for onion breeders for the selection of firm bulbs.

Volbracht and Kuhnke (1956) reported that during the storage period about 10 to 12 percent of potatoes are lost through shrinkage or rotting. A major part of this loss is said to be attributed to mechanical damage. In addition to visible internal or external damages which would lower the eating quality of the potato, shock and impact during mechanical handling can also stimulate sprouting which would reduce viability of the tubers as seed potato (Klapp, 1945). Also, as in the case of other agricultural commodities, any opening on the surface of the tubers would encourage the entrance of microorganisms, fermentation, and in severe cases, eventual spoilage of the potatoes. If open wounds are moderate, formation of wound periderm will close the wound and further advance of infection will be prevented. Ambient conditions such as relative humidity, temperature, and illumination are all factors influencing the life cycle of the infecting organism or the resistance of the host. For example, in the case of flesh wounds, the presence of oxygen and optimum temperature and relative humidity encourage the formation of a layer of wound periderm below the damaged cell thus sealing off the wound. In the absence of these optimum conditions, formation of wound periderm is less likely to occur.

In a study conducted on 13 farms in the Red River Valley of Minnesota and North Dakota, Nylund *et al.*, (1955) analyzed mechanical damage due to field harvesting and subsequent handling operations as follows:

Digging	10.1%
Picking into baskets	0.5%
Filling field sacks	4.9%
Loading and hauling to warehouse	5.7%
Unloading in storage	5.2%
Total	26.4%

The damage was evaluated as cuts, bruises, and cracks computed as percent of total weight selected. Skinning was another major type of damage which was estimated as 15 percent of the skin removed by the time the potatoes reached storage bins. Although of the 26.4 percent total damage only 4.5 percent were serious enough at the beginning of storage to score against grade, at the end of a six-month storage at 36°–40°F, bruises which decreased grade increased from 5.9 to 9.2 percent. In addition to the above injuries, warehouse operations in dry grading resulted in 11.8 percent mechanical injury before the potatoes were sacked. Wet-grading, in which washed potatoes were handled, resulted in only 3 percent injury.

In a more recent study in six fields in California, Zahara *et al.*, (1961) found that as much as 40 to 50 percent of potatoes were damaged before reaching the packing shed. Tubers were stored in a shed for a week before they were examined for injuries. Cracks between ½ inch and 1½ inches long and those with "medium" sized bruises were classified as moderate injury. Tubers with cracks longer than 1½ inches and those with "severe" bruises were classified as severely injured. Table 8.3 shows cumulative percentages of injuries at different stages of harvesting.

In another study in Sweden (Loow, 1964) data from 60 major growers in the country showed that about 50 percent of the damage occurred with the operations in the field, 20 percent during storage and packing, and 30 percent during distribution. Severe damage, referred to injuries greater than 10 percent of total potato's length or surface area was found in about 16 percent of the potatoes handled.

Statistics are available for many agricultural products to illustrate the losses involved as a result of mechanical injuries. For example, the breakage of eggs in the United States during processing was reported approximately 3 percent at a cost of $25 million in 1966 (James and Retzer, 1967).

8.2 CAUSES OF MECHANICAL DAMAGE

Mechanical damage in agricultural products are due either to external forces under static or dynamic conditions or internal forces. Examples of mechanical damage due to external forces are mechanical injuries in fruits and vegetables, seeds, grains, eggshell, bones in dressed poultry, etc. Mechanical damage due to internal forces can be the result of physical changes such as variation in temperature and moisture content or chemical and biological changes. Examples for this type of damage are stress cracks in

Table 8.3 Cumulative percentages of potato tubers damaged at the different stages of harvesting in six fields in California (Zahara *et al.*, 1961)

Operation	Injury		Field					
			I	II	III	IV	V	VI
Digging	Moderate		6	18	29	18	29	17
	Severe		12	10	15	2	1	3
		Total	18	28	44	20	30	20
Picking	Moderate		10	24	50			
	Severe		12	16	5			
		Total	22	40	55			
Loading	Moderate		17	28	52	17	29	56
	Severe		15	1·5	8	5	7	3
		Total	32	43	60	22	36	59
At shed	Moderate		18	26	61	18	54	50
	Severe		30	22	4	6	10	7
	Total		48	48	65	24	64	57

corn and rice, thermal checking of eggshell, and skin cracks in sweet cherries, tomato and sweet potato.

How mechanical damage occurs is a problem which has not yet been investigated. In fact, even the failure of engineering materials under more complicated stress conditions has only been investigated in a few exceptional and more common cases. In regards to agricultural products, a further complication arises from the fact that no satisfactory criterion of failure is available for these materials. The limited information available is based on arbitrary standards which have been developed on the basis of a number of subjective tests rather than well-defined physical measurements.

Failure in engineering materials may occur as a result of excessive elastic deformation, inelastic deformation (yielding), and rupture of the material. In intact agricultural products failure is usually manifested through a rupture in the internal or external cellular structure of the material. Even such terms as a bioyield point in fruits and vegetables, as pointed out in Chapter 4, is the initial cell rupture in the cellular structure under the skin. Actually, even in the case of metals, today, the concept of yield point is being questioned. It is said that the detection of an abrupt change in the stress-strain curve of the material is only an indication of the precision with which the mechanical behavior of the material is being determined.

Some of the more common theories developed for failure of engineering materials are the maximum stress theory, the maximum strain theory, the maximum shear theory, and the maximum energy theory. To apply these theories to any material, it is essential to specify first the principal stresses in the body. The maximum stress theory considers either the maximum principal stress (in the case of simple tension) or the minimum principal stress (in the case of simple compression) as the criterion for strength. In the maximum strain theory either the maximum principal strain (tension) or the minimum principal strain (compression) are taken as the criterion for strength. The maximum shear theory assumes that failure begins when the maximum shear stress in the material becomes equal to one-half of the maximum stress at the yield point (in the case of ductile materials) in a simple tension test. Finally in the maximum energy theory the strain energy per unit volume of the material is used as the criterion for failure. For further details on these failure theories, textbooks such as Timoshenko (1956) and Seely and Smith (1955) should be consulted.

In order to determine the maximum allowable stresses for various agricultural products under the stress conditions which occur in practical conditions, one must first know the mechanical properties of these materials under simple stress conditions such as tension, compression and shear as a result of static or dynamic loading. Knowing the behavior under simple stress conditions, one may then apply one of the theories of failure to predict when failure will occur under combined stress conditions. This, as will be discussed later, has been attempted by Hamann (1967) and Hammerle (1967, 1968) with some degree of success.

8.3 BIOLOGICAL AND CHEMICAL REACTIONS FOLLOWING MECHANICAL DAMAGE

The course of all infectious diseases consists of infection, incubation, and then healing and rehabilitation if the infected part is survived. In the absence of a wound (puncture, crack, etc.) successful infection of the host organism seldom occurs. There are certain types of fungi which secrete toxins that penetrate, disrupt and kill the host cells opening the way for penetration of the fungus. Therefore, it is important to prevent injuries that expose the susceptible internal tissues of plant and animal materials. The degree of infection of mechanically damaged products depends on the nature of the attacking fungi present and the resistance of the product to the invading organism after infection has occurred.

Caldwell and Davies (1957), who studied the development of pathogenic fungi in cereal grains, found that most molds show no signs of penetrating through the living aleurone layer of undamaged wheat. A break in the testa, the aleurone layer, or both, however, provides a point of entry. Such breakage would not necessarily be externally visible. This occurs particularly in grain harvested at high moisture contents. The tentative conclusion was that dry grain being brittle, breaks easily, thus providing a readily available spot for mold growth. Damp grain may suffer compression during which the protective tissues are damaged. Reducing the moisture content of grain to 14 percent (wet basis) would stop the activity of most fungi on grain, though they persist as inactive spores in some cases for a long time. For long-term storage, a moisture level of below 13 percent was recommended.

Mechanical injuries in the form of cutting, peeling, and bruising which result in the crushing of tissues in fresh fruits and vegetables under load produce a darkening of the tissues called browning. The chemistry of browning, despite extensive research, is not yet fully understood. It is known, however, that enzymes are involved in certain types of browning and not in others. Therefore, there are two types of browning. Enzymatic browning, which occurs when injured tissue is exposed to air. Examples are browning of apple, pear, peaches, apricot, banana, cherry, grape, and strawberry. The air supply for enzymatic browning is provided either by intercellular air such as the case of internal browning of apples, or direct exposure to atmosphere. Nonenzymatic browning occurs in processed foods such as fruit juices and in dried fruits. For a discussion of the chemistry of browning, the reader is referred to Meyer (1960).

In potatoes there are two types of internal damage which have by some researchers been attributed to the action of static and dynamic forces in mechanical handling. These internal injuries are black heart and black spot. Finney (1963) found that compression of the whole potato tuber produced an internal injury very similar to black heart. Limited experiments in which tubers were subjected to parallel plate loading until internal cracking was heard, verified Finney's observation. Although no external evidence of injury could be detected on the compressed tubers, cutting of the specimens after a few days showed the breakdown and discoloration of the tissues in the center of the potatoes.

It has been shown that black heart is caused by the deficiency of oxygen in the interior of the potato. It is postulated that crushing of the tissues in the center of potato creates a condition which accelerates cellular respiration and demand for oxygen. It is, therefore, deficiency of oxygen which produces the dark color of the damaged tissues. The fact that discoloration did not occur in specimens cut immediately after compression should support this postulation.

"Black spot" has been observed in potatoes subjected to drop tests (Hopkins, 1953). The affected tissues turned black within 24 hours. In some cases no external evidence of injury could be detected whereas in others a sunken area over the injury was developed due to drying out of the impacted tissues. According to Wiant *et al.*, (1951) black spots in potato is due to mechanical injury to tissues by impact. The grayish black discoloration is the result of formation of melanine which is a black pigment. The reaction involved is believed to be the fermentative oxidation of the tyrosine, contained in the cell sap, to melanine (Talburt and Smith, 1967).

8.4 TERMS AND DEFINITIONS RELATED TO EXTERNAL DAMAGE

The following terms and definitions related to mechanical damage of fruits and vegetables were compiled by the Northeast Regional Research NE-44 Technical Committee. Some of the terms are taken from the publication developed in connection with the NE-44 Project (Mohsenin, ed., 1965). Others were contributed by the members of the Technical Committee but were not given in the publication.

Abrasion abrasive injury, usually occurred in sweet potato and Irish potato, may vary in severity from separation of the periderm or skin to removal of part or all of the pericyclic cortex. Skin abrasions are difficult to detect at time of harvest, but can readily be observed after storage for a week or more at low humidities.

Bruising damage to plant tissue by external forces causing physical change in texture and/or eventual chemical alteration of color, flavor, and texture. Bruising does not break the skin.

Distortion a change in shape of an intact fruit or vegetable which is not characteristic for the variety.

Crack a cleavage without complete separation of the parts.

Cut penetration or division by the sharp edge of an object.

Puncture a small hole or wound on the surface of the fruit made by a pointed object or stem of other fruits.

Shatter cracks one or more tortuous cracks radiating from a point of impact.

Skin break fracture of the periderm or a crack limited to the skin.

Skinning and feathering the separation of periderm from the plant part by scraping, rubbing, etc. Feathering means the same except the separated periderm is still attached to the unseparated periderm.

Split division or separation into parts.

Stem end tearing skin break caused by separation of stem from the fruit.

Swell-cracking cracking due to uptake of water by osmotic pressure.

8.5 DETECTION AND EVALUATION OF MECHANICAL DAMAGE

Detection and evaluation of mechanical damage, particularly if damage is invisible, can become a problem requiring special techniques and instrumentation. The usual methods of evaluation are primarily descriptive. For example, mechanical injury to apples are given by such descriptive terms as hail damage, limb rub, stem puncture, cracks, and bruises. The USDA classification for visible bruises in apples is as follows:

Extra Fancy Firm, flat, shallow bruises only (not over ⅛ inch to deepest point). One bruise ½ inch in diameter or several bruises each less than ½ inch in diameter, and total area of which does not exceed ¾ inch in diameter.

Fancy Firm, flat shallow bruises only (not over 3/16 inch to deepest point). One bruise ¾ inch in diameter or several bruises each less than ¾ inch in diameter, and total area of which does not exceed 1 inch in diameter.

Utility Shallow bruises (not over ⅜ inch to deepest point). One bruise ⅞ inch in diameter or several bruises each less than ⅞ inch in diameter, and total area of which does not exceed 1¼ inch in diameter.

Cull Any apple with a bruise over 1 inch in diameter, or more than ¼ inch deep, or with broken skin, or soft, or with total area of bruises each less than 1 inch in diameter and total area of which exceeds 1¼ inches in diameter.

less than 1 inch in diameter and total area of which exceeds 1-1/4 inches in diameter.

Typical of external symptoms for mechanical damage in tomatoes, as described by McColloch (1962), are fruits being permanently pressed out of shape with flattened sides, *v*-shaped or rounded surface indentations. Internal damage, which invariably accompanied external bruise marks, were water-soaked tissues with signs of varying degrees of damage to the locular tissue surrounding the seeds (See tomato structure, Chapter 2). In mature green tomatoes, the water-soaked condition disappeared during ripening, but serious internal damage remained. Permanent damage to locular tissues resulted in failure of the tissue to change to the typical gelatinous mass of cells that accompany normal ripening. In less damaged locules, the development of the gelatinous tissue was incomplete and had a thick, sometimes stringy, consistency darker than normal. Upon cutting the damaged section showed a bleached, hollow area which resulted from loss of moisture by the crushed tissues.

In addition to size and number, the age of the bruise also affects the quality of the fruit. Older bruises have either darker color which does not "blanch" out in the cooking process or have developed into corkiness. Both of these defects indicate non-usable fruit flesh and must be removed before processing. The fruit variety and the type of process for which the product is being used also have a bearing on evaluation of damage. Therefore, in the absence of accepted standards, it is the trimmer's decision whether a bruise is objectionable or not.

In the absence of any standard, researchers have often designed their own scale for detection and evaluation of mechanical damage. The methods vary from those which are purely subjective, such as minor, moderate and major or severe damage, to those which are based on some physical measurements such as the volume of bruised tissues in fruits and vegetables. In evaluating bruising of apples many workers have expressed bruise damage as mean diameter of bruised area. Austin and Dedolph (1962), however, found that depth and consequently volume of bruised tissue may vary considerably with location of the bruise in some varieties. They concluded that in studies of apple bruising, damage should be estimated by both diameter and depth of bruises. The volume may be calculated using the formula for a segment of a sphere together with measured values of maximum depth and maximum width of the bruise [See Eq. (8.26)].

In many cases injury to commodities cannot be detected by subjective visual methods. The invisible damage may cause either internal defects, such as black spot in fruits and vegetables, or damage to the embryo and lower the quality of the product as seed. In such cases staining to bring out the damaged section (such as *p*-cresol for potato tissues), *x*-ray, light transmittance, and other nondestructive techniques may be used (Johnson, 1962; Birth, 1960; McMaster, 1959; Chung, 1968). Figure 8.2 shows the application of a light-absorption difference method for detecting external and internal damage to shelled corn. Other methods which may have possibilities but their potentials have not yet been explored are wave propagation techniques such as ultrasonic and pulse techniques illustrated in Table 7.2, Chapter 7, and electrical impedance technique, briefly discussed in Chapter 1 (Fig. 8.4). Sonic techniques are used extensively in detecting flaws in fabricated parts or raw engineering materials as well as such biological materials as bones, etc. Preliminary experiments with electrical impedance techniques have indicated that this method has possibilities for detection of bruised tissues in fruits such as apples. Table 8.4 shows damage measurement techniques as summarized by Agness (1968) for corn kernels.

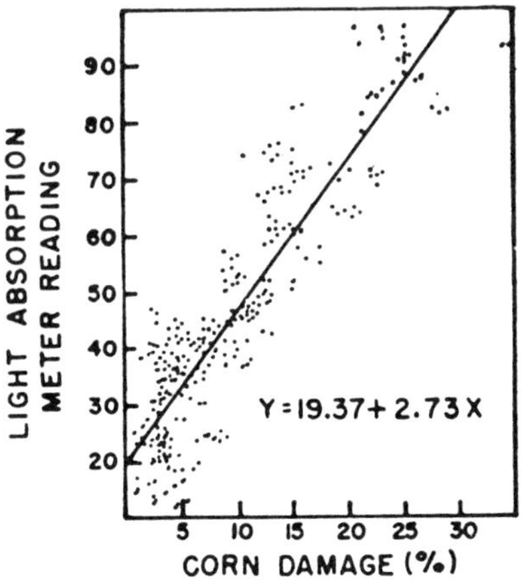

Figure 8.2 Corn damage detected by the method of light absorption (Johnson, 1962)

Table 8.4 Damage measurement techniques for corn kernels (Agness, 1968)

Test description	Functional principle	Change measured	Test mode	Principal advantage	Principal limitations
Carbon dioxide evolution in a sealed system	Damaged kernel support more biological activity	Biological	Bulk	Measured actual deterioration	Very dependent on moisture, temperature and oxygen supply
Temperature rise in an insulated container due to heat evolution	Damaged kernels support more biological activity	Biological	Bulk	Measured actual deterioration	Very dependent on moisture, temperature and oxygen supply
Water absorption by damaged kernels	Damaged seed coat is more permeable to water	Physical	Bulk	No expensive equipment	Differences small compared to system errors
Light absorption by water extract from damaged kernels	Damaged seed coat is more permeable to nutrients	Physical	Bulk	Preferential to germ damage	Differences small compared to system errors
Fines created by standard handling treatment	Quality loss is proportional to structural weakness	Physical	Bulk	Physical breakage is most universal damage problem	Moisture content is critical
Visual separation of damaged kernels	Important damage is visible	Physical	Discrete	Can find even one bad kernel	Human judgment and fatigue
Germination loss due to physical damage	Quality is proportional to viability	Biological	Discrete	Directly measures germ damage	Time consuming
Mold growth in a controlled atmosphere	Damaged kernels support more biological activity	Biological	Discrete	Measures actual deterioration	Very dependent on moisture, temperature and time

Research has shown that in such cases as seeds intended for planting, germination test is the surest way for detecting percent damage, even though there are factors other than mechanical damage affecting viability of seeds. The seed structure and the location as well as protection provided for the embryo varies with different seeds. As has been shown in the case of cottonseed, visually damaged seeds may not necessarily have damaged embryo and vice versa (Clark *et al.*, 1967).

Bulk density has also been used as a quantitative measure of bruising in red tart cherries (LaBelle and Moyer, 1960). Bruising reduces resistance to compression of the individual cherry resulting in greater deformation when cherries are bulked and greater bulk density.

Timofeev (1956) described a method for assessing mechanical damage to fruits and vegetables resulting from impact. The method consisted of replacing the qualitative symbol of a damage by its numerical mechanical equivalent which was derived directly from impact experiments. The mechanical parameters considered were potential energy, energy consumed, and rebound of the impacting product. The qualitative evaluation was based on an arbitrary scale classifying damage into skin damage, bruising, splitting, etc. Energy consumed, E_{ab}, was represented by

$$E_{ab} = (1 - e^2)\, WH \tag{8.1}$$

where e is the coefficient of restitution and W and H are, respectively, weight and height of drop of the impacting fruit or vegetable. Incorporating a weighting factor for each type of the qualitatively assessed damage, an average degree of damage could be computed for the lot of the commodity under consideration. Using data for potato damage, the method was illustrated to evaluate two types of potato harvesters.

Another quantitative method for evaluating mechanical damage is measuring the rate of respiration by the damaged tissue in fruits and vegetables. It has been established that wounding of living plant tissue results in an increased respiration rate (Stiles and Leach, 1961). Respiration rate is expressed usually in terms of volume of CO_2 evolved per unit weight of product per unit time. The reaction equation for glucose sugar, for example, is as follows:

$$C_6H_{12}O_6 + 6\ O_2 \rightarrow 6\ CO_2 + 6\ H_2O + \text{Heat}$$

An established method for measuring the respiratory CO_2 is by infrared gas analysis (IRGA) which involves passing the exhaust air from the living plant tissue through an IRGA unit. Apparatus can be set up to obtain a continuous record of CO_2 evolution versus time. The application of respiration rate technique for measuring mechanical damage to sweet potato roots and shelled corn are shown in Fig. 8.3. The sharply increased rate of CO_2 evolution observed in the case of sweet potato was stated to be in response to the mechanical injury and not as a result of reduced resistance to gaseous exchange.

Since bruising is directly proportional to respiration rate and removal of oxygen from the environment, addition of oxygen has been suggested as a method for reducing scald in bruised red tart cherries when they are dumped into water tanks prior to processing. It has been found that if oxygen level is dropped below two parts per million in the water where the cherries are held, scald development would accelerate on Montmorency cherries (Dekazos, 1966). An aeration system to maintain the level of oxygen above two parts per million should keep scald development at a minimum.

8.6 IMPACT DAMAGE

One of the most common causes of mechanical damage to agricultural products is shock and impact during mechanical handling. Before we consider other causes of mechanical damage, the mechanics of impact and its implications to agricultural products will be discussed.

Mechanics of impact

The concept of impact is differentiated from the case of static rapid loading by the fact that the forces created by the collision are exerted and removed in a very short period of time (duration of impact) and that the collision produces stress waves which travel away from the region of contact. It should be pointed out that, to date, no general impact theory has been developed. The bases of current theories were laid by St. Venant who proposed the wave theory, and Hertz who introduced the contact phenomenon for elastic bodies (See Chapter 6).

St. Venant suggested that the total period of collision is determined by the time required for an elastic compression wave to travel through the solid and be back again. According to this theory the total period of collision will be given by $2\,L/V$, where L is the length of the body and V is the velocity of sound through the body. Recent experiments, however, have shown that for relatively small bodies the collision period is much longer

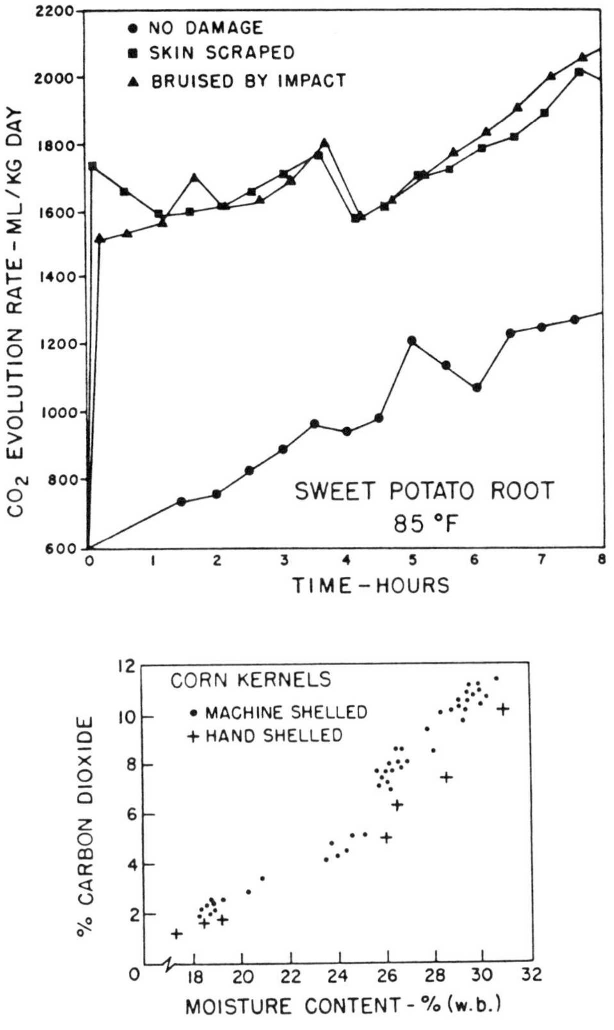

Figure 8.3 Application of the rate of respiration technique for evaluation of mechanical damage to sweet potato (Fluck and Splinter, 1966) and corn kernels (Waelti and Buchele, 1967)

and depends primarily on the deformation occurring at the region of contact. Apparently during the deformation process there is sufficient time for elastic waves to travel to and fro several times before they are dissipated throughout the colliding bodies.

Analysis of numerous experiments have shown that the Hertz theory of contact provides a good description of the collision of two spheres or the impact of a sphere against a thick plate only if the materials are hard and the initial velocity is low (Goldsmith, 1960). For soft materials or higher impact velocities, the Hertz theory must be replaced by an analysis accounting for plastic deformation. In this regard, the hydrodynamic theory and the theory of plastic strain propagation have been considered (See Goldsmith, 1960). There are many uncertainties that still exist concerning the actual contact mechanism and the use of Hertz law during impact. For example, as the first step in the analysis of an impact problem, the deformations are considered elastic and the elastic equations based on Hertz law are employed. Next a plastic yield stress is used to describe the force acting on the body (Tabor, 1951). Despite the inconsistencies, the relative simplicity of elastic solutions, using the Hertz law, and the fact that the method provides good correlation with experimental results have been the main reasons for extensive use of this approach.

Four phases of impact

Bowden and Tabor (1954) divided the impact of colliding bodies into four phases.

1 Initial elastic deformation during which the region of contact will be deformed elastically and will recover fully without residual deformation. The time of impact, mean pressures and deformation for this phase can be found using equations based on the Hertz theory.

2 Onset of plastic deformation during which the mean pressure exceeds the dynamic yield pressure of the material and the resulting deformation will not be fully recovered. It has been shown that this condition occurs at extremely small impact energies (Tabor, 1951). The fact that the indentation produced by a spherical indenter is partially elastic at this stage has been proved by applying the original load several times and observing the recovered indentation which remained essentially unchanged in diameter and in radius of curvature (Tabor, 1948).

3 Full plastic deformation during which the deformation continues from elastic-plastic to fully plastic until the pressure falls below the dynamic yield pressure.

4 Elastic rebound during which a release of elastic stresses stored in both bodies takes place.

These four phases of impact have been demonstrated through numerous collision experiments by dropping spheres of hard and soft materials against another sphere or against flat surfaces (Andrews, 1931; Tabor, 1948; Bowden and Tabor, 1954). Figure 8.4a shows typical oscillograph traces obtained by measuring the electrical conductance between the colliding metal surfaces during their collision. The striking sphere can be freely suspended by a fine wire to complete the electrical circuit. As the bodies come together during the collision, the circuit resistance decreases and conductance increases to a maximum and then falls off as they come apart. Using this technique, Bowden and Tabor (1954) measured the areas of contact of the colliding bodies during a collision. The almost perfectly symmetrical conductance curve of the hard steel ball in Fig. 8.4a indicates that the collision is essentially elastic. The nonsymmetric conductance curve of the mild steel sphere, however, shows plastic deformation in the striking bodies. The part *ACD* gives the time for approach of the surfaces (time of deformation) while part *DEB* gives the time for separation of surfaces (time of restitution). The four phases of impact are clearly demonstrated in the trace for mild steel. At the point *A* contact begins and the trace *AC* corresponds to the plastic phase while *DEB* represents the elastic rebound. Andrew

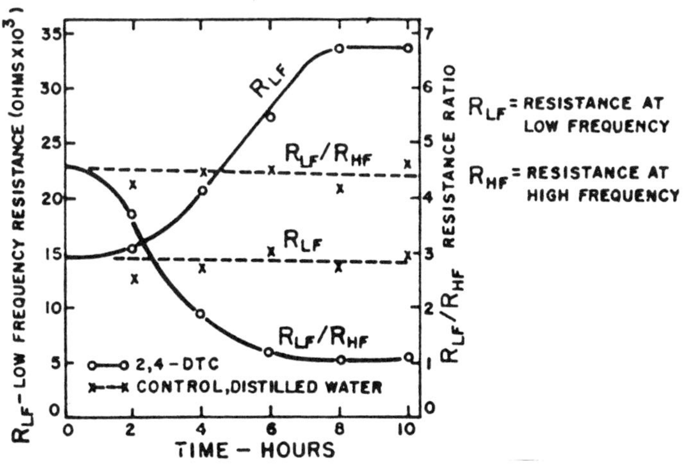

Figure 8.4 Rate of injury to plant tissues can be detected from changes in electrical properties 2,4-DTC=2,4-dichlorophenoxymethyl triethylammonium chloride (Greenham and Cole, 1950)

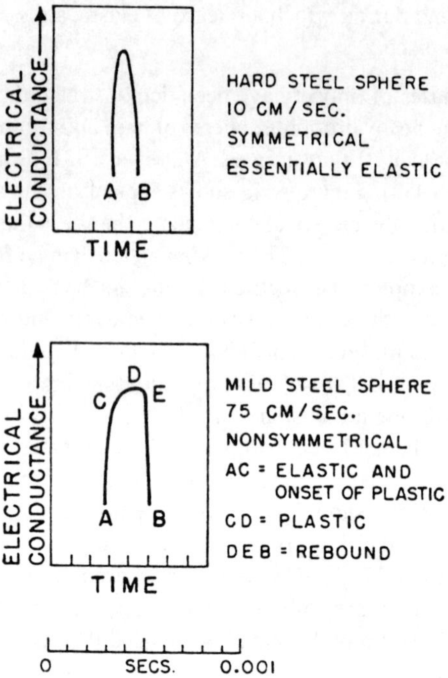

Figure 8.4a Four stages of impact of colliding bodies illustrated by oscillograph traces of conductance versus time (Adopted from Bowden and Tabor, 1954)

(1930) has given an approximate analysis of the four phases of impact with emphasis on the time of collision.

The corresponding deformation-time curves for impact are given in Fig. 8.5. Note that in the case of perfectly elastic impact an axis of symmetry exists about the point of maximum deformation, while an unsymmetrical curve is obtained in the case of elastic-plastic impact. In the case of the perfectly plastic impact, the bodies do not separate. Experiments with falling fruits striking hard flat surfaces have produced results similar to colliding metal spheres (Fig. 8.6).

Elastic impact

The Hertz law for force-deformation relationship of elastic bodies, as discussed in Chapter 6, can be used for the examination of the impact of

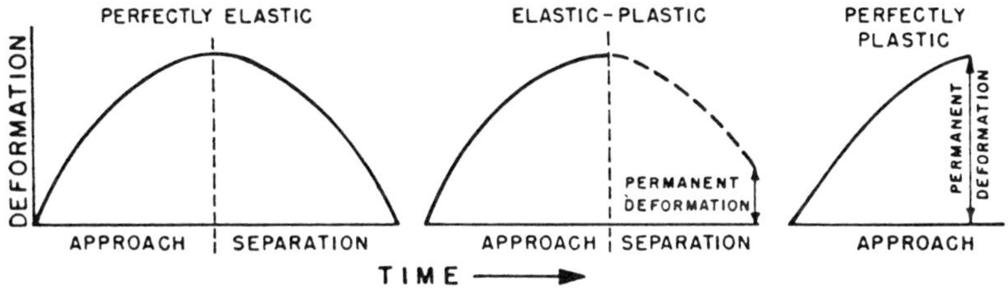

Figure 8.5 Four stages of impact shown as deformation-time curves (Adopted from Goldsmith, 1960)

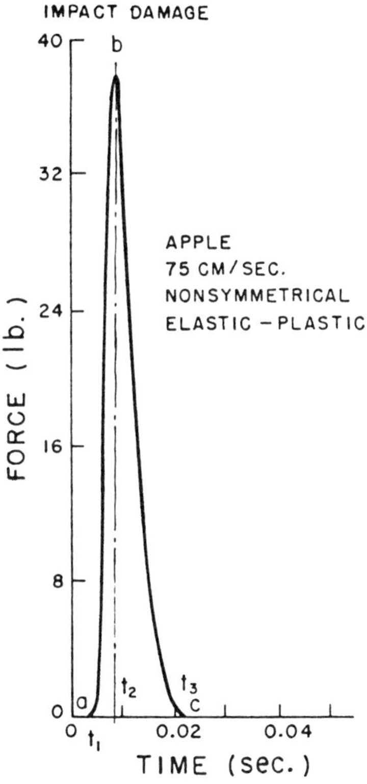

Figure 8.6 Impact of an apple on a rigid plate. The non-symmetrical trace shows the presence of plastic deformation in the apple

elastic bodies if the vibrations produced by the collision can be neglected. Combining Newton's second law of motions with Hertz's force-deformation relationship for contact of elastic bodies [Eq. (6.5)] yields the following relationships as summarized by Goldsmith (1960).

Two spheres of radii R_1 and R_2

$$D_{max} = \left[\frac{15V_1^2 A m_1 m_2}{16\,(m_1 + m_2)}\right]^{2/5} \left[\frac{R_1 + R_2}{R_1 R_2}\right]^{1/5} \tag{8.2}$$

$$t = 4.53\left[\frac{A m_1 m_2}{\pi(m_1 + m_2)}\right]^{2/5} \left[\frac{R_1 + R_2}{V_1 R_1 R_2}\right]^{1/5} \tag{8.3}$$

$$S_{max} = 0.2515\left[\frac{\pi^4 V_1^2}{A^4}\left(\frac{m_1 m_2}{m_1 + m_2}\right)\left(\frac{R_1 + R_2}{R_1 R_2}\right)^3\right]^{1/5} \tag{8.4}$$

Sphere of radius R_1 and a massive plane surface

$$D_{max} = \left[\frac{15V_1^2\, A m_1}{16\sqrt{R_1}}\right]^{2/5} \tag{8.5}$$

$$t = 4.53\left[\frac{A m_1}{\pi\sqrt{(R_1 V_1)}}\right]^{2/5} \tag{8.6}$$

$$S_{max} = 0.2515\left[\frac{\pi^4 V_1^2 m_1}{A^4 R_1^3}\right]^{1/5} \tag{8.7}$$

In the above equations, D_{max} is approach or maximum combined deformation, t is contact time, S_{max} is maximum contact stress or pressure, V_1 is the initial relative velocity, m_1 and m_2 are masses of the two bodies, and A is given by Eq. (6.4). Any consistent units can be used in the above equations. The assumption of perfect elasticity and applicability of Hertz law for the period just before plastic deformation begins was demonstrated by Andrew (1930). He produced experimental curves for the impact velocity of spheres of soft metals fitting the inverse fifth-power law for the duration of contact. As shown in Eq. (8.3) such relationship is derived by assuming perfect elasticity and applying the Hertz theory.

Plastic impact

If the impact is not purely elastic, the kinetic energy is converted into permanent deformation of the material and eventual dissipation of this energy in the form of heat. In such a case the Hertz theory is no longer applicable and the force-deformation relationship must be described by other theories.

In an attempt to replace the Hertz law by a formula applicable for plastic deformation due to impact, several of the relationships established for static conditions have been considered. For example, the Meyer's law for plastic indentation, described in Chapter 7 has been applied to impact of spheres on plane metal surfaces. This empirical law, expressed by Eq. (7.14), relates force and diameter of permanent indentation together. For a sufficiently shallow indentation of predominately plastic character where deformations in the plane surface outside the region of contact are assumed to be absent, the radius of the permanent indentation, *a,* and the central indentation, *D,* can be related by the simple relationship $a^2 \approx 2R_1 D$ as shown in Fig. 8.7. Then the Meyer's law in terms of central indentation, *D,* can be written as follows:

$$F = k' D^{n'} \tag{8.8}$$

where k' and n' are new constants which may be evaluated as shown for k and n in Eq. (7.9). The above relationship applies until the beginning of rebound which occurs elastically. When deformations outside the region of contact are not negligible, the relationship would be the corresponding elastic case from Hertz law where $a^2 = R_1 D$ (Goldsmith, 1960). For application of Meyer's law to dynamic and static indentation of apples see Fig. 8.8.

Andrew (1930) applied a combination of elastic and plastic contact deformations to the impact of two spheres of soft metals which produced permanent deformation when the velocity of approach exceeded a critical velocity. He developed a theory giving duration of impact, diameter of permanent deformation and the coefficient of restitution for the various phases of impact. A simplified method, where the forces of impact are considered was later proposed by Tabor (1948).

Example

Apples of certain stage of maturity were subjected to impact tests using a ½-inch steel ball weighing 0.019 pounds and the pendulum setup of Fig. 7.26. Data were obtained for one series of tests to include height of drop, height of rebound, bruise diameter and bruise depth. Assuming that a value of 0.003 1/psi can be taken for the elastic constant *A* of the contacting bodies, that bruise width be equal to the chordal diameter of permanent

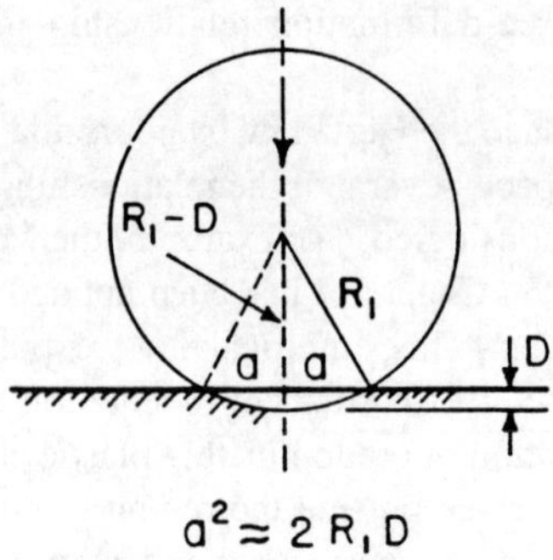

Figure 8.7 Approximate indentation relationship for a hard sphere on a plane surface assuming very shallow indentation

deformation, $2a$, and that depth of bruise be equal to central indentation D, determine (1) maximum deformation D_{max} and contact time for the elastic phase of impact at 4-inch drop height, and (2) the values of k' and n' in Eq. (8.8) for the plastic phase of impact.

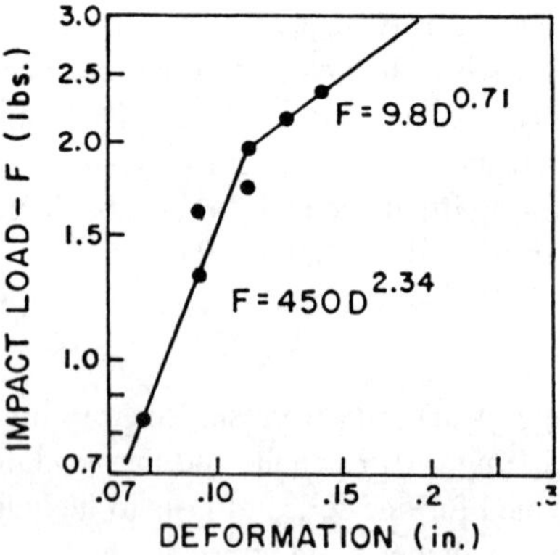

Figure 8.8 Meyer's law as applied to deformation of apple as impacted by a steel ball

Assuming that the surface of the rigidly mounted apple at the point of impact is a massive plane surface, the maximum deformation according to (8.2) is calculated to be

$$V_1 = \sqrt{2gh_1} = \sqrt{2 \times 32.2 \times 12 \times 4} = 55.4 \text{ in/sec}$$

$$m_1 = \frac{0.019}{32.2 \times 12} = 4.94 \times 10^{-5} \frac{\text{lb} - \text{sec}^2}{\text{in}}$$

$$D_{\text{max}} = \left[\frac{15(55.4)^2 \times 3 \times 10^{-3} \times 4.94 \times 10^{-5}}{16\sqrt{0.25}} \right]^{2/5} = 0.06 \text{ in}$$

Likewise the contact time according to (8.3)

$$t = 4.53 \left[\frac{3 \times 10^{-3} \times 4.94 \times 10^{-5}}{3.14\sqrt{0.25} \times 55.4} \right]^{2/5} = 3.17 \times 10^{-3} \text{ sec}$$

Using the experimental data and Eq. (8.10), the force of impact corresponding to each drop height can be calculated as given in the following table.

Height of drop h_1 (in)	Rebound height h_2 (in)	Bruise width $2a$ (in)	Bruise depth D (in)	Impact force F (lbs)
2	0.375	0.172	0.078	0.83
4	0.750	0.219	0.094	1.31
5	1.000	0.250	0.094	1.62
6	1.125	0.250	0.109	1.72
7	1.375	0.266	0.109	1.96
8	1.625	0.266	0.125	2.14
10	1.875	0.281	0.141	2.36

To determine k' and n', force of impact is plotted against depth of bruising on a log-log paper. Using the method of slope and intercept the values of k' and n' may be determined as shown in Fig. 8.8.

When apples, hung from a string, were impacted on a massive plate, it was interesting to note that the measured depth of bruise was about twice the maximum depth of elastic deformation as calculated by Eq. (8.5). The results of these tests for a sample of 50 McIntosh apples taken from cold storage two months after harvest are shown in the following table.

Drop height (in)	Rebound height (in)	Max. def. at impact (in)			Max. bruise depth (in)		
		Mean	σ_e	cv	Mean	σ_e	cv
6	1.600	0.137	0.0011	5.6	0.265	—	—
12	2.945	0.181	0.0014	5.6	0.376	—	—
18	4.170	0.212	0.0017	5.6	0.439	0.0062	10
24	5.462	0.239	0.0019	5.6	0.507	0.0056	7.9
30	6.604	0.241	0.0028	8.1	0.595	0.0065	7.7

σ_e = Standard error
cv = Coefficient of variability

Dynamic yield pressure

Whenever the pressure during impact reaches the value of dynamic yield pressure P_d, plastic flow occurs and so long as plastic flow continues the pressure remains constant at this value. Dynamic yield pressure is not necessarily the same as the static pressure required to cause plastic flow. Assuming that the elastic waves set up in the impacting materials absorb a negligible amount of energy, the work done in producing a permanent indentation of volume V_r is

$$W_3 = W_1 - W_2 = P_d V_r = P_d \pi a^4/4r_2 \qquad (8.9)$$

where W_3, W_1, W_2 are, respectively, work for plastic deformation, total impact work and rebound work, a is one half the chordal diameter d_2 of the permanent indentation, and r_2 is the radius of curvature of the permanent indentation.

After the impact has occurred, there is a release of elastic stresses in the indentation which accounts for the fact that the radius of curvature of the indentation, r_2, is always larger than the radius of the impacting sphere, R_1. It has been shown that for hard metals the radius of curvature of the permanent indentation may be as much as three times as large as that of the indenting sphere. Tabor (1951) showed that the total elastic energy, which is equal to the rebound energy W_2, can be given by the following relationship

$$W_2 = m_1 g h_2 = 3/10(F^2/a)A \qquad (8.10)$$

where h_2 is the height of rebound, m_1g is the weight of the free falling sphere, F is the impact load, and A is given by Eq. (6.4). Assuming that at the end of plastic deformation F is equal to $P_d \pi a^2$ and introducing V_a as the volume of indentation which would be obtained if the indentation were considered to have the same radius of curvature as the indenter ($r_2 = r_1$), Tabor (1951) derived expressions for dynamic yield pressure P_d given as dynamic hardness in Eqs. (7.11) and (7.12) in Chapter 7. For ease of reference these equations will be repeated here as follows:

$$P_d = \frac{m_1 g\,[h_1 - (3/8)\,h_2]}{V_a} \tag{7.11}$$

$$P_d^5 = 6.3 \times 10^{-3}\,(m_1 g/R_1^3)\left[\frac{1}{h_1 - (3/8)\,h_2}\right]^3\left[\frac{h_2}{A}\right]^4 \tag{7.12}$$

The constant A is a function of elastic properties of the two impacting bodies given by Eq. (6.4). In this derivation the value of P_d was assumed to be constant during the collision, an assumption which is not absolutely true. Tabor (1951), however, estimated that the difference will never be more than about 10 percent.

It can be shown that Eq. (7.12) is valid right up to the 100 percent rebound or the condition of perfectly elastic collision. In that case, the equation for determining the final average pressure under the elastic condition would be obtained by simply setting $h_2 = h_1$ in Eq. (7.12) which will give

$$P_d^5\ (\text{elastic}) = 25.8 \times 10^{-3}\left(\frac{m_1 g h_1}{R_1^3}\right)(1/A)^4 \tag{8.11}$$

Example

Calculate the dynamic yield pressure for the steel ball impacting an apple as given in the previous example. Assume a drop height of 6 inches and a rebound height of 1.5 inches.

Direct substitution in Eq. (7.12) gives the dynamic yield pressure for the 6-inch drop as follows:

$$P_d^5 = 6.3 \times 10^{-3}\left(\frac{0.019}{(0.25)^3}\right)\left[\frac{1}{6 - (3/8)\,(1.5)}\right]^3\left[\frac{1.5}{0.003}\right]^4$$

$$P_d = 19.75 \text{ psi}$$

The static load required to produce the onset of plastic deformation can be determined from the mean contact pressure and Eq. (6.7) as follows:

$$P_m = \frac{F}{\pi a^2} \quad \text{or} \quad a = \left(\frac{F}{P_m \pi}\right)^{1/2}$$

Substituting a in Eq. (6.7) and simplifying yields

$$F = 17.24 P_m^3 R_1^2 A^2 \tag{8.12}$$

Experimental data shows that it takes extremely light impacts to start plastic deformation in a material as hard as an alloy steel. The reason for this is the extremely short duration of impact which creates very high impulsive forces.

Example

Assume a 4-gram steel ball of diameter 0.4 inch is dropped on a massive plate of alloy steel with a static yield stress, y=178,500 psi (see Fig. 7.23). If the static mean pressure, P_m, required to produce the onset of plastic deformation is $1.1Y$, find the static load corresponding to P_m.

$$A = \frac{1 - (0.3)^2}{30 \times 10^6} + \frac{1 - (0.3)^2}{30 \times 10^6} = 6 \times 10^{-8} \text{ in}^2/\text{lb}$$

$$P_m = 1.1Y = 1.1 \times 178{,}500 = 196{,}000 \text{ lb/in}^2$$

Direct substitution in Eq. (8.12) yields

$$F = 17.24(196 \times 10^3)^3 (0.2)^2 (6 \times 10^{-8})^2 = 18.7 \text{ lbs}$$

To compare the magnitude of this static load with that required to produce the onset of plastic deformation in a material such as fruits and vegetables, the following data were obtained by loading a whole apple with a ½-inch steel ball and determining the mean values of force and deformation for five sets of tests.

Load F (lb)	Deformation D (in)	Chordal dia. $2a$ (in)	Mean pressure P_m (psi)
0.25	0.0050	0.109	26.93
0.50	0.0114	0.149	28.76
0.75	0.01745	0.1955	28.44
1.00	0.0228	0.2085	29.34
1.25	0.0282	0.2325	29.48

Figure 8.9 shows the plot of mean pressure versus load from the onset of plastic deformation to full plastic flow. Assuming that the onset of plastic deformation in the apple is about $P_m = 10$ lbs, the corresponding load, according to Eq. (8.12) would be

$$A = \frac{1 - 0.33^2}{300} + \frac{1 - 0.3^2}{30 \times 10^6} \cong 3 \times 10^{-3}$$

$$F = 17.24(10)^3 \, (0.25)^2 \, (3 \times 10^{-3})^2 = 0.0097 \text{ lb}$$

In his dynamic tests, Nelson (1967) employed the pendulum apparatus of Fig. 7.26 and the concept of dynamic yield pressure to determine maximum allowable impact load that McIntosh apples at 40°F and 90°F can withstand. In this apparatus, instead of the steel ball, the apple was suspended from the end of the string. The release mechanism held the apple until the catch was released and the fruit was allowed to fall against the massive rigid block fastened to the wall. A simple radial chart enabled readings of the drop height, h_1, and rebound height, h_2. A piece of white paper affixed to the block recorded the contact area of the falling fruit which was coated with Prussian blue at the point to be impacted. The average radius of curvature of the impacting point was also determined prior to each test using the radius meter of Fig. 6.3. To avoid extra bruising after each fall the fruit was allowed to bounce only once before it was caught. After each impact test the same fruit was tested for determination of its mechanical properties under quasistatic conditions.

The impact loads were expressed in terms of energy absorbed, $E' = w(h_1 - h_2)$ and momentum absorbed, $M = w/g(V_1 - V_2)$. The impact and rebound velocities, V_1 and V_2, were calculated from drop and rebound

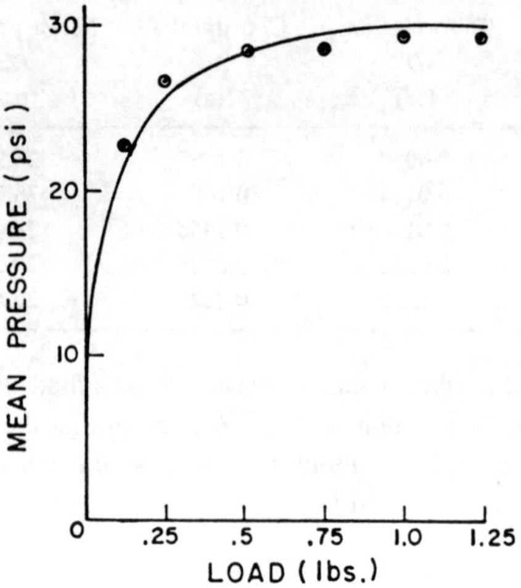

Figure 8.9 The onset of plastic deformation and full plastic flow in McIntosh apple loaded with a 1/2-inch diameter steel ball

heights h_1 and h_2. Also the coefficient of restitution and dynamic yield pressure were calculated.

For determining the dynamic yield pressure for the apple impacting the rigid surface of the block, Nelson employed the following equation which was developed by Tabor (1951) by setting the total elastic energy stored in the surfaces equal to the energy of rebound.

$$P_r = \left(\frac{Wh_2}{2.96a^3} \frac{Ea}{1 - \mu^2} \right)^{1/2} \tag{8.13}$$

In this equation P_r is dynamic yield pressure based on the rebound height, E_a and μ are apparent elastic modulus and Poisson's ratio of the fruit, W is weight of the falling fruit, and a is the radius of the area of maximum contact during the impact. Since by definition dynamic yield pressure is the pressure at which plastic or non recoverable deformation occurs under impact, this property should be studied further to find out if it is a mechanical property analogous to the bio-yield point in fruits demonstrated under quasistatic tests. If so, one may use the dynamic yield pressure as a criterion

criterion for maximum allowable pressure that fruits can withstand under dynamic loading.

Time of plastic impact

Consider an undeformable sphere of mass m_1 and radius R_1 impacting a surface which has an average dynamic yield pressure P_d. If the impact is predominantly plastic, at any instant when the sphere has penetrated a distance D, the indentation has a radius a and the decelerating force on the sphere is $P_d \pi a^2$. Since according to Fig. 8.7 for shallow indentations $a^2=2R_1 D$, the decelerating force becomes $2\pi R_1 D P_d$ and the equation of motion will be

$$2\pi R_1 D P_d = -m_1 \frac{d^2 D}{dt^2}$$

or

$$\frac{d^2 D}{dt^2} + \frac{2\pi R_1 P_d}{m_1} D = 0$$

For the case when the impacting sphere is brought to rest, $dD/dt=0$ and the solution of the above differential equation gives the following expression for the time of plastic impact.

$$t = \pi/2 \left(\frac{m_1}{2\pi P_d R_1} \right)^{1/2} \tag{8.14}$$

Both Tabor (1951) and Andrew (1930) have given the above expression for time of plastic collisions and point out the conclusion that the time of plastic collision is independent of the velocity of impact. As seen from Eq. (8.3), in the case of elastic impact the time is dependent on the impact velocity, although the rate of variation with velocity is extremely small. With reference to Fig. 8.4a, the time given in (8.14) corresponds approximately to the time from A to D.

Example

A hard steel ball of 0.4 inch diameter is dropped on a massive plate of mild steel with a dynamic yield pressure of 290,000 psi. If the ball weighs 4 grams, find the time for plastic collision.

Using Eq. (8.14)

$$t = \pi/2\left(\frac{(4/452)/386 \text{ lb} - \text{sec}^2/\text{in}}{2\pi \times 290 \times 10^3 \times 0.2 \text{ lb} - \text{in/in}^2}\right)^{1/2} = 1.3 \times 10^{-5} \text{ sec}$$

Similar calculations for other metals have shown that the time of plastic collision is of the order of several 10^{-5} seconds. Comparing results with an identical elastic collision using Eq. (8.3) will show that it is only for very small velocities of impact that the time of collision for elastic impact will be appreciably different than that of plastic collision.

Viscoelastic impact

A viscoelastic solution to the problem of impacting spherical fruits has been given by Hamann (1967). Using a sinusoidal compression test based on the same principle as Fig. 5.43, complex axial compressive moduli for apples between frequencies of 50 to 365 cps were determined. A Maxwell model was found to be the best representative of the material's stress-strain behavior with a calculated elastic element constant of 3000 to 750 psi and a viscous element constant of 20 to 4.5 psi/sec, with low values corresponding to riper and softer apples. Relaxation moduli calculated by transformation of complex compression moduli along with the initial values of Poisson's ratio (Chappell, 1967) were then used for the solution of the impact problem of two Maxwell spheres. The solutions were given for the case of one apple falling onto another supported underneath and an apple impacting a rigid plate. The approach of the apple centers was found from the computer solution of a nonlinear second order differential equation. Surface indentations, surface pressures, and internal stresses from time of initial contact to the time of maximum indentation were also given. An illustrated example involved the following values as inputs to the computer. Apple weight=0.342 pound, apple diameter at the contact point=2.88 inches, drop height=2 inches and 12 inches, elastic and viscous constants of the Maxwell body representing the apple, E_0=2627 psi and η=17.93 psi/sec, respectively, and the apple Poisson's ratio μ=0.219.

Results of this viscoelastic analysis showed that the maximum surface pressure for the 2-inch drop height was about 204 psi which was approximately 2/3 of the maximum surface pressure for the 12-inch drop height. Furthermore, comparison of the internal stresses showed that even though the value of stress for the 12-inch drop is only 1½ times that of the 2-inch drop, stress is large to a considerably greater depth for the 12-inch drop than for the 2-inch drop. This shows that in the case of 12-inch drop, the extent of bruising in the apple must be considerably greater than that of the 2-inch drop.

Coefficient of restitution

This coefficient is usually defined as the ratio of final to initial relative velocity components of the striking bodies in the direction normal to the contact surfaces. Denoted by e, the ratio can be expressed as

$$e = V_2/V_1 = (h_2/h_1)^{1/2} \tag{8.15}$$

where subscripts 1 and 2 represent the initial and final states and h_2 and h_1 denote height of rebound and height of drop in free fall, respectively. If the principle of a single pendulum is applied to determine coefficient of restitution (See Fig. 7.26), the angles of the pendulum deflection can be used directly as seen from the following relationship:

$$e = \frac{\sin \beta/2}{\sin \alpha/2}$$

where α and β are initial and rebound angles of the pendulum.

If the striking bodies are made of different materials, the coefficient of restitution is given by Hodgkinson (cited in Goldsmith, 1960) as follows:

$$e_{12} = \frac{e_{11} E_2 + e_{22} E_1}{E_1 + E_2} \tag{8.16}$$

where E is the modulus of elasticity of the two materials involved.

Although some authorities question the fundamental significance of the coefficient of restitution, it is widely employed in analysis of engineering problems. For example, the value of e indicates the degree of elasticity or plasticity in an impact. Values of $e=1$ and $e=0$ denote the idealized concepts of perfectly elastic and perfectly plastic impact, respectively. It is also agreed and well proven by experiments that the coefficient of restitution is not a constant and varies with the velocity of impact. At very low velocities of impact the collision process is primarily elastic and the coefficient of restitution approaches unity. As the impact velocity increases, the amount of plastic deformation increases, there will be a decrease in rebound velocity and coefficient of restitution. This dynamic behavior is illustrated in Fig. 8.10, both for metals and for soft fruits.

If coefficient of restitution can be assumed to be constant for a given range of impact velocities, the energy dissipated in impact can be expressed in terms of coefficient of restitution. If the second body is rigid, the change of kinetic energy $\Delta E = \frac{1}{2} m_1 V_1^2 (1 - e^2)$. If the second body with mass m_2 also has a velocity V_1', then the change of kinetic energy

$$\Delta E = \frac{1 - e^2}{2} \frac{m_1 m_2}{m_1 + m_2} (V_1 - V_1')^2 \tag{8.17}$$

Impulse-momentum law for rigid bodies

The concept of impulse-momentum forms the classical theory of impact. The linear impulse-momentum law is given by

$$\Delta mv = mv_1 - mv_2 = \int_0^t F \, dt \tag{8.18}$$

where m is the mass, v is linear velocity, F is compressive force, and t is time. The objection to this law is that in an impact process not all elements of each body will be instantaneously subjected to the same change of motion. Examination of the theory of wave phenomena reveals that the disturbance generated at the contact point propagates into the interior of the bodies. When this wave propagation accounts for appreciable amount of energy loss, the predictions based primarily on impulse-momentum theory will be seriously in error.

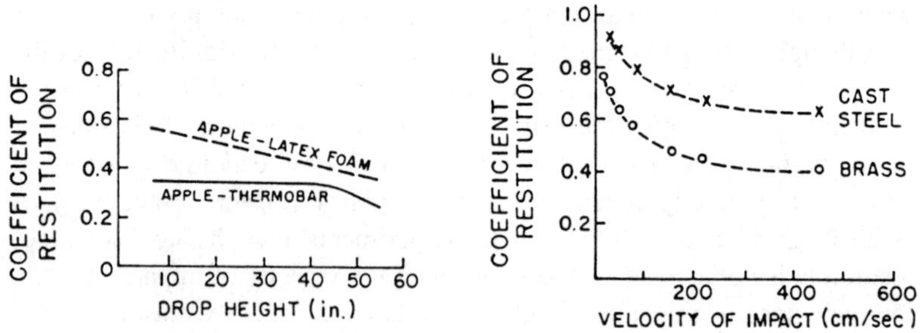

Figure 8.10 Variation of coefficient of restitution with impact velocity: Right (Tabor, 1941), Left (Hammerle, 1966)

Figure 8.6 shows a record of impact force versus time obtained using an optical oscillograph in a study of impact of fruits striking rigid surfaces as well as surfaces covered with cushioning materials. A force transducer was mounted under the rigid plate to sense the impact force. For measurement of rebound height, two successive impacts were recorded and the time of rebound was taken as one-half of the time from the end of the first impact until the beginning of the second impact. Knowing the time of rebound, the rebound height, h_2, was determined from the relationship

$$h_2 = 1/2\, gt^2$$

where g is the acceleration of gravity and t is the time of rebound (Fig. 8.11).

The fruit, weighing 0.339 pound, was dropped on the rigid plate from a height h_1=2.5 feet and the rebound height h_2=0.74 feet was determined from the time of rebound. At point a, the falling fruit with a momentum mv_a contacted the plate. During the time (t_2-t_1)=3.8 milliseconds, this momentum was changed to impulse $\int_{t_1}^{t_2} F\,dt$, which resulted in approximately 38 pounds of force on the steel plate. Assuming that no energy was transmitted, and that the energy losses due to wave propagation were zero, the remaining momentum of the apple was absorbed by the fruit itself causing the resulting damage. The experimental value of impulse $\int_{t_1}^{t_2} F\,dt$ was determined by measuring the area under the portion of the curve from point a to point b. From point b to point c, at which the fruit left the plate for rebound, the impulse $\int_{t_2}^{t_3} F\,dt$ changed to momentum mv_2 which lifted the fruit to a rebound height h_2.

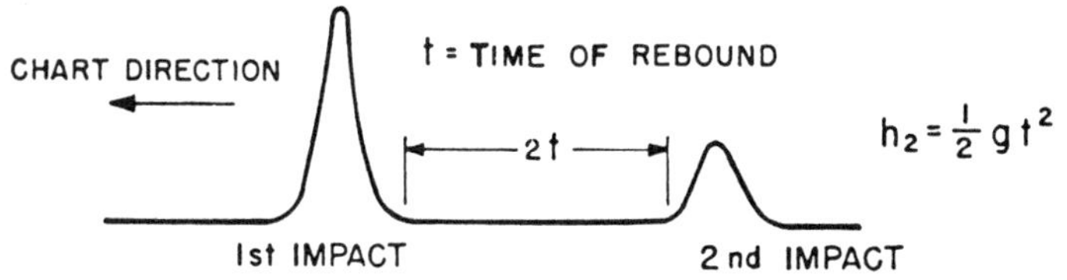

Figure 8.11 Recording of two successive impacts for measurement of rebound height h_2

Having the initial momentum, $mV_a = m\sqrt{2gh_1}$, and the momentum for rebound, $mV_c = m\sqrt{2gh_2}$, by the law of conservation of momentum the energy absorbed by the fruit resulting in mechanical damage can be determined. Numerical analysis of the data given in Fig. 8.6, however, as expected, showed some errors which were either due to the lack of rigidity in the system or unaccounted energy losses due to wave propagation.

For the left hand side of the curve:

$$\int_{t_1}^{t_2} F\,dt = 0.0736 \text{ lb} - \text{sec (measured by planimeter)}$$

$$Va = \sqrt{2gh_1} = \sqrt{2 \times 32.2 \times 2.5} = 12.7 \text{ ft/sec}$$

$$m(V_a - V_b) = 0.339/32.2\,(12.7 - 0) = 0.134 \text{ lb-sec}$$

Note that initial momentum, 0.134 lb-sec, is greater than the recorded impulse $\int_{t_1}^{t_2} F\,dt = 0.0736$ lb-sec. For the right hand side of the curve

$$\int_{t_2}^{t_3} F\,dt = 0.1296 \text{ lb-sec (measured)}$$

$$Vc = \sqrt{2gh_2} = \sqrt{2 \times 32.2 \times 0.74} = 6.9 \text{ ft/sec}$$

$$m\,[V_b - (-V_c)] = 0.339/32.2\,[0 + 6.9] = 0.0725 \text{ lb-sec}$$

Note that the momentum given to rebound, 0.0725, is less than the recorded impulse $\int_{t_2}^{t_3} F\,dt = 0.1296$ lb-sec.

The above analysis assumed that at t_2 velocity of the fruit $V_b = 0$. Note that if the fruit were considered viscoelastic, this assumption would no longer be valid. In that case, computation on the basis of total impulse and total change in momentum yields

total impulse: 0.0736 + 0.1296 = 0.202232 lb-sec
(measured)

total momentum change: 0.134 − (−0.0725) = 0.2065 lb-sec
(calculated)

which shows a good agreement between the measured and calculated values.

Energy-absorbing capacity of cushioning materials

In studying the protective value of cushioning materials, Hammerle and Mohsenin (1966) made use of the coefficient of restitution and found that this coefficient remained fairly constant for most of the cushioning materials tested with a higher coefficient associated with the harder, denser materials. As in the case of metals, it was found that coefficient of restitution varies with drop heights or impact velocity. This is illustrated in Fig. 8.10 where the effect of bottoming in the case of apple impacting a rigid surface covered with thermobar is shown by the sharp drop in the curve.

In this work with cushioning materials, Hammerle and Mohsenin also introduced a new parameter called the fragility factor denoted by G. Fragility factor was defined as the ratio of peak force, F, occurring during impact to the weight, W, of the impacting body so that

$$G = F/W \tag{8.19}$$

Note that G is dimensionless and is a multiple of acceleration due to gravity. It was found that fragility factor, taken as the protective value of a cushioning material, varies with the impacting specimen as well as the loading rate or drop height. In the case of apples impacting a rigid surface covered with a given cushioning material, in general the fragility factor increased with drop height with a smaller G indicating a better cushioning material (Fig. 8.12). Note the difference between latex foam (No. 10) and polyethylene (No. 4). The former had the lowest G protective value corresponding to the highest G value (next to the bare plate). Note also the effect of density on fragility factor. In general the more dense the cushioning material, the better would be its ability to reduce deceleration.

The dynamic testing apparatus employed in these tests is illustrated in Fig. 8.13. On the slider of the apparatus, carrying an accelerometer, could be mounted either a bare rigid plate, a plate covered with a cushioning material, a metal hemisphere, or a rigid plunger. These loading devices could be used to impact a specimen, which could be a fruit or other products, mounted on top of a dynamic load cell. The pneumatic hold-down ring was used to hold a piece of cushioning material for the case when a rigid sphere was dropped to compare its energy losses with those fruits. The data from the load cell gave the rebound time according to Fig. 8.11. The accelerometer was used to determine the impact force according to $F=ma$.

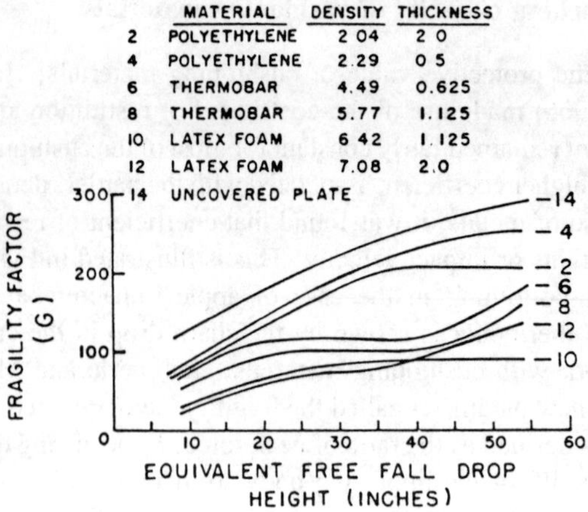

Figure 8.12 Fragility factor of cushioning materials supported by a rigid plate and impacted by an apple (Hammerle and Mohsenin, 1966)

While the drop testing apparatus of Fig. 8.13 offered means by which impact force and displacement could be measured during the tests, it lacked the requirements of absolute rigidity and frictionless guide tracks which were essential in an energy balance study followed by Bittner *et al.*, (1968). These investigators used the pendulum apparatus of Fig. 7.26, which offered a simple means of measuring height of drop and height of rebound when either a fruit or a rigid ball was allowed to impact a piece of cushioning material affixed to the rigid impacting surface on the wall.

The concept of energy balance employed by Bittner *et al.*, (1968) can be expressed by the following relationship.

$$E_{\text{impact}} = E_{\substack{\text{absorbed by}\\ \text{cushion}}} + E_{\substack{\text{absorbed by}\\ \text{specimen}}} + E_{\text{rebound}}$$

It can be shown (Hammerle and Mohsenin, 1966) that if a non-deformable wooden ball of approximately the same density and equivalent diameter as an energy absorbing object, such as an apple, were used to compare the protective value of cushioning materials, the above expression can be reduced to the following form

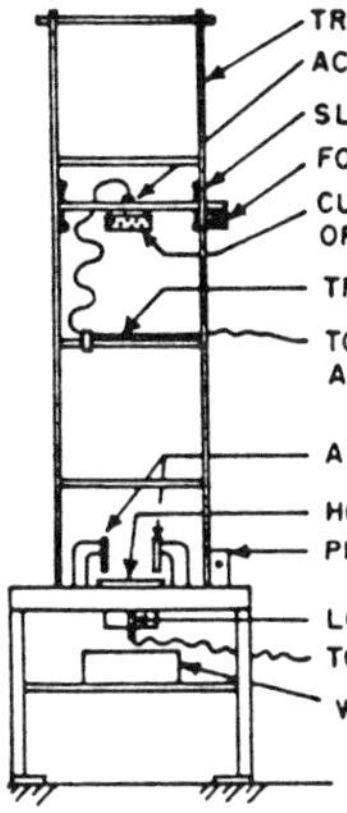

Figure 8.13 A drop test apparatus for impact test of foods and agricultural products (Hammerle and Mohsenin, 1966)

$$E_{\substack{\text{absorbed by}\\ \text{fruit}}} = H_{\substack{\text{rebound}\\ \text{wooden ball}}} - H_{\substack{\text{rebound}\\ \text{fruit}}}$$

where E is energy in inch-pound per pound falling weight and H is height in inches. Using the pendulum apparatus and the above expression any cushioning material could be tested for its ability to reduce energy transferred to the fruit as well as giving a measure of rebound of the fruit after impact. It should be noted that both of these factors are important characteristics to be considered in selecting cushioning materials for machines and equipment to be used in handling of fruits and vegetables.

In conducting the tests, the wooden ball was first dropped several times against the cushioning material until a maximum rebound height was obtained. The fruit drop tests were then followed and height of drop and height of rebound were recorded for each test. By measuring the effective diameter of the fruit at the point of impact, a comparison could be made with values for a wooden ball of approximately the same diameter. In this way the rebound height of the rigid body (the wooden ball), which by assumption did not absorb any energy, was compared with the rebound height of the non-rigid body (the fruit) and by subtraction the energy absorbed by the fruit was determined. Employing this technique Bittner *et al.*, (1968) have reported results of their tests for two cushioning materials in terms of energy absorbed by the cushion, energy absorbed by the impacting objects, rebound of impacting object as well as the effects of cushion temperature and size of the impacting bodies.

Rehkugler (1964) gave the relationship between the thickness of several padding materials and ratio of impact to compression energy of failure of eggshell (Fig. 8.35). The ratio, being greater than unity, increased as the material thickness increased. Also the difference between different types of cushioning materials became greater as the material thickness increased.

Fridley *et al.*, (1964) also studied some of the factors affecting suitability of cushioning materials for mechanical handling of fruits and vegetables. The following shows the amount of energy found required for yielding or initiation of bruising in Halford Cling peaches.

Method of load application	Yield energy (in–lb)
Compression test with plunger	0.264
Impact with plunger	0.786
Compression tests on 0.361-in sponge rubber	5.500
Compression tests on 0.727-in sponge rubber	11.600
Impact test on 0.361-in sponge rubber	12.700
Impact test on 0.727-in sponge rubber	23.400

Comparing compression and impact energies the fruit required several times more energy to bruise under impact than quasistatic compression. Note that the ratio of impact to compression ratio without any padding material is about three compared to a ratio of about two for several varieties of apples found by Mohsenin and Goehlich (1962). When the fruit was impacted or compressed onto the sponge rubber, breakdown of the total energy values shown in the above table into energy absorbed by the rubber and that absorbed by the fruit showed that the fruit bruising energy increased several times while the ratio of impact to compression bruising energies increased only slightly. It was concluded that damage caused by impact on hard surfaces can be minimized by the use of cushioning materials which can absorb the kinetic energy of the impacting fruit without allowing the bruising energy of the fruit to be exceeded. Furthermore, stress-strain curves of padding materials can provide information on energy absorbing capacity of each material and thus materials with high hysteresis can be selected over those which momentarily store energy and then release it causing a high rebound and repeated impact of the fruit.

Example

To determine the energy absorbed by a fruit dropped from a height of 50 cm on a rigid steel plate, a trace of time vs height on an oscilloscope shows that the fruit bounces off the plate at $t=0$ second and strikes the plate again at after 0.5 seconds. If the fruit weighs 250 gms, find the percent of total energy absorbed by the fruit.

From Fig. 8.11 and the given information rebound time $t=0.5/2$ or 0.25 seconds. Therefore rebound height

$$h = 1/2\, gt^2 = 30 \text{ cm}$$

Assuming transmitted energy is negligible, from the energy balance, percent of energy absorbed by the fruit is

$$Ea + 250 \times 30 = 250 \times 50$$

or

$$Ea = 250(50 - 30)/(250 \times 50)$$

$$= 40\%$$

Loading rate and mechanical damage

As discussed in the previous chapters, it has well been established that mechanical properties of viscoelastic materials are strain rate dependent. This means that each property of a material may vary as a function of the rate of loading to which it is subjected under field conditions. Therefore, it is essential to characterize the material over a broad range of test speeds from creep to fast rates of impact loading as well as at discrete points over the complete range simulating the practical field conditions.

It has been shown that plotting of either force or deformation at failure as a function of rate of loading yields a material characteristic curve which may reveal a critical velocity at which some property of the material changes drastically. In the examples given in Figs. 8.14 and 8.15, a critical velocity as well as a change in the fracture geometry of the material are shown at higher rates of loading. When several materials are compared, as in Fig. 8.15, a material such as amorphous polyethylene can be assumed to be the strongest of all the plastic films tested if loading rates were limited to below 100 percent per minute. Higher rates of loading, however, showed that the same material is the weakest if the loading rates were extended beyond the 100 percent per minute. The practical implications of loading rate sensitivity for engineering materials are quite obvious.

Investigations for agricultural products have shown that these materials may also exhibit critical loading rates. By introducing a third variable such as moisture content, maturity, chemical composition, etc., a three-dimensional representation of, for example, maximum breaking force-loading rate-moisture contour for a given grain can be obtained which may prove to be quite useful in milling and other mechanical processing of that grain.

Fletcher *et al.*, (1965) and Hammerle and Mohsenin (1966) have obtained data for several mechanical properties of fruits as a function of rates of loading varying from quasistatic speeds to speeds equivalent to several feet of free fall. Figure 8.16 shows the impact apparatus employed by Fletcher *et al.*, (1965) for studies at fast rates of loading. Data for intermediate and slow rates of loading tests were obtained using the testing machine illustrated in Fig. 4.22 and another version of this machine utilizing

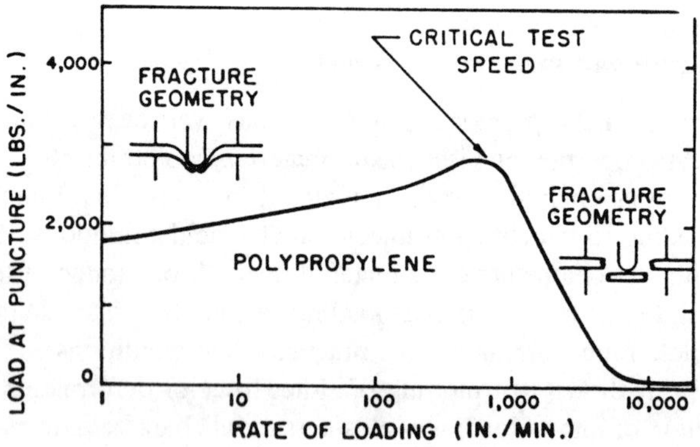

Figure 8.14 Characteristic load-speed curve for puncturing of a general purpose polypropylene (Furno *et al.*, 1962)

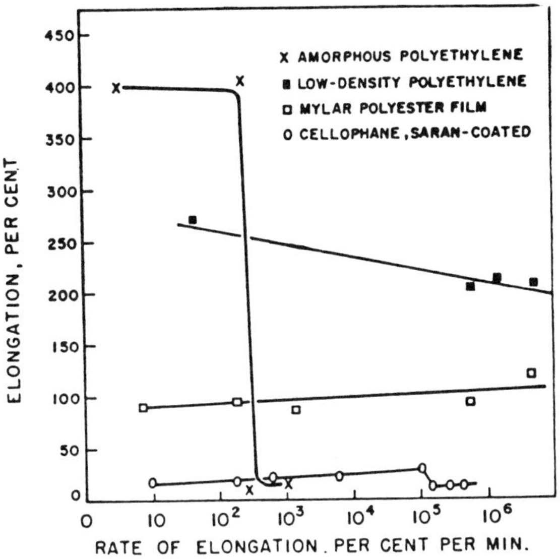

Figure 8.15 Characteristic elongation-speed curves for several plastic films (Supnik, 1962)

a higher speed pneumatic cylinder. The data included variation of force, deformation and energy to yield point and skin rupture point for five varieties of apples at pre-harvest, harvest, and post-harvest stages as well as limited data on loading rate sensitivity of peaches, pears, and potatoes.

Figure 8.17 shows a sample of force-time and deformation-time curves for apples obtained using the pneumatic impact machine of Fig. 8.16. The stroke of the machine was adjusted so that the required data were recorded before the piston rod was intercepted by the stop. As shown, the velocity of the piston for the duration of the test was constant. This constant velocity, obtained from the slope of the displacement-time curve, was also assumed to be the loading rate which was used to obtain the following parameters.

Deformation to yield = AE (sec) × piston velocity (in/sec)

Deformation to rupture = AD (sec) × piston velocity (in/sec)

Energy to yield = area under ABE (lb-sec) × piston velocity (in/sec)

Energy to rupture = area under $ABCD$ (lb-sec) × piston velocity (in/sec)

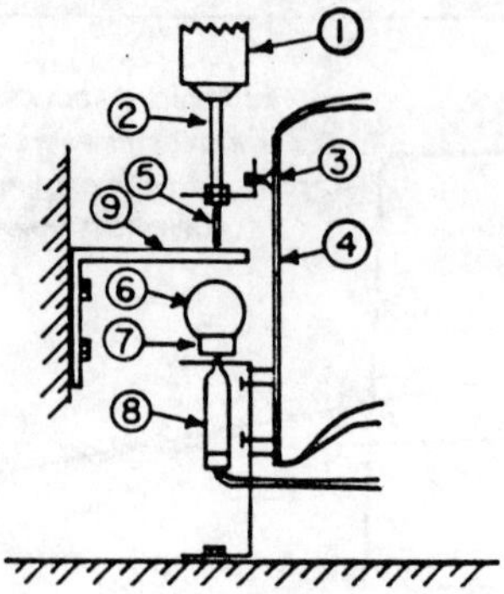

Figure 8.16 A high speed pneumatic apparatus for impact test of foods and agricultural products (Fletcher *et al.*, 1963). 1—high speed air cylinder; 2—piston rod; 3—slide wire wiper; 4—slide wires; 5—1/4-inch plunger; 6—fruit specimen; 7—specimen holder; 8—load cell; 9—crosshead stop

Examples of loading rate characteristic curves for one variety of apple is shown in Fig. 8.18. Results of this study showed that the bioyield properties of apples do not follow any consistent trend with increased rate of loading. The bioyield point actually disappeared at higher rates of loading as shown in more detail in Fig. 8.19. In this illustration the higher numbers correspond to higher loading rates. Tests with a magnetic tape recorder

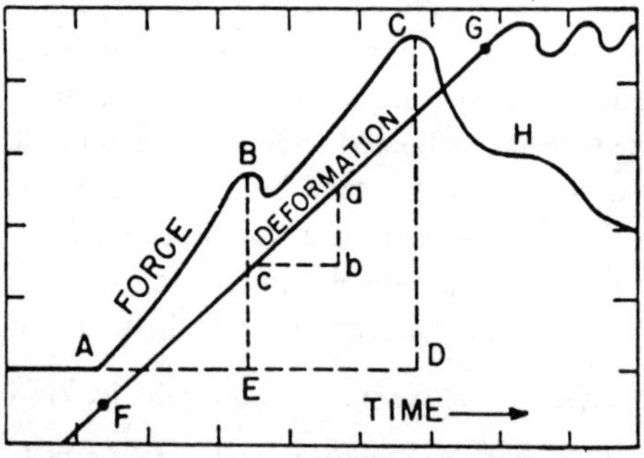

Figure 8.17 Dynamic behavior of apple using impact apparatus of Fig. 8.16 (Fletcher *et al.*, 1965). *AE*: deformation to yield; *AD*: deformation to rupture; *G* and *H*: stop of crosshead; *ab*/*ac*: piston velocity

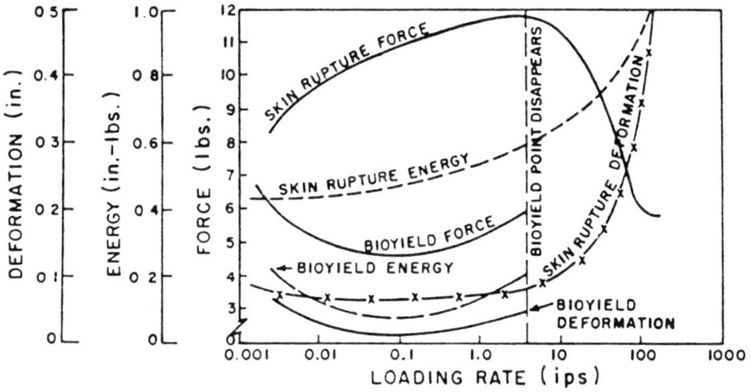

Figure 8.18 Some mechanical properties of golden delicious apples as a function of loading rate. Specimens were tested at harvest maturity using apparatus of Fig. 8.16 (Fletcher *et al.*, 1965)

expanding the strip chart of the optical recorder sixteen times to provide a more accurate examination of the data still did not show any break in the curve at fast rates of loading. Apparently at fast rates of loading the plunger moved into the specimen with such a high velocity that the initial cell rupture and the complete rupture of the skin and flesh could not be differentiated. Unlike the bioyield point the skin rupture force, deformation and energy initially increased with increased loading rate. The force for skin rupture, however, showed a sudden drop beyond a certain speed which varied with different varieties and stage of maturity and then leveled off.

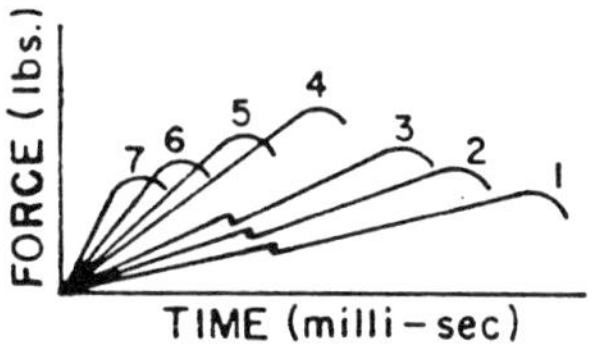

Figure 8.19 Change of bio-yield point in apples at higher rates of loading (Fletcher *et al.*, 1965)

This peculiar phenomenon was similar to the critical loading rate for some high polymers which were mentioned earlier in this section.

The reason for the apparent continued increase of deformation and energy at skin rupture beyond the "critical loading rate" was not explained. It is possible that the sudden decrease of rupture force and the increase of rupture deformation and energy are due to the effect of the inertia of the masses involved between the point of force application (the tip of the plunger) and the point of force measurement (the load cell). This inertia could be absorbing an increasing amount of the applied force as the loading rate was increased. In other words, the impact force was divided as follows

$$\text{Total impact force} = \text{reaction force sensed by transducer} + \text{force accelerating the mass}$$

or

$$F_{\text{total}} = F_{\text{measured}} + ma_{\text{inertia}}$$

If this were the case, then a significant part of the applied force may have never reached as far as the load cell to be measured resulting in a dropoff of the curve force versus rate of loading. At the same time, if the apple and the holder exhibited some displacement at higher rates of loading, the crosshead had to travel a farther distance before rupturing the fruit skin. This error would show a displacement greater than the actual deformation of the fruit tissues under the plunger.

If this were the case, the data could be corrected providing that the masses, the spring factors, and the natural frequencies of the apple and the holder at higher rates of loading were known. With this information one would make a rough estimate of the acceleration of the system from

$$a = 4\pi^2 f^2 A \tag{8.20}$$

where f is the natural frequency and A is the amplitude of vibration. Having acceleration a, the inertia could be calculated from $F = ma$. The acceleration could also be measured directly if suitable accelerometers were attached on the apple and the holder.

To minimize such errors as may have occurred in Fletcher's apparatus, one would either measure the impact force directly at the point of impact or use an accelerometer in a drop test apparatus and determine the force from the product of acceleration and the dropping mass. The latter technique, which is the preferred method, was used by Hammerle and Mohsenin (1966) in the apparatus illustrated in Fig. 8.13. In using this apparatus a rigid plate attached to the slider was dropped on the apple which was fixed on top of the load cell. Impact force was calculated from the acceleration. Deformation during impact was calculated from the product of the average impact velocity ($\frac{1}{2}\sqrt{2gh_1}$) and the time of impact which was given by the load cell according to Fig. 8.11. The impact energy was found from the product of one half the deformation and the impact force. The results of tests on Golden Delicious apples are shown in Fig. 8.20 and the resulting mechanical damage, as a function of impact energy, is given in Fig. 8.21.

The value of a complete force-deformation curve of a material for all rates of loading experienced under the field conditions cannot be overemphasized. In breeding and development of new varieties of products, a force-deformation curve of the varieties under consideration at the rate of loading to which the materials will be subjected when in use would serve as a guide for selection of the most suitable variety. As a hypothetical example, assume that the two curves in Fig. 8.22 represent the force-deformation curve of two materials under a given rate of loading applicable to field conditions. If we consider the energy to rupture as the criterion for selection, this energy, being the area under the curves up to rupture points, is the same for both materials. Therefore, the conventional methods of

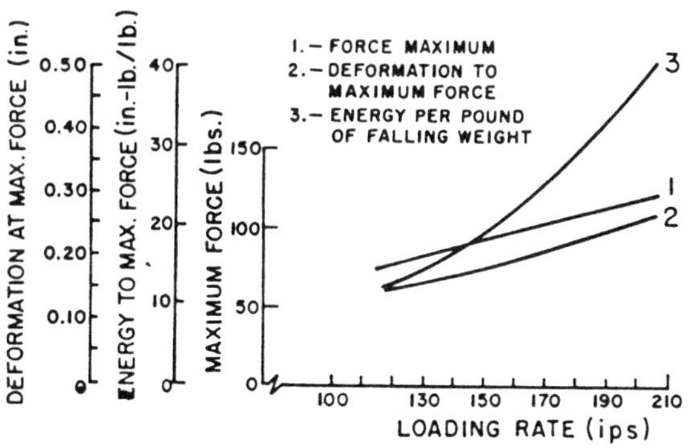

Figure 8.20 Impact of Golden Delicious apples (six months after harvest) and a rigid plate (Hammerle and Mohsenin, 1966)

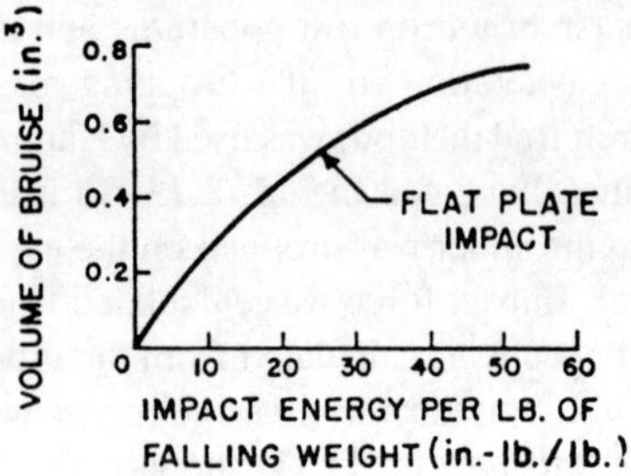

Figure 8.21 Bruise volume in Golden Delicious apples as a function of impact energy (Hammerle and Mohsenin, 1966)

impact tests giving energy to rupture do not offer any help in the selection. However, if any of the other parameters such as energy to yield, load-bearing capacity for a given deformation, or maximum deformation for a given load were taken as the criterion for selection, considerable difference exists between the two curves and the selection would not be as difficult.

Kinematics and Dynamics of Impact Tests

Different kinds of impact tests have been utilized by researchers in investigating the phenomenon of fruit impact. The five common types of impact tests generally used in the past are dropping the fruit on a surface, impacting the fruit with a falling mass, two types of pendulum tests, and impact test by a ram (spring, pneumatic, or electric) striking the fruit. Fig. 8.22a shows, schematically, these different impact tests. The specimens used for these impact tests can be either intact fruits, fruit halves, or cylindrical specimens. Table 8.4a summarizes the basic kinematics and dynamics for the different impact tests shown in Fig. 8.22a. Table 8.4b summarizes the different methods of impact tests and specimen preparations used by researchers for fruit impact, utilizing the geometries shown in Fig. 8.22a. A critical analysis of these techniques is given by Manor (1978). In the following pages, these techniques are compared to show that, in general, the information from different tests is not available on a common basis for comparison and synthesis. To test the validity of his critical analysis and interpretation, Manor (1978) conducted new tests, analyzed the results and related them to the previously reported work.

Table 8.4a Basic Kinematics and Dynamics for Different Methods of Impact Tests (Manor, 1978)

	Drop Test	Falling Mass	Simple Pendulum	Compound Pendulum	Impact Ram
1. Kinematical Description.	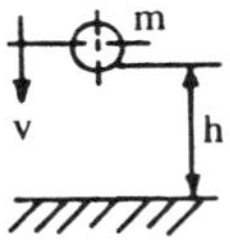	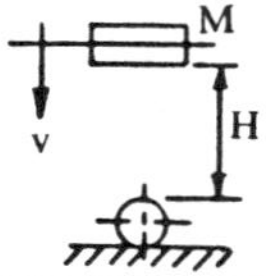	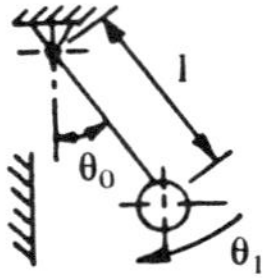	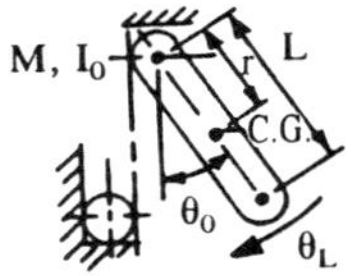	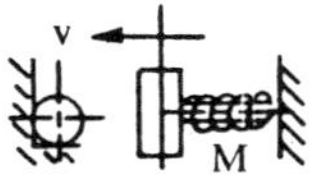
2. Kinetic Energy of Object at Start of Impact.	mgh	"zero"	$mgl(l-\cos\theta_0)$ [1]	"zero" [2]	"zero"
3. Initial Position for Equivalent[3] of: (a) Initial Energy, (b) Initial Velocity.	$h; E_i=mgh$ $h; v=(2gh)^{1/2}$	$H=m/M\,h$ $H=h$	[4] $\theta_0=\cos^{-1}(l-h/l)$ $\theta_0=\sin^{-1}(K\sqrt{2h/l})$	[5] $\theta_0=\cos^{-1}(l-mh/Mr)$ $\theta=\sin^{-1}(K/L\sqrt{2hR})$	$v=(2mgh/M)^{1/2}$ $v=(2gh)^{1/2}$
4. Contact Forces During Impact (FBD).	 1 force	 2 forces	 1 major force	 2 major forces	 2 major forces
5. Equivalent Inertial Forces at Start of Impact.		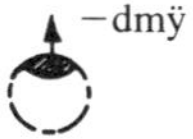	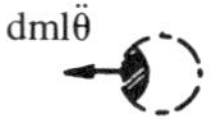		

1—not including the case when a ball attached to the string is striking the stationary object.
2—not including the case when the object is attached to the pendulum arm.
3—corresponding to the free fall drop height, h, of the drop test.
4—$K=[1+0.250\sin^2(\theta_0/2)+0.141\sin^4(\theta_0/2)+0.0976\sin^6(\theta_0/2)+\ldots]$; correction factor for large oscillations.
5—$R=I_0/(Mr)$; equivalent length of simple pendulum.

Table 8.4b Methods of Impact Test for Apples and Peachs and Specimen Preparation (Manor, 1978)

Specimen and Product	Impact Test	Parameter Measured	Results Obtained	Reference
Whole fruits	Drop tests on cushioned surfaces.	Drop height, bruise volume, fruit mass, fruit diameter, duration of impact.	1. Maximum drop height for certain cushioning material. 2. Minimum thickness of a cushioning material for certain drop height.	Mohsenin et al. (1978)
Whole apples and peaches	Portable drop tester.	Rebound time, drop height, bruise volume.	1. Coefficient of restitution vs. maturity. 2. Bruise volume vs. coefficient of restitution. 3. Bruise volume vs. drop height.	Diener et al. (1977)
Apple halves	Spring ram.	Displacement and acceleration of the projectile, impact and rebound velocities of the projectile, bruise volume.	1. Bruise volume vs. absorbed energy. 2. Bruise volume vs. stiffness.	Holt and Schoorl (1977)
Whole apples	Drop tests on hard surface.	Drop height, peak force.	Peak force vs. drop height.	Sarig (1976)
Whole fruits (apples)	Compound pendulum.	Fruit dimensions, peak force, duration of impact, area of contact, volume of indentation.	1. Maximum surface and internal shear stress. 2. Modulus of elasticity. 3. Energy absorbed. 4. Dynamic hardness.	Jindal and Mohsenin (1976)

Table 8.4b (Continued)

Specimen and Product	Impact Test	Parameter Measured	Results Obtained	Reference
Whole peaches	1. Drop tests on hard and cushioned surfaces. 2. Falling plunger. 3. A peach dropped on another peach.	Fruit dimensions and mass, drop height, impact energy, bruise depth and radius.	1. Percent of bruised fruits vs. drop height on hard and 2. Depth and location of bruise.	Kunze et al. (1975)
Whole fruits	Review of: 1. Drop tests. 2. Falling mass tests. 3. Pendulum tests.	Drop height, different measures of damage, peak force.	1. Damage vs. drop height. 2. Peak force vs. drop height.	Fluck (1975)
Whole apple	Drop test.	Numerical values for finite-element program were taken from Hamann (1970).	Contact stresses, internal stresses, principle stresses, indentation vs. time, shear relaxation modulus vs. time.	Rumsey (1974)
Pseudo-fruit	Drop test.	Maximum acceleration and duration of impact.	1. A pseudo-fruit as a means of obtaining quantitative evaluation of equipment. 2. Acceleration vs. time. 3. Hertzian shear stress during impact.	Rider et al. (1973)
Whole fruits, apple halves, cylindrical speci-	Falling mass.	Drop height, duration of impact, peak force, im-	1. Coefficient of restitution. 2. Force time curves.	Fluck and Ahmed (1973)

Table 8.4b (Continued)

Specimen and Product	Impact Test	Parameter Measured	Results Obtained	Reference
men of potato		pulse, attenuation.	3. Comparing curves for whole and halves of summer squash. 4. Energy absorbed by fruit. 5. Bruise vs. impulse and duration of impact.	
Peach halves	Compound pendulum.	Drop height, bruise volume, fruit dimensions and mass, elastic modulus of fruit and impacting surface.	Internal maximum shear stress vs. statistical pressure test.	Horsfield et al. (1972)
Apple halves	Compound pendulum.	Fruit dimensions and mass, drop height, bruise diameter, contact diameter, force time curve.	1. Peak force vs. drop height for different cushioned surfaces. 2. Maximum deformation vs. drop height for different cushioned surfaces. 3. Hertzian contact stress vs. drop height for different cushioned surfaces.	Simpson (1971) Simpson and Rehkugler (1972)
Whole apples	Dropping fruits on limbs and rods.	Mass of fruit, drop height, bruise dimension.	Bruise diameter vs. drop height for different cushioned surfaces of padded limbs and foam-protected steel rods.	Rehkugler and Burkwardt (1971)

Table 8.4b (Continued)

Specimen and Product	Impact Test	Parameter Measured	Results Obtained	Reference
a. Whole fruits (apples)	Simple pendulum of steel ball.	Drop and rebound heights, bruise depth and width.	Calculated maximum force, duration of impact, and maximum deformation vs. drop height, energy and momentum absorbed.	Plastic Impact of this section of this book
b. Whole apples	Simple pendulum.	Drop and rebound height, bruise depth maximum deformation.	Calculated dynamic yield pressure.	
Whole apples	Falling mass with or without cushioned surface.	Drop height, bruise volume, rebound height, peak force, weight of fruit.	1. Coefficient of restitution. 2. Energy absorbed. 3. Fragility factor. 4. Impact efficiency.	Hammerle and Mohsenin (1966) Mohsenin (1970)
Apples	Apple-limb impact.	Fruit diameter and mass, drop height, motion of fruit-limb impact by high-speed camera.	1. Deformation of apple vs. time. 2. Calculated force vs. time. 3. Calculated elastic modulus for apple.	Davis (1969) Davis and Rehkugler (1969)
Whole apples	Compound pendulum.	Dimensions and mass of fruit, drop and rebound height, bruise volume, radius of contact area. Experiments of two different temperatures of fruits.	Bruise volume, contact area, and coefficient of restitution for two different temperatures of	Nelson (1967) Nelson and Mohsenin (1968)

Table 8.4b (Continued)

Specimen and Product	Impact Test	Parameter Measured	Results Obtained	Reference
Whole apples	Simple pendulum.	Fruit diameter and mass, drop height, rebound height.	1. Energy absorbed by cushions and by impacting objects for two cushioning materials. 2. Rebound energy of fruit compared to that of wooden ball.	Bittner et al. (1967)
Peach halves	Falling mass as pendulum	Potential energy of the arm. index.	1. Initial kinetic energy to pendulum arm, bruise fruit. 2. Comparisons between cushioning materials.	Fridley et al (1964) cause yield deformation in

There is a difference in the rigid body motion between the free fall test and the other tests such as pendulum and impacting mass. In the free fall, all the elements of the falling object have the same velocity; the rigid body velocity at the start of impact. After the impact starts, the particle velocity is reduced in the process of momentum transfer. In contrast, a stationary material subjected to impact of a falling mass or pendulum has no particle velocity at the start of the impact. This stationary material behaves like a medium that is used for transferring the momentum of the falling mass to the base under the material. It follows that in the free fall only one contact force is generated, while in the impacting mass or pendulum two contact forces are generated, one between the falling mass and the impacted material and the second between the material and the surface on which the material lies. The magnitudes and distributions of the stresses in these two zones are dependent on the ratios between the impedances, "ρc" (density multiplied by the stress wave velocity) of the impacting mass and material being impacted. In terms of fruit damage, the free falling fruit may get only one

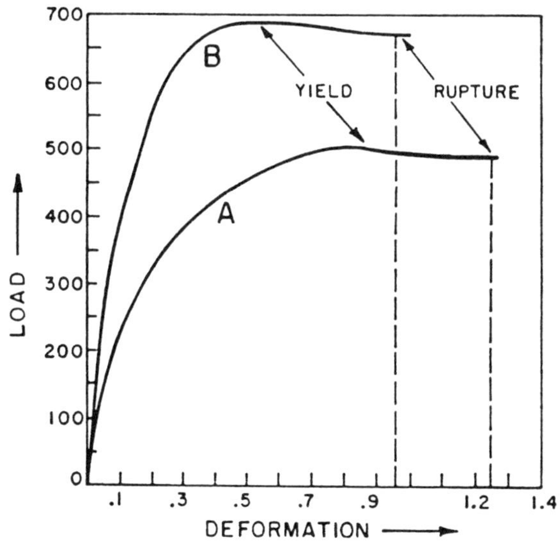

Figure 8.22 Hypothetical force-deformation curves at a given impact rate of loading for two agricultural products

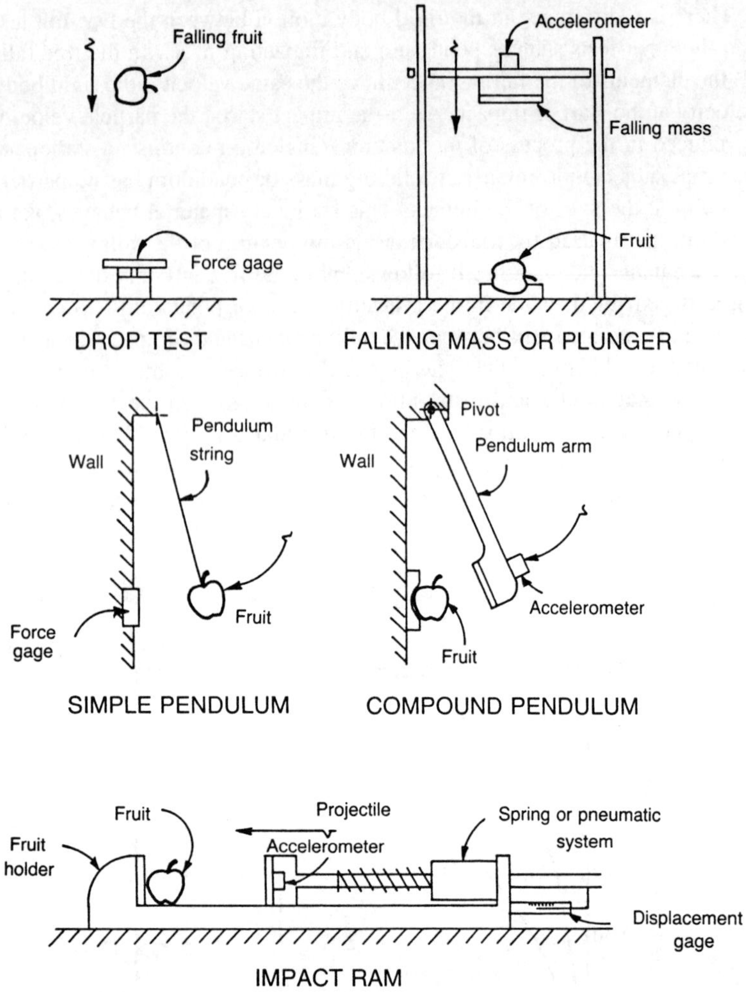

Figure 8.22a Different methods of impact tests used for fruits (Manor, 1978)

zone of damage at the vicinity of the contact, while in the other cases, two zones of damage can be created.

The simple drop tests, either of the product upon a hard or cushioned surface or of a mass upon the product, and also the pendulum test, have been widely used for fruit impact tests. Since the impulse measurements have been considered as very important parameters, most of the fruit impact testers have been instrumented to measure the impulse during the impact. It can be shown (Manor, 1978) that the impulse momentum law cannot correctly represent the phenomenon of the fruit impact. It is expected that the measured impulse, P, be always bigger than the rigid body momentum change, Δmv. That is:

$$P = \int_0^t F dt \geqslant \Delta mv \tag{8.20a}$$

where F = the measured contact force
t = the duration of impact

This inequality can be explained by the fact that each element of the impacted product, as a part of a continuous medium, transfers its initial momentum to adjacent elements which finally contribute to the measured impulse independently of the rigid body velocity. During the impact, the rigid body velocity (i.e., at the center of gravity) could be zero, or even in an upward direction, while the contact force still exists. This means that the elements at the bottom of the impacting body during contact are still exchanging momentum, and their particle velocities are in the downward directions. Consequently, there should be a difference between the measured impulse and the momentum change in the case of fruit impact and this difference might have a significant meaning. Nelson and Mohsenin (1968) called this difference momentum absorbed responsible for fruit damage.

One of the questions involved in any impact problem is the effect of inertia forces generated in the impacting body during impact. In this regard, the difference between rigid body mechanics and stress wave theory should be examined to evaluate the significance and magnitude of inertia forces in impacting of soft fruits.

While the rigid body motion represents only the whole mass of a body concentrated in one point (center of gravity), the continuum medium motion represents the directional motions of each small element in the body. It is possible, therefore, that in certain cases the kinetic energy represented by the rigid body is zero, but the body is exhibiting internal vibration, which means that the body has a kinetic energy. It is also possible that in certain moments the rigid body velocity at the center of gravity, having a contact with another body, is zero but at the contact surface there exists a motion (particle velocity) of the contacting elements. The differences in particle velocities of each element are the actual cause of the inertial force among the elements of a body in a dynamic situation.

It is common to assume that a quasistatic condition exists for certain dynamic-loading cases in which the duration of loading (or impact) is long enough. Goldsmith (1960), for instance, has stated that when the duration of impact is large, compared to the period of the lowest resonance frequency of the impacted sphere, the vibrational effects can be neglected. According to Goldsmith, in these cases of relatively long duration of impact, several reflections of waves are occurring during the contact period allowing the stress waves to travel back and forth in the impacting body. Under such conditions, the impacted bodies may be considered to be in a state of quasistatic equilibrium.

Engle (1976) has explained the justification for neglecting the vibrational effects in certain impact cases by another approach. According to Engle, the vibrational effects (stress-wave) can be neglected where the contact duration time is far longer than the natural periods of vibration of the system, thus avoiding amplification. Accordingly, the following inequality should be satisfied:

$$t^* \ll t \tag{8.20b}$$

where t^* is the period for the lowest natural frequency of the body and t is the duration of impact.

For the case of impact of fruits, there exist enough experimental data so that one can check the validity of neglecting the vibrational effects during the apple's impact. For example, experimental data reported by Manor (1978) have shown that the duration of impacts of apples ranged from 3.7 [m sec] for hard surfaces, up to 4.9 [m sec] for cushioned surfaces. These results are on the same order found by many other researchers such as Simpson (1971) who measured durations of 2 [m sec] up to 8 [m sec] for impacts of apples.

Furthermore, the experimental data of Abbott *et al.* (1968) have shown that the lowest resonant frequency of apples is of the order of 80 Hz. The second lowest resonant frequency, which has been associated with flexural vibration and has been strongly influenced by the size and firmness of apples, was on the order of 370 to 1488 Hz. The lowest frequency found by Finney (1970) was in the 300 to 400 Hz range. Cooke (1972) has considered the above first mode as the torsional class vibration (no radial motion and constant volume), and the second mode as the spheroidal class of vibration (radial motion).

If we consider the torsional first mode as the significant mode for the impact, then the period of this lowest frequency is:

$$t_1^* = \frac{1}{f_1} = \frac{1}{80} = 12.5 \times 10^{-3} \text{ sec}$$

It follows that in this consideration, the duration of impact is not longer than the period of the lowest resonance. Only if we consider the spheroidal mode, the mode that exhibits radial motion, and take the lowest frequency, 300 Hz, as the lowest limit, we find that t_2^*=3.3 [m sec] which means that $t^* \leqslant t$.

The instantaneous velocity, v_i, of a moving point within the medium is called the particle velocity of that point. So a particle velocity can be assigned to every point of a disturbed medium. In an elastic body transmitting a wave, the particle velocity at any point in the wave is linearly related to the instantaneous stress at that point (Rinehart, 1975). The relationship between stress, particle velocity and the material parameters can be derived by applying the principle of impulse momentum, to the stressed region. These relationships for one dimensional waves become:

$$\sigma_{ii} = \rho c_1 v_i, \quad c_1 = (E/\rho)^{1/2} \tag{8.20c}$$

for longitudinal waves, where v_i is particle velocity directed along the direction of the wave propagation, c_1 is velocity of wave propagation, ρ is density, E is elastic modulus, and σ_{ii} is a tensile or compression stress. Similarly, it can be shown that for distortion waves:

$$\sigma_{ji} = \rho c_2 v_j, \quad c_2 = (G/\rho)^{1/2} \tag{8.20d}$$

where the particle velocity, v_j, is perpendicular to the direction of the wave propagation, c_2 is wave propagation velocity, ρ is density, G is shear modulus, and σ_{ji} is a shear stress.

The proportionality constant "ρc" is called the "impedance" of the material. If, for example, c_1 of the apple material is 150 [m/sec], and the apple's mass density is ρ=800 [kg/m^3], then the impedance of an apple is ρc_1=1.20×10^5 [kg/m^{2}sec]. That is, for a particle velocity of v_1=1 [m/sec], the stress will be σ_{11}=1.20×10^5 [N/m^2] or (17.4 PSI). For comparison, the same particle velocity in steel (c_1=5800 [m/sec], ρ=7800 [kg/m^3]) will create a stress of $\sigma_{11}=4.52\times10^7$ [N/m^2], which is about 380 times the stress in the apple.

The use of stress wave velocities in fruits is only a specific use of equations taken from stress wave theory. It provides a basis for a complete analysis of fruit impact based on stress wave theory. However, care must be exercised when assuming the fruit material as a continuous medium. The effect of any discontinuity, such as the air spaces in the tissue, is to attentuate severely the amount of momentum transmitted to an element of material (Rinehart, 1975). Indeed, experimental results of Garret (1970) have shown severe attenuation of pulses propagating through the material of apples. The plastic and shock wave theories, which consider the discontinuity in density, pressure and temperature, have, therefore, a better potential to analyze the fruit impact phenomenon.

To test the validity of the above theories, Manor (1978) used the drop test apparatus shown in Fig. 8.22b. It consisted of the following items: a vacuum fixture releasing the fruit to drop on its cheek (without rotation), a dropping fixture which enabled the drop height to be varied, different

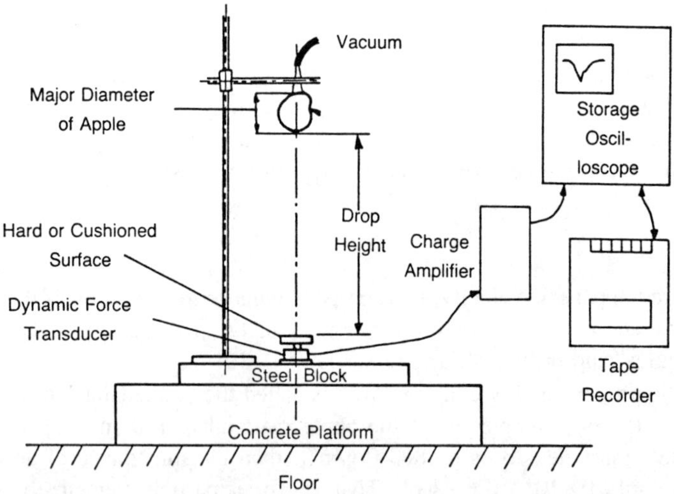

Figure 8.22b Drop test apparatus and the associated instrumentations (Manor, 1978)

kinds of surfaces (cushioning materials) placed on the plate of the force gauge, a heavy block of cast iron (50 kg) on which the force gauge was mounted, and a heavy concrete block (300 kg) as a base for the drop test apparatus. This heavy base, made of concrete, ensured that the momentum transferred by a light (200 grams of apple compared to 300kg of the block) falling product would not cause a significant movement of the base during impacts. It was possible, therefore, to assume that the momentum change of the base is negligible in terms of the rigid body theory. The electrical setup was designed to fill the requirements of fruit impact measurements.

The dynamic force gauge (Endevco, Model 2103-100) had a sensitivity of 29.44 [pcb/N] (131 pcb/lb) and a natural frequency of 15,000 Hz. This force gauge had a satisfactory response for forcing frequencies up to 1500 Hz, which was higher than the equivalent frequency of fruit impact, 300 Hz.

The charge amplifier (Kistler, Dual Mode Amplifier, model 504E) had a very high input impedance which allowed for a dynamic response from an input frequency of up to 100 kHz, without distorting the output.

The oscilloscope (Nicolete 1090A Explorer Digital Oscilloscope) had a built-in amplifier with a band width of 5 MHz (-3 db). The oscilloscope also had a built-in triggering system which used the input signals, provided they were bigger than 0.25 volts. The response of the trigger was fast enough to generate recording of most of the starting signal.

Each apple was measured, weighed, and dropped on its cheek transverse to the major diameter. The apple was caught before a second impact occurred to avoid accumulating bruising. The fruit was then measured for firmness by a Magness-Taylor pressure tester and left for 24 hours at room temperature for bruise measurements. Bruise was measured by sectioning the fruit at the center of the damaged area and then measuring the depth and average diameter of the browned tissue by a vernier caliper.

Fig. 8.22c shows the measured parameters obtained from the trace on the oscilloscope. The measured parameters and those evaluated from the recorded information were as follows:

(a) Drop height, h, was the vertical distance between the bottom of the apple and the impacted surface [cm].
(b) Duration of impact, t, was the time between beginning and end of contact [m sec].
(c) Maximum force, F, the peak force during the impact shown on the force-time trace [N].

(d) Impulse, P, was the integral of the force along time, $P=\int_0^t F\,dt$ [N-sec]. It is equivalent to the area under the force-time curve. The area was measured from the oscilloscope trace by using a planimeter.
(e) Initial momentum, mv, the momentum of the dropped apple at the beginning of impact [N-sec]. The relevant velocity, v, was calculated from the drop height, $v=(2gh)^{1/2}$ [m-sec], and m [kg] is the measured mass of the apple.
(f) Bruise volume, B, the volume of the damaged tissue [cm^3], evaluated as given by Eq. (8.26).

In conducting tests, first apples were dropped on other apples. Both the dropped and the stationary apples were impacted on their cheeks whereas the stationary apple was set on the hard surface, the plate of the dynamic force gauge. The bruise volume was determined in the same way as described earlier for fruits dropped on hard or cushioned surfaces. Next, two different tests were conducted to determine the source of variability in the bruise volume measurement. (1) One apple was dropped on a cushioned surface from a drop height of 1.0 m. Then it was cut into halves at the center of the bruise. The depth and width of the bruise were measured, ten replications on each half. (2) Nineteen apples were dropped on a hard surface, and then each one was measured for the depth and width of the bruise.

A rubber ball, considered as a sphere made of elastic material, and a molded clay ball, considered as a sphere-like object made of plastic material, were dropped on a hard surface from the drop height of $h=0.86$ m. The balls had the same mass, $m=8$ grams, but due to different mass densities, their diameters were different; the diameter of the rubber ball was 2.54 cm, while the clay ball had a diameter of 2.21 cm. The recorded force-time curves obtained from the drop tests are presented on a common scale in Fig. 8.22d. The rebound height of the rubber ball was measured, while in the case of the molding clay ball, no rebound height was detected.

The drop test method was used to supply numerical data on the duration of impact, t, maximum impact force, F, the measured impulse (area under the force-time curve), P, and the resulting bruise volume. Table 8.4c presents the means and standard deviation of each variable of eight different treatments.

Table 8.4d presents the measured values obtained from apple-on-apple impact tests. It can be seen from the table that the bruise volume created in the stationary apple was always much bigger than the bruise volume of

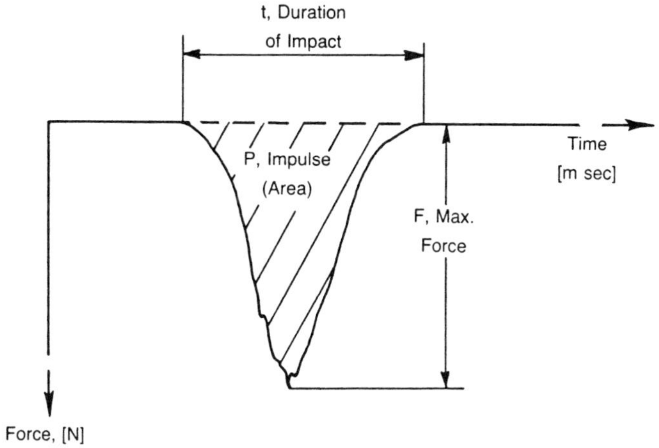

Figure 8.22c Typical force-time curve recorded from drop test of apples with the obtained parameters (Manor, 1978)

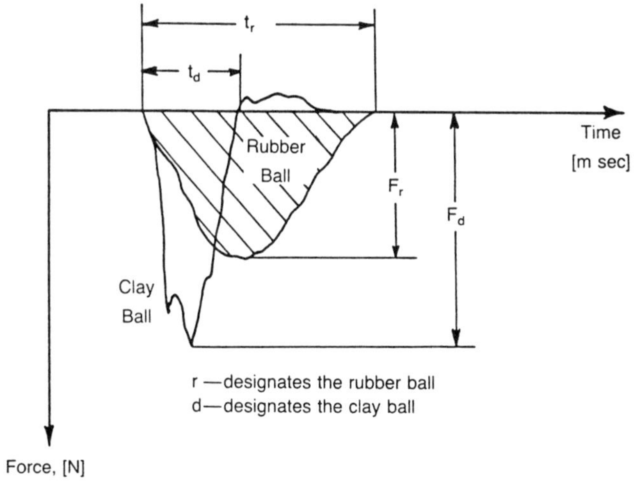

Figure 8.22d Force-time curves for rubber and molding-clay balls of the same mass dropped from the same drop height. t—duration of impact; F—maximum force (Manor, 1978)

Table 8.4c Means and S.D. of Eight Different Drop Experiments of Rome Beauty Apples on Hard and Cushioned Surfaces (Manor, 1978)

Treatment	Number of Apples	Mass of the Apple (kg) –m–	Major Diameter (cm) –d–	Height of Drop (cm) –h–	Surface of Impact	Initial Momentum (N-sec) –mv–	Max. Force (N) –F–	Duration of Impact (m sec) –t–	Measured Impulse (N-sec) –P–	Bruise Volume (cm^3) –B–
1	8	0.236 (0.042)	8.76 (0.49)	25.4	Hard*	0.526 (0.094)	394 (66)	4.72 (0.42)	0.983 (0.191)	3.680 (1.487)
2	6	0.227 (0.016)	8.78 (0.24)	25.4	Cushioned**	0.506 (0.037)	180 (14)	13.71 (0.68)	1.047 (0.080)	0
3	6	0.203 (0.023)	8.56 (0.58)	19.05	Hard*	0.392 (0.045)	289 (16)	5.26 (0.61)	0.733 (0.084)	2.710 (0.791)
4	6	0.203 (0.023)	8.21 (0.40)	12.7	Hard*	0.320 (0.037)	243 (28)	5.20 (0.51)	0.622 (0.079	1.691 (0.404)
5	4	0.206 (0.008)	8.57 (0.14)	6.35	Hard*	0.230 (0.009)	182 (16)	5.39 (0.90)	0.452 (0.018)	1.073 (0.326)
6	5	0.189 (0.015)	8.41 (0.25)	45.72	Cushioned**	0.594 (0.044)	266 (17)	10.23 (0.18)	1.281 (0.115)	0
7	6	0.216 (0.038)	8.48 (0.41)	85.00	Cushioned**	0.882 (0.155)	504 (89)	8.99 (0.54)	1.838 (0.303)	0.841 (1.048)
8	8	0.248 (0.022)	8.99 (0.34)	65.00	Cushioned***	0.885 (0.080)	627 (61)	7.01 (0.39)	1.660 (0.152)	3.576 (1.694)

* Plastic plate, thickness 1.2 cm, with material's impedance of $\rho c = 8 \times 10^5$ kg/(m^2–sec).

** Polyvinyl closed cell 1.27 cm (½ in) thick, 64.4 kg/m^3 (402 lb/ft^3) density.

*** Same as above but one-half the thickness.

Table 8.4d Measured Values and Parameters Evaluated for Apple-on-Apple Impact Experiments (Manor, 1978)

Number	Mass of the Apple (kg) –m–	Major Diameter (cm) –d–	Drop Height (cm) –h–	Initial Momentum (N-sec) –mv–	Duration of Impact (m sec) –t–	Max. Force (N) –F–	Measured Impulse (N-sec) –P–	Bruise Volume (cm^3) –B–
1*	.189	3.16	45	0.4932	8.70	319	0.8884	3.2329
2**	.166	3.04	45					0.2177
3*	.200	3.38	45	0.5170	7.65	382	0.9549	3.9745
4**	.174	2.96	45					2.6739
5*	.194	3.34	45	0.5319	7.80	361	0.9217	4.6032
6**	.179	3.20	45					0.00
7*	.201	3.10	45	0.5611	7.60	418	1.0545	7.8778
8**	.189	3.15	45					0.1854
9*	.201	3.37	45	0.5200	7.10	363	0.9382	2.1719
10**	.202	3.20	45					1.3518
11*	.180	3.14	45	0.5170	6.80	385	0.9715	3.1857
12**	.174	3.08	45					0.9985
13*	.196	3.21	45	0.5140	8.00	351	0.9134	4.5313
14**	.173	3.14	45					0.0989

* Stationary fruit, set on the plate of force gauge.
** Dropped fruit on the stationary one.

the dropped apple. Furthermore, the dropped fruit got only one bruise, at the vicinity of the contact force, while the stationary fruit got two bruises, in the vicinity of the two contact forces. It was observed that the falling apple behaved like an apple dropped on a cushioned surface, while the stationary apple responded like an apple subjected to an impact of a falling mass. The measured duration of impact for apple-on-apple was shorter from the values obtained for apples dropped on cushioned surfaces (from the same drop height), but it was longer than the duration of impact obtained from apples dropped on a hard surface (see Table 8.4c). These observations showed that an apple-on-apple impact is a special case. It is not similar to the simple drop test.

The following data shows the means, standard deviations, and coefficients of variation for the depths, diameters of bruise, and the calculated bruise volumes, based on 20 measurements of a single apple.

Parameter	Mean	S.D.	c.v.
Depth of bruise [cm]	0.6216	±0.0678	10.9%
Diameter of bruise [cm]	2.882	±0.1365	4.7%
Calculated bruise volume [cm^3]	2.164	±0.3548	16.4%

In addition, analysis of variance showed that no significant differences were obtained between the two groups of ten measurements on each apple half. Actually, the coefficient of variation obtained for the bruise volume is a combination of the variability of both the depth and diameter of bruise, as expressed by the formula for bruise volume. However, the variability obtained for the depth measurement (c.v.=10.9%) is surprisingly high, considering that the measurements were conducted in a short time period, by the same person, and with the same vernier caliper.

The means, standard deviations, and coefficients of variation for the measurements of 19 apples dropped on a hard surface from a drop height of 25 cm were as follows:

Parameter	Mean	S.D.	c.v.
Mass of apple [kg]	0.158	±0.0083	5.3%
Firmness, M-T [N]	84.35	±5.84	7.0%
Depth of bruise [cm]	1.205	±0.112	9.3%
Diameter of bruise [cm]	2.989	±0.338	11.3%
Bruise volume [cm^3]	5.257	±1.38	26.3%

The results showed that the variations among apples were different from those obtained for measurements on a single apple. The fact that c.v. among apples for depth of bruise is lower than the value obtained for a single apple can indicate that the variability of bruise measurements is a function of the bruise shape and characteristic. That is, for a cushioned surface and a high drop height (1 m), the shape of the bruise is different than for a hard surface and a lower drop height (25 cm). The accuracy in measuring the bruise dimensions and calculating the bruise volume is, therefore, dependent on the bruise appearance. Nevertheless, it can be seen that even if the dimensions of the bruise were bigger (as on the hard surface) and, therefore, easier to measure, the c.v. of the calculated bruise volume was higher. The above experiments were conducted only to show the possible variability of using small samples, as found in most of the past work.

In addition to the force-time curves recorded during the impacts, (Fig. 8.22d), the coefficients of restitution were also obtained. For a rubber ball the coefficient of restitution was found to be $e=0.87$. In the case of the clay ball, no rebound height was detected. The maximum force, F, was higher by 57 percent for the clay ball, while the duration of impact, t, was longer by 57 percent for the rubber ball. The momentum to impulse ratio of the rubber ball was 46 percent, while for the clay ball it was 77 percent.

These results showed that even for elastic material, the measured coefficient of restitution was $e=0.87$. This led to the conclusion that one cannot consider the coefficient of restitution, measured from fruit impact, as a direct measure of the amount of energy required to bruise an impacted fruit. On the other hand, the fact that for the plastic material $e=0.0$ would indicate that some relationship does exist between the coefficient of restitution and plastic deformation during impact. Apparently, a function that includes the initial momentum to impulse ratio has the potential to indicate plastic deformation of fruits during impact.

8.7 DAMAGE UNDER DEAD LOAD

Economic conditions in recent years have made it imperative that producers use bulk transportation and storage of agricultural products. As a result of this bulk handling, the material may be subjected to excessive pressures resulting in external and internal injuries in the form of distortion, cracks, internal bruising and other injuries affecting the quality and grade of the material.

Because of the time element together with dead loads involved in bulk handling, the mechanics of the force-deformation relationship is usually a creep phenomenon. In cases where the material is packed in containers with a tight lid, analysis of the action of the forces may lead to a stress relaxation phenomenon. The analysis of this problem from the rheological point of view has already been discussed for several agricultural products in Chapter 5.

In an attempt to derive readily used relationships which would be applicable to practical situations, Nelson (1967) investigated the dead loads required over a period of time to cause undesirable distortion or internal browning in McIntosh apples. Fruits were subjected to dead loads for a period of 24 hours at 90°F and 100 hours at 40°F. Figure 8.23 shows the fixture used for measuring deformation of fruits under dead load. Three fruits of the same approximate diameter were grouped together under three equal weights at the three corners of an equilateral triangle. To avoid rolling of the apples under load, they were affixed on a small nail projecting from the base of the testing shelf. The final contact area under the dead load was obtained from a record of contact on white paper affixed to the bottom of the triangular plate. With this arrangement the arithmetic average deformation of three specimens could be read from the dial gage mounted at the center of the triangle. The free swinging arrangement of the fixture enabled several readings to be taken at all the test shelves without disturbing the fruits under dead loads. An automatic readout system for this type of

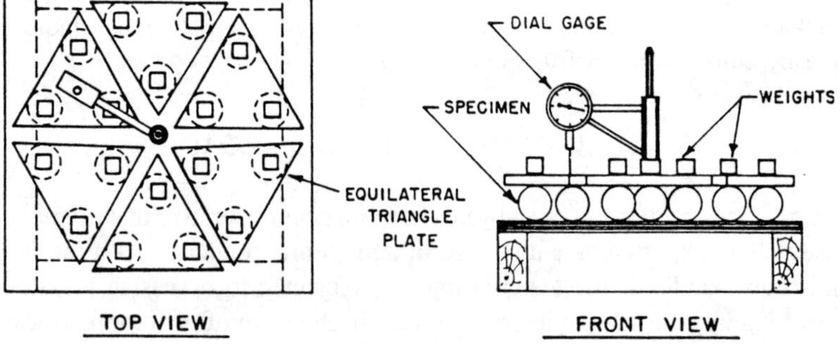

Figure 8.23 Setup for measuring distortion of fruits under dead load (Nelson, 1967)

test employing *LVDT* displacement transducers was employed by Mohsenin (1963).

The results of these tests gave the following regression equations for predicting bruise volume, V, or surface distortion, D, from force, F, or pressure, P, as a result of dead loads on the fruit at 40°F.

$$V = 0.00028F - 0.00055 \quad \text{or} \quad V = 0.00005P - 0.0006 \tag{8.21}$$

$$D = 0.00415\,F + 0.0023 \quad \text{or} \quad D = 0.00077\,P + 0.001 \tag{8.22}$$

Due to short duration of dead load tests at 90°F, the results showed no definite correlation.

Example

Assume that tests of a sample of McIntosh apples to be handled by pallet boxes shows 166 apples per cubic foot, an average diameter of equivalent sphere of 2.3 inches, and an average weight of 0.28 pound per apple. If a bruise volume of not more than 0.5×10^{-3} cubic inches is specified for these apples as the maximum allowable bruise, find the acceptable depth for the pallet box in which these apples are to be stored for a minimum of 100 hours at 40°F.

An analysis similar to that used by Ross and Isaacs (1961) for forces acting on stacks of granular materials may be used to estimate the forces acting on a fruit at the floor of a bulk bin. Figure 8.24 shows the rhombic stacking model assumed for this analysis and the forces acting on an apple somewhere in the stack. Assuming that apples are represented by spheres of an average characteristic diameter d with a stack arrangement which depend on their characteristic diameter and their number per unit volume, the following equation has been given for the arrangement shown in Fig. 8.24

$$N = \frac{1}{4d^3 \cos^2\theta \sin\theta} \tag{8.23}$$

where N is assumed to be the number of apples per unit volume and θ is the angle between a horizontal plane and the θ-axis. Furthermore, Ross and Isaacs (1961) give the following equation for total vertical force F_v acting on a particle at any layer n on the stack.

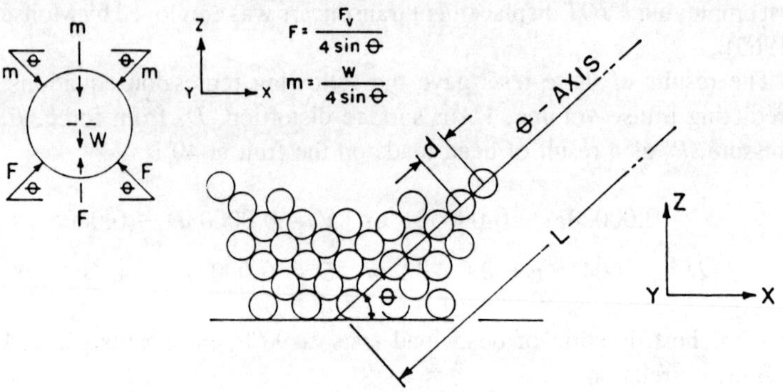

Figure 8.24 An elevation view through the stack of apples in a rhombic arrangement

$$F_{v_{(n)}} = L_\theta W/d \tag{8.24}$$

where L_θ is the average length of the four θ-axis and W is the weight of the particle.

Using Eq. (8.21) the maximum allowable force on an apple at the bottom layer can be estimated as follows.

$$0.5 \times 10^{-3} = 0.00028F_v - 0.00055$$

$$F_v = \frac{0.00105}{0.00028} = 3.75 \text{ lbs}$$

substituting F_v, d, and W in Eq. (8.24) for the bottom layer of apples,

$$L_\theta = F_v d/W = \frac{3.75 \times 2.3}{0.128} = 30.8 \text{ inches}$$

And angle θ can be solved from Eq. (8.23)

$$166 = \frac{1}{4(2.3/12)^3 \cos^2 \theta \sin \theta}$$

or

$$\theta=60.2 \text{ degrees}$$

Therefore, the maximum depth to limit mechanical damage to the specified value is, by reference to Fig. 8.24

$$\text{depth}=L_{\theta}\ \sigma\iota\nu\ \theta=30.8 \sin 60.2 =26.7 \text{ inches}$$

Experiments with 1.5-inch diameter solid plastic spheres stacked in a one-foot square box with a force transducer in its false floor have shown that Eq. (8.24) is valid for cubic and cubic tetrahedral packing arrays of spheres. Other types of packing arrays deviated to some extent from the straight line relationship given by Eq. (8.24). In all types of the packing arrays, however, the theoretical relationship was true only to the point at which the θ-axis intersected the bin wall. Additional layers showed a leveling off of the curve force versus number of layers above the load cell.

This experiment indicated that in designing a bulk bin after the maximum allowable depth is determined, the angle of the θ-axis may be used to calculate the floor dimensions of the bin such that the θ-axis of an apple in the center of the floor would intersect a wall before the allowable depth is exceeded. Using these dimensions the container can then be filled as high as desired without any increase in the force on the fruit at the bottom layer.

8.8 COMPARISON OF STATIC AND DYNAMIC RESISTANCE TO MECHANICAL DAMAGE

Experiments with engineering materials over the years have shown that analysis designed for cases of static loading usually do not apply to the corresponding impact process and discrepancies between results from the two sets of tests are often quite large.

Andrews (1952) has reported, however, that attempts to correlate the dynamic and static data on the basis of an approximate method in viscoelasticity have been reasonably successful. In this work dynamic complex modulus of rubberlike materials measured with sinusoidal vibrations was correlated with static modulus of the same materials measured in experiments of relaxation of stress at constant strain.

To compare static and dynamic hardness of metals, Bowden and Tabor (1954) dropped hard steel balls onto massive anvils of various metals and calculated a dynamic yield pressure from height of drop, height of rebound, and the chordal diameter of indentation. Then static experiments were carried out on the same specimens to determine the static yield pressure required to produce the impressions of the same diameter as in the corresponding impact experiments. Although these investigators did not attempt to correlate the results of their static and dynamic tests, their technique should be applicable to several foods and agricultural products. Whether or not a satisfactory correlation can be established for any of these biological materials is a question which needs to be investigated.

Published work with agricultural materials includes both static and dynamic test data with little attempt to compare the two types of behavior. Few examples where such comparison has been made are found in connection with strength of eggshell, resistance to bruising of fruits and vegetables, and breaking strength of seed grains.

Mueller (1957) reported the following correlation coefficients for eggs subjected to puncture by pressure and by impact.

	Puncture by impact	Shell thickness	Specific gravity
Puncture by pressure	0.563	0.787	0.703
Puncture by impact	—	0.353	0.365
Shell thickness	—	—	0.756

Correlation coefficients larger than 0.321 were significant at 0.1% level.

In comparing strength of eggshell under impact and compression, Tyler (1963) reported that under compression when the effect of thickness is taken out, results with thousands of eggs have shown lowest breaking strength at the minor diameter and highest at the narrow pole with the broad pole intermediate. Under impact the results are reversed with the broad pole again intermediate. Then he comments that "evidently these two methods are measuring different things." It should be pointed out that the impact methods that Tyler speaks of are all based on increasing either heights of fall or weight of the impacting device which, as will be discussed later, is not a valid measurement of resistance to impact.

Sluka *et al.* (1965) compared the results of their hydrostatic tests of egg (Fig. 5.24a) with impact tests employing a standard drop testing apparatus.

In impact tests the eggs were held in position with a cushioning pillow on one pole of each egg while the other pole was imbedded in "silly putty". A correlation coefficient of 0.45 was obtained between the hydrostatic compression *P* and the acceleration ratio *G* required to cause shell failure. The correlation was significant at the 5% level.

Mohsenin and Goehlich (1962) compared the results of their quasistatic tests on apple with results of impact tests using an apparatus which consisted basically of an impact arm and an *LVDT* displacement transducer. The energy of impact was calculated from the height of fall and the effective mass of the falling fruit-impact arm system. The maximum deformation due to impact and the permanent deformation due to single impact were also found from the output of the *LVDT* (linear variable differential transformer). It was shown that energy required for bruising segments of five varieties of apples was 1.5 to 2.7 times that required under quasistatic compression tests. When the plunger as the impacting surface was replaced with either a flat plate or another apple segment of the same fruit, the required bruising energy under impact was again about twice that under the corresponding compression energy as shown in Fig. 8.25. As will be seen later, other investigators have shown that the same trend seems to exist for several other agricultural products. This behavior is also in agreement with that found for such non-biological materials as metals, as discussed earlier in this chapter.

By means of another approach and a different method of experimentation the same relationship was found to exist between compression and impact energies required to cause injury in apples. A series of compression and impact tests, using a ½-inch steel ball, were conducted on McIntosh apples

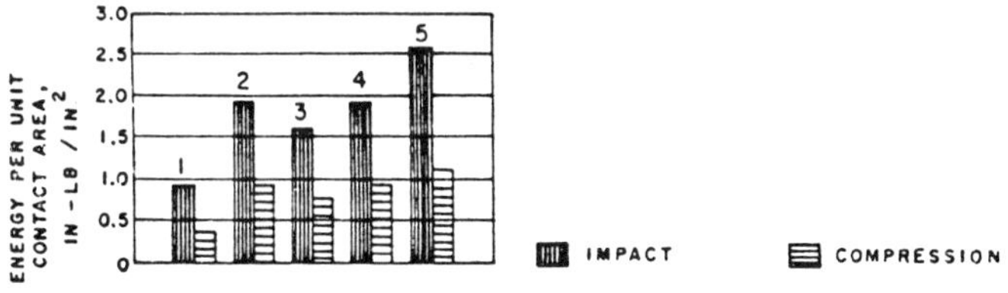

Figure 8.25 Energy requirement to bruise apples at the same stage of maturity (Mohsenin and Goehlich, 1962). 1—McIntosh, 2—Golden Delicious, 3—Delicious, 4—Stayman, 5—Rome Beauty

using the testing machine of Fig. 4.22 and impact apparatus of Fig. 7.26. Compression tests were quasistatic under loads of approximately 2, 3, 4 and 5 lbs. These loads were applied on different spots on each of the 20 selected apples. Bruise volume was calculated from maximum depth and average width of each bruise. The bruising energy was taken as the total compression energy input determined from the area under the loading part of the force-deformation curve.

Impacts tests consisted of striking the apple with a ½-inch steel ball from the heights of 4″, 6″, 8″, and 10″. Bruise volume, in a manner similar to the compression tests was determined on the same 20 apples. Total energy input was determined from the weight of the ball and the height of drop.

Figure 8.26 shows the relationship between bruise volume and total input energy for both compression and impact tests. It shows that the same input energy the apple bruises more severely under compression than under impact. Furthermore, for the same amount of bruising the ratio of impact energy to compression energy is about 1.5.

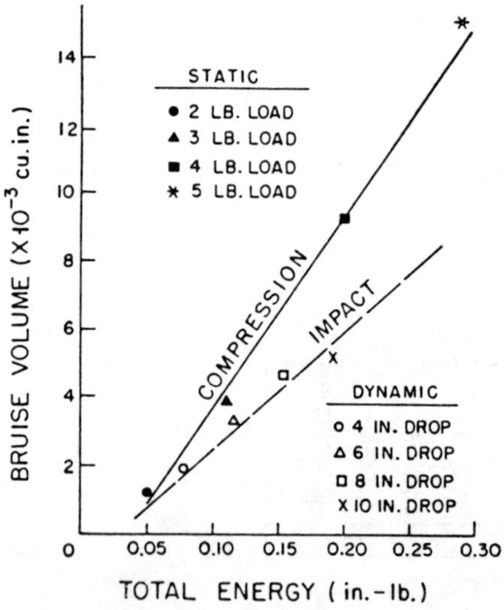

Figure 8.26 Bruise volume of apples at various energy levels of compression and impact (Each point is the average of 20 tests by means of a 1/2-inch steel ball)

Witz (1954) in his experiments with a number of different varieties of potatoes tested for resistance to static penetration of a 0.155-inch plunger and impact indentation by a ¾-inch steel ball found no significant correlation between static and dynamic tests.

Figure 8.27 shows comparison of impact energy and compression energy required to break corn kernels and pea beans by shear. In these tests the grain held in a vice was broken off by a combination of bending and shearing force resulting from either a quasistatic compression or the impact of a simple pendulum. Note that beyond a certain moisture content the grains are more resistant to impact shear than compression shear. Figure 8.28 shows the effect of moisture on mechanical damage to wheat by impact shear. According to these data, energy required to break the grain by impact shear increases with increase of moisture content.

In comparing the static and dynamic strength of corn kernels subjected to high temperatures in drying, Chizhikov (1960) reported that for some varieties, at temperatures about 60°C, greater force was required to rupture the kernel under static conditions of parallel plate loading than under hammer-like action of dynamic loading. Considering the rather crude method of test and analysis in the dynamic tests, the above conclusion can be subject to question.

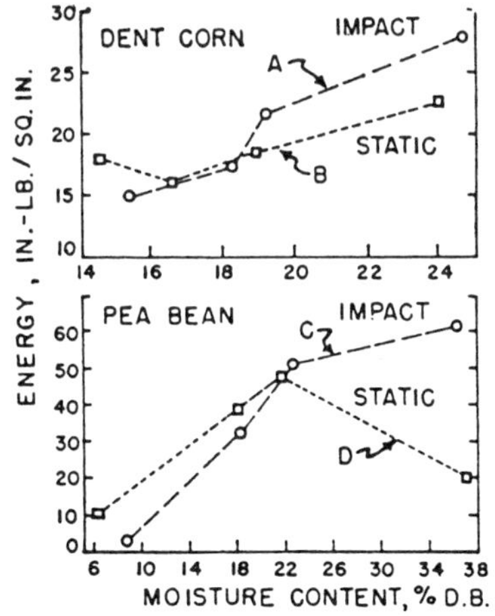

Figure 8.27 Comparison of static and dynamic shear energies required to rupture grains (Zoerb and Hall, 1960)

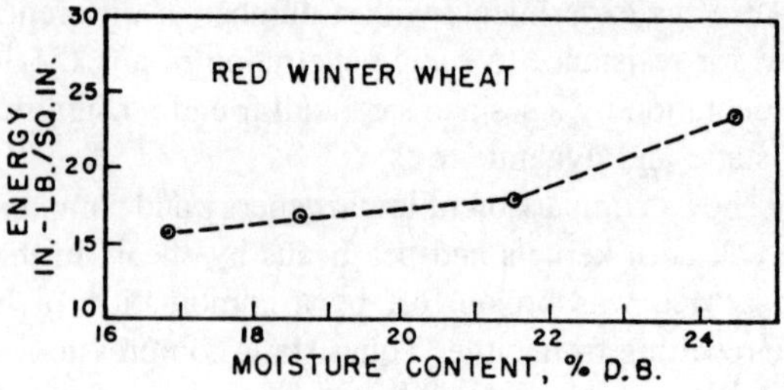

Figure 8.28 Energy required to rupture wheat kernels by impact shear (Zoerb and Hall, 1960)

Clark *et al.* (1967) compared the effect of slowly applied loads and impact loads on energy absorbed versus germination of cottonseed. As shown in Fig. 8.29, beyond 3⁄16 in-lbs of energy, for the same energy level, quasistatic loading caused greater reduction in germination than impact loads. This is another experimental evidence in support of the hypothesis that agricultural products can withstand greater mechanical energy under dynamic loading than under slowly applied quasistatic loading.

From the data reported by Bilanski (1966) on strength of five grains, a comparison of breaking energies under impact and quasistatic loading is not possible. However, from the data reported on energy required to break barley and oats under high velocity and medium velocity loading tests, it is evident, as seen from Table 8.5, that barley takes more energy to break as loading rate increases. This, of course, is expected as it has been shown in the case of other products. However, the trend is reversed for oats. The reason for this change of usual trend cannot be explained without further tests with consideration of the differences in structure of the oats kernel as compared with other grains. The oats kernel is known to have a thicker hull which could take more deformation at slower rate of loading resulting in greater energy to break.

The fact that greater amount of energy is required to produce a deformation of given volume under impact than under static loading was also reported by Bowden and Tabor (1954) in connection with their hardness tests of metals. These authors suggested that the displacement of larger volumes of metals under dynamic loading is the result of viscous flow of

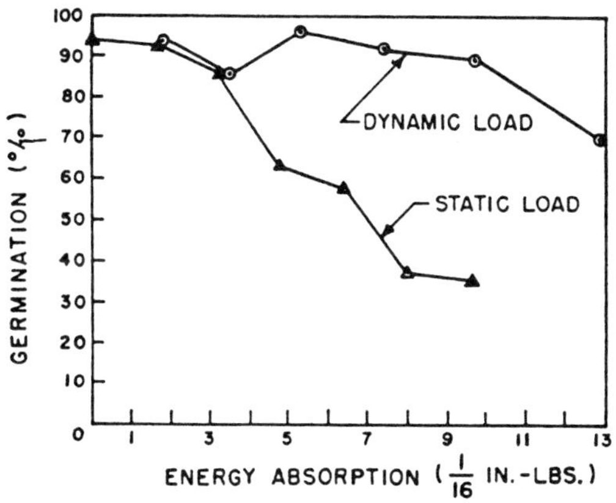

Figure 8.29 Comparison of static and dynamic loads on cottonseed germination (Clark *et al.*, 1967)

Table 8.5 Resistance to breakage of barley and oats under high velocity and low velocity impacts (Data summarized from Bilanski, 1966)

Grain	Moisture %	Minimum and maximum values of breaking energy (in-lbs)	
		Low velocity impact[1]	High velocity impact[2]
Barley	1.0	0.080–0.238	0.077–0.308
	10.0	0.116–0.365	0.148–0.442
	17.0	0.225–0.462	0.395–0.812
Oats	1.0	0.150–0.315	0.0037–0.149
	10.0	0.340–0.634	0.089–0.193
	17.0	0.476–0.718	0.100–0.355

[1] Impact of a pendulum of a known weight dropped at a known height.

[2] Rotating paddle.

the deformed material surrounding the indentation and, therefore, forces of "quasi-viscous" nature must be present under these conditions. This conclusion was substantiated by the fact that the dynamic yield pressures calculated from the rebound height were closer to the corresponding static yield pressures than those calculated on the basis of total energy part of which was used to produce plastic flow. If this is true, then for biological materials which are viscoelastic in nature, the difference between the energy values required to produce a given amount of deformation must be even greater. Furthermore, this trend should be stronger as moisture content of the material increases. The examples given from work by Zoerb and Hall (1960) on grain and others should support this hypothesis.

8.9 VIBRATION DAMAGE

Little work has been reported in relation to vibration damage during transport of agricultural products. Transit injury to fruits has been investigated by O'Brien *et al.*, (1960, 1965). According to these investigators transport damage in fruits, referred to as "roller bruising," is an important factor affecting the quality of fresh and processed fruit (see Fig. 1.6). The cause of the damage is stated to be fatigue due to repeated forces of vibration on the fruit resulting in cell rupture beneath the skin.

The intensity and duration of vibration will determine the severity of damage. In an attempt to determine the causes of in-transit fruit damage, accelerometers and the appropriate readout and recording system were employed in simulated transport tests and laboratory vibration tests (Fig. 8.30). Since vibration damage is due to the motion of the fruit in the pack (bin or lug boxes), the magnitude of acceleration measured in *g* was considered as the criterion for evaluating the intensity of vibration. It was shown that the magnitude of acceleration in the upper layers of fruit where most of the damage occurred (Fig. 8.31) depended on (1) the depth of fruit in the container (O'Brien *et al.*, 1963), (2) the tightness of fill (Guillou, 1963), (3) the type of suspension system used in the truck, (4) the magnitude of forced vibration from the road bed, and (5) the vibrating characteristics of the fruit species.

The importance of the fruit natural frequency in design of the suspension system for fruit-transport trucks in an attempt to keep the resonance frequency of the truck away from the range of fruit resonance frequency has already been discussed in Chapter 5 (see Fig. 5.37). The curves given in Fig. 5.37 can be used as a guide for selecting spring factor or stiffness

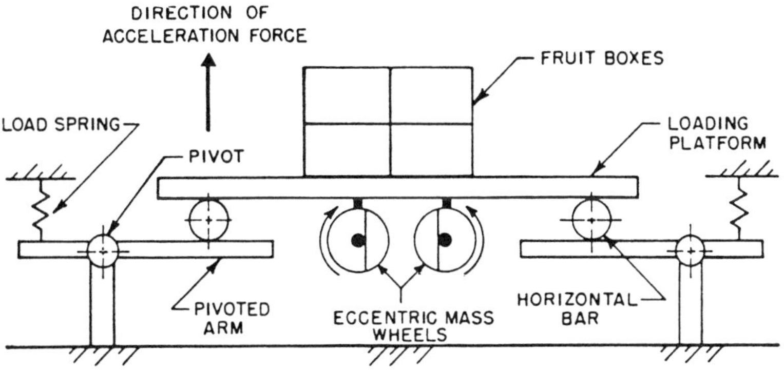

Figure 8.30 Simulated laboratory test of fruit vibration in transport (O'Brien *et al.*, 1960)

values for several fruits from which the resonance frequency can be calculated. In regards to other factors, insofar as design of the fruit-hauling trucks is concerned, following are some of the recommendations offered (O'Brien *et al.*, 1960):

1. A more rigid bulk bin so as to raise bin natural frequency.
2. Softer truck tires to lower truck's natural frequency.
3. Softer air suspension or softer springs to isolate the load from the truck tires.
4. A system of forced spring lubrication to insure independent movement of leaves in springs.

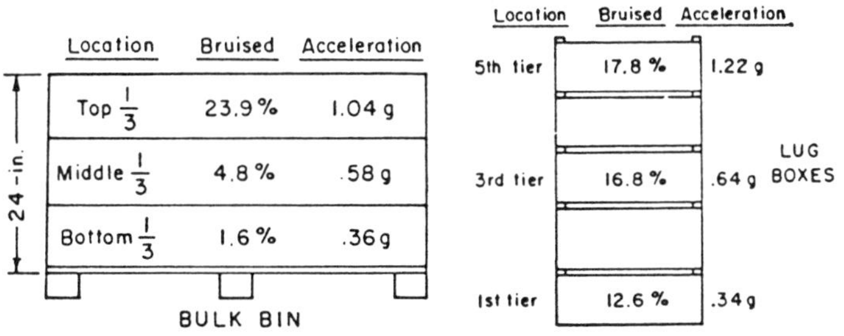

Figure 8.31 A 100-mile simulated severe transit test of damage to cling peaches (O'Brien *et al.*, 1965)

Figure 8.31 shows the magnitude of acceleration and resulting damage with cling peaches in a 100-mile severe simulated transit test. The total amount of bruised fruit was about 10 percent in bins and 16 percent in lug boxes. The reason for greater damage in lug boxes is that these containers have a higher percentage of their fruit in top layers than do bulk bins.

To minimize transit injury, Guillou (1963) developed a method of packing known as "tight-fill" pack. In this method the fruit container was vibrated for 5 to 10 seconds in a vertical plane at a frequency of 600 to 900 cycles per minute with a stroke adjusted to give a maximum acceleration of 2*g*. A closing pressure of about 10 pounds per square foot was also applied on top of the pad of the container. To arrive at the 2*g* acceleration the relationship for sinusoidal vibration, $a = r\omega^2$, was reduced to the following form

$$(\text{acceleration}, g) = \frac{(\text{stroke, inches})\,(\text{frequency, cpm})^2}{70{,}400} \tag{8.25}$$

Such an acceleration can be obtained by mounting a counterweight directly on a shaft of a 1750 rpm electric motor to produce a stroke of about $3/32$ inch. Note that the stroke in Eq. (8.25) is 2 × amplitude of vibration.

The "tight-fill" method was later applied by Gentry *et al.* (1965) to packing of peaches, nectarines and plums. The fruit was graded to size, random filled into a container, settled by vibration at 1100 cpm and displacement of 3/16 inch for 5 seconds. The container was top padded and closed tightly enough to exert light pressure on the fruit inside. Results of tight-fill packing have shown reduction of in-transit bruising for pears and plums while peaches and nectarines received as much bruising as under the hand-placing method of packing. It should be noted that in addition to reduction of bruising, tight-fill method of packing also increases the total volume of fruit that can be packed in a container. Guillou (1963) reported a 10 percent reduction in the volume of the fruit pack prepared by the tight-fill method.

Laboratory tests in which apples were subjected to vibration and repeated impact on a rigid surface were conducted to determine the relationship between amplitude, frequency, and duration of vibration and the extent of mechanical damage to the fruit. The individual fruits were tied down with rubber bands in such a manner that repeated impact could occur only in the vertical plane. Using Eq. (8.25) and setting the amplitude for a fixed value, the fruit was subjected to different levels of "vibration intensity" expressed in terms of g, by varying the frequency. Increase in duration of vibration increased width of surface bruise while depth of bruise showed to be dependent upon amplitude of vibration. For a ⅛-inch amplitude for example, the increase in depth of bruising leveled off after a given duration. From the practical point of view, this experiment showed that prolonged vibration would result in greated surface damage to apple while depth of bruising is apparently limited by the amplitude of vibration. Furthermore, there seemed to be a rather sudden increase of bruising as intensity of vibration increased from 1.0 g to 1.5 g.

8.10 STRESS-CRACKING

As applied to agricultural products, stress cracking is referred to a type of damage which is due primarily to internal forces of expansion and contraction resulting from changes in moisture and temperature. The problem of stress-cracking is of major economic importance in drying and subsequent processing of cereal grains such as corn, rice and wheat. The external cracks observed in several fruits and vegetables such as sweet cherries (Levin *et al.*, 1959), tomatoes (Voisey and Lyall, 1965), white and sweet potatoes (Werner and Dutt, 1941) can also be traced to the effects of moisture and temperature gradients established within these products. Levin *et al.* (1959) observed greater increase in diameter of sweet cherries due to increase of soak water temperature (skin expansion) than due to absorption of water. Thermal checking of eggs, reported in cases where temperature differential between wash water and the egg is excessive, may also be due to the same phenomenon causing stress cracks in other products.

While some research has been done on the subject of stress cracks in grains, skin cracks in fruits and vegetables has only been mentioned casually and any research in this area has apparently been incidental. It is known that, due to the uptake of water by osmotic pressure, cracking of the skin and the underlying tissue occurs in some fruits and vegetables. Cracking of sweet cherries and tomatoes in tank-hauling of these crops to the processing plant and cracking of sweet cherries and grapes following a rainfall are examples of this type of crack. Apparently this so-called "swell-cracking" occurs when combined stresses in the skin due to internal pressure exceeds the tensile strength of the skin and the underlying tissue.

Levin *et al.* (1959) subjected sweet cherries to artificial rainfall and then drying them in still air and in air streams equivalent to 5, 10 and 20 miles per hour. During the rainfall, cracking continued at 0, 5 and 10 miles per hour air flow but at 20 miles per hour the rate of transpiration was apparently great enough to keep the cherry skin from excessive expansion and cracking. In these experiments the modulus of elasticity of the cherry skin and its relationship to the direction of skin cracking was also determined. No attempt was made, however, to relate the mechanical properties of the skin to environmental conditions, stress evaluation and the resulting stress cracks. In some grains such as rice and corn, stress crack may develop during growth and ripening before any mechanical forces come into play. Bermistrova *et al.* (1956) have reported that in some Russian varieties of rice cracks may affect as much as 40 percent of the standing crop. Also considerable cracking took place while rice was drying in the windrow.

Stress crack in rice due to environmental factors during harvest operations in the field and subsequent stages of storage, handling and processing is an old problem and considerable research has been reported in the literature (Henderson, 1954; Stahel, 1935; Kunz and Hall, 1964, 1966; and Stermer, 1966). By employing an observation and inspection chamber where samples of rice could be subjected to different relative humidities at various temperatures, Kunz and Hall (1964, 1966) found that moisture gradients were more effective in producing stress cracks than were temperature gradients. They also found that about 50 percent of the rice samples initially at storage moisture content and then subjected to 100 percent relative humidity at 92°F fissured within a period ranging between 0.8 and 1.6 hours depending on the variety of rice.

Using *x*-ray and freeze-sectioning techniques, Grosh and Milner (1959) demonstrated the presence of internal fissures in wetted wheat grains (Fig. 8.1). To condition the wheat to that of the tempering process before milling, the grain was wetted to 17 percent moisture then sectioned and studied. Cracks radial and transverse to the crease were observed within the first hours after wetting. The movement of moisture along these cracks could be seen penetrating the endosperm. It was postulated that the forces causing cracks might be the result of a residual stress set up within the wheat endosperm during the kernel development or swelling stresses produced when a moisture gradient is established within the wheat kernel. These authors concluded that studies of the mechanical properties of wheat endosperm and the mechanism causing cracking would be of great value in the tempering and milling process of wheat.

Stress cracks in corn grain are fissures in the endosperm. The seed coat is not ruptured, which indicates that stress crack does not extent to the surface of the kernel. It has not yet been established whether or not viability and germination capacity of the kernel as seed is affected by internal stress crack. It is the increased susceptibility of the kernel to subsequent mechanical damage in handling as well as susceptibility to mold and insect damage which may affect the quality of the grain as seed as well as its milling quality and the quality of the final products.

In the milling of corn grain, production of large grits and reduction of fine materials are two major objectives. Corn with stress cracks will produce small grits and excessive amount of fines which must be removed as foreign material. Chizhikov (1960) considered the unequal distribution of moisture along the cross section of the kernel responsible for rising of internal stresses and the resulting cracks. To determine the influence of drying conditions on development of stress cracks in corn, this investigator measured the rupture force of the corn kernel subjected to drying by both static and dynamic methods. Results showed that there is a considerable reduction in both static and dynamic strength of kernels subjected to adverse conditions of drying for a temperature range of 70° to 244°F. For example, for one particular variety, there was about three times reduction in dynamic force required to rupture the grain as drying temperature increased from 104° to 284°F.

Thompson and Foster (1963) also reported two to three times reduction in the kernel strength as drying temperature increased from 140° to 240°F. The breakage evaluation employed by these investigators measured the amount of fine material that passed through a given size sieve after being impacted in various types of machines. They found that drying-air temperature of 140°, 190°, and 240°F resulted in 20.2, 29.6, and 33.9 percent stress-cracked kernels. Rate of drying expressed in percent moisture removed per hour was the most significant factor in stress crack development. Rapid heating through increased drying temperature and increased air flow rate both increased rate of drying and consequently stress cracks. Likewise, rapid cooling of the dried corn also increased stress cracks.

While rapid drying with heated air and forced ventilation is accepted to be a cause of stress crack in grains, no basic work has been conducted to combine the environmental factors with mechanical properties of the grain to determine the nature, magnitude, and distribution of failure stresses resulting from temperature and moisture gradients.

The rising of the temperature and the lowering of the moisture content of corn kernels during the drying process produce a temperature gradient and a moisture gradient within the horny endosperm. On the basis of this observation, one may conclude that stress crack in corn is the result of an increase in temperature, a decrease in moisture, or a combination of the two. To establish whether or not temperature gradient alone can cause the stress crack in corn, Ekstrom *et al.* (1966) first determined the coefficient of cubical thermal expansion and some tensile properties of individual corn kernels. Having these properties, a simplified model was considered for the corn kernel and an estimate was made of the temperature difference required to cause either a strain equal to the strain at which the kernel would rupture or a bending stress at which the inside surface of the kernel would fail. It was found that in order for stress crack to occur exclusively by temperature gradient, there must be a temperature rise of at least 175°F. Considering the corn temperature going into the drier being 50°F, the drying air must have a minimum temperature of 225°F to produce stress cracks in corn. Since most driers are operated with drying air at 180° to 190°F, the moisture gradient must also play a role in producing stress cracks in corn.

A more basic approach to the problem of stress crack in agricultural products was undertaken by Hammerle (1968) who studied the mechanics of stress crack formation in carefully prepared slabs of corn horny endosperm. In this work, hydro-stresses, resulting from moisture gradient, were superimposed on thermal stresses, resulting from variation in temperature, on slabs of horny endosperm considered as viscoelastic discs. Knowing the geometry of the specimen and the applied gradients, the various theories of failure and fracture available in mechanics were applied to determine the stresses at any point within the body. With this approach, a failure equation was developed which expressed failure stresses in terms of temperature, T, moisture content, M, time, t, uniaxial tensile relaxation modulus, $E(t)$, shift factors for temperature, a_T, and moisture, a_M, and a preselected position, z, within the material.

Such a basic approach should be of value in the analysis of any stress crack problem arising from expansion and contraction of the material resulting from changes in the ambient conditions or changes in material's temperature and moisture content.

8.11 MAXIMUM ALLOWABLE LOAD FOR AGRICULTURAL PRODUCTS

Which of the mechanical parameters, force, deformation, or energy causes mechanical damage in plant and animal materials? Nelson (1967) showed that dead loads as high as seven pounds applied for 100 hours on individual apples in a refrigerated room caused appreciable amount of distortion of the fruit surface with little or no browning of the flesh tissues. The impact produced sizable bruises with considerable browning when dropped on rigid surfaces from heights as low as two inches. When measurable bruises were developed under dead loads, the depth to diameter ratio tended to be smaller than for impact bruises. These comparisons indicate that undesirable impact bruises are much more likely to occur than undesirable dead load bruises. Also, the deeper impact bruises would be more difficult and costly to remove in processing operations than the more shallow dead load bruises. In this case the maximum allowable force, resulting from impact inertia, would probably be the parameter to be selected as the criterion for damage. Other materials, such as eggshell, can stand little deformation before shell fracture occurs. In mechanical handling of eggs then, the maximum allowable shell deformation should probably be taken as the criterion for damage.

Strength of eggshell

The work on eggshell strength has concentrated primarily on attempts to predict from static tests resistance to cracking and crushing of the shell under impact conditions. Tyler (1961) has reviewed the literature on measurement of shell strength and its relationship not only to shell thickness, to which it is highly correlated, but also to other factors as shell curvature, shell chemical composition, shell membranes, shell pores, breeds, diet, season, as well as interrelation of all these factors with shell thickness. He divided the method of measurements into crushing, puncturing and impact methods using whole specimens as well as pieces of shell.

Crushing, puncturing and dropping tests

Examination of the testing methods show that, like many other common physical tests of agricultural products, the test is not usually defined in such terms which would lend itself to analysis employing the laws of mechanics. The crushing has been done by usually loading the egg between two parallel plates. In puncturing usually a rod, flat-ended, rounded, or pointed has been employed to produce a puncture in the shell. The impact method in general has relied on a variation of the falling ball technique. However, there have been cases where the egg was allowed to fall. Table 8.6 gives a summary of results for shell strength compiled by Tyler (1961). Examination of this table shows the lack of consistency of units of measurement as well as variation in the methods of tests in regard to contact area, point of contact, supporting surfaces, position, size and shape of puncturing device, etc. Despite such variation in values, Heuser and Norris (1946) claim that normal handling created little difficulty until "shell strength," as described in Table 8.6, falls below 3.0–3.5 kg.

It is claimed that such factors as number and size of pores on the outer surface of the shell, shell protein, membrane thickness, surface curvature, color, size, season of the year, variations in breeds and feeding will affect the strength of eggshell. However, when many of these effects were corrected for shell thickness, their correlation with shell strength was non-significant. Tyler, who has published considerably in this area, stated that ". . . there is a multiplicity of methods, an overwhelming amount of data, but no clear idea of what is really being studied." Tyler also recommended a fundamental study of the actual mechanics of egg cracking in an attempt to find out the initiation of crack with respect to surface pores inside or outside of the shell and the point of load application.

Other interesting findings based on these less precisely defined tests can be summarized as follows:

1 Removal of the contents from the egg and removal of the membrane does not affect the strength of the shell, neither is there any change in strength when water replaces the egg content (Tyler and Geake, 1964).
2 Eggs appear to crack more readily along a longitudinal rather than along a latitudinal line. Using the method of a falling ball from increasing heights showed that more blows are required to produce a dent than a crack (Tyler, 1964).
3 Water weakens eggshells if given the chance to wet the outer surface thoroughly. However, on drying out the shell recovers its strength.
4 Eggs are weakest immediately after laying. Aging strengthens the shell.
5 Translucent areas of an egg are weaker than opaque areas of the same shell.

Dependence on shell thickness

It has been established that strength of eggshell is directly related to thickness of the shell. Some of the factors influencing the shell thickness, as summarized by James and Retzer (1967), are genetic strain, rate of egg production, diet and age of the hen, and the environmental temperature. Higher rate of production, older hens, and higher temperature cause thinner eggshell. James and Retzer also reported on a radiation technique which can be used to classify eggs non-destructively for shell strength in terms

Table 8.6 Summary of results for shell strength[12] (For references to specific tests, see Tyler, 1961)

Methods	Shell strength (kg)		Contact		Point of
	Range	Mean	Upper	Lower	pressure
Crushing methods	1.0–8.5[2]	4.5	Flat[9]	Flat	Poles
	3.6–5.5		Flat	Flat	Poles
	1.3–7.8[1]	4.6	Flat	Flat	Poles
	3.9–4.7	4.3	Flat	Flat	Poles
	2.3–3.6	2.9[3]	Flat	Flat	Poles
	1.7–7.2[1]	4.7[3]	Flat	Ring[11]	Poles
		5.6[4]	Flat	Ring	Poles
	7.4–9.8	8.9	Flat (padded)	Flat (padded)	Poles
	15.0–17.7	17.0	Socket[10] (padded)	Socket (padded)	Poles
		3.7	Flat	Flat	Side
	1.8–5.5	3.7	?	?	Side
	3.5–3.9	3.7	Flat	Flat	Side
		4.0	Socket	Socket	Poles
	2.8–4.7	4.0	Flat	Socket	Poles
	2.7–3.8	3.0	Flat	Flat	Poles
	3.9–4.2	4.1	Socket (padded)	Socket (padded)	Poles
	0.7–21.1[2]		Rubber diaphragm	Ring	Poles
	51.0–70.3		Flat	Socket	Poles
Puncturing methods	4.7–8.0[5]	6.4	?	?	?
	3.2–3.6[2]	3.4	1 mm pin	Socket	Side
	0.9–2.7		1 mm pin	Socket (padded)	?
	0.78–0.93	0.87[3]	Needle	?	Poles
	0.89–0.99	0.94[4]	Needle	?	Poles
Impact methods	569–606[6]	588	Falling ball	(3.53 g.)	Side
	343–347[6]	345	Falling ball	(1.48 g.)	Side
	2.1–2.3[7]	2.1	Falling egg		Side
Miscellaneous methods	10.6–11.0[8]		Piece of shell	broken	
	0.22–0.72	0.47	Piece of shell from inside	punctured	

[1] Range for individual eggs: all other values are group means. [2] kg/cm^2. [3] Values refer to broad pole. [4] Values refer to narrow pole. [5] Unknown units. [6] Dynes. [7] Unit not specified clearly (possibly kg-cm). [8] g/mm of break. [9] Flat surface making "point" contact with egg. [10] Socket or depression making large surface of contact with egg. [11] Ring maing contact with egg around one latitude. [12] Tyler (1961).

of the relative magnitude of thickness and density of the shell. The technique is called beta backscatter technique which employs a source of beta particle emitter. It makes use of the principle that when beta particle faces a thick and/or dense barrier such as eggshell, it reverses its initial direction of travel and backscatters. The amount of backscatter, which can be measured by means of a Geiger-Mueller tube, can be related to thickness and/or density of the shell. It was claimed that a better correlation was obtained with impact resistance than with the thickness of the shell indicating that strength of eggshell is dependent on density as well as shell thickness. Thick shells can sometimes be porous which would reduce their resistance to impact (see Fig. 8.32).

Resistance to static and impact loads

Interest of some agricultural engineers in recent years in measurement of strength of eggshell has led to the development of several new techniques which lend themselves to analyses similar to those employed for engineering materials.

Load and deformation required to rupture the eggshell by three different methods of loading were studied. The eggs were embedded both in vertical and horizontal positions in plaster of Paris until hardened. The testing consisted of loading-unloading of one or two points on each egg with a gradually increasing load starting from 0.3 pound and up to rupture of the shell. Under each load the recovered and residual deformations were recorded. Table 8.7 shows the results of these tests which give the effect of the loading methods, a measure of elasticity of the shell, as well as rupture strength in terms of force and deformation. With these data and the information on the radii of curvature of the eggs at the point of loading, it is possible to calculate modulus of elasticity and surface pressure at rupture (see Chapter 6).

Approximate values for ultimate stress of eggshell both in tension and in compression as well as modulus of elasticity of the shell are reported by Rehkugler (1963). The specimens consisted of narrow rings of the shell prepared by means of a small grinding wheel. Having the dimensions and shape of the ring's cross section, mean diameter of the ring, and load and deflection required to fracture the ring, ultimate tensile and compressive stresses and modulus of elasticity were calculated using the equations given for a ring specimen in Table 7.2. The values reported for full ring and semicircular ring specimens are shown in Table 8.8. Note that the values of failure stress for the section of the ring under tension is greater than

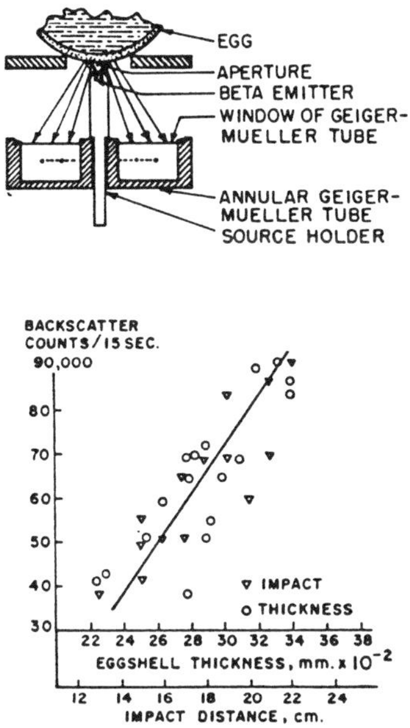

Figure 8.32 Application of beta backscatter technique for measurement of strength of eggshell (James and Retzer, 1967)

Table 8.7 Strength of eggshell measured by three methods of loading[1]

Loading method	Linear limit		Rupture		Elastic deform.	Radii of curvature		per cent
	force (lb)	deform. (10^{-3} in)	force (lb)	deform. (10^{-3} in)	at 1 lb. (10^{-3} in)	R_1 (in)	R_1' (in)	elasticity
Spherical indenter (0.0625″)	1.92	1.33	2.88	2.41	0.66	0.87	1.44	87
Parallel plates	—	—	—	—	0.55	0.88	1.42	90
Cylindrical die (0.016″)	2.18	1.63	3.10	2.97	0.60	—	—	71

[1] Each value is the mean of 6 to 12 samples. Loading rate 0.020 in/min. Egg in flat position, in a mold of plaster of Paris.

that under compression. This shows that the shell material shows less resistance to fracture under tension than under compression. Also in Table 8.8 are shown the values for limestone, $CaCO_3$, which is basically the material of the eggshell.

Rehkugler (1964) also studied maximum static and dynamic load-carrying capacity of whole eggs with and without cushioning materials and related his findings to design of egg-handling equipment. Static tests were conducted by loading the egg across its minor diameter and recording the force-deformation curve until fracture. Impact resistance was investigated by dropping one egg from successively greater heights onto another egg resting on a steel plate with or without padding materials. The same padding materials were used on steel plates employed in static tests.

Figure 8.33 shows the static force and energy required to fracture the egg resting on a rigid plate padded with cushioning materials with the given modulus of elasticity. Note that in each case the sample mean force or mean energy at failure as well as the sample mean minus twice the standard deviation are given so that prediction for selecting a suitable material be based on the acceptance of a maximum of 2.5 percent egg failure which will probably be allowable in design of egg handling equipment. These data are for a given thickness of the padding materials. Increasing material thickness would increase the maximum allowable force or energy that the egg would withstand.

Table 8.8 Mechanical properties of eggshell[1] (Original data from Rehkugler, 1963)

Material	Modulus of elasticity (10^6 psi)	Failure stress (10^3 psi)			
		Full ring specimen		Semi-circular specimen	
		Compression	Tension	Compression	Tension
Eggshell					
Mean	2.16	2.7	3.5	4.0	2.9
Range	1.5–2.9	2.0–3.8	2.9–5.3	2.9–4.7	2.2–3.5
Coef. of var.	19.9%	26.7%	10.8%	24%	22.8%

[1] Statistical summary for 10 tests. For comparison see the following values for limestone

Modulus of elasticity	$4.91 + 10^6 - 8.96 + 10^6$ (lb/in^2)
Compressive failure stress	3313 − 20,620 (lb/in^2)
Tensile failure stress	231 − 910 (lb/in^2)

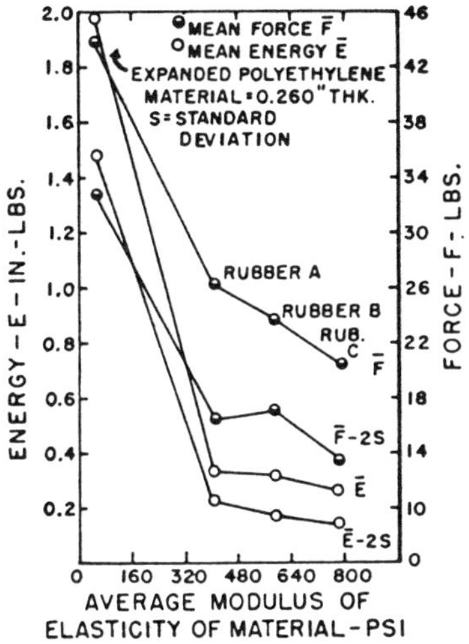

Figure 8.33 Energy and force at failure of eggshell compressed between padding materials with given modulus of elasticity (Rehkugler, 1964)

Figure 8.34 shows the energy required to fracture eggs dropped on different materials of given modulus of elasticity. Note that the stiffer the padding material, the lower would be the energy level causing egg damage. As in the case of static loading (Fig. 8.33), for design purposes one would use the lower curve to predict failure energy for a given padding material. The relationship between energy requirements to fracture egg under static or impact conditions is given in Fig. 8.35. Note that as the material thickness increases, the difference in their cushioning ability become more pronounced.

Voisey and Hunt (1967a) have published several papers on the subject of physical properties of eggshell with emphasis on shell strength. The results of their quasistatic tests when an egg was pressed between flat parallel surfaces showed that location of the moving surface influenced the percentage of shells fractured (Fig. 8.36). No explanation, however, could be given for these results. Also, from the force deformation curves, a stiffness value (ratio of force to deformation) was calculated and on that basis an egg spring factor was suggested as follows:

Loading point	Per cent stiffness	Energy to failure (in/lb)	Spring factor (lb/in)
Equator	100	0.0288	885
Poles	164	0.0189	1450

From the force-deformation curves they also concluded that eggshell can be considered as a Hookean body with a limited energy capacity. Although no unloading data were reported, based on percent of elasticity of eggshell given in Table 8.7 for loading the egg under a flat plate, as the first approximation and for the purpose of machine design, this is probably a valid assumption.

Voisey and Hunt (1967b) also reported on an impact device to measure the resistance of eggshell to shock loading. The apparatus incorporated a falling steel rod with flat or rounded end striking the egg mounted in a cradle. The force of impact was measured by two piezoelectric transducers mounted at the end of the rod. One transducer was calibrated in force units while the other was calibrated in acceleration units. A peak-shock-meter together with a digital voltmeter recorded the force or acceleration. For each test the correlation between shell thickness and impact force at failure was determined as summarized in Table 8.9.

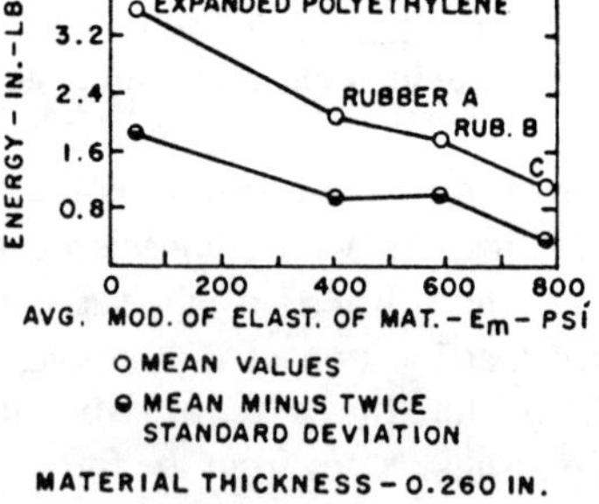

Figure 8.34 Energy required to fracture eggs dropped on cushioning materials of given modulus of elasticity (Rehkugler, 1964)

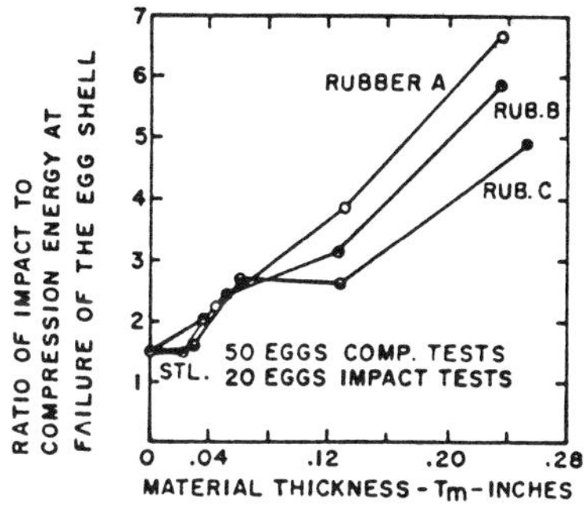

Figure 8.35 Ratio of mean values of impact to compression energies to fracture eggshell (Rehkugler, 1964)

The results of this investigation showed that variations both in shell thickness and in impact force to failure on a between-bird basis were about four times that of a within-bird basis. Furthermore, the shape of the striking surface as well as the weight of the striker affected the results. The striker with a spherical end fractures the egg at less impact load than one with a

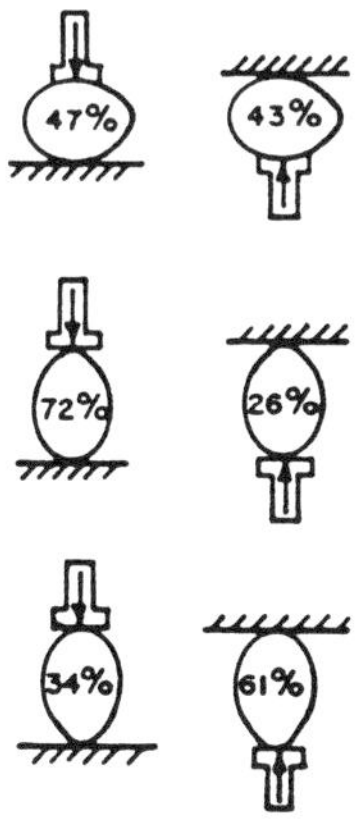

Figure 8.36 Per cent shell fracture and location of force application (Voisey and Hunt, 1967a)

Table 8.9 Correlation of impact force and eggshell thickness (adopted from Voisey and Hunt, 1967b)

Test No.	I	II	III
Shape of striking surface	flat	spherical	flat
Measurement recorded	force	deceleration	deceleration
Weight of striker (kg)	0.0745	0.0853	0.0843
Distance of fall (cm)	2.8	2.8	2.8
Number of eggs tested	1457	1670	1393
Mean shell thickness (cm)	0.0332	0.0332	0.0332
Mean force to fracture (kg)	5.033	3.966	5.756
Correlation coefficient	0.466	0.643	0.606

flat end. This may be due to the fact that a spherical contact presents smaller contact area and thus larger stress as compared with flat plate contact. The effect of weight of the striker in tests I and III may be explained on the basis of impulse-momentum theory. Since the impact force was measured at the point of impact, it would be difficult to account any energy losses for the lower value of measured force in test I as compared to test III. However, the heavier striker in tests III will produce greater momentum which will deform the shell at a higher rate and thus shorter time than that of test I. According to impulse-momentum theory, this would result in a larger force in the case of the heavier striker. With such an analysis Voisey and Hunt conclude that impact tests which alter either the weight or falling distance to achieve different forces on the shell are not valid measures of strength. It should be noted that most impact tests reported in literature for egg or some fruits and vegetables have employed the method of varying either the weight or the height of the falling object to evaluate resistance to mechanical damage.

Application of shell theory

Sluka *et al.* (1965) employed hydrostatic pressure in an attempt to compare shell failure under hydrostatic pressure with that under impact drop tests. Their technique led to the conclusion that eggshell was weaker in the latitudinal direction. Accordingly, it was assumed that when stresses in this direction were to exceed the ultimate value S_t given by Eq. (5.14), shell failure should occur in the form of a single longitudinal crack. In a later paper, Sluka *et al.* (1967) gave an analysis of impact stresses in eggshell leading to another equation for latitudinal stress which involved certain physical characteristics of the egg as well as deceleration ratio G defined as ag, with a being deceleration due to impact force and g, gravitational acceleration. Using hydrostatic pressure along with the apparatus of Fig. 5.24a and dynamic tests employing a drop tester to measure deceleration, data were obtained by testing eggs from single combed White Leghorns.

When the eggs were subjected to an acceleration ratio of $G = 478$, results were mean values of latitudinal failure stress equal to 2140 psi for hydrostatic tests and 2460 psi for impact tests. The fact that the results from dynamic tests were very close to those of the static tests should be an indication of some errors either in the experimental results or the analytical approach to the problem. One major source of error could be the assumption that the bending moments on the eggshell were negligible. Voisey and Hunt (1967c) have shown that when the outer surface of the shell is subjected to compression, the inner surface is subjected to tension and the bending stresses can be quite significant in both meridional and latitudinal directions. In fact, according to shell theory, loading of a spherical shell can produce severe bending moments (Flugge, 1960). If the egg is packed in a material which deforms easily, such as a foam rubber, the load on the egg will be distributed over a larger area resulting in less unit pressure, reduced bending moments on the shell, and greater load-carrying capacity of the egg.

It should also be pointed out that failure of the eggshell under internal pressure cannot be due to only one of the principal stresses but rather a combined stress condition may be involved. In that case the analysis of the failure stress as proposed by Sluka *et al.*, is not complete and the effect of other stresses should be taken into consideration.

In another attempt to apply the shell theory to determine failure stresses in eggshell, instead of one, two of the principal stresses at failure were determined (Hammerle and Mohsenin, 1967). To achieve failure of eggshell, the egg was subjected to internal pressure as in the experiment by Sluka *et al.* (1965). The experimental technique was modified, however, to assure that pressure was applied at a continually increasing and known rate rather than at an unknown rate step-loading as was the case with the water-filled hypodermic syringe of Fig. 5.24a. Carefully selected eggs from White Leghorn pullets were subjected to internal air pressure through a hypodermic needle. The applied air pressure was sensed by a piezoelectric transducer and recorded as a function of time on the chart of an optical oscillograph. The rate of pressure application to the shell could be changed by means of a pressure control valve and determined from the slope of the recorded pressure versus time curve. The mounting of the egg in the testing fixture was such that no axial force was applied so that the internal pressure resulted in pure tension in the shell. Pressure was increased uniformly at a constant rate until egg fracture occurred.

Having the internal pressure required to fracture an egg specimen, two principal stresses, latitudinal stress, S_t, and meridional stress, S_a, were calculated for 11 points along the major axis x as shown in Fig. 8.37. The calculating equations were those given in (5.14) and (5.15). Shell thickness t was determined for each point after egg fracture using a micrometer. The radii r_1 and r_2 were determined also for each point using the following relationship (Marks, 1941).

$$r_1 = \left[\frac{1 + (\dot{y})^2}{\ddot{y}}\right]^{3/2}$$

$$r_2 = \left[\frac{(\dot{y})^2}{y} + y^2\right]^{1/2}$$

where $\dot{y} = dy/dx$ and $\ddot{y} = d^2y/dx^2$. This required a knowledge of the equation of the egg profile $y = f(x)$. The radius meter of Fig. 6.3 could

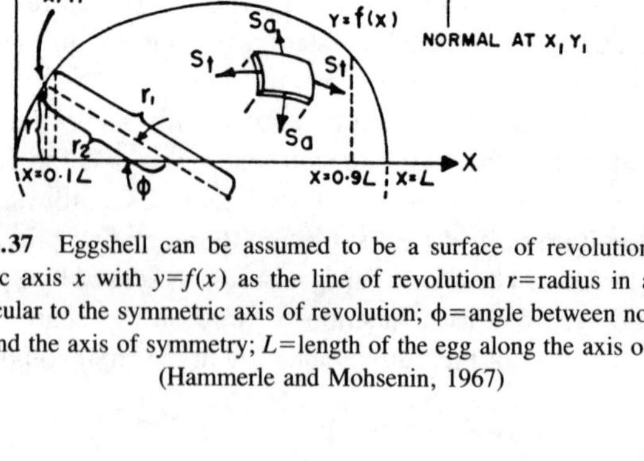

Figure 8.37 Eggshell can be assumed to be a surface of revolution about the symmetric axis x with $y=f(x)$ as the line of revolution r=radius in any section perpendicular to the symmetric axis of revolution; ϕ=angle between normal to the surface and the axis of symmetry; L=length of the egg along the axis of symmetry (Hammerle and Mohsenin, 1967)

be used only for the two ends of the egg where the same curvature existed on all sides. For this purpose, the egg profile was first obtained by projection utilizing a photographic enlarger. With the axis of symmetry aligned with the x axis, values of y were measured for 0.25 inch increments of x. With these data and the use of a computer a fourth degree polynomial equation was derived by the method of least squares (Cziffra and Moraveski, 1959). Table 8.9a shows such an equation and other calculated parameters including the shell failure stress at each point for one egg. Table 8.10 shows the shell failure stress as affected by the egg content, and the hens producing the eggs. Results of statistical analysis showed that there was no significant difference between the failure stresses when the contents of the eggs were removed or when they remained in the egg. This was in agreement with the findings of Tyler and Geake (1964) which was mentioned earlier. Also no significant difference was found among the results on a hen to hen basis except for the rate sensitivity of the shells with contents removed. In general, the eggs produced by Hen *B* showed more variations in shell

Table 8.9a Physical characteristics and failure stress at various points on the shell of old egg[1] (Hammerle and Mohsenin, 1967)

Projection data			Shell Parameters				
			Thickness	Radii of curvature		Failure stresses	
x (in)	y (in)	x	t (in)	r_1 (in)	r_2 (in)	S_a (psi)	S_t (psi)
0.12	0.32	0.0 0.015	0.54	0.54	551	551	
0.25	0.48	0.1L	0.015	1.58	2.15	1859	791
0.50	0.66	0.2L	0.015	1.37	1.19	957	1001
0.75	0.78	0.3L	0.014	1.37	0.91	787	1052
1.00	0.84	0.4L	0.014	1.44	0.84	744	1099
1.25	0.87	0.5L	0.014	1.39	0.83	742	1105
1.50	0.85	0.6L	0.014	1.22	0.84	743	1009
1.75	0.77	0.7L	0.013	1.07	0.88	822	926
2.00	0.61	0.8L	0.013	1.08	1.13	1007	859
2.18	0.41	0.9L	0.013	1.40	2.07	1864	777
		L	0.013	0.96	0.96	734	734

Derived polynomial

$$y = 0.15456701 + 1.5801119x - 1.4232465x^2 + 0.68869045x^3 - 0.15764632x^4$$

[1] Loading rate 13.3 psi per second, failure pressure 24 psi, egg contents removed, L = 2.3 in. For symbols see Figure 8.37.

failure stress than the eggs produced by Hen *A*. However, the variation was not to the extent found by Voisey and Hunt (1967). The fact that little variation was found for strength of shells produced on very similar cycles by hens placed side by side and fed the same ration indicate that the great variations usually observed in tests of physical properties of biological materials can be reduced considerably if the measurement is well defined in terms of physical or engineering parameters. In this case the failure stress of the shell material, as defined in the shell theory, was taken as the physical parameter to be used for comparison of strength of eggshell produced by different birds (see Table 8.10).

One difficulty with the method of polynomials used for deriving the equation of the eggshell was that the resulting equation could not be used to fit the egg profile at the two poles. To avoid this problem, a more accurate and simpler method for determining the radii of curvature r_1 and r_2 of the egg profile was suggested. Once the egg profile is obtained by projection, a transparent mirror can be used to obtain slope of the curvilinear profile and the angle ϕ at any point as shown in Fig. 8.39. By aligning the portion of the curve on one side of the mirror with the portion on the other side, as visible through the mirror, the mirror will be normal to the curve and the angle ϕ can be read directly. Values of ϕ are then plotted with respect to s, the distance along the surface from the end of the egg, and r_1 is calculated (Fig. 8.38a). To find r_2, the value $r_1 \cos \phi$ is plotted as a function of ϕ (Fig. 8.38b). Using a planimeter, the area between the curve and the ϕ-axis is determined for any value of ϕ. From this area, r_2 can be calculated as given in Fig. 8.38a.

Table 8.10 Failure stresses in egg shells (Hammerle and Mohsenin, 1967)

	Contents removed				Contents in			
	Low rate		High rate		Low rate		High rate	
Test	Hen *B*	Hen *A*	Hen *B*	Hen *A*	Hen *B*	Hen *A*	Hen *B*	Hen *A*
1	1805	2467	3719	3289	2928	2999	2300	2678
2	2516	1864	3714	3629	3111	2795	2569	3122
3	1925	1986	2548	3060	2546	2632	2204	3541
4	2336	2347	3165	2830	3703	2286	2964	3256
Mean	2146	2166	3286	3202	3072	2678	2509	3149
S. D.	332	287	613	373	482	301	496	360
Coef. ov var. %	15.5	13.3	18.7	11.6	15.7	11.2	19.8	11.4

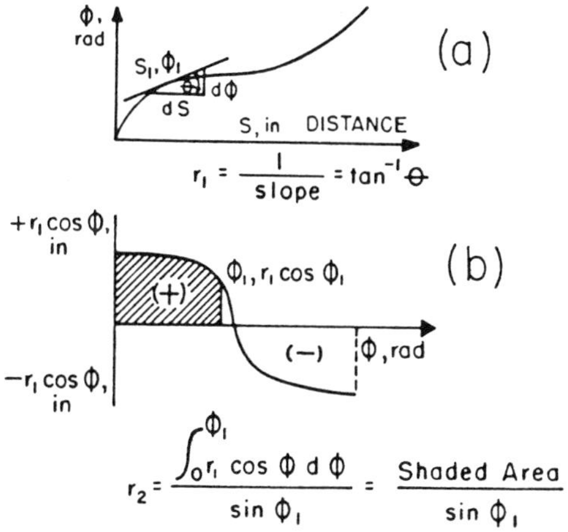

Figure 8.38 Apparatus for determining slope of the egg profile at any point (Hammerle and Mohsenin, 1967)

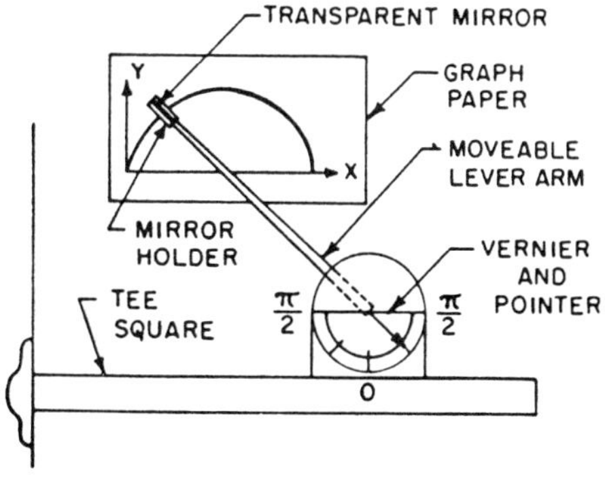

Figure 8.39 Graphical method for determining radii of curvature of the egg profile at any point (Hammerle and Mohsenin, 1967)

Application of stress-coat technique

To understand the eggshell's reaction to external forces, Voisey and Hunt (1967c) studied stress distribution in the shell employing the "stress-coat" technique common for stress analysis of machine elements (Fig. 8.40). In this technique, also called brittle lacquer method, the test specimen is covered with a special resin-like material which, when dry, cracks perpendicular to the axis of principal strains. With known loads on a cantilever steel beam covered with the same brittle coating, the threshold tensile strain of the coating can be calculated. Knowing the strain and the elastic modulus of the specimen, stress can be calculated from Hooke's law $\sigma = E\varepsilon$. The experimental ultimate stress of the egg obtained by this method can be utilized to calculate the force at fracture employing the analysis of a concentrated force on a flat plate given by Timoshenko and Wornowsky-Krieger (1959). When compared with experimental values of load required to fracture the shell, Voisey and Hunt (1967c) found that the calculated values were about 20 times smaller. The explanation given for this large order of error was the larger area of contact which probably existed between the experimental load and the shell surface which in turn supported greater loads. This was supported by the observation that the principal stresses in the force zone extended a considerable distance into a theoretical transition zone indicating that failure at the point of contact resulted in a contact area much greater that that assumed in the calculating formula.

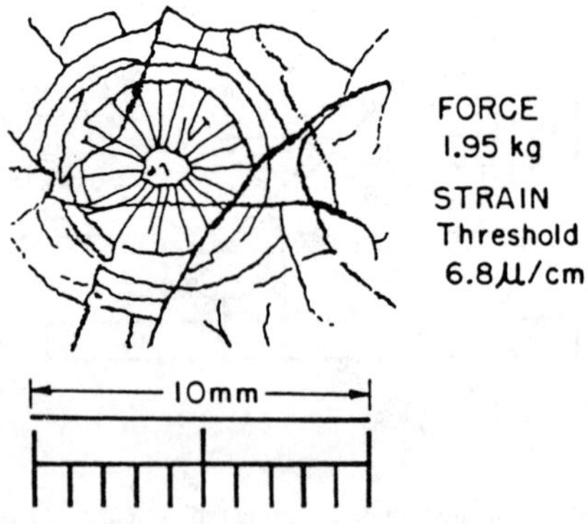

Figure 8.40 Stress-coat technique applied to eggshell (Voisey and Hunt, 1967c)

Other significant results from this study were the confirmation of some of the previously reported findings that the large pole of the egg is weaker than the small pole (Romanoff and Romanoff, 1949; Tyler and Moore, 1965; Voisey and Hunt, 1967a) and that the shell would more likely fail in the meridional direction (Tyler and Moore, 1965; Voisey and Hunt, 1967a; Sluka *et al.*, 1965). It was also found that the initial fracture occurs under the point of contact at the inner surface while the shell material is under tension. This led to the conclusion that the strength of the shell depends not only on its thickness and radii of curvature but also on the tensile strength of the shell material. Figure 8.41 shows the variation of thickness in eggshell.

Strength of seeds and grains

Resistance to static and impact loads

The effect of different combinations of drum speeds and concave clearance on damage to wheat and peas in threshing operations has been reported by King and Riddolls (1959) (see Table 1.2). The conclusions reached were that when wheat or peas are harvested for seed, even at fairly low moisture contents, visible and invisible damage can be kept to low levels by avoiding excessively high drum speeds. Since close concave clearance appears to have little effect except at very high drum speeds, total damage can be reduced to a minimum by keeping the drum speed low and getting

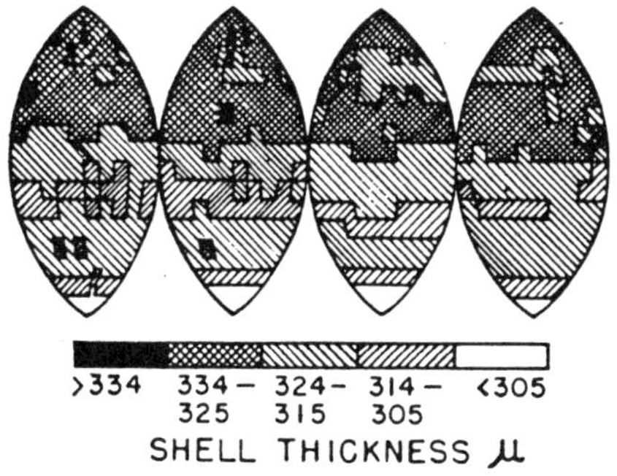

Figure 8.41 Variation in thickness of eggshell from pole to pole (Tyler, 1958)

the required degree of threshing by adjusting the clearance between drum and concave.

The work of Zoerb and Hall (1960) on mechanical properties of dent corn, wheat and pea beans has already been discussed. Figure 8.27 and 8.43 show resistance of these grains to impact and compression shear and ultimate strength in comparison as related to the grain moisture content.

Mitchell and Rounthwaite (1964) studied the resistance of two varieties of wheat to impact induced by striking the individual grains with a rotating hammer. The grains were hit with broad side down. Germination tests of visually undamaged grains showed that the speed of impact was the prime cause of damage. Beyond 3280 ft/min of hammer speed damage increased with about ⅔ of the grains damaged at speeds above 7000 ft/min (Table 8.11) At lower levels of moisture, increasing impact speed increased the amount of visually damaged grain, as expected, due to brittleness of the grain. Increasing impact speed also reduced germination capacity of the apparently undamaged grain with a stronger effect on grains with higher moisture content. The relationship between percentage of undamaged grains (variety Koga II) and impact speed was given by the following approximate equation

$$y = 94 + 3.9 - 1.6S^2$$

Table 8.11 Effect of impact velocity on damage to one variety of wheat (Koga II) (Summarized from Mitchell and Rounthwaite, 1964)

Moisture (%)	Hammer speed (ft/min)						
	Visually undamaged grains (%)						
	0	3280	4180	5170	6090	7080	Mean
15	—	91.0	87.3	93.0	80.7	75.3	85.5
20	—	95.3	94.0	92.7	94.3	92.3	93.7
25	—	98.0	94.3	93.7	95.3	92.7	94.8
Mean		94.8	91.9	93.1	90.1	86.8	91.3
	Germination of visually undamaged grains (%)						
15	96.7	95.2	92.1	83.2	71.0	65.8	81.5
20	95.3	96.1	86.9	80.1	59.6	52.2	75.0
25	90.3	91.8	90.8	83.3	50.6	37.9	70.9
Mean	94.1	94.4	90.0	82.2	60.4	52.0	75.8

where y is percent of undamaged grains and S is impact speed in ft/min $\times 10^{-3}$. Note that the above equation accounts for impact speed only. Statistical analysis showed that impact speed accounts for 97.3 percent of the variation between the number of undamaged grain as compared to only 93.5 percent when moisture levels alone were included in the analysis.

Bilanski (1965) also made an attempt to express breaking strength and damage to seed grains in terms of impact forces and energy. The grain was dropped, one at a time, into the path of a rotating paddle revolving at speeds similar to a threshing cylinder. The energy imparted to the grain was calculated from $KE = \frac{1}{2}m(r\omega)^2$. Where m is the mass of the grain, r is the distance between the center of rotation and the point of impact with the grain, and ω is the angular velocity of the paddle in radians per second. Seeds were also subjected to lower rate of impact by holding individual grains on an anvil and striking them with a pendulum of known potential energy. By increasing or decreasing the levels of impact energy until all of the grains broke at the upper level and remained unbroken at the lower level, the upper and the lower limits of strength were determined. Twelve settings with 20 samples tested at each setting were tried for each of the five different grains investigated. Criteria for damage were visible cracks of the seed, cracks on the seed coat of the soybeans, cracked hull in oats, or the excessive deformation in the case of high moisture grain.

Figure 8.42 shows the effect of moisture content on energy requirements to damage three types of grains. Note that greater energy is required to break by impact the grains having a higher moisture content than those having a lower moisture content. Figure 8.43 shows the amount of deformation and load that grains with higher and lower moisture contents can withstand under quasi-static type of loading. A high-moisture grain which takes a greater amount of energy to break (Fig. 8.42) is not as resistant to deformation as a low-moisture grain (Fig. 8.43). If hardness of grain is defined as resistance to deformation, addition of moisture reduces hardness. On the other hand, if breaking energy is to be taken as a criterion for strength and hardness, addition of moisture increases resistance to breakage.

At lower impact velocities where the kernel could be positioned on the anvil, Bilanski (1965) found that orientation also influenced the breaking strength. The effect of orientation is also reported by Zoerb (1960) and Arnold and Roberts (1967) but results are not conclusive. While each type of grain reacted somewhat differently to the various tests, in general larger grains or grains with higher moisture contents required greater amount of energy to damage. However, it should be pointed out that expenditure of a large amount of energy to break a grain does not necessarily mean high load as energy is a function of both force and deformation. Zoerb and Hall (1960) also found that for yellow dent corn and wheat, the energy required to rupture the kernel by impact increased with increase of moisture content (Fig. 8.27). In this work energy was expressed in terms of in-lb per square inch of kernel cross section. As pointed out by Bilanski, in actual threshing operations, similar grains could probably withstand greater impact energies since the chaff and straw also would act as absorbers of energy, leaving less energy to be absorbed by grain.

Damage to bulk grain under compression was reported by Narayan and Bilanski (1966). Figure 8.44 shows the extent of damage for bulk of wheat at different moisture levels being compressed in a cylinder. Damage was

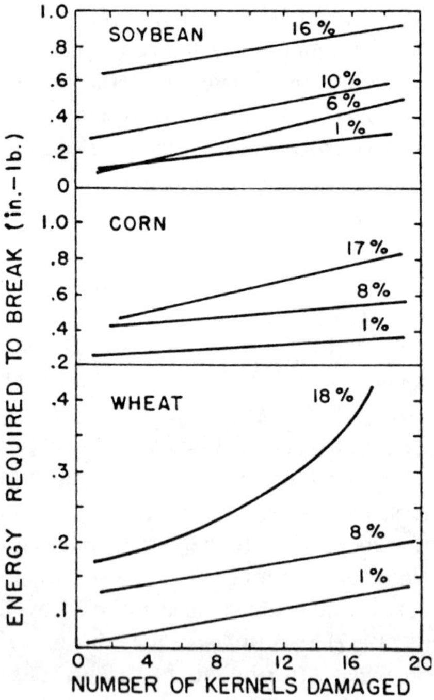

Figure 8.42 Influence of moisture on breaking resistance to grains to impact loading (Bilanski, 1966)

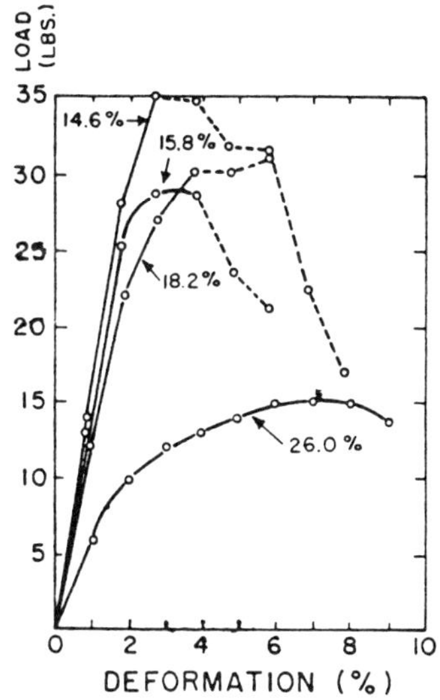

Figure 8.43 Load-deformation behavior of Yellow-Dent corn at different levels of moisture content (Zoerb, 1960)

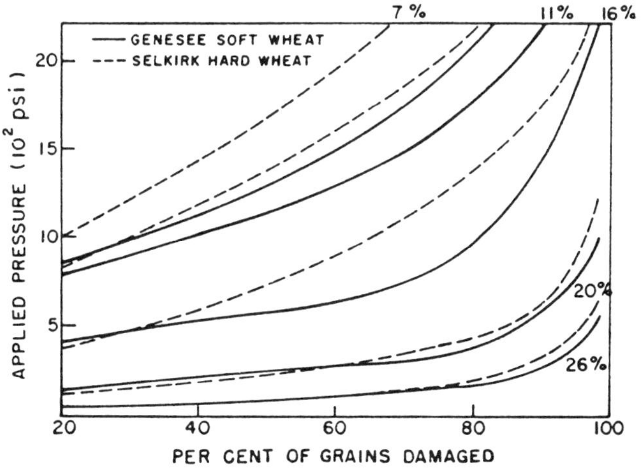

Figure 8.44 Damage to bulk wheat at different moisture contents under pressure (Narayan and Bilanski, 1966)

evaluated by subjecting the grain to a pressure equal to ⅛ of that required for total damage and then releasing the pressure and counting the damaged grain in a sample of 100 kernels. The cylinder was then reloaded with a new sample to be compressed at a higher pressure and again checked for damage. This procedure was repeated until the total-damage pressure was reached. As expected, the higher moisture content grain required less pressure for total damage. A vertically applied pressure of 3000 psi damaged about 90 percent of the two varieties of wheat tested. In other word, contrary to its behavior under impact shear, the grain with lower moisture is more resistant to breakage under compression.

Leonhardt *et al.* (1961) made an attempt to investigate the relationship between the impact energy absorbed by sorghum seed and resulting mechanical damage to the seed. In their tests the kernel was shot with a spring loaded gun at a known velocity against a cantilever beam with strain gages mounted at its base. After impact, the seeds were planted and the extent of damage was evaluated by means of germination tests. Figure 8.45 shows the extent of damage at different impact velocities to sorghum seeds at different moisture contents. As seen from these graphs, the number of damaged seeds increased with either an increase in the impact velocity or a decrease in the moisture content of the kernel or both. This trend increased rapidly at velocities greater than 2700 ft/min. The relationship between energy absorbed and corresponding seed damage, however, is not given.

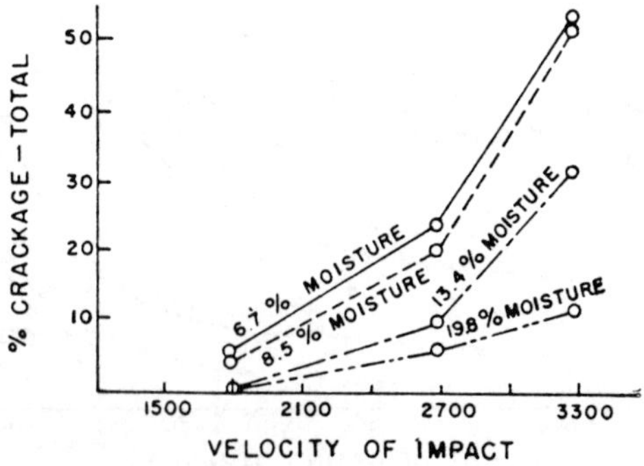

Figure 8.45 Damage to sorghum seeds as affected by moisture content and impact velocity (Leonhardt *et al.*, 1961)

It was only reported that at an impact velocity of 3300 ft/min, seeds initially absorbed from 93 to 96 percent of their initial kinetic energy. This energy was calculated from the known mass and initial velocity of the seed using equation $KE = \frac{1}{2}mv^2$. Considering the experimental set up and the calibration procedure followed for the cantilever beam, it is questionable if the actual initially absorbed energy by the seed was of this order of magnitude.

Turner *et al.* (1965) used an impact machine having a rotating arm to evaluate impact damage to peanuts. A flat metal plate attached to one end of the arm impacted the peanut kernels carried to its path by a conveyor. The effect of moisture content, orientation and velocity of impact on both hull and kernels were evaluated. They found that hull damage did not vary greatly with orientation, the apical end of the hull (See Chapter 2) was most volunerable to damage; and samples at intermediate moisture levels showed the least damage. The kernel damage was evaluated microscopically after a special staining. As expected the apical end of the hull which was least resistant to damage transmitted more injury to the kernel than the other parts. The damage increased with increase in moisture content. At high velocities, however, kernels tended to split particularly at lower moisture levels. This is a trend observed also in the case of cereal grains. Therefore, the impact velocity and moisture content are the two important parameters influencing the peanut injury in mechanical processes. Based

Table 8.12 Seedling development from apical kernels of peanuts impacted on the apical end (Turner *et al.*, 1965)

Kernel moisture at impact (%)	Impact velocity (ft/sec)	Emergence (%)	Irregular roots (%)	Irregular tops (%)
	0	100.0	6.2	0.0
42.0	20	81.2	84.6	15.4
	40	12.5	100.0	50.0
	0	87.5	14.3	0.0
10.2	20	100.0	62.5	6.2
	40	56.2	77.8	33.3
	0	100.0	12.5	0.0
5.6	20	100.0	18.8	0.0
	40	30.8	100.0	0.0

on their studies, Turner *et al.* (1965) suggested that for higher impact velocities, kernels at 20 percent moisture would probably receive the minimum damage. For low velocity impact, a low moisture level was recommended. These conclusions were based on not only damage evaluation but also on laboratory germination tests (Table 8.12).

Using the concept of impulse momentum, Turner *et al.* showed how the impact force on the kernel could be determined. From prints of high speed photographs of peanuts being impacted, the time of contact between the peanut and impact head was determined. Knowing the contact time, Δt, and the peanut velocity, V_p, the average value of impulse force, F_{av}, on the nut was derived to be

$$F_{av} = m_p V_p / \Delta t$$

where m_p is mass of the peanut. Also from maximum rate of change of velocity the maximum acceleration, $a_{\max}$, experienced by the peanut and thus the maximum force on the peanut during impact was estimated as

$$F_{\max} = m_p a_{\max}$$

The values calculated for two independent runs were given as

$$F_{av} = 5.55 \text{ and } 4.11 \text{ lbs} \quad F_{\max} = 10.0 \text{ and } 8.25 \text{ lbs.}$$

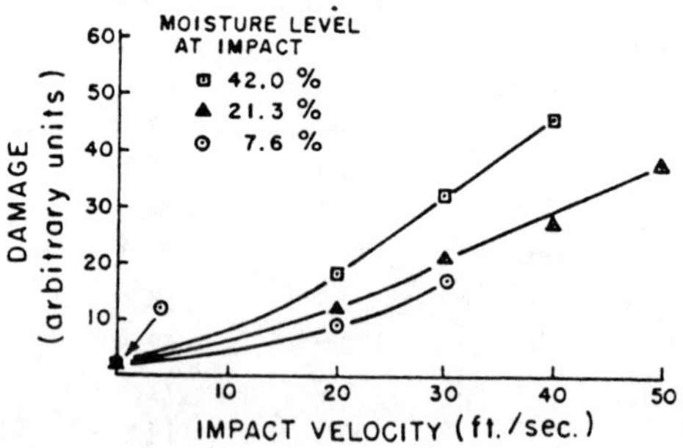

Figure 8.46 Damage to apical end of peanuts impacted on the apical end (Turner *et al.*, 1965)

The relationship between the impact force, moisture content and mechanical damage, however, was not given.

Kirk and McLeod (1967) reported on mechanical damage to cottonseed and its relationship with impact velocity. The impact was achieved by blowing the seed by means of compressed air out of a blowpipe against a steel plate. The initial velocity of the seed was obtained by means of stroboscopic pictures using a 35 mm camera. Results showed that energy required to rupture cottonseed under either quasistatic loading or high velocity impact was independent of moisture content within the limits indicated in Table 8.13. The relationship between the percent seed rupture and seed velocity was a logarithmic function as given in Fig. 8.47. To predict percent seed damage in such mechanical handling as pneumatic conveying, one needs to know the relationship between seed velocity and air velocity. This relationship for the experiment reported by Kirk and McLeod was seed velocity=0.71 air velocity. The authors did not, however, recommend using this ratio for practical situations.

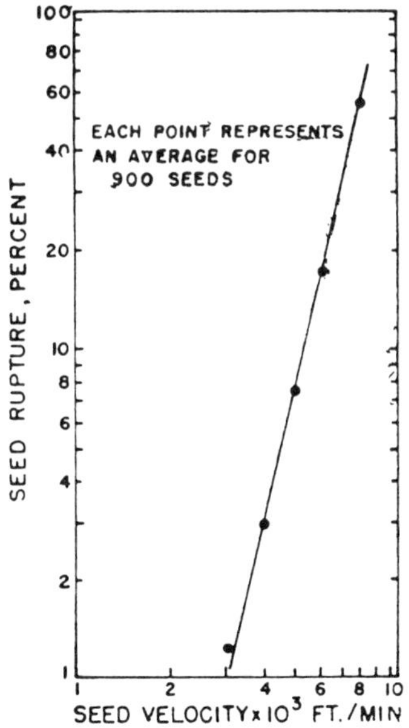

Figure 8.47 Relationship between cottonseed damage and seed impact velocity (Kirk and McLeod, 1967)

Table 8.13 Resistance of cottonseed to rupture under quasi-static loading[1] (Kirk and McLeod, 1967)

Moisture % (d.b.)	Rupture force (lb)	Rupture deformation (lbs)	Rupture energy
6	18.8	0.060	0.714
10	15.6	0.075	0.696
14	12.9	0.099	0.691

[1] Loading rate not given.

Cooke and Dickens (1967) have reported on the mathematical analysis of a centrifugal gun for imparting known, uniform impact velocities to seeds for testing. The gun consisted of a 10-inch long straight rigid tube with ½-inch diameter rotating in a horizontal plane concentric with a 12-inch diameter rigid ring. The seeds were metered into the centroid of the tube and upon discharge impacted on the surface of the surrounding ring. Tests with corn seeds showed that a wide range of seed breakage can be obtained by simply varying the rotational speed of the gun. To test the mathematical analysis of the equation of motions, the path of white-painted rubber and steel balls, ⅝-inch in diameter dropped on a rotating impeller, was recorded photographically. The impeller, painted black, consisted of a 12-inch diameter disc with four equally spaced vanes radiating outward to the edge of the disc.

The insight provided by the mathematical analysis in this work revealed a number of factors which may prove to be useful in establishing design criteria for centrifugal distribution of seeds, chopped forage, and granular materials as well as high speed rotating devices for sorting and grading of food materials subject to mechanical injury. For example, the following were some of the more practical findings for the frictionless case: (1) the discharge velocity is a linear function of the angular velocity and is independent of the seed mass or little affected by the frictional forces involved as illustrated by the following equation:

$$|V| = \omega[2r^2 - r_0^2]^{1/2}$$

where $|V|$ is the velocity vector, ω is the angular velocity, t is the radial distance from the origin to the center of mass and r_0 is the initial radial displacement. (2) The direction of discharge velocity for a radial tube will be at least 45 degrees ahead of the axis of the tube if started from the center. This is shown by the following equation derived for the discharge angle β

$$\beta = \tan^{-1}[r_1/(r_1^2 - r_0^2)^{1/2}]$$

This means that the impact velocity will not be perpendicular to the impact surface (the concentric ring in this case). If such condition is desired, a target may be mounted to the tip of the discharge tube for particle impaction. (3) The kinetic energy of a seed at discharge is a function of the seed mass, the square of the angular velocity, and the square of the tube length as seen from the following equation

$$E = m\omega^2(r_1^2 - r_0^2)$$

where E and m are the energy and the mass, respectively. This information on energy requirement should be useful for estimating power requirements in a specific design. (4) The maximum normal horizontal force, N, depends upon the particle mass, the square of the angular velocity, and the first power of the tube length. In other words, if the normal force for large velocities is to be held within an allowable level, if possible, this should be done by increasing the tube length rather than increasing the angular velocity, as depicted in the following equation

$$N = 2m\omega^2(r^2 - r_0^2)^{1/2}$$

Important physical properties of the seed such as aerodynamic drag, specific heat and thermal conductivity can be taken into consideration for a more extensive study of this problem. For example, the seed tube should be constructed of a material having high thermal conductivity to minimize heat damage to the seed. Cooke and Dickens (1967) suggested that if the tube is made out of steel, the heat effect would be negligible. The effect of frictional properties of the seeds upon discharge conditions has also been discussed by these authors.

A more recent work by Clark *et al.* (1967) on cottonseed has made use of a mechanical system for impacting the seed with a given orientation. They found that orientation of the seed is an important factor in the extent of damage that the seed will receive upon impact. It was shown that beyond a velocity of 5000 ft/min the seed is more susceptible to damage from impacts on the side (seed coat) than on the radical end where the embryo is located. The percents of visible damage on the side reported at 6000 and 7000 ft/min were much greater than the average percentages shown in Fig. 8.47. Germination tests showed that visible damage and germination reduction are not necessarily directly related in the cottonseed. Considering the structure of the seed, it was stated that percent damage is an evaluation of the amount of energy absorbed by the embryo. These investigators also found that contrary to the finding of Kirk and McLeod (1967), moisture content has an effect on the extent of impact damage but no direct relationship could be established.

Mechanical damage to castor bean by compression and impact is reported by Burmistrova *et al.* (1956). It was found that resistance to compression of individual beans was from 2.2 to 20 pounds. When metal surfaces used in loading the bean were covered with rubber, the average crushing resistance increased from 14.7 to 23.8 pounds. Under impact using a pendulum apparatus, no seeds were damaged at an impact velocity of 17 ft/sec but almost 100 percent of seeds were damaged at impact velocity of 50 ft/sec.

Bouse *et al.* (1964) have also reported on mechanical damage to castor seed as they are handled by screw conveyors. Effect of auger speed, entrance opening height, shear plane orientation on seed damage, are given.

Perry and Hall (1966) have reported on mechanical damage to pea beans under impact by dropping individual beans on a layer of pea beans from three heights 11½, 22½, and 45 feet. Drop test was used in these tests because previous studied had shown that 30 to 40 percent of losses due to splitting and other mechanical injuries seemed to occur when beans were dropped into deep storage bins. The drop tests by Perry and Hall showed that moisture and drop height as well as temperature have considerable influence on extent of injury to pea beans. The effect of drop height and temperature for beans at 12 percent moisture content is shown in Fig. 8.48. The increase of moisture content up to 18 percent decreased mechanical damage.

Resistance of sugar beet seed to impact was investigated by Kunz *et al.* (1965) who impacted the seed with a small sphere or pneumatically against a steel plate. Impact energy of 25 gm-cm by the sphere reduced shoot growth 13 percent and root growth 6 percent. Energy of pneumatic impact on the order of 26 gm-cm reduced root growth by 15 percent and shoot growth by 22 percent.

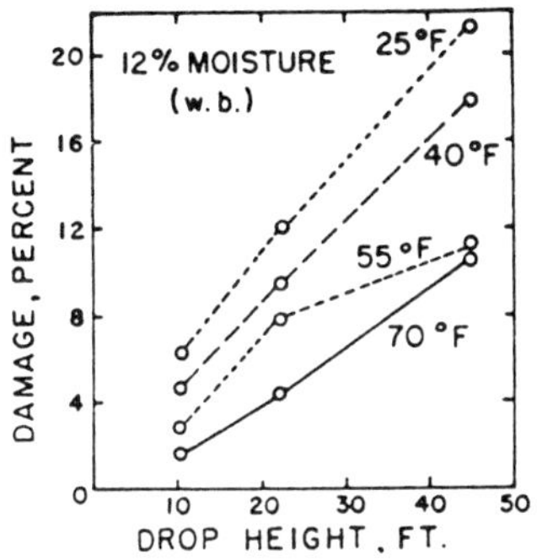

Figure 8.48 The effect of temperature and drop height on mechanical damage to pea beans (Perry and Hall, 1966)

In light of the foregoing discussion the effect of moisture content on mechanical damage to seeds and grains apparently depends on the type of mechanical damage. If visible damage, in the form of splits, cracks and failure of the seed coat is considered, the dryer grain will show greater mechanical damage than the grain with higher moisture content (Figs. 8.27, 8.28, 8.42 and 8.49). On the other hand, if the invisible damage which affects the seed viability and germination capacity is to be considered, the drier grain is more resistant to this type of injury (Figs. 8.46, 8.50). Another factor to consider is the type of forces to which the grain is subjected to in mechanical handling. In the threshing of wheat, the principle type of force is probably impact shear under which, as discussed earlier, the grain with less moisture content takes less energy to fail and shows more damage than the grain with higher moisture content. Research results show this to be true both for wheat kernels tested in laboratory (Fig. 8.28) and in actual threshing operation (Fig. 8.49). In shelling of corn, however, the principal type of force is probably compressive forces under which kernels with less moisture content appear to be stronger and show less damage than kernels with higher moisture content. Waelti and Agness have demonstrated this with both laboratory and field shelling of corn as shown in Fig. 8.3 and 8.51. Another factor contributing to increased damage in shelling of high moisture corn is the increase of force required to separate the kernels form the cob with the increase of moisture content. Waelti and Buchele (1967) found that the most important plant physical property

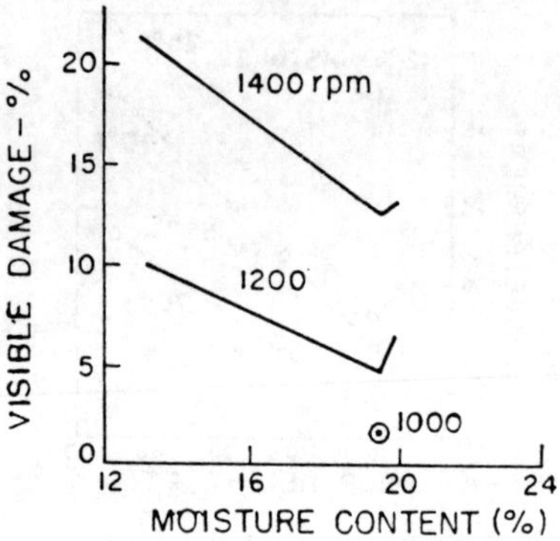

Figure 8.49 Effect of moisture content on visible damage to wheat grain during threshing (King and Riddolls, 1962)

influencing mechanical damage in shelling of corn, were in the order of importance, kernel detachment force, compressive force required to cause rupture of seed coat, kernel deformation under compressive forces, and cob compressive strength. They found that low kernel damage at any given moisture content was associated with low detachment force, high kernel

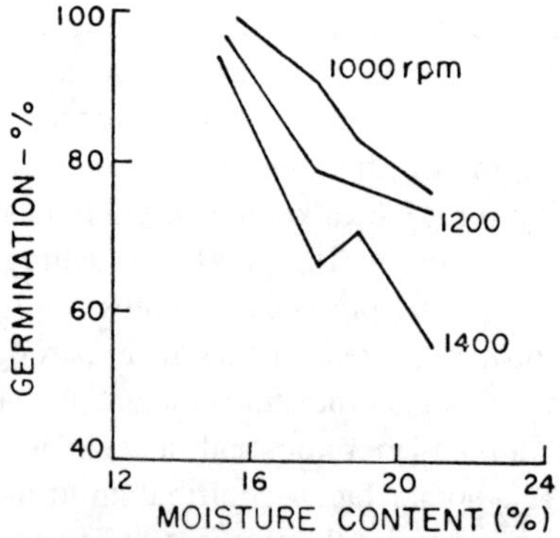

Figure 8.50 Effect of moisture content on invisible damage to wheat grain during threshing (King and Riddolls, 1962)

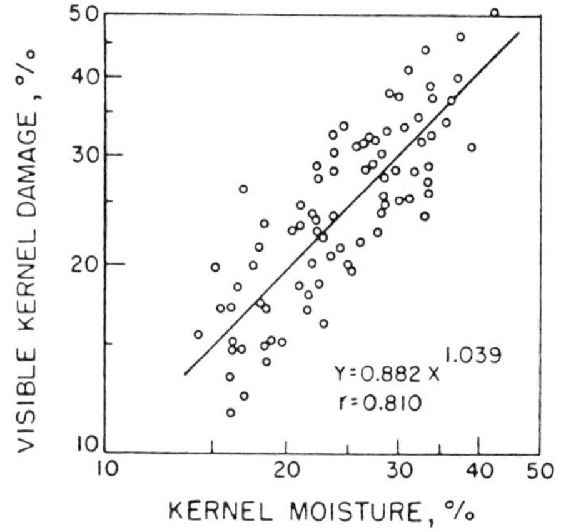

Figure 8.51 Effect of moisture content on kernel damage due to mechanical shelling for four varieties of corn (Waelti and Buchele, 1967)

strength, low kernel deformation, and low cob strength. Apparently the greater resistance of the cob to failure and the kernels to separate at higher moisture contents required longer time for shelling which subsequently caused greater damage to the kernels.

Stress distribution using photo-elastic technique

In an attempt to understand the mechanism of failure in wheat grains, Arnold and Roberts (1969) investigated the stress distribution in wheat specimens under light and heavy loads. They found that for light load conditions the theory of elasticity making use of Hertz contact problem (see Chapter 3) can be applied for stress analysis. For heavy loads causing rupture, however, stress distribution became too complex to analyze theoretically. Under these conditions, these investigators employed photo-elastic models of wheat grain and microscopic deformation studies on actual grains to make a qualitative analysis of the problem of grain failure.

Large sectional models of the wheat grain, ¼-inch thick, were cast and the changing patterns of isochromatics were studied with the aid of a polariscope. As seen from Fig. 1.3, the areas of high stress (high fringe density) occur at the points of contact and around the crease. As the load increased, the stressed region extended from top to bottom leaving the outer regions of the grain unstressed. Grain failure was considered the breaking away of the outer regions from the stressed center section.

In connection with resistance of grains to mechanical damage, Arnold and Roberts (1967) also suggested the concept of load index. Load index was defined as the ratio of load to specific deformation, where specific deformation is the ratio of the actual deformation to the original size in the direction of loading or in terms of symbols

$$\text{Load index} = (F/D)(L)$$

where F/D is proportional to stiffness of the grain and L is the length of the grain in the direction of loading. It was found, as expected, that harder grains sustained greater loads and had larger load index values. The measurement of grain hardness was based on some of the standard hardness tests available to the grain industry (pearling index, etc.).

Conditioning of grains to resist greater forces of impact

In comparing energy requirements for damage to yellow dent corn, soft winter wheat, and pea beans, Zoerb and Hall (1960) found that kernels at lower moisture contents required less breaking energy under impact than under compression. At higher moisture contents, however, grains required more energy to fracture under impact than under compression. In other words, the addition of moisture toughened the grain enabling the material to withstand greater mechanical energy before failure.

Research on several other grains has also shown a range of moisture content which render the material more resistant to dynamic forces of impact. For example, pea beans at 17 to 18 percent moisture (wet basis) are more resistant to damage than beans at less than 15 percent moisture. To reduce the susceptibility of pea beans to this type of damage, it has been proposed to subject the product at lower moisture content to air flow of such temperature and relative humidity conditions that will increase the moisture content of the material to its equilibrium moisture content. This happens to be about 17 to 18 percent at air relative humidities and temperatures between 70 and 80 percent and 70° and 80°F, respectively. Bakker Arkema and Bickert (1967) have given some theoretical and experimental data to explore the feasibility of this idea for toughening the skins of pea beans through environmental control during storage. This concept of conditioning of the crop to resist greater forces in mechanical handling may be applicable to other seeds and grains.

Strength of fruits and vegetables

Resistance to static and impact loads

Maximum drop height and maximum compression loads that apples of different sizes can be subjected to without bruising were reported in terms of height of drop and diameter of bruise (Gaston and Levin, 1951). Mohsenin and Goehlich (1962) and Mohsenin *et al.* (1965) have reported data on maximum allowable load for several varieties of apples at harvest maturity. Resistance to bruising under dead load and quasistatic loading in terms of surface pressure, deformation, and energy and under impact loading in terms of impact energy and deformation, as well as resistance of skin to punch shear are among the data reported in these works (Tables 7.3, 7.4).

Fridley and Adrian (1964) have reported data on resistance to mechanical injuries for apples, pears, apricots, and peaches employing the compression tests and impact tests which were basically similar to those used by Mohsenin and Goehlich (1962). Their findings indicated that pears, at the time of commercial harvesting in California, can withstand a large compression force before bruising and thus present little problem in bulk handling of the fruit in containers filled to large depths. On the other hand, apricots are bruised by a relatively small compressive force and thus their bulk handling is more critical than pears. On the basis of relatively large force and deformations that pears can take (Fig. 8.52) without bruising, these fruits should also present the least problem when shipped in a confined space such as lidded boxes.

It should be noted that stress relaxation characteristics of the fruit under a fixed deformation can also be a factor in the resulting mechanical damage. Fruits which relax their internal stresses more readily than others will not be subjected to prolonged effect of stresses and should show greater resistance to such packing methods as lidding the shipping container. The presence of such relaxation phenomenon and the magnitude of static force remaining on the product has been demonstrated in a case where forces imposed on crisp head lettuce during packing the material in cartons were measured (Harriott *et al.*, 1967).

Relative resistance of four fruits to impact damage is shown in Fig. 8.53. According to these data, for the time of commercial harvesting in California, for which the fruits firmness readings are given, peaches were most resistant to injury by impact while apples were the least resistant. From the standpoint of potential problems in mechanical harvesting and

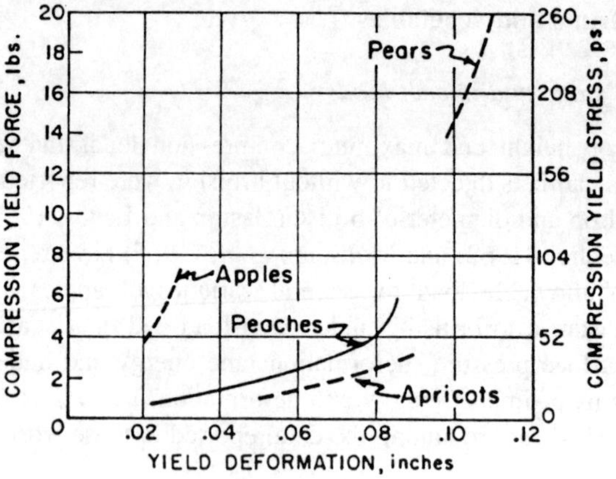

Figure 8.52 Resistance of some fruits to force and deformation under quasi-static loading (Fridley and Adrian, 1964)

handling, Fridley and Adrian conclude that peaches have the best potential for systems involving impacts while apples have the least potential. The past and current research activities and the degree of success which has been achieved for mechanical harvesting of these two crops as well as

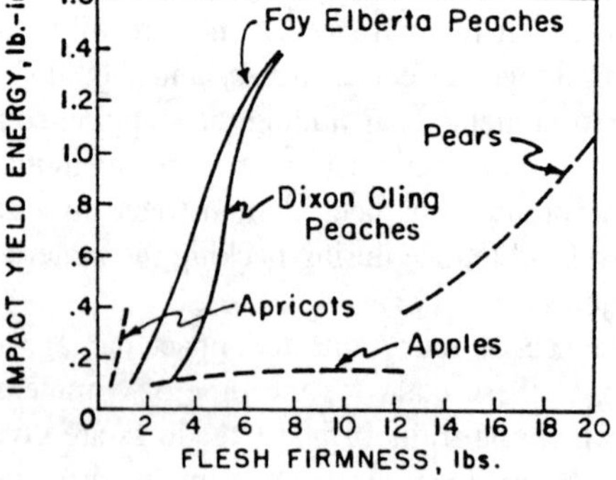

Figure 8.53 Resistance of some fruits to impact (Fridley and Adrian, 1964)

findings by others (Krogenberg, 1964) substantiate the prediction of these researchers made on the basis of their analysis of mechanical properties of the fruits.

Tabachuck (1953) found that potato tuber damage begins at absorbed kinetic energies of as small as 1.5-inch-pounds when the tuber was striking a flat metal surface. Maximum allowable drop height was found to be 10 to 20 inches on a flat surface but only 4 to 6 inches on steel rods of the type used in potato harvesters. The absorbed impact energy by the tuber, E_{ab}, was found from the following equation which can be derived from the principle of conservation of momentum.

$$E_{ab} = (1 - e^2)(M_1\ M_2/(M_1 + M_2))\ H$$

where e is the coefficient of restitution, M_1 and M_2 are, respectively, masses of the potato and the surface being impacted, and H is the height of drop. Since the quantity $(M_2/(M_1+M_2))$ is approximately equal to unity, energy absorbed will actually be a function of e, H, and W, the weight of the potato, as given by Eq. (8.1).

Lamp (1959) has given a rather comprehensive review of literature on evaluating mechanical damage to potatoes. His own work included investigation of mechanical injury to potato tuber as influenced by test location on the tuber, weight and radius of curvature of the tuber, effect of variety, climate, storage period, soil fertility, chemical composition, and position in the hill. Potatoes from the same batch tested by penetration of a plunger under quasistatic loading were subjected to impact and drops in a stationary potato harvester. It was stated that potatoes which showed less resistance to puncture under compression tests were also more susceptible to damage in the harvester.

Dropping potatoes from different heights onto steel rods of the shape and size as those used in a potato harvester revealed that below 6-inch height potatoes are not damaged, at 20-inch height potatoes were lightly damaged or wounded with surface cracks, and at 40 inches, extensive crack wounds as well as flesh wounds occurred (Volbracht and Kuhnke, 1956). When weight of the tubers and the kinetic energy of impact were taken into consideration, it was found that for the same impact energy the large potatoes suffered less bruising than the smaller potatoes which were dropped from a greater height. Although no explanation was given for these results, as pointed out in the discussion of eggshell strength, the reason is the greater momentum and thus greater impulse to which the smaller potatoes dropped from a greater height are subjected. In an actual potato harvester, the small and large potatoes would likely drop from the same height. In this case the larger and heavier potatoes should receive more damage than the smaller ones. The data presented by Volbracht and Kuhnke (1956) support this prediction. Volbracht and Kuhnke also have reported on the influence of radius of curvature, location of tuber in the soil, maturity, storage duration, variety, and fertilization on the extent of mechanical damage to potatoes. With respect to the depth of the tuber in the ground and maturity, Volbracht and Kuhnke (1956) found that resistance of tubers to mechanical damage decreases with both depth in the ground and maturity. Lamp (1959) also found that potatoes located deeper in the ground had less resistance to damage. However, the observation on maturity is not in agreement with reports from others (*Anon.*, 1963) who found that under most conditions, potato skin toughens as the tubers mature and thus resistance to damage increases.

Parke (1963) determined the relationship between impact energy absorbed by potato tubers and the resulting damage in terms of number of split or bruised tubers as well as the volume of tissue bruised. In these experiments the potato was suspended by a pair of threads in the path of a swinging pendulum. The pendulum was released from a known height and the maximum height to which the potato swung was observed. The amount of energy absorbed by the potato was calculated from the principle of conservation of momentum employing the following equation

$$E_{ab} = {}^1/_2\left[Mm/(M + m)\, V_1^2\left(1 - \frac{U_2 - V_2}{V_1}\right)^2\right]$$

where M and m are masses of pendulum and potato, respectively, V_1 is velocity of pendulum just before impact, and V_2 and U_2 are, respectively, velocities of pendulum and potato immediately after impact. The energy absorbed by potato was related to bruise volume by the following equation.

$$V = 2.93 \times 10^{-4} E_{ab} - 382$$

where V is bruise volume in cubic millimeters and E_{ab} is energy absorbed in ergs. It was stated that of the methods chosen for expressing damage, bruise volume appeared to be the most sensitive because it expressed the damage in a quantitative term.

Finney (1964) has reported on the influence of variety and time upon resistance of potatoes to mechanical damage (Fig. 8.54). Mohsenin (1965) has reported on a technique for evaluating the resistance of potatoes to abrasion and skinning against self or against wood or metal (see Chapter 10).

Techniques for measuring resistance of different varieties of potatoes to puncture, abrasion, and impact have been given by Witz (1954). Potato varieties were ranked for resistance to compression in terms of pressure that they can withstand without skin puncture, for resistance to skin abrasion in terms of inch-pounds of torque required for skinning of tubers at a given normal pressure, and for resistance to impact in terms of penetration of a ¾-inch steel ball dropped from different heights.

Edgerly (1951) and Isenberg (1955) conducted impact tests on onions of different sizes dropped from various heights. Bruising was evaluated as that damage where the outer fleshy sheath appeared soft and flattened. They recommended drop heights not greater than 5 feet for onions smaller than 2¼ inches in diameter and not greater than 3 feet for those larger than 2¼ inches.

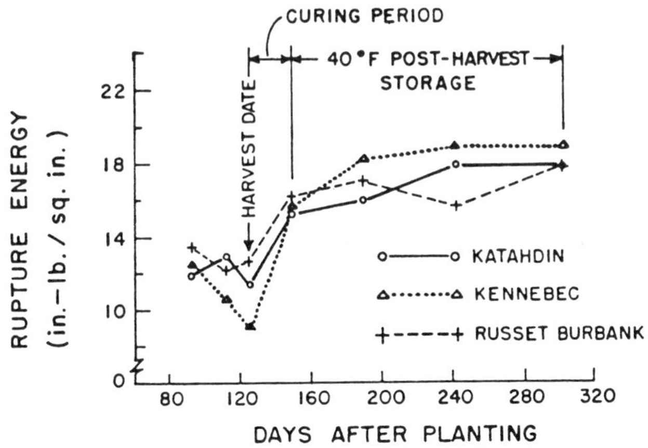

Figure 8.54 Effect of aging on rupture energy of potato loaded with a rigid die (Finney, 1964)

Mechanical damage to tobacco leaves has been reported by Splinter *et al.* (1962). They reported force to shear leaf midrib 26 pounds, energy to break midrib by impact 1 foot-pound, allowable pressure on lamina without bruising 10 psi, and allowable impact energy without bruising 0.22 inch-pounds/psi.

Despite extensive development in mechanical harvesting of tomatoes in recent years, research on resistance of this product to mechanical damage has been quite limited. Mechanical damage to tomato is manifested by water-soaked cellular breakdown of the cross-wall and seed cavity (locular area) which is not necessarily accompanied by external evidence of injury. McColloch (1962) reported on hidden and cumulative bruising injury on tomatoes due to dead load and impact corresponding to the manner in which tomatoes are bruised during commerical handling. Although the work does not quantify the maximum allowable loads in well defined physical parameters, it does describe and illustrate the external and internal symptoms of bruising.

Tomatoes were subjected to a specified load using a ½-inch diameter steel bar and a dial scale. The pressed area was marked for later observation. Dead loads were applied for five to six days either by pressing the tomatoes between two boards or by packing them in a lidded lug box. The magnitude of load on the tomatoes was not measured. Impact tests were simply drop tests on a hard surface or on surfaces padded with foam rubber.

Compression loads above 30 and 35 pounds on green tomatoes produced dry and stringy gel tissue in the affect locules (McColloch, 1962). Loads of 10 to 20 pounds produced gel tissue which became darker than normal and appeared cloudy. Impact damage resulted generally in water-soaked tissues, cloudy gel, and in severe cases, the gel became more liquid and the seeds were detached from the placenta. The extent of damage under different static loads and the external and internal symptom of damage to tissue structure are given in Table 8.14.

Additional data on resistance of fruits and vegetables and seeds and grains to mechanical damage are given in tables in Chapters 5, 6, 7 and in the Appendix.

Influence of temperature

Influence of temperature on impact resistance of pea beans, given in Fig. 8.14 bruising of cherries, and breakage of eggshell has already been dis-

Table 8.14 Influence of maturity and static load on appearance of external and internal injury of tomatoes (McColloch, 1962)

Maturity	Load (lbs)	Total fruits	External condition (% of fruits affected)			Internal condition (% of fruits affected)							
						Parts watersoaked			Condition of gelatinous tissues				Hollow pockets in outer wall
			Immediately water-soaked	Water-soaked condition remaining	Pressure mark remaining visible	Outer wall	Cross wall	Pla-centa	Cloudy	Cloudy and slimy	Stringy	Dry	
25 to 40% colored	15	20	60	20	15	60	40	90	50	50	0	0	0
20% colored	15	14	50	21	7	36	0	50	78	21	0	0	0
5 to 10% colored	15	35	14	0	0	0	0	34	91	9	0	0	0
Mature green	10	20	0	—	100	0	0	0	75		0	0	0
Do—	20	20	5	0	65	0	0	0	50	0	10	40	0
Do—	30	20	80	0	65	0	0	0	5	0	30	65	60
Do—	35	36	100	0	33	0	0	0	0	0	0	100	8

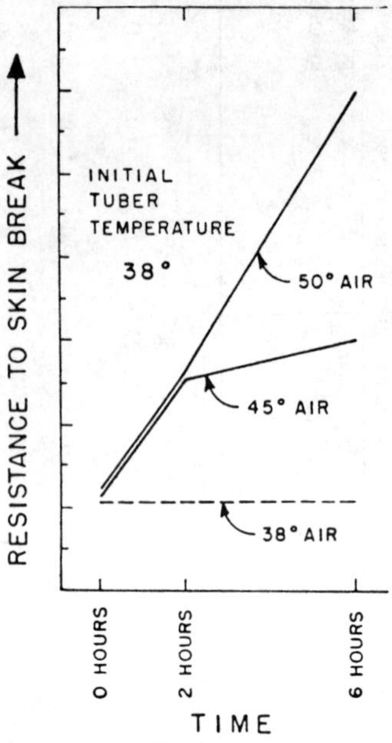

Figure 8.55 Effect of temperature on resistance of potato to impact damage (Johnston and Toko, 1963)

cussed. Some research conducted on potatoes and apples and other fruits will be discussed in the following.

Hopkins (1953) conducted field and laboratory tests to evaluate the effectiveness of padding materials to be used with potato barrels used in New England States for handling potatoes. Potatoes of the same weight were dropped from the same distance on the same padding material covering the bottom of the barrels. About 6 per cent more external injuries were detected on the potatoes when temperature dropped from a range of 53°–56°F to a range of 42°–49°F. Johnston and Toko (1963) have verified this behavior as shown in Fig. 8.55. Resistance to bruising was evaluated by measuring the drop height of a given weight required to break the surface of the tuber. Figure 8.55 shows the increase of resistance to bruising as the internal temperature of the tuber increased. In handling potatoes taken out of cold storage, such relationships may be quite important in reducing tuber damage.

Hawkins and Sando (1920) reported increased resistance to epidermis rupture for decrease in temperature of strawberries, raspberries, blackberries and cherries. Rose *et al.* (1934) also reported increased resistance to puncture at lower temperatures. Hartman (1924) reported increased firmness in pears at 51°F compared to 97°F. Haller (1941) reported no significant change in Magness firmness reading of several varieties of apples from 34°F to 77°F. However, Mattus *et al.* (1960) reported that cold apples were more resistant to bruising than warm apples under impact. For Bartlett pears, which are subjected to internal brown spot when dropped, the opposite trend was reported.

Quasistatic tests conducted by Nelson (1967) employing a ¼-inch die on intact McIntosh apples showed higher values for bioyield point for apples at 90°F than at 40°F (Table 8.15). The dynamic tests in which the apple was used as a pendulum striking a rigid plate showed greater resistance to bruising at 40°F than at 90°F. In other words under static conditions of loading, the apples were more resistant to damage at the higher temperature.

Note that in Table 8.15 while apples at lower temperature are "firmer," they are less resistent to internal cell rupture as manifested through their lower values of bioyield force. The higher yield force exhibit at the higher temperature could be due to the reduced turgidity which enabled the plant cells to undergo a greater deformation without rupturing.

Table 8.15 Effect of temperature on "firmness" and resistance to yield and skin rupture of McIntosh apples (Nelson, 1967)

Property	Number in sample	Temp. °F	Range	Mean	Standard deviation
Bioyield force					
(lbs)	53	40	2.4–6.2	4.08	0.75
	54	90	3.5–6.2	4.50	0.56
Rupture force					
(lbs)	54	40	6.7–9.9	7.95	0.71
	54	90	6.5–9.5	7.87	0.61
Firmness[1]	52	40	250–414	326	43
	53	90	226–353	287	36

[1] Firmness $= E_a/(1 - \mu^2) = F/D\ (1/2a)$

The results of dynamic tests by Nelson (1967) on another sample of apples are given in Table 8.16. Note that in this work the criterion for mechanical injury was the volume of the bruise calculated using the following formula.

$$V = \pi h/24\ (3d^2 + 4h^2) \tag{8.26}$$

where V is volume in cubic inches of the bruised section, assumed to be a segment of a sphere, h is maximum depth of the bruise in inches, and d is average diameter of the bruise in inches. It was found that there was a significant correlation between dynamic yield pressure and resistance to bruising. The dynamic yield pressure was significantly higher at 40°F than at 90°F, indicating, based on these data, that apples are subject to greater impact bruising at 90°F than at 40°F. The same trend is shown in comparing the bruise volume per unit energy absorbed or bruise volume per unit mementum absorbed.

Although this finding is contrary to the results of the static tests shown in Table 8.15, it should be noted, that the impact-tested fruits were taken from apples stored several months and are not the same fruits tested in fall after harvest. When we consider the static tests of these apples (footnote Table 8.16), we find that both bioyield and rupture values were higher for cooler apples, indicating greater resistance to mechanical damage, which is in agreement with results of the impact tests of the same apples. It should be noted that in actual practice reportedly apples bruise more easily from impact when they are cold than when they are warm. This observation, which is contrary to results of laboratory impact tests suggests the need for further investigation in this area. One way to improve the experiment would be to do both the static tests and impact tests at the same time and on a much larger sample than those reported in Table 8.16.

Nelson (1967) also established the correlation between bruise volume and both energy absorbed and momentum absorbed for the two temperatures as shown by the following regression equations.

$$V = 0.08E_{ab} - 0.02 \text{ and } V = 6.91M_{ab} - 0.10 \text{ (at 40°F)} \tag{8.27}$$

$$V = 0.09E_{ab} + 0.00 \text{ and } V = 7.37M_{ab} - 0.07 \text{ (at 90°F)} \tag{8.28}$$

If such relationships are established for fruits and vegetables which bruise easily under impact, when the maximum allowable bruise is specified

Table 8.16 Some dynamic properties of McIntosh apples as affected by temperature and drop height (Condensed from Nelson, 1967)[1]

Number in sample	Apple temp. (°F)	Drop height (in)	Rebound height (in)	Apple weight (lb)	Radius of contact	Coeff. of restit.	Energy absorbed (in-lb)	Momentum absorbed (lb-sec)	Dynamic yld. pres. (psi)	Bruise volume (in^3)	[2]	[3]
5	40	2	0.60 (0.14)[2]	0.39 (0.04)	0.36 (0.03)	0.30 (0.07)	0.54 (0.05)	0.02 (0.003)	21.0 (1.0)	0.02 (0.02)	0.037	1.00
5	90	2	0.75 (0.18)	0.38 (0.02)	0.38 (0.03)	0.38 (0.04)	0.47 (0.07)	0.02 (0.004)	22.3 (2.0)	0.03 (0.01)	0.064	1.50
5	40	4	1.35 (0.22)	0.39 (0.02)	0.43 (0.02)	0.34 (0.06)	1.05 (0.12)	0.02 (0.003)	26.97 (2.31)	0.07 (0.03)	0.067	3.50
5	90	4	1.50 (0.0)	0.39 (0.02)	0.45 (0.02)	0.38 (0.0)	0.97 (0.06)	0.02 (0.002)	25.1 (3.29)	0.10 (0.06)	0.103	5.00
5	40	6	1.70 (0.21)	0.41 (0.04)	0.46 (0.05)	0.28 (0.03)	1.95 (0.14)	0.03 (0.003)	25.4 (2.34)	0.12 (0.03)	0.062	4.00
5	90	6	1.95 (0.21)	0.38 (0.02)	0.50 (0.04)	0.32 (0.04)	1.55 (0.23)	0.03 (0.01)	23.4 (1.66)	0.14 (0.10)	0.090	4.67
5	40	8	2.45 (0.27)	0.38 (0.06)	0.53 (0.02)	0.30 (0.01)	2.11 (0.25)	0.03 (0.01)	24.0 (2.60)	0.16 (0.04)	0.076	5330
5	90	8	2.45 (0.27)	0.39 (0.04)	0.56 (0.01)	0.30 (0.03)	2.18 (0.29)	0.04 (0.01)	20.1 (1.60)	0.19 (0.01)	0.087	4.75
5	40	10	2.80 (0.21)	0.40 (0.04)	0.55 (0.02)	0.28 (0.02)	2.91 (0.14)	0.04 (0.004)	25.2 (3.88)	0.19 (0.07)	0.065	4.75
5	90	10	3.05 (0.21)	0.38 (0.01)	0.58 (0.02)	0.30 (0.02)	2.66 (0.18)	0.04 (0.004)	22.1 (2.84)	0.21 (0.04)	0.079	1.25

Number in parentheses gives standard deviation. [1] The mean values of bioyield force of all these apples were 2.7(0.6) lb. and 2.5(0.5) lb. and 40° and 90°F, respectively. The mean values of skin rupture force for all the apples were 6.1(0.6) and 5.3(0.5) at 40° and 90°F, respectively. [2] Bruise volume per unit energy absorbed. [3] Bruise volume per unit momentum absorbed.

in terms of maximum depth and average width, substitution of calculated bruise volume in the regression equations would give an estimate of maximum drop height or impact velocity which should not be exceeded.

8.12 LIST OF SYMBOLS FOR CHAPTER 8

A —elastic constant $= \frac{1 - \mu_1^2}{E_1} + \frac{1 - \mu_2^2}{E_2}$ (Eq. 6.4)
a —radius of permanent indentation or deceleration
D —deformation
d_2 —chordal diameter of indentation
E_{ab} —energy adsorbed
e —coefficient of restitution
F —impact force
G —fragility factor or deceleration ratio $G = \frac{a}{g}$
g —gravitation acceleration
h_1 —height of drop
h_2 —height of rebound
M_{ab} —momentum absorbed
m —mass
P —hydrostatic pressure
P_d —dynamic yield pressure
R —radius of sphere
S_{max} —maximum contact stress
t —time or thickness
V —velocity, volume of damaged tissue

Table 8.17 Summary of tests for maximum allowable static and dynamic loads for agricultural products

Product	Type of test	Information reported	Ref. no.	Fig. no.	Table no.
Apple	Drop test and compression test	Maximum drop height or load to avoid bruising	1		
	Quasistatic compression test (die loading and plate loading)	Force, deformation, energy to initiate bruise	2		
	Impact of apple segment attached to striking arm and displacement transducer	Energy of impact to initiate bruising, ratio of impact to compression energy	2	8.25	
	Quasistatic dead load compression, and impact of whole apple as a pendulum	Bioyield force, rupture force, bruise volume per unit impact energy, dynamic yield pressure at 40°F and 90°F	3	8.23	8.15 8.16
Castor bean	Mechanical handling in screw conveyor	Effect of various design parameters in screw conveyor handling seeds, resistance to compression	4		
	Plate compression and pendulum impact	Resistance to breakage with or without cushioning	5		
Cottonseed	Pneumatic impact of seed against steel plate	Effect of seed velocity	6	8.47	
	Quasistatic compression of single seeds	Rupture force, deformation, energy, moisture effect	6		8.13
	Impacting seed with known orientation	Effect of velocity, orientation, moisture, comparison of static and dynamic loads	7	8.29	
Eggs	Crushing, puncturing, dropping of whole egg	Force of shell failure, loading device, part of egg under load	8		8.6

Table 8.17 (Continued)

Product	Type of test	Information reported	Ref. no.	Fig. no.	Table no.
Eggs	Compression of whole egg between parallel plates	Correlation coefficient of 0.76 between breaking load and shell thickness	9		
	Measurement of shell thickness and density employing beta backscatter technique	Correlation of drop height and beta back-scatter count, claimed related to both shell density and thickness	10	8.32	
	Compression point-loading of whole egg embedded in plaster of Paris	Linear limit, rupture force, elastic deformation, percent elasticity, modulus of elasticity, failure stress in tension and compression	11		8.7
	Compression of narrow rings of shell	Modulus of elasticity, failure stress in tension and compression	12		8.8
	Compression and drop of whole egg against padding materials with known modulus of elasticity	Mean force and mean energy at failure. Comparison of static and dynamic failure energy	13	8.33 8.34 8.35	
	Compression of whole egg between parallel plates	Effect of location of loading point, energy to failure, stiffness or spring factor	14	8.36	Text
	Impact of whole egg with steel rod incorporating force transducer and accelerometer	Correlation between impact force at failure and shell thickness	15		8.9
	Stress coat technique with whole egg under plate compression	Force to fracture from stress analysis	16	8.40	
	Hydrostatic internal compression using water and drop tests of whole eggs	Latitudinal failure stress	17		

Table 8.17 (Continued)

Product	Type of test	Information reported	Ref. no.	Fig. no.	Table no.
Eggs	Hydrostatic internal compression using air pressure, continually increasing at a known constant rate	Latitudinal and meridional failure stresses at various points of shell with known thickness	18	8.37 8.38	8.9 8.10
Fruits	Compression and impact tests of fruit segments (apple, apricot, pear, peaches)	Force-deformation and energy to bruise under compression and impact.,	19	8.52 8.53	
	Simulated transit tests	Magnitude of acceleration at different layers of fruit in transit, natural frequency of fruits	20	8.30	
Grains	Quasistatic compression of whole grain at different position (barley, corn, oats, soybeans, wheat)	Force to fracture	21		Appendix
	Single grain dropped into path of rotating paddle (corn, soybean, wheat)	Energy requirement to damage grain, effect of moisture	21	8.42	
	Uniaxial compression, parallel shear of grains (corn, pea, beans, wheat)	Comparison of compression and impact shear, ultimate strength at different moisture	22	8.27 8.43	
Onions	Compression tests and drop tests	Distortion, pressure bruising of bulbs of various sizes	23		
Pea beans	Drop test on a layer of pea beans	Effect of moisture, temperature, drop height	24	8.48	
Peanut	Impact of peanut on flat plate of a rotating arm	Effect of impact velocity, moisture, seed orientation on damage	25	8.46	8.12
Potato	Drop test on flat steel plate and steel rods	Absorbed impact energy in terms of mass and coefficient of restitution	26		

Table 8.17 (Continued)

Product	Type of test	Information reported	Ref. no.	Fig. no.	Table no.
Potato	Quasistatic compression using plate and plunger, impact in a potato harvester	Influence of variety, maturity, size, weight, location in soil, etc.	27		
	Swinging pendulum striking suspended tuber	Energy absorbed, bruise volume	28		
	Abrasion, penetration by compression and impact	Resistance of different varieties	29		
	Drop tests on padding materials	Effectiveness of padding materials and temperature effect	30		
	Die-loading compression	Effect of variety and aging on rupture energy	31	8.54	
Sorghum	Impact by shooting grain with a spring loaded gun against rigid surface	Energy absorbed and resulting damage, effect of moisture	32	8.45	
Sugar beet/ seed	Impacting seed with a sphere or pneumatically against a steel plate	Reduction in shoot and root growth	33		
Tobacco leaf	Compression and impact	Force and energy to break midrib, allowable impact energy and compression force	34		
Tomatoes	Drop tests and compression tests	Symptoms of external and internal damage	35		8.14
Wheat	Compression of bulk grain	Damage to bulk grain at different moisture	36	8.44	
	Effect of threshing drum speed and concave clearance	Optimum drum speed and concave clearance	37		1.2
	Impact by striking grain with rotating hammer, broad side down	Effect of impact velocity, effect of moisture content	38		8.11

Table 8.17 (Continued)

Product	Type of test	Information reported	Ref. no.	Fig. no.	Table no.
Wheat	Stress analysis of photoelastic models of single grain under compression	Applicability to Hertz contact problem	39		

(1) Gaston and Levin, 1951. (2) Mohsenin and Goehlich, 1962. (3) Nelson, 1967. (4) Bouse *et al.*, 1964. (5) Burmistrova *et al.*, 1956. (6) Kirk and McLeod, 1967. (7) Clark *et al.*, 1967. (8) Tyler, 1961. (9) Brooks and Hale, 1955. (10) James and Retzer, 1967. (11) Mohsenin and Shelef (unpublished). (12) Rehkugler, 1963. (13) Rehkugler, 1964. (14) Voisey and Hunt, 1967a. (15) Voisey and Hunt, 1967b. (16) Voisey and Hunt, 1967c. (17) Sluka *et al.*, 1965. (18) Hammerle and Mohsenin, 1967. (19) Fridley and Adrian, 1966. (20) O'Brien *et al.*, 1963, 1965. (21) Bilanski, 1965. (22) Zoerb and Hall, 1960. (23) Ang *et al.*, 1960. (24) Perry and Hall, 1966. (25) Turner *et al.*, 1965. (26) Tabachuk, 1953. (27) Lamp, 1959. (28) Parke, 1963. (29) Witz, 1954. (30) Hopkins, 1953. (31) Finney, 1964. (32) Leonhardt *et al.*, 1961. (33) Kunz *et al.*, 1965. (34) Splinter *et al.*, 1962. (35) McColloch, 1962. (36) Narayan and Bilanski, 1966. (37) King and Riddolls, 1960. (38) Mitchell and Rounthwaite, 1964. (39) Arnold and Roberts, 1966.

9

AERO- AND HYDRODYNAMIC CHARACTERISTICS

IN HANDLING and processing of agricultural products often air or water is used as a carrier for transport or for separating the desirable product from the unwanted materials. The pneumatic separation and conveying has been in use in agricultural machinery and food processing equipment for many years. Use of water, however, as a carrier for more economical transport or less injury to such products as fruits and vegetables, is a relatively new idea in the agricultural industry. In either case, fluid flow occurs around the solids and the problem involves the action of the forces exerted by the fluid on these solids. As the principles of this subject, known as fluid mechanics, find increasingly wide applications in handling and processing of agricultural products, it becomes necessary to have a knowledge of those physical properties which affect the aero- and hydrodynamic behavior of agricultural products. In this chapter we will discuss particle dynamics in terms of drag coefficient and terminal velocity and their applications in handling and processing of agricultural products.

9.1 DRAG COEFFICIENT

When fluid flow occurs about immersed objects, the action of the forces involved can be illustrated by diagrams such as the classic example shown in Fig. 9.1. The pressure on the upper side of the object is less than and that on the lower side is greater than the pressure P in the undisturbed fluid stream. This results in a decrease of pressure, $-P$, on the upper side indicated by arrows drawn away from the surface, and an increase of pressure, $+P$, shown by arrows drawn toward the object. In addition to these forces normal to the surface of the object, there are shear stresses, τ, acting tangential to the surface in the direction of flow and resulting from frictional effects.

The resultant force F_r may be resolved into components, F_D, the drag and F_L, the lift. The equations for calculating drag and lift have been derived by dimensional analysis assuming the smooth object having a projected area, A_p, moving through a fluid of mass density, ρ_f, viscosity, η, and modulus of elasticity, E, with a velocity, V. Therefore,

$$F_D = f_1(A_p, \rho_f, \eta, E, V)$$

$$F_L = f_2(A_p, \rho_f, \eta, E, V)$$

Employing the methods of dimensional analysis, the following equations have been established for drag and lift

$$F_D = C_D A_p \frac{\rho_f V^2}{2} \tag{9.1}$$

$$F_L = C_L A_p \frac{\rho_f V^2}{2} \tag{9.2}$$

where C_D and C_L are the (dimensionless) drag coefficient and lift coefficient of the object, respectively.

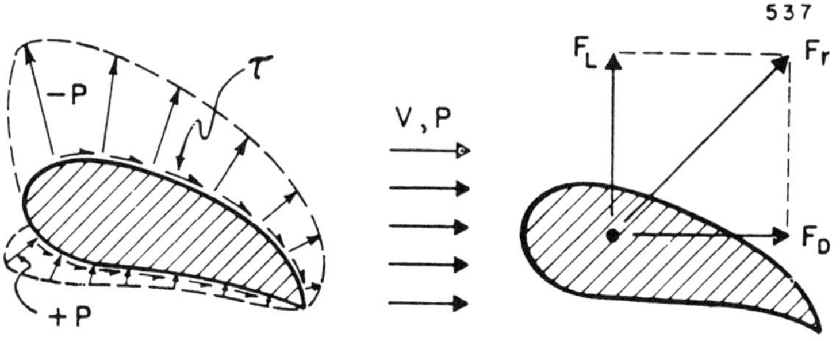

Figure 9.1 Flow about immersed object

In most agricultural engineering applications the moving object is usually free to assume its own random orientation. For this reason the net resistance force F_r can be given in terms of an overall drag coefficient C as follows

$$F_r = \tfrac{1}{2} C A_p \rho_f V^2 \tag{9.3}$$

where

F_r = resistance drag force (lbs) = weight of particle at terminal velocity
C = overall drag coefficient (dimensionless)
A_p = projected area normal to direction of motion $\cdots$ ft^2
ρ_f = mass density of the fluid $\cdots$ slugs/ft^3 = (lb$-$sec^2)(ft$-$ft^3)
V = relative velocity between main body of fluid and object $\cdots$ ft/sec

However, in certain cases it is desirable to resolve the resultant force into the components of frictional drag due to tangential forces on the body surface and profile drag due to pressure distribution around the body. In laminar or low velocity flow where variation in fluid density is small and viscous action governs the flow, the profile or pressure drag is negligible. In turbulent or high velocity flow where fluid compression and not viscous action governs the flow, the frictional drag is negligible. The classic example of frictional drag is the drag force exerted on one side of a smooth flat plate aligned with the flow. An example of profile drag is the drag force on a blunt object, such as a sphere.

Frictional drag

The following equations have been given for frictional drag coefficient, C_f, (Vennard, 1961; Prandtl, 1932). For flat plate with a laminar boundary layer

$$C_f = \frac{1.328}{(N_R)^{0.5}} \tag{9.4}$$

For flat plate with a turbulent boundary layer

$$C_f = \frac{0.455}{(\log N_R)^{2.58}} \tag{9.5}$$

In these equations N_R is the Reynolds number defined by

$$N_R = \frac{Vd\rho_f}{\eta} \tag{9.6}$$

where d is the effective dimension of the object such as length of a plate or diameter of a sphere and η is the absolute viscosity of the fluid (lb-sec/ft^2).

The equation for the transition region, where laminar flow changes to turbulent flow is given by Prandtl (1932) as follows

$$C_f = \frac{0.455}{(\log N_R)^{2.58}} - \frac{1700}{N_R} \tag{9.7}$$

Having the coefficient C_f, the drag force can be calculated using Eq. (9.1). In the case of plate-like materials, it should be noted that the drag force calculated from (9.1) needs to be multiplied by 2 to account for the two sides of the plate. Also note that the above equations are for frictional drag of smooth flat plates aligned with the flow. If the plate or a circular disk is placed normal to the flow, the total drag will contain negligible frictional drag and does not change with Reynolds number (see Fig. 9.2). Experiments have shown that this is usually true for all brusque or very rough objects in a fluid flow. This should have applications in fluid handling of agricultural materials.

Profile or pressure drag

When a blunt object, such as a sphere, is placed in a fluid flow, the frictional drag, usually but not always, can be neglected because of the small surface area on which frictional effects can act. The exception is the case of flow at very low Reynolds numbers where Stokes law is applicable. Stokes has proven that at Reynolds numbers less than unit, where the inertia forces may be neglected and those of viscosity alone considered, the flow closes behind a sphere-like object and the profile drag is composed primarily of frictional drag. For a sphere of diameter d_p, moving at a velocity V through a fluid of viscosity η, Stokes law gives drag force by

$$F_D = 3\pi\eta V d_p \tag{9.8}$$

Equating the above equation with Eq. (9.1) and taking A_p to be the frontal area and equal to $(\pi/4)d_p^2$, the profile drag coefficient is found to be

$$C_D = \frac{24}{N_R} \tag{9.9}$$

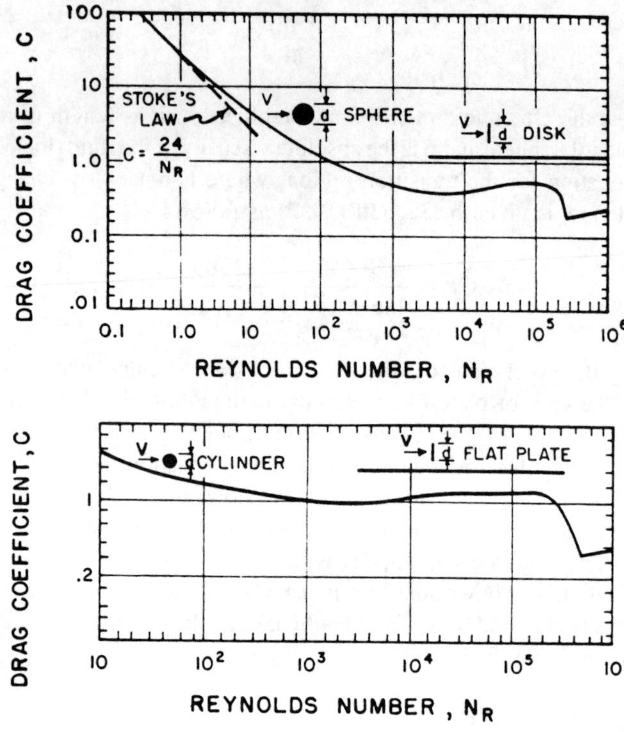

Figure 9.2 Drag coefficient for particles with regular geometric shapes (Redrawn from Vennard, 1961)

As Reynolds number exceeds unity, the Stokes law is no longer applicable because flow opens up behind the blunt object and the drag coefficient is a combination of frictional drag as well as pressure drag in a range up to $N_R = 1000$. Above this figure the frictional effects may be neglected. The variation of drag coefficient of spheres of many sizes with Reynolds number is shown in Fig. 9.2 which has been confirmed for flow in many fluids. As shown in this Figure, the value of C_D is roughly constant for the range of $2.5 \times 10^5 > N_R > 1 \times 10^3$. At about $N_R = 2.5 \times 10^5$, there is a sudden drop of C_D followed by a gradual increase.

9.2 TERMINAL VELOCITY

In free fall, the object will attain a constant terminal velocity, V_t, at which the net gravitational accelerating force, F_g, equals the resisting upward drag force, F_r. A man jumping from a plane after some uninterrupted fall will be moving so fast that the drag of the air will be as great as his weight, resulting in a constant velocity (terminal velocity) and zero acceleration. If the man does not have his parachute open, he will reach his terminal velocity at about 120 miles per hour. If he opens his parachute, his terminal velocity is reduced to about 14 miles per hour, which is equivalent to a jump of about 7 feet.

Under the steady state condition, where terminal velocity has been achieved, if the particle density is greater than the fluid density, the particle motion will be downward. If particle density is smaller than the fluid density, the particle will rise. When air stream is used for separation of a product such as wheat from its associated foreign materials such as straw and chaff, a knowledge of terminal velocity of all the particles involved would define the range of air velocities affecting good separation of the grain from foreign materials. For these reasons, terminal velocity has been used as an important aerodynamic characteristic of materials in such applications as pneumatic conveying and separation from foreign materials.

To derive a general expression for terminal velocity, set the gravitational force, F_g, (corrected for buoyancy) equal to the resisting drag force, F_r, and velocity, V, equal to terminal velocity, V_t. Therefore,

$$F_g = F_r \text{ when } V = V_t$$

Substituting for F_g and F_r, the expression for terminal velocity will be

$$m_p g \left[\frac{(\rho_p - \rho_f)}{\rho_p} \right] = \tfrac{1}{2} C A_p \rho_f V_t^2 \tag{9.10}$$

or

$$V_t = \left[\frac{2W(\rho_p - \rho_f)}{\rho_p \rho_f A_p C} \right]^{1/2}$$

and

$$C = \frac{2W(\rho_p - \rho_f)}{V_t^2 A_p \rho_p \rho_f} \tag{9.11}$$

where

g = acceleration due to gravity—ft/sec^2
m_p = mass of the particle—slugs = w/g = lb-sec^2/ft
ρ_p = mass density of the particle—slugs/ft^3 = (lb-sec^2)/(ft^4)
ρ_f = mass density of the fluid—slugs/ft^3 = (lb-sec^2)/(ft^4)
A_p = projected area of the particle normal to the motion—ft^2
W = weight of particle—lbs

Note that the drag coefficient is given by C which is considered to be an overall drag coefficient such that $C = C_f + C_D$. Where flow is laminar, C_f is generally negligible. For turbulent flow, C_f is usually negligible except for streamlined bodies. The expressions for terminal velocity of objects to various shapes are given in the following.

Spherical bodies

For a sphere of diameter d_p, substitution for $A_p=(\pi/4)d_p^2$ and $W=(\pi/6)\rho_p g d_p^3$ yields the following expression for terminal velocity.

$$V_t = [4gd_p(\rho_p - \rho_f)/3\rho_f C]^{1/2} \tag{9.12}$$

For conditions of laminar flow, the values of the drag coefficient, C, are calculated from (9.9) for Reynolds numbers of less than 1.0. Substitution for C and N_R in (9.12), gives the equation of terminal velocity (also called settling velocity) according to Stokes law.

$$V_t = gd_p^2(\rho_p - \rho_f)/18\eta \tag{9.13}$$

Even for Reynolds number of 2, Eq. (9.9) gives a good approximation of drag coefficient (Lapple, 1956). For Reynolds numbers greater than 2, the values of drag coefficient can be found from Fig. 9.2.

For conditions of turbulent flow in the region where $10^3<N_R<2\times10^5$ and C is equal to approximately 0.44, the following equation has been given for terminal velocity (Lapple, 1956).

$$V_t = 1.74[gd_p(\rho_p - \rho_f)/\rho_f]^{1/2} \tag{9.14}$$

Finally for an intermediate region of $2<N_R<10^3$, the drag coefficient is given by

$$C = \frac{18.5}{(N_R)^{0.6}} \tag{9.15}$$

and the terminal velocity by

$$V_t = \frac{0.153g^{0.714}d_p^{0.142}(\rho_p - \rho_f)^{0.714}}{\rho_f^{0.286}\eta^{0.428}} \tag{9.16}$$

Non-spherical bodies

In Table 9.1 comparative equations for spheres and other objects with regular geometric shapes are given.

Very little work has been done on irregular shapes which are particularly complicated by their random orientation and the variety of methods of expressing their size and dimensions to be used in calculation of Reynolds numbers and frontal area. Schiller (1932) has presented data on drag coefficient in terms of Reynolds number for some irregular particles. His graphical presentation, shown in Fig. 9.3, is based on the work of several investigators using such materials as sand grains, gravel, coal, quartz, dust particles, with shapes not well-defined geometrically, as well as such geometric shapes as spheres and cubes. Drag coefficients and Reynolds numbers were obtained by determining the suspension velocity of the experimental particles under free fall in air, liquid paraffin and water.

As seen in Fig. 9.3, for Reynolds numbers of less than 50, the assumption of sphericity for irregular shapes holds fairly well if the particle shapes are not extremes and if the diameter is taken as the diameter of an equivalent sphere. For Reynolds numbers greater than 50, however, the drag coefficient curves level off and the assumption of sphericity will result in considerable error. For this reason, an attempt should be made to obtain a more reliable estimate of Reynolds number by assuming shapes other than sphere and applying the methods described previously, or actually determining the drag coefficient and terminal velocity experimentally.

Terminal velocity from drag coefficient-Reynold's number relationship

Since both drag coefficient and Reynolds number equations include a velocity term, calculation of terminal velocity from Reynolds number and drag coefficient relationship require a trial-and-error solution. To eliminate a trial-and-error solution, the terms CN_R^2 or C/N_R are first calculated and plotted against N_R. Since these two terms do not include velocity, V_t, and

Table 9.1 Comparative summary of equations of motion of spheres, disks and circular cylinders (Adopted from Lapple, 1956)

	Sphere (any direction[1])	Thin disk (normal to face[1])	Thin disk (parallel to face[1])	Infinite circular cylinder (normal to axis[1])
Reynolds No. equ.	$d_p V \rho_f / \eta$	$d_p V \rho_f / \eta$	$2 L V \rho_f / \eta$	$d_p V \rho_f / \eta$
Frontal area A_p	$(\pi/4) d_p^2$	$(\pi/4) d_p^2$	$(d_p) L$	$(d_p) K$
Mass m_p	$\rho_p (\pi/6) d_p^3$	$\rho_p (\pi/4) d_p^2 L$	$\rho_p (\pi/4) d_p^2 L$	$\rho_p (\pi/4) d_p^2 L$
Drag relationships				
streamline flow				
$N_R < 0.2, F_D =$	$3\pi\eta V d_p$	$8\eta V d_p$	$(16/3)\eta V d_p$	$(4\pi/K)\eta V L$
$C_D N_R =$	24	$64/\pi$	64/3	$8\pi/K$
turbulent flow				
C_D (average)	0.44	1.12	—	1.2
N_R (range)	$1\times10^3 - 2\times10^5$	>1000	—	$1\times10^2 - 2\times10^5$
Terminal velocity V_t^2	$\dfrac{4 g d_p (\rho_p - \rho_f)}{3 C \rho_f}$	$\dfrac{2 g L (\rho_p - \rho_f)}{C \rho_f}$	$\dfrac{g d_p (\rho_p - \rho_f)}{2 C \rho_f}$	$\dfrac{g d_p (\rho_p - \rho_f)}{2 C \rho_f}$

[1] Direction of flow or motion.

L = Thickness of disk, length of rod or cylinder, length of flat plate along direction of flow or motion

$K = 2.002 \ln N_R$

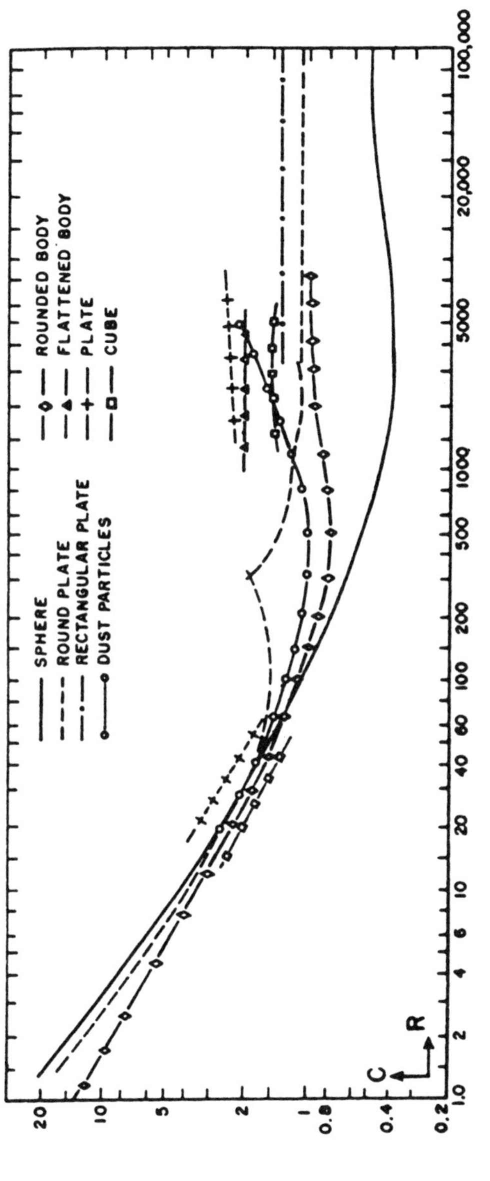

Figure 9.3 Drag coefficient for geometric and non-geometric shapes (Redrawn from Schiller, 1932)

particle diameter, d_p, respectively, depending on whether V_t or d_p is unknown, the value of CN_R^2 or C/N_R can be calculated from the given information and by reference to the appropriate plot, the corresponding value of N_R can be obtained.

For spherical particles the terms

$$CN_R^2 = \frac{4g\rho_f d_p^3(\rho_p - \rho_f)}{3\eta^2} \tag{9.17}$$

and

$$C/N_R = \frac{4g\eta(\rho_p - \rho_f)}{3\rho d_f^2 V_t^3} \tag{9.18}$$

are obtained by combining Eqs. (9.6) and (9.12). If Eqs. (9.11) and (9.6) are combined, the term CN_R^2 will not include d_p but will require a knowledge of weight and mass density of the object, ρ_p, as seen from the following resulting equation

$$CN_R^2 = \frac{8W\rho_f(\rho_p - \rho_f)}{\pi\eta^2\rho_p} \tag{9.19}$$

The plot of CN_R^2 versus N_R for spherical objects is shown in Fig. 9.4.

The values used to obtain the graphs of Fig. 9.4 are taken from tables such as given by Lapple (1956) and Henderson and Perry (1952). Table 9.2 gives absolute viscosity and mass density of air and water which are normally used as the fluid medium in aero- and hydrodynamic problems in handling of agricultural products.

Example

Consider the following data taken from a study of aerodynamic properties of beans for the purpose of separation from foreign materials (Tiwari, 1962). For one particular bean

Major diameter = 0.584 in.
Minor diameter = 0.252 in.
Intermediate diameter = 0.313 in.
Geometric mean of the three diameters = $(0.584 \times 0.252 \times 0.313)^{1/3}$ = 0.358 in.

Frontal area = $\pi/4(0.358)^2 = 0.101$ in^2
Weight = 1.12×10^{-3} lb.
Volume = 1.52×10^{-5} ft.^{3}, diameter of equivalent sphere = 0.372 in.
Weight density = 73.7 lb/ft^3
Mass density $\rho_p = 73.7/32.2 = 2.22$ slug/ft^3
Mass density of the air $\rho_f = 2.38 \times 10^{-3}$ slug/ft^3
Absolute viscosity of the air $\eta = 3.74 \times 10^{-7}$ lb-sec/ft^2

Assuming a spherical shape for the bean with the geometric mean of its three diameters equal to the diameter of the equivalent sphere, from Eq. (9.19).

$$CN_R^2 = \frac{8 \times 1.12 \times 10^{-3} \times 2.38 \times 10^{-3}\,(2.22 - 2.38 \times 10^{-3})}{\pi(3.74 \times 10^{-7})^2 \times 2.22} = 48.7 \times 10^6$$

And, from plot of CN_R^2 versus N_R, N_R=10,700. At this Reynolds number the bean will theoretically attain its terminal velocity which can be found from Eq. (9.6).

$$10{,}700 = \frac{V_t \times 0.358/12 \times 2.38 \times 10^{-3}}{3.74 \times 10^{-7}}$$

Table 9.2 Absolute viscosity and mass density of air and water at various temperatures

Dry Air at 14.7 psi[1]			Water[2]		
Temp. (°F)	Abs. viscosity $\eta \times 10^7\ \frac{\text{lb-sec}}{\text{ft}^2}$	Mass density $\rho_f \times 10^3$ slug/ft^3	Temp. (°F)	Abs. viscosity $\eta \times 10^7\ \frac{\text{lb-sec}}{\text{ft}^2}$	Mass density ρ_f slug/ft^3
0	3.43	2.69	32	3.746	1.940
32	3.55	2.51	40	3.229	1.940
59	3.74	2.38	50	2.735	1.940
70	3.78	2.33	60	2.359	1.938
100	3.98	2.2	70	2.050	1.936
			80	1.799	1.934
			90	1.595	1.931
			100	1.424	1.927

[1] From various sources
[2] From Vennard (1961)

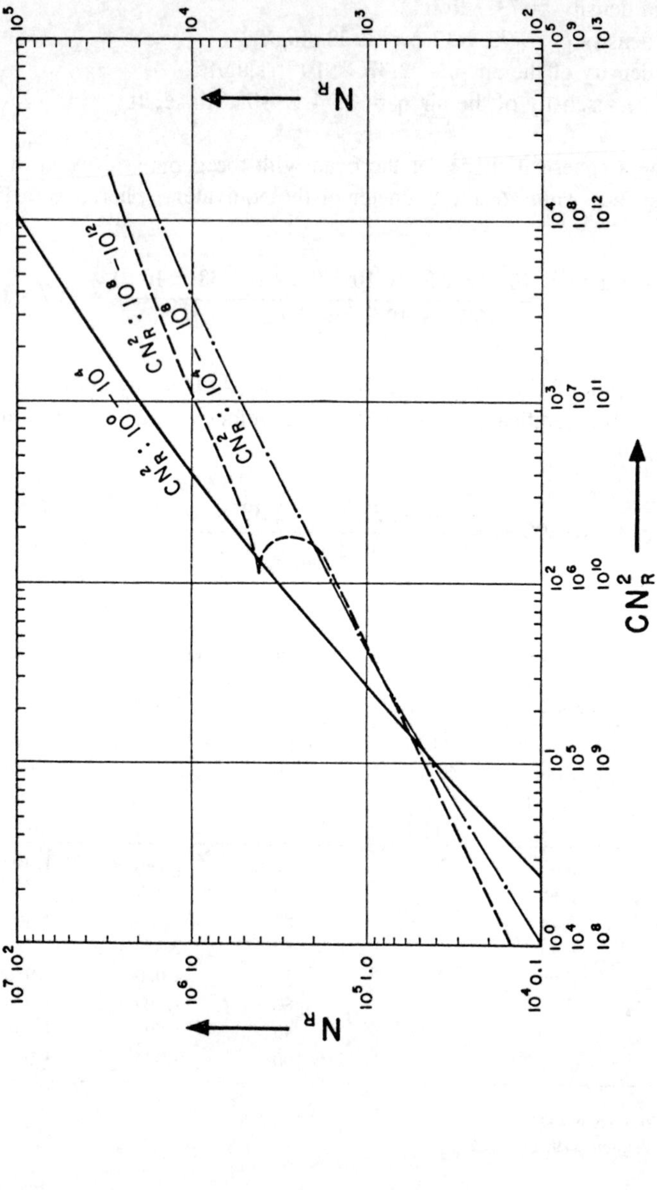

Figure 9.3 Reynolds Number, N_R, *versus* CN_R^2 for spheres

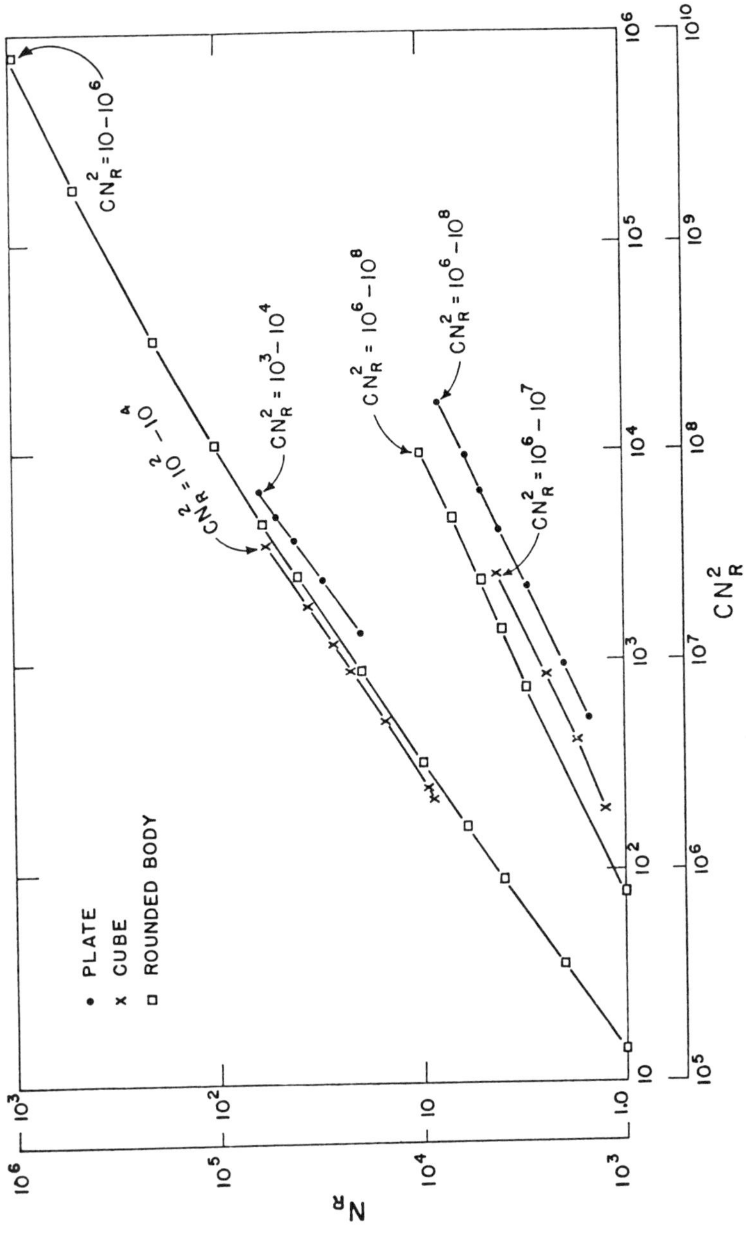

Figure 9.4a Reynolds number, N_R *versus* CN_R^2 for plate-like, cube and rounded bodies

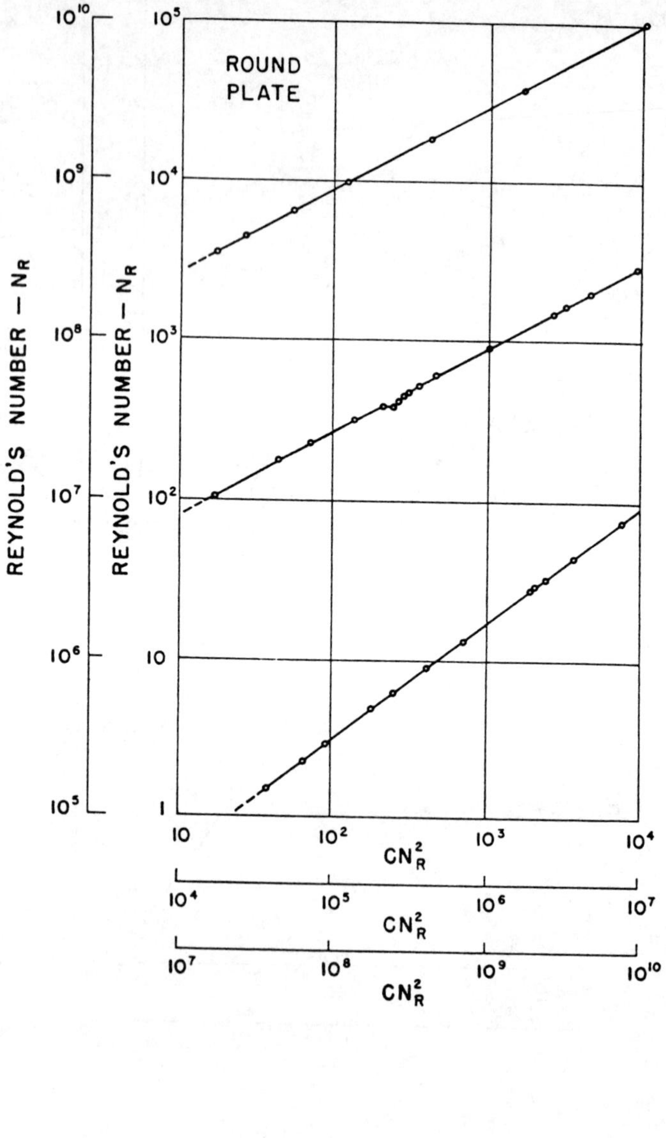

Figure 9.4b Reynolds number *versus* CN_R^2 for round plate

or

$$V_t = 56.5 \text{ ft/sec}$$

The actual terminal velocity of this particular bean, found experimentally by a pneumatic separator, was about 47.8 ft/sec (Tiwari, 1962). Using this actual terminal velocity in Eq. (9.11), the drag coefficient will be

$$C = \frac{2 \times 1.12 \times 10^{-3} (2.22 - 2.38 \times 10^{-3})}{(47.8)^2 (0.101/144)(2.22)(2.38 \times 10^{-3})} = 0.59$$

The Reynolds number based on the actual terminal velocity from Eq. (9.6) will be

$$N_R \text{ (actual)} = N_R \text{ (calculated)} \left[\frac{V_t \text{ (actual)}}{V_t \text{ (calculated)}} \right]$$

$$= 10{,}700 \, [47.8/56.5]$$

$$= 9000$$

According to Fig. 9.3, the shape of a body with $C=0.59$ and $N_R=9000$ falls between the shape of a sphere and that of a "flattened body". This departure of the shape of this particular bean under experiment from that of a sphere can numerically be evaluated using Eq. 10 Chapter 3.

$$\text{Sphericity} = \left[\frac{0.313 \times 0.252}{(0.584)^2} \right]^{1/3} = 61\,\%$$

Considering this deviation of the shape of the object from the assumed spherical shape, the agreement between the experimental and calculated terminal velocities and Reynolds numbers appears much better than one would expect.

Example

A single wheat grain with a geometric mean diameter of 0.157 in. and a density of 87.4 lb/ft^3 is dropped to fall freely into a column of water at 70°F, determine the terminal velocity of the grain in ft/sec.

From Table A-19 (Appendix), a good estimate for the weight of this wheat grain is 10.37×10^{-5} lb. Assuming for water at 70°F, $\rho_f = 1.936$ lb-sec^2/ft^4 and viscosity $M = 2.05\times10^{-5}$ lb-sec/ft^2,

$$CN_R^2 = \frac{8 \times 10.37 \times 10^{-5} \times 1.936\ (87.4 - 62.4)}{\pi(2.05 \times 10^{-5})^2 \times 87.4} = 3.5 \times 10^5$$

From Fig. 9.4a, $N_R^2 \cong 800$

$$\begin{aligned} V_t &= R_N \eta / D\rho \\ &= 800 \times 2.05 \times 10^{-5}/0.0/29 \times 1.936 \\ &= 0.66\ \text{ft/sec} \end{aligned}$$

Example

To estimate the diameter of starch granules, their settling velocity in water at 70°F was found to be 0.1 mm/sec. If the granules' density were 93.63 lb/ft^3, determine the average diameter of the granules in mm. If these granules were falling in air, what would be their terminal velocity?

a. Terminal velocity in water $V_t = 0.\times10^{-3}$ m/sec. This being a laminar flow, use Stoke's law eq. 9.13 and solve for diameter dp

$$(dp)^2 = \frac{V_t}{g(P_p - P_f)/18M} = \frac{0.1 \times 10^{-3}}{9.8(1500 - 997/18M}$$

Assuming $\eta = 0.98 \times 10^{-3}$ Kg/sec m

$$\begin{aligned} dp &= (2.028 \times 10 \times 0.98 \times 10^{-3} \times 18)^{1/2} \\ &= 1.9 \times 10^{-5}\ \text{m or } 0.019\ \text{mm} \end{aligned}$$

b. Terminal velocity in air:
From eq. 9.13

$$\begin{aligned} V_t &= 9.8 \times 3.58 \times 10^{-10}(1500 - 1.2)/18 \times 18.5 \times 10^{-6} \\ &= 0.016\ \text{m/sec} \end{aligned}$$

Terminal velocity from time-distance relationship

In determining terminal velocities by experiment it is desirable to have a relationship between time and displacement of the object. Such relationship can be obtained by solving the differential equation of the motion as shown in the following.

In free fall of an object in still air, the net force on the object is the difference in the force of gravity, mg, and the resultant frictional or drag force, kV^2, given by Eq. (9.3). This net force is also equal to $m\frac{dV}{dt}$ or mass times acceleration. From these basic principles the equation of motion can be set up as the following differential equation:

$$m\frac{dV}{dt} = mg - kV^2$$

or

$$\frac{dV}{dt} = g\left(1 - \frac{k}{mg}V^2\right) \qquad (9.20)$$

where k is the constant equal to $1/2\ CA\rho_f$ of Eq. 9.3). Now let $k/mg=a^2$. Integrating

$$\int\frac{dV}{(1 - a^2V^2)} = g\int dt$$

yields

$$\frac{1}{2a}\ln\frac{1 + aV}{1 - aV} = gt + c$$

For the initial conditions of $t=0$, $V=0$, $c=0$, the integrated expression then simplifies to

$$\frac{1 + aV}{1 - aV} = e^{2agt}$$

or

$$V = \left(\frac{1}{a}\right)\frac{(e^{2agt} - 1)}{(e^{2agt} + 1)}$$

or

$$V = \frac{ds}{dt} = \frac{1}{a}\left(\frac{e^{agt} - e^{-agt}}{e^{agt} + e^{-agt}}\right) = \frac{1}{a}\tanh agt$$

or

$$ds = \frac{1}{a(ag)} \tanh agt\,(ag\,dt)$$

where s and t are displacement and time, respectively. Integrating the last expression yields

$$s = \frac{1}{a^2 g}\ln\cosh agt + c_2 \tag{9.21}$$

Again for the initial conditions, $t=0$, $s=0$ and, therefore, $c_2=0$.

In aerodynamic applications where air is the fluid being used and the air mass density, ρ_f is small enough in comparison with mass density of the product, ρ_p, mass density of air can be assumed zero and Eq. (9.10) can be written in the simplified form

$$V_t = \left[\frac{2W}{\rho_f AC}\right]^{1/2} \tag{9.22}$$

Combining the above equation with expression $k=\frac{1}{2}CA\rho_f$, a^2 in Eq. (9.21) can be evaluated as follows

$$V_t = \left[\frac{W}{\frac{1}{2}CA\rho_f}\right]^{1/2} = \left[\frac{W}{k}\right]^{1/2}$$

or

$$k = \frac{W}{V_t^2}$$

therefore

$$a^2 = \frac{k}{mg} = \frac{W/V_t^2}{mg} = \frac{1}{V_t^2}$$

Substituting the value of a^2 in (9.21), the relationship between displacement, time, and terminal velocity for objects considerably heavier than air will be as follows

$$s = \frac{V_t^2}{g} \ln \cosh \frac{g}{V_t} t \qquad (9.23)$$

To determine terminal velocity, V_t, experimental data on time versus distance of free fall of the object in still air are plotted to obtain a time-displacement curve. If the height of fall is sufficient for the object to reach its terminal velocity, the time-displacement curve becomes linear when such velocity is attained. In such cases, the slope of the linear portion of the curve can be taken as the terminal velocity. For some particles, the height of fall at which V_t is attained is considerable (see Fig. 9.5). In such cases, Eq. (9.23) can be used to calculate terminal velocity from a given time-displacement data.

To show the validity of Eq. (9.23) Bilanski *et al.* (1962) measured the time of fall of various agricultural grains through still air from zero velocity. The apparatus consisted of a 2-foot diameter guard tube, a glass tube as the dropping unit, a photocell circuit as the timer-starter, a striking plate and microphone as the timer-stopper, and an electronic timer. The data produced for various grains are shown in Fig. 9.5. As seen, the experimental points of each displacement-time curve follow very closely the theoretical curves fitted to them by means of Eq. (9.23). Actual terminal velocity for each grain was found from the slope of the linear portion of the plotted curve for that grain. When this terminal velocity was inserted into Eq. (9.23), a function was produced which fitted the heavier grains which did not reach their terminal velocities for the height of fall available.

Having the terminal velocity from the slope of the curve or from Eq. (9.23), the drag coefficient and Reynolds number were calculated by means of Eqs. (9.22) and (9.6), respectively. The projected or frontal area assumed for calculation of drag coefficient was based on the two largest dimensions L_1 and L_2 used in the following expression

$$A_p = (\pi/4) L_1 L_2$$

The diameter chosen for calculation of Reynolds number was the average of the three linear dimensions. Table 9.3 shows the physical measurements and aerodynamic characteristics of the grains tested. A plastic sphere, for

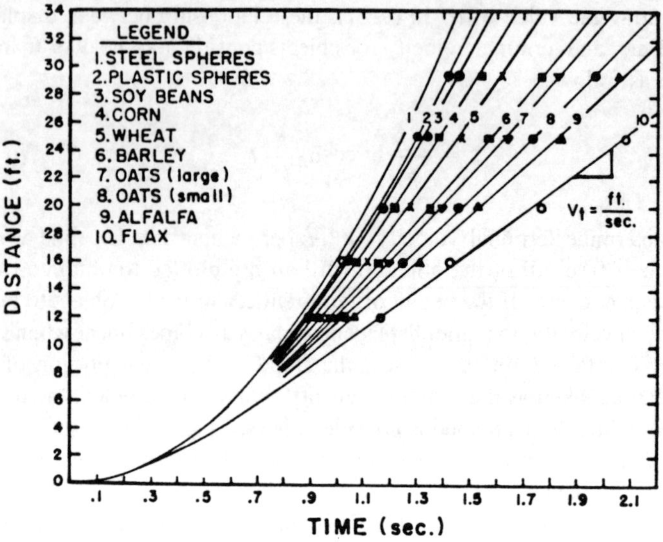

Figure 9.5 Displacement-time curves for spheres and agricultural seed grains—Experimental curves are fitted to analytical curves based on equation 5.22 (Bilanski *et al.*, 1962)

which published data were available, was included as a check for confirming the values of drag coefficients found for agricultural grains. The value of C=0.43 for the plastic sphere is in agreement with C=0.425 reported by others. Also the values of Reynolds numbers are within a region where the drag coefficient is roughly constant (see Fig. 9.2 and 9.3). The values reported in Table 9.3 all lie in a narrow range (0.45 to 0.56) which can be considered nearly constant.

9.3 APPLICATIONS TO AGRICULTURAL PRODUCTS

The principles outlined in the previous sections have been employed by several investigators in applications such as separation of foreign materials from seeds, grains, potato, blueberry; pneumatic conveying and handling

Table 9.3 Aerodynamic properties of seed grains (Bilanski *et al.*, 1962)

	Length $L_1 \times 10^2$ ft	Width $L_2 \times 10^2$ ft	Depth $L_3 \times 10^2$ ft	Weight $W \times 10^4$ lb	Terminal velocity V_t fps	$CA_p \times 10^4$ ft^2	C	$N_R \times 10^{-2}$
Alfalfa	0.77	0.47	0.35	0.053	17.9	0.141	0.50	6.01
Flax	1.42	0.74	0.36	0.118	15.3	0.432	0.52	8.36
Wheat	2.28	1.10	0.97	1.00	29.5	0.99	0.50	27.2
Barley	2.89	1.05	0.78	0.73	23.0	1.18	0.50	22.8
Small oats	3.15	0.80	0.68	0.40	19.3	0.93	0.47	19.0
Large oats	4.01	0.92	0.71	0.74	20.8	1.47	0.51	24.8
Corn	3.82	2.63	1.36	6.30	34.9	4.44	0.56	57.7
Soybeans	2.55	2.22	1.93	4.54	44.3	1.98	0.45	62.8
Plastic spheres	3.12	3.12	3.12	11.6	55.0	3.30	0.43	108.5

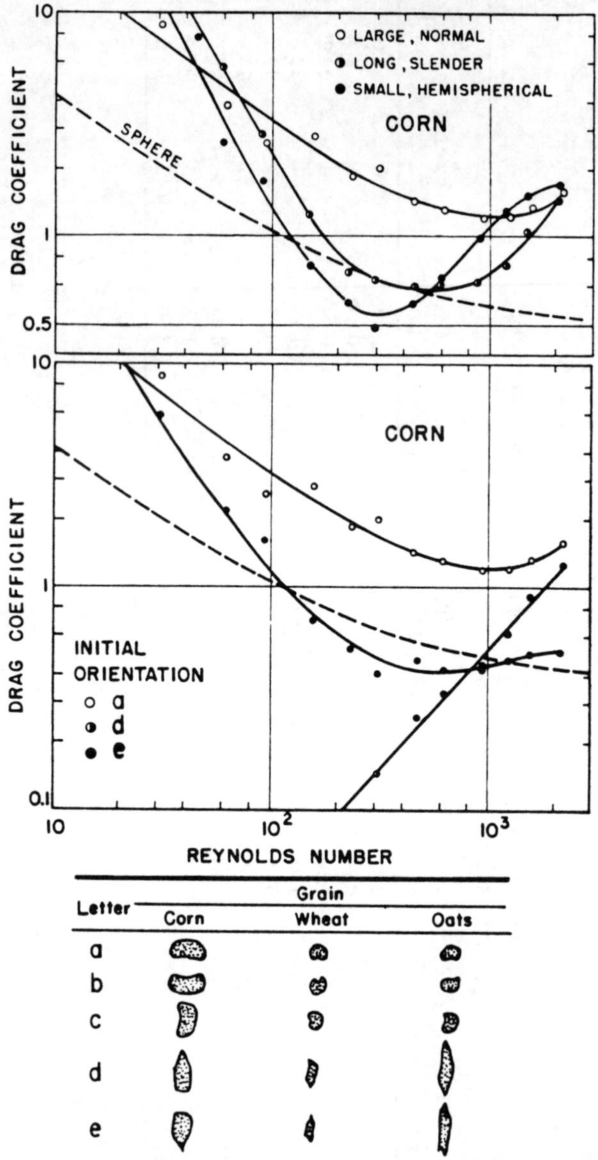

Figure 9.6 Effect of shape, size, initial orientation on aerodynamic drag of agricultural grains (Condensed from Garrett and Brooker, 1965)

of grains, chopped forage, silage and small and large fruits; and hydraulic handling of apples, cherries, and potatoes.

Frontal area

In applying the aerodynamic principles to agricultural products, one must choose a method for expressing the projected or frontal area of the particle. Table 1.1 in Chapter 1 gives some of the methods used by various investigators. In most cases, the object has been assumed to be spherical in shape with the projected area equal to $(\pi/4)d_p^2$, where d_p is the diameter of the sphere of the same volume as the object. In cases where the volume of the object is not known or is difficult to evaluate due to the porous nature of the material (feed pellets and wafers), the geometric mean of the three axial dimensions is a good approximation of the diameter of the equivalent sphere providing that the shape factor is close to unity. Frontal area for non-spherical geometric shapes are defined in Table 9.1.

Particle of sieve size dimensions are usually considered spherical in shape and of such a dimension as the sieve analysis indicates.

Particle orientation

In streamline or laminar motion a particle will generally retain its initial orientation while settling. Some authorities take the minimum cross section area as the frontal area for the laminar region. In the turbulent region the particle will theoretically assume a position of maximum resistance (Davies, 1947). However, in aerodynamic studies of agricultural grains, Bilanski *et al.* (1963) and Bilanski and Lal (1964) found that grains and wheat straw have random orientation at any instant and actually rotated about a vertical axis with their longest dimension tending toward the horizontal plane. This rotation and inclination was explained to be due to the fact that the aerodynamic force on the unsymmetrical grain will neither pass through the center of gravity nor will be parallel to the air velocity. This force can be considered as the sum of a vertical force through the center of gravity, a horizontal force which tends to rotate the grain about a vertical axis, and a couple in the vertical plane which tends to change the inclination of the long axis of the grain. An interesting observation was that when straw or kernels began to rotate, they would also rise in the wind tunnel into a region of lower air velocity. In other words, rotation caused a higher drag and lower terminal velocity.

Garrett and Brooker (1965), in their study of aerodynamic drag of farm grains, also observed the tumbling of grains falling in still air. Figure 9.7 shows the effect of shape and size and initial orientation on drag coefficient of corn kernels falling in a tube. The diameter of the sphere of equal volume was used in calculation of Reynolds number and the characteristic frontal area. The variations in the three curves were believed to be due to the changes in the orientation of the grains during the fall. Since the projected area changes continually, the characteristic area used in calculations cannot be expected to give the true drag coefficient. With a large number of tests, a range of drag coefficients might be established which can be used in calculations.

Separation from foreign materials

Bilanski and Lal (1964) have reported on the aerodynamic behavior of threshed materials in a vertical wind tunnel. The wind tunnel was calibrated so that the velocity of the air at any given height could be determined from the fan speed. To measure the terminal velocity of a particle, the fan speed was varied until the particle was observed to float without appreciable vertical movement.

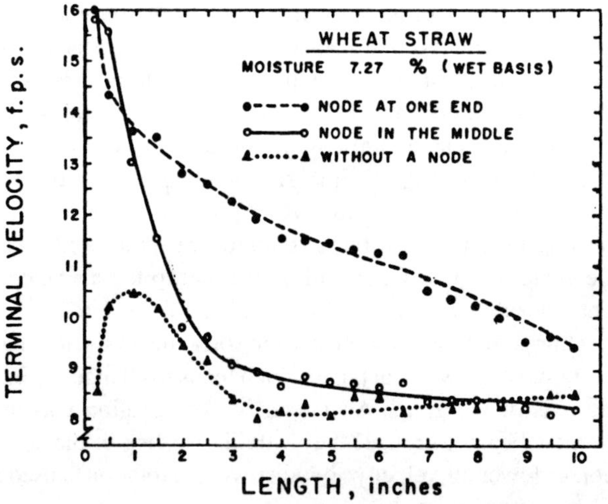

Figure 9.7 Terminal velocity of wheat straw as affected by length (Bilanski and Lal, 1964)

The aerodynamic drag of the particles was presented both in terms of drag coefficient and another coefficient termed resistance coefficient indicated by k in Eq. (9.20). Since at terminal velocity $\frac{dv}{dt} = 0$ and $F_r = F_g = mg$ (neglecting buoyancy of the air), from Eq. (9.20)

$$k = \frac{mg}{V_t^2} \tag{9.24}$$

and from Eq. (9.3)

$$mg = \tfrac{1}{2} C A_p \rho_f V_t^2$$

or

$$C = \frac{2mg}{A_p \rho_f V_t^2} \tag{9.25}$$

where mg=weight of the particle at terminal velocity.

Table 9.4 shows the resistance coefficient and drag coefficients calculated for various types of particles in threshed materials. The resistance coefficient k, being a parameter free from the assumption of the effective frontal area necessary when calculating drag coefficient, appears to be a better criterion for separation of various particles than the drag coefficient.

The plot of resistance coefficients against the straw sample length showed linear relationship given by the following equations

$k=(1.003L)\,10^{-6}$ for straw pieces with node in the middle or without a node

and

$k=(0.786L)\,10^{-6}$ for straw pieces with node at one end

Calculation of drag coefficients for 21 different lengths of straw with node in the middle showed little variation. The calculations were based on an effective frontal area equal to the length times the average diameter of the straw sample. The orientation of the straw pieces was not taken into consideration. Omitting the values around 0.60 which occurred only for three of the 21 samples, the drag coefficients for straw pieces were in the

Table 9.4 Aerodynamic properties of threshold materials (Data compiled from Bilanski and Lal, 1964)

Material	Weight $W \times 10^5$ lb	Terminal velocity ft/sec	Resistance coefficient $k \times 10^6 \frac{\text{lb-sec}^2}{\text{ft}^2}$	Drag coefficient C
Wheat chaff	0.7	4.12	0.412	
Wheat heads	45.01	7.09	8.954	
Wheat grains	9.3	28.8	0.112	
1/4-in. straw*	6.52	15.75	0.20	0.84
1-in. straw*	11.70	13	0.687	0.66
3-in. straw*	24.57	9	3.03	0.90
10-in. straw*	74.66	8.25	11	0.91

* Node in the middle

order of that of a circular cylinder and within the same range of Reynolds number. Also, the terminal velocity of the straw pieces without a node followed a pattern similar to terminal velocity of circular cylinders reported by Rice (1960).

Figure 9.7 shows the change of terminal velocity of wheat straw with length and presence or absence of nodes. The rapid decrease of terminal velocity was explained to be due to the effect of spinning and orientation. The data on terminal velocity of wheat straw in this work does not agree with similar data reported by Uhl and Lamp (1966). In fact the plotted data on straw length versus terminal velocity give trends opposite to those shown in Fig. 9.7. Bilanski and Lal (1964) have pointed out this lack of agreement and claim that it is due to the fact that Uhl and Lamp did not consider the node position in the straws tested.

Also the drag coefficients for various grains reported by Bilanski *et al.*, (Table 9.3) are lower than drag coefficients reported for similar grain by Uhl and Lamp (Fig. 9.7). The primary reason for this could be the difference in the assumptions made for the frontal area to be substituted in Eq. (9.25). Bilanski and Lal, as previously indicated, based their calculation of the frontal area on the two largest dimensions while Uhl and Lamp considered the least cross section area as the frontal area. The values of terminal velocity of the various grains reported by one group, however, lie within the range of terminal velocities reported for that grain by the other group (compare Tables 9.3 and 9.5).

Uhl and Lamp (1966) also found percent separation of grain, straw and chaff at various air velocities (Fig. 9.8). Some of the physical characteristics of various grains and the range of air velocities required to airborn these grains, as reported by these investigators, are shown in Table 9.5. It is interesting to note that the value of the terminal velocity for 100 percent separations of wheat grain (30 ft/sec, Fig. 9.8) is about the same as that reported in Table 9.3. Other values of terminal velocities in Table 9.3 are within the range given in Table 9.5 for the specific product.

The work reported by Uhl and Lamp followed very closely the procedure followed earlier by Persson (1957). In both cases, a sample of unwinnowed grain was taken from a combine harvester operating under normal field conditions and was then analyzed in a wind tunnel.

Bilanski and Manzies (1967) conducted aerodynamic tests on alfalfa particles and reported terminal velocities in terms of particle weight so that the effect of moisture content can be taken into consideration. Having terminal velocity from experimental data, the parameters drag coefficient and resistance coefficient were found from plots of velocity squared versus weight/area Eq. (9.25) and velocity squared versus weight Eq. (9.24). Reynold's number for the alfalfa particles was determined by taking length as the characteristic dimension and experimental suspension velocity as the terminal velocity Eq. (9.6). Particle density was determined by using a density gradient column similar to that in Fig. 3.9.

Table 9.5 Range of physical characteristics of grains and air velocity required to airborne (Condensed from Uhl and Lamp, 1966)

Grain	Kernel weight lb $\times 10^6$	Unit density lb/ft^3	Least cross section area in$^2 \times 10^4$	Bulk density lb/ft^3	Terminal velocity ft/sec
Wheat	50.84	62.4–77.4	62–91	35.6–49.2	19–30
Rye	39.53	72.4–76.1	51–66	39.3–44.3	20–27
Oats	29.58	46.1–60.5	34–95	23.0–31.8	17–26
Corn	569–692	71.1–74.9	500–610	42.7–46.2	26–42
Soybeans	233–419	74.3–72.4	240–470	43.6–43.1	30–60

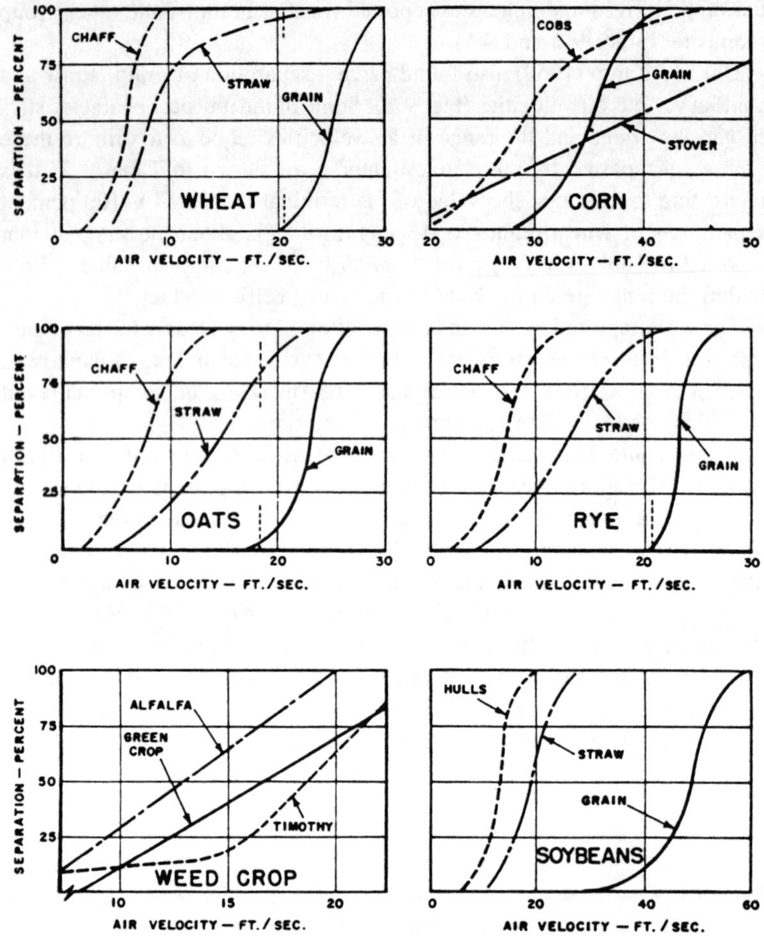

Figure 9.8 Percent separation versus air velocity for several grains and associated materials (Uhl and Lamp, 1966)

The results of this work are shown in Table 9.6 and Fig. 9.10. According to these results the behavior of alfalfa particles in an air stream depends on length, nodal position, and length to diameter ratio.

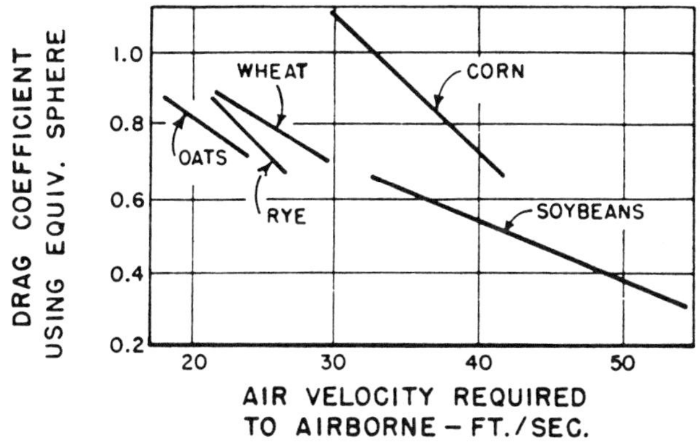

Figure 9.9 Drag coefficient *versus* terminal velocity for various grains (Uhl and Lamp, 1966)

Tiwari (1962) employed the aerodynamic principles to investigate the possibilities of pneumatic separation of good beans from samples of threshed dry edible bean crop containing undesirable materials such as damaged beans, stones, leaves, stems, roots, etc. Actual suspension velocities for individual beans were found to be about 4 to 6 ft/sec less than the calculated values (Fig. 9.11). The lower actual suspension velocities were probably because of the observed spinning and rotation of the beans

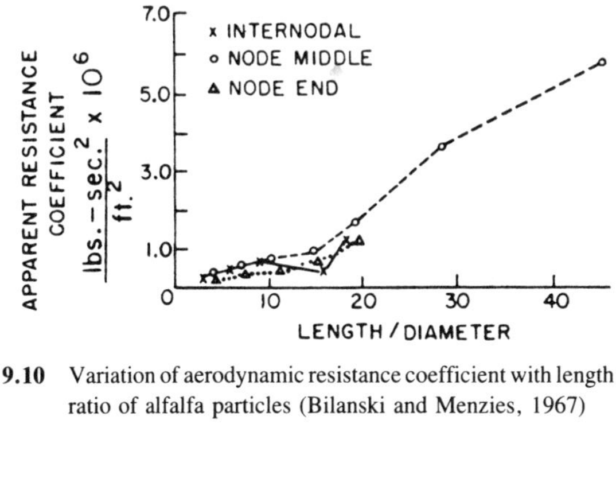

Figure 9.10 Variation of aerodynamic resistance coefficient with length to diameter ratio of alfalfa particles (Bilanski and Menzies, 1967)

Table 9.6 Some aerodynamic properties of alfalfa particles (Bilanski and Menzies, 1967)

Configuration and length (in)	Reynolds number $\times 10^{-4}$	Suspension velocity[2] V_t^2 (ft^2/sec^2)	Drag coefficient C	Resistance coefficient k_2 ($(lb\text{-}sec^2)/ft^2$)
0.25[1]	0.251	$V_t^2 = 99.59 + 4.38 \times 10^6 W$	0.71	0.230
0.25[2]	0.240	$V_t^2 = 97.65 + 4.06 \times 10^6 W$	0.75	0.246
0.25[3]	0.232	$V_t^2 = 107.56 + 3.48 \times 10^6 W$	0.75	0.287
0.50[1]	0.506	$V_t^2 = 114.40 + 2.56 \times 10^6 W$	0.66	0.391
0.50[2]	0.480	$V_t^2 = 118.29 + 2.36 \times 10^6 W$	0.70	0.423
0.50[3]	0.451	$V_t^2 = 91.91 + 2.53 \times 10^6 W$	0.57	0.395
1.0[1]	0.976	$V_t^2 = 113.38 + 1.47 \times 10^6 W$	0.64	0.680
1.0[2]	0.964	$V_t^2 = 137.23 + 1.34 \times 10^6 W$	0.62	0.747
1.0[3]	0.879	$V_t^2 = 68.85 + 1.67 \times 10^6 W$	0.50	0.600
1.50[1]	1.217	$V_t^2 = 32.43 + 1.35 \times 10^6 W$	0.44	0.742
1.50[2]	1.355	$V_t^2 = 74.65 + 1.06 \times 10^6 W$	0.47	0.947
1.50[3]	1.266	$V_t^2 = 28.02 + 1.29 \times 10^6 W$	0.44	0.776
2.0[1]	1.915	$V_t^2 = 65.28 + 0.65 \times 10^6 W$	0.64	1.529
2.0[2]	1.565	$V_t^2 = 48.03 + 0.56 \times 10^6 W$	0.74	1.776
2.0[3]	1.607	$V_t^2 = 42.15 + 0.70 \times 10^6 W$	0.61	1.422
3.0[2]	2.215	$V_t^2 = 71.35 + 0.27 \times 10^6 W$	1.22	3.731
4.0[2]	2.567	$V_t^2 = 82.82 + 0.18 \times 10^6 W$	1.52	5.714

[1] Internodal (no node)
[2] Node middle
[3] Node end

in the air stream. Based on actual terminal velocities and projected areas, using geometric means of the three diameters, drag coefficients for four varieties of beans were found to vary between 0.45 to 0.65. Values of terminal velocities for all the four varieties were found to be between 40 to 65 ft/sec. Within this range about 80% of damaged beans could be separated without significant loss of whole beans. Light materials such as leaves, stems and roots were completely separated at terminal velocities of about 21 ft/sec.

Since suspension velocities of good beans were within the velocity range of 26 to 80 ft/sec required for stone separation, it proved difficult to separate the stones from beans employing only a pneumatic device. A separator

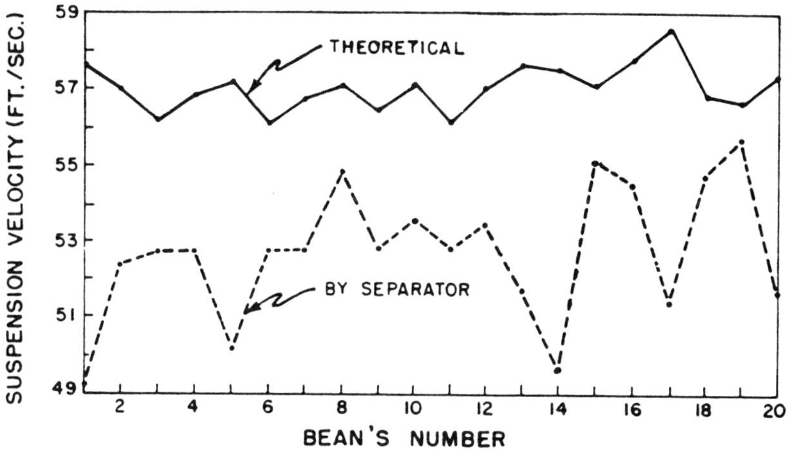

Figure 9.11 Comparison of theoretical and experimental terminal velocities for yellow eye beans (Tiwari, 1962)

based on combination of pneumatic and other principles, such as rolling resistance, was suggested to be more suitable for this purpose.

Keck and Goss (1965) and Garrett and Brooker (1965) also determined time-displacement curves of grains and seeds in free fall. The apparatus used by Keck and Goss is shown in Fig. 9.12. A selected seed was picked off by a "vacuum drop" device in the dropping head and was released into a drop tube by breaking the vacuum. A phototube mechanism at the top of the tube started an electronic timer and a microphone circuit stopped the timer upon the impact of the seed on the diaphragm of a receiving cup. The dropping head and the receiving cup could be moved from one drop tube to another to cover a range of 2 to 30 feet of drop height for each seed.

Table 9.7 shows the results of tests dropping a small sphere as well as seeds of alfalfa and clover. Due to the limited length of the tubes, terminal velocities were not achieved but a theoretical check using Eq. (9.14) showed that the 1/8-inch sphere and both seeds tested reached about 94% of their terminal velocities at 30 feet. For effective diameter, d_p, in Eq. (9.14) the geometric mean of the three mutually perpendicular measured seed dimensions was used. Using the diameter of the equivalent sphere gives a lower terminal velocity which indicates that the geometric mean diameter may vary considerably for irregular objects such as seeds.

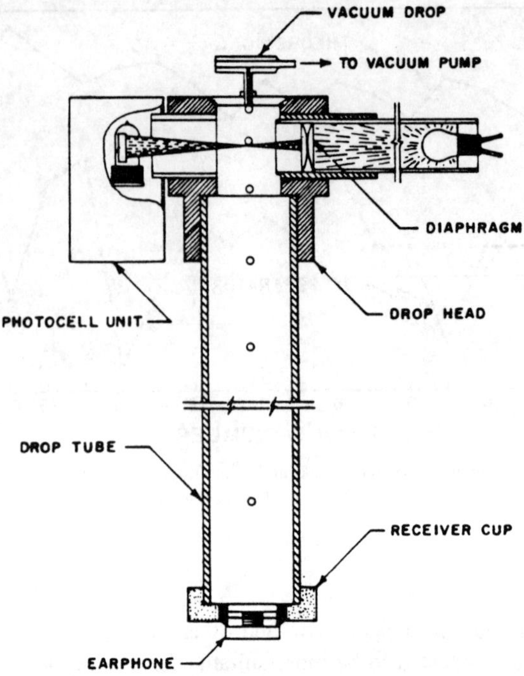

Figure 9.12 Drop head and receiving cup and section of drop tube used for determining time-displacement curves of seeds and grains (Keck and Goss, 1965)

For this reason, Keck and Goss (1965) suggested the determination of a shape factor, defined as the ratio of geometric mean diameter to the diameter of equivalent sphere, for different seeds. If these shape factors are then plotted against the geometric mean diameters, an average curve can be obtained which might be used to pick off the shape factor of any seed with a particular geometric mean diameter. Note that this is another method of finding the diameter of the equivalent sphere.

The shape factor for clover seed (1.04) was closer to that of sphere (1.00) than the shape factor of alfalfa seed (1.16). This greater sphericity of clover seed was also demonstrated by its drag coefficient and terminal velocity values which were in closer agreement with those of the sphere than the values for alfalfa seed. This trend is also seen in the time-displacement data of Table 9.7.

Table 9.7 Free fall of spheres and seeds in still air (Keck and Goss, 1965)

Height of fall (ft)	Time (seconds)			
	⅛″ theoretical sphere	⅛″ nylon[1] sphere	Rose[2] clover	Alfalfa[3]
2	0.3581	0.350	0.358	0.363
4	0.5123	0.502	0.520	0.532
6	0.6343	0.624	0.655	0.681
8	0.7402	0.729	0.776	0.806
10	0.8362	0.826	0.888	0.920
12	0.9255	0.915	0.996	1.050
14	1.0097	1.000	1.103	1.155
16	1.0902	1.078	1.201	1.290
18	1.1678	1.158	1.295	1.381
20	1.2430	1.232	1.396	1.479
22	1.3162	1.308	1.494	1.601
24	1.3879	1.379	1.591	1.712
26	1.4582	1.449	1.683	1.806
28	1.5274	1.517	1.766	1.889
30	1.5957	1.586	1.867	2.028

[1] Average of ten spheres dropped through each height of fall.
[2] Average of ten seeds, each seed dropped ten times for each height of fall.
[3] Average of twenty seeds droped once through each height of fall.

The technique used by Garrett and Brooker (1965) was unique in that the actual time-displacement curves of grains falling in still air were recorded photographically. A camera was focused on the path of a grain falling in a transparent tube. The shutter was held open so that the image of the falling grain formed a vertical trace on the line. At the same time the film, mounted on a rotating drum, moved horizontally at a constant speed. Thus, a photographic time-displacement curve was obtained from which a velocity-time curve was constructed by means of tangents to the curve drawn with the aid of a prismatic derivator. Using acceleration, $\frac{dv}{dt}$ at convenient points taken from the velocity-time curve, the following equation was used to calculate the aerodynamic drag, C, of corn, oats and wheat falling in still air:

$$C = \frac{2W}{A_p \varrho_f V^2}\left[\left(\frac{\varrho_p - \varrho_f}{\varrho_p}\right) - \left(\frac{\mathrm{d}V/\mathrm{d}t}{g}\right)\right] \tag{9.26}$$

where W is the weight of the particle. Note tht the above equation is the same as Eq. (9.20) except that the buoyant effect of the air has been taken into consideration.

Aerodynamic characteristics of potato and blueberries have also been investigated in an attempt to explore their possibilities in separating the product from the associated foreign materials. Gilfillan and Crowther (1959) used a vertical air blast with velocities given in Fig. 9.13, to separate stones and clods of soil from potatoes. In this figure are shown a series of curves of terminal velocity versus diameter for spheres and flat plates in the size and specific gravity range as potatoes, stones, and soil clods. Spheres were in the range of 0.5- to 5-inch diameter. Flat plates were 1-inch thick. Terminal velocities were calculated using Eq. (9.11), (9.6) and (9.3). Theoretical drag coefficients were obtained from published data for sphere. In Eq. (9.3), as usual the drag force was set equal to mg, the

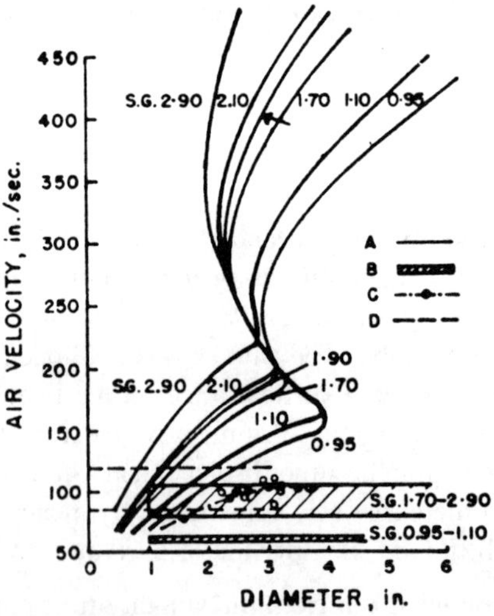

Figure 9.13 Air velocities required to lift potatoes, stones and clods—A spheres; B plates (1-in. thick); C potato; D range for potato samples. A & B calculated, C & D measured (Gilfillan and Crowther, 1959)

weight of the materials being lifted. This graphical presentation showed to what extent the behavior of the objects tested in the wind tunnel agreed with that to be expected from aerodynamic consideration. The inflexion of the terminal velocity curves for spheres is due to the change of flow conditions from laminar to turbulent flow which occurs at Reynolds number between 10^5 and 10^6. The actual air velocities at which the potatoes and stones were lifted are shown at the bottom of Fig. 9.13. The experimental observations lie between the values calculated for spheres and flat plates having the same specific gravities as potatoes and stones.

Figure 9.14 shows the percent of potatoes and stones lifted at various air velocities. As shown, an air velocity of about 115 ft/sec was required to lift all the potatoes while at the same air velocity as much as 60 percent of both stones and soil clods were also lifted. In other words, the vertical air stream was no more effective in separating potatoes from clods than in separating potatoes from stones. This is a case where the shape of the object proves to be more important than the density inasmuch as pneumatic separation is concerned. The authors observed that flat and near-flat stones were lifted readily at the air velocities necessary to lift potatoes even though the stones were heavier.

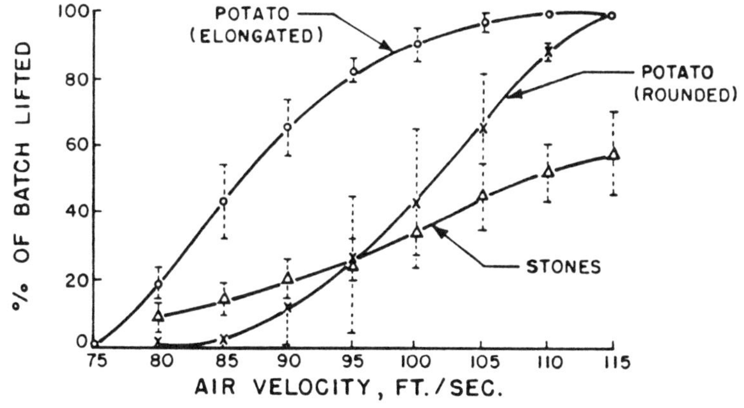

Figure 9.14 Effect of vertical air stream in separation of potatoes and stones (Gilfillan and Crowther, 1959)

The range and mean values of some of the physical characteristics and aerodynamic properties of the potatoes and stones tested are shown in Table 9.8.

Other attempts to separate potatoes from stones and soil clods include brine solutions of a specific gravity greater than that of potato and several methods involving separation in vertical air streams (Maak, 1957) and combined pneumatic and mechanical (oscillatory motion) separation (Kolchin, 1957).

Some physical characteristics and terminal velocities of blueberries have also been investigated to explore their possibilities in cleaning the blueberry from sand and other foreign materials. The plate-like sand, found in blueberry areas of Maine, proved to be very difficult to separate from blueberries. The terminal velocity of the sand was found to be about 50 percent of that of blueberry, indicating possibilities for aerodynamic separation.

Table 9.9 shows some aerodynamic properties of Lowbush blueberries. The relationship between actual drag coefficient and the drag coefficient based on the assumption of the particle being a sphere is shown in Fig. 9.15. The difference in the two drag coefficients demonstrated that the

Table 9.8 Some physical characteristics and aerodynamic properties of potatoes and stones (Condensed from Gilfillan and Crowther, 1959)

Material (No. of specimens	Weight (lb)	Volume cu. in.	Specific gravity	Max. cross sectional area-in^2	Terminal velocity ft/sec
Potatoes	0.06–0.80	6.66–19.4	1.12–1.15	2.0–11.7	76–108
(15)	(0.36)	(11.2)	(1.12)	(6.49)	(98)
Round stones	0.15–0.89	4.32–9.18	2.17–2.97	1.9–8.0	115–122
(19)	(0.53)	(6.74)	(2.65)	(4.50)	(120)
Flat stones	0.22–0.71	2.64–9.42	2.04–2.97	3.5–9.9	90–116
(10)	(0.36)	(4.404)	(2.59)	(5.98)	(99)
Triangular	0.31–1.06	8.94–12.78	2.27–2.91	2.3–5.7	118–122
(5)	(0.84)	(9.99)	(2.67)	(3.93)	(120)
Rectangular	0.32	3.60	2.45	4.92	99
(1)	(0.32)	(3.60)	(2.45)	(4.92)	(99)

Mean values are given in parentheses under the range of values for each property.
Drag coefficient equals 0.64 for potato and 0.6–1 for others.

Table 9.9 Some aerodynamic properties of Lowbush blueberries (Soule, 1968)

Physical property	Mean	Std. deviation
Average diameter (cm)	0.835	0.586
Volume (cm^3)	0.288	0.140
Projected area (cm^2)	0.550	0.172
(Weight/area) ratio (g/cm^2)	0.511	0.082
Reynolds Number	6732	1713
Spherical drag coefficient	0.533	0.058
Actual drag coefficient	0.494	0.053
Terminal velocity (cm/sec)	1320	125

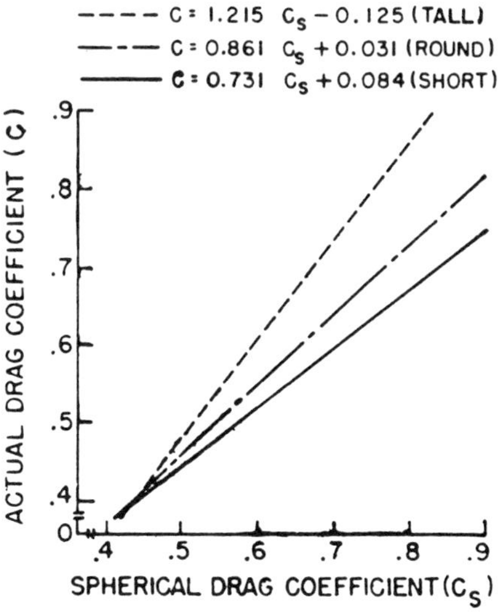

Figure 9.15 Relationship between actual drag coefficient and spherical drag coefficient for three different shapes of blueberries (Soule, 1968)

behavior of blueberries in a turbulent air stream differs significantly from that of spheres under the same conditions. The ratio of height to average diameter, which caused variation in projected area for tall, round, or short blueberries of a given volume, was considered to be a factor in producing this difference.

Schmidt and Levin (1963) have reported on terminal velocities of selected small fruits determined by allowing the free-falling fruit to interrupt light beams directed at a series of phototubes located along a 23-foot drop chute. Velocity of fall was determined from distance between phototube pips on a recording chart, the chart speed, and the distance between the phototubes. Some physical characteristics along with calculated and experimental terminal velocities for blueberries, grapes, cherry and cranberry were reported. In order that the theoretical terminal velocities, using an equation similar to (9.10) for sphere, come close to the experimental terminal velocities, a drag coefficient of 1.0 had to be assumed. This value of drag coefficient is about twice as much as the drag coefficient for sphere in turbulent flow.

Mueller *et al.* (1967) found that aerodynamic and weight characteristic of black walnuts make it possible to design an aerodynamic separator which will effectively separate good from bad walnuts.

To determine the extent of separation that could be obtained aerodynamically, two differential equations defining the motion of walnuts released into a horizontally-moving air stream were developed and solved by an analog computer. The differential equations were in terms of the nut velocity, air velocity, the vertical and horizontal distances traveled by the nut, and an expression of terminal velocity given by Eq. (9.25). The analog solution predicted the distance traveled by each nut which served as the criterion for design of an experimental separator as well as the desirable air velocity and initial nut velocity.

The possibilities of using cyclones for separating grain from chaff in combine harvesters were investigated by Hassebrauck (1962). Cyclones have no mechanically operated part and are easy to manufacture. They are used in pneumatic installations where air, which is used as a carrier, is to be separated from the solid particles such as grain, chaff, saw dust, wood chips, etc. The air, filled with solid particles, enters tangentially at the top into the cylindrical separation chamber. Here a rotating motion is formed in which the centrifugal force acts on the solid particles and forces them to the outside while the force of gravity tends to draw the particle downward. Under the influence of these forces, the particles move spirally downward into a collector while the air is discharged at the top.

The action of the forces involved in a cyclone separator is shown in Fig. 9.16. Under the condition of steady state, the centrifugal force, F_c, is equal to the air resistance force, F_r.

$$F_c = F_r$$

$$\frac{WV_c^2}{gr} = 1/2\, CA_p\varrho_f V_r^2$$

where W is the weight of the particle, r is the radius of rotation, and V_c and V_r are the tangential or peripheral and radial velocities, respectively. When $F_r = F_c$, the particle retains the same circle of rotation. When $F_c > F_r$, the particle would follow the centrifugal force and would settle downward due to the resultant force of net outward force and the force of gravity. The rate of settling is determined by terminal velocity of the particle. Finally, if $F_c < F_r$, the particle would follow the air resistance force and move with the air to the cyclone axis and is discharged. From the condition of equilibrium stated above, the following velocity ratio can be deducted:

$$\frac{V_c}{V_r} = \left(\frac{CA_p\varrho_f gr}{2W}\right)^{1/2} \tag{9.27}$$

Substitution of experimental data for chaff and wheat kernels in the square root quantity of Eq. (9.27) indicated that the difference in velocity ratios between these two particles is sufficiently large to make separation by cyclone possible. However, the actual range of velocity ratios in a cyclone of conventional design is considerably higher than that which is required for equilibrium of the forces. In other words, the centrifugal force considerably exceeds the force of air resistance so that only separation of the grain and chaff particles from the air can be expected and not separation of one solid particle from another.

Pneumatic transport and handling

Terminal velocity and drag coefficients are often among the parameters seen in mathematical expressions giving the relationships between the moving air and solid particles. In studying the air-solid interactions in the vertical pipe of forage blowers Chancellor (1960) has shown the influence of air velocity on the particle exit velocity from the pipe. Using the following derived expression

$$H_t = t(V_a) - \frac{V_s}{g}\left[\frac{V_s}{2}\ln\frac{V_s^2 + V_r^2}{V_s^2 + V_{ro}^2}\right] \qquad (9.28)$$

where

H_t = total height of the particle above any fixed reference point
t = time, V_a=air velocity, V_s=suspension or terminal velocity
V_r = relative velocity of the particle in the air stream
V_{r0} = V_r at time zero

Chancellor illustrated that the primary function of air movement in the pipe is to reduce the effect of aii resistance on the solid material. For example when V_s=17 ft/sec, the material impelled upward into still air at 117 ft/sec will rise only 15 feet.

Kiker and Ross (1965) investigated gravitational flow of granular materials in enclosed vertical pipes. Using a high speed movie camera, velocity and position of the particles in the pipe were obtained. A specially mounted mirror gave a view of the reflected image of the pipe that was at a right angle to the view of the real image. This technique enabled plotting of a

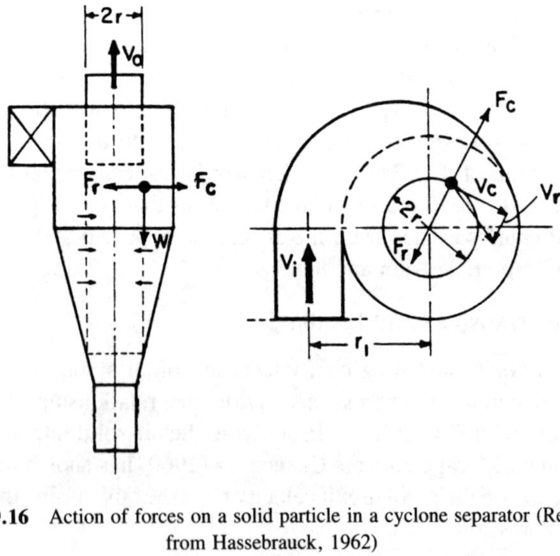

Figure 9.16 Action of forces on a solid particle in a cyclone separator (Redrawn from Hassebrauck, 1962)

velocity profile of the particle flow. Results of experiments with small seeds, falling vertically in a 2-inch diameter flexiglass pipe at a flow rate of about 0.03 ft^3/sec, showed that the position of the seeds in the cross section of the pipe had no effect on the velocity of the seed. This indicated that the velocity profile is blunt and an average particle velocity can be used in the equation of motion. Furthermore, since the particles fell in the center of the pipe and made only occasional contact with the wall and with each other, the frictional effects to be considered in the equation of motion could be neglected.

With the frictional force acting on the falling particle neglected, an equation of motion was then developed by equating the summation of forces acting on the particle with inertia forces according to Newton's law as was previously given in (9.20). In this case, however, the buoyant force, $m_p g \left(\frac{\rho_f}{\rho_p}\right)$, was substracted from the weight of the particle to obtain the effective weight. The resulting equation of motion was given by

$$m_p V_p \left(\frac{\mathrm{d}V_p}{\mathrm{d}L}\right) = m_p g \left(\frac{\rho_p - \rho_f}{\rho_p}\right) - \frac{C A_p \rho_f (V_p - V_a)^2}{2} \qquad (9.29)$$

where $\frac{V_p \mathrm{d}V_p}{dL}$ is replacing $\frac{\mathrm{d}V_p}{dT}$ of Eq. (9.20) to show the change in velocity with respect to distance moved, and the term $(V_p - V_a)^2$ is replacing V_p^2 in (9.20) to account for the velocity of air, V_a.

From Eq. (9.29) the changes in velocities of particles were calculated and plotted against distance as shown in Fig. 9.17. As seen the velocity of the downward moving air in the top six feet of the pipe was greater than the particle velocity, causing the particle to be dragged down. Beyond the six feet distance, the particle velocity exceeded the air velocity and the particles dragged the air through the pipe. The engineering implications of this air-solid velocity relationship is given by Chancellor (1960) who found that under most conditions the pressure at the outlet of forage blowers was less than that required to overcome the pipe friction resulting in a slow down of particle movement. In other words, the solid material was imparting energy to the air stream instead of the air stream imparting energy to the solid material. To overcome this situation Chancellor suggests providing more air inlet area for the blower.

Kiker and Ross (1965) also give the relationship between the number of particles N in a section of the pipe and the velocity of the particle, V_p, at that point by equating the weight of the particles as follows:

$$NV_p \rho_p = AL \rho_{dp} \tag{9.30}$$

where A and L are, respectively, cross section area and length of the pipe at the point of consideration, v_p is the volume of an average particle, and ρ_{dp} is weight density of dispersed particles given by

$$\rho_{dp} = \frac{Q_B \rho_B}{AV_p}$$

with Q_B and ρ_B being bulk flow (ft^3/sec) and bulk density (lb/ft^3) of the particles, respectively. Substituting for ρ_{dp} in Eq. (9.30)

$$N = \frac{Q_B \rho_B L}{V_p v_p \rho_p} \tag{9.31}$$

From the above relationship, as the velocity of the particles increases down the length of a pipe, the number of particles in a given volume of air will decrease. Multiple particles act as obstructions to the air flow in much the same way as a screen placed perpendicular to the flow. Therefore, a decrease in the number of particles will tend to decrease the free-stream turbulence intensity which causes the drag coefficient to decrease beyond a certain length of the pipe. This change of drag coefficient with length of the duct was theoretically determined from Eq. (9.29) and is shown plotted against N_R and falling distance in Fig. 9.18. Based on this analysis, Kiker and Ross concluded that the drag coefficients obtained from standard drag curves are only suitable for single particles falling in a very large body of still fluid and not for multiple-particle flow in a confined tube. The treatment of this subject to a great depth is found in a series of articles by Torobin and Gauvin.

Most of the published work on pneumatic conveying of solids deals with flow in horizontal or vertical pipes (Hariu and Molstad, 1949; Vogt and White, 1958, and others). Crane and Carleton (1957) adapted the theoretical analysis of Pinkus (1952) to agricultural grains to include any pipe inclination from 0 to 90 degrees.

Figure 9.19 shows the forces acting on a single particle being conveyed by air in a pipe with an angle of θ. From Newton's law the summation of forces in the tangential direction is equated to the product of mass times

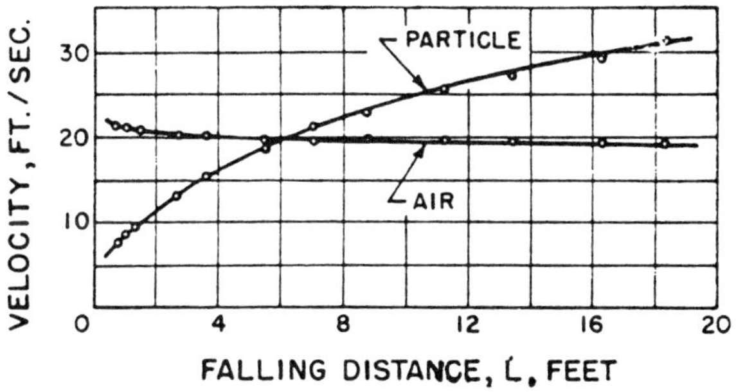

Figure 9.17 Velocity relationship of air and solid particle in free fall (Kiker and Ross, 1965)

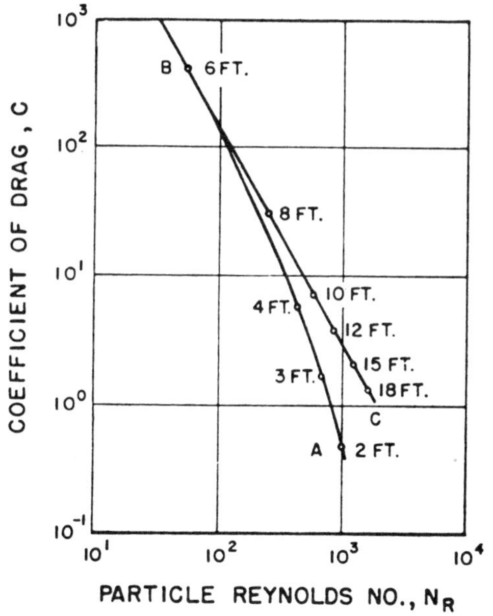

Figure 9.18 Variation of drag coefficient with Reynolds number and distance along the conveying duct as found from equation 5.27 (Kiker and Ross, 1965)

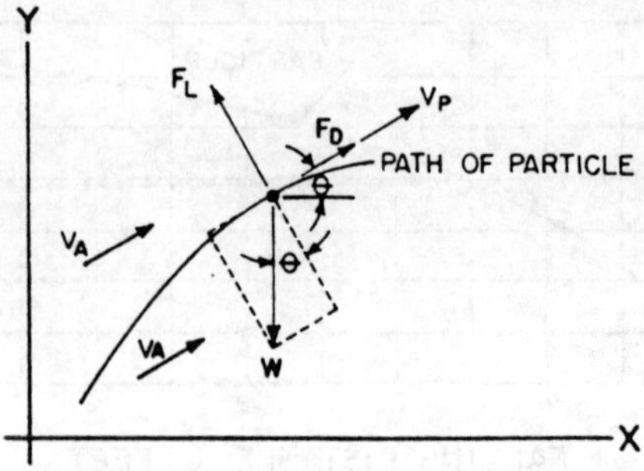

Figure 9.19 Particle conveyed by air in a pipe with angle θ (Crane and Carleton, 1957)

acceleration as in Eq. (9.20). However, because of the multiple-particle flow in a closed pipe, another term, to account for friction of particles due to collision with other particles and with the wall of the pipe, must be added to Eq. (9.20). Based on the work of Pinkus (1952), several investigators have used the Darcy-Weisback friction equation which can be given by

$$F = \frac{f_s V_p^2 m_p}{2D} \tag{9.32}$$

where f_s is the solid factor (dimensionless) and D is the diameter of the pipe. With consideration of the friction term, F, and the angle of the pipe θ, the complete equation of motion based on the free-body diagram of Fig. 9.18 and for the condition of steady state where $dV/dt=0$ can be written as follows:

$$\frac{CA_p\rho_f(V_a - V_p)^2}{2} - m_p g \frac{(\rho_p - \rho_f)}{\rho_p} \sin\theta - \frac{f_s V_p^2 m_p}{2D} = 0 \tag{9.33}$$

For the case of agricultural grains, where the term $(\rho_p - \rho_f)/\rho_p$ is very close to unity and thus can be neglected, Crane and Carleton derived the following expression for predicting solid friction factor, f_s, for all the particles.

$$f_s = \frac{D[\rho_f CA_p(V_a - V_p)^2 - 2gv_p\rho_p \sin\theta]}{V_p^2 v_p \rho_p} \tag{9.34}$$

where v_p is the volume of individual particles which is related to the total number of particles present through Eq. (9.31).

Converting each of the terms in Eq. (9.33) into equivalent expression for pressure drop per unit of mass flow given by $G_s = \left(\frac{Q_B \rho_B}{A}\right)$ (lb/sec-ft^2), results in the following expression for pressure drop ΔH in feet of water.

$$\Delta H = \frac{f_s V_p L G_s}{2 D g \rho_{H_2O}} + \frac{G_s L}{V_p \rho_{H_2O}} \sin\theta + \frac{f_a L V_a^2 \rho_f}{2 D g \rho_{H_2O}} \tag{9.35}$$

where ρ_{H_2O} in this equation is in lb/ft^3, L and D are length and diameter of the pipe in feet, respectively, and f_a is the value of Darcy-Weisback friction factor for the flow of air in a pipe.

If the solid friction factor f_s has experimentally been determined for a material to be transported pneumatically, calculation of pressure drop in the conveying pipe from Eq. (9.35) also requires a knowledge of solid velocity, V_p. For the condition of steady state, where terminal velocity has been achieved, Crane and Carleton give the following equation:

$$V_p = \frac{-2C_3}{C_2 - (C_2^2 - 4C_1C_3)^{1/2}} \tag{9.36}$$

where

$$C1 = \frac{0.2 A_p \rho_f}{m_p} - \frac{f_s}{2D}$$

$$C_2 = \frac{-0.4 A_p \rho_f V_a}{m_p} - \frac{20 A_p \eta}{m_p d_p}$$

$$C_3 = \frac{0.2 A_p \rho_f V_a^2}{m_p} + \frac{20 A_p \eta V_a}{m_p d_p} - g \sin\theta$$

From Eqs. (9.33), (9.35), (9.36) and the experimental pressure drops, the curves shown in Fig. 9.20 and 9.21 were developed using the following procedure:

1 The drag coefficient was evaluated from

$$C = 0.4 + 40/N_R = 0.4 + 40/79.9(V_a - V_p)$$

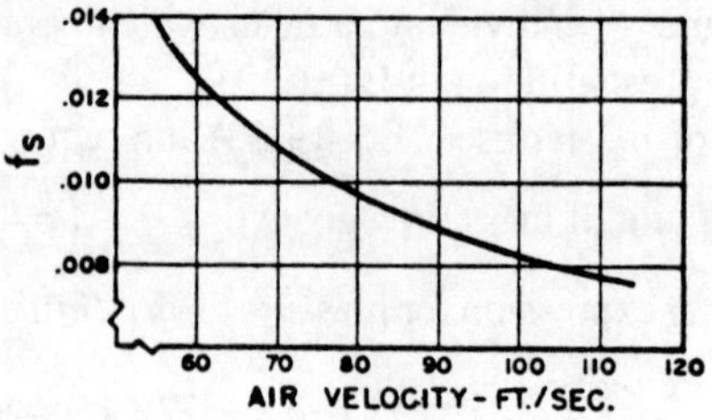

Figure 9.20 Friction factor of wheat for pipe angles 0° to 90° (Crane and Carleton, 1957)

2 A series of best fit lines were obtained for experimental pressure drops versus air velocities at various flow rates of the grain.

3 A series of curves of air velocity versus particle velocity were obtained for each flow rate by omitting f_s between Eq. (9.34) and (9.35) and solving for V_p.

4 An average curve, calculated from the series of air velocity versus particle velocity curves, obtained in the preceding steps, was used to calculate solid friction factor curve shown in Fig. 9.20.

5 The f_s values of Fig. 9.20 were used to determine the calculated curves of Fig. 9.21.

Examination of the pressure drop curves shows that the pressure drop due to solid particles increases with increase of pipe angle, with increase of air velocity, and with increase of flow rates.

A more recent work on the mechanics of pneumatic conveying systems for agricultural materials is reported by Cornish and Charity (1966). These investigators applied the Buckingham Pi Theorem for dimensional analysis of the various system parameters affecting elbow pressure losses. The parameters considered were sweep radius, R, conveying air velocity, V, solid flow rate, M, fluid density, ρ_f, fluid viscosity, η, particle density, ρ_p, particle diameter representative length, d, pipe diameter, D, and pressure drop due to solid phase, P_s. The resulting dimensionless equation was found to be of the following form

$$\frac{P_s}{\rho_f V^2} = f\left(\frac{R}{D}, \frac{M}{\rho_f V D^2}, \frac{D}{d}, \frac{\rho_p}{\rho_\phi}, \frac{\rho_f V D}{\eta}\right)$$

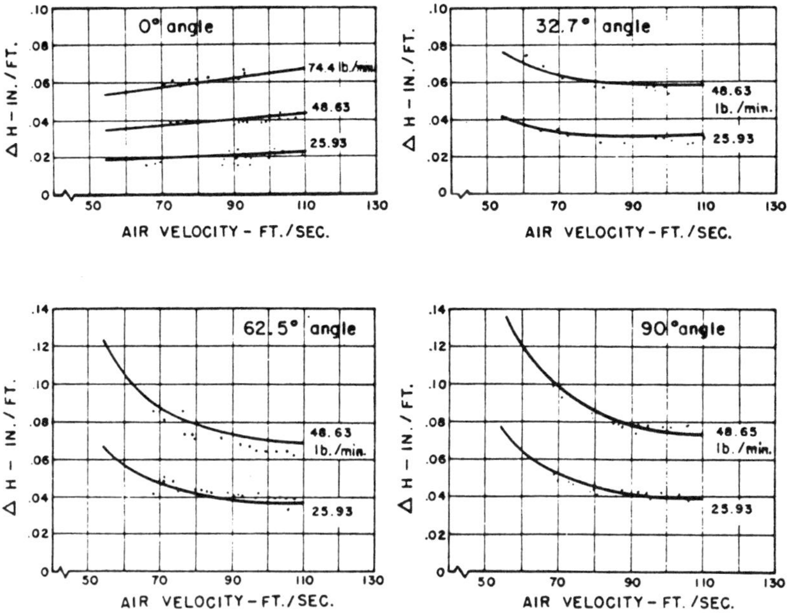

Figure 9.21 Calculated curves and experimental points for pressure drop caused by wheat (air drop excluded) in A 4-in. pipe at various angles (Condensed from Crane and Carleton, 1957)

Using peas, grain sorghum, and rape seed, experimental data were then obtained by determining the response of the dependent *Pi* term, $P_s/(\mathrm{r})\,\rho_f V^2$, to variations of one of the independent *Pi* terms, while remaining four *Pi* terms were held constant. Results of this experiment showed that to minimize pressure losses due to sweep elbows, the ratio R/D should be greater than six. It was also found that the static pressure loss, P_s, through a pneumatic conveying system sweep elbow is a linear function of M, the solid rate phase of flow.

To investigate the effectiveness of an energy-dissipating backstop for particle separation in a pneumatic conveying system, the concept of coefficient of restitution has been employed by Whitney and Porterfield. A sketch of the separator placed in-line with the conveying system is given in Fig. 9.22. Depending on the physical properties of the material being

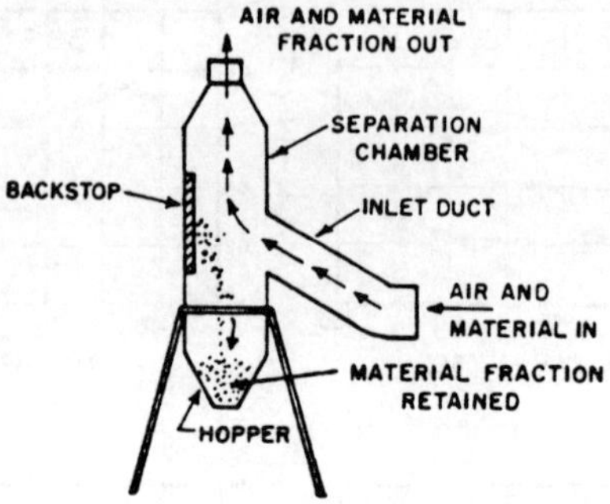

Figure 9.22 A pneumatic separator with a backstop (Whitney and Porterfield, 1968)

conveyed, the light particles will be carried upward and out of the separator while the heavy particles will collide with the back stop, losing energy, and become dependent on the drag force of the air to accelerate and convey them upward. If the drag force is insufficient to overcome the gravitational forces, the heavier particles will drop to the hopper below.

The padding materials used as the back stop were canvas, carpet, denim, gasket material, polyurethane foam, and vinyl sponge. The particles tested were plastic balls, soybeans, wheat, and sorghum. The energy dissipating characteristics of the padding materials was determined by determining the coefficient of restitution of the plastic balls dropped against the back stop padded with a given material. A photographic method, far more complicated than that shown in Fig. 8.11, was employed to determine the height of rebound.

Results, as expected, showed that as energy absorbing capacity of the back stop increased, more particles were separated out of the air stream. Of all the materials tested, denim as shown in Fig. 9.23, had the least coefficient of restitution or the greatest energy absorbing capacity. This material was then selected for the remainder of the tests to develop a prediction equation for determination of the effect of the system parameters on separation performance, given in terms of a "separation coefficient." The system parameters considered were particle diameter, density and flow rate, air density, viscosity and flow rate, and the geometry of separation chamber and other parts of the conveying system. It was found that with

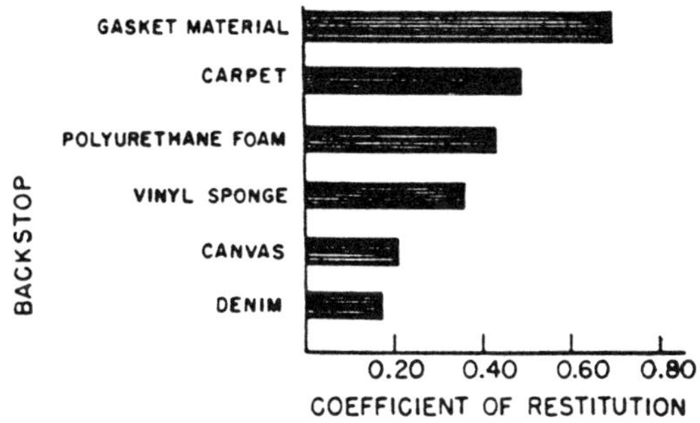

Figure 9.23 Coefficient of restitution of a plastic ball impacting various materials used on the backstop of the separator in Fig. 9.22 (Whitney and Porterfield, 1968)

soybeans, sorghum, and wheat, feed rate and concentration of a given size class had little effect on separation. Increase in the number of size classes, in the mixture of grains, however, resulted in larger prediction error.

Air flow through products and drag coefficient of materials

In aeration and drying of agricultural crops such as grains and hay, the static pressure against which the fan must work depends on air flow rate, depth of the material, and the type, shape, size, orientation and packing of the material. The last set of variables affect the drag coefficient of the material and its resistance to air flow.

Air flow through a bed of granular material has been studied rather extensively for beds of granular chemicals usually with regular geometric shapes such as spheres. Most investigators have considered the problem as analogous to flow in a pipe of varying diameter and surface roughness. Pressure losses are then related to expansion and contraction of air and to surface friction (Burke and Plummer, 1928; Carman, 1937; Ranz, 1952). Under the assumption that the drag forces for individual particles in a free stream should be additive and appear as pressure loss across a packed bed of the particles, Ranz (1952) proposed the following equation

$$P/L = \frac{KCA_pV^2N}{2g} \tag{9.37}$$

where

P/L = pressure drop inches of water per foot depth of the packed material
K = in. water per lb/ft^3
N = number of particles per cu. ft.

This equation is simply the sum of the drag forces Eq. (9.3) on N particles per unit volume of the packed material. The number of particles can be estimated from porosity factor, h, weight of each particle, W, and weight density of each particle, γ, as shown below

$$h = 1 - NW/\gamma \tag{9.38}$$

or

$$N = \frac{\gamma}{W}(1 - h) \tag{9.39}$$

Garrett and Brooker (1965) used Eq. (9.37) to predict pressure losses due to air flow through a grain mass employing the drag coefficients found experimentally (see Fig. 9.6). For the value of V, velocity of air passing between particles, a relationship was to be found between V and the air bulk velocity, V_0, measured across the packed bed. For this purpose two methods were considered. First, they assumed a velocity ratio

$$V/V_0 = 10.73 \tag{9.40}$$

proposed by Ranz (1952) for uniformly packed spherical bodies and found by a geometrical analysis of rhombohedral packing. This method resulted in prediction curves on the high side of the experimental air flow versus pressure drop curves reported by others. Second, the method proposed by Carman (1937), who assumed that the open area of a layer of granular material of differential thickness is the same as the porosity of the packed bed and proposed the following velocity ratio

$$V/V_0 = 1/h \tag{9.41}$$

where h is the porosity. Using the Carman's velocity ratio, displaced the prediction curves for the three grains tested to the low side of the experimental flow rate versus pressure drop curves (see Fig. 9.23a). This showed that if velocity ratio is the only uncertain parameter in the procedure of predicting pressure drop from aerodynamic data, a relationship which places V/V_0 between 10.73 and $1/h$ must be found. Further calculations showed that for closer agreement with measured air flow versus pressure drop data, the relative velocities should be 9.1 for corn, 7.8 for oats, and 4.2 for wheat. Assuming that equation (9.37) was valid and the porosity values taken from published data were suitable, it was suggested that the lack of closer agreement between the measured and predicted curves was due to the incorrect velocity ratio. It should be added that the values of drag coefficients, which were found for single grains falling in a large body of air, might also be different under the multiple particle and packed condition of flow, as discussed previously.

Day (1963) in his study of the resistance to hay to air flow has made an attempt to make use of dimensional analysis and the Buckingham *Pi* theorem to express pressure drop, *P*, as a function of air velocity, *V*, control length, *L*, diameter of air passage, *d*, density and viscosity of air,

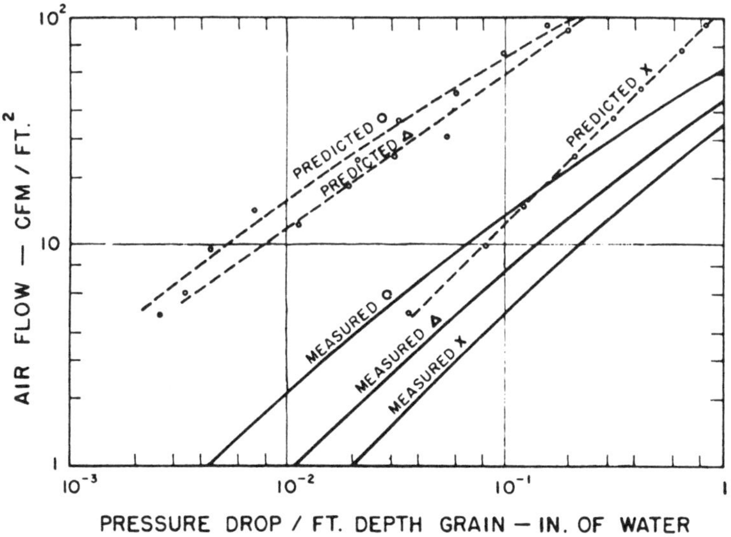

Figure 9.23a Comparison of measured and predicted pressure drop for various grains (Garrett and Brocker, 1965)

ρ_f and η, porosity, h, and a roughness factor, r, for the material. However, the dimensional analysis of the function written as

$$P = f(V, L, \rho_f, \eta, d, h, r)$$

was not carried out. The reason for precluding the dimensional analysis was that even with such a relationship known for a given batch of materials under a given set of conditions, it may be impossible to reproduce the same set of conditions for another batch of the same material. Instead, the effects of certain variables such as mechanical treatment of the material, moisture content, and bulk weight density on resistance to air flow was studied. It was found that the resistance of hay to air flow is increased by chopping and crushing, but the increases due to mechanical conditioning were minor when compared to those resulting from an increase in bulk density.

Other studies on the resistance of hay to air flow, based on derived equations, considering the drag coefficient are reported by Guillou (1946) and Bickel (1963).

In studying pneumatic transport of chopped forage Collins *et al.* (1965) also made an attempt to develop a prediction equation for pressure drop in a horizontal pneumatic system using the technique of dimensional analysis. Pressure drop due to drag of the material was not, however, considered a variable. After grouping the various parameters into dimensionless units, it proved difficult to control the variables to obtain the dimensionless relationships. Instead, the following formula for predicting pressure drop in horizontal pneumatic system conveying grain (Madison, 1949), was used for chopped hay.

$$m = a(R/K + 1)$$

where

m = pressure drop when handling material
a = pressure drop with air alone
R = ratio of weight of materials to weight of air
K = a function of velocity (dimensionless)

The Darcy-Weisback friction equation (9.32) was employed to estimate pressure drop with air alone. For the velocity term in this equation, the air velocity when conveying both air and material was substituted. Results were close to the measured pressure drops for the air velocities used. The formula, however, has not been verified over a wide range of air velocities and material flow rates.

Other applications of aerodynamic properties

To investigate the possibilities of using a high velocity airstream to reduce impact forces received by fruits in mechanical harvesting, Quackenbush *et al.* (1962) conducted laboratory tests and determined some of the physical properties and terminal velocities of several fruits (see Table 9.10).

In calculating the terminal velocities, a spherical shape was assumed for the fruit. A value of $C=0.44$, approximate value for sphere at $400<N_R<10^5$, was used for drag coefficient. And, the buoyant effect of air was neglected by assuming the term $(\rho_p-\rho_f)/\rho_p$ being equal to unity. With these assumptions and under the standard air conditions of 70°F and 50 per cent relative humidity, Eq. (9.10) was reduced to the following simple relationship for terminal velocity

$$V_t = 31{,}800\,(W/A)^{1/2}$$

where V_t is terminal velocity in ft/min and W/A is the ratio of weight to projected area in lb/in^2 based on an average diameter. To further simplify the calculation of terminal velocity, a regression equation was established between weights W, and weight to projected area ratio, W/A for each fruit. For example, the regression equation for the Red Haven peaches tested was of the following form

$$Y = 0.0438 + 0.05338X \pm 0.00205$$

where $Y = W/A$ and $X = W$.

The high correlation between V_t and W/A has also been demonstrated by Soule (1968).

Some fruits, such as McIntosh apples, will damage when dropped on a hard surface from a height as low as one inch. Therefore, results of this study showed that air velocities close to terminal velocities will be required if air cushioning is to be used in mechanical harvesting. Since the required terminal velocities are about 8000 ft/min (Table 9.10), the power requirement to drive the fans is excessive and under the existing conditions would be quite costly and impractical.

Table 9.10 Calculated terminal velocities and Reynolds numbers for several fruits (Quackenbush, Stout and Ries, 1962)

Fruit	Weight lb	Ratio of weight to projected area, lb/in^2	Reynolds number	Terminal velocities ft/min
Apples				
McIntosh	0.371	0.063	1.83×10^5	7950
Cortland	0.392	0.066	1.89×10^5	8130
Jonathan	0.288	0.056	1.63×10^5	7580
Northern Spy	0.403	0.066	1.91×10^5	8100
Apricots				
Montgamet	0.064	0.039	7.60×10^4	6250
Blueberries				
Jersey	0.001	0.012	1.02×10^4	3450
Cherries				
Montmorency	0.010	0.019	3.12×10^4	4440
Peaches				
Red Haven	0.396	0.066	1.89×10^5	8080
Elberta	0.404	0.066	1.92×10^5	8160
Plums				
Stanley Prune	0.078	0.039	8.45×14^4	6290

Hydraulic transport and handling

Hydraulic transport and handling of solids through pipes is a handling process used for many years by civil, chemical and mining engineers. Examples are dredging and filling work, oil well drilling, transport of pulp in paper making, water and sewage treatment, disposal of waste materials, flow of underground transport and subsequent hoisting of coal for power plants, ore extracting and transport, hydraulic transport of muds, sludges and suspensions, and many other applications involving particles from few microns to few inches in size.

Recently considerable interest is being shown in transport, separation, and handling of fruits and vegetables including potatoes, apples, cherries, tomatoes, and citrus fruits by water at the packing stations and the processing plants. (Edgar and Claycome, 1957; Pflug and Levin, 1961; Stout *et al.*, 1966). For transport of potato over long distances which may include vertical rises and where open channel transport may not be economical, pumping of the mixture of water and the product has also been considered. In addition to the advantages of hydraulic transport claimed for non-biological solid materials, water handling of fruits and vegetables is particularly advantageous in that mechanical damages can be minimized, cooling and quality preservation can be affected, and in the case of potatos, the material is washed while being transported. Depending upon the mode of transport (open channel or pipeline) and the size and concentration of the material in water, resulting in a homogeneous or a heterogeneous mixture, certain questions must be answered in order that specific design characteristics of the system can be determined. These characteristics are such things as dimensional and physical characteristics of the product as well as the pipe or channel, maximum allowable velocity to avoid mechanical injury to the product, minimum transport velocity (in the case of small particles) to avoid settling of the solid, solids concentration, head losses, and power requirements.

The flow characteristics and subsequent design of the pumping and transport systems for the homogeneous single-phase mixtures has been discussed in Chapter 5 under Flow of Organic Fluid. The application of the same principles for hydraulic transport of muds, sludges and suspensions under both laminar and turbulent flow is given by Caldwell and Babbit (1941). Heterogeneous mixtures which are discussed in this section consist of two-phase systems in which the carrying fluid retains its own individuality and the solid particles move with the flow either by means of *suspension* or *saltation*. Suspension occurs when particles are small and the flow velocity is high. Saltation, or moving along in a series of short intermittent jumps, takes place when the particles are large and the flow velocity is low. In either case, the terminal velocity of the particle in the fluid will determine the transition from a nondeposit regime to a deposit regime. Despite the industrial importance of hydraulic transport of heterogeneous solid-water mixtures, very little basic work has been published in this area.

Settling velocity and apparent drag coefficient

As in the case of pneumatic transport the two important parameters for hydraulic transport of heterogeneous materials are drag coefficient and terminal velocity. When a particle falls through a liquid at rest, its maximum settling velocity is reached when the apparent weight of the particle due to gravitational forces and buoyancy equals the drag forces. Thus, the equations presented previously are applicable for the entire range of

Reynolds numbers associated with the particle movements. Assuming a spherical shape for the particle, under the conditions of laminar flow and $N_R<1$, the settling velocity is found from Stokes law, Eq. (9.13). For $10^3>N_R>2$, Eq. (9.16), and for $2\times10^5>N_R>10^3$ Eq. (9.14) can be employed, if a spherical shape can be assumed.

For settling velocity of non-spherical and irregular shapes that present their largest cross section area to the flow of water, Condolios and Chapus (1963) give the following equation

$$V_t = \left[\left(\frac{4}{3}\right) g d_e \frac{(\varrho_p - \varrho_f)}{\varrho_f} \left(\frac{\psi}{C}\right)\right]^{1/2} \tag{9.42}$$

where

d_e = diameter of equivalent sphere
ρ_p, ρ_f = specific gravity of the particle and the fluid (water), respectively
C = overall drag coefficient
ψ = a shape factor always less than unit and defined by

$$\psi = \frac{A_e}{A_1} \tag{9.43}$$

where A_e and A_1 being the equivalent sphere and the largest cross section areas, respectively. Note that Eq. (9.42) is exactly the same as Eq. (9.12) with the exception of the shape factor ψ which is introduced to account for deviation of the shape of the particle from that of a true sphere. Experience with mixtures of sand and water with different concentration, different pipe diameters, and various sizes of particles have shown that the true criterion characterizing materials being transported hydraulically is not the drag coefficient alone but the parameter

$$(C_a)^{1/2} = (C/\psi)^{1/2}$$

which may be called the *apparent drag coefficient* (Condolios and Chapus, 1963). These authors also pointed out that when a wide range of particle sizes in the solid-water mixture is present, the weighted average of the drag coefficients given by

$$(C_a)^{1/2} = p_1(C_1)^{1/2} + p_2(C_2)^{1/2} + \ldots + p_n(C_n)^{1/2} \tag{9.44}$$

should be used when $p_1, p_2, \ldots, p_n$ represent fractions of the particles with the corresponding drag coefficients.

In pumping heterogeneous mixtures of water and solid particles, the prediction of head losses for calculation of power requirements becomes an important engineering problem. Condolios and Chapus (1963) have given the following empirical equation for the non-deposit regime

$$\Phi = 180\left[\frac{V^2}{gd}(C_a)^{1/2}\right]^{-3/2} \tag{9.45}$$

and

$$\Phi = 180\left[\frac{(V_e)^2}{4gR_h}(C_a)^{1/2}\right]^{-3/2} \tag{9.46}$$

for the deposit regime. In these expressions

$$\Phi = (J - J_e)/J_e C_t$$

$$V_e = V_t\left(\frac{4}{d}R_h\right)^{1/2}$$

where

V = velocity of the mixture
J = head-loss gradient of the mixture
J_e = head-loss gradient of clear water
C_t = concentration of solids
Φ = a function of head-losses gradient and concentration
d = diameter of the pipe
R_h = hydraulic radius of the free-flow section of the pipe
V_e = flow velocity in the free section of the pipe

As seen from the above velocity expression, as height of deposit approaches zero, R_h approaches the hydraulic radius of the pipe, $(d/4)$, and V_e approaches the settling velocity V_t. Note that under these conditions the mean flow velocity, V, cannot be simply related to the cross-sectional area of the pipe, A, and the flow discharge, q, through $q=AV$.

Based on the above expression and laboratory tests the power requirements increase directly with an increase in size of the particle. For example, everything else being equal, the hydraulic transport of sand requires approxiamtely 20 times as much energy as the transport of cement slurry, and the transport of gravels 50 times more energy than the slurry.

Application of hydraulic transport and handling of agricultural products has been reported for potato transport to the packing line, apple transport and sorting before storage, at the processing plant, and apricots in preparation for drying.

Fluming of potato is accomplished by directing the high pressure water on potato pile in storage and washing them into a flume which conveys them to the vertical elevator and the packing line (Fig. 9.24). Comparing the cost of flumes or pipes with the cost of mechanical conveyors, it has been found that beyond a distance of 100 feet the cost of flumes and pipes is distinctly lower than the cost of mechanical conveyors (Edgar, 1957). When the total costs of equipment, power requirements, labor and maintenance are considered, it is claimed that pumping is probably more economical than any other bulk handling method for moving potatoes to the packing line. No data, however, are available to substantiate this claim. The advantages of pumping in pipes are that the material can be conveyed up an incline and in any direction with pipes crossing obstacles, taking less space, easier installation and maintenance, easier to operate with greater possibilities for automatic control.

To minimize bruising and stem puncture of soft-fleshed fruits such as McIntosh apples during unloading from bulk boxes, water flotation dumpers were developed and put in use as far back as 1960 (Pflug and Levin, 1961). The bulk box of fruit is submerged in a tank of water allowing the apples to float away from the box and spread out into a single layer at the water surface. The water in the tank is circulated to carry the fruit from the box unloading end to a mechanical conveyor at the opposite end. This system of handling has been reported to result in little or no bruising or stem puncture.

The success with water flotation dumpers suggested water as a likely medium to be used in handling of apples for sizing, sorting and filling in boxes. Research on water handling of apples was then continued to provide basic data essential for development of equipment for sorting and sizing of apples destined for storage in pallet boxes (Matthews *et al.*, 1965; Dewey *et al.*, 1966). Such behavior as fruit orientation (stem up, stem down, or cheek up), the upward terminal velocity, the angle of repose of submerged apples, the effect of hydrostatic pressure on bruising, and the depth of water needed to cushion the falling fruit were investigated and

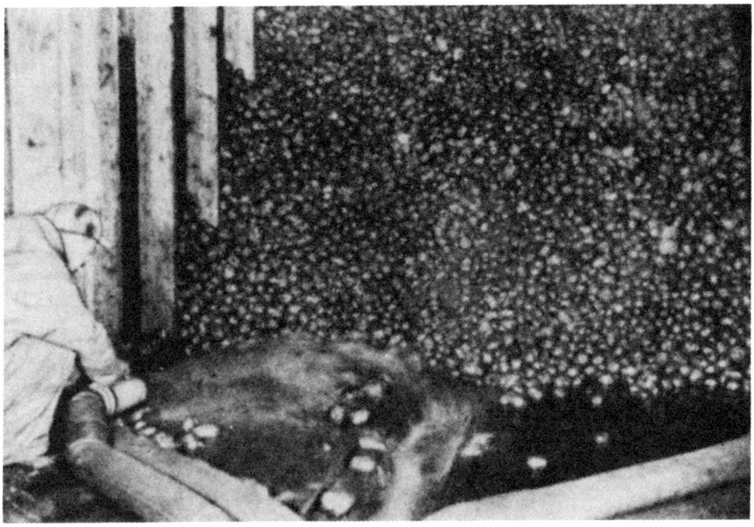

Figure 9.24 Fluming of potatoes (Hunter and Wilson, 1966)

their relationship to design of the handling equipment were determined.

The terminal velocity of the fruit was determined by releasing the fruit from rest at the bottom of a 15-inch diameter container and recording the motion on a 16 mm movie film. From this film, a graph of displacement versus time was plotted which enabled determination of terminal velocity (Fig. 9.25). As shown in Table 9.11, the experimental values of terminal velocities were lower than the calculated values. The wall effects of the small container were stated to be the reason for lower values of experimental terminal velocities. However, due to the limited depth of the container, it is possible that the fruits did not reach their terminal velocities. The corrected terminal velocities shown in Table 9.11 are experimental velocities multiplied by 1.06, a correction factor to account for wall effects as determined from experiments with a rubber ball. Since a turbulent condition of flow existed, a drag coefficient of 0.44, appropriate for a true sphere, was used for calculation of terminal velocity. Obviously, the stem and the non-spherical shape of the apple resulted in a higher drag coefficient. In order to make the calculated and experimental terminal velocities equal,

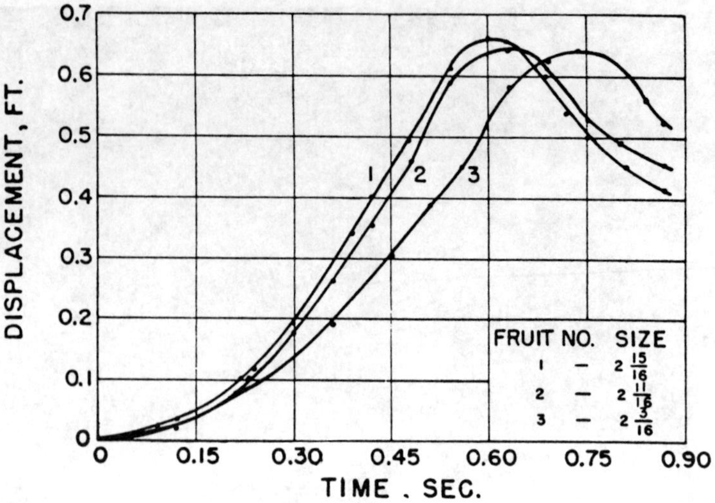

Figure 9.25 Graphs of displacement *vs* time for determination of terminal velocity of apples in water (Matthews *et al.*, 1965)

Table 9.11 Some physical characteristics of apples under water (Taken from Matthews *et al.*, 1965; Dewey *et al.*, 1966).

Apply variety	Floating orientation (percent)			Angle of repose deg.	Terminal buoyant velocity				
	Stem up	Stem down	Cheek up		Ave. maj. dia. in.	Sp. gr.	Expl. vel. fps.	Corr. vel. fps.	Cal. vel. fps.
Jonathan	95	1	4	30	2.94	0.76	1.83	1.94	2.34
McIntosh	64	34	2	33	—	—	—	—	—
Delicious	71	2	27	36	2.84	0.82	1.59	1.68	1.94

a drag coefficient of 0.68 was required. It seems, therefore, that experiments with a deeper and wider water container and assumption of a larger drag coefficient should have brought the calculated and experimental terminal velocities closer together.

According to the results reported by these investigators, apples reach their terminal buoyant velocity in water after 2 to 3 inches of travel. Since this is the highest velocity the fruit would attain, a knowledge of its magnitude is important in predicting bruise damage resulting from contact with solid objects under water. The maximum corrected terminal velocity of 1.94 given in Table 9.11, is equivalent to about 0.7-inch fall in air which can cause bruising in soft-fleshed fruits such as McIntosh apples. Therefore, to minimize impact under water, it was concluded that the areas of equipment that fruit may contact after floating up more than 2 inches should be covered with a cushioning material.

Also, fruit dropped in water from 2.5 inches sank to an average depth of 7 inches as compared to 18 inches when dropped from a height of 36 inches. This indicated that if the depth of water is not adequate to completely decelerate the fruit before it contacts another object, cushioning material should be used to reduce impact forces.

The angles of repose of submerged apples were also determined by observing the pyramid of fruit resulting from the accumulation of apples under water. This indicated that the design of an under water filling device must provide adequate space for fruit pyramiding or utilize a distributing device.

Submersion of fruits causes water uptake and if the hydrostatic pressure is excessive, injury may occur on the cell structure. The extent of water uptake and tissue injury increases with depth and holding time under water. Submergence to a depth of 6 feet of water for up to 15 min. however, was found to be permissible for water handling of apples.

Based on these hydrodynamic characteristics of apples and other observations related to design and performance of water handling equipment, first a pilot model filler of one bushel capacity was built and tested. The success achieved with this small scale pilot model warranted the design and construction of a complete prototype of the system for sorting and sizing apples up to 600 bushels per hour (Stout *et al.*, 1966). In this prototype, apples are received from pallet boxes of 16- to 20-bushel capacity, undersize fruits are eliminated, the culls are removed by a sorting device, and after sizing (chain sizer) the apples are returned to pallet boxes.

The possibilities of hydraulic handling of fruits has also been investigated in connection with several operations involved in preparing apricots for drying (Lorenzen and Lamouria, 1964). The operations involved in the process are aligning, metering, orienting, cutting, pitting, realigning, re-orienting, and spreading. The inability of the experimental mechanical methods to accommodate soft fruits led to the use of water as the means for transporting the product and controlling the various operations.

Separation of stemmed cherries from stemless fruit by fluming method have proved to be difficult because of the different flow patterns in the flume. In an attempt to determine the theoretical depth for a water column to separate stemmed sweet cherries from stemless fruit by water settling, Tennes *et al.* (1968) determined several physical properties including terminal velocities for two varieties of sweet cherries. Using the experimental method employed by Mathews *et al.* (1965), described earlier in this Chapter, they found considerable difference between experimental values and the theoretical values of terminal velocities of the sweet cherries tested. One reason for this lack of agreement was the assumption of sphericity for the cherries. According to the experimental data, however, the sphericity of the fruits tested did not exceed 76.6%.

To consider the actual degree of sphericity of the fruit, the following relationship, given by Pettyjohn and Christiansen (1948), was used to determine a resistance coefficient, C_r, for calculating the terminal velocity.

$$C_r = 5.31 - 4.88S$$

where S is sphericity. Substituting the value of 0.766 for sphericity in the above expression yields a coefficient of resistance equal to 1.57. Using this value for drag coefficient in the terminal velocity equation resulted in terminal velocities which were still about twice as much as that found experimentally. This shows that, assuming experimental data are reasonably correct, other factors much be present which increase the drag coefficient considerably. It was interesting to note that experimental values for terminal velocity of cherries with stem were greater than those of stemless cherries, as indicated by Fig. 9.26. According to the authors, the cherries with stem inscribed a "snake-like" path down the water column, apparently reducing the drag. The same fruit, without the stem, showed a tumbling action during the fall, which apparently increased the drag.

Example

To effectively flume a given variety of sweet cherries into different tanks, a 24-inch separation between cherry fruits with and without stems is required. If the terminal velocities of the cherries with and without stems are respectively 5.24 and 4.54 ft/sec, find the required depth of water to separate the two lots.

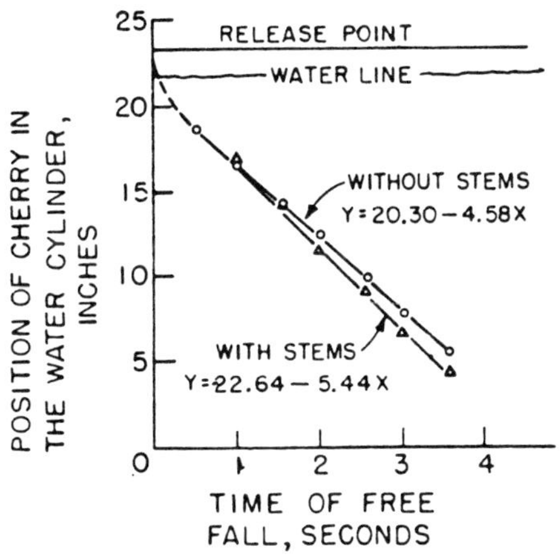

Figure 9.26 Time-displacement relationship for free fall of napoleon cherry fruit in water (Tennes *et al.*, 1968)

Time required to achieve the 2-ft. spread

$$= \frac{2 \text{ ft}}{5.24 \ -4.54 \text{ ft/sec}}$$

$$= 2.86 \text{ sec}$$

If the fruits with stems are traveling at 5.24 ft/sec, they would require a column $5.24 \times 2.86 = 15$ feet deep to produce the 24-inch spread.

Example

To separate a mixture of peas into two separate sizes, a sample of the peas is dropped into a column of water at 70°F. The average diameter of one group of peas is 6 mm with a density of 65.54 lb/ft^3 and that of a second group is 7.5 mm with a density of 67.41 lb/ft^3. Determine how far the larger peas have traveled when the smaller peas have reached a distance of 1.64 ft.

If design restrictions require a minimum of 2 ft for fluming the larger peas away from the smaller peas, what should the length of water column be? From Eq. (9.19)

$$CN_R^2 = 8W\rho_f(\rho_p - \rho_i)/\pi\eta^2\rho_p$$

$$\rho_p = 65.54\ \mathrm{lb/ft^3} \qquad d = 6\times10^{-3}\,\mathrm{m}$$

$$\rho_f = 62.4\ \mathrm{lb/ft^3} \qquad = 19.8\times10^{-3}\,\mathrm{ft}$$

$$W = (65.54\ \mathrm{lb/ft^3})(\pi/6)(19.8\times10^{-3})^3\,\mathrm{ft^3}$$

$$= 2.7\times10^{-4}$$

$$\eta_{H_2O} = 2\times10^{-5}\ \mathrm{lb\ sec/ft^2}$$

$$CN_R^2 = \frac{(8\times2.7\times10^{-4})\,(62.4/32.2)\,(65.54-62.4)}{\pi(2.05\times10^{-5})^2\,(65.54)}$$

$$= 1.7\times10^{-5} \text{ or } N_R = 600$$

$$V_t = N_R\times\eta/\rho D$$

$$= 600\times2\times10^{-5}/1.94\times19.8\times10^{-3}$$

$$= 0.31\ \mathrm{ft/sec}\ \text{(smaller peas)}$$

Similar calculation for larger peas yields:

$$V_t = 0.42\ \mathrm{ft/sec}$$

$$\frac{1.64\ \mathrm{ft}}{0.31} = 5.29\ \text{sec. for small peas to travel 1.64 ft}$$

$$0.42\times5.29 = 2.2\ \text{ft distance traveled by larger peas}$$

$$\frac{2\ \text{ft spread}}{(0.42-0.31)\ \mathrm{ft/sec}} = 18.18\ \mathrm{sec}$$

Length of water column: 18.18 sec × 0.42 ft/sec − 7.6 ft

9.4 LIST OF SYMBOLS FOR CHAPTER 9

A_p —projected area
C —overall drag coefficient
C_d —drag coefficient
C_f —frictional drag coefficient
C_L —lift coefficient
d —effective length of particle
d_e —diameter of equivalent sphere
d_p —particle diameter
F_D —drag force
F_g —gravitational force
F_L —lift force
F_r —resistance drag force

g —acceleration due to gravity
h —porosity (per cent)
m_f —mass of the fluid
m_p —mass of the particle
N_R —Reynold's number
P —fluid pressure
ρ_f —fluid mass density
P_B —bulk density
Q_B —flow rate
V —relative velocity between fluid and particle
V_f —fluid velocity
V_p —particle velocity
V_p —particle volume
V_t —terminal velocity
W —weight, lbs.
w, γ —weight density
τ —fluid shear stress
ΔH —pressure drop, ft.
η —viscosity
ψ —shape factor

10

FRICTION

10.1 SOME BASIC CONCEPTS OF SOLID FRICTION

The need for a knowledge of coefficient of friction of agricultural materials on various surfaces has long been recognized by engineers concerned with rational design of grain bins, silos and other storage structures. In design of agricultural machinery, however, the need for this information has been recognized rather recently. For example, in the design of a chopping and impelling unit, the engineers of a major manufacturer needed some information on the sliding coefficient of friction of chopped alfalfa and corn on steel. Not finding this information in any handbook or published data, it became necessary to set up a friction test apparatus and obtain the information needed (Hintz and Schinke, 1952). Obviously, before granular or unconsolidated materials can flow from a bin by gravity or a loaded auger can be started by a power source, the forces of static friction must be overcome. Likewise, once the forced flow has begun, the dynamic coefficient of friction is needed before the power requirement for continued flow can be estimated.

In this chapter some of the basic concepts of friction will first be introduced. Next, in light of these concepts, the friction of agricultural materials and the effects of various parameters such as moisture content, load, surface-roughness, and sliding speed on static and sliding coefficient of friction will be discussed. Finally, the applications of this important physical property in design of storage structures and materials handling and processing equipment will be given.

The laws of friction

Whenever a body is pressed against another by a force equal to its own weight, W, the first body will not move in the transverse direction until the frictional force between the two bodies is overcome. The frictional forces acting between surfaces at rest with respect to each other are called forces of *static friction*. The static force of friction is that necessary to start motion. Once the motion is started, the frictional forces usually decrease so that a smaller force is required to maintain motion. The friction forces existing between the surfaces in relative motion are called forces of *kinetic friction*.

The ratio between the force of friction, F, and the force normal to the surface of contact, W, is given by the well-known relationship

$$f = F/W \tag{10.1}$$

where f is called the *coefficient of friction*. In this book we denote static and kinetic coefficients of friction by f_s and f_k, respectively.

Coefficient of friction is also given by the tangent of the angle of the inclined surface upon which the friction force tangential to the surface and the component of the weight normal to the surface are acting. This angle is also called the angle of repose for granular materials. It is the angle with the horizontal at which the material will stand when piled.

Based on Eq. (10.1), the laws of friction, which were first stated by Amontons and later verified by Coulomb, are usually stated as follows:

1 Frictional force is proportional to the normal load.
2 Frictional force is independent of the area of the sliding surfaces.
3 Frictional force is largely independent of sliding velocity.
4 Frictional force depends upon the nature of the materials in contact.

Recent investigations using improved measuring techniques have shown that some of these time-honored concepts of friction can no longer be explained by the simple Coulomb's law of coefficient of friction. With this in mind, Sherwood (1951), summarized the more commonly accepted concepts of friction as follows:

1 The friction force may be defined as the force acting in a plane containing the contact point or points and in such a manner as to resist relative motion of the contact surfaces.
2 The friction force may be regarded as being composed of two main components, a force required to deform and sometimes shear the asperities of the contacting surfaces, and a force required to overcome adhesion or cohesion between the surfaces.
3 The friction force is directly proportional to the *actual* contact area.
4 The friction force depends on the sliding velocity of the contacting

surfaces because of the effect of the velocity on the temperature of the contacting materials.

5 The friction force depends on the nature of the materials in contact.

6 The friction force is not dependent on the surface roughness except in the extremes on very fine and very rough surfaces.

As his final conclusion, Sherwood (1951) states that despite these new findings, the Coulomb's law of friction continues to be useful even though it does not explain fully many of the observed events. As the mechanism of friction and the influence of various factors on the coefficient of friction are studied, the reasons for the above conclusions will be seen.

Effect of load and properties of contacting bodies

Although the basic laws of friction have been known for a long time, a partially accepted explanation of the mechanism of friction resulting in these relationships was given only recently by Bowden and Tabor (1950, 1964), who introduced the concept of asperity contact. These investigators made an experimental study of the physical and chemical processes involved during the contact of sliding solid surfaces. Their experiments showed that even the most carefully prepared surfaces contain hills and valleys which are large compared with molecular dimensions.

Figure 10.1 shows the irregularities on the surface of a finely turned copper surface. The solid sliding on such a surface will be supported on the peaks of the highest of these irregularities. This situation makes the real area of contact very small and almost independent of the apparent area of the sliding surfaces. The real area is directly proportional to the applied load because under the intense pressure at the localized points of contact, plastic deformation occurs until the area is large enough to support the load. An accurate determination of the real area of contact is not easy. Determining the electrical resistance between the touching surfaces is one technique used by several investigators (Holm, 1950; Bowden and Tabor, 1950).

In metals it is shown that when sliding takes place as a result of friction and rise of temperature over the area of the asperities, localized softening occurs which causes adhesion at the points of contact. The frictional force is in a large measure the force required to shear these "welded" junctions. The increase in friction is caused by the shearing, deformation, and plucking away of the softer material at the points of junction. The "welding" of the junctions is produced by the intense pressure in the regions of contact

Figure 10.1 Asperities on a cross-section of finely turned copper surface. Irregularities are about 2×10^{-4} in. high (Traced from a photomicrograph in Bowden and Tabor, 1956)

under static conditions. Upon motion, this welding process is accentuated by the rise in temperature and the softening of the material.

Therefore, friction phenomenon cannot be considered purely as a surface effect since the frictional force is greatly influenced by the properties of the two sliding bodies. The exact physical processes involved during sliding are too complex for a quantitative treatment. However, if we accept the concept of asperity contact, it is clear that during sliding points of junctions are formed and sheared. Furthermore, if one surface is harder than the other, the asperities of the harder surface "plow out" the softer material. To obtain an approximate expression for friction in terms of the physical properties of the sliding materials, Bowden and Tabor (1950), expressed frictional resistance, F, as the sum of two terms. The first term represents the shearing force, S, and the second term represents the plowing force, P, so that

$$F = S + P \tag{10.2}$$

If the real area of contact is given by A and the mean yield pressure of the softer material under the load, W, acting at the point junctions is given by p_m, then

$$A = W/p_m \tag{10.3}$$

and shear term S is given by

$$S = As \tag{10.4}$$

or from (10.3)

$$S = Ws/p_m \tag{10.5}$$

where s is the shear stress of the softer material acting in a direction tangential to the interface. Likewise, the plowing term, P, can be found by the product of the cross section of the grooved tract A' and the mean pressure, p_d, required to displace the softer material, so that

$$P = A' p_d \tag{10.6}$$

The terms mean yield pressure, p_m, and mean displacement pressure, p_d, are assumed to be equal in metals. The mean yield pressure, p_m, is the mean pressure when a hard sphere is pressed into a flat surface and is expressed by the ratio of load, W, to the area of the circle of contact, πa^2.

$$p_m = W/\pi a^2$$

The radius of the circle of contact, a, is given by the Hertz's Eq. (6.7) in Chapter 6. The relationship between p_m load and, W, may be presented by a curve similar to Fig. 4.3 in Chapter 4. The initial deformation up to the elastic limit is elastic. Plastic deformation commences at point LL and continues until the whole of the indentation is flowing plastically. In the case of metallic contacts, the tips of the asperities are considered as the spherical indenters deforming the material around them well beyond the elastic limit and into the range of full plasticity so that $p_m \simeq 3y_e$, where y_e is the elastic limit (Bowden and Tabor, 1950). If the asperities are softer than the surface against which they press, the asperities themselves are deformed plastically and similar considerations apply (O'Neill, 1934).

Substituting in (10.2) for S and for p from (10.5) and (10.6), respectively, yields

$$F = Ws/p_m + A' p_d \tag{10.7}$$

Equation (10.7) states that frictional resistance is proportional to the load and is independent of the apparent area of surface contact. Furthermore, since frictional force, F, is directly proportional to load, W, the coefficient of static friction f_s should, according to (10.1), be virtually independent of the load. This has been shown to be true experimentally. For example, friction of steel sliding on polished aluminum has shown that the coefficient of friction is almost constant for loads ranging from about 10 mg to 10 kg (Bowden and Tabor, 1950, p. 99).

Under certain conditions the plowing term in Eq. (10.7) is very small and can be neglected so that the coefficient of friction, f, may be expressed in terms of mechanical properties of the materials in contact. This is evident if we neglect the plowing term in (10.7) and combine this equation with (10.1) which yields

$$f = \frac{s}{p_m} = \frac{\text{shear strength of softer material}}{\text{yield pressure of softer material}} \tag{10.8}$$

In other words, since the shearing usually occurs within the softer material, coefficient of friction becomes a function of the physical properties of the softer of the two sliding bodies. Since s and p_m are strength properties of the same material, they vary together and their ratio is roughly the same. This conclusion, again, has been substantiated by observations in metals. Coefficient of friction for a very wide range of metals lies between 0.6 and 1.2 which shows that the coefficient does not vary by a very large factor.

The above conclusion that coefficient of friction is dependent on strength properties of the two materials in contact is also substantiated by the work reported by Karelitz (1938), who derived an expression giving coefficient of friction in terms of the modulus of elasticity in compression, E, and the modulus of rigidity (shear), G, as follows:

$$f = 1.07 \times 10^4 (H_1 + H_2)^{2/3} \tag{10.9}$$

where $H = \dfrac{3E + 4G}{G(3E + G)}$, constant 1.07×10^4 has the units of $\dfrac{1}{(\text{lb/in}^2)^{2/3}}$, and subscripts 1 and 2 refer to the two materials in contact.

Those experts who hold to the concept of asperities contact advocate that in addition to shearing and deforming of asperities, adhesion and cohesion between the surface in contact are also involved in the forces of friction. Adhesive and cohesive forces can be defined as those which cause surface attraction between dissimilar and similar materials, respectively. Two very smooth and very clean metallic surfaces may be held together in such a way that a large force will be needed to induce relative motion. If the surfaces are not clean or are not intimately associated, the force needed to induce relative motion will be much smaller.

To conclude this section, it is proposed to consider that friction force is the result of concurrent action of two types of forces: (1) shearing and deforming asperities, and (2) adhesion and cohesion.

Effect of sliding velocity and contact surface temperature

Under many conditions the sliding between surfaces may not be a continuous process but may proceed in a series of intermittent jerks referred to as "stick-slip." The friction builds up to a maximum during the "stick" and falls rapidly during the "slip" immediately following (Fig. 10.2). The physical and mechanical properties of the friction measuring system will usually determine the nature of the motion observed. In metals, for example, at extremely slow speeds of sliding, the strength of metallic junctions is often greater than that under higher speeds of sliding. This results in static friction being higher than kinetic friction. If one of the sliding surfaces has a certain degree of elastic freedom, the motion may not be continuous and proceed by a process of "stick-slip". As the speed increases, the time during which the surfaces can remain together becomes smaller and the motion becomes relatively smooth.

As the sliding velocity increases, frictional energy will usually increase and is released as heat energy. As a result of this, the temperature of the real contact surfaces increases at a rate depending on the ability of the materials to dissipate heat energy (thermal conductivity), sliding speed, and load. In dry friction of engineering materials, it is frequently observed that the friction force decreases as the velocity increases. This has been traced to the increase of temperature. The effect of temperature change in turn has been related to the change of physical properties of the materials. In metals, for example, the strength decreases with increase of temperature resulting in a decrease of friction with speed (Fig. 10.3). In general, it has been observed that at very low speeds of sliding, the coefficient of kinetic friction increases with speed while at high speeds of sliding, friction either decreases or is little influenced by sliding velocity.

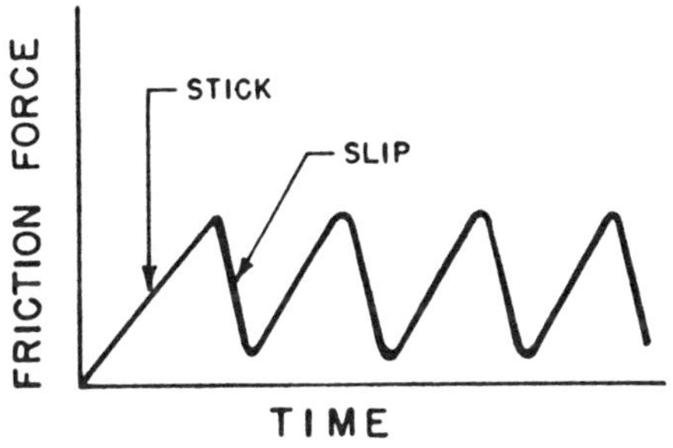

Figure 10.2 Stick-slip at low speeds

Effect of water film (friction and adhesion)

An adsorbed layer of water will considerably influence the friction coefficient. The formation of surface layers of water molecules or other liquids depends on the wetting power of the liquid. The wetting power of a liquid depends upon the attraction between the liquid and the surface, as measured by the ratio of adhesion tension, T_{ad}, and surface tension, T_s, of that liquid. The contact angle, δ, between liquid and solid (Fig. 10.4) is given by the equation

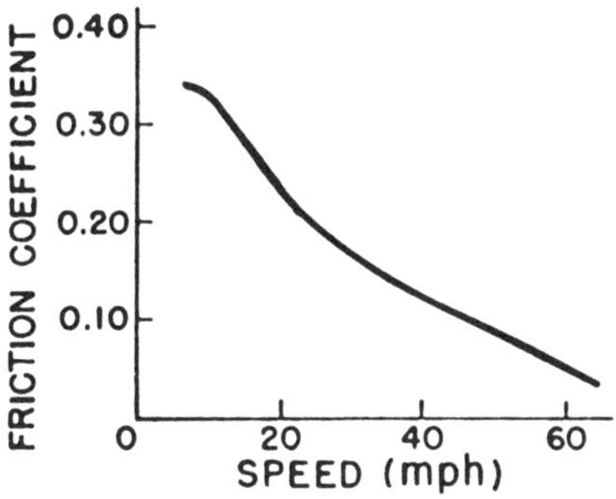

Figure 10.3 Friction of cast iron on steel as affected by speed (Gemant, 1950)

$$\cos \delta = \left(\frac{T_{ad}}{T_s} \right) - 1 \tag{10.10}$$

Wetting becomes complete when $T_{ad}=2T_s$. As long as angle δ (counterclockwise notation) remains acute, i.e., $2T_s > T_{ad} > T_s$, the liquid wets the surface. For conditions when $T_{ad} < T_s$, angle δ=90° and the liquid does not wet the surface. For conditions when $T_{ad} > 2T_s$, the equation is not valid.

In certain hard materials, the adhesion between surfaces is very small but increases when a film of liquid such as water is present between the surfaces. One example is the addition of a drop of water between two glass surfaces. The increase of adhesion has been shown to be due essentially to the surface tension forces acting on the thin film of water adsorbed on the surfaces (Bowden and Tabor, 1950). High humidity in the atmosphere has shown to produce the same effect.

It has been shown that the increase of adhesion force is accompanied by an increase in friction force. If the coefficient of adhesion, v, is defined as the ratio of adhesion force to the normal force, the theoretical relationship

$$f^2 = 0.3(v^2 - 1) \tag{10.11}$$

has been derived for plastic flow under combined normal and tangential stresses (Bowden and Tabor, 1950). Experimental data for friction and adhesion of steel on indium has shown the validity of the above relationship. In other words, as one surface begins to move over another surface, there is a steady rise both in the tangential and in the adhesive forces. Introducing a film of lubricant between two rubbing metals diminishes the amount of

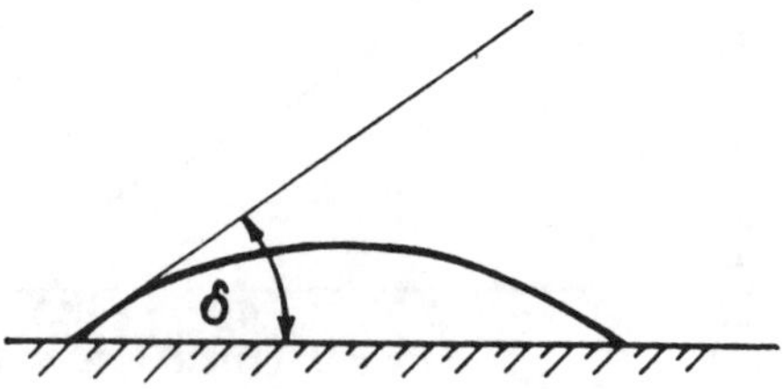

Figure 10.4 The contact angle between a liquid and a solid

intimate metallic contact which in turn reduces both the friction and adhesion between the surfaces.

Based on the above considerations, under certain conditions increase of moisture on the surfaces may cause an increase of adhesion and a subsequent increase in friction. On the other hand, it has been shown that with many metals the lubrication provided by fatty acids is really due to the formation of metal soap as a result of chemical reaction with the metal surface (Bowden and Tabor, 1950). It is also shown that with dilute solutions of fatty acids, the chemical attack is favored by the presence of water as well as oxygen. From this point of view, the presence of moisture together with fatty acids reduces the coefficient of friction. Consideration of these two opposing effects of moisture on friction phenomenon will probably provide some explanation for the friction data of organic material with various moisture contents on various surfaces.

Effect of surface roughness (directional friction coefficient)

Except for extremes of very fine and very rough surfaces, the friction force has been shown experimentally to be independent of surface roughness. This rather surprising observation has been shown by Bowden and Tabor (1950), by extending the "ratchet" theory (Makinson, 1948) to the effect of surface roughness on friction. As will be seen later, an understanding of this aspect of friction is particularly important in biological materials which may have surface texture with scales, hairs, etc., providing directional frictional properties. The following analysis has been applied to friction studies on wool fiber which has a very fine scale-like structure at the surface. Due to this phenomenon, the frictional properties for motion with the scales is different than that against the scales.

Suppose the scales or surface roughness of one body can be assumed as inclined faces making a small angle θ with the general surface as shown in Fig. 10.5a, and the other body rests at a number of points on these unclined faces. If the true coefficient of friction of the two faces at the points of contact is f, on account of the inclination of the faces the apparent coefficient of friction when sliding with the direction of roughness (left to right) will be

$$f_1 = f + \tan \theta \tag{10.12}$$

for numerous contacts of type I. For motion against the scales or roughness, where contacts of type II between the steep face of the incline and an asperity of the upper surface occur, the apparent coefficient of friction, f_2, contains not only a friction term but also a plowing or tearing term as follows:

$$f_2 = (1 - m)(f - \tan\theta) + F/W \quad (10.13)$$

where m is the fraction of the load, W, borne by all the contacts of type II and F is the total force required to tear off the tips of the scales. From microscopic examination, θ is approximated. Apparent coefficient, f_1, is determined experimentally by knowing normal load and measuring horizontal friction force. Knowing f_1 and θ the true coefficient f is calculated from (10.12). The value of m is very small, and it has been shown that it has little effect on the value of tearing force F (Makinson, 1948). Therefore, if f_2 is determined, by knowing the normal force and measuring the horizontal force, the value of F can be calculated.

Bowden and Tabor (1950) applied the above experimental and theoretical observations to study the effect of surface irregularities in friction of metals. In Fig. 10.5b, if a soft metal is assumed to rest on a harder metal with asperities tilted to the right at an angle of θ, in order to allow sliding of the upper surface from left to right, we must also raise the upper surface. The apparent coefficient of friction will then be the sum of the true coefficient f and the component $\tan\theta$ of the weight being lifted on the inclined surface. If angle θ is small and f is not too large, f_1 can be estimated from (10.12). If the upper soft material flows so that the irregularities of the lower surface are completely covered (Fig. 10.5b), upon sliding from left to right, shearing will occur in the soft material in the plane AA' and the upper surface will not be raised along the asperities. In such a case, friction will be independent of the surface roughness and $f_1 = f$. The same arguments apply for sliding from right to left, i.e., $f_2 = f$. For this reason, it has been shown that over a wide range of surface finish, the friction of metals is independent of surface roughness (Ernest and Merchant, 1940).

10.2 SOLID FRICTION IN AGRICULTURAL MATERIALS

Limited data on static coefficient of friction of grains have been available for many years (*Agricultural Engineering Yearbook*, 1966). In recent years additional data were made available on both static and kinetic friction of grains as well as chopped forage, corn silage, straw, wool, and other agricultural materials. While most of these investigators have undertaken friction studies for a particular use, a few have studied the subject in order to gain an understanding of the mechanism involved and the factors affecting the resulting friction data. As a result of the latter investigations, the validity of some of the published data on friction coefficients of agricultural materials are being seriously questioned. Furthermore, these better controlled recent experiments are providing data which in some cases are in conformity with the basic concepts of friction discussed earlier in this chapter.

Measurement of static and kinetic coefficient of friction

The methods used by various investigators to determine static and kinetic coefficients of friction of agricultural materials have usually been designed to suit the particular conditions of the material. The usual methods include the tilting of an inclined plane or moving of a given surface against the material. The method of inclined plane has been used for rough rice (Kramer, 1944), cereal grains (Burmistrova *et al.*, 1956), coffee fruit and coffee bean (Eschenwald, 1959), tobacco (Suggs and Splinter, 1960), and fruits (Cooper, 1962). In the method which places the material in contact with a positively driven surface, the surface is either mounted on a revolving circular disk or on a horizontal table. The former methods has been used with chopped forage, straw, silage and grains (Hintz and Schinke, 1952; Richter, 1954; Balis, 1961; Lorenzen, 1957; Stewart *et al.*, 1967). The latter method was employed for friction studies of grains on various surfaces (Buelow, 1961; Brubaker, 1965; Snyder *et al.*, 1967). In either method, the surface was driven at a known velocity against the material which was held in a container loaded with dead weights. The horizontal force of friction was measured either with a spring scale or a system of beams and linkages with strain gages or transducers as the force sensing element (Figs. 10.6, 10.7).

Two other methods of friction measurement are shown in Figs. 10.8, 10.9 and 10.10. The apparatus in Fig. 10.8 which was developed for determining the resistance of potato to skinning, can also be used for determining static and kinetic coefficients of friction of one convex body, such as fruits and vegetables, against another (Mohsenin, 1965). In the case of potato, the friction values were obtained while a skinning test was conducted.

Referring to Fig. 10.8, the pivoting plate (7) held one-half of the potato while the other half of the same potato was fixed on three nails projecting from the bed of the carriage (5). The carriage, carrying one-half of the

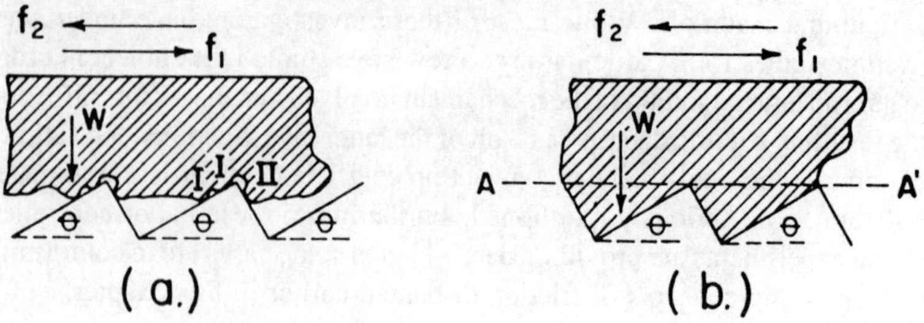

Figure 10.5 Effect of surface roughness on friction (Bowden and Tabor, 1950)

specimen, was pulled at a known speed passing under the stationary plate and skinning the upperhalf of the specimen, if the skinning pressure was exceeded. The dial indicator gave the "peak-to-peak" position of the specimens which was marked automatically on the recorder. If a given normal weight resulted in skinning the "peak-to-peak" position was re-established by backing up the carriage, the contact area was printed on a sheet of paper, and the normal pressure causing skinning was calculated.

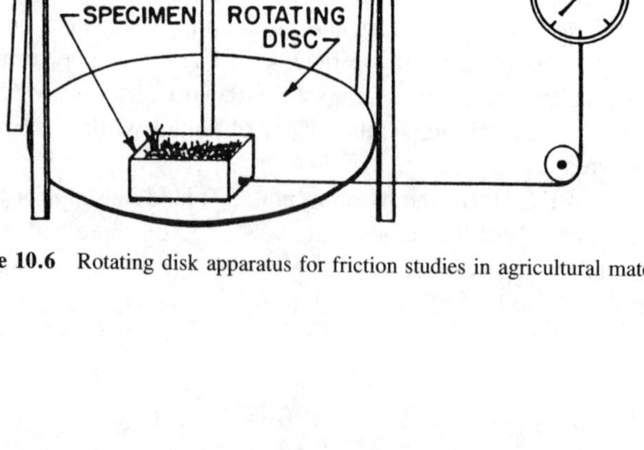

Figure 10.6 Rotating disk apparatus for friction studies in agricultural materials

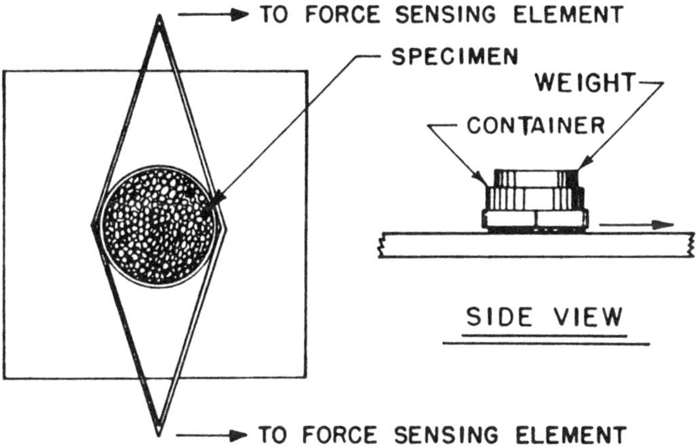

Figure 10.7 Horizontal moving surface for friction studies (Bickert and Buelow, 1966)

In the apparatus of Fig. 10.9, a single fiber of wool or a stalk of straw or grass was placed on the rotating drum, *B*, covered with the friction surface under study, and force F_2 for a given F_1 and a given drum speed was determined. The coefficient of friction was calculated using the following expression

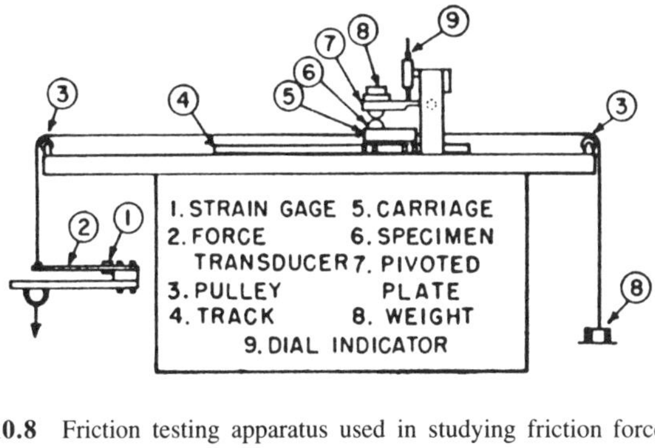

Figure 10.8 Friction testing apparatus used in studying friction forces causing skinning of potatoes (Mohsenin, 1965)

$$f = \frac{1}{\alpha} \ln \frac{F_2}{F_1} \tag{10.14}$$

where α is the angle of contact of the material with the friction drum. The derivation of Eq. (10.14) is found in Sloane (1947).

Effect of normal pressure

In determining the coefficient of friction of shelled corn on sheet steel, Buelow (1961), found that over a range of values from very low pressures to those equivalent to a 2-foot head of grain, the effect of load on coefficient of friction was not statistically significant. Also, the work reported by Wieneke (1956) on friction coefficient of straw, grass, and wool on various surfaces was inconclusive insofar as the dependence of friction coefficient on pressure is concerned. Moreover, Richter (1954) reported that there was no change in static coefficient of friction for chopped dry hay and straw on highly polished galvanized steel for normal pressures up to 70 lb/ft^2 (Fig. 10.11). For chopped grass and corn silage with about 72% moisture, however, a decrease of static coefficient with pressure was observed. More recent work by Snyder *et al.* (1967), also reported no significant change in kinetic coefficient of friction of wheat on steel when normal pressure was varied in the range of 50 to 288 lb/ft^2.

From these investigations, it appears that the coefficient of friction of organic materials is independent of the normal pressure. Note that this finding is in agreement with what has been found for dry friction of engineering materials.

Effect of sliding velocity

Preliminary friction tests of shelled corn at 19% and 12% (w.b.) moisture content on sheet have shown (Fig. 10.12) a slight increase of coefficient of friction with speed for a range of 3 to 15 inches per second (Buelow, 1961). At high velocities, Hintz and Schinke (1952) and Richter (1954) found no significant change of kinetic coefficient of friction for high moisture chopped grass and silage as well as dry chopped straw on steel for a range of speeds up to 6000 ft/min (Fig. 10.12). Another investigator (Wieneke, 1956), has reported an increase of kinetic coefficient of friction of dry straw and hay up to about 800 ft/min and then remaining constant for speeds up to 2000 ft/min. For freshly cut grass, this same investigator reported a decrease of the coefficient from 100 ft/min to 600 ft/min.

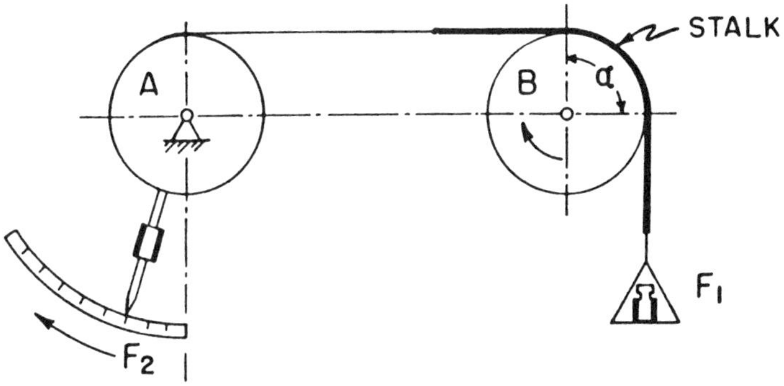

Figure 10.9 Apparatus used for determining friction of single fibres of wool and stalks of plant materials (Wieneke, 1956)

Wieneke (1956), who also studied the influence of pressure, contact time, moisture, and roughness on coefficient of friction of wool and several plant materials, found little change of friction force with speed for dry hay and straw but a decrease of friction for freshly cut grass. He considered bending of the leaf hairs on the fresh materials as having an effect on the decrease of friction at higher velocities.

Effect of surface conditioning

To obtain reproducible results, several investigators have found that the surface must be conditioned by making repeated passes before recording

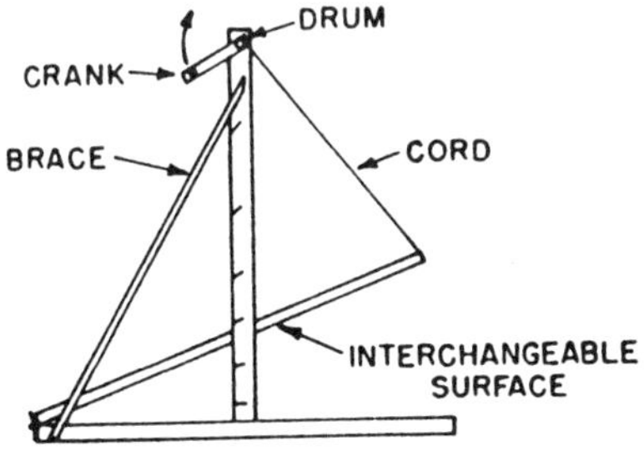

Figure 10.10 An inclined-plane apparatus for determining coefficient of friction and rolling resistance of fruits (Cooper, 1962)

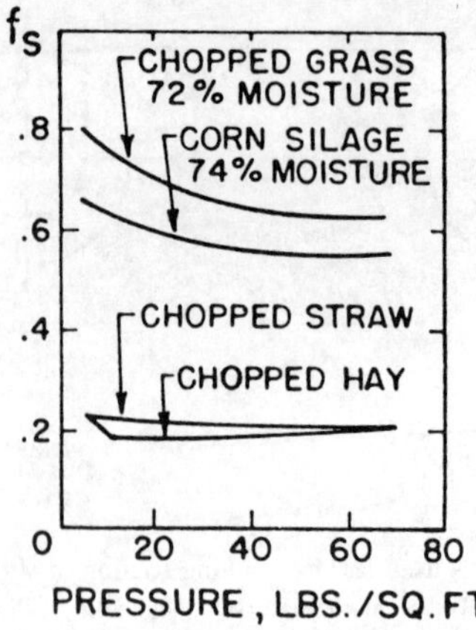

Figure 10.11 Influence of normal pressure on static coefficient of friction. Note the insignificant effect for dry material but a decrease of coefficient for moist material (Richter, 1954)

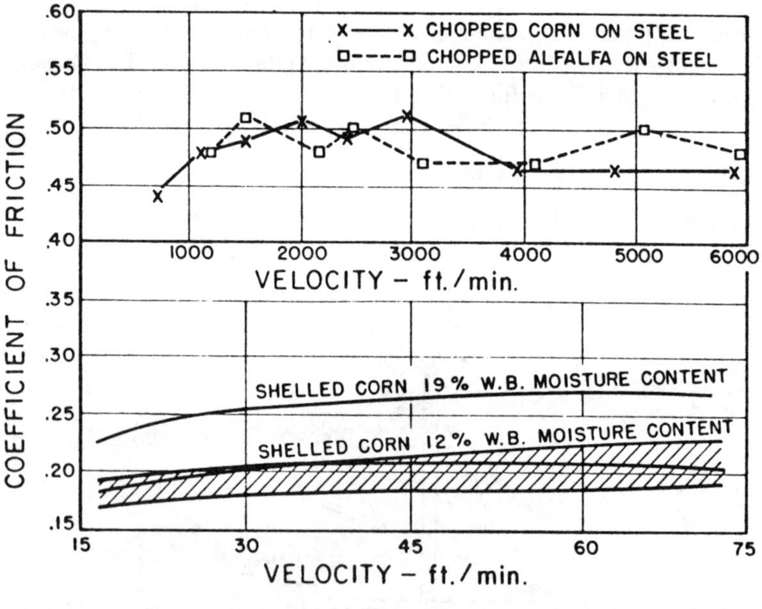

Figure 10.12 Influence of sliding speed on kinetic coefficient of friction (Below): Buelow, 1961; Top; Hintz and Schinke, 1952)

friction data. Bickert and Buelow (1966) found as much as 50 per cent increase in friction with shelled corn on a glass surface during the conditioning. On sheet steel approximately 400 to 500 cycles of repeated passes were necessary before the kinetic friction leveled off. They concluded that the transfer of oils, fats and waxes from the grain to the surface caused an increase in friction because washing the surface with carbon tetrachloride reduced the friction to the original value. Snyder *et al.* (1967), who made the same observation, allowed the carbon tetrachloride to evaporate and collected the remaining substance. Checking with a spectrophotometer, they found that the remaining substance was in fact cutin which is a waxlike substance containing numerous fatty acids and is found on the surface of the corn kernel. This increase of friction force during conditioning was also observed by Wieneke (1956) for straw and wool when rubbed repeatedly over a steel surface. Richter (1954), on the other hand, reported a decrease of friction of chopped straw and hay on galvanized steel with increase of contact time. It was stated, however, that the progressive treatments of the surface required nearly three days without giving the details of moisture content of the materials, number of passes, changes in relative humidity, and other factors which were found by the more recent investigators to influence the observed friction. The data reported by Richter (*Agricultural Engineering Yearbook*, 1966), were obtained, however, after the friction coefficient was stabilized with respect to time (Fig. 10.13).

Bickert and Buelow (1966) related the increase of friction during conditioning to the concept of contact of asperities and the rise of temperature at the interface as proposed by Bowden and Tabor (1950). The latter investigators have found that the friction force is greater when fatty acids at the interface are in liquid form rather than solid form. Both the time of contact of the asperities and the temperature rise at the points of contact affect the magnitude of shearing and adhesion and the resulting friction force. The deposit of a thin layer of cutin with low melting point on the rubbing surface apparently creates the same situation as has been found in lubrication of metals.

Wieneke (1956), also considered the influence of surface roughness in his friction studies with wool, straw and grass. On the drum of his friction apparatus (Fig. 10.9), he produced polished or rough surfaces. He also covered the friction drum with emery cloth of various fineness. He found no significant difference in friction for surfaces with variable roughness. This result, which is in agreement with what was discussed under surface roughness, was explained to be due to the fact that there were more points

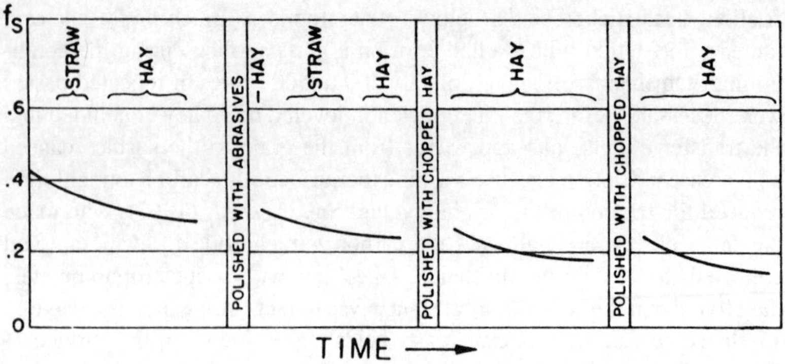

Figure 10.13 Effect of surface treatment (Richter, 1954)

of contact on the polished surface than on the rough surface. To determine the surface roughness, a thin layer of varnish was first used to coat the surface. After the solvent was evaporated, the coat of varnish was peeled off, sectioned and examined with a microscope. Lorenzen (1957) stated that according to his observations, grain on steel does not necessarily have a coefficient of friction less than grain on wood. Others have reported the same thing without offering an explanation. Perhaps the effect can also be explained by the concept of asperities in friction and the effect of roughness.

Richter (1954) in his friction work with chopped hay and straw, polished the surface of the rough galvanized steel with abrasive and found a decrease in friction. Subsequent polishing of the surface with chopped hay, however, did not decrease the coefficient of friction beyond a minimum value. Although no data were given as to the values of surface roughness, it is possible that the range of roughness from the galvanized steel in its original condition to partially polished by hay was sufficiently great to cause a decrease in static friction. As discussed previously, a decrease with roughness may occur only in the extremes of very fine and very rough surfaces. Tests by Stewart *et al.* (1967) has shown that considerable decrease in friction occurs with sorghum on Teflon, little affected by change in moisture content of the grain.

Effect of moisture

As discussed previously, the presence of moisture on the rubbing surfaces may cause an increase in friction due to increasing adhesion. Osmak (1954) reported an increase of 30 percent in static coefficient of friction of corn stalk on steel when moisture content increased from 25 to 77 percent. The same increase in moisture content of the corn stalk produced 15 percent increase in friction against rubber. The same trend was found in kinetic coefficient of friction of corn stalk on sheet steel. The kinetic coefficient of friction was increased from 0.32 to 0.54 when moisture content was increased from 13 to 88 percent, respectively. Wieneke (1956) found an increase of coefficient of friction with increase of moisture content of chopped hay and straw on steel. Richter reported the same trend for dry and wet chopped straw, grass and silage (Fig. 10.11). Brubaker and Pos (1965) found an increase in static coefficient of friction of various grains on various surfaces when moisture of the grain was increased. Bickert and Buelow (1966) found that kinetic coefficient of friction of shelled corn and barley on sheet steel and plywood begins to increase when certain levels of moisture content were exceeded. These levels of moisture content were found to be 19% (w.b.) for shelled corn and 17.5% (w.b.) for barley. This is shown in Fig. 10.14 and Table 10.1. The broken lines show extrapolation into the moisture range for which no data were available. The normal loads were varied but no definite effect was found due to variation in normal pressure. Snyder, *et al.* also reported an increase of kinetic coefficient of friction of wheat on sheet metals with increase of moisture content (Fig. 10.15). Lorenzen (1957), also reported that static coefficient of friction for wheat, corn, milo, barley, rice, and sawdust on wood and steel surfaces increased with increase of moisture content beyond the 13 percent level. This effect was less evident, however, for dynamic coefficient of friction. Stewart *et al.* (1967) reported the same trend observed for sorghum grain. Lorenzen (1957), also found that grains with hull intact were less affected by moisture than grains without hulls. Apparently, the cohesion part of apparent friction force is less influenced in the presence of rough hulls.

Since addition of moisture affects the adhesion and cohesion properties of the material, it would be desirable to separate the effects of cohesion from that of friction. Spangler (1951) shows a method based on Mohr's theory of failure for separating the effects of cohesion and friction (see discussion of Mohr's circle in flow of solid granular materials).

Effect of environment

In determination of the coefficient of friction, the need for a controlled environment to prevent changes in the moisture content of the organic material and the surfaces being tested was pointed out by the work of Brubaker and Pos (1965) and Snyder *et al.* (1967). In both of these works, the friction experiments were conducted under controlled atmosphere and the grains as well as the surfaces to be tested were allowed to come into equilibrium with the surrounding atmosphere. It was found that coefficient of friction of grains on various surfaces changes rapidly under non-equilibrium moisture content test conditions as illustrated in Fig. 10.16. This

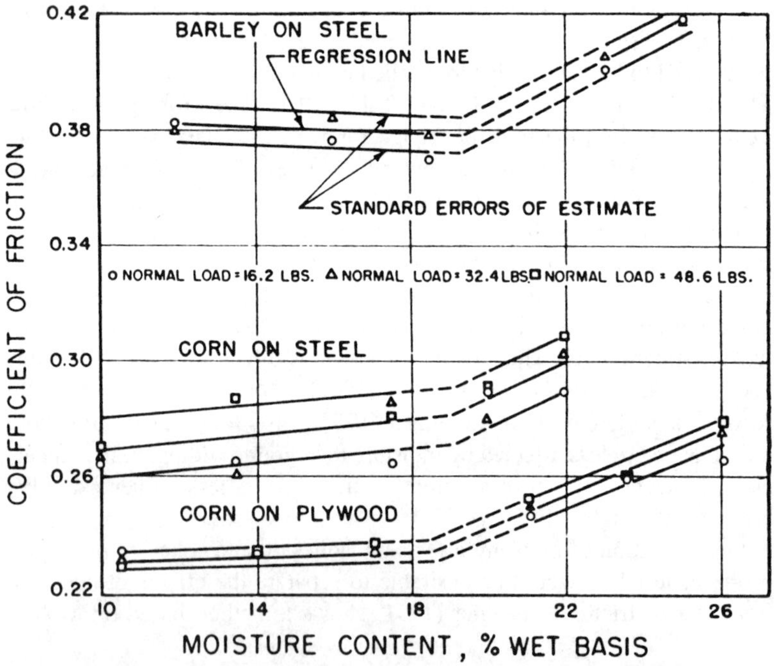

Figure 10.14 Coefficient of kinetic friction of grains on steel and plywood (Bickert and Buelow, 1966)

Table 10.1 Relationship between kinetic coefficient of friction (f_k) and moisture content (MC) of grains on various surfaces (Bickert and Buelow, 1966)

Shelled corn on sheet steel

Moisture range (%, w.b.)	Regression equation	Standard error of estimate
10–17.5	$f_k = 0.256 + (1.34 \times 10^{-3})\,MC$	0.01
20–22	$f_k = 0.153 + (6.67 \times 10^{-3})\,MC$	0.008

Shelled corn on plywood

Moisture range (%, w.b)	Regression equation	Standard error of estimate
10.5–17	$f_k = 0.225 + (4.5 \times 10^{-4})\,MC$	0.003
21–26	$f_k = 0.137 + (5.33 \times 10^{-3})\,MC$	0.004

Barley on sheet steel

Moisture range (%, w.b.)	Regression equation	Standard error of estimate
10–16.5	$f_k = 0.388 - (5.8 \times 10^{-4})\,MC$	0.007
21–23	$f_k = 0.248 + (7.5 \times 10^{-3})\,MC$	0.005

observation is also supported by experiments by Stewart *et al.* (1967) for sorghum grains on sheet steel and concrete.

10.3 ROLLING RESISTANCE

In some material handling applications, rolling resistance or maximum angle of stability in rolling of agricultural materials with rounded shapes may serve as useful design information. One example is gravity conveying of fruits and vegetables.

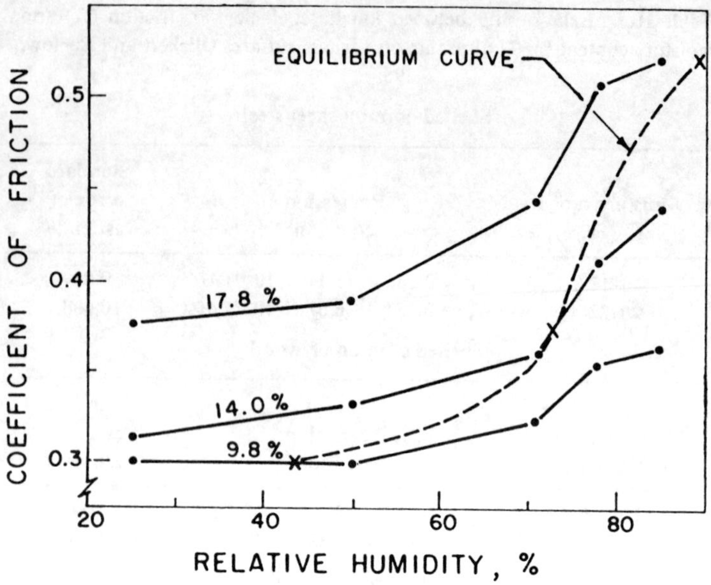

Figure 10.15 Effect of wheat kernel surface drying upon kinetic coefficient of friction (Snyder *et al.*, 1967)

When a ball or cylinder rolls over a horizontal surface with a force, F, if the surface deforms, there will be a resultant force, R, which the surface exerts on the body as illustrated in Fig. 10.17a. If the point of application of R is taken as a moment axis, neglecting accelerating forces,

$$\sum M_B = Fb - Wc = 0$$

For small deformation of the surface, $b \cong r$ so that

$$c = \frac{Fr}{W} \quad \text{or} \quad F = \frac{cW}{r} \tag{10.15}$$

The term "c" can be called coefficient of rolling resistance while F is the rolling resistance. From Fig. 10.17a and Eq. (10.15), it can be seen that the more rigid a surface is, the smaller will be c which results in a smaller

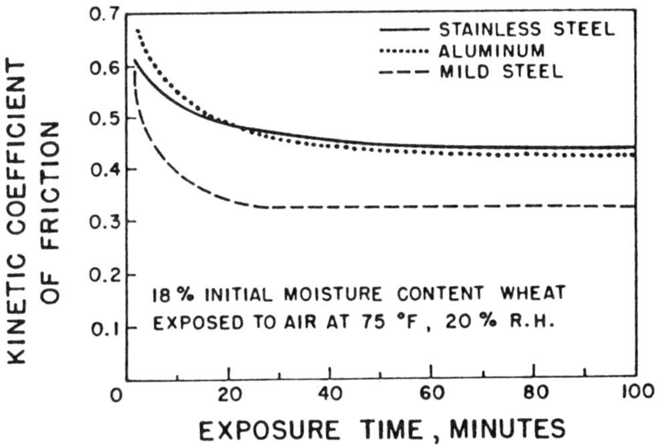

Figure 10.16 Changes of coefficient of friction of wheat under nonequilibrium moisture-content test conditions (Brubaker and Pos, 1965)

rolling resistance. Therefore, rolling resistance is directly proportional to the weight of the rolling object, indirectly proportional to the effective radius of the rolling object, and directly proportional to the coefficient of rolling resistance which depends upon the rigidity of the supporting surface.

The laws of rolling resistance are not thoroughly understood but Drutowski (1959), Bowden and Tabor (1964), and others have shown that the work lost in rolling a ball or a cylinder on an elastic plate is dissipated in hysteresis losses of the elastically strained surface. According to this theory, rolling resistance, F, is numerically equal to the product of elastic input energy per unit distance of rolling denoted by Φ and the hysteresis loss fraction, α, of this total energy.

$$F = \alpha\Phi \tag{10.16}$$

The elastic input energy per unit distance of rolling can be calculated by considering the couple exerted on the rolling element by the pressure acting over the front half of the region of contact in the process of rolling (Fig. 10.17b). For a rigid sphere of radius r with weight W over an elastic surface such as a rubber plate, the pressure distribution at a point distance r from the center is given by Hertz's problem of contact stresses as follows:

$$p = \frac{3W}{2\pi a^2}\left(1 - \frac{r^2}{a^2}\right)^{1/2} \tag{10.17}$$

where a is the radius of the circle of contact given by Eq. (6.7). The vertical force f on any strip perpendicular to the direction of rolling with an elemental area dx dy is p dx dy. This force exerts a couple fx on the ball, and the total couple, G, exerted by all elements on the front half of

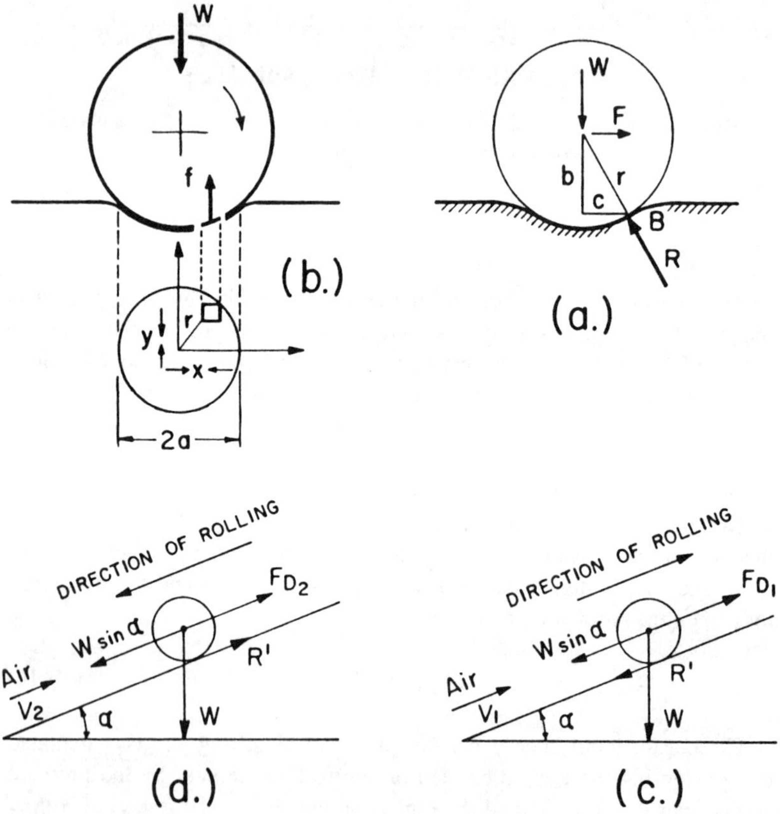

Figure 10.17 Rolling resistance. a—A rolling ball on a deformable surface; b—The couple exerted on the rolling element; c,d—Determination of rolling resistance of potatoes and stones using air stream and an inclined plane (Maack, 1957)

the circle of contact is the integral of fx between $x=0$ and $x=a$ as shown by Bowden and Tabor (1964).

$$G = \frac{3W}{4a^3} \int_0^a (a^2 - x^2)\, x dx = \frac{3Wa}{16} \tag{10.18}$$

From the above equation and consideration of the fact that work done by the ball on the elastic surface in rolling forward a unit distance is G/r, the elastic input energy per unit distance of rolling will be

$$\Phi_b = \frac{3}{16} \frac{Wa}{r}. \tag{10.19}$$

Substituting for a from Eq. (6.7) yields

$$\Phi_b = \frac{3}{16} \left(\frac{3}{4}\right)^{1/3} \frac{(W)^{4/3}}{(r)^{2/3}} \left(\frac{1 - \mu_1^2}{E_1} + \frac{1 - \mu_2^2}{E_2}\right)^{1/3} \tag{10.20}$$

where subscripts 1 and 2 refer to the surface and the ball, respectively.

Following the same analysis, Bowden and Tabor (1964) give the equation for elastic input energy per unit distance of rolling of a rigid cylinder over a deformable plate by

$$\Phi_c = \frac{(W)^{3/2}}{(r)^{1/2}} \left(\frac{16}{9\pi^3} \frac{1 - \mu^2}{E}\right)^{1/2} \tag{10.21}$$

where E and μ are Young's modulus and Poisson's ratio of the plate. Experimental data for rolling resistance of rigid spheres and cylinders on rubber surfaces with known hysteresis losses have shown that rolling resistance can be predicted from hysteresis losses if it be assumed that in rolling of a rigid cylinder about three times and in rolling of a rigid sphere about two times as much energy is lost as in their respective hysteresis tests. Having data on simple loading-unloading hysteresis test of the elastic material, with the above information one can make an estimate of hysteresis loss factor α and calculate the rolling resistance using Eq. (10.20) or (10.21) and (10.16).

Drutowski (1959), who confirmed the above theory for balls rolling on elastic plates, concludes that rolling energy measurements can be translated into materials hysteresis losses and that the equilibrium rolling resistance increases more rapidly with load in the range where plastic deformation occurs than it does in the load range where only elastic rolling occurs.

The application of these theories advanced for rolling resistance of balls and cylinders on deformable surfaces has not been reported for agricultural materials. Cooper (1962) employed the apparatus illustrated in Fig. 10.10 to study rolling resistance of apples on various surfaces. He reported maximum stability angles in rolling and no record of weight and diameter of the fruits were obtained. Allshouse (unpublished data) followed Cooper's technique and obtained the same type of data for tomatoes. Some of these data are given in Table 10.2. In these tests the platform was raised until the fruit began to roll from its initial equilibrium position. This initial equilibrium position was selected by noting the position at which a moving fruit came to rest. For most varieties this was the calyx end. For dynamic angle of stability in rolling, the inclined plane was adjusted until the fruit rolled down the incline at a uniform velocity. It was noted that surface condition, shape, size, and mass of the fruit as well as the amount of

Table 10.2 Coefficient of friction and rolling resistance (maximum stability angle in rolling) for fruits and vegetables

	Coefficient of friction		Rolling resistance (degrees)	
Surface	Static (f_s)	Kinetic (f_k)	Static	Kinetic
	Apples (six different varieties)[1]			
Plywood	0.32–0.44	0.24–0.33	12–18	2.5–4.5
Galvanized steel	0.38–0.46	0.28–0.36	13–18	2.5–4.0
Rigid foam	0.34–0.44	0.28–0.38	13–18	2.5–4.0
Soft foam	0.72–0.93	0.55–0.75	11–16	4.0–5.0
Canvas	0.36–0.44	0.25–0.36	12–16	4.0–5.0
	Tomatoes (four different varieties)[2]			
Sheet aluminum	0.33–0.52	0.28–0.40	7–11	3.6–4.8
Plywood	0.41–0.60	0.41–0.56	9–14	3.6–4.8
Rigid foam	0.44–0.56	0.48–0.56	11–13	4.2–4.8
Soft foam	0.77–0.83	0.68–0.79	11–13	4.8–4.8
Canvas	0.48–0.75	0.49–0.67	13–14	4.8–7.0

[1] For specific varieties see Cooper, 1962.
[2] Allshouse, G. W. (Unpublished data).

indentation that fruit caused on the test surface influenced the stability angle in rolling (Cooper, 1962).

Coefficient of friction of apples on several surfaces has also been measured by Vis *et al.* (1968). These investigators held the apple in a special mechanism equipped with a load cell and moved the test surface, mounted on a horizontal platform driven by an electric motor, at a given speed. The coefficient of friction was calculated from the effective normal weight of the fruit assembly and the horizontal friction force. Although the surfaces were conditioned by repeated runs before each test, as indicated by the authors, the frictional forces existing under actual field conditions would probably vary considerably. These variations would occur as a result of changes in temperature, relative humidity, surface dirt and fruit conditions. The values of friction found by Vis *et al.* (1968) are given in Table 10.2a.

One of the numerous techniques which have been employed for separating potatoes and stones was making use of the difference in rolling resistance of these two materials. To determine the rolling resistance, samples of potato or stone were placed on an inclined surface against an adjustable air stream using the following analysis (Maack, 1957).

Figure 10.17 (c) and (d) shows the free body diagram for the two conditions of the object rolling up or rolling down the inclined plane. With reference to this figure,

$$F_{D_1} = W \sin \alpha + R'$$

$$F_{D_2} = W \sin \alpha - R'$$

or

$$R' = W \sin \alpha \left(\frac{F_{D_1} - F_{D_2}}{F_{D_1} + F_{D_2}} \right) \tag{10.22}$$

where R' is the rolling resistance. From Eq. (9.1)

$$F_{D_1} = C_D A_p \varrho_f \frac{V_1^2}{2}$$

$$F_{D_2} = C_D A_p \varrho_f \frac{V_2^2}{2}$$

Substituting for F_{D_1} and F_{D_2} in (10.22) yields

Table 10.2a Dynamic coefficient of friction between apple fruit and various surfaces (Vis *et al.*, 1968)

	Surface							
	Teflon		Stainless steel		Rigid foam		Food belt[1]	
Varieties	Mean	S.D.	Mean	S.D.	Mean	S.D.	Mean	S.D.
Michigan McIntosh	0.11	(0.02)	0.60	(0.09)	0.49	(0.03)	1.97	(0.18)
Michigan Red Delicious	0.13	(0.01)	0.39	(0.08)	0.63	(0.12)	2.03	(0.33)
New York McIntosh	0.12	(0.02)	0.67	(0.12)	0.86	(0.10)	1.83	(0.20)
Washington Red Delicious	0.18	(0.03)	0.57	(0.10)	1.04	(0.21)	2.17	(0.58)
Washington Red Romes	0.17	(0.02)	1.04	(0.25)	1.22	(0.08)	2.78	(0.51)

[1] A canvas-backed belt with a rubber sealed surface.

$$R' = W \sin \alpha \frac{V_1^2 - V_2^2}{V_1^2 + V_2^2} \qquad (10.23)$$

For a horizontal plane, rolling resistance will be

$$R = R' \cos \alpha$$

The results of the measurements showed that the rolling resistance of stones differs from that of potatoes. Equal rolling resistance was observed only for very large potatoes and very small stones. This observation indicated that stones and potatoes may be separated according to their rolling resistances. Several mechanisms designed on the basis of difference in rolling resistance of potatoes and stones are described by Maack (1957). (See Table 3.1).

10.4 ANGLE OF INTERNAL FRICTION AND ANGLE OF REPOSE

Frictional properties of granular materials such as seeds and grains are important in design of equipment for solid flow and structures for storage of these materials.

Coefficient of friction between granular materials is equal to the tangent of the angle of internal friction for that material. The angle of repose is the angle with the horizontal at which the material will stand when piled. The size, shape, moisture content and orientation of the particles have a decided influence on the angle of repose. Although engineers assume generally that the angle of repose and the angle of internal friction are approximately the same, Stewart (1968) has shown that at least for one granular material that he tested (sorghum grain), the two angles are different and the use of one in place of the other in design may introduce errors. Lorenzen (1957) also found a difference between the two angles as will be discussed later.

Angle of repose

There are two angles of repose, i.e., static angle of repose and dynamic angle of repose. The static angle of repose is the angle of friction taken up by a granular solid to about slide to upon itself. The dynamic angle of repose is more important than the static angle as it arises in all cases where the bulk of the material is in motion such as the movement of solids discharging from bins and hoppers.

Kramer (1944) studied the angle of repose for rice on rice using a wooden frame full of rice mounted on a tilting top drafting table. The table top was tilted until the rice began to move leaving an inclined surface. The angle of the inclined frame was then measured as the angle of repose for that particular sample (Fig. 10.18). It was found that the angle increased very rapidly when moisture content exceeded 16 to 17 percent.

The effect of moisture content on the angle of repose has been reported by Fowler and Wyatt (1960) and Stewart (1968). The former investigators used an apparatus similar to Fig. 10.19 which consisted of a circular platform immersed in a box filled with the granular solids and with a glass window in one side. The platform was supported by three adjustable screwed legs and was surrounded by a metal funnel leading to a discharge hole. The solid was allowed to escape from the box, leaving a free-standing cone of solids on the platform. A cathetometer (traveling microscope) was

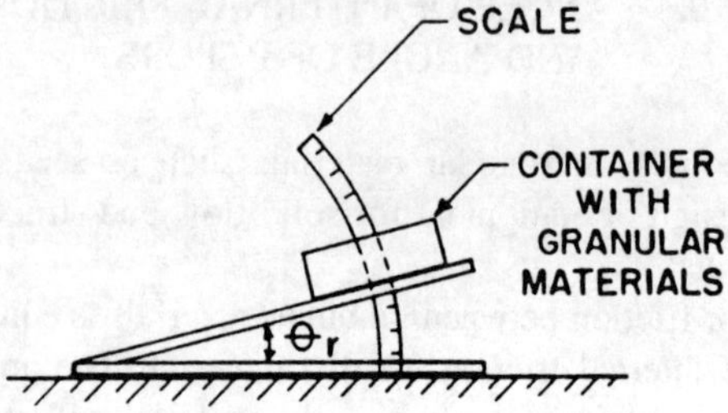

Figure 10.18 Simple device for measuring angle of repose of granular materials

used to measure the heights indicated in Fig. 10.19. The angle of repose, Φ_r, was obtained from the geometry of the cone as follows:

$$\Phi_r = \tan^{-1} \frac{2(H_c - H_p)}{D_p}$$

where H_c, H_p and D_p are height of the cone, height of the platform, and diameter of the platform, respectively.

When moisture content of the granular solids was varied, the experimental values of the angle of repose were highly correlated with the calculated values using the following empirical equation.

$$\tan \Phi_r = an^2 + b\left(\frac{M}{D_{av}}\right) + cs + d \qquad (10.24)$$

where Φ_r = angle of repose

n = shape factor = $\dfrac{\text{specific surface of solid*}}{\text{specific surface of sphere}}$

M = percent moisture content

D_{av} = average screen particle diameter

s = specific gravity

a, b, c, d = constants

*See Chapter 3 for determination of specific surface.

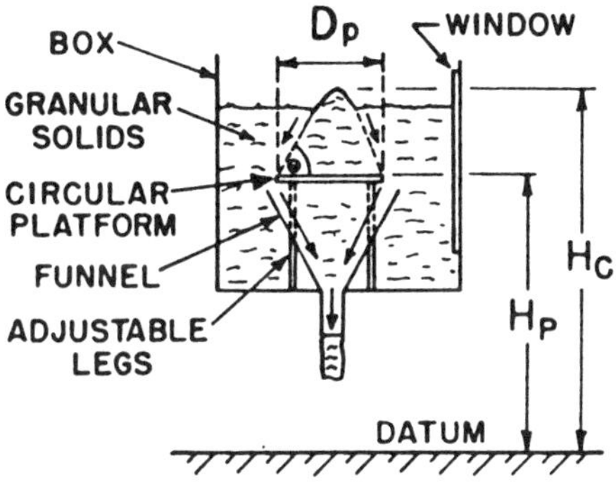

Figure 10.19 Apparatus for measurement of angle of repose—Platform immersed in a container of granular materials (Fowler and Wyatt, 1960)

Using the values of a=0.4621, b=0.0342, c=−0.0898, d=0.0978, in Eq. (10.24) a correlation coefficient of 97 percent was reported for a plot of experimental versus calculated angles of repose for wheat, sand, canary seed and a few other solids. For example, the following data were given for a wheat sample:

$$s = 1.376; \quad D_{av} = 0.3162 \text{ cm}; \quad n = 1.12$$

Using the above data and the given values of the constants in Eq. (10.24), the calculated angles of repose at two different moisture contents will compare with the experimental values of the angles as follows:

Moisture content M(%)	Angle of repose Φ_r (degrees) Calculated	Experimental
0	33.7	33.5
3.46	39.7	40

The result of added moisture, as can be expected, was an increase in the angle of repose. It was suggested that the variation of angle of repose with moisture content is due to the surface layer of moisture that surrounds each particle and that surface tension effects become predominant in holding aggregates of solids together. The increase of the repose angle with increase of moisture content for several granular materials is shown in Fig. 10.20.

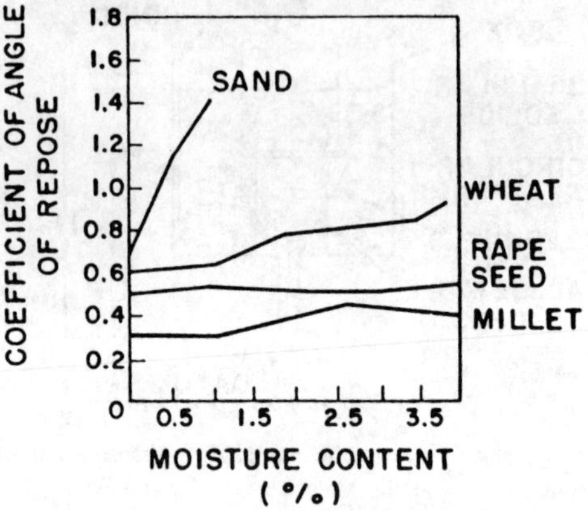

Figure 10.20 Variation of angle of repose with moisture content (Fowler and Wyatt, 1960)

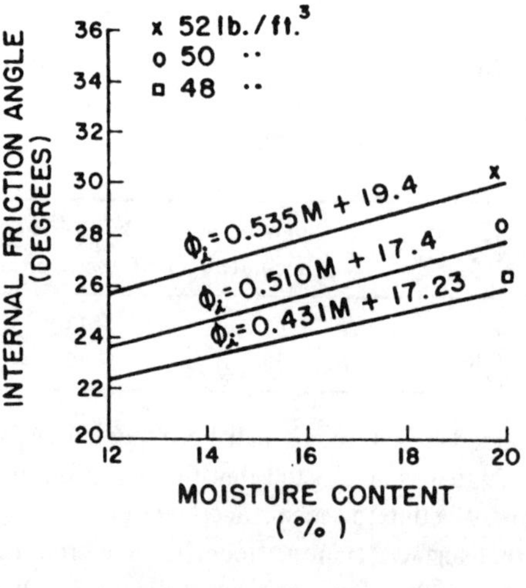

Figure 10.21 Variation of internal friction angle with moisture content and density (Stewart, 1968)

It is interesting to note that in the case of canary seed (millet), which has an elongated ellipsoidal shape with a shape factor of n=1.78, Eq. (10.24) predicts a higher angle of repose while the experimental values showed the angles being less than those of wheat at any moisture content. The authors suggested that since derivation of Eq. (10.24) was by statistical methods based on random ordering, the shape of canary seeds resulted in an "ordered" orientation which invalidated the use of the equation for this material.

Angle of internal friction

In predicting the lateral pressure on a retaining wall in storage bins or design of bins and hoppers for gravity flow the coefficients of friction between granular materials is needed as a design parameter.

For example in design of shallow bins, Rankine equation

$$\sigma_3 = wy \tan^2 (45 - \Phi_i/2) \tag{10.25}$$

is used where σ_3 is lateral pressure against the wall at a point y feet below the top of the wall, w is weight density of the material in lb/ft^3, and Φ_i is the angle of internal friction.

In design of deep bins and other similar storage structures, the pressure ratio k, referred to the ratio of lateral pressure σ_3 to vertical pressure σ_1 at a given point in the material ($k=\sigma_3/\sigma_1$) is needed which can also be found from the angle of internal friction as seen from the following expression

$$k = \frac{1 - \sin \Phi_i}{1 + \sin \Phi_i} \tag{10.26}$$

Knowing the value of k, the horizontal pressure against the wall can be estimated for any given vertical pressure. The vertical pressure causes a column action while the lateral pressure causes a bending action on the wall. In grain bins when the height of material in the bin exceeds a certain limit (2 to 2.5 times the bin diameter), no increase in bottom pressure can be detected with increasing depth of grain. This indicates that the wall must be supporting the additional weight. Furthermore, Ketchum (1919) and others have stated that k is not constant but varies with the type of the material and the geometry of the bin, as well as depth, friction and cohesion properties and moisture content of the material. The influence of these various factors on the pressure ratio is best illustrated by the well-known Janssen's equation given for lateral pressure, σ_3, in deep bins.

$$\sigma_3 = \frac{wR}{f_s}\left(1 - e^{\frac{-kf_sh}{R}}\right) \tag{10.27}$$

where

R = hydraulic radius or the ratio of cross section area to circumference $\left(\frac{\text{ft}^2}{\text{ft}}\right)$
w = weight density of material (lb/ft^3)
f_s = static coefficient of friction of material against wall
h = depth of material (ft)

Lorenzen (1957) gives in the following table the effect of moisture on various components of Janssen's equation calculated on the basis of $\sigma_1=3$ psi for wheat grain.

Pressure ratio k is obtained either directly by pressure measurements in full-sized or model bins or by the use of a triaxial compression chamber and the Mohr's circle. Williams and Ross (1968) determined k for dried citrus pulp by measuring the frictional forces exerted by the pulp on a vertical and horizontal steel blade inserted in a model bin. According to Mohr's theory a material under stress fails along the plane at which a certain combination of normal stress and shear stress occurs. Fig. 10.22 (b) shows the three principal planes—major, intermediate and minor—in

Table 10.3 Effect of moisture content of wheat on various components of Janssen's equation (Lorenzen, 1957)

Moisture M(%)	Pressure ratio k	Static coeff. f_s	Weight density w	Repose angle Φ_r	Angle of internal friction Φ_i
7.3	0.463	0.453	49.3	29.6	23.5
11.0	0.420	0.432	49.3	29.3	24.5
14.1	0.357	0.433	47.2	31.0	26.5
17.1	0.280	0.471	45.4	35.6	27.3
19.3	0.310	0.592	43.9	41.0	23.2

the order of decreasing magnitude of direct stress through a point 0 in the granular mass. On any plane normal to the figure and with an angle of θ with the major principal plane *AO*, here will be present both direct stress σ and shearing stress τ. If the length *AB* is designated by *L*, the forces acting on the sides of the element will be as shown in Fig. 10.22 (c). Equilibrium of these forces in directions normal and parallel to *AB* yields the equations for normal direct stress σ and shear stress τ as follows:

$$\sigma = \sigma_1 \cos^2 \theta + \sigma_3 \sin^2 \theta \tag{10.28}$$

$$\tau = (\sigma_1 - \sigma_3) \sin \theta \cos \theta \tag{10.29}$$

Graphical presentation of stresses at any point in the granular mass can be given by a Mohr circle shown in Fig. 10.23 from which the following relationships can be deducted.

$$\sin \Phi_i = \frac{\sigma_1 - \sigma_3}{\sigma_1 + \sigma_3} \tag{10.30}$$

$$\frac{\sigma_3}{\sigma_1} = \frac{1 - \sin \Phi_i}{1 + \sin \Phi_i} \tag{10.31}$$

The values of σ_1 and σ_3 can be obtained by a triaxial compression test where the minor principal stress σ_3 will be equal to the chamber pressure and the major principal stress σ_1 will be the chamber pressure plus the intensity of axial thrust (see Taylor, 1948).

If the angle of internal friction, θ_i, is assumed to be a constant, the Mohr envelopes shown in Fig. 10.23 will represent the shear strength. If the Mohr circle representing the stress conditions at a given point within the granular material falls below these Mohr envelopes, the shearing stress of the point is smaller than the shearing strength. Under this stress condition, there is no possibility of failure or flow of the granular material. In other words, any Mohr circle within the Mohr envelope represents a stable condition, whereas any circle tangent to the envelope represents an unstable condition at which failure on the plane represented by the point of tangency may occur. Accordingly, *AC′* in Fig. 10.23 shows the orientation of a plane of failure while *AC* shows the orientation of a stable plane.

In establishing the Mohr envelope at least two sets of data are needed from which semicircles given in Fig. 10.23 can be described and the tangent line established. From this graphical method shear strength of the granular

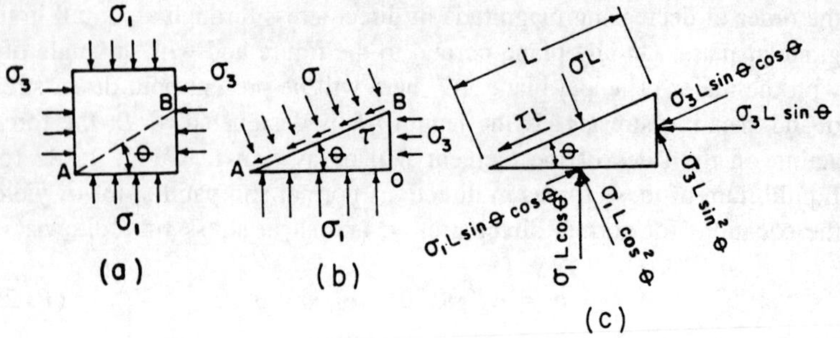

Figure 10.22 (a, b) Stresses acting on a cubical unit in a granular mass and a shear plane ab from this unit; (c) equilibrium of forces in directions normal and parallel to ab giving Eqs. (6.29) and (6.30)

solid as well as its angle of internal friction can be determined. These two parameters may also be calculated directly from the equation of Mohr's circle. From Fig. 10.23

$$\csc \Phi_i = \frac{C' + \sigma_3 + \dfrac{\sigma_1 - \sigma_3}{2}}{\dfrac{\sigma_1 - \sigma_3}{2}}$$

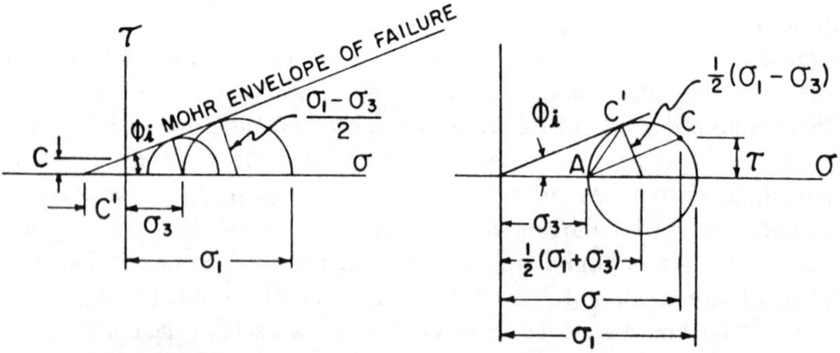

Figure 10.23 Mohr circle and mohr envelope of failure

After simplification, the above expression yields the equation of Mohr's circle as follows:

$$\sigma_3\left(\frac{\sigma_1}{\sigma_3} - 1\right)\csc \Phi_i - \sigma_3\left(\frac{\sigma_1}{\sigma_3} + 1\right) - 2C' = 0 \qquad (10.32)$$

Having two sets of data, Eq. (10.32) can be written twice and solved simultaneously to obtain Φ_i and C'. From these two parameters the value of C in Fig. 10.23 can be determined

$$C = C' \tan \Phi_i$$

The parameter C is usually taken as cohesion of the granular material. However, some authorities consider C as the experimental error and not a property of the material.

If $\sigma_3/\sigma_1=k$ is substituted in Eq. (10.32) and the granular material is assumed cohesionless, i.e., $C=0$, Eq. (10.26) would result.

Lorenzen (1957) attempted to relate internal friction angle, Φ_i, and repose angle, Φ_r, with the hope that a simple test of repose angle determination would yield the value of Φ_i from which k could be determined. Results showed that the values of the two angles run almost parallel to each other for various moisture contents (Fig. 10.24) but no simple relationship existed whereby Φ_i could be estimated from Φ_r with reasonable accuracy.

Because of this difficulty of predicting the angle of internal friction from the angle of repose, other investigators have obtained these data using the more elaborate technique of triaxial tests. Stewart (1968) applied a triaxial compression test, which was developed for studying the shear properties of soils, to the study of sorghum grain and the effect of density as well as moisture content on internal angle of friction. Lorenzen (1957), who investigated the effects of the same factors on internal friction angle of wheat, rice, corn, milo, barley, and sawdust, also employed a triaxial compression apparatus.

To adopt a triaxial soil testing apparatus to shear test of grains, Stewart prepared specimens in the form of grain cylinders four inches in diameter enclosed in a rubber membrane. After forming the grain cylinders, the membrane was sealed using rubber o-rings to clamp the membrane to the loading caps. An internal vacuum maintained the cylinder shape after the forming jacket was removed. The grain cylinder was then placed inside the test chamber where a pressure up to a predetermined σ_3 was built up gradually as the vacuum was released. Once the grain cylinder was positioned in the test chamber, the standard procedure used in triaxial compression test of soils was followed to obtain data on σ_1 and σ_3 for construction of Mohr circles and Mohr failure envelope as discussed previously. In this procedure, the axial load is applied through the loading piston either by a universal testing machine or other loading devices and

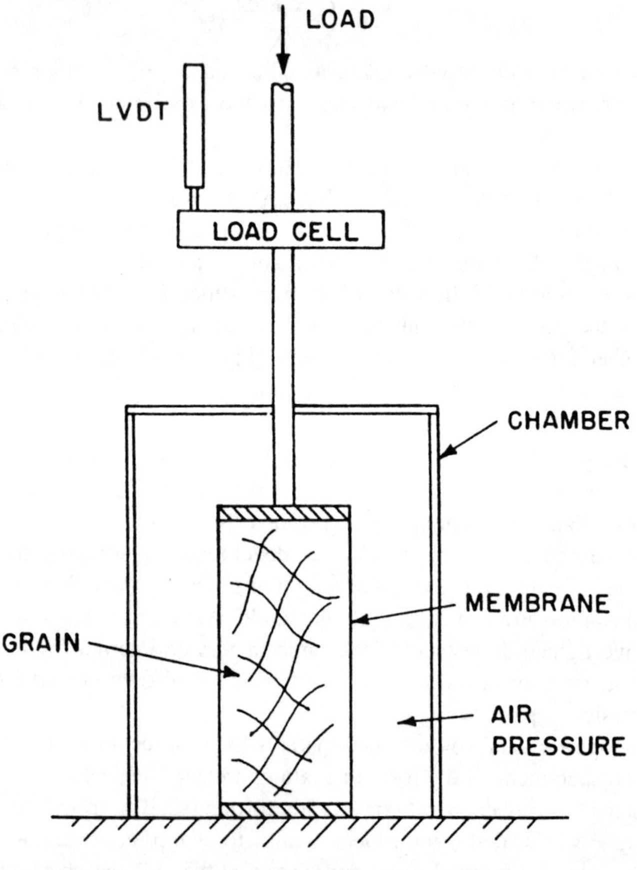

Figure 10.23a Schematic for application of triaxial compression test to agricultural grains

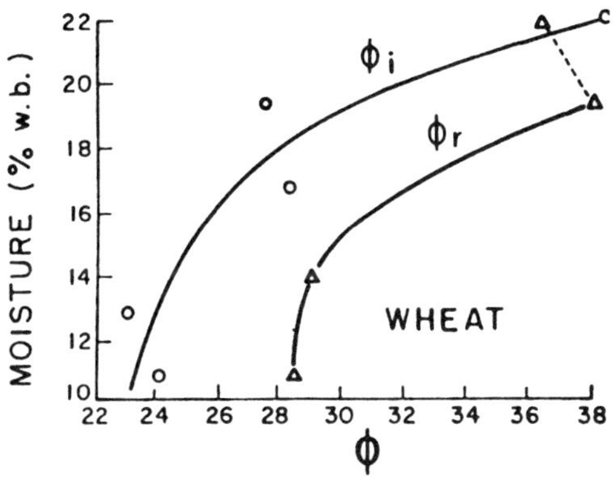

Figure 10.24 Variation of angle of repose and angle of internal friction with moisture content (Lorenzen, 1957)

a stress-strain diagram is obtained for evaluation of axial pressure σ_1. If the stress-strain curve levels off after a certain load, σ_1 is taken as the point of maximum stress on the stress-strain diagram. For those materials which do not develop a peak on the curve, σ_1 is measured for some limiting strain.

Stewart in his triaxial compression study of sorghum grain constructed the Mohr circles from the stress condition at a total deformation of one inch. A typical stress-strain diagram for a minor stress of 5 psi and a series of Mohr circles obtained from three such stress-strain diagrams at chamber pressures of 2.5, 5.0 and 7.5 psi are shown in Fig. 10.25. Note the presence of an apparent cohesion which could be attributed to moisture in the grain.

Following this procedure, the internal angle of friction was determined for grain at various moisture contents with various densities (Fig. 10.21). The correlation coefficients for the regression equations in the order of decreasing density were 0.84, 0.87 and 0.81, respectively.

When the values of internal angles of friction found by the above method were compared with the angle of repose determined by means of a simple repose apparatus, the angles of repose were consistently higher for grains of approximately the same moisture content and density. The values reported were 22.5 to 35 degrees for internal friction compared to 32 to 38 degrees for angles of repose.

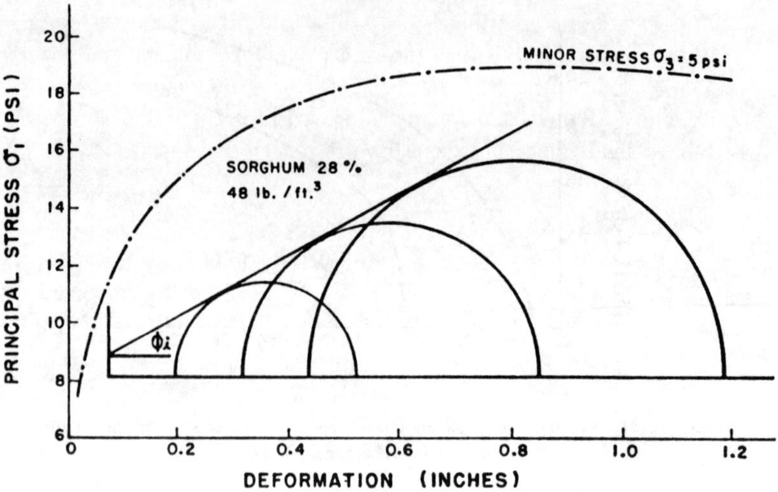

Figure 10.25 Typical stress-strain diagram and the resulting mohr circles from such diagrams in triaxial compression test of sorghum grain (Stewart, 1964)

10.5 FLOW OF BULK GRANULAR MATERIALS

Many of the concepts discussed in the previous section have applications in problems of flow of granular materials encountered in the design of gravity and forced flow equipment. For example, an important flow property of solids known as "critical flow factor" for conical hoppers is dependent on the angle of friction between the solid and the wall of the hopper, the hopper slope, the outlet diameter, the internal angle of friction of the solid, as well as the solid's size distribution and moisture content.

Gravity flow in bins and hoppers

Bins and hoppers constitute major items of equipment in the handling of granular materials. Properly designed bins are to hold a known volume of bulk solid and feed it at a prescribed rate at the required time. An improperly designed bin may cause obstruction to flow, erratic flow, development of dead zones resulting in degradation of the solid, segregation and several other problems. One of the most recent works in the area of gravity flow of bulk solids is the contribution by Jenike and his associates (Jenike *et al.*, 1960, Johanson and Colijn, 1964). In this work a fundamental theory, based on plasticity principles, has been suggested for analysis of the flow of bulk solids. Methods and apparatus for measurement of those properties which have significance in design have also been developed.

Shear apparatus for determination of flow properties

A direct shear test apparatus was developed by Jenike (1961, 1964) for study of bulk solids. Figure 10.26 shows the three basic components of the machine, namely, a shear cell, a controlled loading device and a recorder. The shear cell is composed of a base, a shear ring, and a cover with a loading bracket. The normal load is applied by weights acting vertically. The shearing action is provided by means of an electrical or mechanical drive with a load cell or dynamometer in line for force measurement. The shearing force acts in the plane of contact between the base and the ring. The shear cell shown in Fig. 10.26 is supposed to assure a uniform stress distribution across the specimen and shear of solid in the plane between the ring and base.

In conducting a test, a uniform and representative sample of the solid with known moisture content and size distribution is packed into the shear cell using the mold ring and the twisting top illustrated in Fig. 10.27. The shearing ring is offset approximately 0.10 inches. A pre-selected consolidating load W_c is placed on top of the twisting cap and by means of a special wrench a number of oscillating twists are applied to the cover. After this so-called pre-consolidation step, the twisting top and mold ring are removed, the excess material scraped off level with the top of the shear ring, and the cover is placed in position. The next step is consolidation which involves a shearing operation causing the material to flow under consolidating stresses until a steady state is approached. With the consolidating load W_c in place and the shearing ring offset, a horizontal shearing is applied on the loading pin until a constant shearing force is reached. The loading pin is then retracted to release the shearing force.

Knowing the load W_c and the corresponding shearing force S_c, a point can be established in a shear force versus normal load diagram. To obtain other points sufficient for drawing of a smooth curve, the consolidation and actual shear tests must be repeated for several loads less than the limiting load W_c. The resulting curve was named by Jenike (1964) the "yield locus" denoted by *YL*. For each actual shear test, with the load $W<W_c$ in place, the loading pin is driven forward until the shearing force

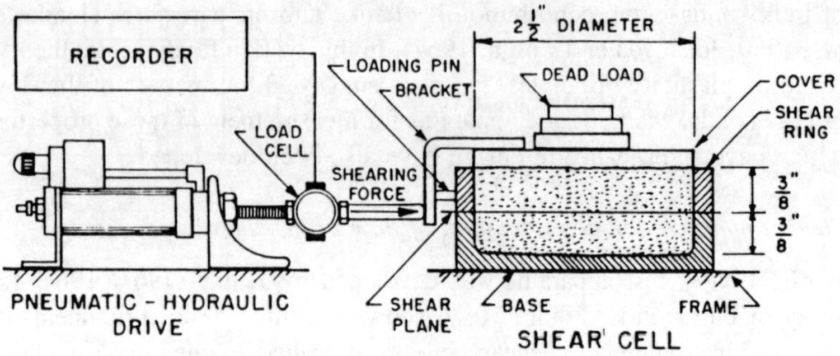

Figure 10.26 Shear apparatus for determination of flow properties of granular materials

reaches a maximum and then drops off indicating the actual shearing of the specimen. The purpose of the consolidation step is to assure that a uniformly packed specimen at approximately the same stress condition exists prior to each actual shear test.

Yield locus, time yield locus, and effective yield locus

Suppose flow properties of a granular material are studied using the shear apparatus given in Fig. (10.26). The consolidating load prior to each shear test is W_c=8.9 lb. The recorded chart given in Fig. (10.28) shows the sequence of three consolidation and actual shear tests. The three sets of data on shearing and one set of data on consolidation and shearing are plotted on a shearing force versus normal load diagram (Fig. 10.28). A

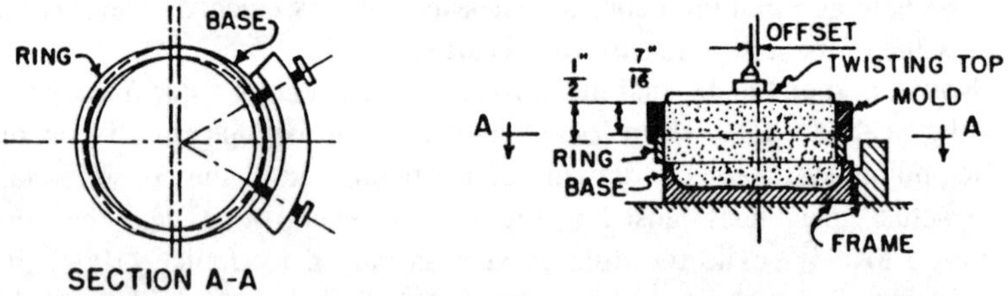

Figure 10.27 Packing mold for preparation of a uniform specimen (Jenike, 1960)

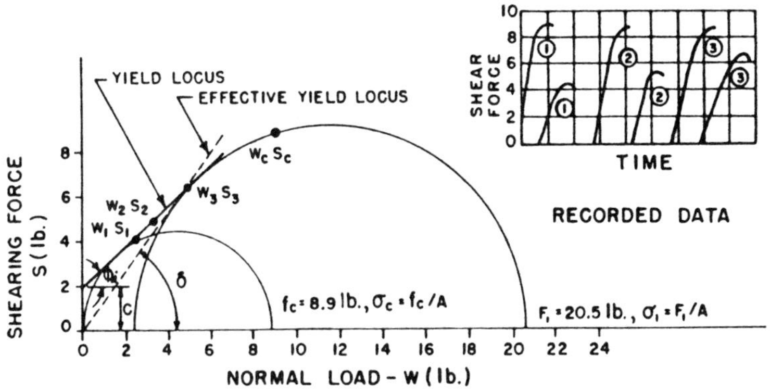

Figure 10.28 Plot of shear force versus normal load showing flow properties of a granular material (Data from Jenike, 1964)

smooth curve is drawn through the three points corresponding to the shearing tests. The resulting curve, which is supposed to be slightly concave for granular materials, is the *"yield locus," YL*. The major principal stress σ_1 can be obtained by dividing the major principal factor F_1 by the shearing area A. The major principal force F_1 is read at the point where a Mohr semi-circle tangential to the yield locus and passing through the point (W_c, S_c) intercepts the horizontal axis. Drawing another Mohr semi-circle through the origin and tangential to the yield locus will define the unconfined yield pressure $\sigma_c = f_c/A$ as shown in Fig. (10.28). Both σ_1 and σ_c are important parameters in design problems. They represent one coordinate point of the flow function curves shown in Fig. (10.30) for different solids. The major consolidating principal pressure σ_1 corresponds to the pressure applied to the solid if it were compressed in a cylinder with frictionless rigid wall. The unconfined yield pressure σ_c is a measure of the solid's strength at a free surface. It corresponds to the pressure necessary to cause the unconfined consolidated cylinder of solid (taken out of the rigid cylinder) to collapse.

The angle of the internal friction of the solid, Φ_i, is the slope of the yield locus. This is a kinetic angle of internal friction and varies from point to point due to the curvature of the yield locus.

The yield locus defines the stress conditions (shear stress and principal stress) for which flow occurs. The bulk solid is stable for any stress condition represented by a Mohr semi-circle not touching the yield locus.

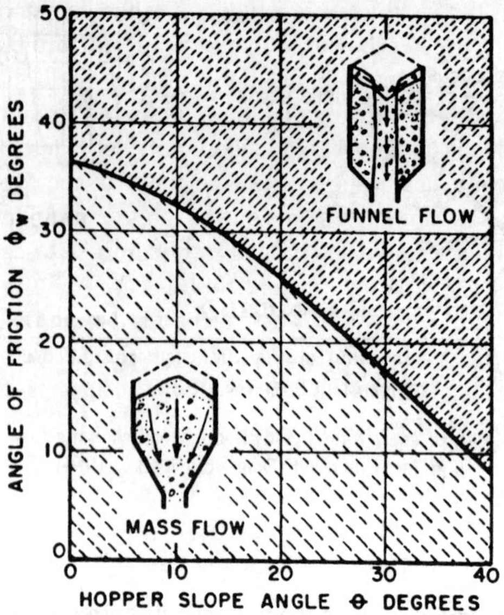

Figure 10.29 Types of flow as affected by hopper slope angle and friction angle of solid in hopper wall (Courtesy Jenike and Johanson, Inc.)

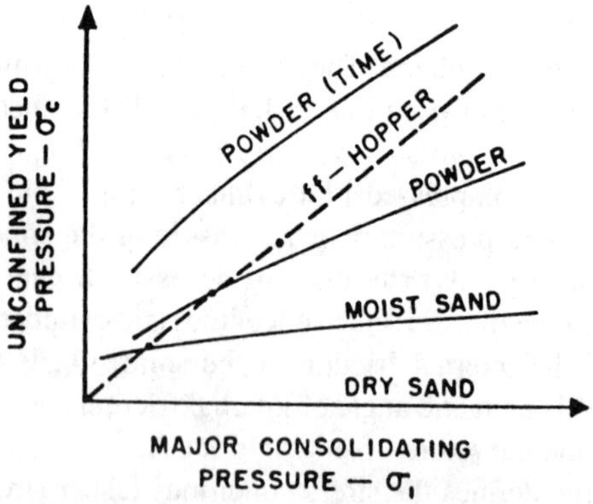

Figure 10.30 Critical flow factor ff of a hopper superimposed on flow functions of several solids (Courtesy Jenike and Johanson, Inc.)

The position of the yield locus is a function of the degree of consolidation of the solid. The greater the consolidation load, the further the yield locus expands. For a loose, unconsolidated material the yield locus may reduce to a point at the origin showing that the solid has no strength. On the other hand the yield locus does not extend indefinitely with increasing consolidating load but terminates at the point of tangency of the Mohr semi-circle.

Reduction in flowability of solids due to consolidation during the storage period in the hopper can be determined by obtaining a so-called "*time yield locus*," TYL, for any particular solid. In this case, the same procedure as for getting a yield locus is followed except that before any shearing takes place, the sample is left under static loads for a prescribed period of time. After this time, the samples are sheared in the usual manner.

Repeated tests on a solid with various consolidating loads will result in a range of yield locus curves with each having a definite Mohr semi-circle. Jenike (1964) referred to the envelope of these Mohr semi-circles as the "*effective yield locus*" denoted by EYL. Experimental data with granular material have shown that a straight line passing through the origin is approximately tangential to these Mohr semi-circles and thus defines the effective yield locus for that material. Note that this effective yield locus and the Mohr envelope of failure shown in Fig. (10.23) are the same for a cohesionless material where $C=0$. Accordingly, the slope of the effective yield locus must be the tangent of the angle of internal friction for materials with zero cohesion. This angle is denoted by δ in Fig. 10.28. As the first approximation for the angle δ, Jenike (1960) suggested drawing a straight line through the origin and tangential to the Mohr semi-circle defining major and minor principal stresses in the solid (Fig. 10.28). Angle δ is a constant for a particular granular material. Therefore, if we refer to δ as the *effective angle of internal friction*, the shear apparatus described above should be a convenient method to determine this angle. Note that the ratio of lateral pressure (minor principal stress) to vertical pressure (major principal stress), given as k in previously stated Eq. (10.26), can be calculated using δ in place of Φ_i. Therefore,

$$k = \frac{1 - \sin\delta}{1 + \sin\delta} \tag{10.26}$$

In cases where the coefficient of friction, yield locus and other flow properties of the granular material along the bin or hopper wall are needed, in place of the base of the shear cell in Fig. 10.26, a sample of the wall material is used as the friction surface. The yield locus obtained in this manner is referred to as wall yield locus. The slope of the wall yield locus is given by the kinetic angle of friction, Φ_w, between the material and the wall. The relationship between the angle of friction Φ_w of a stored solid against the wall of a conical hopper and the half included angle of the cone for two types of flow is shown in Fig. 10.29. According to this figure, if the cone is sufficiently steep and the surface coefficient is small, the channel expands from the outlet upward along the walls of the bin and all the solid is in motion. This type of flow is known as *mass-flow*. Since in mass-flow the flowing channel of the granular material coincides with the walls of the hopper, the shape and frictional effects of the walls have a great influence on flow. In an ideal mass-flow there is no dead zone and all the solid in the hopper is in motion whenever any of it is drawn out of the outlet. In contrast to this situation is *funnel-flow* in which the solid flows toward the outlet in a channel formed within the solid itself. The solid surrounding this channel is at rest (dead zone) and has a tendency to spoil, cake or oxidize if it is not emptied out. This situation develops if the hopper slope angle is large or the hopper wall is not sufficiently smooth.

Flow function and flow factor

During flow of a granular material from a hopper, as an element of the solid flows down, it becomes consolidated under the pressure prevailing in the hopper. To each value of consolidating pressure there corresponds a strength of the solid referred to as unconfined yield pressure of unconfined yield strength. Some of the problems associated with gravity flow of granular materials from a bin or hopper are erratic feeding, flooding, sticking to the wall, and the formation of obstructions such as "piping" and "arching." When the granular material has sufficient strength to support its own weight it may form an arch across the outlet of a hopper and thus cause an obstruction to the flow. The strength of the solid depends on its moisture content, surface roughness, and degree of consolidation. Greater consolidation results in greater strength.

The relationship between the unconfined yield pressure of a solid and the consolidation pressure is called "*flow function*" $\sigma_c = f(\sigma_1)$ denoted by $FF = d\sigma_1/d\sigma_c$. To obtain the flow function curve of a solid, shear tests are conducted for several consolidating loads and the resulting σ_1 and σ_c are plotted as shown in Fig. 10.30. Jenike (1964) showed that for a given hopper there exists a critical line such that as long as the flow-function curve of the solid lies below this line, the strength of the solid is insufficient to support an arch and there would be no obstruction to flow. This critical line was referred to as the "*critical flow factor*" of the hopper. This factor, denoted by *ff*, is the ratio of the critical major consolidating pressure σ_1 to the critical unconfined yield pressure σ_c determined experimentally.

$$ff = (\sigma_1/\sigma_c \text{ critical} = (F_1/F_c) \text{ critical}$$

This is the condition at which the bulk material is just on the point of forming an obstruction to flow. In other words, in a free-flowing solid

$$\sigma_c = f(\sigma_1) < \sigma_{c\,\text{critical}} = (1/ff)\,\sigma_1$$

or

$$\sigma_1/\sigma_c > ff$$

While flow function *FF* depends on the material only, the flow factor *ff* depends on both the material and the hopper geometry, surface wall characteristics, etc. If a solid with certain flow properties, represented by its *FF* curve, is placed in a hopper with a certain critical flow factor *ff*, represented by a straight line on the σ_c versus σ_1 coordinates, then the critical values of σ_c and σ_1 are given by the intersection of the curve and the line. For different size hopper outlets, this point of intersection will determine the minimum size of the outlet required to avoid arching or piping. Critical flow factors for arching and for piping for a number of outlet channel configurations have been reported by Jenike (1964).

Jenike (1964) also classified solids according to their limiting flow function, given by letters *FF*, as follows

$$FF < 2 \text{ very cohesive and non-flowing}$$

$$4 > FF > 2 \text{ cohesive}$$

$$10 > FF > 4 \text{ easy-flowing}$$

$$FF > 10 \text{ free-flowing}$$

Figure 10.30 shows the flow functions for several solids and the critical flow factor of a hopper. The flow function of a dry sand coincides with the consolidating pressure axis because the unconfined yield pressure for dry sand is zero and it cannot be consolidated. However, when the moisture is added, the sand gains strength. The effect of time consolidation is also shown in the case of a powder which flow freely but is capable of arching and doming if stored at rest for 24 hours to give the time flow-function.

Example

A shear test, using the shear apparatus of Fig. 10.25 was used to obtain the following data on pelleted feed. The cell used was 5 inches in diameter and the consolidation load was 14 lbs., sheared at 17 lbs.

Normal load W (lb)	Shearing force S (lb)
11.0	15.0
7.5	11.0
6.0	9.5
4.5	7.0
3.0	5.5
1.5	4.0

Using the Mohr's semi circle method of analysis outlined in the text, find the following information:

1. Yield Locus of the materials.
2. Φ_i = the internal angle of friction.
3. F_1 = major principal force.
4. σ_1 = major principal stress.
5. f_c = unconfined yield force.
6. σ_c = unconfined yield pressure.
7. FF = Flow Function and classification of the solid.
8. C = Cohesion of the material.

From the graph of shearing force vs. normal load, shown in Fig. 10.30a:

$$\Phi_i = 50^\circ$$
$$C = 2\text{ lb}$$
$$f_c = 11\text{ lb}$$
$$F_1 = 44.75\text{ lb.}$$

$$A_{\text{cell}} = \frac{\pi D^2}{4} = \frac{\pi(5)^2}{4} = 19.6\text{ in}^2$$

$$\sigma_1 = \frac{F_1}{A_{\text{cell}}} = 44.75/19.6 = 2.28\text{ psi}$$

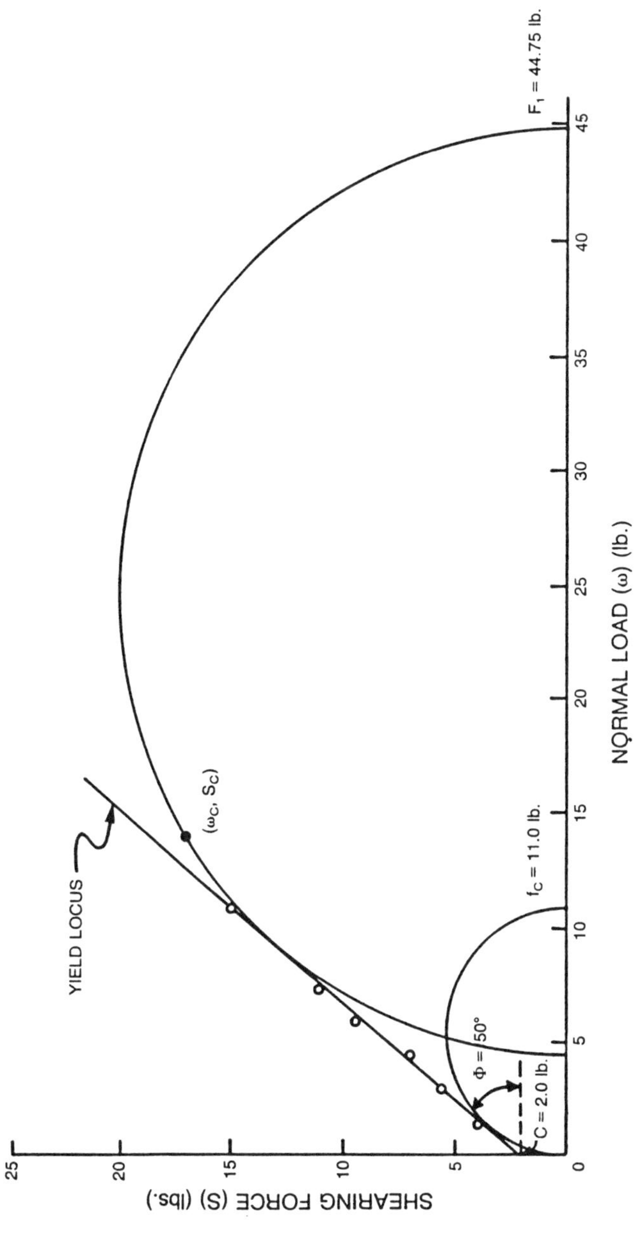

Figure 10.30a Sheating force vs. normal load (Feed Pellets)

$$\sigma_c = \frac{F_c}{A_{\text{cell}}} = 11.0/19.6 = 0.56 \text{ psi}$$

$$\text{FF} = \sigma_1/\sigma_c = 2.28/0.56 = 4.07$$

The material is cohesive.

Example

In a triaxial test of wheat grain at 7% moisture content, the following data were obtained.

Displacement (in)	Load (lb)	Air Pressure (psi)
0.0123	363	12
0.0216	643	22
0.0285	853	32
0.0320	953	42

Assuming a 4-inch cell for the test specimen, use the graphical method of the Mohr's semicircles and determine:

1. Angle of internal friction.
2. Cohesion
3. Pressure ratio σ_3/σ_1

$$\text{Axial pressure} = P = \frac{F}{A}$$

$$A = \frac{\pi}{4}d^2 = \frac{\pi}{4}(4)^2 = 12.56 \text{ in.}^2$$

The values of axial pressure are tabulated as the following

σ_3 (air pressure) (psi)	Displacement (in)	Load (lb)	P axial (psi)	$\sigma_1 = \sigma_3 + P$ axial (psi)
12	0.0123	363	28.9	40.9
22	0.0216	643	51.2	73.2
32	0.0285	853	67.8	99.8
42	0.0320	953	75.8	117.8

σ_3 = minor principal stress
σ_1 = major principal stress

Plotting the data yields: Fig. 10.30b from which the following properties of wheat grain are determined

1. Angle of internal friction = 25°
2. Cohension $C = 4.5$ psi
3. Pressure ratio $K = \left(\frac{\sigma_3}{\sigma_1}\right)_{\text{average}} = 0.325$

Critical dimensions of hopper openings

To determine the critical dimension of the hopper openings, failure conditions must be established for two basic obstructions; namely, "arching" where no flow may take place, and "piping" where flow may be reduced or limited. Figure 10.31 shows the free-body diagram of a granular material with bulk density, w, forming an arch with a uniform thickness, T. Let B denote the diameter of a circular hole or the width of a slot with length L. For small arcs, the equilibrium of forces, resulting from the weight of the mass acting downward and the vertical component of force due to compressive pressure P in the arch acting upward, yields:

$$wBLT = 2PLT \cos \alpha \sin \alpha$$

or

$$B = (P/w) \sin 2\alpha \text{ (for slot)}$$

and,

$$w(\pi/4)\, B^2T = P\pi BT \cos \alpha \sin \alpha$$

or

$$B = (2P/w) \sin 2\alpha \text{(for circle)}$$

Giving the above analysis, Johanson and Colijn (1964) suggested that in order for failure to occur, the major compressive pressure P should be equal to the unconfined yield strength σ_c. Substituting σ_c for P in the above expressions, and assuming $\sin 2\alpha - 1$, which considers the strongest possible arch that may form, the critical opening dimension B becomes

$$B \geqq \sigma_c/w \text{ (for slot opening)}$$

$$B \geqq 2\sigma_c/w \text{ (for circular opening)}$$

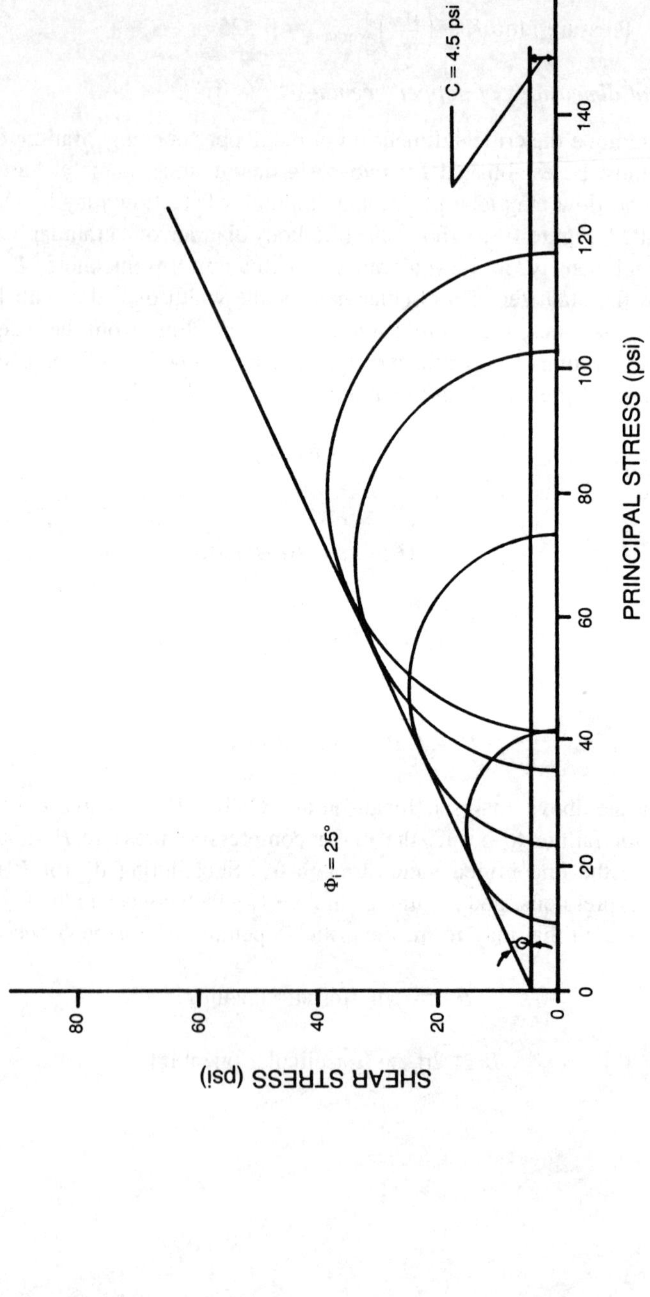

Figure 10.30b Mohr's semicircles for triaxial test of wheat at 7% M.C.

The critical opening dimension to prevent "piping" has been derived from the stability analysis of vertical pipes and is given in terms of σ_c, w and angle of internal friction Φ (Jenike, 1961).

Example

Using the data provided in Fig. 10.32 and 10.33, calculate the critical width B for arching of the slot opening of a wedge shaped, mild steel hopper with $\theta=30$ degrees. Figure 10.33 shows that for a mild steel hopper with the wall friction angle $\Phi_w=35°$, the maximum effective angle of friction $\delta=55°$. Figure 10.32 shows that for $\delta=55°$ and $\Phi_w=35°$, the critical flow factor $ff=1.25$. Superimposing $ff=1.25$ on Fig. 10.33, shows no intersection with the instantaneous flow function curve, indicating that no arching problems would develop if the material were not to wait in the hopper for any length of time. However, the intersection with the 24-hour flow function curve gives $\sigma_1=65$ lb/ft^2, $\sigma_c=50$ lb/ft^2, $w=90$ lb/ft^3 and $\delta=55$ degrees. Using these data in Eq. 10.33 yields

$$B \geqq 50/90 \geqq 0.6 \text{ ft}$$

Therefore, the critical slot width for arching in the hopper is about seven inches.

Base pressure in circular hoppers

A small opening at the bottom of a circular hopper and insufficient static pressure at that point often result in arching over the opening and unsatisfactory discharge rate. In a straight-sided hopper with a flat bottom, the

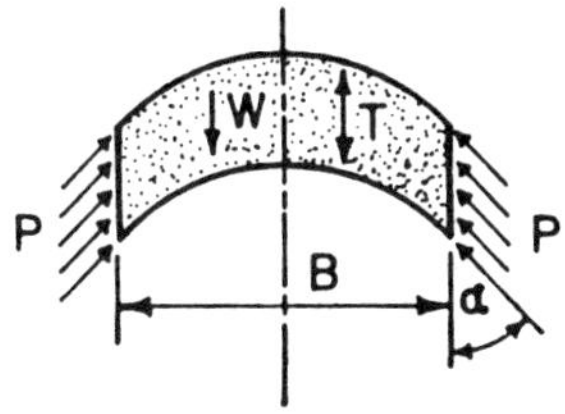

Figure 10.31 Free-body diagram of a mass of granular material forming an arch

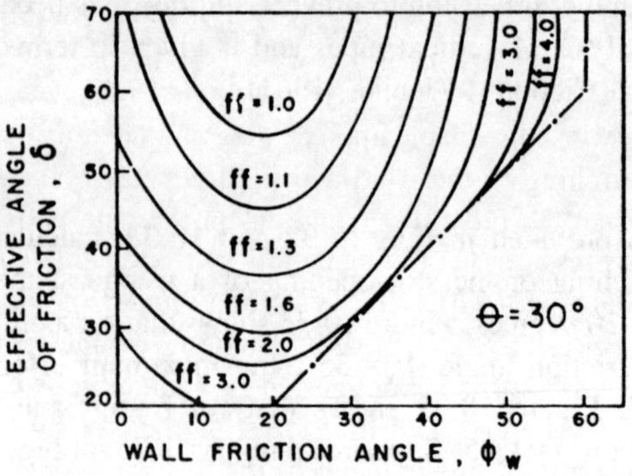

Figure 10.32 The critical flow for mass-flow hoppers with w=30° (Johanson and Colijn, 1964)

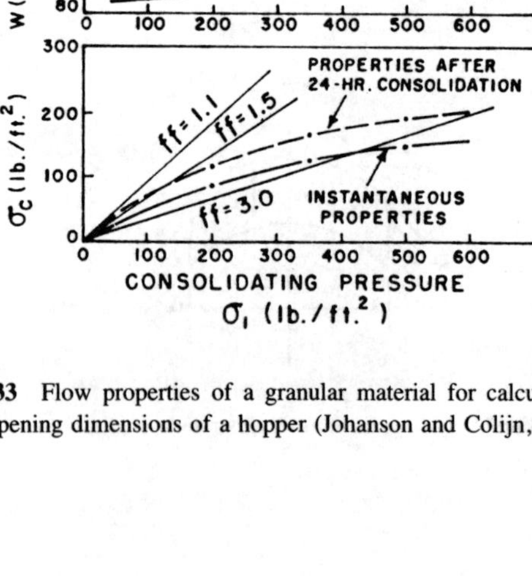

Figure 10.33 Flow properties of a granular material for calculation of critical opening dimensions of a hopper (Johanson and Colijn, 1964)

rate of flow varies as the cube of the bottom opening diameter (Ketchum, 1919) and the base pressure, estimated by Eq. (10.27), is little affected by the motion of the solid. In a slope-sided hopper, however, the change in base pressure may be appreciable upon closing and opening of the discharge gate. To avoid arching of the solid across the opening, the base pressure must be larger than the pressure change when the gate is opened.

For predicting pressures at varying depths in a slope-sided hopper, Smith (1955) applied Janssen's Eq. (10.27) to the cylinder of supported solids and made allowance for the increased frictional support of the sloping walls. Using this analysis he has suggested two equations, one for an angle of inclination of the hopper wall $\theta=5°$ and another for $\theta=20°$. He found that with coarse solids, if the diameter of the discharge opening is at least eight to ten times the diameter of the largest particle, a base pressure of 10 to 20 lb/ft^2 will provide satisfactory flow. With fine solids, with a pressure ratio, K, of 0.1 to 0.2, a minimum value of 50 lb/ft^2 base pressure was recommended.

Gravity flow through orifices

A knowledge of the laws governing flow of solids through orifices would be of value in design of such field machines as seed drills and fertilizer distributors as well as installations such as storage bins, automatic filling machines, etc.

The law of hydrodynamics do not apply to the flow of solid granular materials through orifices for the following reasons:

1. Pressure is not distributed equally in all directions due to the development of arches and to frictional forces between the granules.
2. The rate of flow is not proportional to the head, except at heads smaller than the container diameter.
3. No provision is made in hydrodynamics for size and shape of particles, which greatly influence the flow rate.

Because of these factors a number of researchers have attempted to derive mathematical expressions governing the flow of solid particles. It is generally agreed that the flow rate through orifices is a function of the geometry of the container and the orifice as well as certain physical properties of the granular material. The effect of head, container shape and size, orifice shape, diameter and location on the container, and head to orifice diameter ratio are some of the equipment factors which have been studied. Shape, size, surface roughness, density, porosity, specific surface, angle of repose, angle of internal friction, and moisture content are some of the solid's physical properties investigated.

A review of the published work shows that despite considerable effort in this area, there is still a need for a reliable general formula for predicting the flow rate of granular solids of different physical characteristics. Some reserachers have considered several of the possible physical factors which may influence the flow rate and through dimensional analysis and experimental data have developed formulas for estimating flow rates of granular materials of given characteristics. Others, admitting that certain physical parameters such as frictional effects are difficult to evaluate experimentally, resorted to experimentation only and developed empirical relationships for estimating the rate of flow. Among the first category is the work of Deming and Mehring (1929), Franklin and Johanson (1955), Brown and Richards (1959) and Fowler and Glastonbury (1959). In the second category, the work of Ewalt and Buelow (1963) and Beverloo *et al.*, (1961) are to be mentioned.

Deming and Mehring (1929) applied dimensional analysis to the flow of solid particles through funnels and found that when the coefficient of internal friction (assumed equal to angle of repose Φ_r) and the cone included angle, θ', were kept constant, the flow rate could be defined by the following function.

$$tB^{2.5}w = f(d/B)$$

where t is the time in minutes for flow of 100 grams of the solid, B is diameter of the orifice in millimeters, w is bulk density in grams per cubic centimeter, and d is the average diameter of the assumed spherical particle in millimeters. Experimental data using various materials gave the form of the function as follows

$$Q = \frac{100B^{2.5}w}{\tan\Phi_r[(34.6 + (67.4 + 444\sin {}^1/_2\theta')(d/B + 0.130 - 0.161\tan\Phi_r)]} \tag{10.33}$$

where Q is the flow rate in grams per minute and $\tan\Phi_r$ is the tangent of the angle of repose measured between the sloping side of the material and its base when poured into a heap upon a level surface. In the case of nonspherical particles, in the above expression d was calculated from the average major diameter d_2 and the average minor diameter d_1 of the particles using the following derived equation.

$$d = 0.8(d_2^5 - d_1^5)/(d_2^4 - d_1^4)$$

However, when $d_2/d_1 < 2$, it was found that the arithmetic average of d_1 and d_2 was sufficient to obtain good agreement between calculated and experimental flow rates.

Based on their experimental data presented in Table 10.4, Deming and Mehring concluded that Eq. (10.33) was valid for estimating the rate of flow of solid particles through funnels with included angles between 20 and 110 degrees. Apparently the equation is valid for particles of any density, shape, and size provided that the size is not small enough to introduce cohesion as a factor.

Franklin and Johanson (1955) developed the following empirical equation for flow rate of granular materials through horizontal orifices

$$Q = \frac{w_s B^{2.93}}{(6.29 \tan \Phi_i + 23.16)(d + 1.89) - 44.9} \qquad (10.34)$$

where Q is flow rate in pounds per minute, B is diameter of orifice in inches, d is the average diameter of particles in inches, w_s is solid density of the particles in pounds per cubic foot, Φ_i is the internal angle of kinetic friction. To modify the above expression for inclined orifices, the following correction formula was proposed.

$$Q_\alpha = Q \frac{\cos \Phi_r + \cos \alpha}{\cos \Phi_r + 1} \qquad (10.35)$$

where Q_α is the corrected flow for the orifice inclined at an angle of α with the horizontal, and Φ_r is the kinetic angle of repose.

For cone-bottomed containers terminating in a horizontal orifice, Franklin and Johanson (1955) suggested the use of the Eq. (10.33) up to cone angles of $(180\text{-}2\Phi_r)$. For cone angles greater than this, Eq. (10.34) was suggested to be satisfactory. The experimental data reported by Franklin and Johanson contained particles with diameters ranging from 0.03 to 0.21 inches, orifice diameters from 0.24 to 2.28 inches, and bulk density from 7.3 to 676 pounds per cubic foot. The experimental and calculated data were in agreement within ± 7 per cent. They suggested that the degree of accuracy will be reduced for orifice to particle diameters ratio less than 5, column diameter to orifice diameter less than 6.3 and bed height less than one column diameter.

Table 10.4 Observed and predicted flow rates of granular materials through horizontal circular orifices in containers (Sources of data are the same as those of the predicting equations)

Material	Orifice diam. B mm	Particle diam. d mm	Bulk density w g/cc	Solid density w_s g/cc	Repose angle Φ_r deg	Wall angle from vertical θ (deg)	Flow rate Q g/min obsd.	Flow rate Q g/min calcd.	Predicting equation
Lead shot #10	10.1	1.78	6.55		19.5	15	7200	7350	10.33
						30	5560	5520	10.33
Lead shot #10	10.70	1.25		10.8	20.2	0	4990	5450	10.34
Glass beads	10.1	3.54	1.57		27.4	15	1295	1315	10.33
						30	961	961	10.33
Glass beads	10.7	0.788		2.77	25	0	271	220	10.34
Marbles	73	13.5	1.32		27	30	129000	137000	10.33
Crushed rock	34.3	4.01		3.3	37	0	17400	18600	10.34
Sand	50.7	0.92	1.6			0	80,000	87500	10.36
Phosphate rock	5	0.43	1.25		34.6	30	147	154	10.33
						45	133	135	10.33
Urea crystals	10.1	0.66	0.44		45	15	385	333	10.33
						30	263	313	10.33
Puffed Rice	33.7	5.1		0.12	31.5	0	750	690	10.34
Sugar	25	0.9	0.83			0	8050	8415	10.37
Vetch seed	16	3.4	0.82		22.5	45	1400	1450	10.33
Kale seed	10.1	1.7	0.69		22.5	15	675	690	10.33
Mustard seed	10.1	2.2	0.75		25.1	15	621	699	10.33
Green peas	40	7.0	0.84			0	14340	12590	10.37
Soybeans	40	6.6	0.75			0	13690	13540	10.37
Lupin	40	6.5	0.70			0	12920	14220	10.37
Wheat	25.1	5.0	0.8			0	3830	3820	10.37
Wheat	25.3	4.1	0.86			0	4440	5980	10.36
Rice	25.3	3.2	0.88			0	5640	6920	10.36

Fowler and Glastonbury (1959), based on their previous work, considered that the coefficient of friction is a function of particle shape, size, roughness and void friction of the bed of packed material. On that basis, instead of evaluating the frictional properties directly and incorporating them in their flow equation, they included those parameters which would influence angle of repose and angle of internal friction. Their dimensional analysis and experimental data resulted in the following equation.

$$Q = 0.236\,(B'/d_s)^{0.185}\,[wA\sqrt{2gB'}] \qquad (10.36)$$

where Q is the rate of flow in pound per second, B' is the effective diameter of the orifice in feet given by 4 × area/perimeter of orifice, d_s is the spherical diameter of the particle in feet, w is bulk density in pound per cubic foot, A is the orifice area in square feet, and g is the gravitational constant in feet per second per second. The constant 0.236 is the weighted average for 347 tests on seven different shapes of orifices. The variation of the flow velocity due to the shape of the orifice is shown in Fig. 10.34.

Beverloo *et al.*, (1961) also developed an equation for flow of solids and compared it with those published previously. Their equation was given as

$$Q = 35w\sqrt{g}\,(B - 1.4d)^{2.5} \qquad (10.37)$$

where Q is in grams per second, w is in grams per cubic centimeter, g is in centimeter per second per second, and B and d are in centimeters. Note that in this equation d is the average screen size of particles and B is the orifice diameter. Comparison of observed and calculated data for several seeds showed an average deviation of 5 percent and a maximum deviation of 12.5 percent. When the results based on the above equations were compared with those reported by others, the authors claimed that while Eq. (10.37) agrees best with (10.36), it gives values as much as 10 percent lower than those reported by several other investigators.

In Table 10.4 is shown a summary of the observed and calculated flow rates reported by several investigators for various materials. As seen from this table the little data which are available for agricultural materials pertain to flow through horizontal circular orifices only without any reference to the moisture content of the material. Stahl (1950), reporting on flow of grains from horizontal and vertical orifices, indicated that the flow from a vertical orifice was one-third that from a horizontal opening. Ewalt and

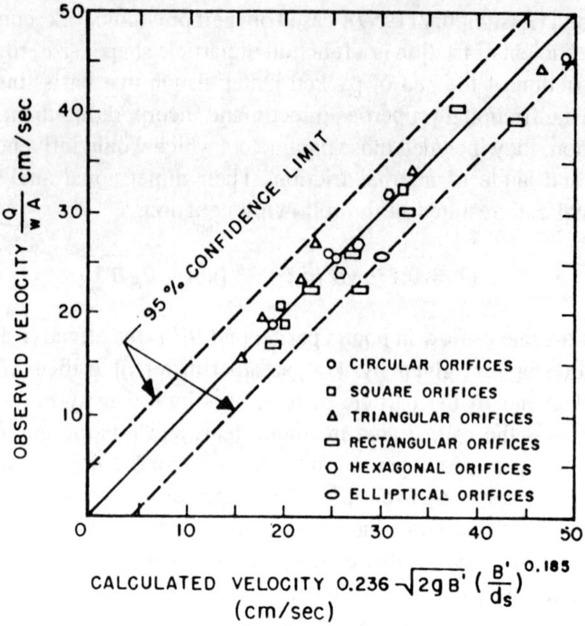

Figure 10.34 Relationship between observed and calculated velocities for different orifice shapes (Fowler and Glastonbury, 1959)

Buelow (1963), who determined the relationship between the flow of shelled corn from bins and the size and location of bin orifice, found that there was no direct correlation between the flow through a vertical and a horizontal opening. Furthermore, they found that there was a slight decrease in flow rate as moisture content of the grain increased. Following are the flow rate equations reported by Ewalt and Buelow determined by measuring flow from straight-sided wooden bins equipped with test orifices:

Horizontal openings, circular orifice (moisture content 8.4% db)

$$Q = 0.1196B^{3.1}$$

Horizontal openings, rectangular orifice (moisture content 12.1% db)

$$Q = 0.1531\, W^{1.62} L^{1.4}$$

Vertical openings, circular orifice (moisture content 12.7% db)

$$Q = 0.0351\, B^{3.3}$$

Vertical openings, rectangular orifice (moisture content 12.4% db)

$$Q = 0.0573\, W^{1.75} L^{1.5}$$

In these equations, the flow rate Q is in bushels per minute and B, W and L are respectively diameter, width and length of the orifices in inches.

As seen from the above relationships the flow rate in each case is, as found by others, an exponential function of the form $Q = KW^n$, where K and n are two constants which can be found either by substituting experimental data from two sets of tests and solving the two equations simultaneously or by determining them directly from the slope and intercepts of the straight line plot of Q versus one of the dimensions on a log-log graph paper. Apparently these constants contain all of the other physical parameters, such as frictional properties, etc., which influence the flow of solids through orifices.

Rausch (1949), in his theoretical and experimental study of flow of granular materials through orifices, derived a general expression which was used to determine the effects of head of solid, ratio of orifice diameter to head of solid, the ratio of particle diameter to the orifice diameter, and the flow of solid into mediums other than air. For the usual case of flow in air where the ratio of bin diameter to orifice diameter is large and also the ratio of head to orifice diameter is large, Rausch simplified his general expression to yield the following

$$Q = Cw\,(\pi/4\, B^2)\sqrt{\frac{gB/2}{k \tan \Phi_r}} \tag{10.38}$$

Constant C in this equation is the ratio of bulk density of particles in flow through orifice to bulk density of particles at rest. In passing through an orifice, the bulk density of the solid may differ from the normal bulk density of the material depending upon the ratio of the orifice diameter to the particle diameter. As this ratio increases, the constant C approaches unity. Other symbols in Eq. (10.38) are the same as those given previously. Note that the flow rate in this equation is proportional to the power 2.5 of the orifice diameter as found by Deming and Mehring in Eq. (10.33).

Equation (10.38) does not provide for the effect of solid particle size upon flow rate. However, experimental data showed that particle size is important in some way which is not accounted for by the factor *C*. To include this effect, Rausch assumed that *k* was approximately of the same value for all solids and employed dimensional analysis which yielded the two following dimensionless groups for solid flow in air.

$$(Do/Dp)^n \quad \text{and} \quad \frac{Q\sqrt{\tan \Phi}}{C\sqrt{g(w)}\,D_p^{2.5}}$$

Where D*o* and D*p* are orifice diameter and particle diameter, respectively. It was then surmised that if the theory was correct the log-log plot of these two dimensionless factors should yield a straight line. Experimental results showed that the log-log plot was an approximately straight line but it was different for each angle of approach or cone angle. This was expected since the cone angle was not considered in the derivation of flow rate formula.

At this time it should be pointed out that the angles of repose appearing in various equations of flow are either kinetic repose angle or static repose angle. Various authors have referred to these two types of repose angles as, respectively, angle of repose in filling or piling and angle of repose in emptying or funneling (Stahl, 1950) Caughey *et al.*, 1951). As expected, the former is smaller than the latter, as seen from a set of data given by Stahl in Table 10.5.

Lorenzen (1957) also reported two types of repose angles, one representing the angle of the pile at rest referred to as repose angle and the other representing that angle of the solid's surface with horizontal at which the flow begins. The latter, which was called the flow angle, is shown with the corresponding repose angle for several grains in Fig. 10.35.

The following expression is used for flow of low-density powders through an orifice in the bottom of a storage container:

$$Q = c(\pi/4)w(gD^5/2 \tan \Phi_r) \tag{10.38a}$$

where Q is flow rate in lb/min, c is the discharge coefficient, usually between 0.5 and 0.7, w is weight density in lb/ft^3, and D is diameter of the orifice in ft. This equation is applicable when d/D, or particle size diameter to orifice diameter, is less than 0.1.

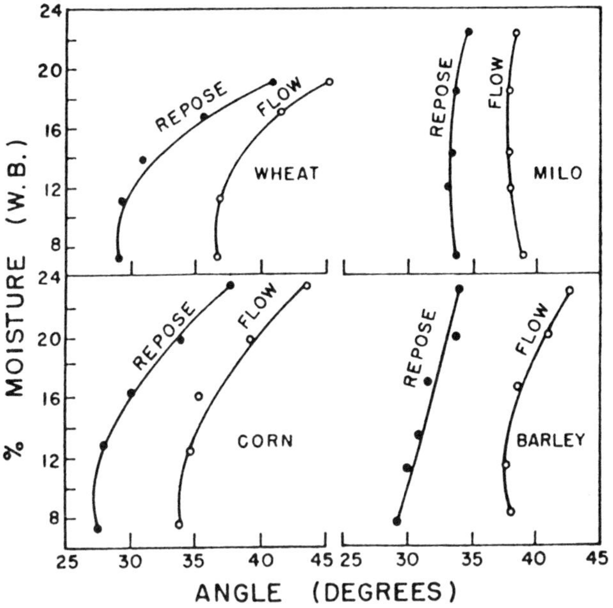

Figure 10.35 Angle of repose and angle of flow at various moisture contents of grains (Lorenzen, 1957)

Example

A 3-in diameter hole in the bottom of a 40-in diameter container holding a dry food powder is to discharge the powder by gravity. The density of the powder is 35 lb/ft^3 and as the powder is flowing through the hole, a 30-in high mound if formed in the container, estimate rate flow for this material.

From the geometry of the powder mound

$$\tan \Phi_r = 30/20 = 1.5$$

Using Eq. (10.38a), assuming $c = 0.6$

Table 10.5 Angle of repose for several seeds and grains (Stahl, 1950)

Material	Repose angle (degrees)	
	Filling or piling	Emptying or funneling
Barley	16	28
Corn (shelled)	16	27
Flax seed	14	25
Oats	18	32
Rice (rough)	20	36
Rye	17	26
Sorghum (grain)	20	33
Soybeans	16	29
Vetch	14	25
Wheat	16	27

$$Q = 0.6(\pi/4)\ 35\ (32.2(3/12)^5/2 \times 1.5)$$
$$= 0.173 \text{ lb/sec}$$

Gravity flow through chutes

In handling granular materials through chutes, the relationship between the volume of flow, stream thickness, and chute inclination angle is an important design criterion. For fast flow conditions in a fully enclosed chute the stream thickness should be less than the depth of the chute to avoid choking. While in an open chute an optimum stream thickness is necessary in order to prevent overflow. Undesirable buildup will also be avoided if the chute is designed with correct inclination angle to handle a given flow rate.

The relationship between these parameters has been reported by Roberts (1967). Of the chute shapes studied, the straight inclined chute was reported as being one of the most satisfactory types. The following relationship was given for the surface profile of wheat grain flowing through straight inclined chutes:

$$H = \frac{Q}{b\sqrt{V_0^2 \sin^2 \beta + 2gs(\sin \beta - f_E \cos \beta)}}$$

where H is the stream thickness in feet, Q is the volume flow rate in cubic feet per second, V_0 is the entry velocity of grain to chute in feet per second, b is chute width in feet, β is the chute inclination angle, g is acceleration due to gravity, s is distance down the chute in feet, and f_E is the combined coefficient of friction of grain against the bottom and side walls of the chute.

A comparison of experimental and calculated points using the above expression is given in Fig. 10.36. The effect of coefficient of friction f_E on design parameters for optimum performance and efficiency is also illustrated in Fig. 10.36. In this illustration the optimum chute cutoff angle is plotted against the radius of curvature for curved chutes. The cutoff angle corresponds to the point where the stream thickness is at its minimum and flow velocity is at its maximum.

Forced flow of solids

Broadly speaking, there are two types of forced flow systems available today for transport of solid agricultural materials. The first method is hydraulic handling of the material where a pump is used to transport the

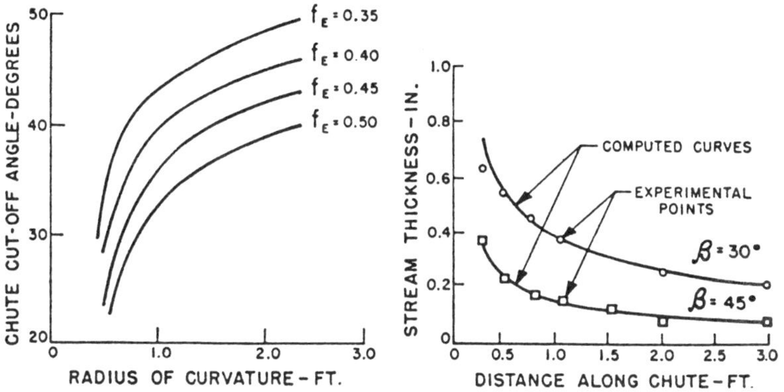

Figure 10.36 (right) Comparison of experimental data and computed stream thickness through straight chutes at two angles of inclination. (left) Effect of friction coefficient on the "optimum" cutoff angle for curved chutes (Roberts, 1966)

solid immersed in a carrying medium such as water. This subject is briefly discussed in Chapter 9 under hydrodynamic properties. The second method is pneumatic handling of the solid where again a pump or blower is used to transport the solid but air is used as the carrying medium. In this method, depending on the solid-air ratio, the conveying system can be classified as (1) lean-phase transport or conventional *pneumatic conveying*, and (2) dense-phase transport or *fluidized conveying*. In the former, the solid is conveyed as single particles in an air stream and not as a mass. In the latter, the solid is made to flow as aerated slugs that form, move, dissipate, and reform in a cyclical manner thoughout the transport system. The subject of pneumatic conveying and its application to handling of agricultural materials is covered under the aerodynamic properties. Since difficulties in fluidized conveying of a solid are due largely to its frictional properties, this application of physical properties was reserved for the chapter on friction.

Fluidizing of a solid is a process by which a mass of solid particles is permeated with air giving them fluid-like properties. When air is introduced through a contained mass of solid particles, the mass expands, the upper surface becomes level, air bubbles rise and burst at the surface and the surface permits objects to sink in as in a liquid. Fluidized conveying is characterized by high solid-air ratio, low solids velocity, and a large degree of fluidity and mobility. This is in contrast to ordinary pneumatic conveying where particle and air velocities are relatively high and concentration of solid particles in air is low. In certain applications such as conveying of seeds and grains the high velocity in pneumatic conveying is actually objectionable because of possible mechanical damage to the conveying material (Segler, 1951). In fluidized conveying, this mechanical damage may be reduced as a result of lower velocities. Furthermore, an almost perfect cleanout of the conveying lines is possible by simply increasing the air flow and purging the system. Thorough cleanout of the system is essential in such applications as handling of certified seeds and chemicals where high purity is a prime requirement.

Roberts (1966) has reported some experiments with forced flow of grains which he has referred to as "grain pumping" (Fig. 10.37). In this work, first the force required to lift a column of grain through a circular tube without fluidization was theoretically and experimentally determined. In the theoretical analysis it was assumed that the coefficient of friction, f, of the grain against the tube wall and the pressure ratio k were constant, that the pressure distribution over the surface of the piston was constant,

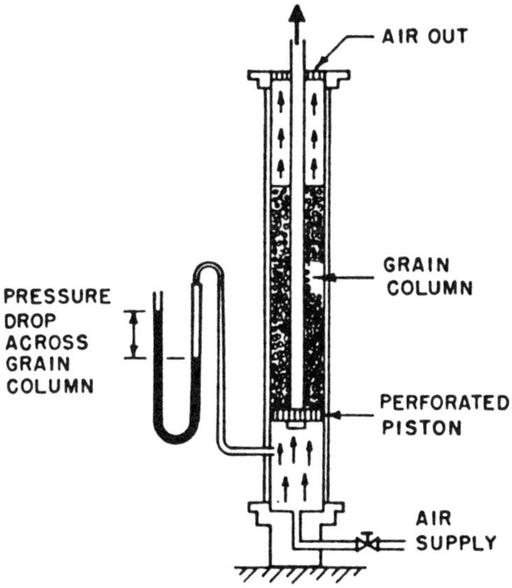

Figure 10.37 Schematic for combination of mechanical and pneumatic pumping of granular materials

and that the grain was incompressible and cohesionless. With these assumptions, from the equilibrium of forces in the tube, the following relationship was derived.

$$F = w\pi R^3/2fk\left[e\frac{2fkL}{P} - 1\right] \qquad (10.39)$$

where F is the force in pounds required to move the column of the grain L inches high in the tube against wall friction, w is bulk density in pounds per cubic inch, R is radius of the tube in inches, and P is pressure acting on the piston in inches. Roberts (1966) produced curves plotting forces versus tube length and showed that considering the simplifying assumptions made, there was a fair agreement between experimental and predicted results. However, the magnitude of forces required to lift the grain, as expected, was considerable.

To reduce the force requirement, the grain was next fluidized by drilling holes in the moving piston and the fixed flange on the cap of the tube and mixing measured quantities of air with the column of grain (Fig. 10.37). The amount of air introduced was up to and slightly exceeding the amount required for fluidization. The point of fluidization was determined by observing the pressure drop across the grain column. Just before the point of fluidization, the pressure drop was steady. At the point of fluidization, due to "bubbling," there was variation in pressure drop. Beyond the point of fluidization, due to "slugging," pressure drop varied widely in a pulsating manner. Figure 10.38 shows that the effect of fluidization in reduction of the conveying force is appreciable. In terms of air horsepower required for conveying, experiments by Roberts (1966) showed that to elevate millet a height of 8 feet through a 4-inch vertical tube at a velocity of 2 feet per second, only 7.53 cubic feet per minute of airs is required which would result in a 3.5 psi pressure drop and 1.53 air horsepower.

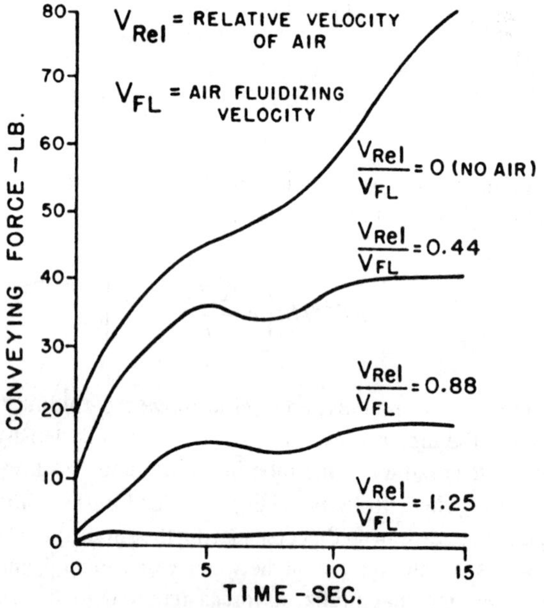

Figure 10.38 Effect of fluidization on force requirement to lift a 15.5-inch column of millet (Roberts, 1966)

When compared with ordinary pneumatic conveying, the power requirements are higher for the dense-phase operation. This was shown by Brandenburg and Harmond (1964) who applied fluidization to handling of seeds. They have reported on the development of a fluidized system for handling of crimson clover and bean seeds considering low velocities for minimizing seed damage. Figure 10.39 shows a schematic of the fluidized seed conveyor system used for seed fluidization studies. The auxiliary airflow, passing the air lock and supplying 70 percent or more of the conveying air, proved valuable in lowering the operating pressures and power requirements. The horsepower requirements for seed transport in the dense-phase was about 3.5 compared with 2.5 horsepower in the lean-phase operation. The solid to air ratio tried by these investigators was as high as 28:1. The major advantage claimed for the dense-phase transport was low velocity of seed flow through flexible, small pipes resulting in minimum seed damage and with a capacity large enough to meet the requirements of the seed industry.

Daffin Industrial Products Division has compiled data on a long list of materials giving those physical properties which are important in design of fluidized transport systems. Table 10.6 shows selected materials from that list. In this table, the weights per cubic foot are in loose dry state, coefficient of friction refers to tangent of the angle of slide on a clean steel plate, and the numbers giving fluidization characteristics are defined in the footnote of the table. In addition to the information given in the table, a measure of the hardness of the material is also required. This information will supposedly aid in determining the abrasiveness of the material and its resistance to mechanical damage.

10.6 PRESSURE DISTRIBUTION IN STORAGE STRUCTURES

The resistance to lateral displacement of granular materials and silage when placed in bins or silos is supplied by the retaining walls of these structures. Due to insufficient knowledge of the physical properties of the materials to be stored in these structures, estimation of lateral and vertical pressures and floor loads, under both static and dynamic conditions, continues to be a critical task for design engineers.

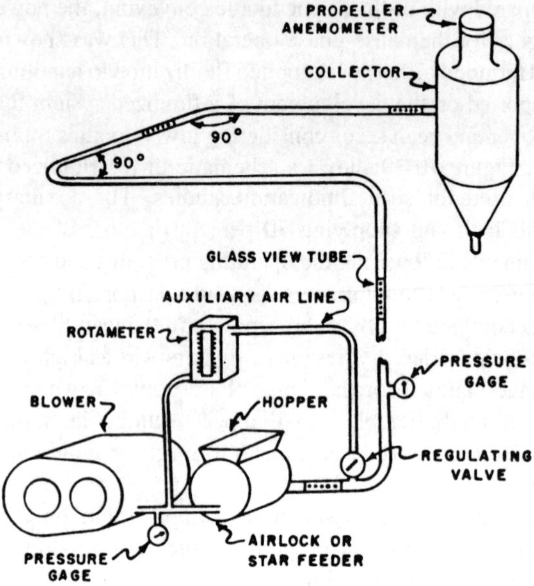

Figure 10.39 USDA fluidized seed conveyor (Brandenburg and Harmond, 1964)

Definitions of shallow bins and deep bins

There are several definitions given for shallow and deep bins. A grain bin is referred to as a shallow bin when the depth of grain is less than or equal to equivalent diameter. In a deep bin, depth of grain is greater than the equivalent diameter. The equivalent diameter is taken as four times the hydraulic radius of the bin.

Another method to determine the deepness of a bin is to draw a line at an angle equal to the angle of repose of the granular material from the intersection of the bin wall and floor to the opposite bin wall. In a deep bin, this line intersects the opposite wall before passing through the upper surface of the granular material. In a shallow bin, the line meets the opposite wall at or above the surface of the granular material (see Fig. 10.40).

Table 10.6 Physical properties of selected materials required for design of fluidized transport systems (From data supplied by Farmhand-Industrial Products Division)

Material	Bulk density lb/ft^3	Repose angle (deg)	Coeff. of friction	Fluidization characteristics*
Alfalfa, ground	15.0	45°	0.6	
Alfalfa, seed	48		0.7	
Barley (48 lb/bu)	38.5	30°	0.6	1, 4, 8
Beans, cocoa	67	30°–45°	0.5	1, 3
Beans, soy, whole	50	30°	0.4	1, 3, 8
Beans, white, navy	50		0.4	1, 4
Bluegrass seed	11.0		0.7	
Buckwheat (46 lb/bu)	42	30°	0.5	1, 4, 8
Buckwheat, hulls	13		0.6	
Coffee beans, green	42	45° plus	0.5	1, 4
Coffee, roasted bean	23	30°	0.5	1, 4
Corn, ear	28		0.4	
Corn shelled (56 lb/bu)	45	30°	0.4	1, 4, 8
Cotton seed	25	30°–45°	0.6	1, 4
Flax seed (56 lb/bu)	45	30°	0.4	1, 4, 8
Hay, loose	5			
Millet	40		0.5	
Mustard seed	45	30°	0.7	1, 4, 8
Oats (32 lb/bu)	26	30°	0.4	1, 4, 8
Oat hulls	8		0.7	
Peanuts, hulled	30–45	30°–45°	0.8	1, 4
Peas, dried (64 lb/bu)	50	30°	0.6	1, 4, 8
Rape seed	48		0.7	
Rice, clean	48	30°–45°	0.4	1, 4
Rice, hulls	20		0.6	
Rice, rough	32	30°–45°	0.5	1, 4, 8
Rye (56 lb/bu)	45	30°	0.4	1, 4, 8
Sand, dry & loose	90–106	34°	0.8	2, 3
Sawdust, dry	10–30		0.8	
Sorghum seed	32	30°–45°	0.7	1, 3
Sugar, granulated	50	30°–45°	0.7	1, 4
Timothy seed (45 lb/bu)	36	30°–45°	0.6	1, 4, 8, 9
Tobacco stems	15	45° plus	0.5	2, 4, 11
Wheat (60 lb/bu)	48	28°	0.5	1, 4, 8
Wood chips	10–30	45° plus	0.4	2, 4, 9, 11

* 1—Free flowing; 2—sluggish; 3—abrasive; 4—non-abrasive; 8—contains explosive dust; 9—light and fluffy; 11—interlocks.

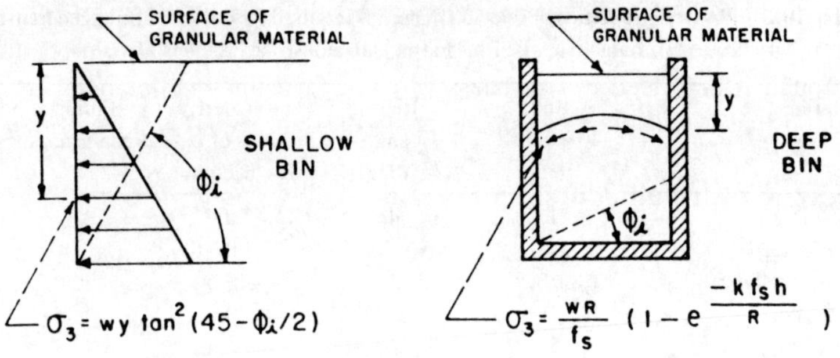

Figure 10.40 Lateral pressure in shallow and deep bins

The most recent definition for a deep bin, applicable to considerably tall structures, is given by Isaacson and Boyd (1965). According to their definition, in a deep, bin

$$H_d/D \geqq 0.75\,(1/f_s k) \qquad (10.40)$$

where H_d is the depth of grain and D is the diameter of the circular bin. For example, if for a certain grain the coefficient of friction f_s=0.45 and the pressure ratio k=0.33, according to the above definition a circular cylindrical bin for storage of this grain is deep if

$$H_d/D \geqq 5$$

Pressure distribution in shallow bins

In shallow bins, the Rankine equation is usually used. In this equation, which is given in (10.25), the unit lateral pressure, σ_3, is expressed in terms of the angle of internal friction Φ_i. In this type of bin, the pressure diagram is a triangle with a resultant load, L, acting through a point two-third of the wall height, h, below the top of the grain surface. This resultant load in pound per foot is given by

$$L = \tfrac{1}{2} wh^2 \tan^2 (45 - \Phi_i/2) \qquad (10.41)$$

In addition to shallow bins, Eq. (10.41) may also be used in design of retaining walls for soil and similar materials. To obtain a rough estimate of earth pressures on a vertical retaining wall, if Φ_i is not known, the equivalent fluid density for average soil conditions (30–35 lb/ft^3) is used for w and lateral pressure is estimated from $L=wh^2/2$. This procedure, although it may lead to over design of the structure, is defended on the ground that it will place the designer on the safe side.

Pressure distribution in deep bins

For design of deep bins, Janssen's equation (10.27) is now generally used. Before Janssen's equation, the estimate of grain pressure on walls of deep bins was made by assuming the grain being a semi-liquid of the same density as the grain and applying the principle of hydrodynamics which states that the lateral pressure at any depth h would be the product of h and the bulk density of the grain. In this analysis both the internal friction within the granular material and the friction of the material against the walls were completely neglected and the whole weight of the grain was assumed to be transmitted directly to the bottom of the bin. As a result of this simplification, many structures failed either under the lateral pressure due to internal friction of the grain Eq. (10.26) or under the vertical pressure on the wall due to wall friction (product of coefficient of friction and lateral pressure).

In recent years several investigators have challenged the generality of Janssen's equation particularly for cases involving dynamic loading and storage of grains in deep bins. It is recognized, however, that the Janssen's equation is still valid for static loads or pressures exerted during the storage period. For dynamic loads, occurring during charging and discharging, it has been reported that Janssen's equation cannot predict certain critical pressure increases resulting from this type of loading.

Isaacson and Boyd (1965) have reviewed the recent work in this area and have formulated a general solution for grain pressure in deep bins as a prototype of other solutions arising from different characteristic functions and loading conditions. The characteristic functions are in terms of density, friction and pressure-ratio k. The value of these functions for solution of the equation is proposed to be obtained by experimentation.

Recognizing the basic deficiency of Janssen's equation, namely its static nature, Collins (1963) conducted some experiments involving dynamic effects in cylindrical structures and compared the results with those predicted by Janssen's equation. Figure 10.41 shows such a comparison for horizontal and vertical stresses versus height of dry sand in the cylindrical tank. Values for coefficient of wall friction and pressure ratio were obtained using a shear test machine based on the principles illustrated in Fig. 10.26. Stresses were calculated from strains, measured by strain gages, using the equation for a biaxial state of stress Eq. (10.44). Considering the possible experimental errors involved in the measurement of strains, the agreement of predicted and test data is rather good in the upper portion of the tank as seen from Fig. 10.41. Near the bottom of the tank, however, test data deviate considerably from the prediction curve. To make some qualitative observations as to the dynamic effects, paper cylinders were filled with grain to various depths and discharged at various rates. Wrinkles or indentations on the wall of the paper cylinders were considered buckling. Results showed that the rate of unloading had no effect but the vertical loads on the cylinders were apparently greater during unloading because in every case buckling occurred at the lower part of the cylinder when grain exceeded a certain depth.

Since the results of research on the question of pressure distribution in grain storage structures are still inconclusive, the Janssen's equation remains to be the only simple equation available to the designer.

A technical committee of the American Society of Agricultural Engineer has attempted to pull together data on loads imposed by stored grains. So far no recommendation has been made by this committee. The major problem seems to be the need for reliable data on physical properties of grains. The available data are not complete, and often give conflicting values and lack probability distribution. Until the question of physical properties is settled, the problem would perhaps continue to remain vague and too complex for a simple solution. With this consideration, the following procedure is currently being practiced for the purpose of estimating various types of static loads in deep storage bins.

Three types of pressure are generally considered. These pressures are the lateral pressure and vertical pressure on the bin walls and the vertical load on the bin floor. For estimate of static lateral pressure on vertical walls Eq. (10.27) and the appropriate constants as given in Table 6.3 and the Appendix may be used. The vertical pressures on vertical walls of the bin would be the product of lateral pressure and the coefficient of friction between the grain and the wall material. Tables in the Appendix may be used to find an appropriate friction coefficient. Vertical pressure on the bin floor may be obtained either from the product of pressure ratio k and the lateral pressure or from the following equation given by Barre and Sammet (1950).

$$P = W - \sum LCf_s \qquad (10.42)$$

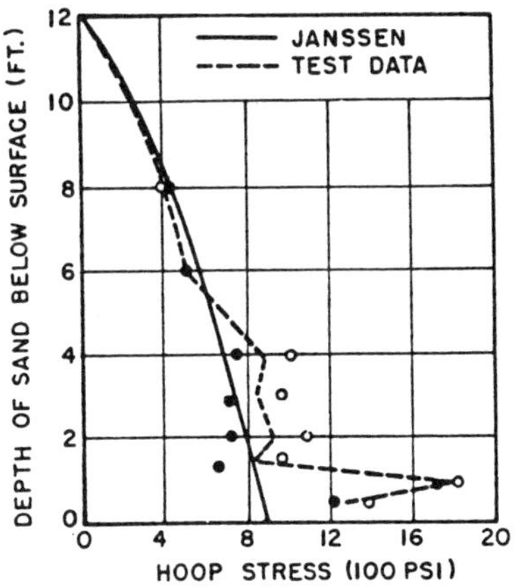

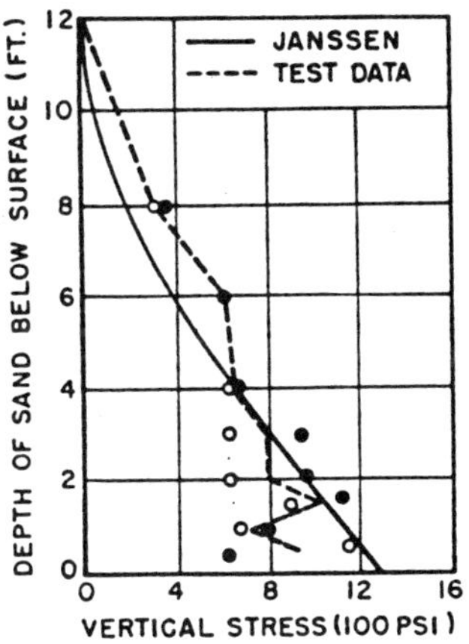

Figure 10.41 Horizontal and vertical stresses versus height of dry sand in a cylindrical tank (Collins, 1963)

where P is the floor load in pounds, W is the weight of the stored material in pounds, C is the circumference of the bin in feet, and L is given by Eq. (10.41). Equation (10.42) gives simply the difference between the total weight of the material and the portion of this weight which is supported by friction on the bin walls. The latter is given as the sum of the product of friction coefficients f_s and the resultant load LC acting through the circumference of the bin on successive increments of wall height. Some observed pressures for ear corn in storage bins and ensilage in silos are given by McCalmont *et al.* (1934, 1946) and the *Agricultural Engineering Yearbook* (1967).

Example

To determine the effect of moisture content of grain on lateral pressures in a wooden bin, the following data were obtained:

Moisture content % (w.b.)	Bulk density lb/cu ft (w)	Coeff. of friction (grain on wood) (f_s)	Pressure ratio (k)
7.3	49.3	0.45	0.44
19.3	43.9	0.59	0.34

Assuming a round wooden bin with a diameter of 12 feet and a height of 30 feet, calculate the percent increase in lateral pressure against the bin wall as the grain dries out from 19.3% to 7.3% moisture content.

The larger repose angle of the grain for 19.3% moisture from Table 10.3 is 41 degrees. With this repose angle and the 12-foot diameter of the bin, the geometry given in Fig. 10.40 shows that this is a deep bin. Equation (10.27) for deep bins and the given data yield lateral pressure for the dry wheat as follows:

Hydraulic radius $R = D/4 = 12/4 = 3$ ft.

$$\sigma_{3\,\mathrm{dry}} = \frac{49.3 \times 3}{0.45}\left(1 - e^{\dfrac{-0.44 \times 0.45 \times 30}{3}}\right) = 283\ \mathrm{lb/ft^2}$$

Similarly lateral pressure for the wet wheat

$$\sigma_{3\,\mathrm{wet}} = \frac{43.9 \times 3}{0.59}\left(1 - e^{\dfrac{-0.34 \times 0.59 \times 30}{3}}\right) = 193\ \mathrm{lb/ft^2}$$

Increase in pressure due to drying

$$\frac{283 - 193}{193} \times 100 = 46.7\ \%$$

While the problem of pressure distribution in grain storage structures remains unsolved, the problem of pressure distribution in large silos for silage has not even gone beyond the empirical stage. Recent developments in production, handling and feeding of silage have necessitated the construction of large silos. Having these structures designed on the basis of experience and extrapolation of available data rather than on a rational basis considering the physical properties of the materials has resulted in numerous failures of large silos during recent years.

The available design formulas are purely empirical and consider only such factors as diameter of silo and depth of the silage (McCalmont *et al.*, 1946; Besley and McCalmont, 1941), diameter, depth and moisture content (Neubauer, 1966), or just depth of silage alone assuming that other variables are absorbed in the constant of the formula (American Concrete Institute, 1946).

Yu *et al.* (1963) have made an attempt to consider bulk density w, coefficient of friction f_s, and pressure ratio k in derivation of an equation for pressure prediction in silos. In this derivation, each of the above parameters were expressed as different functions of depth. Assuming that at any elevation the silage was uniformly distributed over a small increment of depth and free from arch action, the Janssen's formula Eq. (10.27) was applied to derive the following silo formula

$$H_i = \frac{f_{1i}(z)\,R}{f_{2i}(z)}\left(1 - e^{\frac{-f_{3i}(z)\,f_{2i}(z)}{R}}\right) \tag{10.43}$$

where H_i with $i=1, 2, 3$ is lst, 2nd or 3rd degree polynomial approximation of the horizontal pressure, z is depth, R is hydraulic radius, and f_{1i}, f_{2i} and f_{3i} are functions of density, friction coefficient and K ratio, respectively. The values of each of these functions in terms of polynomial equations were found using experimental data and the least square curve fitting techniques on a digital computer. Limited data showed fair agreement between predicted and measured values of pressure. Further work in this area is considering also the effect of vertical forces.

10.7 PRESSURE DISTRIBUTION IN COMPRESSION CHAMBERS

When a bulk of such unconsolidated materials as forage and grain is being compressed in a rigid container, a knowledge of the coefficient of friction of the material against the wall of the container will enable determination of pressure distribution. Consider the cylindrical container illustrated in Fig. 10.42. If the unconsolidated material being compressed in the cylinder were considered elastic, the relationship between the lateral and vertical pressures, according to Timoshinko (1956) will be

$$P_L = \frac{\mu}{1 - \mu} P_x \tag{10.44}$$

where P_L is the lateral pressure, P_x is the vertical pressure at any point below the surface and μ is the Poisson's ratio of the compressed material.

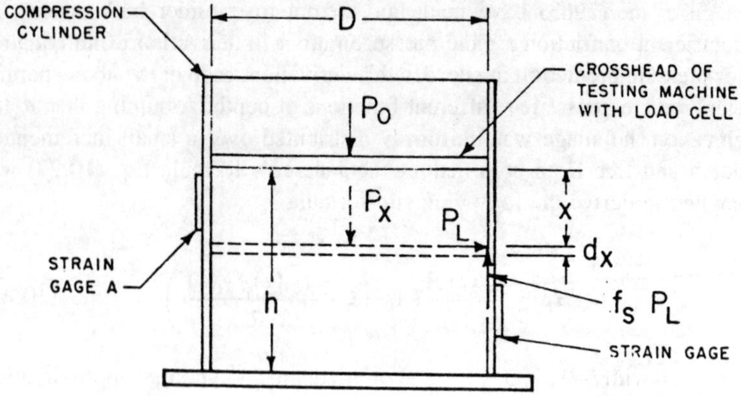

Figure 10.42 Pressure distribution in compression chambers

At point x below the surface of the compressed material, the resistance to axial force will be the sum of the bottom force and the wall friction at that point. The wall friction will be the product of lateral force at that point and the coefficient of friction between the material and the container wall.

If the distance x increases by an amount d_x, the axial pressure decreases by an amount dP_x so that the decrease in axial force is $dP_x(\pi/4)D^2$. This decrease in axial force is balanced by an increase in frictional force given by $(f_sP_L)\ (\pi D)d_x$. The balanced equation is

$$dP_x(\pi/4)\,D^2 = (f_sP_L)(\pi D)\,d_x \tag{10.45}$$

where f_s is the coefficient of friction. Substituting for P_x from Eq. (10.44) and integrating will yield the lateral pressures at any distance x from the surface of the compressed material as follows:

$$P_{Lx} = \frac{\mu}{1-\mu}\,P_0\,e^{-f_s\frac{\mu}{1-\mu}\frac{4}{D}x} \tag{10.46}$$

The technique illustrated in Fig. 10.42 is a method for determining the parameters in Eq. (10.46). Two three-element rectangular rosette strain gages are mounted on the compression cylinder wall with two other sets of gages mounted on another cylinder for temperature compensation. By means of a strain gage indicator, hoop strain, ε_x, and vertical strain, ε_y, of the cylinder wall can be determined. The strain values can then be converted to hoop stress, σ_x, at any point x, using the following expression given by Dally (1965) for a biaxial state of stress.

$$\sigma_x = \frac{E}{1-\mu'^2}(\varepsilon_x - \mu'\varepsilon_y) \tag{10.47}$$

where E and μ' are Young's modulus and Poisson's ratio of the cylinder wall. Knowing the hoop stress σ_x the lateral pressure P_L can be determined from the following relationship given by Marin (1963)

$$P_L = 2t\sigma_x/D \tag{10.48}$$

where t is the thickness of the cylinder wall, and D is the diameter of the cylinder.

Once the lateral pressures for the two locations A and B are determined, two equations can be established by substituting the values of P_L, axial pressure P_0, and the corresponding values of x in Eq. (10.46). Solving these two equations simultaneously, the values of Poisson's ratio and coefficient of friction can be determined.

Note that for $x=0$ and $x=h$, where h is the depth of the compressed material, Eq. (10.46) yields lateral pressures at the surface, P_{L0}, and lateral pressure at the bottom, P_{Lb}, as follows:

$$P_{L0} = \frac{\mu}{1-\mu} P_0 \tag{10.49}$$

$$P_{Lb} = \frac{\mu}{1-\mu} P_0 e^{-f_s \frac{\mu}{1-\mu} \frac{4}{D} h} \tag{10.50}$$

If the pressure ratio $\mu/1-\mu$ in Eq. (10.50) is set equal to the pressure ratio k in Eq. (10.27), Eq. (10.50) becomes Eq. (10.27), which is the familiar Janssen's equation.

Example

Dry shelled corn with a moisture content of 2.9% (w.b.) is compressed in a cylinder to a bulk density of 63 lb/ft^3. The inside wall of the compression cylinder is coated with teflon. The pressure is applied by means of a pneumatic-hydraulic testing machine. Strain measurement gave negligible values for vertical strains. Hoop strains are shown along with other pertinent data in the following table.

Hoop strain (10^{-6} in/in)		Wall thickness (in)	Surface pressure (psi)	Cylinder diameter (in)	Depth of compd. material (in)	Distance of gage below surface (in)	
ε_{xA}	ε_{xB}	t	P_0	D		X_A	X_B
51	34	0.24	272	5.73	6.75	2.8	5.6

Using these data, determine the lateral pressures at points A and B, Poisson's ratio of bulk material, and the corresponding coefficient of friction of corn on teflon wall of the cylinder.

Assuming a Poisson's ratio of $\mu'=0.3$ and a Young's modulus of $E=30\times10^6$ for the steel wall of the compression cylinder, hoop stress in the wall at points A and B are found using

$$\sigma_{xA} = \frac{30 \times 10^6}{1 - (0.3)^2}(51 \times 10^{-6} - 0) = 1680 \text{ psi}$$

$$\sigma_{xB} = \frac{30 \times 10^6}{1 - (0.3)^2}(34 \times 10^{-6} - 0) = 1122 \text{ psi}$$

Using the above hoop stresses in Eq. (10.48), the lateral pressure at A and B are found

$$P_{LA} = \frac{2\,(0.24)\,(1680)}{5.73} = 147 \text{ psi}$$

$$P_{LB} = \frac{2\,(0.24)\,(1122)}{5.73} = 98 \text{ psi}$$

Having the lateral pressures for the two points along the cylinder wall and the information given in the above table, the following two equations can be established by writing Eq. (10.50) twice:

$$147 = \frac{\mu}{1 - \mu} 272\, e^{-f_s \frac{\mu}{1-\mu} \frac{4}{5.73} (2.8)}$$

$$98 = \frac{\mu}{1-\mu} 272\, e^{-f_s \frac{\mu}{1-\mu} \frac{4}{5.73}(5.6)}$$

Solving the above equations simultaneously the value of μ and f_s are found to be $\mu=0.447$ and $f_s=0.26$.

In order to confirm the above coefficient of friction, direct friction tests were conducted using a flat plate coated with teflon and samples from the same dry shelled corn. The coefficients of friction were found to vary from 0.17 to 0.23 with the increase being in the direction of increased normal load. Considering the higher normal pressures involved in compressing the corn in the cylinder, the value of $f_s=0.26$ is not out of line.

In studying the compressibility of straw in a compression chamber with rectangular cross section, Mewes (1958) followed a theoretical analysis similar to that given above. The sidewall and bottom pressures were measured by means of mechanical deflection of springs. As expected, it was found that the coefficient of friction of the straw against the wall of the compression chamber influences the pressure gradient as well as the ratio of vertical pressure to lateral pressure. Figure 10.43 shows changes in lateral pressure as a function of piston pressure for straw arranged vertically to the axis of compression.

10.8 APPLICATIONS IN DESIGN OF HANDLING AND PROCESSING MACHINES

The theoretical performance of machines and mechanisms used in mechanical handling and processing of materials cannot be fully appreciated without a knowledge of the physical properties of the material.

In this section a few examples where the frictional properties constitute important parameters in the derivation of the theoretical and applied equations will be given.

Power losses due to friction

In design of material handling equipment such as mechanical and pneumatic conveying machines, the material comes in direct contact with the trough, casing or other components of the machine over which it must slide. The total power required to drive these machines is composed of several components. Power consumed to overcome friction is one of the components

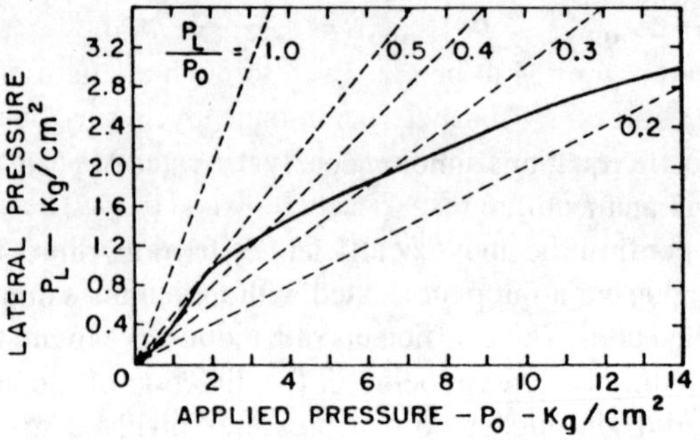

Figure 10.43 Lateral pressure versus piston pressure in compressing straw (Mewes, 1958)

of total power requirements and its rational estimate requires a knowledge of frictional properties of the material to be handled.

For example, a general rational equation for theoretical horsepower for belt conveyors can be derived by considering the algebraic sum of all the gravitational and frictional forces acting along the conveyor at a given speed. The resulting rational equation would be of the following form:

$$\text{Total } HP = HP(\text{empty}) + HP(\text{Loaded}) \pm HP(\text{elevate or lower})$$

Henderson and Perry (1966) have given the following special case of the above expression for flight conveyors.

$$HP = \frac{2VL_{uc}W_cf_c + Q(L_{1c}f_m + H)}{33000} \tag{10.51}$$

where

V = speed of conveyor, ft/min
L_{uc} = horizontal projected length of unloaded conveyor, ft
W_c = weight of flights and chain, lb/ft
f_c = coefficient of friction for chains and flight
Q = weight rate of material to be handled, lb/min
L_{lc} = horizontal projected length of loaded conveyor, ft
f_m = coefficient of friction for material
H = height of lift, ft

Example

Assume that corn grain is to be conveyed through a length of 10 feet and up 4 feet at a rate of 10 bushels per minute by means of a drag-chain conveyor. Coefficients of friction of the grain against steel at 7.3% and 19.3% moisture contents are respectively 0.46 and 0.56. The grain weights 61.5 pounds per bushel when dry and 54.7 pounds per bushel when wet. Excluding the power required for running of the empty conveyor, determine the effect of moisture content on horsepower requirement to lift the grain

$$HP = \frac{Q\,(L_{1c} f_m + H)}{33000} = \frac{10 \times 61.5\,(10 \times 0.46 + 4)}{33000} = 0.160$$

for wet grain

$$HP = \frac{10 \times 54.7\,(10 \times 0.56 + 4)}{33000} = 0.159$$

Note that as friction increases due to increase of moisture content, weight per bushel decreases. This results in little change in power requirements.

Screw conveyors

Although considerable theoretical and experimental work are reported in literature on the performance, efficiencies and power requirements of this type of conveyor, the design procedures are still empirical with frictional properties of the material either absorbed in the constants of the equations or introduced as a "material factor" in the design procedure.

Roberts (1962), in his attempt to determine the conditions for most efficient operation of grain augers, applied dimensional analysis to predict the performance of a full-scale prototype machine. The frictional parameters used were internal coefficients of friction of grain and friction coefficient against casing. Since the frictional parameters were absorbed in the constants of the derived equations, the design equations and charts were recommended for wheat grain only.

For estimating the power requirements for screw conveyors, a common practice among the design engineers is the use of a "material factor" which accounts for several of the material's physical characteristics including the frictional properties. Such material factors appear in *Materials Handling Handbook* (Bolz, 1958), and the *Engineering Manual of the Industrial*

Machinery Company. Table 10.7 shows the material factors along with other physical characteristics of several foods and agricultural materials selected from the latter source. The horsepower formula given by the Industrial Machinery Company for transport of materials only by screw and drag-type conveyors is as follows

$$MHP = \frac{CP \times MF \times L}{1{,}000{,}000} \tag{10.52}$$

where *MHP* is material horsepower, *CP* is capacity in pounds per hour, *MF* is material factor, and *L* is conveyor length in feet. It is also recommended that if the calculated material horsepower was less than 5, the horsepower should be corrected by the scale shown in Fig. 10.44.

Example

For conveying dry undelinted cottonseed at a rate of 10 tons per hour, a 10-inch screw conveyor 35-feet long operating at 45 rpm has been selected. Calculate the horsepower required to move the material.

From Table 10.7 horsepower factor for dry undelinted cottonseed is 0.8. Therefore,

$$MHP = \frac{10 \times 2000 \times 0.8 \times 35}{1{,}000{,}000} = 0.56 < 5$$

Reference to correction scale in Fig. 6.41 gives the corrected horsepower as 1.32.

In conveying granular materials in vertical screw conveyors of given geometrical proportions there is a critical speed of operation at which reverse flow in the central region ceases. Several investigators have studied such a condition in screw conveyors and have developed equations to give the critical velocity (Gutyar, 1956, Baks and Schmid, 1960, both cited in Roberts, 1963, 1964; Vierling and Sinha, 1960; Ross and Isaacs, 1961). These equations along with a knowledge of the geometrical characteristics of screw conveyors as well as the frictional properties (angle of repose and coefficient of friction) of the material to be conveyed would enable theoretical prediction for conveying capacity and critical velocity. Figure 10.45 shows a screw conveyor with forces acting on a particle at the critical speed of rotation. Roberts (1963, 1964) made a critical analysis of work in this area and showed that while some of these equations may hold for

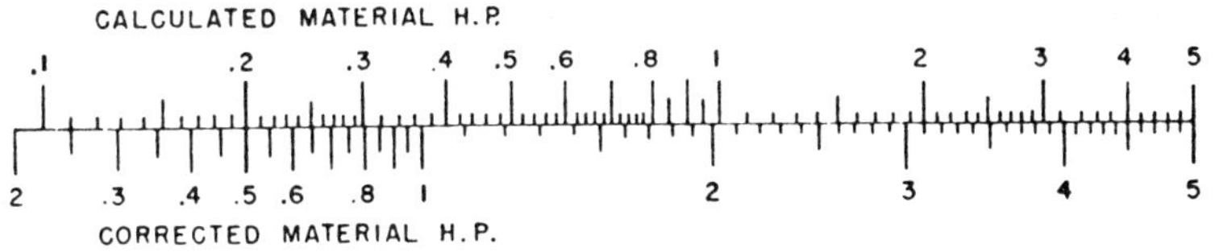

Figure 10.44 Scale for correcting material horsepower (Courtesy Industrial Machinery Company)

a single grain particle, it does not hold for a mass of grain passing through an auger. In turn he developed an expression for critical velocity which involves several constants, presumably incorporating the physical properties of the material.

Oscillating conveyors

Berry (1959) showed that the forward motion of materials such as grains on an oscillating conveyor may proceed by either "stick-slip" or continuous slipping throughout the cycle. On the surface of the conveyor deck, slipping will not occur unless the instantaneous frictional force necessary to accelerate the material exceeds $f_s W$, where f_s is the static coefficients of friction of the material on the deck surface and W is the instantaneous effective weight of the particle. After setting up mathematically the conditions for stick-slip and pure slipping, the following relationships were derived to give the efficiency of the conveying mechanism.

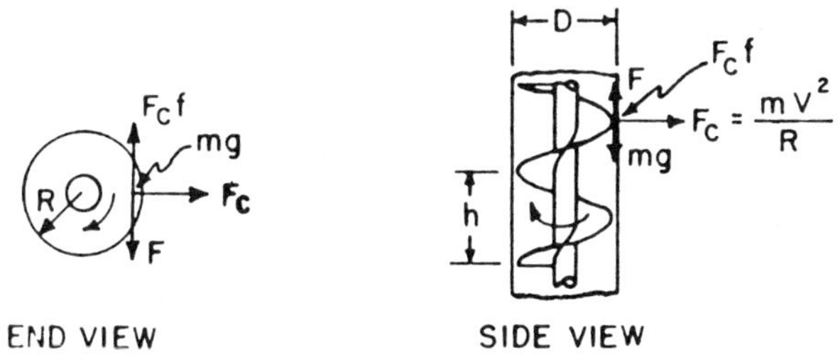

Figure 10.45 A screw conveyor with forces acting at the critical speed of rotation (Vierling and Sinha, 1960)

Table 10.7 Maximum particle size, bulk density and material horsepower factor for several foods and agricultural products (adopted from Engineering Manual, Industrial Machinery Company)

Material	Maximum particle size in	Bulk density lb/ft^3	Horse-power factor
Alfalfa Meal	—1/8	17	0.6
Alfalfa Seed	—1/8	48	0.5
Almonds	—1/2	28–30	0.9
Barley	—1/8	37–48	0.4
Beans, Castor	—1/2	36	0.5
Beans, navy	—1/2	48	0.5
Bran	—1/8	10–20	0.4
Clover, seed	—1/8	48	0.4
Cocoa, beans	—1/2	30–45	0.4
Cocoa, powdered	—100*M*	30–35	0.9
Cocoanut	shred	20–22	1.0
Coffee, green bean	—1/2	32–45	0.5
Coffee, ground	—1/8	25	0.6
Coffee, roasted bean	—1/2	22–26	0.4
Corn, cracked	—1/2	40–50	0.7
Corn, shelled	—1/4	45	0.4
Corn, meal	—1/8	32–40	0.5
Cottonseed, dry, delinted	—1/4	22–40	0.9
Cottonseed, dry, undelinted	—1/4	18–25	0.8
Cottonseed, cake	+1/2	40–45	1.0
Cottonseed, meal	—1/8	35–40	0.4
Flaxseed	—1/8	43–45	0.4
Flour, wheat	—100*M*	30–46	0.6
Grass seed	—1/8	10–32	0.4
Milk, whole, dried	—100*M*	20	0.4
Milo	—1/4	56	0.4
Mustard seed	—1/8	45	0.4
Oats	—1/2	25–35	0.4
Peanuts, shelled	—1/4	35–45	0.4
Peanuts, unshelled	+1/2	15–24	0.6
Peas, dried	—1/2	45–50	0.5
Rice, hulled or polished	—1/8	45–48	0.4
Rice, rough	—1/8	32–36	0.4
Rice bran	—1/8	16–20	0.4
Rye	—1/8	44–48	0.4
Sawdust, dry	—1/8	10–30	0.5
Sorghum, seed	—1/8	32–52	0.5

Table 10.7 (continued)

Material	Maximum particle size in	Bulk density lb/ft^3	Horse-power factor
Soybeans, whole	—1/4	45–50	0.4
Sugar, granulated	—1/8	50–55	0.7
Sugar, raw, cane	—1/8	55–65	1.0
Tobacco, scraps	+1/2	15–25	0.5
Tobacco, snuff	—100*M*	30	0.9
Walnut shells, crushed	—1/8	35–40	1.0
Wheat	—1/4	45–48	0.4
Wheat, cracked	—1/8	35–45	0.4
Wood bark	+1/2	10–20	1.2
Wood chips	+1/2	10–30	0.6

$$\underset{\text{(slipping)}}{\text{Eff.}} = \left(\frac{\Delta x}{x_0}\right)\left(\frac{f_k g}{\pi a_1}\right)$$

$$\underset{\text{(stick-slip)}}{\text{Eff.}} = (1 + f_k \tan\theta)$$

where Δx is displacement of the particle, a_1 is the harmonic coefficient of the first cosine component, f_k is kinetic coefficient of friction on the conveyor surface, and $\tan\theta$ is the ratio of the conveyors vertical amplitude y_0 to horizontal amplitude x_0 (Fig. 10.46). The efficiency in these equations refers to the ratio of useful work to move the particle a distance Δx to the work done by the conveyor to move a distance x_0. When the efficiency for pure slipping is computed in terms of the quantity ($f_k \tan\theta$) and is compared with the efficiency of stick-slip condition, as shown in Fig. 10.47, it appears that oscillating conveyors operating in the stick-slip region will always be more efficient than any other type of conveyor operating in the pure slipping region. Experimental work, using a shallow layer of wheat, substantiated this theory (Berry, 1958, 1959). Figure 10.47 is drawn for the case where $y_0\omega^2/g = 1$, with ω being the angular frequency of vibration. Thus in this example, the derived efficiency equations give the

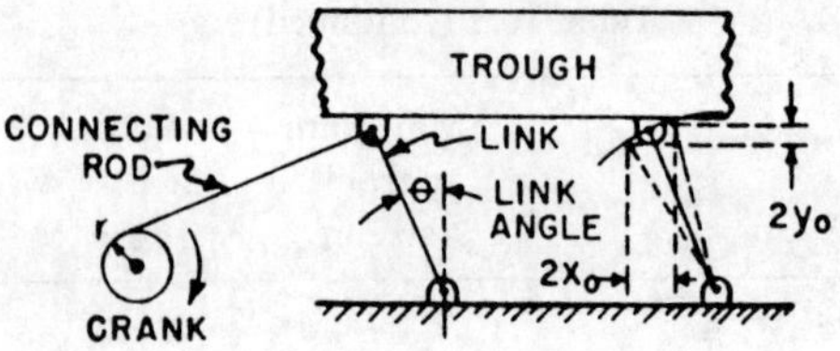

Figure 10.46 Oscillating conveyor showing link angle θ, and horizontal and vertical displacements (Berry, 1959)

preferred link angle θ and frequencies for the most efficient operation of the conveyor.

Schertz and Hazen (1963) developed the predicting equations for performance of oscillating conveyors considering four types of motion, namely free fall, sliding negatively, sliding positively, and riding. As expected, the predicting equations for the two types of sliding motions involved the kinetic coefficient of friction of the material against the conveyors surface. The equation for riding motion involved the static coefficient of friction. The theory was applied to shelled corn to predict the rate of material

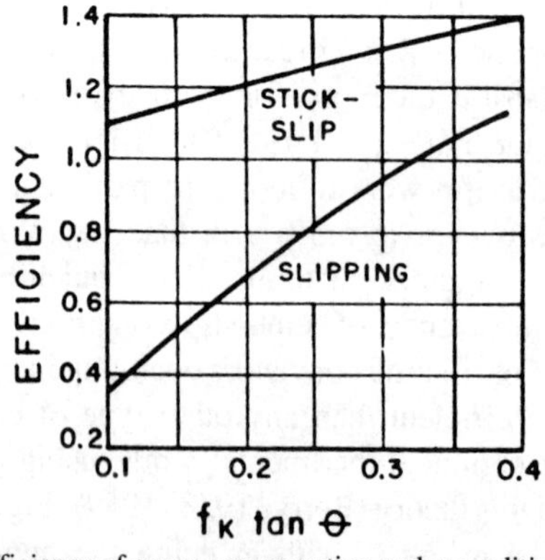

Figure 10.47 Efficiency of conveyors operating under conditions of stick-slip and pure slipping (Berry, 1959)

movement. The results indicated that the assumption of mean particle velocity being independent of depth of grain layer was not valid. The observed increase of mean velocity with increase in layer depth was felt to be due in part to the difference between coefficients of friction of the corn on the conveyor surface and friction coefficient of corn on corn.

Forage harvesters

Blevins (1954) tested several commercial forage harvesters with rotary chopping and conveying elements for determining the horsepower losses due to friction. The expression for friction horsepower was derived from the frictional force and peripheral speed of the chopped forage against the fan housing.

The frictional force F_f would be the product of coefficient of friction f_s and the normal force F_n, which was taken equal to the centrifugal force of the particles impacting against the housing.

$$F_f = f_s F_n = f_s\left(\frac{W}{g}\frac{V^2}{R}\right) \tag{10.53}$$

where g is acceleration of gravity, R is radius of the cutting unit in feet, V is tangential velocity of particles in feet per minute, and W is the total weight of the material against the housing at any instant in pounds. In terms of machine capacity Q in pound per minute, W can be given by the following expression

$$W = QR\alpha/V$$

where α is the angle over which the total weight W is uniformly distributed on the housing in radians. Substituting for W in (10.53) and employing the horsepower formula yields

$$\begin{aligned} HP_{\text{friction}} &= F_f V/33000 \\ &= \frac{f_s(QR\alpha/Vg\ V^2/R)\ V}{33000\,(60)^2} \end{aligned}$$

which after simplification gives

$$HP_{\text{friction}} = 25 \times 10^{-11}\ Q\alpha f_s V^2 \tag{10.54}$$

Example

Alfalfa at 74% (w.b.) moisture content is chopped and blown into a wagon using a machine with a 7.25-inch reel operating at 888 rpm. The machine is set for a 2-inch cut at an average output of 1190 pounds per minute. If the coefficient of friction of alfalfa on steel is 0.6 and the angle α in this machine is approximated to be 1.57 radians, find the theoretical power loss due to friction of the material.

From the speed and radius of the chopper,

$$V = 2\pi\, 7.25/12 \times 88 = 3330 \text{ ft/min}$$

Theoretical power loss due to friction is estimated from (10.54)

$$HP_{\text{friction}} = 25 \times 10^{-11} \times 1190 \times 1.57 \times 0.6(3330)^2 = 3.2$$

In cases where α could not be determined, Blevins (1954) conducted tests with and without the fan housing in place and assuming that the air horsepower remained unchanged, the friction horsepower was determined from the difference of power requirements in the two tests. Increased speeds and shorter length of cut were found to increase frictional losses. Other factors being equal, the power loss due to friction of chopped corn in comparison with chopped alfalfa, as expected, was directly proportional to the ratio of the friction coefficients for the two crops times the friction loss in alfalfa. Since power loss due to friction in forage harvesters of conventional designs (particularly the flywheel type) is considerable, it was suggested that an ideal machine would be one which would cut the material and discharge it from the cutting unit, allowing no time for frictional drag or excessive velocity increases. Once the material is cut and discharged, a more efficient conveying mechanism may be designed for elevating and transport. The main advantage of the rotary cutting and conveying machines, is however, the simplicity in design and relatively low cost of manufacturing.

Sifting on an oscillating riddle

Baader (1961) investigated the movement of a mass of material on an oscillating riddle in an attempt to formulate theoretically the conditions for most efficient sifting. He found that the condition of persistent slipping and sticking can be defined by the following derived expression

$$P_t/P_n = \tan\beta - \frac{\ddot{x}}{(g + \ddot{y} - \ddot{x}\tan\beta)\cos^2\beta} = f$$

where P_t and P_n are tangential and normal components of the force exerted by riddle base on material, β is the angle of inclination of the riddle surface with horizontal, $\ddot{x}$ and $\ddot{y}$ are components of acceleration of a particle on the riddle, g is acceleration due to gravity, and f is coefficient of friction. Baader then concluded that if the left-hand side of the above equation is less than f throughout the period of oscillation, the particle is at rest in relation to the riddle base. If it intermittently exceeds f, there is an alternation between sticking and slipping. Finally, if it is steadily greater than f, the material slips continuously over the riddle surface.

Grain threshing

The concept and the problems of a threshing cone instead of the conventional threshing cylinder in a combine has been reported by several workers (Lamp and Buchele, 1960; Lalor and Buchele, 1963; Buchanan and Johnson, 1964; Hamdy *et al.*, 1967). The system consists of two coaxial truncated cones, one within the other with adjustable clearance. In one system, the outer cone is made of perforated sheet metal and is stationary, while the inner cone is made of rubber-covered angle-bar beaters and is rotating. The mechanism makes use of the centrifugal force for threshing of grain heads and as the material moves from the small end of the cone to the large end, more time is made available for the grain to separate from chaff and straw.

Lalor and Buchele (1963) determined the optimum angle of the cones by an analysis of sliding friction of the material being threshed with the inner surface of the outer cone. In Fig. 10.48 the material is resting on an inclined plane representing the inner surface of the perforated cone. F_c is the centrifugal force resulting from circular motion of the material within the cone. The components of this force, $F_c \sin\theta$ and $F_c \cos\theta$, are, respectively, the force which tends to cause the material to slide down the plane and the normal force. The friction force $fF_c \cos\theta$ prevents the material from sliding down the plane. This analysis shows that when slipping is impending,

$$fF_c \cos\theta = F_c \sin\theta \text{ or } \tan\theta = f$$

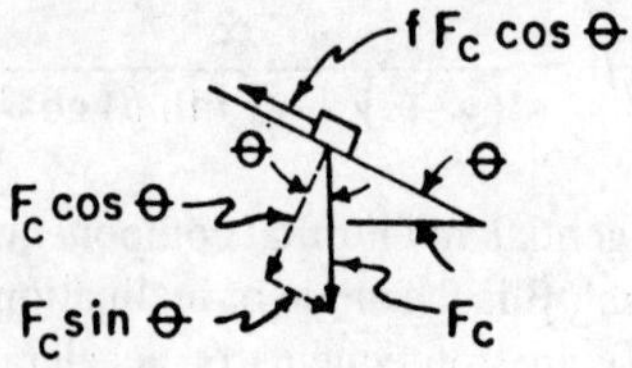

Figure 10.48 Determination of the optimum cone angle in a threshing cone from the analyses of sliding friction (Lalor and Buchele, 1963)

Knowing the coefficient of friction f, the cone angle $= 2\theta$ can be determined. For the material to slip on the surface, θ must be greater than the $\tan^{-1} f$.

Fertilizer spreaders

Cuningham (1963) reported theoretical and experimental investigations on the performance of granular fertilizer spreaders. He considered four types of spinner configurations for the spreading mechanism. Analysis of particle motion along the blades for each of the four types are given in Fig. 10.49. In the force diagram shown, W is weight of particle, a is acceleration of fertilizer particle with respect to spinner blade, M is particle mass, V is velocity of particle with respect to blade, ω is angular velocity of spinner, R is spinner radius, P is radius of blade curvature, r is radial coordinate on spinner, r_0 is radial position at which fertilizer is delivered to spinner, Φ is angle of blade with respect to spinner radius at r, δ is angle of blade with respect to spinner radius at r_0, and α is spinner cone angle.

To show the effect of various accelerations on fertilizer blade friction force, F, the reverse effective force associated with each acceleration is shown in the force diagrams by multiplying the acceleration by the particle mass. Setting the vector sum of the reverse effective forces along with the friction and gravity forces equal to zero, an expression for force F is derived for each type of blade configuration. The force expression in each case is actually the product of reverse effective normal forces and the coefficient of blade-fertilizer friction, f.

Applicability of this theory was checked by experiments including high speed photography. It was found that theoretical equations were capable of predicting departure velocity of fertilizer particle and angular location of distribution patterns.

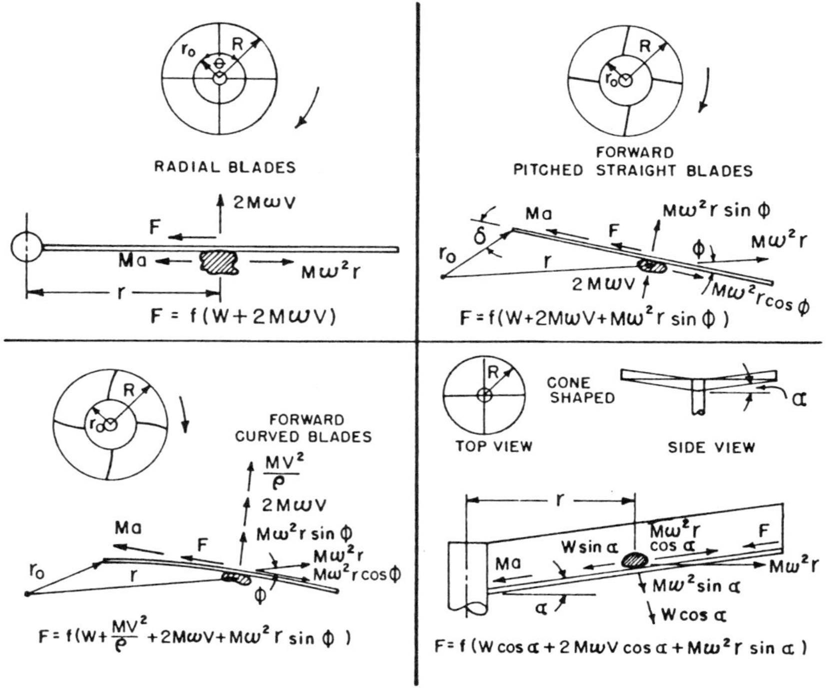

Figure 10.49 Force diagrams for four types of spinner configurations in fertilizer speaders (Cunningham, 1963)

Machine design criteria for minimum mechanical injury

One example where physical properties of the product being handled has been used as a guide in establishing the design criteria of the machine components is that reported by Wang (1963). Poha berry is an approximately spherical fruit covered by a leafy husk. It is used in making jams and is gaining importance in the economy of Hawaii. The problem was to develop a husking machine incorporating two frictional rollers to apply enough force on the husk to separate it from the berry without damage to the berry itself. The theoretical approach was to develop the equations giving the coefficient of friction of roller surface, the rotational speed, the size of the rollers, and the forces to which the berry will be subjected.

Figure 10.50 (left) shows two husking rollers with a single poha berry ready to be husked. On the right is shown the free-body diagram and the forces acting on the berry. Taking the summation of vertical forces at equilibrium,

$$W + 2F \sin \theta - 2N \cos \theta = 0$$

$$W/\sin \theta + 2F(1 - \cot \theta/f) = 0$$

where $f = F/N$, the coefficient of friction of the berry in its husk and the surface material of the rollers. Since W, the weight of the berry, cannot be negative, for the above expression to hold, $\cot \theta/f > 1$ or $\cot \theta > f$.

$$\cot \theta = \frac{\sqrt{(R_r + R_b)^2 - R_r^2}}{R_r} = \sqrt{\frac{R_b}{R_r}\left(2 + \frac{R_b}{R_r}\right)} > f \qquad (10.55)$$

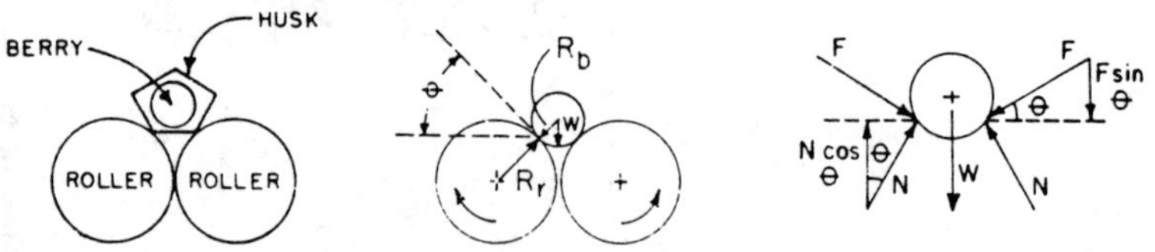

Figure 10.50 Establishment of design criteria for a husking machine. Left: A poha berry between two husking rollers. Right: Freebody diagram and the forces acting on the berry (Wang, 1963)

Thus the above expression establishes the maximum value for the radii of the rollers.

Another expression which gives the net horizontal force, F_h, on the berry can be derived as follows:

$$F_h = 2\,(F \cos\theta + N \sin\theta) = \frac{W(f \cos\theta + N \sin\theta)}{\cos\theta - f \sin\theta} \qquad (10.56)$$

If the maximum horizontal force that poha berry can withstand is denoted $F_{h_{max}}$, then

$$\frac{W(f \cos\theta + N \sin\theta)}{\cos\theta - f \sin\theta} < F_{h_{max}}$$

Substitution of physical properties of poha berries and the surface material of the husking rollers in the theoretical expression satisfies the theoretical considerations and shows that the larger the size of the rollers, the greater damage should be expected in the husking process.

Example

The crushing resistance of poha berries is given as 2.5 pounds. Sizes of these berries vary from R_b=0.187 to R_b=0.338 inch. Coefficient of friction of unhusked poha berry over rollers with rubberized surface (Durometer 60) is 0.625. If two sets of rollers, R_{r1}=0.625 inch and R_{r2}=0.75, are selected as the basic components of a husking machine and only the 0.187-inch berries are considered, find the maximum horizontal forces that one may expect from each of the two given roller sizes in terms of the berry weight W.

For $R_b = 0.187$ inch and $f = 0.625$ from Eq. 10.55

$$\cot\theta = 0.832 \text{ or } \theta = 50 \text{ deg } 14 \text{ min}$$

and from (10.56)

$$F_h = 7.29W \text{ lb for } R_{r1} = 0.625 \text{ in and}$$

$$F_h = 14.5W \text{ lb for } R_{r2} = 0.75 \text{ in}$$

Therefore, the weight of the berries, W, should be such that $F_h = 14.5W < 2.5$ lb. Apparently, the range of weights found in poha berries satisfies the above conditions.

10.9 LIST OF SYMBOLS FOR CHAPTER 10

a —radius of the circle of contact
A —area of contact
c —coefficient of rolling resistance
C —cohesion of granular materials
E —modulus of elasticity
EYL—effective yield locus
f —coefficient of friction
ff —critical flow factor
f_k —kinetic coefficient of friction
f_s —static coefficient of friction
F —frictional resistance, rolling resistance
FF —flow function
G —modulus of rigidity
$k = \dfrac{\sigma_3}{\sigma_1}$ = pressure ratio
P_d —mean displacement pressure of softer material
P_m —mean yield pressure of softer material
s —specific surface or shear stress of softer material
S —shearing force
w —weight density or bulk density
W —normal weight
YL —yield locus
δ —effective angle of internal friction
μ —Poisson's ratio
σ_1 —vertical pressure or major principal stress
σ_3 —lateral pressure or minor principal stress
σ_c —unconfined yield pressure
Φ_i —angle of internal friction
Φ_r —angle of repose
Φ_w —wall internal friction angle

APPENDIX

AS IT has been emphasized throughout this volume, the subject of physical properties of biological materials has not yet developed to the point when design standards based on mechancial properties of the material and analytical considerations of the mechanisms involved can be established. However, a rather extensive amount of data are available, at least for some materials, which may well serve as the beginning of the eventual delineation of design criteria.

Some of these data have been given along with discussion of the subject matter. Some other available data on physical properties of materials as well as some useful data for conversion of units, etc. are given in this Appendix. Data on physical properties of of agricultural products may also be found in the following sources:

Agricultural Engineering Yearbook. American Society of Agricultural Engineers, Saint Joseph, Michigan.

Agricultural Engineering Handbook, by C. B. Richey, P. Jacobson, and C. W. Hall. McGraw-Hill Book Company, Inc., New York.

Drying of Farm Crops, by C. W. Hall. Agricultural Consulting Associates, Inc., Box 330, Wooster, Ohio.

Principles, Equipment and Systems for Corn Harvesting, by W. H. Johnson and B. Lamp. Agricultural Consulting Associates, Inc. Box 330, Wooster, Ohio.

Agricultural Process Engineering by S. M. Henderson and R. L. Perry. Edwards Brothers, Inc., Ann Arbor, Michigan.

Physicomechanical Properties of Agricultural Crops by M. F. Burmistrova *et al.* Translated from Russian and published for the National Science Foundation by the Israel Program for Scientific Translations. Available from Office of Technical Services, U.S. Department of Commerce, Washington, D.C.

In addition, the Engineering Parameters Subcommittee of the Physical Properties Committee of the American Society of Agricultural Engineers, has been active for the past several years compiling information on agricultural products. The data sheet planned for each product is to give definition of terms, range and mean of observed values, and references. It is expected that these data will be published periodically until such time when a handbook can be developed for physical properties of agricultural materials.

Table A–1 Modulus conversion factors

To convert from	to	multiply by
psi	dynes/sq cm	6.895×10^4
dynes/sq cm	psi	1.450×10^{-5}
psi	kg/sq mm	7.03×10^{-4}
kg/sq mm	psi	1.422×10^3
gm/cm	psi	1.422×10^{-2}
dynes/sq cm	kg/sq mm	1.02×10^{-8}
kg/sq mm	dynes/sq cm	9.806×10^7

Table A–2 Ratio of lateral to vertical pressures in cylindrical structures for various grains (Lorenzen, 1957)*

Material	Pressure ratio with moisture content (% wet basis) in parenthesis				
Barley	0.41 (7.9)	0.43 (10.8)	0.41 (13.3)	0.39 (16.6)	0.43 (19.5)
Corn (shelled)	0.53 (7.3)	0.49 (13)	0.42 (16.2)	0.43 (19.5)	0.39 (23)
Milo	0.44 (7.8)	0.41 (12)	0.36 (14.3)	0.38 (18.6)	0.32 (22.1)
Rice (paddy)	0.38 (9.6)	0.27 (12.1)	0.28 (15.4)	0.23 (17.6)	0.31 (25)
Wheat	0.44 (7.3)	0.42 (11.0)	0.38 (14)	0.33 (17.1)	0.34 (19.3)

*The following additional data are given in the ASAE Yearbook.

Material	Pressure ratio	Reference
Wheat	0.3–0.5	Ketchum, 1911
	0.6	Jamieson, 1905
Rye	0.23–0.45	Ketchum, 1911

Table A–3 Bulk density (lb/ft^3) of grains at various moisture contents (Lorenzen, 1957)

Material	Moisture content (% wet basis) in parenthesis				
Barley	36.5 (7.9)	37 (10.8)	37 (13.3)	36 (16.6)	35.5 (19.5)
Corn (shelled)	47 (7.3)	46 (13)	45 (16.2)	43 (19.5)	41 (24.9)
Milo	47 (6.8)	47 (12)	47 (14.3)	46 (18.6)	45 (22.1)
Rice (paddy)	37.5 (9.6)	38 (12.1)	38 (15.4)	38.5 (17.6)	38 (25)
Wheat	49 (7.3)	49 (11)	47 (14.1)	45 (17.1)	44 (19.3)

Table A–4 Physical properties of water (Adopted from Vennard, 1963)

Temperature, °F	Vapor pressure lb/in^2 abs.	Weight density lb/ft^3	Mass density $\frac{\text{lb-sec}}{\text{ft}^4}$	Viscosity $10^{-5}\frac{\text{lb-sec}}{\text{ft}^2}$	Surface tension lb/ft	Modulus of elasticity 10^3 lb/in^2
32	0.09	62.42	1.940	3.746	0.00518	287
40	0.12	62.43	1.940	3.229	0.00514	296
50	0.18	62.41	1.940	2.735	0.00509	305
60	0.26	62.37	1.938	2.349	0.00504	313
70	0.36	62.30	1.936	2.050	0.00498	319
80	0.51	62.22	1.934	1.799	0.00492	324
90	0.70	62.11	1.931	1.595	0.00486	328
100	0.95	62.00	1.927	1.424	0.00480	331
110	1.27	61.86	1.923	1.284	0.00473	332
120	1.69	61.71	1.918	1.168	0.00467	332
130	2.22	61.55	1.913	1.069	0.00460	331
140	2.89	61.38	1.908	0.981	0.00454	330
150	3.72	61.20	1.902	0.905	0.00447	328
160	4.74	61.00	1.896	0.838	0.00441	326
170	5.99	60.80	1.890	0.780	0.00434	322
180	7.51	60.58	1.883	0.726	0.00427	318
190	9.34	60.36	1.876	0.678	0.00420	313
200	11.52	60.12	1.868	0.637	0.00413	308
212	14.70	59.83	1.860	0.593	0.00404	300

Table A–5 Elastic properties of materials[1]

Materials (metals)	*E*	*G*	Sp. Gr.
Aluminum, pure	9.0	3.7	2.7
Aluminum, alloys	10.0	4.0	2.78
Antimony	11.3	2.8	6.62
Beryllium	42.0		1.85
Brass	13.5	5.0	8.4
Bronze	13.0	5	8.6
Cadmium	10.0	3.5	8.6
Columbium	13.5	5.4	8.57
Copper, Hard drawn	13.0	6.0	8.9
Copper, Beryllium	18.0	6.9	8.23
Gold	11.5	3.7	19.3
Hastelloy	29.5		8.2–8.9
Incoloy	28.0		8.05
Inconel	30.0		8.2–8.4

Table A–5 (continued)

Materials (metals)	*E*	*G*	Sp. Gr.
Iron, Ferritic Malleable	25.0	10.0	7.3
Iron, Pearlitic Malleable	26.5		7.3
Iron, gray cast #20	11.8	4.8	7.2
Iron, gray cast #25	13.2	5.3	7.2
Iron, gray cast #30	14.7	5.9	7.2
Iron, gray cast #35	15.3	6.35	7.2
Iron, gray cast #40	18.0	7.1	7.2
Iron, gray cast #50	20.8	7.6	7.2
Iron, gray cast #60	22.0	8.2	7.2
Iridium	76.0		22.5
Lead	2.3	0.8	11.3
Magnesium	6.5	2.4	1.8
Molybdenum	46.0		10.2
Monel	26.0	9.6	8.6
Nickel, whrought	30.0	10.0	8.9
Osmium	80.0		22.6
Palladium	17.0	6.4	12.0
Phosphor bronze	16.0	6.0	8.89
Platinum	23.0	9.1	21.5
Rene 41	32.0	12.0	8.3
Rhodium	48.0		12.4
Ruthenium	60.0		12.2
Silver	11.0	3.8	10.5
Steel, low alloy	29.0	12.0	7.85
Steel, stainless	29.0	12.0	7.8
Tantalum	27.0	10.0	16.6
Tin	6.8	2.4	7.3
Titanium	16.0	6.5	4.5
Tungsten	50.0	19.6	19.3
Vanadium	19.0	6.73	6.1
Zinc (die cast Zamak)	12.0	5.0	7.1

Material (non metals)	*E*	*G*	Sp. Gr.
Acrylic (Plexiglass)	0.44		1.2
Alkyd—Fibrous Filler	2.2		
Alkyd—Granular Filler	2.1		
Cellulose Acetate Butyrate	0.12		1.2
Diallyl Phthalate	0.5		1.3
Epoxy—Unfilled	0.35		1.25
Epoxy—60% glass fiber	3.0		1.8
Ethyl Cellulose	0.25		1.1
Melamine & Glass Fabric	1.6		1.9

Table A–5 (continued)

Materials (non metals)	E	G	Sp. Gr.
Melamine & Alpha Cellulose	1.35		1.5
Melamine & Wood Floor	1.0		1.5
Melamine & Rag	1.4		1.5
Melamine & Asbestos	1.95		1.5
Nylon 6	0.38		1.14
Nylon 6/6	0.41		1.14
Nylon 6/10	0.28		1.14
Phenolic—Molded	1.3		1.4
Phenolic & Glass Fabric	1.4		1.65
Phenolic—Cotton Fabric	0.9		1.33
Phenoxy	0.39		1.25
Polycarbonate	0.34		1.2
Polycarbonate—40% glass fiber	1.7		1.52
Polyester—rigid	6.0–7.0		1.25
Polyester—resilient	5.2		1.12
Polyester—30% glass cloth	1.8		1.6
Polyester—30% glass mat	1.5		1.5
Polyimides	0.46		1.43
Polypropylene	0.16		0.91
Polystyrene	0.45		1.06
Silicone—glass fabric	1.3		1.65
Silicone—fiber	2.5		1.88
Silicone—granular silica	1.6		1.86
Urea—alpha cellulose	1.35		1.5
Vinyl—rigid	0.43		1.4
Brick, soft	1.8		1.87
Brick, hard	3.5		2.24
Concrete	3.2		2.34
Glass	8.7[2]	2.0[3]	2.7
Ice	1.4[3]		
Limestone	8.4[2]	3.3[2]	2.58
Fired structural clay	10.0		2.5
Rubber, soft	0.0005		1.0
Rubber, hard	0.2		1.2
Silk	1.6[3]	0.14[3]	
Wood, common varieties	1.3		0.34
parallel to grain, dry			0.78

[1] Courtesy University Precision Measurement Co.
[2] Nielson, 1962.
[3] Scott Blair and Reiner, 1957.

E = Young's modulus in 10^6 psi
G = shear modulus in 10^6 psi
Sp. Gr. = specific gravity

Table A–6 Energy and force required to initiate fracture in grains under quasi-static loading at a crosshead speed of 0.050 in/min (Bilanski, 1965)

Type of grain	Moisture %	Kernel position	Energy (in-lbs)		Force (lbs)	
			Ave.	Min.	Ave.	Min.
Soybeans	1.0	Hilum horizontal	0.034	0.025	13	11
		Hilum vertical	0.029	0.021	12	7
	6.0	Hilum horizontal	0.038	0.033	13	11
		Hilum vertical	0.047	0.042	12	11
	10.0	Hilum horizontal	0.083	0.050	12	8
		Hilum vertical	0.089	0.063	9	7
	16.0	Hilum horizontal	0.279	0.166	10	9
		Hilum vertical	0.150	0.083	7	6
Corn	1.0	Germ side down	0.200	0.167	32	28
		Kernel on edge	0.023	0.017	12	10
	8.0	Germ side down	0.496	0.334	90	72
		Kernel on edge	0.043	0.033	43	28
	17.0	Germ side down	0.606	0.508	71	64
		Kernel on edge	0.043	0.033	22	18
Wheat	1.0	Crease down	0.046	0.025	13	9
		Crease on side	0.040	0.033	11	9
	8.0	Crease down	0.098	0.076	13	12
		Crease on side	0.058	0.042	12	9
	18.0	Crease down	0.107	0.092	10	8
		Crease on side	0.114	0.077	9	8
Barley	1.0	Crease down	0.035	0.025	14	12
		Crease on side	0.028	0.025	7	6
	10.0	Crease down	0.045	0.042	13	11
		Crease on side	0.042	0.042	14	12
	17.0	Crease down	0.051	0.042	12	11
		Crease on side	0.041	0.029	10	8
Oats	1.0	Crease down	0.077	0.063	9	7
		Crease on side	0.111	0.073	8	5
	10.0	Crease down	0.105	0.083	8	7
		Crease on side	0.151	0.075	9	7
	16.0	Crease down	0.117	0.083	9	8
		Crease on side	0.143	0.121	10	9

Table A–7 Specific gravity of fruits and vegetables

Product	Specific gravity	Reference	Product	Specific gravity	Reference
Apple			Lisbon	0.92	Turrel & Slack, 1948
Delicious	0.83	Cooper, 1962	Marsh Grapefruit	0.81	Turrel & Slack, 1948
Delicious	0.83	Mohsenin *et al.*, 1965			
Golden delicious	0.75	Cooper, 1962	Citrus-Orange		
Golden delicious	0.81	Mohsenin *et al.*, 1965	Valencia	0.93	Turrel & Slack, 1948
Golden delicious	0.79–0.84		Washington Naval	0.95	Turrel & Slack, 1948
	0.81	PSU, Unpublished			
McIntosh	0.74	Cooper, 1962	Coffee	1.01–1.09	
McIntosh	0.81	Mohsenin *et al.*, 1965		1.05	Eschenwald, 1959
McIntosh	0.77–0.80				
	0.78	PSU, Unpublished	Pear		
Melba	0.71	Cooper, 1962	Maxine	0.98–1.00	
Melba	0.79	Mohsenin *et al.*, 1965		0.99	PSU, Unpublished
Melba	0.76–0.81				
	0.80	PSU, Unpublished	Peach		
Rome beauty	0.79	Cooper, 1962	Elberta	0.99–1.01	
Rome beauty	0.82	Mohsenin *et al.*, 1965		0.99	PSU, Unpublished
Rome beauty	0.82–0.87		Red Haven	0.98–1.03	
	0.84	PSU, Unpublished		1.00	PSU, Unpublished
Stayman	0.82	Cooper, 1962			
Stayman	0.86	Mohsenin *et al.*, 1965	Plum	0.99–1.08	
Stayman	0.82–0.89			1.05	PSU, Unpublished
	0.86	PSU, Unpublished			

Table A–7 (continued)

Product	Specific gravity	Reference	Product	Specific gravity	Reference
			Potato		
Blueberry			Golden wonder	1.15	Gilfillan & Crowther, 1959
Jersey	1.01	Sides *et al.*, 1962	Katahdin	1.06–1.08	
—	0.70–1.20			1.07	Sides *et al.*, 1962
	1,00	PSU, Unpublished	Katahdin	1.07	Finney, 1963
			Kennebec	1.07	Finney, 1963
Cherry			Kerr's Pink	1.12	Gilfillan & Crowther, 1959
—	0.97–1.05		Netted Gem	1.07–1.11	
	1.02	PSU, Unpublished		1.08	Timbers, 1964
			Onaway	1.06	Finney, 1963
Citrus Lemon			Russet Burbank	1.07	Finney, 1963
Eureka	0.93	Turrel & Slack, 1948	Sebago	1.07	Finney, 1963

Table A–8 Physical characteristics of fruits and vegetables[1]

Product	a (in)	b (in)	c (in)	GMD (in)	½ $(a \times c)$	W × 10^2 (lb)	Volume (in^3)	d_s	Reference
Apple									
Cortland					2.75	39			Quackenbush, 1961
Golden delicious	2.6–3.0	2.5–2.9	2.0–2.4			27–41	8.9–13.9		
	2.76	2.66	2.22	2.54		32	11.0	50.0	PSU, Unpublished
Jonathan					2.55	29			Quackenbush, 1961
McIntosh	2.4–3.0	2.4–2.9	1.7–2.3			29–32	6.7–11.5		
	2.70	2.59	1.94	2.38		25	8.8	49.3	PSU, Unpublished
McIntosh					2.73	37			Quackenbush, 1961
Melba	1.9–2.5	1.8–2.5	1.6–2.1			30–31	3.1–7.0		
	2.34	2.23	1.93	2.16		38	6.1	51.0	PSU, Unpublished
Northern spy					2.79	40			Quackenbush, 1961
Red delicious	2.7–3.3	2.6–3.3	2.0–3.0			30–63	9.5–21.3		
	3.02	2.82	2.50	2.77		46.4	14.9	54.0	PSU, Unpublished
Rome beauty	2.8–3.5	2.7–3.4	2.0–2.8			29–65	9.5–21.8		
	3.13	2.97	2.33	2.78		43	14.3	52.0	PSU, Unpublished
Stayman	2.6–3.1	2.4–3.0	1.9–2.5			32–46	7.9–14.6		
	2.90	2.76	2.22	2.61		36	11.7	53.0	PSU, Unpublished
Apricot									
Henderson						4			Abu-Gheida, 1961
Montgamet						6			Quackenbush, 1961
Blueberry									
Jersey	0.50		0.46			0.17		48.4	Schmidt & Levin, 1963

[1] a = major diameter; b = intermediate diameter; c = minor diameter; GMD = geometric mean diameter; ½$(a + c)$ = average diameter; W = weight (lb); d_s = unit density (lb/ft^3).

Table A–8 (continued)

Product	a (in)	b (in)	c (in)	GMD (in)	½ ($a \times c$)	W × 10^2 (lb)	Volume (in^3)	d_s	Reference
Blueberry (continued)									
Jersey	0.34		0.25			0.07		83.5	Schmidt & Levin, 1963
						0.10			Quackenbush, 1961
—	0.46	0.45	0.34	0.41		0.18	0.05	62.2	PSU, Unpublished
Cherry									
Giant						1.0			Abu-Gheida, 1961
Montmorency	0.75		0.63		0.67	0.82		82.9	Schmidt & Levin, 1963
						0.73			Abu-Gheida, 1961
					0.83	1.00			Quackenbush, 1961
—	0.7–0.8	0.6–0.7	0.6–0.7			0.7–1	0.18–0.24		
	0.76	0.65	0.63	0.67		0.82	0.22	64.5	PSU, Unpublished
Coffee				(0.25–0.375)			(0.019–0.023)	(69–72)	Ghosh, 1966
Cranberry	0.75		0.56		0.66	0.32		37.7	Schmidt & Levin, 1963
Grape									
Delaware	0.53					0.35		81.2	Schmidt & Levin, 1963
	0.66					0.68		79.5	Schmidt & Levin, 1963
Peach									
Cardinal								59.4	Bennett, 1965
Dixie Gem								59.5	Bennett, 1965
Early red free								60.1	Bennett, 1965
Elberta								59.7	Bennett, 1965
Elberta					2.78	40			Schmidt & Levin, 1963
Elberta	2.5–2.9	2.5–2.8	2.3–2.8			30–47	9.2–12.9		
	2.71	2.63	2.57	2.60		39	10.8	62.0	PSU, Unpublished

Table A–8 (continued)

Product	a (in)	b (in)	c (in)	GMD (in)	½ ($a \times c$)	W × 10^2 (lb)	Volume (in^3)	d_s	Reference
Peach (continued)									
Hale haven								59.5	Bennett, 1965
Red haven								59.8	Bennett, 1965
Red haven					2.76	40			Schmidt & Levin, 1963
Red haven	2.2–2.4	2.2–2.4	2.1–2.3			22–29	5.8–7.9		
	2.29	2.25	2.16	2.21		25	6.8	62.5	PSU, Unpublished
Pear									
Maxine	3.1–3.7	2.6–3.2	2.5–3.0			37–63	10.4–17.4		
	3.29	2.88	2.68	2.90		48	13.3	62.4	PSU, Unpublished
Plum									
Stanley prune					1.59	8			Quackenbush, 1961
—	1.7–2.0	1.6–1.9	1.5–1.8			9.5–15	2.7–4.0		
	1.81	1.76	1.61	1.71		11.6	3.1	64.6	PSU, Unpublished
Potatoes									
(Kerr Pink)	3.22	2.83	2.06	2.66	2.64	44.9	11.23	69.8	Gilfillan (1959)
Tomato	1.9–3.4	1.6–3.2	1.5–2.3			12–60	3.2–16.5		
	2.54	2.33	1.86	2.20		28	7.7	62.8	PSU, Unpublished
Walnut									
(black)						1.1–1.37	0.59–1.35	29–69	Mueller *et al.*, (1967)
						1.17	0.88	57	

Table A–9 Mechanical properties of skins of fruits and vegetables in tension[1]

Product	Stage of develop-ment	Rate of loading (ipm)	Rupture stress (psi)	Rupture energy (in-lb)	Initial tangent modulus (psi)	Secant modulus (psi)	Poisson's ratio	Reference
Apple								
Winsesap	[2]	0.21, 0.84	84–232 154	38×10^{-3}	1370	1300 at 0.05 in/in 1230 at 0.1 in/in	0.47	Clevenger, 1966
Golden delicious	[2]	0.21, 0.84	99–183 134	13×10^{-3}	2085	1930 at 0.05 in/in 1450 at 0.1 in/in	0.31	Clevenger, 1966
Red delicious	[2]	0.21, 0.84	134–263 205	41×10^{-3}	2085	1900 at 0.05 in/in 1620 at 0.1 in/in	0.32	Clevenger, 1966
Potato								
Kennebec	[3]	0.0002–15 in/in/sec	100			2400 at 0.05 in/in		Huff, 1966
Cherry (bing)								
Long. section	[2]							
35% (m.c.)					120			Levin *et al.*, 1959
43% (m.c.)					105			Levin *et al.*, 1959
Trans. section	[2]							
35% (m.c.)					218			Levin *et al.*, 1959
43% (m.c.)					141			Levin *et al.*, 1959

[1] All tests were performed at room temperature. Mean values are shown below range.
[2] The stage of maturity and other details were not specified.
[3] Stored at 34°F for one to four months after harvest.

Table A–10 Mechanical properties of flesh of fruits and vegetables[1]

Product	Stage of maturity	Rupture stress (psi)	Apparent elastic modulus (psi)	Secant modulus	Degree of elasticity	Shear stress (psi)	Reference
Apple							
Delicious	*M*20		1320				Cooper, 1962
Delicious	*M*09		486				Mohsenin *et al.*, 1965
Delicious	—	9.6 (0.8)	856			27.4 (2.1)	PSU, Unpublished
Delicious	—				0.75		Mohsenin & Gohlich, 1962
Golden delicious	*M*10		1020				Cooper, 1962
Golden delicious	*M*20		549				Mohsenin *et al.*, 1965
Golden delicious	—	9.8 (1.7)	677			26.3 (4.6)	PSU, Unpublished
Golden delicious	—				0.75		Mohsenin & Gohlich, 1962
McIntosh	*M*18		720				Cooper, 1962
McIntosh	*M*09		693				Mohsenin *et al.*, 1965
McIntosh	—	10.4 (1.3)	879			29.3 (2.2)	PSU, Unpublished
McIntosh					0.75		Mohsenin & Gohlich
Melba	*M*16		1040				
Melba	*M*08		613				
Melba	—	4.7 (1.2)	676			12.6 (4.5)	PSU, Unpublished
Melba	—				0.75		Mohsenin & Gohlich

Table A–10 (continued)

Product	Stage of maturity	Rupture stress (psi)	Apparent elastic modulus (psi)	Secant modulus	Degree of elasticity	Shear stress (psi)	Reference
Rome beauty	*M*19		1519				Cooper, 1962
Rome beauty	*M*11		1031				Mohsenin *et al.*, 1965
Rome beauty	—				0.75		Mohsenin & Gohlich, 1962
Stayman	*M*21		795				Cooper, 1962
Stayman	*M*13		895				Mohsenin *et al.*, 1965
Stayman					0.75		Mohsenin & Gohlich, 1962
Potato							
Kennebec		142		1209 at fail.			Huff, 1966
Kennebec		96		328 at fail.			Huff, 1966
Netted gem			465–652		0.64–0.91		
			579		0.77		Timbers, 1964
Russet burbank			450–650		0.32–0.60		
			543		0.46		Finney, 1967

[1] All tests were performed at room temperature. Single numbers represent means. Numbers in parenthesis are standard deviations. All apple data refer to the commercial harvest date. Numbers preceded by *M* indicate Magness pressure readings. Apple tests were compression tests of ½″ × ½″ cylindrical plugs. Potato tests were tensile tests of specimen taken from center and between skin and center in the order given. Rates of loading were 1 ipm or less.

Table A–11 Aerodynamic properties of fruits and vegetables[1]

Product	Weight (lb)	Ratio of wt. to projected area (psi)	Terminal velocity (fpm)	Reynold's number $\times 10^{-5}$	Drag coefficient	Reference
Apple						
Cortland	0.39	0.066	8130	1.89		Quackenbush, 1961
Jonathan	0.29	0.056	7580	1.63		Quackenbush, 1961
McIntosh	0.37	0.063	7950	1.83		Quackenbush, 1961
Northern spy	0.40	0.066	8100	1.91		Quackenbush, 1961
Apricots						
Montgamet	0.06	0.039	6250	0.76		Quackenbush, 1961
Blueberries						
Jersey	0.001	0.002	3450	0.10		Quackenbush, 1961
	0.001	—	1800	—		Schmidt, 1963
Cherries						
Montmorency	0.010	0.019	4440	0.31		Quackenbush, 1961
	0.008	—	2830	—		Schmidt, 1963
Cranberry	0.003		1910			Schmidt, 1963
Coffee (fruit)			3950			Eschenwald, 1959
Grape						
Delaware	0.006		2990			Schmidt, 1963

[1] Single values are the mean and numbers in parenthesis represent standard deviation.

Table A–11 (continued)

Product	Weight (lb)	Ratio of wt. to projected area (psi)	Terminal velocity (fpm)	Reynold's number $\times 10^{-5}$	Drag coefficient	Reference
Peaches						
Elberta	0.40	0.060	8160	1.92		Quackenbush, 1961
Red haven	0.40	0.066	8080	1.89		Quackenbush, 1961
Plums						
Stanley Prune	0.08	0.039	6290	0.85	0.64	Quackenbush, 1961
Potatoes						
Golden wonder	0.32		4500–6900	1.09–2.17		Gilfillan, 1959
			5280	1.0		
Kerr's pink	0.37		4800–6900	1.09–2.17	0.64	Gilfillan, 1959
				1.0		
Walnut (black)			4096	0.42	0.73	Mueller, 1966
			(499)		(0.07)	

Table A–12 Coefficient of friction of fruits and vegetables

Product	Kinetic					Static					Reference
	wood	metal	etha-foam	plastic foam	canvas	wood	metal	etha-foam	plastic foam	canvas	
Apple											
Delicious	0.33	0.31	0.29	0.60	0.28	0.37	0.40	0.40	0.72	0.36	Cooper, 1962
Golden delicious	0.32	0.34	0.36	0.73	0.31	0.44	0.46	0.40	0.93	0.44	Cooper, 1962
McIntosh	0.29	0.28	0.33	0.64	0.27	0.36	0.38	0.38	0.80	0.38	Cooper, 1962
Melba	0.24	0.36	0.38	0.72	0.25	0.32	0.44	0.44	0.80	0.39	Cooper, 1962
Rome beauty	0.29	0.29	0.28	0.75	0.36	0.33	0.39	0.34	0.88	0.42	Cooper, 1962
Stayman	0.33	0.36	0.34	0.55	0.27	0.42	0.44	0.38	0.74	0.40	Cooper, 1962
Tomato	0.42	0.31	0.48	0.71	0.49	0.43	0.36	0.47	0.74	0.50	PSU, Unpublished

Table A–12a Coefficient of friction of Virginia-type tobacco leaves on surfaces (From Suggs and Splinted, 1960)

Variety	Kinetic			Static		
	Sheet metal	Neoprene canvas	Plywood	Sheet metal	Neoprene canvas	Plywood
Hicks	0.75–1.13	0.73–1.13	0.87–0.93	0.81–1.15	0.93–1.12	0.97–1.23
D.B. 101	0.84–1.38	0.82–1.09	0.73–1.04	0.93–1.42	0.90–1.33	0.92–1.19
McNair 121	0.85–1.38	0.87–1.04	0.73–0.90	1.00–1.60	0.98–1.48	0.95–1.23
N.C. 73	0.84–1.26	0.88–1.17	0.87–1.11	0.88–1.35	1.05–1.51	0.92–1.30

Table A–12b Kinetic coefficient of friction of grains on surfaces (Condensed from Bickert and Buelow, 1967)

Product	Moisture (%w.b.)	Surface	
		sheet metal	plywood
Barley	10–16.5	0.38	
	20–23	0.40–0.42	
Corn (shelled)	10–17.5	0.27	0.23
	20–22	0.29–0.30	0.24–0.25

Table A–13 Part I Static coefficient of friction of seed grains

Product	Moisture content[1]	Concrete: Plastic smooth finish	Concrete: Steel trowel finish	Concrete: Wood float finish	Concrete: (Surface speci-fied)	Self grain	Metal: Galvanized sheet metal	Metal: Mild steel	Metal: Steel	Reference
Barley	10.7	0.23	0.56	0.50			0.20	0.20		Brubaker & Pos, 1965
	12.3	0.25	0.55	0.52			0.17	0.25		Brubaker & Pos, 1965
	14.3	0.24	0.57	0.51			0.20	0.23		Brubaker & Pos, 1965
	16.4	0.33	0.62	0.55			0.34	0.21		Brubaker & Pos, 1965
	—				0.45	0.51			0.38	Airy, 1898
	—	0.45				0.53			0.38	Stahl, 1950
	7.9								0.40	Lorenzen, 1957
	10.8								0.40	Lorenzen, 1957
	13.3								0.40	Lorenzen, 1957
	16.6								0.38	Lorenzen, 1957
	19.5								0.39	Lorenzen, 1957
Beans					0.44	0.62			0.37	Airy, 1898
Corn					0.42	0.52			0.37	Airy, 1898
	7.5	0.27	0.41	0.46			0.20	0.20		Brubaker & Pos, 1965
	9.9	0.25	0.59	0.62			0.24	0.25		Brubaker & Pos, 1965
	12.2	0.33	0.68	0.65			0.25	0.23		Brubaker & Pos, 1965
	13.9	0.35	0.64	0.54			0.34	0.21		Brubaker & Pos, 1965
Corn	7.3								0.53	Lorenzen, 1957
	13.0								0.47	Lorenzen, 1957
	16.2								0.48	Lorenzen, 1957
	19.5								0.59	Lorenzen, 1957
	23.1								0.76	Lorenzen, 1957

Table A–13 Part I (continued)

Product	Moisture content[1]	Concrete				Self grain	Metal			Reference
		Plastic smooth finish	Steel trowel finish	Wood float finish	(Surface speci-fied)		Galvanized sheet metal	Mild steel	Steel	
Flaxseed					0.41	0.46			0.34	Airy, 1898
		0.41				0.47				Stahl, 1950
Milo	6.8								0.35	Lorenzen, 1957
	12.0								0.32	Lorenzen, 1957
	14.3								0.29	Lorenzen, 1957
	18.6								0.39	Lorenzen, 1957
	22.1								0.52	Lorenzen, 1957
Oats					0.47	0.53			0.41	Airy, 1898
		0.47				0.62			0.41	Stahl, 1950
	10.6	0.28	0.40	0.43			0.22	0.20		Brubaker & Pos, 1965
	13.0	0.34	0.44	0.44			0.24	0.26		Brubaker & Pos, 1965
	14.0	0.33	0.51	0.42			0.18	0.21		Brubaker & Pos, 1965
	16.0	0.29	0.46	0.46			0.41	0.20		Brubaker & Pos, 1965
	17.3	0.50	0.65	0.64			0.32	0.44		Brubaker & Pos, 1965
Peas					0.30	0.47			0.26	Airy, 1898
Rice										
	14				0.52	0.73			0.41	Stahl, 1950
Blue rose	14	0.52–0.53				0.73	0.40–0.41			Kramer, 1964
Rexora	14	0.47–0.61				0.68	0.45			Kramer, 1964
Rye					0.85					Ketchum, 1911
		0.35				0.49			0.41	Stahl, 1950

Table A–13 Part I (continued)

Product	Moisture content[1]	Concrete: Plastic smooth finish	Concrete: Steel trowel finish	Concrete: Wood float finish	Concrete: (Surface speci-fied)	Self grain	Metal: Galvanized sheet metal	Metal: Mild steel	Metal: Steel	Reference
Sorghum-grain		0.33				0.65			0.37	Stahl, 1950
Soybeans										
		0.44				0.55			0.37	Stahl, 1950
	7.1	0.25	0.39	0.39			0.21			Brubaker & Pos, 1965
	8.1	0.32	0.55	0.52			0.21			Brubaker & Pos, 1965
	9.8	0.31	0.47	0.37			0.18			Brubaker & Pos, 1965
	12.2	0.36	0.55	0.52			0.20			Brubaker & Pos, 1965
Sugar beet seed		0.82				0.80			0.52	Stahl, 1950
Wheat					0.44	0.47			0.41	Airy, 1898
					0.40–0.43	0.53			0.36–0.46	Jamieson, 1905
		0.42				0.53				Stahl, 1950
	11.2	0.36	0.52	0.51			0.10			Brubaker & Pos, 1965
	13.0	0.46	0.52	0.55			0.14			Brubaker & Pos, 1965
	15.0	0.50	0.55	0.51			0.27			Brubaker & Pos, 1965
	15.7	0.56	0.68	0.69			0.33			Brubaker & Pos, 1965
	7.3								0.37	Lorenzen, 1957
	11.0								0.39	Lorenzen, 1957
	14.1								0.43	Lorenzen, 1957
	17.1								0.44	Lorenzen, 1957
	19.3								0.55	Lorenzen, 1957

[1] Moisture content is given in percent wet basis except where noted otherwise.

Table A–14 Part II Static coefficient of friction of seed grains

Product	Moisture[1] content %	Plastic			Wood							Reference
		Teflon	Poly-ethyl-ene	Plexi-glass	Douglas Fir Grain par.	Douglas Fir Grain perp.	Oak Grain par.	Oak Grain perp.	Rough wood	Smooth wood	Unspeci-fied	
Barley	10.7	0.17	0.23		0.27	0.32	0.23	0.29				Brubaker & Pos, 1965
	12.3	0.15	0.28		0.28	0.31	0.21	0.28				Brubaker & Pos, 1965
	14.3	0.13	0.28		0.30	0.32	0.21	0.28				Brubaker & Pos, 1965
	16.4	0.11	0.35		0.37	0.41	0.30	0.33				Brubaker & Pos, 1965
									0.42	0.33		Airy, 1898
	7.9										0.46	Lorenzen, 1957
	10.8										0.45	Lorenzen, 1957
	13.3										0.47	Lorenzen, 1957
	16.6										0.46	Lorenzen, 1957
	19.5										0.50	Lorenzen, 1957
Beans									0.44	0.32		Airy, 1898
Corn									0.34	0.31		Airy, 1898
	7.5	0.17	0.22		0.27	0.29	0.24	0.25				Brubaker & Pos, 1965
	9.9	0.18	0.27		0.31	0.31	0.28	0.31				Brubaker & Pos, 1965
	12.2	0.16	0.30		0.33	0.33	0.26	0.29				Brubaker & Pos, 1965
	13.9	0.12	0.35		0.37	0.38	0.29	0.36				Brubaker & Pos, 1965
Corn	7.3										0.39	Lorenzen, 1957
	13.0										0.40	Lorenzen, 1957
	16.2										0.42	Lorenzen, 1957
	19.5										0.45	Lorenzen, 1957
	23.1										0.49	Lorenzen, 1957

Table A–14 Part II (continued)

Product	Moisture[1] content %	Plastic			Wood							Reference
		Teflon	Poly-ethyl-ene	Plexi glass	Douglas Fir		Oak		Rough wood	Smooth wood	Unspeci-fied	
					Grain par.	Grain perp.	Grain par.	Grain perp.				
Flaxseed								—	0.41	0.31		Airy, 1898
Milo	6.8										0.46	Lorenzen, 1957
	12.0										0.44	Lorenzen, 1957
	14.3										0.47	Lorenzen, 1957
	18.6										0.48	Lorenzen, 1957
	22.1										0.51	Lorenzen, 1957
Oats									0.45	0.37		Airy, 1898
	10.6	0.13	0.20		0.27	0.29	0.20	0.23				Brubaker & Pos, 1965
	13.0	0.14	0.24		0.29	0.35	0.24	0.25				Brubaker & Pos, 1965
	14.0	0.13	0.38		0.34	0.36	0.23	0.25				Brubaker & Pos, 1965
	16.0	0.11	0.31		0.37	0.37	0.31	0.31				Brubaker & Pos, 1965
	17.3	0.14	0.50		0.48	0.50	0.46	0.48				Brubaker & Pos, 1965
Peas									0.29	0.27		Airy, 1898
Rice												
Blue rose	14				0.44	0.50						Kramer, 1944
Rexora	14				0.50	0.53						Kramer, 1944
Rye											0.38	Stahl, 1950
											0.37–0.55	Ketchum, 1911

Table A–14 Part II (continued)

Product	Moisture[1] content %	Plastic: Teflon	Plastic: Poly-ethyl-ene	Plastic: Plexi glass	Wood: Douglas Fir Grain par.	Wood: Douglas Fir Grain perp.	Wood: Oak Grain par.	Wood: Oak Grain perp.	Wood: Rough wood	Wood: Smooth wood	Wood: Unspeci-fied	Reference
Soybeans	7.1		0.25		0.29	0.31						Brubaker & Pos, 1965
			0.32		0.32	0.37						Brubaker & Pos, 1965
	9.8		0.29		0.33	0.31						Brubaker & Pos, 1965
	12.2		0.43		0.35	0.44						Brubaker & Pos, 1965
Sugar beet seed										0.70		Stahl, 1950
Wheat									0.41	0.36		Airy, 1898
											0.42–0.45	Jamieson, 1905
											0.25–0.45	Ketchum, 1911
Vetch										0.26		Stahl, 1950
Wheat										0.46		Stahl, 1950
	11.2		0.27		0.31	0.35						Brubaker & Pos, 1965
	13.0		0.35		0.35	0.38						Brubaker & Pos, 1965
	15.0		0.39		0.47	0.46						Brubaker & Pos, 1965
	15.7		0.45		0.48	0.50						Brubaker & Pos, 1965
	7.3										0.45	Lorenzen, 1957
	11.0										0.43	Lorenzen, 1957
	14.1										0.43	Lorenzen, 1957
	17.1										0.47	Lorenzen, 1957
	19.3										0.59	Lorenzen, 1957

[1] Moisture content is given in percent wet basis except where noted otherwise.

Table A-15 Coefficient of friction of coffee fruit and coffee beans[1]

Surface	Ripe fruits		Green fruits		Pulped unwashed beans	Pulp	Washed wet beans	Dry beans 12% m.c. (d.b.)
	d.s.[2]	w.s.	d.s.	w.s.				
Concrete (wood floated)	0.64	0.59	0.64	0.62	0.94	0.70	1.03	0.55
Concrete (as taken from forms)	0.68	0.55	0.59	0.59	0.93	0.62	0.98	0.51
Rough wood (with grain)	0.61	0.75	0.77	0.85				
Rough wood (across grain)	0.68	0.74	0.59	0.66				
Planed wood (with grain)	0.61	0.70	0.51	0.67	0.90	0.75	1.07	0.42
Planed wood (across grain)	0.61	0.66	0.59	0.67	0.93	0.81	1.15	0.49
Plywood (with grain)	0.59	0.69	0.50	0.60				
Plywood (across grain)	0.64	0.71	0.55		0.62			0.53
Galvanized iron	0.57	0.47	0.50	0.47	0.54	0.60	0.87	0.40
Aluminum	0.58	0.52	0.54	0.47	0.49	0.58	0.78	0.40
Stainless steel	0.64	0.55	0.49	0.45			0.65	0.34
Steel	0.59	0.49	0.59	0.58	0.65	0.73	0.90	0.53
Steel perforated 1/16″–5/32″ diameter							0.90	
Steel perforated 3/16″ diameter							0.75	
Window screen 14 × 18 mesh	0.78	0.79	0.65	0.65			1.11	
Green fruit				1.33				
Ripe fruit		1.15						

[1] Original data by Eschenwald (1959).

[2] d.s. and w.s. indicate dry surface and wet surface, respectively.

Table A–16 Coefficient of friction of roughage materials on various surfaces*
(From Farm Building Standards, Supplement No. 6 to the National Building Code of Canada)

Material	Moisture content %	Surfaces: Concrete: Plastic smooth finish	Concrete: Steel trowel finish	Concrete: Wood float finish	Wood: Oak: Grain par.	Wood: Oak: Grain perp.	Wood: Douglas Fir: Grain par.	Wood: Douglas Fir: Grain perp.	Plastic: Teflon	Plastic: Poly-ethyl-ene	Metal: Mild steel C.R.	Metal: Galvan-ized sheet metal
Alfalfa	82.0	0.737	0.686	0.775	0.610	0.674	0.697	0.614	0.191	0.610	0.653	0.535
	33.3	0.478	0.562	0.714	0.373	0.478	0.383	0.488	0.179	0.394	0.510	0.374
	22.2	0.328	0.649	0.655	0.310	0.333	0.334	0.374	0.160	0.320	0.458	0.359
Timothy	79.3	0.584	0.598	0.765	0.521	0.532	0.637	0.594	0.226	0.661	0.570	0.526
	30.5	0.373	0.480	0.731	0.441	0.382	0.417	0.522	0.192	0.383	0.388	0.483
	16.7	0.270	0.446	0.629	0.347	0.423	0.398	0.439	0.215	0.213	0.315	0.318
Alfalfa (75%), Timothy (25%)	77.0	0.631	0.677	0.775	0.576	0.603	0.603	0.698	0.266	0.645	0.650	0.644
	26.2	0.269	0.488	0.727	0.306	0.391	0.361	0.423	0.191	0.332	0.363	0.375
	21.3	0.258	0.486	0.616	0.312	0.355	0.315	0.386	0.197	0.194	0.345	0.266
Alfalfa (25%), Timothy (75%)	81.1	0.619	0.686	0.827	0.518	0.638	0.659	0.647	0.225	0.618	0.569	0.591
	49.3	0.514	0.602	0.822	0.437	0.558	0.448	0.585	0.210	0.611	0.431	0.498
	21.6	0.245	0.530	0.663	0.308	0.379	0.368	0.428	0.203	0.226	0.315	0.291
Corn silage	78.4	0.456	0.560	0.699	0.583	0.563	0.567	0.581	0.184	0.401	0.569	0.493
Oat straw	14.95	0.202	0.360	0.454	0.197	0.260	0.222	0.253	0.139	0.219	0.351	0.304
Wood shavings	9.5	0.354	0.699	0.725	0.460	0.528	0.427	0.514	0.200	0.294	0.566	0.384

* Richter (1954) based on his experiments, recommeded the following values for roughage friction on polished galvanized steel.

Material	Coefficient of friction: Static	Coefficient of friction: Kinetic
Grass and silages (chopped)	0.80	0.70
Straw of hay (chapped)	0.35	0.30

Table A–17 Friction force and pressure required to cause skinning of potatoes (Mohsenin, 1965)

Date of test	Variety	No. of runs	Friction force (lbs)		Normal force (lbs)		Contact area $(in)^2$		Normal pressure (lb/in^2)		Coefficient of friction	
			Range	Mean	Range	Mean	Range	Mean	Range	Mean	Range	Mean
(Potato against potato)												
10/12/65	Kennebec	13	2.1–3.2	2.50	3.5–6.0	4.54	0.080–0.195	0.13	23.6–50.0	36.2	0.52–0.68	0.56
10/24/62	Kennebec	20	1.8–2.6	1.97	4.0–6.0	4.50	0.064–0.186	0.11	31.2–67.5	41.0	0.36–0.50	0.44
10/30/62	Kennebec	20	1.9–4.8	3.16	4.0–13.5	7.90	0.079–0.346	0.20	26.0–54.5	40.0	0.27–0.52	0.42
11/27/62	Kennebec	20	1.8–3.4	2.50	4.0–7.0	5.45	0.076–0.216	0.13	26.0–65.5	43.3	0.33–0.60	0.45
2/5/63	Kennebec	20	6.0–15.0	10.87	13.0–22.8	14.50	0.244–0.620	0.34	36.8–58.0	47.4	0.47–0.80	0.68
3/21/63	Kennebec	20	6.0–12.0	9.38	9.0–19.5	15.95	0.186–0.514	0.34	31.0–72.5	44.2	0.55–0.77	0.65
10/26/62	Katahdin	20	0.7–2.0	1.24	2.0–4.0	2.50	0.050–0.100	0.07	25.0–50.0	35.0	0.35–0.60	0.49
10/31/62	Katahdin	20	1.8–3.4	2.50	4.0–7.0	4.95	0.074–0.184	0.11	31.5–66.0	45.4	0.40–0.58	0.50
2/5/63	Katahdin	20	4.0–13.0	7.93	7.0–19.8	13.24	0.140–0.610	0.30	32.5–63.3	45.7	0.40–0.76	0.60
3/21/63	Katahdin	20	6.0–13.0	9.10	9.2–19.5	13.77	0.210–0.486	0.36	33.0–52.0	38.9	0.50–0.72	0.67
(Potato against plywood)												
11/6/62	Katahdin	10	1.7–3.0	2.24	2.0–4.0	2.80	0.075–0.140	0.10	20.5–40.0	27.0	0.75–0.90	0.82
2/6/63	(across	10	2.0–6.0	4.37	4.0–9.0	6.60	0.114–0.284	0.19	28.2–46.0	35.66	0.56–0.78	0.68
3/21/63	grain)	10	3.0–7.0	4.90	4.0–9.0	6.60	0.100–0.300	0.20	21.5–51.7	33.92	0.57–0.86	0.73
11/6/62	Katahdin	10	5.5–11.0	8.95	9.0–22.3	16.60	0.170–0.440	0.31	50.0–68.0	54.0	0.45–0.74	0.57
11/28/62	(along	4	11.5–14.0	12.4	15.5–18.5	16.15	0.335–0.435	0.40	35.5–46.0	41.1	0.75–0.78	0.76
2/6/63	grain)	10	10.0–20.0	16.6	19.0–32.2	28.1	0.290–0.700	0.53	46.0–68.0	54.4	0.59–0.62	0.59

Table A–18 Part I Physical characteristics of seeds and grains[1]

Product[2]	Weight × 10^5 (lb)	Axial dimensions: *a* Major dia. (in)	*b* Interm. dia. (in)	*c* Minor dia. (in)	Volume × 10^4 (in^3)	Specific gravity	Moisture content (% w.b.)	Seeds per lb × 10^{-3}
Alfalfa								
Atlantic	0.44	0.090	0.056	0.040	0.942	1.30	5.80	226
	0.00	0.008	0.005	0.004	0.01			4.5
Africa	0.551	0.098	0.057	0.041	1.180	1.29	6.60	182
	0.009	0.007	0.004	0.003	0.018			2.8
Buffalo	0.450	0.093	0.055	0.040	0.961	1.30	6.35	220
	0.017	0.009	0.004	0.004	0.036			8.4
Ranger	0.400	0.091	0.055	0.038	0.846	1.31	5.80	250
	0.007	0.009	0.006	0.004	0.015			3.6
Barley								
Arivat	10.58	0.417	0.139	0.113	20.68	1.42	7.50	9.46
	0.313	0.039	0.010	0.012	0.612			0.56
Atlas 46	7.97	0.418	0.128	0.102	1596.	1.38	7.90	12.5
	0.239	0.036	0.008	0.008	0.478			0.75
Club	9.94	0.414	0.136	0.102	19.96	1.38	8.20	10.0
Mariout	0.359	0.032	0.011	0.011	0.722			0.74
Hero	11.86	0.430	0.151	0.120	23.43	1.40	7.50	8.43
	0.384	0.036	0.013	0.008	0.759			0.59
Rojo	10.77	0.393	0.142	0.114	21.22	1.40	7.60	9.29
	0.581	0.038	0.008	0.008	1.14			0.97

Table A–18 Part I (continued)

Product[2]	Weight $\times 10^5$ (lb)	Axial dimensions: *a* Major dia. (in)	*b* Interm. dia. (in)	*c* Minor dia. (in)	Volume $\times 10^4$ (in^3)	Specific gravity	Moisture content (% w.b.)	Seeds per lb $\times 10^{-3}$
Tenn. Winter	8.47	0.392	0.125	0.098	17.04	1.38	7.90	11.8
	0.223	0.028	0.010	0.011	0.448			0.62
Clover								
Alsike	0.148	0.051	0.043	0.029	0.309	1.33	6.30	674
	0.002	0.004	0.003	0.003	0.003			7.2
Ladino	0.122	0.045	0.039	0.029	0.255	1.32	8.20	823
	0.003	0.004	0.002	0.003	0.006			19.0
(Red clover)	0.446	0.081	0.059	0.044	0.955	1.29	7.80	224
Kenland	0.008	0.005	0.003	0.004	0.017			4.1
Rose clover	0.722	0.087	0.066	0.055	1.488	1.34	6.40	138
	0.019	0.008	0.006	0.005	0.038			3.6
Corn								
Pfister 347	76.39	0.640	0.798	0.504	163.54	1.29	6.70	1.30
	2.15	0.036	0.042	0.028	4.59			0.04
Flax								
Argentine	1.553	0.206	0.102	0.039	3.767	1.14	6.50	64.4
	0.027	0.008	0.004	0.004	0.065			2.2
B 5128	1.561	0.199	0.093	0.041	3.776	1.14	6.50	64.0
	0.022	0.008	0.005	0.003	0.053			1.8
Imperial	1.463	0.196	0.088	0.040	3.525	1.15	6.30	68.4
	0.017	0.007	0.005	0.002	0.040			1.5

Table A–18 Part I (continued)

Product[2]	Weight × 10^5 (lb)	Axial dimensions: *a* Major dia. (in)	*b* Interm. dia. (in)	*c* Minor dia. (in)	Volume × 10^4 (in^3)	Specific gravity	Moisture content (% w.b.)	Seeds per lb × 10^{-3}
Punjab	1.147	0.175	0.085	0.042	2.742	1.16	6.80	87.2
	0.013	0.006	0.005	0.004	0.031			2.0
Punjab 47	1.658	0.203	0.096	0.043	4.026	1.14	6.30	60.3
	0.035	0.008	0.004	0.003	0.084			2.5
Punjab 53	1.444	0.193	0.091	0.041	3.06	1.14	6.50	69.3
	0.025	0.008	0.004	0.002	0.062			2.4
Redwood	1.410	0.186	0.091	0.042	3.427	1.14	5.90	70.9
	0.021	0.008	0.005	0.005	0.052			2.1
Milo								
D.D.38	6.342	0.170	0.160	0.110	13.43	1.32	9.20	15.8
	0.207	0.010	0.007	0.012	0.435			1.0
Oats								
Calif. Red	7.418	0.510	0.115	0.088	14.88	1.38	8.60	13.5
	0.233	0.047	0.009	0.007	0.466			0.82
Coast black	7.236	0.561	0.115	0.093	14.83	1.35	8.80	13.8
	0.429	0.076	0.017	0.013	0.880			1.7
Kanota	7.443	0.433	0.110	0.089	15.04	1.37	8.80	13.4
	0.448	0.056	0.010		0.904			1.6
Palistine	8.692	0.587	0.122	0.095	17.44	1.38	8.50	11.5
	0.194	0.052	0.011	0.013	0.390			0.52
Ventura	6.722	0.449	0.106	0.083	13.71	1.36	8.60	14.9
	0.336	0.058	0.012	0.009	0.685			1.4

Table A–18 Part I (continued)

Product[2]	Weight × 10^5 (lb)	Axial dimensions: *a* Major dia. (in)	*b* Interm. dia. (in)	*c* Minor dia. (in)	Volume × 10^4 (in^3)	Specific gravity	Moisture content (% w.b.)	Seeds per lb × 10^{-3}
Rice								
Caloro	6.393	0.300	0.143	0.098	13.03	1.36	8.60	15.6
	0.111	0.012	0.006	0.005	0.227			0.54
Calrose	5.934	0.314	0.127	0.089	12.03	1.36	9.20	16.9
	0.087	0.014	0.004	0.004	0.176			0.49
Hy Mix								
Early	5.484	0.385	0.100	0.081	10.9	1.39	8.80	18.2
24–1–1	0.072	0.026	0.003	0.003	0.144			0.48
Sudan grass								
Sudan	2.493	0.230	0.101	0.073	5.08	1.36	8.00	40.1
grass 2e	0.042	0.025	0.005	0.006	0.086			1.4
Piper	2.772	0.219	0.100	0.078	5.66	1.3	8.70	36.1
	0.102	0.013	0.008	0.006	0.208			2.7
Tift	2.366	0.213	0.092	0.070	4.85	1.35	8.30	42.6
	0.070	0.012	0.007	0.007	0.144			2.5
Vetch								
Common	12.47	0.180	0.181	0.155	24.95	1.38	8.00	8.0
	0.575	0.010	0.009	0.011	1.150			0.71
Hairy	6.355	0.140	0.141	0.131	13.01	1.35	8.50	15.7
	0.301	0.009	0.010	0.010	0.617			1.5
Purple	8.302	0.176	0.165	0.134	17.35	1.32	8.10	12.0
	0.280	0.009	0.014	0.011	0.584			0.83

Table A–18 Part I (continued)

Product[2]	Weight $\times 10^5$ (lb)	Axial dimensions *a* Major dia. (in)	*b* Interm. dia. (in)	*c* Minor dia. (in)	Volume $\times 10^4$ (in^3)	Specific gravity	Moisture content (% w.b.)	Seeds per lb $\times 10^{-3}$
Wheat								
Bart 46	10.37	0.288	0.117	0.111	20.36	1.41	8.50	9.6
	0.345	0.017	0.009	0.011	0.678			0.64
Big club	7.420	0.253	0.123	0.120	14.52	1.41	8.30	13.5
43	0.298	0.018	0.012	0.011	0.584			1.1
Onas 53	9.134	0.258	0.128	0.118	17.73	1.43	7.80	10.9
	0.620	0.019	0.013	0.011	1.205			1.5
Pac. blue	9.858	0.263	0.126	0.121	19.16	1.42	8.00	10.1
Stem	0.377	0.018	0.014	0.012	0.734			0.78
Poso	7.777	0.224	0.123	0.129	15.17	1.42	7.80	12.9
	0.154	0.010	0.008	0.010	0.300			0.50
Romona 50	11.21	0.270	0.149	0.138	21.69	1.43	7.50	8.92
	0.366	0.019	0.008	0.008	0.708			0.58
White fed.	10.60	0.253	0.119	0.119	20.65	1.42	6.20	9.43
38	0.231	0.015	0.009	0.006	0.449			0.40

[1] Original data supplied by John R. Goss, University of California, Davis, California.

[2] Under each product variety and down the columns, the two values given are mean and standard deviation, respectively. Single values only represent means.

Table A–19 Part II Physical characteristics of seeds and grains[1]

Product[2]	Average diameter (in) Equiv. sphere $\left(\frac{6}{\pi}V\right)^{1/3}$	Geom. mean $(abc)^{1/3}$	Arith. mean $\left(\frac{a+b+c}{3}\right)$	Percent sphericity $\frac{100(abc)^{1/3}}{a}$	Unit density (lb/ft^3)	Bulk density (lb/ft^3)	Moisture content (% w.b.)	Seeds per lb $\times 10^{-3}$
Alfalfa								
Atlantic	0.056	0.059	0.062	65.6	81.20	52.37 0.122	5.80	227
Africa	0.061	0.062	0.065	63.3	80.70	50.40 0.052	6.60	174
Buffalo	0.057	0.059	0.063	63.4	81.50	49.82 0.223	6.35	220
Ranger	0.055	0.058	0.061	63.7	81.70	50.05	5.80	250
Barley								
Arivat	0.156	0.187	0.223	44.8	88.40	40.63 0.110	7.50	9.35
Atlas 46	0.145	0.176	0.216	42.1	86.30	36.60 0.194	7.90	12.5
Club Mariout	0.156	0.180	0.217	43.5	86.00	36.64 0.262	8.20	10.0
Hero	0.166	0.199	0.234	46.3	87.50	38.54 0.237	7.50	8.50
Rojo	0.159	0.185	0.216	47.1	87.70	39.06 0.271	7.60	9.35
Tenn. Winter	0.148	0.169	0.205	43.1	85.80	35.26 0.251	7.90	11.7

Table A–19 Part II (continued)

Product[2]	Average diameter (in)			Percent sphericity	Unit density	Bulk density	Moisture content	Seeds per lb
	Equiv. sphere $\left(\frac{6}{\pi}V\right)^{1/3}$	Geom. mean $(abc)^{1/3}$	Arith. mean $\left(\frac{a+b+c}{3}\right)$	$\frac{100(abc)^{1/3}}{a}$	(lb/ft³)	(lb/ft³)	(% w.b.)	$\times 10^{-3}$
Clover								
Alsike	0.039	0.040	0.041	78.4	83.00	52.22 0.081	6.30	671
Ladino	0.036	0.037	0.038	82.2	82.20	51.43 0.097	8.20	831
Red Clover Kenland	0.057	0.060	0.061	74.1	80.70	51.03 0.150	7.80	224
Rose Clover	0.066	0.068	0.069	78.2	83.90	52.20 0.097	6.40	138
Corn								
Pfister 347	0.315	0.636	0.647	99.4	80.70	46.48 0.537	6.70	1.30
Flax								
Argentine	0.090	0.093	0.116	45.1	71.20	43.41 0.281	6.50	64.5
B 5128	0.090	0.091	0.111	45.7	71.50	43.73 0.159	6.50	64.1
Imperial	0.088	0.088	0.108	44.9	71.70	44.29 0.181	6.30	68.2

Table A–19 Part II (continued)

Product[2]	Average diameter (in) Equiv. sphere $\left(\frac{6}{\pi}V\right)^{1/3}$	Geom. mean $(abc)^{1/3}$	Arith. mean $\left(\frac{a+b+c}{3}\right)$	Percent sphericity $\frac{100(abc)^{1/3}}{a}$	Unit density (lb/ft^3)	Bulk density (lb/ft^3)	Moisture content (% w.b.)	Seeds per lb $\times 10^{-3}$
Punjab	0.081	0.086	0.101	49.1	72.30	42.82 0.278	6.80	86.9
Punjab 47	0.092	0.094	0.114	46.3	71.20	43.92 0.181	6.30	60.2
Punjab 53	0.087	0.089	0.108	46.1	71.20	43.43 0.134	6.50	69.2
Redwood	0.087	0.089	0.106	47.8	71.10	43.41 0.204	5.90	71.0
Milo								
D. D. 38	0.137	0.144	0.147	84.7	82.20	48.13 0.205	9.20	15.7
Oats								
Calif. red	0.142	0.173	0.283	33.9	86.10	30.12 0.186	8.60	13.5
Coast black	0.141	0.182	0.256	32.4	84.30	22.35 0.217	8.80	13.7
Kanota	0.142	0.162	0.211	37.4	85.50	31.19 0.269	8.80	12.8
Palistine	0.149	0.189	0.268	32.2	86.10	29.34 0.210	8.50	11.5
Ventura	0.138	0.158	0.213	35.2	84.70	31.89 0.130	8.60	14.5

Table A–19 Part II (continued)

Product[2]	Average diameter (in)			Percent sphericity $\frac{100(abc)^{1/3}}{a}$	Unit density (lb/ft^3)	Bulk density (lb/ft^3)	Moisture content (% w.b.)	Seeds per lb $\times 10^{-3}$
	Equiv. sphere $\left(\frac{6}{\pi}V\right)^{1/3}$	Geom. mean $(abc)^{1/3}$	Arith. mean $\left(\frac{a+b+c}{3}\right)$					
Rice								
Caloro	0.136	0.162	0.180	54.0	84.80	35.65 0.106	8.60	15.6
Calrose	0.132	0.152	0.177	48.4	85.20	35.63 0.388	9.20	16.8
Hy mix early 24–1–1	0.128	0.146	0.189	37.8	86.60	36.91 0.583	8.80	22.3
Sorghum			0.134			83.9		
Soybean			0.236			80.8		
Sudan grass								
Sudan grass 23	0.099	0.120	0.135	52.2	84.80	41.17 0.173	8.00	40.1
Piper	0.103	0.119	0.132	54.3	84.60	42.63 0.221	8.70	36.1
Tift	0.097	0.112	0.125	52.6	84.30	41.44 0.153	8.30	42.6
Vetch								
Common	0.168	0.171	0.172	95.0	86.30	52.27 0.401	8.00	8.0
Hairy	0.135	0.137	0.137	97.9	84.40	50.58 0.250	8.50	15.9

Table A–19 Part II (continued)

Product[2]	Average diameter (in) Equiv. sphere $\left(\frac{6}{\pi}V\right)^{1/3}$	Geom. mean $(abc)^{1/3}$	Arith. mean $\left(\frac{a+b+c}{3}\right)$	Percent sphericity $\frac{100(abc)^{1/3}}{a}$	Unit density (lb/ft^3)	Bulk density (lb/ft^3)	Moisture content (% w.b.)	Seeds per lb $\times 10^{-3}$
Purple	0.149	0.157	0.158	89.2	82.70	49.54 0.439	8.10	12.1
Wheat								
Bart 46	0.157	0.157	0.172	53.8	88.00	49.83 0.235	8.50	9.6
Big club 43	0.140	0.155	0.165	61.3	86.30	49.49 0.206	8.30	13.4
Onas 53	0.150	0.157	0.168	60.9	89.00	49.32 0.173	7.80	10.9
Pac. blue Stem	0.154	0.159	0.170	60.5	88.90	49.64 0.158	8.00	10.1
Poso 48	0.143	0.152	0.159	67.9	88.60	51.17 0.176	7.80	12.8
Romona 50	0.161	0.177	0.186	65.6	89.30	50.57 0.106	7.50	8.90
White fed. 38	0.158	0.153	0.164	60.6	88.70	50.07 0.194	6.20	9.44

[1] Original data supplied by John R. Goss, University of California, Davis, California. Values for sorghum and soybean are taken from Whitney and Porterfield (1968).

[2] Under each product variety and down the columns, the values given are mean and standard deviation, respectively. Single values only represent means.

Table A–20 Physical properties of rough rice (Wratten *et al.*, 1968)

Moisture content per cent	Individual grain properties							Bulk properties		
	Length in	Width in	Thickness in	Volume cu in $\times 10^3$	Density lb/cu ft	Specific gravity	Area sq. in.	Density lb/cu ft	Porosity %	Specific gravity
				Medium grain (Saturn)						
12	0.311	0.123	0.077	0.98	82.67	1.374	0.0623	37.35	58.5	0.599
14	0.312	0.123	0.077	1.02	83.47	1.355		38.58	56.5	0.618
16	0.313	0.123	0.078	1.07	84.53	1.350		39.56	55.0	0.630
18	0.314	0.125	0.079	1.17	85.64	1.325	0.0658	40.49	53.1	0.653
				Long grain (Bluebonnet–50)						
12	0.381	0.102	0.075	1.12	85.05	1.384		36.56	59.6	0.586
14	0.384	0.103	0.076	1.13	85.59	1.378		36.72	59.3	0.589
16	0.388	0.104	0.076	1.17	86.00	1.372		37.79	57.9	0.606
18	0.395	0.106	0.078	1.20	86.34	1.358		38.40	56.9	0.615

Table A–21 Physical properties of pineapple fruit during development (Singleton, 1965)

Days from ripe	Eyes per long spiral	Fruit length/width ratio	% of fruit weight			Juice cc/100 g flesh	Cells per eye volume ($\times 10^6$)
			Shell	Core	Flesh		
−109	15.6	1.12	54.4	9.2	36.3	48	—
−105	17.0	1.17	46.7	11.2	42.2	68	16.5
−102	16.8	1.35	47.5	12.2	40.2	57	—
− 98	18.0	1.28	43.0	12.7	44.4	48	11.3
− 91	17.6	1.32	40.6	13.9	45.5	51	9.3
− 84	17.6	1.38	37.9	14.0	48.2	53	8.8
− 77	17.8	1.38	37.0	12.4	50.6	56	8.7
− 70	17.8	1.36	37.4	12.0	50.6	60	7.6
− 63	17.8	1.35	35.4	10.6	54.0	56	7.0
− 56	17.8	1.33	37.8	10.0	52.2	59	6.9
− 49	18.0	1.36	34.8	10.2	55.0	62	7.1
− 42	17.2	1.34	35.8	10.7	53.4	60	7.0
− 35	18.0	1.30	38.0	8.6	53.3	60	7.2
− 28	18.0	1.35	37.8	8.7	53.5	66	7.6
− 21	17.6	1.32	37.7	9.2	53.0	68	7.1
− 14	17.9	1.32	37.4	9.2	53.0	64	6.3
− 7	17.0	1.28	37.3	9.4	53.3	66	7.1
0	18.1	1.34	35.6	10.0	54.5	68	7.7
+ 7	18.4	1.38	35.9	10.0	54.1	73	7.0

Table A–22 Storage modulus and loss modulus for compression and shear tests of flesh of several varieties of apples (Hamann, 1969)

Type of test	Variety	Approximate time in storage (months)	Average mechanical thumb force (lbs)	Equations relating the real and imaginary parts of the complex moduli to ω
Compression	Red delicious	0	19	$E' = 1847 + 0.8193\omega - 0.00020\omega^2$ $E'' = 813.4 - 0.4009\omega + 0.000095\omega^2$
Compression	Winesap	0	23	$E' = 1464 + 1.594\omega - 0.00044\omega^2$ $E'' = 447.2 + 0.1083\omega - 0.000065\omega^2$
Compression	Golden delicious	0	16	$E' = 2072 + 0.2855\omega$ $E'' = 734.3 - 0.3593\omega - 0.000095\omega^2$
Compression	Winesap	4	17	$E' = 1341 + 0.1899\omega$ $E'' = 143.4 - 0.3796\omega - 0.00019\omega^2$
Compression	Golden delicious	4	11	$E' = 629.2 \& 0.1291\omega$ $E'' = 175.6 - 0.0413\omega$
Shear	Winesap	4	17	$G' = 311.3 + 0.1080\omega$ $G'' = 52.54 + 0.02892\omega - 0.000025\omega^2$
Shear	Golden delicious	4	1	$G' = 105.5 + 0.1091\omega$ $G'' = 30.08 + 0.02124\omega - 0.000015\omega^2$

E' = Storage compression modulus—psi
E'' = Loss compression modulus—psi
G' = Storage shear modulus—psi
G' = Loss shear modulus—psi

Table A–23 Separation force for selected plant materials

Material and parent body	Separation force (lbs) Mean	Range	Reference
Corn kernel from cob	3.36	2.57–3.78	Waelti & Buchele, 1967
Cotton fiber from seed	2.15	1.0–5.3	Griffin & Moore, 1964
Fruit from tree or vine[1]			
Apples			
Cortland	5.2		Quackenbush *et al.*, 1962
Golden delicious	3.6		Cooper, 1962
Jonathan	4.4		Quackenbush *et al.*, 1962
McIntosh	4.5		Quackenbush et al., 1962
McIntosh	4.3		Cooper, 1962
Melba	3.9		Cooper, 1962
Northern spy	9.6		Quackenbush, 1962
Red delicious	2.8		Cooper, 1962
Rome beauty	3.9		Cooper, 1962
Stayman	5.3		Cooper, 1962
Apricots			
Montgamet	0.91		Quackenbush, 1962
Blueberries			
Jersey	0.21		Quackenbush, 1962
Cherries			
Montmorency	0.21		Quackenbush, 1962
Grapefruits	17.0		Coppock *et al.*, 1968
Oranges			
Hamlin	16		Coppock *et al.*, 1968
Pineapple	18		Coppock *et al.*, 1968
Valencia	23		Coppock *et al.*, 1968
Peaches			
Red haven	6.0		Quackenbush, 1962
Elberta	3.4		Quackenbush, 1962
Plums			
Stanley pure	1.8		Quackenbush, 1962
Tobacco leaf from stalk			
1959	1.1	1.04–1.26	Splinter *et al.*, 1962
1960	1.2	0.87–1.42	Splinter *et al.*, 1962

[1] Detachment force along the stem at harvest time.

Table A–24 Bulk density of fruits and vegetables[1]

Material	Bulk density (lb/ft^3)
Apples	34–38
Apricots	38
Asparagus	36
Beets	52
Cabbage	28
Carrots	40
Cauliflower	20
Cherries	45
Cucumbers	48
Grapes	23
Green beans	24
Lemons	48
Onions	40–46
Oranges	48
Peaches	38
Pears	40
Plums	45
Spinach	14
Sweet corn	54
Sweet potatoes	44
Tomatoes	42
White potatoes	48

[1] Compiled from various sources.

Table A–25 Some typical values for physical properties of dried citrus pulp (Ross and Kiker, 1967)

Property	Range	Mean
Bulk density (lb/ft^3)	16–21	18
Void space (%)	76–81	79
Moisture content (d.b.)	5.7–15	9.6
Coefficient of friction on		
wood[1]	0.56–0.73	0.64
steel[1]	0.33–0.45	0.38
aluminum[1]	0.34–0.50	0.40
Ratio of lateral to vertical pressure (k)[2]	—	0.62
Modulus of fineness[3]	4.1–4.7	4.4
Modulus of uniformity[3]	4–5–1 to 6–4–0	6–4–0

[1] Moisture equilibrium at 60% RH and 77°F = 13.3.

[2] Williams and Ross, 1967.

[3] See Agricultural Engineering Yearbook.

Table A–26 Poisson's ratio of selected plant materials subjected to compression

Material	Stress or load	Poisson's ratio mean	Poisson's ratio c.v. (%)	Reference
Apple				
Golden delicious	25 psi	0.26	14	PSU[1]
Golden delicious	24 psi	0.24	—	Chappel & Hamann, 1968[2]
McIntosh	25 psi	0.34	12	PSU[1]
Red delicious	24 psi	0.24	16	PSU[1]
Red delicious	21 psi	0.21	—	Chappel & Hamann, 1968[2]
Winesap	24 psi	0.29	—	Chappel & Hamann, 1968[2]
Corn horny endosperm				
(10%) moisture	9500 psi	0.32	9.2	PSU[3]
Corn stalk (dry)				
with nodes and pith	25–100 lbs.	0.23	—	Prince & Bradway, 1969
with nodes without pith	25–100 lbs.	0.34	—	Prince & Bradway, 1969
Potatoes	31 psi	0.45	7	PSU[1]
Potatoes	—	0.49	—	Finney & Hall, 1967[4]

PSU Pennsylvania State University, Agricultural Products Testing Laboratory.
[1] Cylindrical specimens 1″ × 1″.
[2] Cylindrical specimens 15/16″ × 13/16″.
[3] Rectangular slab speciments 0.18″ × 0.14 ″ × 0.03″.
[4] Calculated from uniaxial compression and bulk compression data for whole specimen.

Table A–27 Poisson's ratio for various materials

Material	Poisson's ratio	Source
Cork	0	Reiner (1960)
Concrete	0.19	Schmidt (1948)
Gelatin gel (80%, #20)	0.50	Van Wazer, *et al.*, (1963)
Glass	0.24	Reiner (1960)
Granite	0.21	Jaeger (1962)
Hard rubber	0.39	Schmidt (1948)
Lead	0.43	Nielson (1962)
Polyethylene	0.33	Nielson (1962)
Sandstone	0.10	Jaeger (1962)
Soft rubber	0.49	Schmidt (1948)
Steel	0.30	Reiner (1960)
Vitreous silica	0.14	Nielson (1962)
Wood	0.3–0.5	USDA, Wood Handbook (1955)

Table A–28 Viscosity coefficient for various materials (Also see Table 9.2 in text)

Material	Temperature	Viscosity coefficient (centipoise)	Source
Air	68	0.0186	1
Castor oil	68	720	1
Cotton seed oil	60	91	2
Cream 20% fat	37	6.20	2
Cream 30% fat	37	13.78	2
Glass	68	1×10^{24}	1
Mercury	68	1.56	1
Milk, whole	68	2.12	2
Milk, whole	32	4.28	2
Milk, skim	77	1.37	2
Molasses	70	6600	2
Olive oil	86	84.9	2
SAE No. 30 oil	60	400	3
Soybean oil	86	40.6	2
Sucrose 60% solution	70	60.2	2

[1] Jaeger, 1962.
[2] Agricultural Materials Handling, 1962.
[3] Marks, 1941.

Table A–29 Bulk modulus or incompressibility for various materials

Material	Bulk modulus 10^6 lb/in^2	Source
Clay	1.4	Jaeger (1962)
Ether	0.12	Reiner (1960)
Glass	5.8	Reiner (1960)
Lead	5.5	Jaeger (1962)
Limestone	5.8	Jaeger (1962)
Mercury	7.9	Reiner (1960)
Natural rubber	0.00275	Van Wazer, *et al.* (1963)
Steel	26.0	Reiner (1960)
Water	0.34	Reiner (1960)

A-30 COMPRESSION TEST OF FOOD MATERIALS OF CONVEX SHAPE†

Reviewed by the ASAE Physical Properties of Agricultural Products Committee; approved by the Food Engineering Division Standards Committee; adopted by ASAE December 1973; revised and reclassified as a Standard December 1979.

SECTION 1—PURPOSE

1.1 This standard is intended for use in determining mechanical attributes of food texture, resistance to mechanical injury, and force-deformation behavior of food materials of convex shape, such as fruits and vegetables, seeds and grains, and manufactured food materials.

SECTION 2—SCOPE

2.1 Compression tests of intact biological materials provide an objective method for determining mechanical properties significant in quality evaluation and control, maximum allowable load for minimizing mechanical damage, and minimum energy requirements for size reduction.

2.2 Realizing the shortcomings associated with subjective methods, the use of fully automatic testing machines has become popular in recent years. With the use of testing machines, the need for a recommendation for testing, interpretation of data, and reporting of results has become evident.

2.3 Determination of compressive properties requires the production of a complete force-deformation curve. From the force-deformation curve,

---

†Reprinted with permission by ASAE. Taken from ASAE Standard S368.1 (Revised December 1979).

ASAE Recommendation S-368.1. Compression Tests of Food Materials of Convex Shape. In *Agricultural Engineers Yearbook*. Am. Soc. Agr. Engrs., St. Joseph, Michigan.

Note: For better illustrations and possibly the complete layout, see N. Mohsenin, *Physical Properties of Food and Agricultural Materials*, Gordon and Breach Science Publishers, 1981, p. 55.

stiffness; modulus of elasticity; modulus of deformability; toughness; force and deformation to points of inflection, to bioyield, and to rupture; work to point of inflection, to bioyield, and to rupture, and maximum normal contact stress or a stress index at low levels of deformation can be obtained. Any number of these mechanical properties can, by agreement, be chosen for the purpose of evaluation and control of quality.

SECTION 3—DEFINITIONS

3.1 Bioyield point: A point such as shown in Fig. 1 where an increase in deformation results in a decrease or no change in force.

3.2 Force-deformation curve: A diagram plotted with values of deformation as abscissae and values of force as ordinates.

3.3 Modulus of elasticity: A modulus given by the equations shown in Fig. 2. The equations are based on the Hertz problem of contact stresses in solid mechanics and assume very small and elastic deformations. To

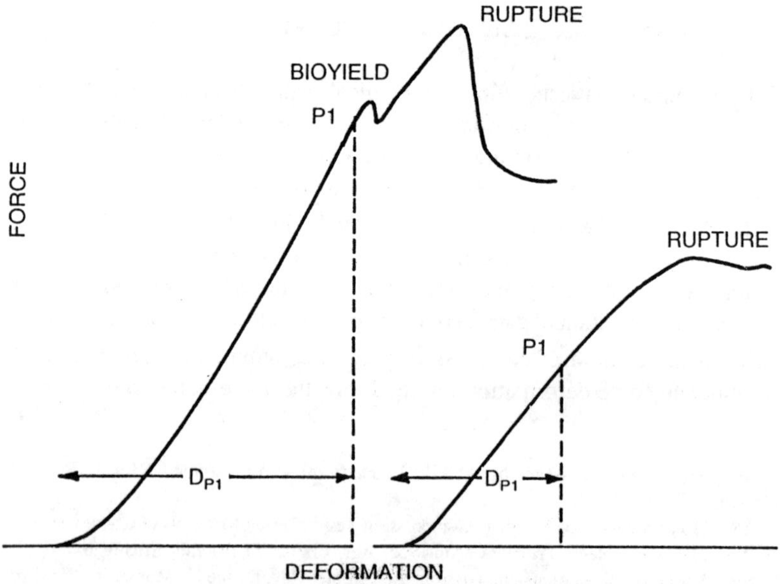

Figure 1 Force-deformation curves for materials with and without bioyield point. PI=point of inflection, D_{PI}=deformation at point of inflection

separate elastic and plastic deformations, an unloading force-deformation curve is required.

3.4 Modulus of deformability: A modulus defined by the same equations as those shown in Fig. 2 for modulus of elasticity except that D in these expressions is the sum of both elastic and plastic deformations. No unloading force-deformation curve is required in this case.

3.5 Point of inflection: A typical force-deformation curve is first concave up and then concave down (Fig. 1). The point at which the rate of change

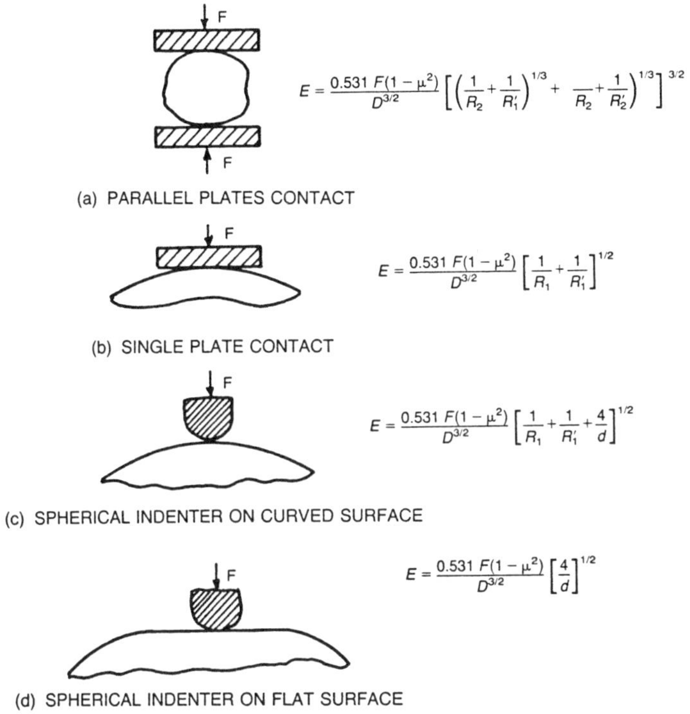

Figure 2 Modulus of elasticity calculated from force and deformation data, E=modulus of elasticity, Pa (psi); F=force in N (lbf); D=elastic deformation at both loading and supporting points of contact, m(in.); μ=Poisson's ratio; R_1, R_1'. R_2, R_2'=radii of curvature of the convex body at the points of contact, m(in.); d=diameter of the spherical indenter, m(in.); 0.531=constant valid for the case where the angle between the normal planes containing the principal curvatures of the convex body is 90 deg and the difference between the curvatures in each plane is small.

of slope (second derivative) of the curve becomes zero is called the point of inflection. This point, designated as PI, can be found by using a straight edge to follow the change of slope of the curve and to determine the point at which the slope begins to decrease.

3.6 Radius of the circle of contact: The radius of the contact area between the compression tool and the convex-shape specimen at any given load and deformation. For a rigid spherical compression tool on convex specimens and very small deformations (Fig. 2c), the radius of the circle of contact can be approximated by the following formula:

$$a^2 = \frac{D}{\frac{1}{R} + \frac{2}{d}}$$

where

a = radius of the circle of contact
D = elastic deformation
d = diameter of spherical indenter
R = average radius of the convex body

For a rigid spherical tool on the flat surface of a specimen (Fig. 2d), the radius of the circle of contact can be approximated by the following formula:

$$a^2 = \frac{dD}{2}$$

3.7 Rupture point: The point on the force-deformation curve at which the loaded specimen shows a visible or invisible failure in the form of breaks or cracks. This point is detected by a continuous decrease of the load in the force-deformation diagram (Fig. 1).

3.8 Stiffness: The ratio of force to deformation at the point of inflection.

3.9 Stress Index: The maximum stress which the material is capable of sustaining before a given deformation can occur. This deformation shall be one-half the deformation corresponding to the point of inflection.

3.9.1 For a convex body loaded with a steel flat plate (Figs. 2a, 2b), stress index is computed from:

$$S_i = 0.365 \left[\frac{F}{A^2} (1/R_1 + 1/R_2)^2 \right]^{1/3}$$

For a convex body loaded with a smooth, steel spherical indenter (Fig. 2c), stress index is computed from:

$$S_i = 0.365 \left[\frac{F}{A^2}(1/R_1 + 1/R_2 + 4/D_2)^2\right]^{1/3}$$

where

S_i = stress index; Pa (psi)
F = force, N (lbf)
R_1, R_1' = radii of curvature of convex body at the point of contact, m(in.)
d_2 = diameter of spherical indenter, m(in.)
A = an elastic constant, 1/Pa (1/psi) given by

$$A = \frac{(1 - \mu^2)}{E}$$

μ = Poisson's ratio
E = modulus of elasticity of material, Pa (psi)*

SECTION 4—APPARATUS

4.1 Testing machine. Any suitable testing machine capable of constant-rate-of-crosshead movement and comprising essentially the following:

4.1.1 A drive mechanism for imparting to the crosshead a constant velocity with respect to the base. The crosshead velocity shall also be repeatable with an accuracy of ± one percent.

4.1.2 A load indicating mechanism capable of showing the total compressive load carried by the test specimen. The mechanism shall be essentially free from inertia-lag at the specified site of testing and shall indicate the load with an accuracy of ± one percent of the maximum anticipated value of the load on the specimen. The accuracy of the testing machine shall be verified at least once a year in accordance with American Society for Testing and Materials Standard E-4, Load Verification of Testing Machines.

---

*For the purpose of this Standard, modulus of deformability shall be used for E.

4.1.3 A suitable instrument capable of determining the change in distance between the point of loading and a fixed member of the base of the testing machine at any time during test. It is desirable that this instrument automatically record the change of distance as a function of the load on the test specimen. The instrument shall be essentially free of inertia-lag at the specified rate of loading and shall conform to the requirements for a class B-2 extensometer as defined in American Society for Testing and Materials Standard E-83, Verification and Classification of Extensometers. Realizing that in the case of test specimens of food materials of convex shape the use of extensometers on the specimen is difficult and in most cases impractical, the second alternative shall be the use of an extensometer on the movable crosshead of the testing machine, sensing the true displacement of the crosshead. The third alternative shall be the use of crosshead velocity as an indirect method of measuring deformation.

4.2 Compression tool. A compression tool with smooth and polished surface is used for applying the load to the specimen. The type of the compression tool used for each test shall depend on the size of the specimen, the nature of the test, and the extent of information expected from that test. Examples of suitable compression tools are shown in Fig. 3. Flat plate and spherical tools may be used for both soft and hard materials. A plunger with end machined to a known radius without rounding the edge (Fig. 3b) is suitable for soft materials such as fruits and vegetables, particularly those which exhibit a bioyield point.

4.2.1 The rounded tip of the plunger in Fig. 3b is the segment of a sphere with a diameter *d* calculated from equations in paragraph 3.6. The radius of the circle of contact, *a,* in this case is one-half of the diameter of the cylindrical portion of the compression tool when full contact is established. Knowing the magnitude of the desired deformation, the appropriate radius of the spherical tool can be computed from the equations. To obtain a distinct bioyield point on the recording chart in case of fruits and vegetables, it is desirable to bring the full diameter of the cylinder in contact with the specimen with as little deformation as possible. For example, if the skin of the fruit is removed and the tool in Fig. 3b is used, the full diameters of the cylindrical portion of the plunger will come in contact with the specimen at about 0.64 mm (0.025 in.) deformation. Such value of deformation for fruits such as apples is below the usual range of deformation reported for bioyield point.

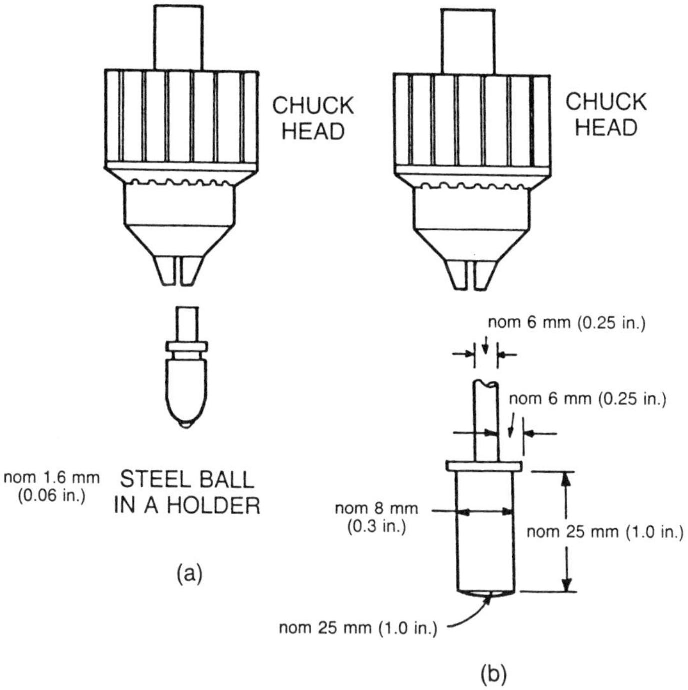

Figure 3 Examples of compression tools with spherical ends

4.2.2 If the modulus of deformability or stress index is to be computed, the force and deformation to be used in the equation shall be those corresponding to the values of one-half the deformation at the point of inflection on the force-deformation curve (Fig. 1). Also, the ratio of the radius of the circle of contact to the radius of the spherical indenter shall not be greater than 1:10.

4.2.3 The 8 mm (5⁄16 in.) diameter cylindrical plunger is recommended for use on fruits because this size corresponds to one of the two plungers of the Magness-Taylor fruit pressure tester commonly used for quality evaluation of such fruits as apples, pears, and peaches. If one of these plungers is used as the loading tool in conducting a compression test on fruits, it is recommended to continue the compression beyond and through the point of rupture. In this manner the force reading at the point of rupture can serve also as the Magness-Taylor pressure reading of the specimen.

4.3 Supporting jig. For small hard specimens such as seeds and grains, a hardened metal plate with smooth surface finish can be used as supporting jig. The test specimens shall be mounted on the plate by using a thin layer of quick hardening cement or glue. The flatter side of the grain is first lightly sanded before being cemented. If the grain specimens have been equilibriated for moisture contents higher than normal storage conditions, it is necessary to use a special glue which will harden at high-humidity atmosphere.

4.3.1 For larger and softer specimens such as fruits and vegetables, it is essential to provide a support to ensure that the deformation of the whole specimen at the support is negligible. This can be accomplished by providing some pre-setting arrangement such as that shown in Fig. 4. It consists of a lower wooden plate and an upper heavy plate, each provided with a rounded circular hole. The specimen shall be placed between the two plates and pre-set by the weight of the upper plate on the top of the specimen. The weight of the upper plate required for pre-setting shall be determined by the maximum load which the specimen can carry without leaving an objectionable mechanical damage where the rounded circular hole of the heavy plate contacts the specimen.

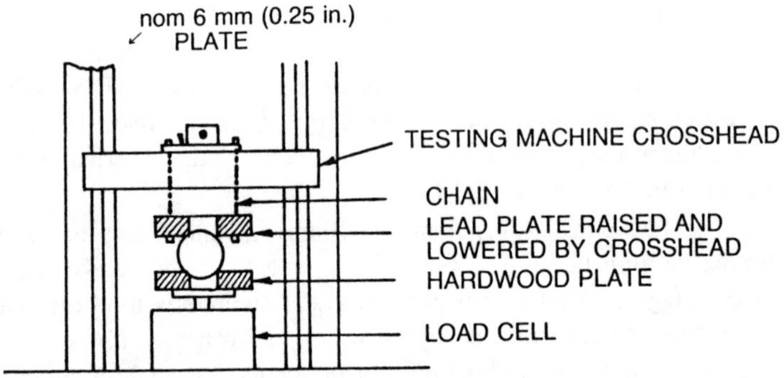

Figure 4 Fixture for pre-setting of the specimen

SECTION 5—TEST SPECIMENS

5.1 Specimens shall be tested in their natural form and size after being placed in their appropriate supporting jigs. For larger convex-bodies such as fruits and vegetables, testing the specimens in its natural form permits taking several readings on the same specimen nondestructively. This would be in contrast with the case where the specimen is cut in the form of a segment of a sphere. If skin is sliced off at the point of compression, the details of test specimen with or without skin shall be given.

5.2 If only relative measurements of force and deformation are to be made, it is desirable to make measurements on shape, size, weight, color and even such constituents as sugar and starch before testing the specimen. For more exact tests where equations given in Fig. 2 are expected to be used, in addition to the above data, the radii of curvature of the convex-shape specimen shall also be determined by either using a radius of curvature meter such as shown in Fig. 5 or by calculation as in Fig. 6. If the loading tool is rounded at the end, the radius of curvature of the tip shall also be specified.

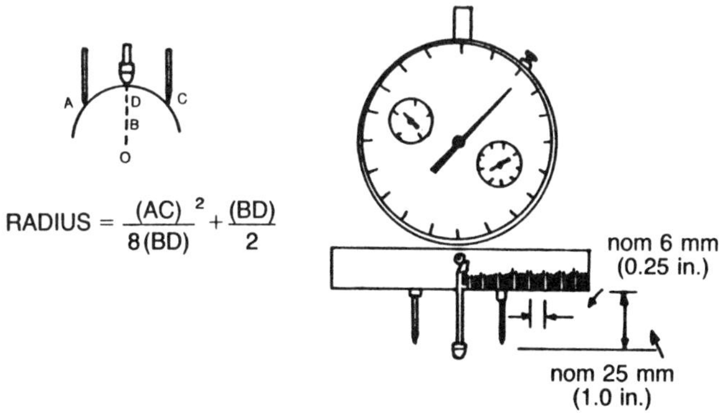

Figure 5 Radius of curvature meter

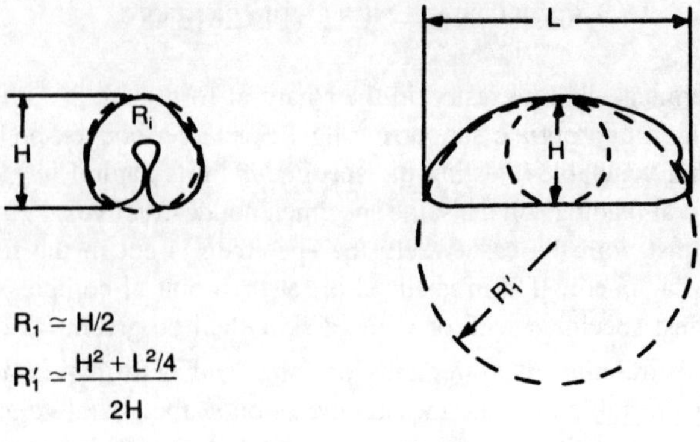

Figure 6 Approximation of R_1 and R_1' for convex bodies

SECTION 6—CONDITIONING

6.1 The test specimen shall be conditioned for the desired temperature and relative humidity before testing. Fruits and vegetables normally require a few hours to adjust their temperatures from field conditions to room temperature. Hard seeds and grains may have to be conditioned in an atmosphere of given relative humidity for several days if they are to be brought up to equilibrium moisture contents other than the ordinary storage conditions. In that case the conditioning chamber shall be brought into the testing room before testing so that moisture gains or losses of the specimen can be minimized.

SECTION 7—NUMBER OF TEST SPECIMENS

7.1 Because of the large variance inherent in biological materials, each experiment shall be statistically designed with sufficient number of replications to result in an acceptable level of confidence insofar as significant differences are concerned. The variation due to shape, size, age and cellular structure are normally such that at least a minimum of twenty specimens is required to be tested for each sample.

SECTION 8—SPEED OF TESTING

8.1 Speed of testing shall be relative rate of motion of the compression tool during the test. Rate of motion of the driven compression tool when the machine is running idle may be used if it can be shown that the resulting speed of testing is within the limits of variations allowed. The recorder response shall be such that the rate of rise of force on the chart shall be at most 1/3 of the maximum theoretical rate of rise of the recorder pen.

8.2 The speed of testing shall be chosen on the basis of the sensitiveness of the specimens to loading rate. For most hard fruits and vegetables the speed of 25 mm per minute ± 20 percent shall be specified as the standard speed. For seeds and grain, the speed of 1.25 mm per minute ± 50 percent shall be specified.

SECTION 9—TESTING PROCEDURE

9.1 Conduct the test under laboratory atmosphere of constant relative humidity and temperature. If possible, tests should be conducted under laboratory conditions of 20°C±5°C and 50 percent relative humidity ± 5 percent.

9.2 Measure the major, minor and intermediate diameters of the specimen to the nearest 10 percent of the dimensions.

9.3 Weigh the specimen and record the weight together with some observation of color and appearance.

9.4 If modulus of deformability is to be calculated, measure radius of curvature at the loading point to the nearest 10 percent.

9.5 Record information on variety, age, maturity and the history of the material prior to testing.

9.6 Select the compression tool according to the requirements given in paragraph 4.3, and install the tool in the testing machine.

9.7 If the compression tool is one with a spherical end, record the radius of the rounded end. Set the speed control at the desired rate and calibrate the recording chart for load and displacement.

9.8 Place the specimen in the testing machine under the compression tool, taking care to align the center of the tool with the peak of the curvature of the test specimen.

9.9 Place the specimen over the support and lower the pre-setting weight as shown in Fig. 4. For specimens of seeds and grains which have been softened in an atmosphere of high relative humidity, a similar pre-setting arrangement can be devised to insure that the deformation of the whole specimen in comparison with the deformation at the point is negligible.

9.10 Start the machine and record the complete force-deformation curve through the point of rupture.

9.10.1 When testing small, hard specimens such as grains, the deflections of most load cells cannot be considered negligible. For this reason either proof shall be given that the load cell deflection is negligible or the deflection is determined and deducted from the recorded deformation.

SECTION 10—CALCULATIONS

10.1 Force and deformation to bioyield and to rupture. These values are read directly from the chart and recorded along with the type of compression tool used.

10.2 Point of inflection. Determine the point of inflection by using a straight edge on the force-deformation curve to locate the point where the slope of the curve begins to decrease.

10.3 Modulus of deformability. Calculate the modulus of deformability by using the appropriate equation in Fig. 2 given for modulus of elasticity, except that for the values of F and D use the force and deformation corresponding to the point of the force-deformation curve at one-half of the total deformation up to the point of inflection, as shown in Fig. 1. Express the results in Pa (psi).

10.4 Stress index. Calculate stress index by using equations in paragraph 3.9.1, depending on the type of compression tool used. For the force value F use the force at one-half of total deformation for the point of inflection. Calculate the elastic constant using the modulus of deformability in place of E. Express the results in Pa (psi).

10.5 For each series of tests, calculate the mean and standard deviation.

SECTION 11—REPORT

11.1 The report shall include the following:

11.1.1 Complete identification of the material tested including source, previous history, time of the day samples were harvested, weight, color and appearance, shape and size (as defined by major, minor, intermediate diameters), and if calculations of mechanical properties are reported, the radii of curvature at the point of testing.

11.1.2 Complete identification of the compression tool used including the diameter of the cylindrical part of the tool and the radius of the spherical end.

11.1.3 Method of preparing and pre-setting the specimen for test.

11.1.4 Conditioning procedure used.

11.1.5 Atmospheric conditions in testing room.

11.1.6 Number of specimens used.

11.1.7 Speed of testing.

11.1.8 Date of test.

11.1.9 Type of testing machine used.

11.1.10 Mean and standard deviation for force and deformation to bioyield and to rupture.

11.1.11 Mean and standard deviation for modulus of deformability and stress index, if of interest.

References:

Last printed in 1977 AGRICULTURAL ENGINEERS YEARBOOK: list available from ASAE Headquarters.

Cited Standards:

ASTM E4, Load Verification of Testing Machines
ASTM E83, Verification and Classification of Extensometers

Selected Conversion Factors (SI) System of Units

To convert from	to	Multiply by[a]
Atmosphere (standard)	pascal (Pa)	1.013 E + 05
bar	pascal (Pa)	1.000 E + 05
Btu (British thermal unit)	joule (J)	1.055 E + 03
Btu ft/h ft^2 °F (thermal conductivity)	watt per meter kelvin (W/m K)	1.73 E + 00
Btu/h	watt (W)	2.930 E − 01
Btu/ft^2	joule per $meter^2$ (J/m^2)	1.136 E + 4
Btu (International Table)/h ft^2 °F (heat transfer coefficient)	watt per $meter^2$ kelvin (W/m^2K)	5.678 E + 00
Btu/lb	joule per kilogram (J/kg)	2.326 E + 03
Btu/lb °F (specific heat)	joule per kilogram kelvin (J/kg K)	4.187 E + 03
calorie	joule (J)	4.187 E + 00
centimeter of mercury (0°C)	pascal (Pa)	1.333 E + 03
centimeter of water (4°C)	pascal (Pa)	9.806 E + 01
centipoise (cP)	pascal second (Pa s)	1.000 E − 03
centistokes	$meter^2$ per second (m^2/s)	1.000 E − 06
degree (angle)	radian (rad)	1.1745 E − 02
degree Celsius	kelvin (K)	$t_k = t_c + 273.15$
degree Fahrenheit	degree Celsius	$t_c = (t_F - 32)/1.8$
degree Fahrenheit	kelvin (K)	$t_k = (t_F + 459.67)/1.8$
degree Rankine	kelvin (K)	$t_k = t_R/1.8$
dyne	newton (N)	1.000 E − 05
dyne cm	newton meter (N m)	1.000 E − 07
dyne/cm^2	pascal (Pa)	1.000 000°E − 01
erg	joule (J)	1.000 E − 07
erg/cm^2 s	watt per $meter^2$ (W/m^2)	1.000 E − 03
erg/s	watt (W)	1.000 E − 07
fluid ounce (U.S.)	$meter^3$ (m^3)	2.957 E − 05
foot	meter (m)	3.048 E − 01
foot of water (39.2°F)	pascal (Pa)	2.989 E + 03
ft^2	$meter^2$ (m^2)	9.290 E − 02
ft^2/h (thermal diffusivity)	$meter^2$ per second (m^2/s)	2.581 E − 05
ft^2/s	$meter^2$ per second (m^2/s)	9.290 E − 05
ft^3	$meter^3$ (m^3)	2.832 E − 02
ft^3/min	$meter^3$ per second (m^3/s)	4.719 E − 04
ft^3/s	$meter^3$ per second (m^3/s)	2.832 E − 02
ft/h	meter per second (m/s)	8.467 E − 05
ft/s	meter per second (m/s)	3.048 E − 01
ft/s^2	meter per $second^2$ (m/s2)	3.048 E − 01
ft lbf	joule (J)	1.356 E + 00
ft lbf/h	watt (W)	3.766 E − 04

Selected Conversion Factors (SI) System of Units (continued)

To convert from	to	Multiply by[a]
ft lbf/min	watt (W)	2.260 E − 02
ft lbf/s	watt (W)	1.356 E + 00
free fall (g)	meter per second^2 (m/s^2)	9.807 E + 00
gallon (U.K. liquid)	meter^3 (m^3)	4.546 E − 03
gallon (U.S. liquid)	meter^3 (m^3)	3.785 E − 03
gal (U.S. liquid)/min	meter^3 per second (m^3/s)	6.309 E − 05
grain (1/7000 lb)	kilogram (kg)	6.480 E − 05
gram	kilogram (kg)	1.000 E − 03
g/cm^3	kilogram per meter^3 (kg/m^3)	1.000 E + 03
gram-force/cm^2	pascal (Pa)	9.807 E + 01
	meter (m^2)	1.000 E + 04
horsepower (550 ft lbf/s	watt (W)	7.457 E + 02
inch	meter (m)	2.540 E − 02
inch of mercury (32°F)	pascal (Pa)	3.386 E + 03
inch of mercury (60°F)	pascal (Pa)	3.377 E + 03
inch of water (39.2°F)	pascal (Pa)	2.491 E + 02
inch of water (60°F)	pascal (Pa)	2.488 E + 02
in^2	meter^2 (m^2)	6.452 E − 04
in^3	meter^3 (m^3)	1.639 E − 05
kelvin	degree Celsius	$t_c = t_k - 273.15$
kilocalorie	joule (J)	4.187 E + 03
kilogram-force (kgf)	newton (N)	9.807 E + 00
kgf/cm^2	pascal (Pa)	9.807 E + 04
km/h	meter per second (m/s)	2.778 E − 01
kWh	joule (J)	3.600 E + 06
liter	meter^3 (m^3)	1.000 E − 03
micron	meter (m)	1.000 E − 06
mil	meter (m)	2.540 E − 05
mile	meter (m)	1.609 E + 03
mi/h	meter per second (m/s)	4.470 E − 01
mi/h	kilometer per hour (km/h)	1.609 E + 00
Millibar	pascal (Pa)	1.000 E + 02
millimeter of mercury (0°C)	pascal (Pa)	1.333 E + 02
minute (angle)	radian (rad)	2.909 E − 04
ounce		
ounce	kilogram (kg)	3.110 E − 02
ounce (U.K. fluid)	meter^3 (m^3)	2.841 E − 05
ounce (U.S. fluid)	meter^3 (m^3)	2.957 E − 01
poise (absolute viscosity)	pascal second (Pa s)	1.000 E − 01
pound	kilogram (kg)	4.536 E − 01
lb/ft h	pascal second (Pa s)	4.134 E − 04

Selected Conversion Factors (SI) System of Units (continued)

To convert from	to	Multiply by[a]
lb/ft s	pascal second (Pa s)	1.488 E + 00
lb/ft^3	kilogram per meter^3 (kg/m^3)	1.602 E + 01
lb/gal (U.K. liquid)	kilogram per meter^3 (kg/m^3)	9.978 E + 01
lb/gal (U.S. liquid)	kilogram per meter^3 (kg/m^3)	1.198 E + 02
lb/h	kilogram per second (kg/s)	1.260 E − 04
lb/in^3	kilogram per meter^3 (kg/m^3)	2.768 E + 04
pound-force (lbf)	newton (N)	4.448 E + 00
lbf ft	newton meter (N m)	1.356 E + 00
lbf s/ft^2	pascal second (Pa s)	4.788 E + 01
lbf/ft^2	pascal (Pa)	4.788 E + 01
lbf/in^2 (psi)	pascal (Pa)	6.895 E + 03
second (angle)	radian (rad)	4.848 E − 06
slug	kilogram (kg)	1.459 E + 01
stokes (kinematic viscosity)	meter^2 per second (m^2/s)	1.000 E − 04
therm	joule (J)	1.055 E + 08
ton (long. 2240 lb)	kilogram (kg)	1.016 E + 03
ton (metric)	kilogram (kg)	1.000 E + 03
ton (refrigeration)	watt (W)	3.517 E + 03
ton (short. 2000 lb)	kilogram (kg)	9.072 E + 02
W h	joule (J)	3.600 E + 03
W s	joule (J)	1.000 E + 00
yard	meter (m)	9.144 E − 01

[a] E denotes a power of 10, e.g. 1 bar = 1.00×10^5 Pa; 1 cP = 1.0×10^{-3} Pa s.

CITED REFERENCES

Abbott, J. A., G. S. Bachman, N. F. Childers, J. V. Fitzgerald, and F. J. Matusik. 1967. *Sonic Techniques for Measuring Texture of Fruits and Vegetables.* Revised Report (August 31, 1966) and Extension Report (February 28, 1967). Nametre Company, Edison, N.J.

Abbot, J. A., C. S. Bachman, R. F. Childers, J. V. Fitzgerald, and F. J. Matusik. 1968. Sonic Techniques for Measuring Texture of Fruit and Vegetables. *Food Technology* 22 (5):101–112.

Abu-Gheida, D. M. 1961. Mechanical Harvesting of Tree Fruits Utilizing a Pulsating Air Stream. M. S. Thesis Michigan State University, East Lansing, Michigan.

Agness, J. B. 1968. Measuring Mechanical Damage to Corn Kernels. ASAE Paper No. 68–620. Am. Soc. Agr. Engrs., Saint Joseph, Michigan.

Agricultural Engineering Yearbook. 1966. American Society of Agricultural Engineers, St. Joseph, Michigan

Agricultural Material Handling (Section 2.5—Liquid Conveyors). 1962. National Committee on Agricultural Engineering, Canada Department of Agriculture, Ottawa, Canada.

Airy, W. 1898. The Pressure of Grain. Institution of Civil Engineers. Minutes of proceedings 121:347.

Akyurt, M., G. E. Zachariah, and C. G. Haugh. 1972. Constitutive relations for plant materials. *Trans. Am. Soc. Agric. Engr.* 15:766–769.

Alfrey, Turner, Jr. 1948. *Mechanical Behavior of High Polymers.* Interscience Publishers, Inc., New York and London.

American Association of Cereal Chemists. 1947. *Cereal Laboratory Methods.* 5th ed., p. 162.

American Concrete Institute. 1946. Recommended Practice for the Construction of Concrete Farm Silos. ACI 714–46 Volume 18, No. 2.

Anderson, D. J. 1956. Measurement of Stress in Mastication, I and II. *J. Dental Res.* 35:664, 671.

Anderson, R. A., D. E. Uhl, W. L. Detheridge, and E. L. Griffin. 1962. Composition of the component parts of two hybrid amylose corns. *Cereal Chem.* 39:282–286.

Andrews, J. P. 1929. Impact of Spheres of Soft Metals. *Phil Mag.* 8(53) 781–800.

Andrews, J. P. 1930. Theory of Collision of Spheres of Soft Metals. *Phil. Mag.* 9(58):593–610.

Andrews, J. P. 1931. Experiments on Impact. *Proc. Phys. Soc. Lond.* 43:8–17.

Andrews, R. D. 1952. Correlation of Static and Dynamic Measurements on Rubberlike Materials. *Industrial and Engineering Chemistry* 44(4):707–715.

Ang, J. K., F. M. Isenberg and J. D. Hartman. 1960. Measurement of Firmness on Onion Bulbs with a Shear Press and a Potentiometric Recorder. *Proc. Am. Soc. Hort. Sci* 75:500.

Anon. 1963. Measuring Potato Bruising. *Agricultural Research*. January.

Anon. 1966. Viscosity. *Lubrication* 52(3):21–48.

Arnold, P. C. and A. W. Roberts. 1966. Stress Distributions in Loaded Wheat Grains. *J. Agric. Engng. Res.* 11(1):38–43.

Arnold, P. C. and A. W. Roberts. 1969. Fundamental Aspects of Load-Deformation Behavior of Wheat Grain. *Trans. of the ASAE* 12(1):104–108.

Arnold, P. C. 1968. An Investigation of Mechanical and Rheological Properties of Wheat Grains. Ph. D. Thesis in Mechanical Engineering, Wollongong University College, The University of New South Wales.

Arnold, P. C. and N. N. Mohsenin. 1970. Recommended techniques for axial compression tests on intact agricultural products of convex shape. *Trans. Am. Soc. Agric. Engr.* 14:78–84.

Atanasoff, J. V. 1937. Measurement of the Viscosity of Eggs by the Use of a Torsion Pendulum. *J. Agr. Research* 54:701–709.

Austin, M. E. and Dedolph. 1962. An Evaluation of Apple Bruise Damage Measurement. *Proc. Am. Soc. Hort. Sci.* 80:125.

Avery, G. S. 1933. Structure and Development of the Tobacco Leaf. *J. An. Botany,* 20:565–592.

Baader, W. 1961. Das Verhalten eines Schüttgutes auf schwingenden Siebrosten. *Grundlagen der Landtechnik* 13:13–20. (Behavior of loose material on oscillating riddles, Translation No. 120, N. I. A. E. Silsoe).

Bailey, C. H. 1912. A Method for the Determination of the Specific Gravity of Wheat and Other Cereals. USDA, Bureau of Plant Industry, Circular No. 99.

Bain, J. M. and R. N. Robertson. 1951. The Physiology of Growth in Apple Fruits. I. Cell Size, Cell Number, and Fruit Development. *Aus. J. Sci. Res.* (Biol. Sci.) 2:75–91.

Bakker-Arkema, F. W., W. G. Bickert and S. T. Dexter. 1968. Environmental Control During Storage to Prevent Cracking of Pea Beans—An Analysis. *Trans. of the ASAE* 11(3):380–383.

Bakker-Arkema, F. W., J. R. Rosenau, and W. H. Clifford. 1971. The Effect of Grain Surface Area on the Heat and Mass Transfer Rates in Fixed and Moving Beds of Biological Products. *Trans. of the ASAE* 14(5):864–867.

Baks, A. and W. L. Schmid. 1960. Vertical Transport met Schroeftransporteu Orgaan van Het Nederlands Institut van Register-Ingenieurs en Afgestudeeren van Hogere Technische Scholen 4–15e Jaarz, February, 196.

Balis, J. S. 1958. Methodology of the Determination of Friction Coefficients of Agricultural Materials. M. S. Thesis in Agricultural Engineering, Purdue University.

Barger, E. L., T. B. Liljedahl, W. W. Gunkel and W. M. Carleton. 1949. Results of Test on Mechanical Loading Devices for Chopped Forage. *Agricultural*

Engineering 30(5):223–225.

Barkas, W. W. 1953. *The Mechanical Properties of Wood and Their Relation to Moisture*. Part A in *Mechanical Properties of Wood and Paper*. R. Meredith, ed., Interscience Publishers, Inc., New York.

Barney, T. E., H. D. Pollock and C. C. Bolze. 1965. *Cereal Chemistry*. 42:215.

Barre, H. J. and L. L. Sammet. 1950. *Farm Structures*. John Wiley and Sons, Inc., New York.

Barrow, G. M. 1973. *Physical Chemistry*. McGraw-Hill Book Company, New York.

Baten, W. D. and R. E. Marshall. 1943. Some Methods for Approximate Prediction of Surface Area of Fruits. *J Agricultural Research* 66(10):357–373.

Beeman, J. F., W. W. Deen, and L. H. Halsey. 1966. Development of a Mechanical Celery Harvester. *Agricultural Engineering* 47(7):376–377.

Bekker, M. G. 1960. *Off-the-Road Locomotion*. The University of Michigan Press, Ann Arbor, Mich.

Bennesen, T. and W. Fenchel. 1948. *Theorie der konvexen Koerper* (Theories of Convex Bodies).

Bennett, E. H. 1950. Kernel Hardness in Corn. I. A. Machine for the Rapid Determination of Kernel Hardness. *Cereal Chemistry* 27:222.

Bennett, E. H. 1950. Kernel Hardness in Corn. II. A Microscopic Examination of Hard and Soft Types of Dent Corn. *Cereal Chemistry* 27:232.

Berry, P. E. 1958. Research on Oscillating Conveyors. *J. Agric. Engng. Res.* 3(3):249.

Berry, P. E. 1959. Basic Theory of Low Acceleration Oscillating Conveyors. *J. Agric. Engng. Res.* 4(3):204–213.

Besch, E. L., S. J. Sluka, and A. H. Smith. 1968. Determination of Surface Area Using Profile Recordings. *Poultry Science*. 47(1):82–85.

Besely, H. E. and J. R. McCalmont. 1941. Observations on the Storage of Grass Silage. *Agricultural Engineering* 22(2):51–53.

Betscher, J. H., I. A. Gould, C. D. Jones, and J. L. Baisdell. 1967. Flow Properties of Fluid Milk Products. *Agricultural Engineering* 48(12):714–715.

Beverloo, W. A., H. A. Leniger, and J. van De Velde. 1961. The Flow of Granular Solids Through Orifices. *Chemical Engineering Science*. 15:260–269.

Bickel, H. 1963. Der Strömungswiderstand von Dürrfutter in Abhängigkeit von der Strömungsrichtung. *Landtechnische Forschung* 13(1):24–26.

Bickert, W. G. and F. H. Buelow. 1966. Kinetic Friction of Grains on Surfaces. *Trans. of the ASAE* 9(1):129–134.

Bilanski, W. K., S. H. Collins, and P. Chu. 1962. Aerodynamic Properties of Seed Grains. *Agricultural Engineering* 43(4):216–219.

Bilanski, W. K. and R. Lal. 1965. The Behavior of Threshed Materials in a Vertical Wind Tunnel. *Trans. of the ASAE* 411–413, 416.

Bilanski, W. K. and D. Menzies. 1967. Aerodynamic Properties of Alfalfa Particles. ASAE Paper No. 67–672, Am. Soc. Agr. Engr., Saint Joseph, Mich.

Birth, G. S. 1960. A Nondestructive Technique for Detecting Internal Discoloration in Potatoes. *Am. Potato Journal* 37(2):53–60.

Bittner, D. R., H. B. Manbeck, and N. N. Mohsenin. 1967. A Method for Evaluating Cushioning Materials Used in Mechanical Harvesting and Handling of Fruits and Vegetables. *Trans. of the ASAE* 19(6):711–714.

Bland, D. R. 1960. *The Theory of Linear Viscoelasticity*. Pergamon Press, New York.

Blevins, F. Z. 1954. Some of the Component Power Requirements of Field-Type Forage Harvesters. M. S. Thesis, Purdue University, E. Lafayette, Indiana.

Bloom, M. 1952. Machine for Testing Jelly Strength of Glues, Gelatins and the Like. U.S. Patent 1,540,979 (June 9).

Bluhm, J. I. 1956. Stresses in Projectile During Penetration. *Proc. Soc. Exp. Stress Analysis*. 13(2):167.

Bockstiegel, G. 1966. The Porosity-pressure Curve and its Relation to the Pore-size Distribution in Iron Powder Compacts in *Modern Developments in Powder Metallurgy Vol. 1*. Fundamentals and Methods (ed., H. H. Hausner) Plenum Press, New York.

Bolz, H. A., ed. 1958. *Materials Handling Handbook*. The Ronald Press Co., New York.

Bond, F. C. 1963. Particle size reduction, theory and practice. *Can. Mining J.* 84:43–48.

Bone, D. 1969. Water Activity: Its Chemistry and Applications. *Food Prod. Dev.* 3:81.

Boner, C. J. 1964. *Gear and Transmission Lubricants*. Reinhold Publishing Corp., New York.

Bourne, M. C. 1965. Studies on Punch Testing of Apples. *Food Technology* 19:113.

Bouse, L. F., Schoenleber, L. G., and J. G. Porterfield, 1964. Screw Conveyor Capacity and Castor Seed Damage. *Trans. of the ASAE* 7(2):152–158.

Boussinesq, J. Applications des Potentiels a l'Etude de l'Equilibre et du Mouvement des Solides Elastiques, Paris (1885).

Bouton, P. E. and P. V. Harris. 1972. The effect of cooling temperature and time on some mechanical properties of meat. *J. Food Sci.* 37:140–144.

Bowden, E. P. and D. Tabor. 1950. *Friction and Lubrication of Solids*. Oxford University Press, London.

Bowden, E. P. and D. Tabor. 1954. *The Friction and Lubrication of Solids*. Oxford University Press, London.

Bowden, E. P. and D. Tabor. 1964. *The Friction and Lubrication of Solids* (Part II). Clarendon Press, London.

Bowden, E. P. and D. Tabor. 1964. *Friction and Lubrication of Solids*. Oxford University Press, London.

Bowen, R. L. 1961. How to Handle Slurries. *Chemical Engineering*, 68(17), 119–122.

Brandenburg, N. R. and J. E. Harmond. 1964. Fluidized Conveying of Seeds.

USDA Technical Bulletin 1315. USDA, Washington, D. C.

Brandt, M. A., E. Z. Skinner and J. A. Coleman. 1963. Texture Profile Method. *J. Food Sci.* 28:404.

Bratzler, L. S. 1932. Measuring the Tenderness of Meat by Means of a Mechanical Shear. M. S. Thesis, Kansas State College Library, Manhattan, Kansas.

Braungart, D. C. and R. H. Arnell, Jr. 1962. *An Introduction to Plant Biology.* The C. V. Mosby Company, St. Louis.

Bright, R. E. and R. W. Kleis. 1964. Mass Strength of Haylage. *Trans. of the ASAE* (72):100–101.

Brook, A. J. *The Living Plant.* 1964. Aldine Publishing Company, Chicago.

Brooks, J. and H. P. Hale. 1955. Strength of the Shell of the Hen's Egg. *Nature* 175(5):848–849.

Browne, D. A. 1962. Variation of Bulk Density of Cereals with Moisture Contents. *J. of Agr. Engr. Res.* 7(4):288–290.

Brown, R. L. and J. C. Richards. 1959. Exploratory Study of the Flow of Granules Through Apertures. Trans. Ins. Chem. Engrs. 37:108–119.

Brubaker, J. E. and J. Pos. 1965. Determining Static Coefficient of Friction of Grains on Structural Surfaces. *Trans. of the ASAE* 8:53.

Brunauer, S. P., H. Emmett and E. Teller. 1938. Adsorption of Gases in Multimolecular Layers. *J. of Am. Chem. Soc.* 60:309–319.

Buchanan, J. C. and W. H. Johnson. 1964. Functional Characteristics and Analysis of a Centrifugal Threshing and Separating Mechanism. *Trans. of the ASAE* 7(4):460–463, 468.

Buelow, F. H. 1961. Determination of Friction Coefficients of Materials Handled on the Farm. ASAE Paper No. 61–822. American Society of Agricultural Engineers, St. Joseph, Michigan.

Burke, S. P. and W. B. Plummer. 1928. Gas Flow Through Packed Columns. *Industrial Engineering Chemistry* 20:1197–1200.

Burkner, P. F. and D. M. Kinch. 1968. Force-Deformation Ratio as an Index of Papaya Maturation. *Trans. of the ASAE* 11(3):437–440.

Burmistrova, M. F. *et al.,* 1956. *Physicomechanical Properties of Agricultural Crops.* Translated from Russian and published for the National Science Foundation by the Israel Program for Scientific Translations. Available from Office of Technical Services, U. S. Department of Commerce, Washington, D. C.

Burrill, L. M., I. Deethadt and R. L. Staffle. 1962. Two Mechanical Devices Compared with Taste Panel Evaluation for Measuring Tenderness. *Food Technology* 16:145.

Bursewitz, G. H. and B. A. Stout. 1970. Modeling the stress-strain behavior of plant materials as an interacting fluid-solid mixture. ASAE Paper No. 70–551.

Burton, L. V. 1938. Quality Separation by Differences of Density. *Food Industries* 10:136–138, 170–175.

Busse, W. 1964. Die theorie auf dem gebiet des verdichtens landwirtschaftliches halmguter. *Landtech. Forschung* 14:6–15.

Caldwell, D. H. and H. E. Babbitt. 1941. The Flow of Muds, Sludges and Suspensions in Circular Pipes. *Am. Inst. Chem. Engrs. Trans.* 37:237–266.

Caldwell, F. and A. C. W. Davies. 1957. The Relation of Development of Pathogenic Fungi to the Mechanical Injury and Storage Moisture Content of Cereal Grains. *J. Agric. Engng. Res.* 2(1):67.

Caldwell, J. S. 1939. Factors Influencing Quality of Sweet Corn. *Canning Trade* 61(42):7–8.

Campbell, L. E. 1938. The Calibration of Jelly-Testers. *J. Soc. Chem. Ind.* 57:413.

Carman, P. C. 1937. Fluid Flow Through Granular Beds. *Inst. of Chem. Engrs. Trans.* 15:150–166.

Casada, J. H., J. N. Walker and E. M. Smith. 1968. A Method of Predicting the Flextural Behavior of Intact Tobacco Stalks. ASAE Paper 68–813. Am. Soc. Agr. Engrs., St. Joseph, Michigan.

Casson, N. 1959. *Rheology of Disperse System.* (C. C. Mill, ed.), Pergamon Press, New York.

Caughey, R. A., Tooles, C. and A. C. Scheer. 1951. Lateral and Vertical Pressures of Granular Materials in Deep Bins. Iowa Exp. Sta. Bulletin 172, Iowa State University, Ames, Iowa.

Chancellor, W. J. 1960. Relation Between Air and Solid Particles Moving Upward in a Vertical Pipe. *Agricultural Engineering* 41(3):168–171, 176.

Chancellor, W. J. 1960. Influence of Particle Movement on Energy Losses in an Impeller Blower. *Agricultural Engineering* 41(2):92–94.

Chancellor, W. J. and G. E. Laduke. 1960. Analysis of Forage Flow in a Deflector Elbow. *Agricultural Engineering* 41(3):234–236, 240.

Chappell, T. W. and D. D. Hamann. 1968. Poisson's Ratio and Young's Modulus for Apple Flesh Under Compressive Loading. *Trans. of the ASAE* 11(5):608–610.

Charm, S. E. 1959. Heat Transfer Coefficients of Pseudoplastic Food Materials in Streamline Flow in Straight Tubes. *Food Research* 24:319.

Charm, S. E. 1960. The Direct Determination of Shear Stress-Shear Rate Behavior of Foods in the Presence of a Yield Stress. *J. of Food Science* 28(1):107–113.

Charm, S. E. 1963. Viscosity of Non-Newtonian Food Materials. *Food Research* 25:351–362.

Charm, S. E. 1963. *Fundamentals of Food Engineering.* AVI Publishing Co., Westport, Conn.

Chen, C. S. and J. T. Clayton. 1970. Equilibrium Moisture Curves of Biological Materials—The Effect of Temperature. Paper No. 70–383, ASAE, St. Joseph, Mich.

Chizhikov, A. 1960. The Influence of Drying Conditions on Fracturing and Strength of a Corn Kernel. Mukomol'no-elevatornaja Promyshlennost. 26(9):15–16.

Chung, Do Sup. 1966. Thermodynamic Factors Influencing Moisture Equilibrium of Cereal Grains and Their Products. Ph. D. Thesis, Kansas State University.

Chung, D. S. and H. B. Pfost. 1967. Adsorption and Desorption of Water Vapor

by Cereal Grains and Their Products. *Trans. ASAE* 10(4):549–557.
Chung, D. S. 1968. Radiographic Examination of Wheat. ASAE Paper No. 68–103. (Mid Central Region). Department of Agricultural Engineering, Kansas State University, Manhattan, Kansas.
Clardy, L., W. D. Phole and V. C. Mehlenbacher. 1952. A Shortening Consistometer. *J. Am. Oil Chemists' Soc.* 29:591.
Clark, R. L., G. B. Welch, and J. H. Anderson. 1967. The Effect of High Velocity Impact on the Germination and Damage of Cottonseed. ASAE Paper No. 67–822. Am. Soc. Agric. Engrs., Saint Joseph, Michigan.
Clark. R. L., W. R. Fox, and G. B. Welch. 1968. Representation of Mechanical Properties of Nonlinear Viscoelastic Materials by Constitutive Equations. ASAE Paper No. 68–811, Am. Soc. Agr. Engrs., St. Joseph, Michigan.
Clevenger, J. T. and D. D. Hamann. 1968. The Behavior of Apple Skin Under Tensile Loading. *Trans. of the ASAE* 11(1):34–37.
Collins, N. E., W. L. Harris, and G. J. Burkhardt. 1965. Pneumatic Conveying of Chopped Forage. *Trans. of the ASAE* 8(2):196–198.
Collins, R. E. 1961. *Flow of Fluids Through Porous Materials*. Reinhold Publishing Corporation, New York.
Collins, R. E. 1963. Determination of Pressures in Cylindrical Storage Structures. *Trans. of the ASAE* 6(2):98–101, 103.
Combs, Y. F. 1944. An Instrument for Determining the Compressibility and Resistance to Shear of Baked Products. *Cereal Chemistry* 21:319.
Condolios, E. and E. E. Chapus. 1963a. Transporting Solid Materials in Pipelines. *Chemical Engineering* 40 (June 24):93–98.
Condolios, E. and E. E. Chapus. 1963b. Designing Solids—Handling Pipelines. *Chemical Engineering* 70(July 8):131–135.
Connell, J. J. 1960. Mechanical Properties of Fish and Fish Products, In *Flow Properties of Blood and Other Biological Systems*. (Copely, A. L. and G. Stainsby, eds.) Pergamon Press, New York.
Connell, J. J. 1964. *Proteins and Their Reactions* (Schultz, H. W. and A. F. Anglemier, eds.) AVI Publishing Co., Westport, Conn.
Cooke, J. R. and J. W. Dickens. 1967. A Centrifugal Gun for Impaction Testing of Seeds. ASAE Paper No. 67–330, Am. Soc. of Agr. Engrs., Saint Joseph, Michigan.
Cooke, J. R. 1972. An Interpretation of the Resonant Behavior of Intact Fruits and Vegetables. *Trans. of the ASAE* 15(6):1075–1080.
Cooper, A. R. and L. E. Eaton. 1962. Compaction Behavior of Several Ceramic Powders. *J. Am. Ceram. Soc.* 45:97–101.
Cooper, H. E. 1962. Influence of Maturation on the Physical and Mechanical Properties of the Apple Fruit. M. S. Thesis, The Pennsylvania State University, University Park, Pennsylvania.
Coppens, R. and H. D. Wei. 1954. Determination of the Hygroscopic Point and Hygroscopicity Curve as a Function of Moisture Content in Agricultural Ma-

terials Which are to be Dried. Translated by J. A. Midgrey. *Annales de Gembloux* 4:225–235. NIAE, Silsoe, England.

Coppock, G. E., S. L. Hedden, and D. H. Lenker. 1968. Biophysical Properties of Citrus Fruit Related to Mechanical Harvesting. ASAE Paper No. 68–117, Am. Soc. Agr. Engrs., Saint Joseph, Michigan.

Cornish, G. K. and L. F. Charity. 1966. Pressure Drop in Elbows of a Pneumatic Conveying Systems. *Trans. of the ASAE* 9(1):29–31.

Crane, J. W. and W. M. Carleton. 1957. Predicting Pressure Drop in Pneumatic Conveying of Grains. *Agricultural Engineering* (3):168–180.

Culbertson, C. C., R. S. Shearer, E. E. Hammond, and J. L. Robinson. 1940. Corn of Different Degrees of Hardness. An. Hus. Leaflet. No. 158, Iowa State College.

Cunningham, F. M. 1963. Performance Characteristics of Bulk Spreaders for Granular Fertilizer. *Trans. of the ASAE* 6(2):108–114.

Cunningham, J. R., I. Hlynka, and J. A. Anderson. 1953. An Improved Relaxometer for Viscoelastic Substances Applied to the Study of Wheat Dough. *Canadian J. of Technology* 31:98–108.

Curray, J. K. 1951. Analysis of Sphericity and Roundness of Quartz Grains. M. S. Thesis in Mineralogy, The Pennsylvania State University, University Park, Pa.

Curtis, L. M. and J. G. Hendrick. 1968. A Study of the Bonding Strength Properties of Cotton Stalks. Paper Presented before the 1968 Southeast Section of the ASAE, February 6, Louisville, Kentucky.

Cziffra, P. and M. J. Moraveski. 1959. A Practical Guide to the Method of Least Squares. University of California. Radiation Laboratory Rev. 8523.

Dally, J. W. 1965. *Experimental Stress Analysis*. McGraw-Hill Book Co., New York.

Daniels, F. 1949. *Outline of Physical Chemistry*. John Wiley and Sons, New York.

Davies, C. N. 1947. Symposium on Particle Size Analysis. Inst. Chem. Engrs. and Soc. Chem. Ind., London.

Davis, D. C. and G. E. Rehkugler. 1969. A Theoretical and Experimental Analysis of the Apple-Limb Impact Problem. ASAE Paper No. 69–879. American Society of Agricultural Engineers, St. Joseph, Michigan.

Davis, J. G. 1937. The Rheology of Cheese, Butter and Other Milk Products. *J Dairy Res*. 8:245.

Davis, R. M. 1962. Tissue Air Space in the Potato: Its Estimation and Relation to Dry Matter and Specific Gravity. *Am. Potato J*. 39:298–305.

Davison, S., A. L. Brody, B. E. Proctor and P. Felsenthal, 1959. A Strain Gage Pea Tenderometer. 1. Instrument Description and Evaluation. *Food Technology* 13:119.

Dawson, R. F. Laboratory Manual in Soil Mechanics. Pitman Publishing Corp., New York.

Day, C. L. 1963. Effects of Conditioning and other Factors on the Resistance of

Hay to Air Flow. *Trans. of the ASAE* 199–201.

Day, C. L. and G. L. Nelson. 1963. Desorption Isotherms for Wheat. ASAE Paper No. 63–424, Am. Soc. Agr. Engrs., St. Joseph, Michigan.

Day, C. L. 1964. Device for Measuring Voids in Porous Materials. *Agricultural Engineering* 45(1):36–37.

Day, C. L., H. H. Panda. 1966. Effect of Moisture Content, Depth of Storage and Length of Cut on Bulk Density of Alfalfa Hay. *Trans. of the ASAE* 9(3):428–432.

Dekazos, E. D. 1966. Relation Between Scald of Montmorency Cherries and Oxygen Content in Soak Tanks. *J. of Food Science* 31(6):956–963.

Deming and Mehring. 1929. The Gravitational Flow of Fertilizers and other Comminuted Solids. *Industrial and Engineering Chemistry* 21(7):661–665.

Dewey, D. H., B. A. Stout, R. H. Matthews, and F. W. Bakker-Arkema. 1966. Development of a Hydrohandling System for Sorting and Sizing Apples for Storage in Pallet Boxes, Marketing Research Report No. 743 STD, UDFS.

Diener, R. G. and D. R. Heldman. 1968. Methods of Determining Rheological Properties of Butter. *Trans. of the ASAE* 11(3):444–447.

Diener, R. G., R. E. Adams, M. Ingle, K. C. Elliott, P. E. Nesselroad, and S. H. Blizzard. 1977. Bruise Energy and Peaches and Apples. ASAE Paper No. 77–1029. American Society of Agricultural Engineers, St. Joseph, Michigan.

Dillewign, C. van. 1952. *Botany of Sugar Cane*. The Chronica Botanica Co. Waltham, Mass.

Dodge, D. W., and A. B. Metzner. 1959. Turbulent flow of non-Newtonian systems. *AICJEJ* 5:189–204.

Dolby, R. M. 1941. The Rheology of Butter. I. Methods of Measuring the Hardness of Butter. *J. Dairy Res.* 12:329–336.

Drake, B. 1962. Automatic Recording of Vibrational Properties of Foodstuffs. *J of Food Sci.* 27:182.

Drake, B. 1963. Food Crushing Sounds—An Introductory Study. *J. Food Sci.* 28:223.

Drake, B. 1968. *The Biorheological Process of Mastication. S. C. I. Monograph* No. 27. Soc. of Chem. Ind. 14 Belgrave Square, London.

Drake, B. 1979. A Fortran Program "Framed" for Simulation of Large Fractoviscoelastic Models. *J. Texture Studies* 10:165–182.

Drow, J. T. and R. S. McBurney. 1946. The Elastic Properties of Wood, Young's Moduli, Moduli of Rigidity and Poisson's Ration of Yellow Poplar. Forest Product Laboratory Publication No. 1528G.

Drutowski, R. C. 1959. Energy Losses of Balls Rolling on Plates. In *Friction and Wear*. Davies, R. ed. Edsevier Publishing Company, N.Y.

Duckworth, R. B. 1971. Differential thermal analysis of frozen food systems. The determination of unfreezable water. *J. Food Tech.* 6:317.

Duffy, J. and R. D. Mindlin. 1957. Stress-Strain Relations and Vibrations of Granular Medium. *Trans. of ASME, J. of Applied Mechanics* 24:585–593.

Duffy, V. 1959. A Differential Stress-Strain Relation for the Hexagonal Close-Packed Array of Elastic Spheres. *Trans. of ASME, J. Applied Mechanics* 26(3):88–94.

Edgar, A. D., R. Claycome and T. Hansen. 1957. Flume Systems for Handling Bulk Stored Potatoes. USDA Marketing Research Report No. 177.

Edgerly, G. O. 1951. Bin Storage of Onions. M. S. Thesis in Agricultural Engineering, Michigan State University, East Lansing, Michigan.

Ekstrom, G. A., J. B. Liljedahl, and R. M. Peart. 1966. Thermal Expansion and Tensile Properties of Corn Kernels and Their Relationship to Cracking During Drying. *Trans. of the ASAE* 9(4):556–561.

Emori, R. and D. Schuring. 1963. Static and Dynamic Penetration Tests of Soils. Proceedings 4th International Congress on Rheology.

Engle, P. A. 1976. *Impact Wear of Materials*. Elsevier Scientific Pub. Co., New York.

Ernest, H. and M. E. Merchant. 1940. Surface Friction of Clean Metals. Summer Conference on Friction and Surface Finish. M. I. T., Cambridge, Mass.

Eschenwald, A. 1959. Some Basic Properties of the Coffee Fruit and Coffee Beans. M. S. Thesis, Michigan State University, East Lansing, Michigan.

Ewalt, D. J. and F. H. Buelow. 1963. Flow of Shelled Corn Through Orifices in Bin Walls. Quarterly Bulletin of the Michigan Agricultural Experiment Station 46(1):92–102. Michigan State University, East Lansing, Mich.

Falk, S., C. H. Hertz and H. I. Virgin. 1958. On the Relation Between Turgor Pressures and Tissue Rigidity. I. Experiments on Resonance Frequency and Tissue Rigidity. *Physiol. Plantarum* 11:802.

Feltham, P. 1955. On the Representation of Rheological Results with Special Reference to Creep and Relaxation. *British J. of Applied Physics*. 6(1):26–31.

Fennema, O. and L. A. Berny. 1974. Proc. 4th Int. Cong. Food Sci. and Tech., Madrid, II, 27.

Ferry, J. D. 1958. Experimental Techniques for Rheological Measurements of Viscoelastic Bodies. *Rheology, Theory and Applications*. Vol. F. R. Eirich, Editor, Academic Press, Inc., New York.

Ferry, J. D. 1961. *Viscoelastic Properties of Polymers.*, John Wiley & Sons, Inc., New York.

Finney, E. E. 1963. The Viscoelastic Behavior of the Potato. Solanum Tubersom, Under Quasi-Static Loading. Ph. D. Thesis, Michigan State University, East Lansing, Michigan.

Finney, E. E., C. W. Hall and G. E. Mase. 1964. Theory of Linear Viscoelasticity Applied to Potato. *J. of Agric. Engng. Res.* 9(4):307–312.

Finney, E. E., C. W. Hall, and N. R. Thompson. 1964. Influence of Variety and Time Upon the Resistance of Potatoes to Mechanical Damage. *Am. Potato J.* 41(6):178–186.

Finney, E. E., and C. W. Hall. 1967. Elastic Properties of Potatoes. *Trans. of the ASAE* 10(1):4–8.

Finney, E. E. 1967. Dynamic Elastic Properties of Some Fruits During Growth and Development. *J. Agric. Engng. Res.* 12(4):249–256.

Finney, E. E. and K. H. Norris. 1968. Instrumentation for Investigating Dynamic Mechanical Properties of Fruits and Vegetables. *Trans. of the ASAE.* 11(1):94–97.

Finney, E. E. 1968. Mechanical Resonance Within Red Delicious Apples and Its Relationship to Fruit Texture. ASAE Paper No. 68–875. Am. Soc. Agr. Engrs., Saint Joseph, Michigan.

Finney, E. E. Jr. 1970. Mechanical Resonance Within Red Delicious Apples and Its Relation to Fruit Texture. *Trans. of the ASAE* 13(2):177–180.

Fitzgerald, E. R. and J. D. Ferry. 1953. Method for Determining the Dynamic Mechanical and Electrical Properties. *J. of Colloid Science* 8:1–34.

Fletcher, S. W., N. Mohsenin, J. R. Hammerle, and L. D. Tukey. 1965. Mechanical Behavior of Fruits Under Fast Rates Loading. *Trans. of the ASAE* 8(3):324–326, 331.

Fliesen, J. A. and C. D. Beal. 1971. Moisture Transfer Rates in Cotton Seed and Lint. Paper No. 71–345, ASAE, St. Joseph, Mich.

Fliesen, J. A. 1974. Predicting Equilibrium Moisture Content Within the Hysteresis Loop. *Trans. ASAE* 17(2):339–341.

Fluck, R. C. and W. E. Splinter. 1966. Infrared Gas Analysis for Measuring Mechanical Damage to Biological Tissues. ASAE Paper No. 66–842, Am. Soc. Agr. Engrs., Saint Joseph, Michigan.

Fluck, R. C., F. S. Wright and W. E. Splinter, 1968. Compression Plunger, Skinning and Friction Properties of Sweet Potatoes. *Trans. of the ASAE* 11(2):167–170, 174.

Fluck, R. C. and E. M. Ahmed. 1973. Impact Testing of Fruits and Vegetables. *Trans. of the ASAE* 16(4):660–666.

Fluck, R. C. 1975. Use of Impact Parameters to Control Handling Damage to Fruits and Vegetables. Design Application of Mechanical Properties of Solid Foods Materials, Proceedings of a Workshop, Aug. 7–8, 1974, Department of Agricultural Engineering, The Pennsylvania State University, University Park, Pa.

Flugge, W. 1960. *Stresses in Shells.* Springer-Verlag, Berlin/Göttingen/Heidelberg.

Fowler, R. T. and Wyatt, F. A. 1960. The Effect of Moisture Content on the Angle of Repose of Granular Solids. *Australian Journal for Chemical Engineers.*

Fowler, R. T. and J. R. Glastonburg. 1959. The Flow of Granular Solids Through Orifices. *Chemical Engineering Science* 10:150–156.

Franklin, F. C. and L. N. Johansen. 1955. Flow of Granular Materials Through a Circular Orifice. *Chemical Engineering Science* 4:119–129.

Frazier, W. A. 1934. A Study of Some Factors Associated with the Occurrence of Cracks in the Tomato Fruit. *Proc. Am. Soc. Hort. Sci.* 32:519.

Frechette, R. J. 1964. Transient Heat Transfer in Apple. M. S. Thesis in Agricultural

Engineering, University of Massachusetts, Amherst, Mass.

Frechette, R. J. and J. W. Zahradnik. 1966. Surface Area-Weight Relationships for McIntosh Apples. *Trans. of the ASAE* 9(4):526–527.

Frederick, D and M. A. Garcia. 1958. The Theoretical Behavior of an Inelastic Material Model. Bulletin of the Virginia Polytechnic Institute Experiment Station. Series No. 85.

Freundlich, H. 1922. *Colloid and Capillary Chemistry*. E. P. Dutton & Co. Inc., New York.

Frey-Wyssling, A. 1952. *Deformation and Flow in Biological Systems*. Interscience Publishers, Inc., New York.

Fridley, R. B., Goehlich, H., Claypool, L. L., and P. A. Adrian. 1964. Factors Affecting Impact Injury to Mechanically-Harvested Fruit. *Trans. of the ASAE* 7(4):409–411.

Fridley, R. B. and P. A. Adrian. 1966. Mechanical Properties of Peaches, Pears, Apricots, and Apples. *Trans. of the ASAE* 9(1):135–142.

Fridley, R. B., Bradley, R. A., Rumsey, J. W. and P. A. Adrian. 1968. Some Aspects of Elastic Behavior of Selected Fruits. *Trans. of the ASAE* 11(1):46–49.

Friedman, H. H., J. Whitney and A. S. Szczesniak. 1963. The Texturometer—A New Instrument for Objective Texture Measurement. *J. Food Sci.* 28:390.

Fung, Y. C. B. 1967. Biomechanics, Its Scope, History, and Some Problems of Continuum Mechanics in Physiology. *Applied Mechanics Review* 21(1):1–20.

Fung, Y. C. B. 1968. Elasticity of Soft Tissues in Simple Elongation. *American Journal of Physiology* 213(6):1532–1544.

Furno, F. J., R. S. Web, and N. P. Cook. 1962. High Speed Puncture Testing of Thermoplastics, in *High Speed Testing,* Vol. IV. Dietz and Eirich, eds., Interscience Publishers, New York.

Gal, S. 1967. *Die Methodik der wasser dampf-Sorptions messungen*. Spring Verlag, Berlin.

Galin, W. A. 1961. *Contact Problems in the Theory of Elasticity*. Translated from Russian by H. Moss, edited by I. N. Sneddon. North Carolina State College, Department of Mathematics and Engineering Research, Raleigh, N. C.

Garrett, A. W., Desrosier, N. W., Kuhn, G. D. and M. C. Fields. 1960. Evaluation of Instruments to Measure Firmness of Tomatoes. *Food Technology* 14:562.

Garrett, R. E. and D. B. Brooker. 1965. Aerodynamic Drag of Farm Grains. *Trans. of the ASAE* 8(1):49–52.

Garrett, R. E., M. Zahara, and R. E. Griffin. 1966. Selector-Component Development for a Head-Lettuce Harvester. *Trans. of the ASAE* 9(1):56–57.

Garrett, R. E., M. Zahara, and F. Zink. 1969. Firmness, Density and Volume of Crisphead Lettuce as Related to Acceptability for Market. *Trans. of the ASAE* 12(4):564–566.

Garrett, R. E. 1970. Velocity of Propagation of Mechanical Disturbance in Selected Fruits and Vegetables. Ph.D. Thesis, Department of Agricultural Engineering,

Cornell University, Ithaca, New York.
Gasnikow, A. 1960. An Increase in the Keeping Quality of Onions by Means of Thermal Treatment. *Sel, sk. Khaz, Sibiri* (8):36–37.
Gaston, H. P. and J. H. Levin. 1951. How to Reduce Apple Bruising. Mich. State College Spec. Bull. 374.
Gemant, A. 1950. *Frictional Phenomena*. Chemical Publishing Company, Inc., Brooklyn, New York.
Gent, A. N. and A. G. Thomas. 1963. Mechanics of Foamed Elastic Materials. Proc. of 7th Annual Technical Conference of the Cellular Plastics Division of the Society of the Plastic Industry, New York.
Gentry, J. P., F. G. Mitchell, and N. F. Sommer. 1965. Engineering and Quality Aspects. of Deciduous Fruits Packed by Volume-Filling and Hand-Placing Methods. *Trans. of the ASAE* 8(4):584–589.
Ghosh, B. N. 1968. Physical Properties of the Different Grades of Arabic Coffee Beans. *Trans. of the ASAE* 9(4):592–593.
Gilfillan, G. and A. J. Crowther. 1959. The Behavior of Potatoes, Stones and Clods in a Vertical Air Stream. *J. of Agric. Engng. Res.* 4:9.
Glassmann, P. 1970. Physical principals of chemical engineering technology (in German). Saver lander, Aarau.
Glasstone, S., K. Lardler and H. Eyring. 1941. *Theory of Rate Processes*. McGraw-Hill Book Company, Inc., New York.
Glover, F. A. and G. W. Scott Blair. 1966. The Effect of Sperm Motility and Viability on the Viscosity of Bull Semen. *Biorheology* 3:189–195.
Goldsmith, W. 1960. *Impact, The Theory and Physical Behavior of Colliding Bodies*, Edward Arnold Publishers, Ltd., London.
Gould, W. A. 1957. Density Can Be a Tool for Control of Quality. *Food Packer* 38(1):30–32.
Goulden, C. H. 1952. *Methods of Statistical Analysis*. John Wiley and Sons, Inc., New York.
Graham, J. R. 1965. Compressive Characteristics of Corn Silage. M. S. Thesis in Agricultural Engineering, The Pennsylvania State University, University Park, Pa.
Green, H. C. 1956. Potato Damage. *J. Agric. Engng.* Res. 1:56–62.
Greenham, C. G. and D. J. Cole. 1950. Studies on the Determination of Dead or Deceased Tissues. *Aust. J. Agric. Res.* 1:103–117.
Greup, D. H. and H. M. R. Hingtzer. 1953. Cereals in *Foodstuffs Their Plasticity, Fluidity and Consistency*. Scott Blair, G. E., ed. p. 66.
Griffin, A. C. and V. P. Moore. 1965. Relation of Physical Properties of Cotton to Commerce and Ginning Research. *Trans. of the ASAE* 8(4):488–490.
Griffin, J. H. and Z. I. Kertesz. 1946. Changes Which Occur in Apple Tissue Upon Treatment with Various Agents and Their Relation to the Natural Mechanism of Softening During Maturation. *Botan. Gaz.* 108:279.
Griffiths, J. C. 1958. Petrography and Porosity of the Cow Run Sand at St. Marys,

West Virginia. *J. Sed Petrology* 28:15–30.

Griffiths, J. C., C. M. Smith. 1964. Relationship Between Volume and Axes of Some Quartzite Pebbles from the Olean Conglomerate Rock City, New York. *Am. J. of Sci.* 262(4):497–512.

Grogg, B. and D. Helms. 1958. A Modification of the Extensograph for Study of the Relaxation of Externally Applied Stress in Wheat Dough. *Cereal Chemistry* 35:189–195.

Grosh, G. M. and M. Milner. 1959. Water Penetration and Internal Cracking in Tempered Wheat Grains. *Cereal Chemistry* 36(3):260–273.

Guillou, R. 1946. Forced Air Flow in Drying Hay. *Agricultural Engineering* 27(11):519–520.

Guillou, R. 1958. Some Engineering Aspects of Cooling Fruits and Vegetables. *Trans. of the ASAE* 1(1):38–39, 42.

Guillou, R. 1963. Settling Packed Fruit by Vibration. *Trans. of the ASAE* 6(3):190–192, and 194.

Gustafson, A. S. and W. L. Kjelgaard. 1963. Hay Pellet: Geometry and Stability. *Agricultural Engineering* 44(8):422–425.

Gustafson, R. J. and G. E. Hall. 1973. Equilibrium Moisture Content of Shelled Corn from 50° to 155°F. Paper No. 73–315, ASAE, St. Joseph, Mich.

Gutyar, E. M. 1956. Elementaryana teoriya verticalnovo vintovovo transportera. Trudy Mosk. Inst. Mekh. i Elekt. Selsk. Khoz. im V. M. Molotov No. 2. 102. (Basic Theory of Vertical Screw Conveyor. Translation No. 84. NIAE, Silsoe, England).

Hadley, W. H. 1966. Surface Penetration Test for Creep Compliance of Viscoelastic Materials. *Materials Research Standards* 6(4):185–187.

Hall, C. W. and Rodriguez-Arias. 1958. Equilibrium Moisture Content of Shelled Corn. *Agricultural Engineering* 39:466–470.

Hall, G. E., R. D. Brazee, and C. W. Hall. 1968. Cross-Sectional Area Measurement of Alfalfa Stems. ASAE Paper No. 68–627, Am. Soc. Ag. Eng., St. Joseph, Michigan.

Haller, M. H. 1941. Fruit Pressure Testers and Their Practical Applications. USDA Circular 627.

Haller, M. H., 1933. The Interrelation of Firmness, Dry Weight, and Respiration in Strawberries. *Proc. Am. Soc. Hort. Sci.* 29:330–334.

Halsey, G. 1947. Non-Linear Viscous Elasticity and the Eyring Shear Model. *J of Applied Physics* Vol. 18.

Halsey, L. H. 1955. Preliminary Studies of Bruising of "Turning" and "Pink" Tomatoes Caused by Handling Practices. *Proc. Fla. State Hort. Soc.* 68: 240–243.

Halsey, L. H. 1963. Studies of Tomato Bruising. *Am. Soc. Hort. Sci. Proc.* 83:291.

Halton, P. and G. W. Scott Blair. 1937. A Study of Some Physical Properties of Flour Dough in Relation to Their Bread-Making Qualities. *Cereal Chemistry* 14:201.

Halyk, R. M. 1962. Tensile and Shear Strength Characteristics of Alfalfa Stems. M. S. Thesis in Ag. Eng., University of Nebraska.

Hamann, D. D. 1967. Some Dynamic Mechanical Properties of Apple Fruits and Their Use in the Solution of an Impacting Problem of Spherical Fruit. Ph. D. Thesis in Engineering Mechanics, Virginia Polytechnic Institute, Blacksburg, Virginia.

Hamann, D. D. 1969. Some Dynamic Mechanical Properties of Apple Fruit Flesh. *Trans. of the ASAE* 12(2):170–174.

Hamann, D. D. 1983. Structural Failure in Solid Foods. In *Physical Properties of Foods,* Peleg and Bagley (Eds.), AVI Publishing Co., Westport, Conn.

Hamdy, M. Y., R. E. Stewart and W. H. Johnson. 1967. Theoretical Analysis of Centrifugal Threshing and Separation. *Trans. of the ASAE* 10(1):87–90.

Hamm, R. and F. E. Dentherage. 1960. Changes in Hydration, Solubility, and Changes of Muscle Proteins During Heating of Meat. *Food Research* 25:587.

Hammerle, J. R. 1966. Some Dynamic Aspects of Fruits Impacting Hard and Soft Materials. M. S. Thesis in Agricultural Engineering, The Pennsylvania State University, University Park, Pa.

Hammerle, J. R. and N. N. Mohsenin. 1966. Some Dynamic Aspects of Fruits Impacting Hard and Soft Materials. *Trans. of the ASAE* 9(4):484–488.

Hammerle, J. R. 1958. Failure in a Thin Viscoelastic Slab Subjected to Temperature and Moisture Gradients. Ph. D. Thesis in Engineering Mechanics. Pennsylvania State University, University Park, Pa.

Hammerle, J. R. and N. N. Mohsenin. 1966. Determination and Analysis of Failure Stresses in Egg Shells. *J Agric. Engng. Res.* 12(1):13–21.

Handleman, A. R., J. F. Conn and J. W. Lyons. 1961. Bubble Mechanics in Thick Foams and Their Effects on Cake Quality. *Cereal Chemistry* 39:294.

Hariu, O. H. and M. C. Molstad. 1949. Pressure Drop in Vertical Tubes in Transport of Solids by Gases. *Industrial and Engineering Chemistry* 41: 1148–1160.

Harmond, J. E., N. R. Brandenburg and D. E. Booster. 1961. Seed Cleaning by Electrostatic Separation. *Agricultural Engineering* 42(1):22–25.

Harper, J. C. 1961. Coaxial-cylinder Viscometer for Non-Newtonian Fluids. *The Review of Scientific Instruments* 32:425.

Harper, J. C. and A. F. ElSahrigi. 1965. Viscometric Behavior of Tomato Concentrates. *J. of Food Science* 30(3):470–476.

Harper, J. C. and K. W. Lebermann. 1964. Rheological Behavior of Pear Purees. First International Congress of Food Science and Technology, London.

Harper, J. C. and I. Kamel. 1966. Viscosity Relationship for Liquid Food Products. ASAE Paper No. 66–837, Am. Soc. Agr. Engrs., Saint Joseph, Michigan.

Harper, R. 1968. Texture and Consistency from the Standpoint of Perception: Some Major Issues. In *Rheology and Texture of Foodstuffs,* S. C. I. Monograph No. 27, Soc. Chem. Ind., 14 Belgrave Square, London.

Harriott, B. L., Oebker, N. F., L. C. Wolf, H. A. Hughes, and R. O. Kuehl.

1967. Physical Characteristics and Damage of Crisphead Lettuce During Harvest Packing Operations. ASAE Paper No. 67–144, Am. Soc. Agric. Engrs., Saint Joseph, Michigan.

Harris, W. L., K. E. Felton, G. E. Burkhart, and N. E. Collins. 1966. Pneumatic Handling of Chopped Alfalfa Hay. University of Maryland, Agricultural Experiment Station Bulletin A-143.

Hartman, H. 1924. Studies Relating to the Harvesting and Storage of Apples and Pears. Oregon Agr. Exp. Sta. Bul. 206

Hassan, A. E., R. M. Halyr, and E. R. Norris, 1973. Energy Cost Consideration in Compacting Wheat. ASAE Paper No. NA73–302. ASAE, St. Joseph, MI 49085.

Hassebrauck, B. 1962. Untersuchungen über die Eignung von Zyklonen zum Trennen von Korn-Strohhäcksel-Gemischen (Investigating on Applicability of Cyclones for Separating Grain-Chaff Mixtures). *Landtechnische Forschung* 12(4):108–112.

Hawk, A. L., D. B. Brooker, and J. J. Cassidy. 1966. Aerodynamic Characteristics of Selected Farm Grains. *Trans. of the ASAE* 9(1):48–51.

Heckel, R. W. 1961. Density-pressure Relationships in Powder Compaction. *Trans. Metallurgical Soc. AIME* 221:671–675.

Heldman, D. R. 1975. *Food Process Engineering*. AVI Publishing Co., Inc., Westport, Conn.

Henderson, S. M. 1952. A Basic Concept of Equilibrium Moisture. *Agricultural Engineering* 33(1):29–33.

Henderson, S. M. 1954. The Causes and Characteristics of Rice Checking. *The Rice Journal* 57(7):16, 18.

Henderson, S. M. and R. L. Perry. 1966. *Agricultural Process Engineering*. Edwards Brothers, Inc., Ann Arbor, Michigan.

Hermans, J. J. 1953. *Flow Properties of Disperse Systems*. Interscience Publishers, Inc., New York.

Herring, H. K., R. G. Cassens, and E. J. Brisky. 1965. Further Studies on Bovine Muscle Tenderness as Influenced by Carcass Position. Sarcomer Length and Fiber Diameter. *J. Food Sci.* 30:1049.

Hertz, H. 1896. *Miscellaneous Paper*. MacMillan and Company, New York.

Herum, F. L., G. W. Isaacs, and R. M. Peart. 1966. Flow Properties of Highly Viscous Organic Pastes and Slurries. *Trans. of the ASAE* 9(1):45.

Heuser, G. F. and L. C. Norris. 1946. Oyster Shells, Calcite Grit, Ground Limestone and Granite Grit in Rations for Hens. *Poultry Science* 4:109–117.

Hill, O. J. and L. S. Palmer. 1938. Study of the Relation of the Feed Consumed by the Cow to the Composition of the Milk Fat and the Properties of Butter. *J. Dairy Sci.* 21:529.

Hill, R. 1950. *Mathematical Theories of Plasticity*. Clarendon Press, Oxford.

Hintz, O. E. and R. Schinke. 1952. Coefficient of Sliding Friction for Corn and Alfalfa on Steel. Advanced Engineering Department, International Harvester

Company. Report 3-A1866-52R1, Chicago, Illinois.
Hodges, L. H. 1949. The Design of a General-Purpose Farm Wagon Rack for Mechanical Unloading. *Agricultural Engineering* 30(3):124–126, 128.
Holm, R. 1946. *Electrical Contacts*. Almquist and Wiksells, Stockholm.
Holt, J. E. and D. Schoorl. 1977. Bruising and Energy Dissipation in Apples. *Journal of Texture Studies* 7(4):421–432.
Hopkins, R. B. 1953. The Reduction of Injuries to Potato Tubers Through the Use of Padding Materials. *Am. Potato Journal* 30(10):247–255.
Horsfield, B. C., R. B. Fridley, and L. L. Claypool. 1972. Application of Theory of Elasticity to the Design of Fruit Harvesting and Handling Equipment for Minimum Bruising. *Trans. of the ASAE* 15(4):746–750, 753.
Houston, R. K. 1957a. Orthogonal Projection as a Criterion of Size for Agricultural Products. M. S. Thesis in Agricultural Engineering, University of California, Davis, California.
Houston, R. K. 1957b. New Criterion of Size for Agricultural Products. *Agricultural Engineering* 38(12):856–858.
Houwink, R. 1958. *Elasticity, Plasticity and Structure of Matter*. Dover Publications, Inc., New York.
Huff, E. R. 1967. Measuring Time-Dependent Mechanical Properties of Potato Tubers, Equipment, Procedure, Results. *Trans. of the ASAE,* 10(3):414–419.
Huff, E. R. 1967. Mechanical Properties of Potatoes-Like Rubber or Like Glass. Maine Farm Research 14(2).
Hunter, J. H. and J. B. Wilson, 1966. Some Aspects of Hydraulic Handling of Potatoes. Maine Agricultural Experiment Station Misc. Rep. No. 120, Orono, Maine.
Hunter, S. C. 1957. Energy Absorbed by Elastic Waves During Impact. *J. Mech. Phys. Solids*. 5:162.
Isaacson, J. D. and J. S. Boyd. 1965. Mathematical Analysis of Lateral Pressures in Flat-Bottomed Deep Grains Bins. *Trans. of the ASAE* 8(3):358–360.
Isenberg, F. M. 1955. The Effect of Height of Fall on Onion Bruising. *Proc. Am. Soc. Hort. Sci*. 65:331–333.
Isherwood, F. A. 1960. Some Factors Involved in the Texture of Plant Tissues. *Texture in Foods,* S. C. I. Monograph No. 7, Soc. Chem. Ind., London.
Jacobs, M. B. 1951. *The Chemistry and Technology of Food and Food Products*. Interscience Publishers, Inc., New York.
Jacobs, M. B. 1958. *The Chemical Analysis of Foods and Food Products*. Van Nostrand, Princeton, N.J.
Jaeger, J. C. 1962. *Elasticity, Fracture and Flow*. John Wiley and Sons, Inc., New York.
James, P. E. and H. T. Retzer. 1967. Measuring Egg Shell Strength by Beta Backscatter Technique. *Poultry Science* 46(5):1200–1203.
James, P. E., D. E. Wilkins and J. R. Menear. 1962. Peaceful Use of Atomic Energy Provides Engineers with Improved Way to Measure Silage Density.

Agricultural Research, August.

Jamieson, J. A. 1905. Grain Pressures in Deep Bins. Canadian Society of Civil Engineers. Transactions, 1093 Vol. 17, Part 2, 554.

Janssen, H. A. 1895. Versuche über Getreidedruck in Silozellen. *Zeitschrift des VDI* 39:1045–1049.

Jenike, A. W., P. J. Elsey, and R. H. Wooley. 1960. Flow Properties of Bulk Solids. *Proceedings of the ASTM* 60:1168–1181.

Jenike, A. W. 1961. Gravity Flow of Bulk Solids. Bulletin No. 108, Utah Engineering Experiment Station, University of Utah, Salt Lake City, Utah.

Jenike, A. W. 1964. Storage and Flow of Solids. Bulletin No. 123, Utah Engineering Experiment Station, University of Utah, Salt Lake City, Utah.

Jenkins, H. V. 1956. Air Flow Planimeter for Measuring the Area of Detached Leaves. *Plant Physiology* 34(5):532.

Jindal, V. K., N. N. Mohsenin, and J. V. Husted. 1974. Surface Area of Agricultural Grains. *Trans. of the ASAE* 17(4):721–725, 728.

Jindal, V. K. 1975. Mechanics of hammermill grinding of corn as related to its mechanical properties. Ph.D. dissertation, Penn State University, University Park, Pa.

Jindal, V. K. and N. N. Mohsenin. 1976. Analysis of a Simple Pendulum Impacting Device for Determining Dynamic Strength of Selected Food Material. *Trans. of the ASAE* 19(4):766–770.

Johannessen, G. A. 1949. Skin Puncture Studies on Red-Ripe Tomatoes. *Proc. Am. Soc. Hort. Sci.* 54:272.

Johanson, J. R. and H. Colijn. 1964. New Design Criteria for Hoppers and Bins. *Iron and Steel Engineer* (10):85:103.

Johnson, R. M. 1962. Determining Damage in Yellow Dent Corn. *Cereal Science Today*. 7(1).

Johnston, E. F. and H. V. Toko. 1963. Resistance of Potatoes to Bruising as Influenced by Air and Tuber Temperatures. A Progress Report. *Maine Farm Research*, January 20–24.

Jones, N. R. 1968. The Texture of Agar Jellies—Measurement and Modification. *Rheology and Texture of Foodstuffs,* Monograph No. 27, Soc. Chem. Ind., 14 Belgrave Square, London.

Jowitt, R. 1977. An Engineering Approach to Some Aspects of Food Texture. In *Food Texture and Rheology*. P. Serman (ed.), Academic Press, New York.

Kaplow, M. 1970. Commercial Development of Intermediate Moisture Foods. *Food Tech.* 24:889.

Kapsalis, J. G. and R. A. Segars. 1972. An instrument for measuring rheological properties by bending. *J. Texture Studies* 3:31–50.

Karel, M., Y. Aikawa, and B. E. Proctor. 1955. *Mod. Packaging* 29(2):153.

Karel, M. and J. T. R. Nickerson. 1964. Effect of relative humidity, air, and vacuum on browning of dehydrated orange juice. *Food Tech.* 18:1215–1218.

Karel, M., O. R. Fennema, and D. B. Lund. 1975. *Physical Principles of Food*

Preservation. Marcel Dekker, Inc., New York.

Karel, M. 1974. In *Advances in Preconcentration and Dehydration of Foods*. Spicer, A. (ed.) p. 45. Applied Science Publishers.

Karelitz, M. E. 1938. "On Mechanism of Dry Friction," in *Contributions to the Mechanics of Solids*. MacMillan Company, New York.

Kattan, A. A. 1957. Changes in Color and Firmness During Ripening of Detached Tomatoes, and the Use of a New Instrument for Measuring Firmness. *Proc. Am. Soc. Hort. Sci.* 70:379.

Katz, R., A. B. Cardwell, N. D. Collins and A. E. Hostetter. 1959. A New Grain Hardness Tester. *Cereal Chemistry* 36:393.

Keck, H. and J. R. Goss. 1965. Determining Aerodynamic Drag and Terminal Velocities of Agronomic Seeds in Free Fall. *Trans. of the ASAE* 8(4):553–554, 557.

Keiper, D. A. 1962. Dynamic Mechanical Properties Tester for Low Audio and Subaudio Frequencies. *The Review of Scientific Instruments* 33(1):1181–1184.

Kelvin, (Sir W. Thomson). 1891. *Popular Lectures and Addresses. Vol. I: Constitution of Matter*, 2nd ed. MacMillan & Co., London, p. 80.

Kertesz, Z. I., M. Eucare and G. Fox. 1959. A Study of Apple Cellulose. *Food Research* 24:14.

Ketchum, M. S. 1919. *The Design of Walls, Bins and Grain Elevators*. The Engineering News Publishing Company, New York.

Kick, F. 1885. The law of proportional resistance and its application (in German), Felix, Leipzig.

Kiker, C. F. and I. J. Ross. 1966. An Equation of Motion for Multiple Granular Particles in Free Fall in Enclosed Vertical Ducts. *Trans. of the ASAE* 9(4):468–473, 479.

King, D. L. and A. W. Riddolls. 1960. Damage to Wheat Seed and Pea Seed in Threshing. *J. of Agric. Engng. Res.* 5(4):387–398.

King, D. L. and A. W. Riddolls. 1962. Damage to Wheat Seed and Pea Seed in Threshing. *J. Agric. Engng. Res.* 7(2):90.

King, N. 1964. The Physical Structure of Butter. *Dairy Science Abstr.* 26:151–162.

Kirk, I. W. and H. E. McLeod. 1967. Cottonseed Rupture from Static Energy and Impact Velocity. *Trans. of the ASAE* 10(2):216–219.

Klapp, E. 1945. *Kartoffelbau*. Stuttgart.

Knipper, N. V. 1956. Use of High Frequency Currents for Grain Drying. Nauch. Trud. Elektrif. Selkhoz. 2. (Translated by E. Harris, NIAE, Silsoe, Befordshire, England).

Kolchin, N. N. 1957. Combined Pneumatic and Mechanical Separation of Potato Tubers from Clods. Translated from Russian by E. Harris. *J. of Agric. Engng. Res.* 2(3):238–240.

Kolganov, K. G. 1956. Mechanical Damage to Grain During Threshing. Translation No. 52. *J. Agric. Engng. Res.* 3(2).

Kosma, A. and H. Cunningham. 1962. Tables for Calculating the Compressive

Surface Stresses and Deflections in the Contact of Two Solid Elastic Bodies Whose Principle Planes of Curvature Do Not Coinside. *J. of Industrial Mathematics* 12(1):31–40.

Kosutany, T. 1907. *Der ungarische Weizen und das ungarische Mehl.* Verlag Molnarck Lapja, Budapest.

Kramer, A. and B. A. Twigg. 1966. *Fundamentals of Quality Control for the Food Industry.* AVI Publishing Co., Westport, Conn.

Kramer, H. A. 1944. Factors Influencing the Design of Bulk Storage Bins for Rough Rice. *Agricultural Engineering* 25:463.

Kranzler, G. A. and R. L. Witz. 1967. Some Mechanical Properties of Frozen High-Moisture Barley. ASAE Paper No. 67–811, Am. Soc. of Agr. Engrs., Saint Joseph, Michigan.

Krogenberg, H. G. 1964. Possibilities for Fruit Mechanical Harvesting. *J. Agric. Engng. Res.* 9(2):194–196.

Kruger, H. W. and R. Flenniken. 1961. Physical Properties of Cells Determined by Microscopic Methods. *Bausch & Lomb Focus* 32:18–22.

Kulwitch, R., R. W. Decker, and R. H. Alsmeyer. 1963. Use of a Slice-Tenderness Evaluation Device with Pork. *Food Technology* 17:83.

Kumar, L. and J. K. Wang. 1969. Response of papaya fruit to dynamic loading. ASAE Paper No. 69–871. Am. Soc. of Agric. Engr., St. Joseph, Mich.

Kumar, M. 1969. Flow Properties of Animal Waste Slurries. M. S. Thesis in Agricultural Engineering. The Pennsylvania State Unviersity, University Park, Pa.

Kumar, M. 1973. Compaction Behavior of Ground Corn. *J. Food Science* 38: 877–878.

Kunkel, R. G., P. F. Edgar and A. M. Binkley. 1952. The Mechanical Separation of Potatoes Into Specific Gravity Groups. *Col. Agr. Exp. Sta. Bul. 422A.*

Kunze, O. R. and C. W. Hall. 1965. Relative Humidity Changes Which Cause Brown Rice to Crack. *Trans. of the ASAE* 8(3):396–399.

Kunze, O. R., F. W. Snyder, and C. W. Hall. 1965. Impact Effects on Germination and Seedling Vigor of Sugarbeets. *J. of the ASSBT* 13(4):341–353. January.

Kunze, O. R. and C. W. Hall. 1967. Moisture Adsoprtion Characteristics of Brown Rice. *Trans. of the ASAE* 10(4):448–450, 453.

Kunze, O. R., W. H. Aldred, and B. D. Reeder. 1975. Bruising Characteristics of Peaches Related to Mechanical Harvesting. *Trans. of the ASAE* 18(5): 939–941, 945.

LaBelle, R. L. 1964. Bulk Density—A Versatile Measure of Food Texture and Bulk. *Food Technology* 18:89.

LaBelle, R. L., E. E. Woodams and M. C. Bourne. 1964. Recovery of Montmorency Cherries from Repeated Bruising. *Proc. Am. Soc. Hort. Sci.* 84:103.

LaBelle, R. L. and J. L. Moyer. 1960. Factors Affecting the Drained Weight and Firmness of Red Tart Cherries. *Food Technology* 14:347.

Labuza, T. P. 1970. Third International Congress on Food Sci. and Tech., SOS

70, p. 618. IFT, Washington.

Labuza, T. P. and H. E. Chou. 1974. Decrease of Linoleate Oxidation Rate Due to Water at Intermediate Water Activity. *J. Food Sci.* 39:112.

Lalor, W. F. and W. F. Buchele. 1963. The Theory of Threshing Cone Design. *J. Agric. Engng. Res.* 8(1):35–40.

Lamb, H. *Hydrodynamics.* 1945. Dover Publications, New York.

Lamp, B. J. and W. F. Buchele. 1960. Centrifugal Threshing of Small Grains. *Trans. of the ASAE* 3(2):25–28.

Lamp, K. 1959. Möglichkeiten zur Messung der Beschädigungsempfindlichkeit von Kartoffelknollen und anderen Früchten. *Landtechnische Forschung* 9(2):50–54.

Lapple, C. E. 1956. *Fluid and Particle Mechanics.* University of Delaware, Newark.

Leaderman, H. 1943. *Elastic and Creep Properties of Filamentous Materials and Other High Polymers.* Textile Foundation, Washington, D.C.

Lebedeva, L. P. 1965. Measurement of the Dynamic Complex Shear Modulus of Animal Tissues. *Soviet Physics Acoustics* 11:163.

Lee, F. A. 1948. Determination of Maturity of Frozen Lima Beans. *N. Y. Agr. Exp. Sta. Bul.* 792.

Lee, H. E. and J. R. M. Radok. 1960a. The Contact Problem for Viscoelastic Bodies. *Trans. of the ASME, J. of Applied Mechanics* 27(9):438–444.

Lee, H. E. and J. R. M. Radok. 1960b. The Contact Problem for Viscoelastic Bodies. Technical Report No. 47, Brown University, Providence, R. I.

Leniger, H. A. and W. A. Beverloo. 1975. *Food Process Engineering.* Reidel Publishing Co., Boston.

Leonhardt, J. L., G. C. Zoerb, and D. D. Hamann. 1961. Investigation of Factors Affecting Impact to Surgham Seeds. ASAE Paper No. 61–125. Am. Soc. of Agr. Engrs., Saint Joseph, Michigan.

Leslie, J. R. 1951. Pulse Technique Applied to Dynamic Testing in "Symposium on Ultrasonic Testing". ASTM Technical Bulletin No. 101:104–116.

Levin, J. H., C. W. Hall and A. P. Deshmukh. 1959. Physical Treatment and Cracking of Sweet Cherries. Quart. Bul. of the Mich. Ag. Exp. Sta. 42:133.

Levine, L. 1982. Estimating Output and Power of Food Extruders. *J. Food Process Engineering* 6(1):1–13.

Lindenmuth, B. E. 1966. Developing a Harvest Prediction Technique from the Physical and Mechanical Properties of Apple Fruit. M. S. Thesis Department of Agricultural Engineering, The Pennsylvania State University, University Park, Pa.

Linke, L. 1969. Flow characteristics of coating processes (in German). *Lebensm. Ind.* 16:452–457.

Lipka, J. L. 1918. *Graphical and Mechanical Computation.* John Wiley and Sons. Inc., New York.

Loewer, O. J., I. J. Ross, D. D. Kratzer, and J. N. Walker. 1977. Properties of

Ground Shelled Corn as Related to Forces in Bulk Storage Structures. *Transactions of the ASAE* 20:155–156.

Loncin, M., J. Bimbenet and J. Lenges. 1968. Influence of Water Activity on the Spoilage of Foods. *J. Food Tech.* 3:131.

Loncin, M. 1975. Basic principles of moisture equilibria. In *Freeze Drying and Advanced Food Technology* (Goldblithns-A, L. R., and Rothmayer, W. W. eds.), pp. 599–617. Academic Press, New York.

Loncin, M., and R. L. Merson. 1979. *Food Engineering and Selected Applications*. Academic Press, New York.

Loow, H. 1964. Mechanical Damage to Potatoes. Bulletin No. 304. Swedish Institute of Agricultural Engineering, Uppsala, Sweden.

Lorenzen, C. and L. H. Lamouria. 1964. Hydraulic Handling of Fruit in Processing Operations. *Agricultural Engineering* 45(5):258–259, 262–263.

Lorenzen, R. T. 1957. Effect of Moisture Content on Mechanical Properties of Small Grains. M. S. Thesis, University of Calif., Davis.

Love, R. M. 1960. Texture Change in Fish and Its Measurement. In *Texture in Foods*. S. C. I. Monograph No. 7, Soc. Chem. Ind., London.

Lynch, L. J. and R. S. Mitchell. 1950. The Physical Measurement of Quality in Canned Peas. Commonwealth Sci. and Ind. Org. (Australia) Bul. 154.

Maack. L. O. 1957. Die mechanische Trennung von Kartoffeln und Steiner "The Mechanical Separation of Potatoes and Stones) Translated by W. E. Klinner. *Landtechnische Forschung* 7(3):71. Translation No. 35, National Institute of Agricultural Engineering, Silsoe, Bedfordshire, England.

Madison, R. D. (editor.) 1949. *Fan Engineering*. Buffalo Forage Co., Buffalo, N. Y.

Magness, F. R. and G. F. Taylor. 1925. An Improved Type of Pressure Tester for Determination of Fruit Maturity. USDA Circular 350.

Makinson, K. R. 1948. *Transactions Faraday Soc*. 44:279.

Manor, A. N. 1978. Critical Analysis of the Mechanics of Fruit Damage Under Impact Conditions. Ph.D. Thesis in Agricultural Engineering, Penn State University, Univ. Park, Pa.

Marin, J. 1963. *Mechanical Behavior of Engineering Materials*. Prentice-Hall, Inc., New Jersey.

Marks, J. D., R. Bernlohr, and J. E. Varner. 1957. Esterification of Phosphate in Ripening Fruit. *Plant Physiology* 32:259–262.

Marks, L. S. 1941. *Mechanical Engineers' Handbook*. McGraw-Hill Book Co., New York.

Marvin, J. W. 1939. The Shape of Compressed Lead Shots and the Relation to Cell Shape. *Am. J. Bot*. 26:280.

Matthews, F. V. and C. W. Hall. 1968. Method of Finite Differences Used to Relate Changes in Thermal and Physical Properties of Potatoes. *Trans. of the ASAE* 11(4):558–562.

Matthews, R. W., B. A. Stout, D. D. Dewey and F. W. Bakker-Arkema. 1965.

Hydrohandling of Apple Fruits. ASAE Paper No. 65–130. Am. Soc. of Agr. Engrs., St. Joseph, Michigan.

Mattus, G. E., L. E. Scott, and L. L. Claypool. 1960. Brown Spot Bruises of Barlett Pears. *Proc. Am. Soc. Hort. Sci.* 75:100–105.

Matz, S. A. 1962. *Food Texture.* The AVI Publishing Co., Westport, Conn.

McCalmont, J. R. and W. Ashby. 1934. Pressures and Loads of Ear Corn in Cribs. *Agricultural Engineering* 15(4):123–125.

McCalmont, J. R., Jrueger, W. C., and C. Eby. 1946. Pressures and Other Factors Affecting Silo Construction. Bulletin 731, New Jersey Agr. Exp. Sta., Rutgers University, New Brunswick, N. J.

McClelland, J. H. and R. E. Spielrein. 1958. An Investigation of Ultimate Bending Strength of Some Common Pasture Plants. *J. of Agric. Engng. Res.* 3:288–292.

McColloch, L. P. 1962. Bruising Injury of Tomatoes. USDA, AMS, Marketing Research. Bulletin No. 513.

McKee, G. W. 1964. A Coefficient for Computing Leaf Area in Hybrid Corn. *Agron. J.* 56:240–241.

McMaster, R. C. ed. 1959. *Nondestructive Testing Handbook.* Vol. I. The Ronald Press Co., New York.

McMasters, M. M. 1965. Kernel morphology and microscopy. *Cereal Chem. Today* 10:268.

McNulty, P. B., R. J. Petrell, D. Y. C. Fung, and N. N. Mohsenin. 1977. Effect of High Pressure of Bulk Corn Kernel on its Microbial Activity. *J. Food Protection* 40:686–688.

McNulty, P. B. and N. N. Mohsenin. 1977. Compaction Behavior of Bulk Corn Kernel at Pressures up to 34.6 MN/m^2. *J. Food Process Engineering* 1:75–95.

McNulty, P. B. and N. N. Mohsenin, 1979. Compaction of Bulk Corn Kernel to Failure. *Trans. of the ASAE* 22(2):264–269.

Meddings, P. J. and O. E. Potter. 1974. The Adsorption of Water by Ground Barley Malt. *Cereal Chemistry* 51(1):1–16.

Menzies, D. R. 1976. An Instrumented Test Chamber to Compact Fodder at High-pressure Conditions. *Can. Agric. Engng.* 18(1):18–20.

Meredith, R. ed. 1956. *The Mechanical Properties of Textile Fibres.* Interscience Publishers, Inc., New York.

Merrill, E. W. 1956. A Coaxial Cylinder Viscometer for Non-Newtonian Fluids. *J. Inst. Soc. Am.* 3(4):124.

Metzner, A. B. 1956. Non-Newtonian Technology in *Advances in Chemical Engineering.* Vol. I, T. B. Drew and J. W. Hooper, Jr., Editors, Academic Press, Inc., New York.

Mewes, E. 1958. Zum Verhalten von Preßgütern in Preßtöpfen. (Behavior of Agricultural Materials in Compression Containers) *Landtechnische Forschung* 8:158–164,

Mewes, E. 1959. Berechnung der druckverteilung an stroh und heupressen. *Landtech. Fourshung* 9:160–169.

Meyer, L. H. 1960. *Food Chemistry*. Reinhold Publishing Corp., New York.

Miles, R. A., R. B. Fridley, and C. Lorenzen. 1969. Strength characteristics of tomatoes subjected to quasi-static loading. *Trans. Am. Soc. Agric. Engr. 12,* 627–630.

Miles, O. A., and G. E. Rehkugler. 1973. A failure criterion for apple flesh. *Trans. Am. Soc. Agric. Engr. 16,* 1148–1153.

Mill, C. C. ed. 1959. *Rheology of Disperse Systems*. Pergamon Press. New York.

Miller, E. E., C. A. Sadbolt and L. Holm. 1956. Use of an Optical Planimeter for Measuring Leaf Area. Plant Physiology 31(6):484–496.

Milloway, W. T., C. C. Surland and I. Skulte. 1961. The Effects of Initial Voids on the Bulk Modulus and Void Formations on Uniaxial Extension. JANAF Bulletin, 20th Meeting.

Mills, J. M. 1956. A Rapid Method of Construction of Linear Density Gradient Columns. *J. of Polymer Science* 19:585.

Mitchell, B. W. and R. M. Peart, 1968. Measuring Apparent Viscosity of Organic Slurries. *Trans. of the ASAE* 11(4):523–525.

Mitchell, F. S. and T. E. Rounthwaite. 1964. Resistance of Two Varieties of Wheat to Mechanical Damage by Impact. *J. Agric. Engng. Res.* 9(4):303.

Mittal, J. P., N. N. Mohsenin, and M. G. Sharma. 1977. Development of constitutive relationship for soft fleshy tissues of apples. Presented at the 48th Annual Meeting for the Society of Rheology, Madison, Wisconsin, October 23–27, 1977.

Mohsenin, N. N. and H. Goehlich. 1962. Techniques for Determination of Mechanical Properties of Fruits and Vegetables as Related to Design and Development of Harvesting and Processing Machinery. *J. of Agric. Engng. Res.* 7:300.

Mohsenin, N., H. Goehlich and L. D. Tukey. 1962. Mechanical Behavior of Apple Fruit as Related to Bruising. *Proc. Am. Soc. Hort. Sci.* 81:61.

Mohsenin, N. 1963a. A Testing Machine for Evaluation of Mechanical and Rheological Properties of Agricultural Products. Pa. Agr. Exp. Sta. Bul. 701.

Mohsenin, N., H. E. Cooper and L. D. Tukey. 1963b. Engineering Approach to Evaluation of Textural Factors in Fruits and Vegetables. *Trans of ASAE* 6(2):85–88, 92.

Mohsenin, N., H. E. Cooper, J. R. Hammerle, S. W. Fletcher and L. D. Tukey. 1965a. "Readiness for Harvest" of Apples as Affected by Phys by Physical and Mechanical Properties of the Fruit. Pa. Agr. Exp. Sta. Bul. 721.

Mohsenin, N., C. T. Morrow and D. L. Tukey. 1965b. The "Yield-Point" Non-Destructive Technique for Predicting Firmness of Golden Delicious Apples. *Proc. Am. Soc. Hort. Sci.* 86:70.

Mohsenin, N. N. 1965c. Friction Force and Pressure Causing "Skinning" of Potatoes. *American Potato Journal*. 42:83.

Mohsenin, N. N. ed. 1965. Terms, Definitions and Measurements Related to Mechanical Harvesting of Selected Fruits and Vegetables. Penna. Agr. Exp. Sta. Progress Report 257.

Mohsenin, N. N., C. T. Morrow, and Y. M. Yang. 1968. On the Spherical Indenter as a Means for Determining the "Firmness" and "Hardness" of Food Materials. Proceedings 5th International Congress of Rheology (in press).

Mohsenin, N. N. 1970. Engineering Techniques for Evaluation of Texture of Solid Food Materials—A Review. *J. Texture Studies* (in press).

Mohsenin, N. N. and J. Zaske. 1976. Stress Relaxation and Energy Requirements in Compaction of Unconsolidated Materials. *J. Agr. Eng. Res.* 21:193–205.

Mohsenin, N. N., V. K. Jindal, and A. Manor. 1978. Mechanics of Impact of Fruit on a Cushioned Surface. *Trans. of the ASAE* Vol. 21:594–600.

Moody, L. F. 1947. An Approximate Formula for Pipe Friction Factor. *Mech. Eng.* 69:1005–1006.

Morris, O. M. 1925. Studies in Apple Storage. Washington Agr. Exp. Sta. Bul. 193.

Morrow, C. T. 1965. Viscoelasticity in a Selected Agricultural Product. M. S. Thesis in Ag. Eng., The Pennsylvania State University, University Park, Pa.

Morrow, C. T. and N. N. Mohsenin. 1966. Consideration of Selected Agricultural Products as Viscoelastic Materials. *J. of Food Science* 31(5):686–698.

Morrow, C. T. and N. N. Mohsenin. 1968. Dynamic Viscoelastic Characterization of Solid Food Materials. *J. Food Science* 33(6):646–651.

Morrow, C. T. 1969. Nonlinear Transient and Dynamic Behavior in Bovine Muscle. Ph. D. Thesis in Engineering Mechanics, the Pennsylvania State University, University Park, Pa.

Morrow, C. T., D. D. Hamann, N. N. Mohsenin, and E. E. Finney. 1971. Mechanical characterization of red delicious apples. Am. Soc. Agric. Engr. Paper No. 71–372.

Mueller, R. A., D. B. Brooker and J. J. Cassidy. 1967. Aerodynamic Properties of Black Walnuts: Application in Separating Good from Bad Walnuts. *Trans. of the ASAE* 11(1):57–61.

Mueller, W. J. 1957. The Relationship Between Selected Egg Characteristics and the Resistance of the Shell Against Puncture by Pressure and Impact. Paper presented at the 46th Annual Meeting of the Poultry Science Association, University, of Missouri, August 6–9.

Muller, H. G. 1968. Some Aspects of Dough Rheology. In *Rheology and Texture of Foodstuffs*. S. C. I. Monograph No. 27, Soc. Chem. Ind., 14 Belgrave Square, London.

Murnaghan, F. D. 1959. *Finite Deformations of an Elastic Solid*. John Wiley and Sons, Inc., New York.

Mustafa, M. A., B. A. Stout and W. A. Bradley. 1967. Theoretical Modeling of the Wheat Plant. *Trans. of the ASAE* 10(6):799–804.

Nakayama, T. and Sato, Yashushi. 1971. Relationship between binding quality of meat and myofibrillar proteins. *J. Texture Studies* 2, 475–488.

Narayan, C. V. and W. K. Bilanski. 1966. Behavior Under High Pressure of Wheat Grains in Bulk. ASAE Paper No. 66–806. Am. Soc. Agr. Engrs., St. Joseph, Michigan.

Naumov, I. A. 1957. Mechanical Properties of the Wheat Grain in Shear (In Russian) *Trudy* 9:10–18.

Nelson, C. W. 1967. Maximum Allowable Static and Dynamic Loads for Mechanical Injury in McIntosh Apples. M. S. Thesis in Agricultural Engineering, The Pennsylvania State University, University Park, Pa.

Nelson, C. W. and N. N. Mohsenin. 1968. Maximum Allowable Static and Dynamic Loads and Effect of Temperature for Mechanical Injury in Apples. *J. Agric. Engng. Res.* 13(4):317–329.

Neubauer, L. W. 1966. A Simplified Equation for Silage Pressures with Moisture Variation *Trans. of the ASAE* 9(2):295–296.

New Departure Engineering Data-Analysis of Stresses and Deflections. General Motors Corporation, Bristol, Connecticut (1946).

Ngoddy, P. O. and F. W. Bakker-Arkema. 1970. A generalized theory of sorption phenomena in biological materials, Part I: The isotherm equation. *Trans. ASAE* 13(5):612–617.

Ngoddy, P. and F. W. Bakker-Arkema. 1972. A generalized theory of sorption phenomena in biological materials. Part II: Adsorption compression on sorbing biological materials—a density function for sorbed water. Trans. ASAE 15(6): 153–1159.

Ngoddy, P. O. and F. W. Bakker-Arkema. 1972. A generalized theory of sorption phenomena in biological materials. Part III—A pore size distribution function of the power low type for biological materials. *Trans. ASAE* 15(6):1160–1164.

Nielsen, L. E. 1962. *Mechanical Properties of Polymers*. Reinhold Publishing Corp., New York.

Nilsson, S. B., C. H. Hertz and S. Flak. 1958. On the Relation Between Turgor Pressure and Tissue Rigidity, II. *Physiologica Plantarum* 11:818.

Nolle, A. W. 1948. Methods for Measuring Dynamic Mechanical Properties of Rubber-Like Materials. *J. Applied Physics* 19:753.

Nolt, I. G. and J. A. Meier. 1965. On the Spherical Indenter as a Means for Determining Viscoelastic Material Functions. Part II: *Proceedings of the Fourth International Congress on Rheology,* E. H. Lee, Editor, Interscience Publishers, New York.

Norris, K. H. Measuring Light Transmittance Properties of Agricultural Commodities. *Agricultural Engineering* 10:640–643, 651.

Nybom, N. 1962. A New Principle for Measuring Firmness of Fruits. *Horticultural Research* 2:1.

Nylund, R. E., P. Hempkill and J. M. Lutz. 1955. Mechanical Damage to Potatoes During Harvesting and Handling in the Red River Operation in the Red River Valley of Minnesota and North Dakota. *Am. Potato Journal* 32(7):237–247.

O'Brien, M., L. L. Claypool, and S. J. Leonard. 1960. Effects of Mechanical Vibrations on Fruit Damage During Transportation. ASAE Paper No. 60–311, Am. Soc. Agr. Engrs., St. Joseph, Michigan.

O'Brien, M., L. L. Claypool, J. S. Leonard, G. K. York, and J. H. MacGillivray.

1963. Causes of Fruit Bruising on Transport Trucks. *Hilgardia* 35(6):113–124.
O'Brien, M., J. P. Gentry, and R. C. Gibson. 1965. Vibrating Characteristics of Fruits as Related to In-Transit Injury. *Trans. of the ASAE* 8(2):241–243.
Okimoto, M. C. 1948. Anatomy and Histology of the Pineapple Inflorescence and Fruit. *Botanical Gazette* 110:217–231.
Oldfield, R. C. 1960. Perception in the Mouth, in *Texture in Foods*. S. C. I. Monograph No. 7 Soc. Chem. Ind., 14 Belgrave Square, London.
Oldroyd, J. G. 1953. The Elastic and Viscous Properties of Emulsions and Suspensions. *Proc. Roy. Soc.,* London. A-218:122.
Oldroyd, J. G. 1956. The Effect of Small Viscous Inclusions on the Mechanical Properties of an Elastic Solid in *Deformation and Flow of Solids*. Grammel, ed., Springer-Verlag, Berlin.
O'Neill, H. 1934. *The Hardness of Metals and Its Measurement*. London.
Orr, C., Jr. 1966. *Particulate Technology*. The McMillan Company, New York.
Osmak, I. T. 1954. Physicomechanical Properties of Corn. (In Russian). Sels, Khozmashina (4):10–15.
Otis, C. K. and J. H. Pomroy. 1957. Density, A Tool in Silo Research. *Agricultural Engineering* 38(11):806–807.
Pall, R. and N. N. Mohsenin. 1978. A Modified Soil Pychnometer for Determination of Porosity and Particle Size Density of Soils. *Trans. of the ASAE* 23(3): 742–745.
Palmer, J. 1961. Electronic Sorting of Potatoes and Clods by Their Reflectance. *J. of Agric. Engng. Res.* 6(2):104.
Park, D. 1963. The Resistance of Potatoes to Mechanical Damage Caused by Impact Loading. *J. Agric. Engng. Res.* 8(2):173–177.
Parker, B. F. and D. E. Wiant. 1954. Use of Chromatic Illumination and Methods of Product Rotation for Sorting Red Tart Cherries. Mich. Agr. Exp. Sta. Quart. Bul. 36(4):435–447.
Parker, M. E., E. H. Harvey and E. S. Stateler. 1952. *Elements of Food Engineering,* Reinhold Publishing Co., New York.
Parker, R. E., J. H. Levin and R. J. Whittenberger. 1966. An Instrument for Measuring Firmness of Red Tart Cherries. Quart. Bull. of Mich. Agr. Exp. Sta. 48(3):471–482.
Patrell, R. J., N. N. Mohsenin, and S. Walner. 1980. Dynamic Mechanical Properties of the Apple Cortex in Relation to Ripening. *J. of Texture Studies* 10:217–229.
Peleg, M. and E. B. Bagley, (eds.). 1983. *Physical Properties of Foods*. AVI Publishing Co., Westport, Conn.
Peleg, M. 1977. Contact and Fracture Elements as Components of the Rheological Memory of Solid Foods. *J. Texture Studies* 8:39–48.
Perry, J. S. and C. W. Hall. 1966. Evaluating and Reducing Mechanical-Handling Damage to Pea Beans. *Trans. of the ASAE* 9(5):696–701.
Persson, S. 1957. Eigenschaften des Reinigungsgutes in Mähdreschern. *Land-*

technische Forschung 7:41–45.

Pettyjohn, E. S. and E. E. Christiansen. 1948. Effect of Particle Shape on Free Settling Rates of Isometric Particles. *Chem. Eng. Prog.* 44:157–172.

Pflug, I. J., F. M. Joffe and R. C. Nicholas. 1960. A Mechanical Recording Pressure Tester. Quart. Bull. Mich. Ag. Exp Sta. 43(1):117–121.

Pflug, I. J. and J. H Levin. 1961. Actual Grower Results with Water Flotation Bulk Box Unloaders for Fruits. *Eastern Fruit Grower* 24(8):6–8.

Pichler, H. J. 1956. Sorption Isotherms for Grain and Rape. Translation No. 27, *Landtechnische Forschung* 6(2):47–52. Translated by W. E. Klinner. NIAE, Silsoe, Bedfordshire, England.

Pickett, G. 1949. Equations for Computing Elastic Constants from Flextural and Torsional Resonant Frequencies of Vibration of Prisms and Cylinders. Proc. Am. Soc. Testing Materials 45:845–865.

Pichler, H. J. 1956. Sorption Isotherms for Grain and Rape. Translation No. 27, *Landtechnische Forschung* 6(2):47–52. Translated by W. E. Klinner. NIAE, Silsoe, Bedfordshire, England.

Pinkus, O. 1952. Pressure Drops in the Pneumatic Conveyance of Solids. *Journal of Applied Mechanics* 19:525–531.

Plateau, M. H. 1872. Bull de 1, Acad. Rov. Belg. 33 (Series 2):376.

Platt, W. and R. Powers. 1940. Compressibility of Bread Crumb. *Cereal Chemistry* 17:601.

Pollack, R. L., C. Riccwti, C. F. Woodward, and H. H. Hills. 1958. Studies on Cherry Scald. I Relation Between Bruising and Respiration in Water *Food Technology* 12(2):

Polya, G. and G. Szezo. 1951. *Isoperimetric Inequalities in Mathematical Physics.*

Potter, N. N. 1970. Intermediate Moisture Foods: Principles and Technology. *Food Prod. Dev.* 4:38.

Powers, J. B., J. T. Gunn and F. C. Jacob. 1953. Electronic Sorting of Fruits and Vegetables. *Agricultural Engineering* 34(3):149–154, 158.

Prantl, L. and O. G. Tietjens. 1934. *Applied Hydro- and Aeromechanics.* Translated by J. P. Den Hartog. McGraw-Hill Book Company, New York.

Prince, R. P. 1961. Measurement of Ultimate Strength of Forage Stalks. *Trans. of the ASAE* 4(2):208–209.

Prince, R. P., D. D. Wolf, and J. W. Bartok. 1965. The Physical Property Measurements of Forage Stalks. Connecticut Ag. Exp. Sta. Bul. 388, Storrs, Conn.

Prince, R. P. and J. W. Bartok, Jr. 1966. A Recording Air-Flow Planimeter. *Agricultural Engineering* 47(9):487.

Prince, R. P. and D. W. Bradway. 1969. Shear Stress and Modulus of Elasticity of Selected Forages. *Trans. of the ASAE* 12(4):426–428.

Proctor, B. E., S. Davison and A. L. Brody. 1956. A Recording Strain Gage Denture Tenderometer for Foods. III. Correlation with Subjective Tests and the Canco Tenderometer. *Food Technology* 10:344.

Pryce-Jones, J. 1953. The Rheology of Honey in *Foodstuffs—Their Plasticity,*

Fluidity, and Consistency. Scott Blair, G. W. ed. Interscience Publishers Inc., New York.

Quackenbush, H. E., B. A. Stout and S. K. Ries. 1962. Pneumatic Tree-Fruit Harvesting. *Agricultural Engineering* 42:388.

Quenoville, M. H. 1952. *Associated Measurements*. Butterworth-Sprinter, Ltd. London.

Quinn, F. C. 1966. How to Select Humidity Measuring and Control Equipment. *Heating Piping and Air Conditioning,* July issue.

Quinn, F. C. 1967. The Quality of Water—The Other Raw Material. *Paper Trade Journal,* September 4 issue.

Rahman, A. R., G. Schafer, G. R. Taylor, and D. E. Westcott. 1969. Studies on Reversible Compression of Dehydrated Vegetables. Technical report 70-36pFL, U.S. Army, Natick Lab. MA.

Rahman, A. R. G. Schafer, P. Prell, and D. E. Westcott. 1971. Non-reversible Compression of Intermediate Moisture Fruit Bars. Technical Report TR-71-60-FL, U.S. Army, Natick Lab, MA.

Rahman, A. R. 1976. Dehydration and Compressed Foods of the future. Paper presented at First International Congress on Engineering and Food held in Boston, August 9–13, 1976.

Ranz, W. E. 1952. Friction and Transfer Coefficients for Single Particles and Packed Beds. *Chemical Engineering Progress* 48:247–253.

Rausch, J. M. 1949. Gravity Flow of Solid Beds in Vertical Towers, Ph. D. Thesis, Princeton, University, Princeton, N. J.

Reeve, R. M. and L. R. Leinbach. 1953. Histological Investigation of Texture in Apples. I. Composition and Influence of Heat on Structure. *Food Research* 18:592.

Reeve, R. M. 1953. Histological Investigations of Texture in Apples. II. Structure and Intercellular Spaces. *Food Research* 18:604.

Reeve, R. M. 1954. Histological Survey of Conditions Influencing Texture in Potatoes. I. Effects of Heat Treatments on Structure. *Food Research* 19:333.

Rehkugler, G. E. 1962. Properties of Eggs Important to the Design and Use of Egg Handling Equipment. *Trans. of the ASAE* 7(2):174–178.

Rehkugler, G. E. 1963. Modulus of Elasticity and Ultimate Strength of the Hen's Egg. *J. Agric. Engng. Res.* 8:352–354.

Rehkugler, G. E. 1964. Egg Handling Equipment Design. *Trans. of the ASAE* 7:174–187.

Rehkugler, G. E. and W. F. Buchele. 1969. Biomechanics of Forage Wafering. *Trans. of the ASAE* 12(1):1–8.

Rehkugler, G. E. and E. D. Markwardt. 1971. An Evaluation of Limb Padding to Reduce Apple Damage in Mechanical Harvesting. *Trans. of the ASAE* 14(4):734–737, 741.

Reiner, M. 1960. *Deformation, Strain and Flow*. H. K. Lewis and Co., Ltd., London.

Reiner, M. 1956. *Rheology, Theory and Applications*. Vol. 3, p. 359. F. R. Eirich, Editor, Academic Press, Inc., New York.

Rice, C. E. 1960. The Effects of Particle Size, Shape and Density of Minimum Suspension and Vertical Transport Velocities. Ph. D. Thesis in Agricultural Engineering, Michigan State University, East Lansing, Mich.

Richter, D. W. 1954. Friction Coefficients of Some Agricultural Materials. *Agricultural Engineering* 35(6):411–413.

Rider, R. C., R. B. Fridley, and M. O'Brian. 1973. Elastic Behavior of a Pseudo Fruit for Determining Bruise Damage to Fruit During Mechanized Handling. *Trans. of the ASAE* 16(2):241–244.

Riedel, L. 1949. Measurement of thermal conductivity of sugar solutions, fruit juices, and milk. *Chem. Eng. Tech.* (in German) 21:340–342.

Riedel, L. 1969. Thermal conductivity measurements on water-rich foods (in German), *Kaltetechnik* 21:315–316.

Rinehart, S. 1975. Stress Transients in Solids. Hyper-Dynamics. Santa Fe, Mexico.

Ritter, H. L. and L. C. Drake. 1945. Pore size distributions in porous materials. *Ind. Eng. Chem. Anal.* 17:782.

Robbins, W. W., T. E. Weier and C. R. Stocking. 1957. *Botany—An Introduction to Plant Science*. John Wiley and Sons, Inc., New York.

Roberts, A. W. and A. H. Willis. 1962. Performance of Grain Augers. *Proc. Instn. Mech. Engrs. Lond.* 178, PT. 176(8):165–194.

Roberts, A. W. 1963–1964. An Investigation of Grain Vortex Motion with Relation to the Performance Within Vertical Grain Augers. *Proc. Instn. Mech. Engrs. Lond.* 178, PT. 1(12):293–310.

Roberts, A. W. 1966. Developments in Material Handling Research at Wollongong University College. Paper presented at the 2nd Annual Conference of the Illawarra Group, the Institution of Engineers, Australia, held at Wollongong University College.

Roberts, A. W. 1967. The Dynamics of Granular Material Flow Through Curved Chutes. Reprint of a Paper presented before a conference of The Institution of Engineers, held in Adelaide, Australia.

Roberts, E. and I. R. Hoener. 1941. Causes of Preference Exhibited by Animals for Certain Inbred Lines of Corn. *J. Am. Soc. Agron.* 33:448.

Robinson, W. L. 1939. Hybrid and Open-Pollinated Corn for Pigs. Bimonthly Bulletin, Ohio Agr. Exp. Sta. 24(201):156.

Rockland, L. B. and G. F. Stewart (eds.). 1981. *Water Activity: Influences on Food Quality*. Academic Press, New York.

Roller, W. L. and H. S. Teague. 1967. Developing a Paste Feeding System for Swine. *Agricultural Engineering* 8(2):82–83.

Romanoff, A. L. and A. J. Romanoff. 1949. *The Avian Egg*. John Wiley and Sons, New York.

Rose, D. H., M. H. Haller and P. L. Harding. 1934. Relation of Temperature of Fruit to Firmness in Strawberries. *Proc. Am. Soc. Hort. Sci.* 32:429.

Rosenbaum, J. and C. E. Sando. 1920. Correlation Between Size of the Fruit and the Resistance to Puncture and its Relation to Infection with Macrosforium Tomato Cook. *Am. J. Botany* 7:78.
Ross, E. 1949. A Quantitative Hardness Tester for Food Products. *Science* 109:204.
Ross, I. J. and G. M. Isaacs. 1961. Forces Acting on Stacks of Granular Materials. *Trans. of the ASAE* 4(1):92–95.
Ross, I. J. and G. M. Isaacs. 1961. Capacity of Enclosed Conveyors Handling Granular Materials (Part II). *Trans. of the ASAE* 4(1):97–100, 104.
Ross, I. J. and Kiker. 1967. Some Physical Properties of Dried Citrus Pulp. *Trans. of the ASAE* 10(4):483–485.
Ross, K. D. 1975. Estimation of Water Activity in Intermediate Moisture Foods. *Food Tech.* 29:26.
Rowan, J. D., K. H. Norris and C. K. Power. 1958. A Method for Measuring the Rheological Properties of Eggs. *Food Research* 23(6):670–676.
Rumsey, T. 1974. Analysis of Viscoelastic Contact Stress in Agricultural Products Using a Finite Element Method. Ph.D. Thesis, University of California, Davis, Calif.
Salwin, H. 1959. Defining minimum moisture contents for dehydrated foods. *Food Tech.* 13:594.
Saravacos, G. D. and J. C. Moyer. 1967. Heating Rates of Fruit Products in an Agitated Kettle. *Food Technology* 21:54A–58A.
Sarig, Y. 1976. Deformation Analysis of Foam Encapsulated Apples Under Impact Loading. Unpublished Ph.D. Thesis, Department of Agricultural Engineering, Michigan State University, East Lansing, Michigan.
Saunders, D. W. 1964. Large Deformations in Amorphous Polymers. In *Biomechanics and Related Bio-Engineering Topics*. Kenedi, R. M., ed., Pergamon Press, New York.
Scheidegger, A. E. 1957. *The Physics of Flow Through Porous Media*. The MacMillan Company, New York.
Schertz, C. E. and T. E. Hazen. 1963. Predicting Motion of Granular Materials on an Oscillating Conveyor. *Trans. of the ASAE* 6(1):6–10.
Schertz, C. E. and T. E. Hazen. 1965. Movement of Shelled Corn on Oscillating Conveyor. *Trans. of the ASAE* 8(4):582–583.
Schiller, L. 1932. *Handbuch der Experimental-Physik*. Vol. IV, Part 2, pp. 337–387. Akademische Verlagsgesellschaft, Leipzig.
Schmidt, A. X. and C. A. Marlies. 1948. *Principles of High Polymer Theory and Practice*. McGraw-Hill Book Company, Inc., New York.
Schmidt, E. D. and J. H. Levin. 1963. Terminal Velocities of Small Fruits. USDA, ARS 42–89.
Schmidt, J. 1962. Zum Problem der sortenbedingten Festigkeit des Fruchtfleisches von Äpfeln (On the Problems of Variety-dependent Firmness of Apple Pulp). *Die Gartenbauwissenschaft* 27(9) Band Heft 3:303–358.
Schomer, H. A. and K. L. Olsen. 1962. A Mechanical Thumb for Determining

Firmness of Apples. *Proc. Am. Soc. Hort. Sci.* 81:61.

S.C.I. Monograph No. 7. 1960. *Texture in Foods*. Society of Chemical Industry, 14 Belgrave Square, London.

Scott Blair, G. W. 1938. An Apparatus for Measuring the Elastic and Plastic Properties of Cheese Curd. *J. Dairy Res.* 9:347.

Scott Blair, G. W., ed. 1953. *Foodstuffs, Their Plasticity, Fluidity, and Consistency*. Interscience Publishers, Inc., New York.

Scott Blair, G. W. and F. M. V. Coppen. 1940. The Subjective Judgment of Elastic and Plastic Properties of Soft Bodies. *Brit. J. Psychol.* 31:61.

Scott, Blair, G. W. and M. Baron. 1949. Constant-Stress Measurements of the Hardening of Soft Materials. *Nature* 164(7):148.

Scott Blair, G. W. and M. Reiner. 1957. *Agricultural Rheology*. Routledge and Kegan Paul, London.

Seely, F. B. and J. O. Smith. 1961. *Advanced Mechanics of Materials*. 2nd ed., John Wiley & Sons, Inc., New York.

Segars, R. A., H. A. Nordstrom, and J. G. Lapsalis. 1974. Textural characteristics of beef muscles. *J. Texture Studies* 5, 283–297.

Segler, G. 1951. Pneumatic Grain Conveying. National Inst. Agr. Engr., Braunschweig, Germany.

Setterfield, G. and S. T. Bayley. 1961. Structure and Physiology of Cell Walls. *Ann. Rev. Plant Physiol.* 12:35–62.

Shafizadeh, F. and W. T. Nearn. 1966. Composition of Wood and the Origin of Its Anisotropic Properties. *Materials Research and Standards* 6:593.

Shama, F. and P. Sherman. 1966. The Texture of Ice Cream—2. Rheological Properties of Frozen Ice Cream. *J. Food Science* 31(5):699–706.

Sharma, M. G. 1965. Theories of Phenomenological Viscoelasticity Underlying Mechanical Testing. Chapter 4, J. V. Schmitz, ed. *Testing of Polymers*. Interscience Publishers, New York.

Sharma, M. G. 1964. Viscoelasticity and Mechanical Properties of Polymers. Mimeographed notes, Summer Institute on Applied Mechanics and Materials Science, Department of Engineering Mechanics, The Pennsylvania State University, University Park, Pa.

Sharma, M. C. and N. N. Mohsenin. 1970. Mechanics of deformation of a fruit subjected to hydrostatic pressure. *J. Agric. Eng. Res. 15* (1):65–74.

Sharrah, N., M. S. Kunze and R. M. Pangborn. 1965. Beef Tenderness; Comparison of Sensory Methods with the Warner Bratzler and L. E. E.-Kramer Shear Presses. *Food Technology* 19:136.

Shedd, C. K. 1953. Resistance of Grains and Seeds to Air Flow. *Agricultural Engineering* 34(9):614–619.

Shelef, L. and N. N. Mohsenin. 1966. Moisture Relations in Germ-Endosperm and Whole Corn Kernel. *Cereal Chemistry* 43(3):347–353.

Shelef, L. and N. N. Mohsenin. 1967. Evaluation of the Modulus of Elasticity of Wheat Grain. *Cereal Chemistry* 44(4):393–402.

Shelef, L. and N. N. Mohsenin. 1968. Effects of Moisture Content on Mechanical Properties of Corn Horny Endosperm. *Cereal Chemistry* 46(3):242–253.

Sherif, S. M. 1976. The quasi-static contact problem for nearly incompressible agricultural products. Ph.D. Thesis, Michigan State University, East Lansing.

Sherwood, R. S. 1951. The Mechanism of Dry Friction. Engineering Report No. 6 of the Iowa Eng. Exp. Sta., Ames, Iowa.

Shimizu, T., H. Fukawa, and A. Ichiba. 1958. Physical Properties of Noodles. *Cereal Chemistry* 35:35.

Shoefield, R. K. and G. W. Scott-Blair. 1933. The Relationship Between Viscosity, Elasticity and Plastic Strength of a Soft Material as Illustrated by Some Mechanical Properties of Flour Dough—III. *Proceedings of the Royal Society of London* 141:72–85.

Shpolyanskaya, A. L. 1952. Structural Mechanical Properties of Wheat Grain. *Colloid Journal* (USSR) 14(1):137–148. Translated by Consultant Bureau, New York.

Simpson, J. B. 1971. Forces, Pressures, and Apple Damage During Cushioned Impact. Unpublished Ph.D. Thesis, Cornell University, Ithaca, New York.

Simpson, J. B. and G. E. Rehkulger. 1972. Forces and Apple Damage During Cushioned Impact. ASAE Paper No. 72–307. American Society of Agricultural Engineers, St. Joseph, Michigan.

Singleton, V. L. 1965. Chemical and Physical Development of the Pineapple Fruit. I. Weight per Fruitlet and Other Physical Attributes. *J. Food Science* 39: 98–104.

Sink, J. D. 1965. Specific Biophysical Features of Post-Mortem Changes in Porcine Muscle. Paper presented at the 18th Reciprocal Meat Conference sponsored by the American Meat Science Association at Kansas State University, June 15–17.

Sloane, A. 1947. *Fundamentals of Engineering Mechanics*. Prentice Hall Inc., New York.

Sluka, S. J., Besch, E. L. and A. H. Smith. 1965. A Hydrostatic Tester for Egg Shell Strength. *Poultry Science* 24(6):1494–1500.

Sluka, S. J., Besch, E. L. and A. H. Smith. 1967. Stress in Impacted Egg Shells. *Trans. of the ASAE* 10(3):364–369.

Smith, I. E., S. D. Robert, R. E. Stokes, and R. J. Hansford. 1977. The Briquetting of Wheat Straw. *J. Agr. Eng. Res.* 22:105–111.

Smith, J. C. 1955. Design a Hopper That Won't Arch. *Chemical Engineering* (9):167–168.

Smith, R. E. and G. L. Nelson. 1969. Transient heat transfer in solids: theory vs. experiment. *Trans. ASAE* 12:823.

Smith, S. E. 1947. The Sorption of Water Vapor by High Polymers. *J. of Am. Chem. Soc.* 69:646–651.

Smith, T. L., J. D. Ferry and F. W. Schremp. 1949. Measurements of Mechanical Properties of Polymer Solutions by Electromagnetic Transducers. *J. of Appliec*

Physics. Vol. 20, 144–153.

Snyder, L. H., W. L. Roller, and G. E. Hall. 1967. Coefficients of Kinetic Friction of Wheat on Various Metal Surfaces. *Trans. of the ASAE* 10(3):411–413, 419.

Somers, G. F. 1965a. Citric-Acid Induced Loss of Rigidity in Potato Tuber Slices. *Plant Physiology* 40:388.

Somers, G. F. 1965b. Viscoelastic Properties of Storage Tissues from Potato, Apple and Pear. *J. of Food Science* 30(6):922–929.

Soule, H. M. 1968. Investigation of Some Aerodynamic Properties of Lowbush Blueberries. ASAE Paper No. 68–847. Am. Soc. Agr. Engrs., Saint Joseph, Michigan.

Spangler, M. G. 1951. *Soil Engineering*. International Textbook Company, Scranton, Pa.

Spinner, S. and W. E. Tefft. 1961. A Method for Determining Mechanical Resonance Frequencies and for Calculating Elastic Moduli for These Frequencies. *Proc. Soc. for Testing Materials*. 61–1221.

Splinter, W. E., C. W. Suggs, and J. R. Beeman. 1962. Physical Properties of Green Virginia Tobacco Leaves. IV. Kinetic Energy Adsorption and Force Relationships During Separation from the Stalk. *Tobacco Science* 6:62–70.

Stahel, G. 1935. Breaking of Rice in Milling in Relation to the Condition of the Paddy. *Tropical Agriculture* 12(10):255–260.

Stahl, B. M. 1950. Grain Bin Requirements. USDA Circular 835, Washington, D.C.

Stambaugh, R. B. 1952. Electrical Analog Method for Studying Elastromer Behavior. *Industrial and Engineering Chemistry* 44(7):1590–1594.

Stamm, A. J. 1962. Wood and Cellulose-Liquid Relationships. North Carolina Ag. Exp. Sta. Techn. Bull. No. 150.

Stamm, A. J. 1964. *Wood and Cellulose Science*. The Ronald Press Company, New York.

Stamm, A. J. and R. M. Seborg. 1935. Adsorption Compression. *J. of Physical Chemistry* 39:133–142.

Stanley, D. W., L. W. McKnight, W. G. S. Hines, W. R. Osborne, and J. M. DeMan. 1972. Predicting meat tenderness from muscle tensile properties. *J. Texture Studies* 3, 51–68.

Stawinski, J. and B. Szot. 1976. The use of the mercury porosimeter for determination of pore size distribution in wheat grain. *Hod. Rosi. Aki. INas* 20(3): 321–327.

Steiner, E. H. 1959. *The Rheology of Disperse Systems*. Pergamon Press, New York.

Stermer, R. A. 1965. A Fast Method for Determining Grass Seed Purity. *Agricultural Marketing*. January.

Stevens, S. S. 1957. On the Psychophysical Law. *Psychol. Rev.* 64:153.

Stevens, S. S. and M. Guirao. 1964. Scaling of Apparent Viscosity. *Science* 144:1157–1158.

Stewart, B. R. 1968. Effect of Moisture Content and Specific Weight on Internal-Friction Properties of Sorghum Grain. *Trans. of the ASAE* 11(2):260–266.
Stewart, B. R., Q. A. Husain, and O. R. Kunze. 1968. Friction Coefficients of Sorghum Grain on Steel, Teflon and Concrete Surfaces. *Trans. of the ASAE* 12(4):415–418.
Stiles, W. and W. Leach. 1961. *Respiration in Plants.* Methuen Monograph on Biological Subjects, 4th Ed. John Wiley and Sons, New York.
Stout, B. A., D. H. Dewey, E. G. Vis, and T. F. Herrick, Jr. 1966. A Prototype Hydrohandling System for Sorting and Sizing Apples Before Storage. ARS 52–14, USDA.
Strohman, R. D. and R. R. Yoerger. 1967. A new equilibrium moisture content equation. *Trans. ASAE* 10(5):675–677.
Suggs, C. W., J. F. Beeman and W. E. Splinter. 1960. Physical Properties of Green Virginia-Type Tobacco Leaves, Part III, Relation of Leaf Length and Width of Leaf Area. *Tobacco Science* 4:194–197.
Suggs, C. W., J. F. Beeman and W. E. Splinter. 1962. Physical Properties of Tobacco Stalks. Part I. Cantilever Beam Properties. *Tobacco Science* 6: 146–151.
Suggs, C. W., J. F. Beeman and W. E. Splinter. 1962. Physical Properties of Green Virginia-Type Tobacco Leaves. Part V. Critical Radius of Curvature. *Tobacco Science* 6:71–77.
Suggs, C. W. and W. E. Splinter. 1960. Physical Characteristics of Green Virginia-Type Tobacco Leaves. Part II: Frictional Characteristics. *Tobacco Science* 4(April 22):84–88.
Supnik, R. H. 1962. Rate Sensitivity: Its Measurement and Significance. *Materials Research and Standards* 2(6):498–500.
Surland, C. C. 1960. Thermodynamic Effects of Hydrostatically-Stressed Composites and Determination of Specific Heat: JANAF Bulletin, 19 Meeting.
Swanson, C. O. 1946. *Physical Properties of Dough.* Burgess Publishing Co., Minneapolis.
Symposium on Grain Damage. 1968. Department of Agricultural Engineering, Iowa State University, Ames, Iowa.
Szczesniak, A. S., M. A. Brant, and H. H. Friedman. 1963. Development of Standard Rating Scales for Mechanical Parameters of Texture and Correlation Between the Objective and the Sensory Methods of Texture Evaluation. *J. of Food Science* 28(4):397–403.
Szczesniak, A. S. 1963. Objective Measurements of Food Texture. *Food Technology* 28:410.
Szczesniak, A. S. 1966. Texture Measurements. *Food Technology.* 20:52.
Tabachuk, V. I. 1953. An Investigation into the Susceptibility of the Potato Tuber to Damage by Impact (USSR). Leningradskii selkhoz. Inst. Zap. 7:90–99. Translation No. 130. NIAE, Silsoe, Bedford, England.
Tabor, D. 1948. A Simple Theory of Static and Dynamic Hardness. *Proc. Roy.*

Soc. A192:247–274.

Tabor, D. 1951. *Hardness of Metals*. Oxford University Press, London.

Talburt, W. F. and O. Smith. 1967. *Potato Processing*. The AVI Publishing Company, Inc., Westport, Conn.

Tamiya H. 1938. Zur Theorie der Turgordehnung und über den funktionellen Zusammenhang der einzelnen osmotischen Zustandsgrößen. *Cytologica*. 8:542.

Taylor, A. A. 1961. Determination of moisture equilibria in dehydrated foods. *Food Technol*. 15:536.

Taylor, D. W. 1948. *Fundamentals of Soil Mechanics*. John Wiley & Sons, Inc., New York.

Tennes, B. R., R. G. Diener, J. H. Levin, and R. T. Whittenberger. 1967. Firmness and Pitter Loss Studies on Tart Cherries. ASAE Paper No. 67–333. Am. Soc. of Agr. Engrs., Saint Joseph, Mich.

Tennes, B. R., J. H. Levin, and B. A. Stout. 1968. Sweet Cherry Properties Useful for Designing Harvesting and Handling Equipment. ASAE Paper No. 68–348, Am. Soc. Agr. Engrs., Saint Joseph, Mich.

Terepka, A. R. 1963. Structural and Classication in Avian Egg Shell. *Experimental Cell Research* 30:171–182.

Thomas, H. R. and V. A. Hoersch. 1930. Stresses Due to the Pressure of One Elastic Solid Upon Another. Eng. Exp. Sta., Univ. of Ill., Bull. 212.

Thompson, R. A. and G. H. Foster. 1963. Stress Cracks and Breakage in Artificially Dried Corn. USDA, AMS, Marketing Research Report No. 631.

Thompson, R. A. and G. W. Isaacs. 1967. Porosity Determinations of Grains and Seeds with an Air-Comparison Pycnometer. *Trans. of the ASAE* 10(5):693–696.

Timbers, G. E. 1964. Some Mechanical and Rheological Properties of the Netted Gem Potato, M. S. Thesis, University of British Columbia.

Timbers, G. E., L. M. Stately and E. L. Watson. 1965. Determining Modulus of Elasticity in Agricultural Products by Loaded Plungers. *Agricultural Engineering* 46(5):274–275.

Timofeev, A. N. 1956. A Method for Determining the Dependence of Potato Damage on Mechanical Factors. Sbor. Trud. Zemled. Mekhan. vsesoyuz. (USSR) Translation No. 131, NIAE, Silsoe, Bedford, England.

Timoshenko, S. 1956. *Strength of Materials*. Part I. Elementary Theory and Problems. D. Van Nostrand Company, Inc., Princeton, N.J.

Timoshenko, S. 1956. *Strength of Materials*. Part II 3rd ed. D. Van Nostrand Co., Inc., New York.

Timoshenko, S. and L. N. Goddier. 1951. *Theory of Elasticity*. McGraw-Hill Book Company, Inc., New York.

Timoshenko, S. and S. Wornowski-Krieger. 1959. *Theory of Plates and Shells*. McGraw-Hill Book Co., Inc., New York, pp. 67–72.

Tiwari, S. N. 1962. Aerodynamic Behavior of Dry Edible Beans and Associated Materials in Pneumatic Separation. M. S. Thesis in Agricultural Engineering.

University of Maine, Orono, Maine.

Toledo, R. T. 1973. Proc. Meat Ind. Res. Conf. Am. Meat Institute Foundation, Chicago, Ill. p. 85.

Torobin, L. B. and W. H. Gauvin. 1959. Fundamental Aspects of Solids—Gas Flow. Part III: Accelerated Motion of a Particle in a Fluid. *Canadian Journal of Chemical Engineering* 37:224–236.

Torobin, L. B. and W. H. Gauvin. 1960a. Fundamental Aspects of Solids—Gas Flow. Part V: The Effect of Fluid Turbulence on the Particle Drag Coefficient. *Canadian Journal of Chemical Engineering* 38:180–200.

Torobin, L. B. and W. H. Gauvin. 1960b. Fundamental Aspects of Solids—Gas Flow. Part IV: The Effects of Particle Rotation, Roughness and Shape. *Canadian Journal of Chemical Engineering* 38:142–153.

Torobin, L. B. and W. H. Gauvin. 1961a. Fundamental Aspects of Solids—Gas Flow. Part VI: Multiparticle Behavior of Turbulent Fluids. *Canadian Journal of Chemical Engineering* 39:113–130.

Torobin, L. B. and W. H. Gauvin. 1961b. Drag Coefficients of Spheres Moving in Steady and Accelerated Motion in a Turbulent Fluid. *American Institute of Chemical Engineering Journal* 7:615–619.

Townsend, C. T. 1956. A Laboratory Manual for the Canning Industry. 22:28.

Treitel, Otto. 1944. Elasticity of Plant Tissues. *Kansas Academy of Science* 47(2):219–239.

Treloar, L. R. G. 1949. *The Physics of Rubber Elasticity*. Clarendon Press, Oxford.

Tressler, S., N. Woodberry and H. Mark. 1946. Application of the Density-Gradient Tube in Fiber Research. *J. of Polymer Science* 1:437.

Troller, J. A. and J. H. B. Christian. 1978. *Water Activity and Food*. Academic Press, New York.

Tuckey, L. D. and N. N. Mohsenin. 1966. Visual representation of apple bruising. Motion picture proceedings of the 17th Int'l. Horticulture Conf. *1*, 125.

Tuomy, J. M., R. J. Lechnir and T. Miller. 1963. Effect of Cooking Temperatures and Time in the Tenderness of Beef. *Food Technology* 17:1457.

Turner, W. K., Suggs, C. W. and J. W. Dickens. 1967. Impact Damage to Peanuts and Its Effects on Germination, Seedling Development, and Milling Quality. *Trans. of the ASAE* 10(2):248–249.

Turrell, F. B. and D. L. Slack. Specific Gravity of Citrus Fruits. *Am. Soc. of Hort. Sci.* 52:245.

Tyler, C. 1961. Shell Strength: Its Measurement and Its Relationship to Other Factors. *British Poultry Science* 2(1):3–19.

Tyler, C. 1963. Egg Shell Structure and Strength. *Chemistry and Industry* June 22:1012–1013.

Tyler, C. and F. H. Geake. 1964. The Testing of Methods for Cracking Egg Shells, Based on Paired Readings From Individual Eggs and the Measurement of Some Effects of Various Treatments. *British Poultry Science* 5(1):19–28. and (3):277–284.

Tyler, C. and D. Moore, 1965. Types of Damage Caused by Various Cracking and Crushing Methods Used for Measuring Egg Shell Strength. *British Poultry Science* 6(2):175–182.

Uhl, J. B. and B. J. Lamp. 1966. Pneumatic Separation of Grain and Straw Mixtures. *Trans. of the ASAE* 9(2):244–246.

USDA, Forest Service, The Forest Products Laboratory. 1955. Agriculture Handbook No. 72, *Wood Handbook,* U.S. Government Printing Office, Washington, D.C.

Usenko, V. V. 1952. Mechanical Damage to Seed. Selektsiya i Semenovodstvo Translated by E. Harris, NIAE, Silsoe, Bedfordshire, England.

Van Buren, J. P., J. C. Moyer, W. B. Robinson and D. B. Hand. 1960. Control of Firmness in Canned Snap Beans. *Farm Research* 26:9.

Van den Berg, C. and H. A. Leniger. 1976. First Int. Congress on Engineering and Food. August 9–13, Boston.

Van den Berg, C. and S. Bruin. 1978. Water Activity and Its Estimation in Food Systems: Theoretical Aspects. Second Int. Symposium on Properties of Water in Relation to Food Quality and Stability (ISOPOW-II), Sept. 10–16, Osaka, Japan.

Van Gilst, Carl, R. M. Peart, T. W. Perry, and R. A. Pickett. 1966. An Automatic Slurry Feeding System for Swine. *Agricultural Engineering* 47(1):24–25, 31.

Van Wazer, J. R., J. W. Lyons, K. Y. Kim and R. E. Colwell. 1963. *Viscosity and Flow Measurement*. Interscience Publishers, Inc., New York.

Vasic, I. and L. M. DeMan. 1968. Effect of Mechanical Treatment on Some Rheological Properties of Butter. *Rheology and Texture of Foodstuff*. Monograph No. 27, Soc. Chem. Ind., 14 Belgrave Square, London.

Vennard, J. K. 1961. *Fluid Mechanics*. 4th Ed. John Wiley and Sons, Inc., New York.

Vierling, A. and G. L. Sinha. 1960. Untersuchungen zum Fördervorgang beim senkrechten Schneckenförderer. *Foerdersau. Heven*. 10(8):587. (Investigations into the process of conveying by vertical screw conveyor, translation no. 95, N.I.A. E. Silsoe).

Vincent, J. F. and L. Z. Szabo. 1947. *Anal. Chem*. 19:655.

Virgin, H. I. 1955. A New Method for Determination of the Turgor of Plant Tissues. *Physiol. Plantarum* 8:954.

Vis, E., D. Wolf, B. A. Stout, and D. H. Dewey. 1968. Physical Properties of Apple Fruit Pertaining to Orientation. ASAE Paper 68–333, Am. Soc. Agr. Engrs., Saint Joseph, Michigan.

Vogt, E. G. and R. R. White. 1948. Friction in the Flow of Suspensions. *Industrial and Engineering Chemistry* 40:1731–1738.

Voisey, P. W. and D. C. McDonald. 1964. An Instrument for Measuring the Puncture Resistance of Fruits and Vegetables. *Am. Soc. Hort. Sci. Proc.* 84:557.

Voisey, P. W. and L. H. Lyall. 1965. Methods for Determining the Strength of

Tomato Skins in Relation to Fruit Cracking. *Proc. Am. Soc. Hort. Sci.* 86: 597–609.

Voisey, P. W. and H. Hansen. 1967. A Shear Apparatus for Meat Tenderness Evaluation. *Food Technology* 21:37A.

Voisey, P. W. and J. R. Hunt. 1967a. Relationship Between Applied Force, Deformation of Egg Shells and Fracture Force. *J. Agric. Engng. Res.* 12(1): 1–4.

Voisey, P. W. and J. R. Hunt. 1967b. Physical Properties of Egg Shells, 3: An Apparatus for Estimating Impact Resistance of the Shell. *British Poultry Science* 8(4):259–263.

Voisey, P. W. and J. R. Hunt. 1967c. Physical Properties of Egg Shells, 4: Stress Distribution in the Shell. *British Poultry Science* 8(4):263–271.

Volbracht, O. and V. Kuhnke. 1956. Mechanische Beschädigungen an Kartoffeln (Mechanical Injuries of the Potatoes). *Der Kartoffel-Bau,* Heft No. 4 and 5, April and May. Translation No. 125, NIAE, Silsoe, Bedford, England.

von Rittinger, P. R. 1967. *Textbook of Mineral Dressing Loss* (in German). Ernst & Korn, Berlin.

Wadell, H. 1934. The Coefficient of Resistance as a Function of Reynolds Number for Solids of Various Shapes. *Journal of the Franklin Institute*. 217:459–490.

Waelti, H. and W. F. Buchele. 1969. Factors Affecting Corn Kernel Damage in Combine Cylinders. *Trans. of the ASAE* 12(1):55–59.

Walls, E. P. and W. B. Kemp. 1940. Relationship Between Tenderometer Readings and Alcohol Insoluble Solids of Alaska Peas. *Proc. Am. Soc. Hort. Sci.* 37:279.

Wang, J. K. 1963. Design of a Ground Berry Husking Machine. *Trans. of the ASAE* 6(4):211–212.

Wang, J. K. and H. S. Chang. 1970. Mechanical properties of papaya and their dependence on maturity. *Trans. Am. Soc. Agric. Engr.* 13:369–371.

Watson, E. L. 1967. Rheology of Apricot and Apple Purees. ASAE Paper No. 67–306, Am. Soc. Agr. Engrs., St. Joseph, Michigan.

Way, S. 1940. Some Observations on the Theory of Contact Pressures. *Trans. of ASME, J. of Applied Mechanics* 7(12):A147–A157.

Weltmann, R. N. 1960. Rheology of Pastes and Paints in *Rheology, Theory and Applications,* Vol. III, F. R. Eirich, Editor, Academic Press, Inc., New York.

Wen, P. R. and N. N. Mohsenin. 1970. Application of Pulse Technique for Determination of Elastic Modulus of Yellow Poplar. *Material Research and Standards* (in press).

Wennergren, E. B. and W. A. Lee. 1961. Economic Aspects of Bruising in the Apple-Processing Industry. Penna. Agr. Exp. Sta. Bulletin No. 675.

Werner, H. O. and J. O. Dutt, 1941. Reduction of Cracking of Late Crop Potatoes at Harvest Time by Root Cutting or Vine Killing. *Am. Potato Journal.* 18: 189–209.

Whipple, H. E. 1965. Forms of Water in Biological Systems. *Ann. N. Y. Acad. Sci.* 125:1.

White, R. K. 1966. Swelling Stress in Corn Kernel as Influenced by Moisture Sorption. M. S. Thesis in Agricultural Engineering, The Pennsylvania State University, University Park, Pa.

White, R. K. and N. N. Mohsenin. 1967. Apparatus for Determination of Bulk Modulus and Compressibility of Materials. *Trans. of the ASAE* 10(5):670–671.

Whitehead, J. B. 1935. *Impregnated Paper Insulation*. John Wiley and Sons, Inc., New York.

Whitney, R. W. and J. G. Porterfield. 1968. Particle Separation in a Pneumatic Conveying System. *Trans. of the ASAE* 11(4):477–479.

Whittenberger, R. T. and C. H. Hills. 1953. Studies on the Processing of Red Cherries. I. Changes in Fresh Red Cherries Caused by Bruising, Cooling and Soaking. *Food Technology* 7:29.

Whittenberger, R. T. 1958. Bruising of Red Cherries. *The Canner and Freezer* May, 12.

Whittenberger, R. T., H. P. Gaston, and J. H. Levin. 1964. Effect of Recurrent Bruising on the Processing of Red Tart Cherries. Research Report No. 4, Michigan Agr. Exp. Sta.

Wiant, J. S. Findlen, H. and J. Kaufman. 1951. Effect of Temperature on Black Spot in Long Island and Red River Valley Potatoes. *Am Potato Journal* 28: 753–756.

Wichser, W. R. 1961. The World of Corn Processing. *American Millers and Processors,* April, 29–31.

Wieneke, F. 1956. Reibungswerte von Pflanzen und Faserstoffen. *Landtechnische Forschung* 6:146.

Wilder, H. K. 1948. Instruction for Use of the Fibrometer in the Measurement of Fiber Content in Canned Asparagus. *Nat'l. Canners' Assoc. Res.* Lab. Rept. No. 12313-C. San Francisco, California.

Wiley, R. C. and Stembridge. 1961. Factors Influencing Apple Texture. *Proc. Amer. Soc. Hort. Sci.* 77:60.

Wiley, R. E. 1962. The Density Gradient Technique. *Plastic Technology* 8(3):31.

Wilhelm, R. H. and S. Valentine. 1951. The Fluidized Bed: Transition State in the Vertical Pneumatic Transport of Particles. *Industrial and Engineering Chemistry* 43:1199–1203.

Wilke, H. L. 1936. An External Measure of Egg Viscosity. Iowa State Ag. Exp. Sta. Bul. 194.

Wilkinson, W. K. 1960. *Non-Newtonian Fluids*. Pergamon Press, New York.

Williams, E. J. and I. J. Ross. 1968. Vertical Bottom Pressures of Confined Stack of Dried Citrus Pulp. *Trans. of the ASAE* 11(6):868–870.

Winslow, N. M. and J. J. Shapiro. 1959. An instrument for the measurement of pore size distribution by mercury penetration. *ASTM Bull. 49*. Am. Soc. Testing and Materials, Phil., Pa.

Winton, A. L. and K. B. Winton. *The Structure and Composition of Foods*. John Wiley & Sons, Inc., New York (Vol. I, 1932, Vol. II, 1935; Vol. III, 1937).

Witz, R. C. 1954. Measuring the Resistance of Potatoes to Bruising. *Agricultural Engineering* 35(4):241–244.

Wood, F. W. 1968. Psychophysical Studies of the Consistency of Liquid Foods in *Rheology and Texture of Foodstuffs*. S.C.I. Monograph No. 27, Soc. Chem. Ind., 14 Belgrave Square, London.

Wratten, F. T., D. P. Wiley, J. L. Chesness, S. Bal, and V. Ramarao. 1968. Physical and Thermal Properties of Rough Rice. ASAE Paper No. 68–809, Am. Soc. of Agr. Engrs., Saint Joseph, Michigan.

Yang, Y. M. 1973. Analysis of the mechanics of deformation of a fruit under a rigid die with spherical end. Ph.D. Thesis. Penn State University, University Park, Pa.

Yang, Y. M. and N. N. Mohsenin. 1974. Analysis of the mechanics of the fruit pressure tester. *J. Texture Studies* 5:213–238.

Yaremenko, M. K. 1956. Mechanical Properties of Silage Material and Choice of a Method of Compressing It. Mekbanizarsiya i Electrikifatsia Sots. Selsk. Khozyaistva (5):22. Translated by E. Harris, NIAE, Silsoe, Bedfordshire, England.

Yeatman, J. N. and K. H. Norris. 1965. Evaluating Internal Quality of Apples with New Automatic Fruit Sorter. *Food Tech.* 19(3):123–125.

Young, J. H., J. M. Bunn and W. H. Henson. 1967. Analytical Description of Moisture Sorption of Cured Tobacco. *Trans. of the ASAE* 10(1):31–37.

Young, J. H. and G. L. Nelson. 1967. Theory of Hystersis Between Sorption and Desorption Isotherms in Biological Materials. *Trans. of the ASAE* 10(2): 260–263.

Yu, W. W., J. S. Boyd, and J. Menear. 1963. Silage Pressure in Large Diameter Silo. ASAE Paper 63–427. American Society of Agricultural Engineers, Saint Joseph, Michigan.

Zahara, M., J. G. McLean, and D. N. Wright. 1961. Mechanical Injury to Potato Tubers During Harvesting. *California Agriculture,* August.

Zoerb, G. C. and C. W. Hall. 1960. Some Mechanical and Rheological Properties of Grains. *J. Agric. Engng. Res.* 5:83.

Zoerb, G. C. 1967. Instrumentation and Measurement Techniques for Determination of Physical Properties of Agricultural Products. *Trans. of the ASAE* 10(1):100–109.

SUBJECT INDEX